A Course in Mathematical Statistics
Second Edition

George G. Roussas

Intercollege Division of Statistics
University of California
Davis, California

ACADEMIC PRESS

San Diego · London · Boston
New York · Sydney · Tokyo · Toronto

This book is printed on acid-free paper. ∞

Copyright © 1997 by Academic Press

All rights reserved.
No part of this publication may be reproduced or transmitted in any form or by any means, electronic or mechanical, including photocopy, recording, or any information storage and retrieval system, without permission in writing from the publisher.

ACADEMIC PRESS
525 B Street, Suite 1900, San Diego, CA 92101-4495, USA
1300 Boylston Street, Chestnut Hill, MA 02167, USA
http://www.apnet.com

ACADEMIC PRESS LIMITED
24–28 Oval Road, London NW1 7DX, UK
http://www.hbuk.co.uk/ap/

Library of Congress Cataloging-in-Publication Data

Roussas, George G.
    A course in mathematical statistics / George G. Roussas.—2nd ed.
      p. cm.
    Rev. ed. of: A first course in mathematical statistics. 1973.
    Includes index.
    ISBN 0-12-599315-3
    1. Mathematical statistics.   I. Roussas, George G. First course in mathematical statistics. II. Title.
QA276.R687   1997   96-42115
519.5—dc20   CIP

Printed in the United States of America
96 97 98 99 00 EB 9 8 7 6 5 4 3 2 1

Table of Selected Discrete and Continuous Distributions and Some of Their Characteristics

Distribution	Probability density function	Mean	Variance
Negative Exponential	$f(x) = \lambda \exp(-\lambda x)$, $x > 0$; $\lambda > 0$	$\dfrac{1}{\lambda}$	$\dfrac{1}{\lambda^2}$
Uniform, U(α, β)	$f(x) = \dfrac{1}{\beta - \alpha}$, $\alpha \le x \le \beta$; $-\infty < \alpha < \beta < \infty$	$\dfrac{\alpha + \beta}{2}$	$\dfrac{(\alpha - \beta)^2}{12}$
Beta	$f(x) = \dfrac{\Gamma(\alpha + \beta)}{\Gamma(\alpha)\Gamma(\beta)} x^{\alpha-1}(1-x)^{\beta-1}$, $0 < x < 1$; $\alpha, \beta > 0$	$\dfrac{\alpha}{\alpha + \beta}$	$\dfrac{\alpha\beta}{(\alpha+\beta)^2(\alpha+\beta+1)}$
Cauchy	$f(x) = \dfrac{\sigma}{\pi} \cdot \dfrac{1}{\sigma^2 + (x-\mu)^2}$, $x \in \mathbb{R}$; $\mu \in \mathbb{R}$, $\sigma > 0$	Does not exist	Does not exist
Bivariate Normal	$f(x_1, x_2) = \dfrac{1}{2\pi\sigma_1\sigma_2\sqrt{1-\rho^2}} \exp\left(-\dfrac{q}{2}\right)$, $q = \dfrac{1}{1-\rho^2}\left[\left(\dfrac{x_1-\mu_1}{\sigma_1}\right)^2 - 2\rho\left(\dfrac{x_1-\mu_1}{\sigma_1}\right) \times \left(\dfrac{x_2-\mu_2}{\sigma_2}\right) + \left(\dfrac{x_2-\mu_2}{\sigma_2}\right)^2\right]$, $x_1, x_2 \in \mathbb{R}$; $\mu_1, \mu_2 \in \mathbb{R}$, $\sigma_1, \sigma_2 > 0$, $-1 \le \rho \le 1$	vector of expectations: $(\mu_1, \mu_2)'$	vector of variances: $(\sigma_1^2, \sigma_2^2)'$.
k-Variate Normal, $N(\mu, \Sigma)$	$f(\mathbf{x}) = (2\pi)^{-k/2}\|\Sigma\|^{-1/2} \times \exp\left[-\dfrac{1}{2}(\mathbf{x}-\boldsymbol{\mu})'\Sigma^{-1}(\mathbf{x}-\boldsymbol{\mu})\right]$, $\mathbf{x} \in \mathbb{R}^k$; $\boldsymbol{\mu} \in \mathbb{R}$, Σ: $k \times k$ non-singular symmetric matrix	mean vector: $\boldsymbol{\mu}$	covariance matrix: Σ

A Course in Mathematical Statistics
Second Edition

To my wife and sons

Contents

Preface to the Second Edition xv

Preface to the First Edition xviii

Chapter 1   **Basic Concepts of Set Theory**   1

     1.1   **Some Definitions and Notation**   1
            Exercises   5
     1.2*   **Fields and σ-Fields**   8

Chapter 2   **Some Probabilistic Concepts and Results**   14

     2.1   **Probability Functions and Some Basic Properties and Results**   14
            Exercises   20
     2.2   **Conditional Probability**   21
            Exercises   25
     2.3   **Independence**   27
            Exercises   33
     2.4   **Combinatorial Results**   34
            Exercises   40
     2.5*   **Product Probability Spaces**   45
            Exercises   47

2.6* The Probability of Matchings 47
Exercises 52

Chapter 3 On Random Variables and Their Distributions 53

3.1 Some General Concepts 53
3.2 Discrete Random Variables (and Random Vectors) 55
Exercises 61
3.3 Continuous Random Variables (and Random Vectors) 65
Exercises 76
3.4 The Poisson Distribution as an Approximation to the Binomial Distribution and the Binomial Distribution as an Approximation to the Hypergeometric Distribution 79
Exercises 82
3.5* Random Variables as Measurable Functions and Related Results 82
Exercises 84

Chapter 4 Distribution Functions, Probability Densities, and Their Relationship 85

4.1 The Cumulative Distribution Function (c.d.f. or d.f.) of a Random Vector—Basic Properties of the d.f. of a Random Variable 85
Exercises 89
4.2 The d.f. of a Random Vector and Its Properties—Marginal and Conditional d.f.'s and p.d.f.'s 91
Exercises 97
4.3 Quantiles and Modes of a Distribution 99
Exercises 102
4.4* Justification of Statements 1 and 2 102
Exercises 105

Chapter 5 Moments of Random Variables—Some Moment and Probability Inequalities 106

5.1 Moments of Random Variables 106
Exercises 111
5.2 Expectations and Variances of Some R.V.'s 114
Exercises 119
5.3 Conditional Moments of Random Variables 122
Exercises 124

Chapter 6 — Characteristic Functions, Moment Generating Functions and Related Theorems 138

- 5.4 Some Important Applications: Probability and Moment Inequalities 125
  - Exercises 128
- 5.5 Covariance, Correlation Coefficient and Its Interpretation 129
  - Exercises 133
- 5.6* Justification of Relation (2) in Chapter 2 134

Chapter 6 Characteristic Functions, Moment Generating Functions and Related Theorems 138

- 6.1 Preliminaries 138
- 6.2 Definitions and Basic Theorems—The One-Dimensional Case 140
  - Exercises 145
- 6.3 The Characteristic Functions of Some Random Variables 146
  - Exercises 149
- 6.4 Definitions and Basic Theorems—The Multidimensional Case 150
  - Exercises 152
- 6.5 The Moment Generating Function and Factorial Moment Generating Function of a Random Variable 153
  - Exercises 160

Chapter 7 Stochastic Independence with Some Applications 164

- 7.1 Stochastic Independence: Criteria of Independence 164
  - Exercises 168
- 7.2 Proof of Lemma 2 and Related Results 170
  - Exercises 172
- 7.3 Some Consequences of Independence 173
  - Exercises 176
- 7.4* Independence of Classes of Events and Related Results 177
  - Exercise 179

Chapter 8 Basic Limit Theorems 180

- 8.1 Some Modes of Convergence 180
  - Exercises 182
- 8.2 Relationships Among the Various Modes of Convergence 182
  - Exercises 187
- 8.3 The Central Limit Theorem 187
  - Exercises 194
- 8.4 Laws of Large Numbers 196
  - Exercises 198

	8.5 **Further Limit Theorems** 199
	Exercises 206
	8.6* **Pólya's Lemma and Alternative Proof of the WLLN** 206
	Exercises 211

Chapter 9 — Transformations of Random Variables and Random Vectors 212

9.1 **The Univariate Case** 212
Exercises 218
9.2 **The Multivariate Case** 219
Exercises 233
9.3 **Linear Transformations of Random Vectors** 235
Exercises 240
9.4 **The Probability Integral Transform** 242
Exercise 244

Chapter 10 — Order Statistics and Related Theorems 245

10.1 **Order Statistics and Related Distributions** 245
Exercises 252
10.2 **Further Distribution Theory: Probability of Coverage of a Population Quantile** 256
Exercise 258

Chapter 11 — Sufficiency and Related Theorems 259

11.1 **Sufficiency: Definition and Some Basic Results** 260
Exercises 269
11.2 **Completeness** 271
Exercises 273
11.3 **Unbiasedness—Uniqueness** 274
Exercises 276
11.4 **The Exponential Family of p.d.f.'s: One-Dimensional Parameter Case** 276
Exercises 280
11.5 **Some Multiparameter Generalizations** 281
Exercises 282

Chapter 12 — Point Estimation 284

12.1 **Introduction** 284
Exercise 284

12.2 **Criteria for Selecting an Estimator: Unbiasedness, Minimum Variance** 285
Exercises 286
12.3 **The Case of Availability of Complete Sufficient Statistics** 287
Exercises 292
12.4 **The Case Where Complete Sufficient Statistics Are Not Available or May Not Exist: Cramér-Rao Inequality** 293
Exercises 301
12.5 **Criteria for Selecting an Estimator: The Maximum Likelihood Principle** 302
Exercises 308
12.6 **Criteria for Selecting an Estimator: The Decision-Theoretic Approach** 309
12.7 **Finding Bayes Estimators** 312
Exercises 317
12.8 **Finding Minimax Estimators** 318
Exercise 320
12.9 **Other Methods of Estimation** 320
Exercises 322
12.10 **Asymptotically Optimal Properties of Estimators** 322
Exercise 325
12.11 **Closing Remarks** 325
Exercises 326

Chapter 13 — Testing Hypotheses 327

13.1 **General Concepts of the Neyman-Pearson Testing Hypotheses Theory** 327
Exercise 329
13.2 **Testing a Simple Hypothesis Against a Simple Alternative** 329
Exercises 336
13.3 **UMP Tests for Testing Certain Composite Hypotheses** 337
Exercises 347
13.4 **UMPU Tests for Testing Certain Composite Hypotheses** 349
Exercises 353
13.5 **Testing the Parameters of a Normal Distribution** 353
Exercises 356
13.6 **Comparing the Parameters of Two Normal Distributions** 357
Exercises 360
13.7 **Likelihood Ratio Tests** 361
Exercises 369
13.8 **Applications of LR Tests: Contingency Tables, Goodness-of-Fit Tests** 370
Exercises 373

13.9 Decision-Theoretic Viewpoint of Testing Hypotheses 375

Chapter 14 Sequential Procedures 382

14.1 Some Basic Theorems of Sequential Sampling 382
Exercises 388
14.2 Sequential Probability Ratio Test 388
Exercise 392
14.3 Optimality of the SPRT-Expected Sample Size 393
Exercises 394
14.4 Some Examples 394

Chapter 15 Confidence Regions—Tolerance Intervals 397

15.1 Confidence Intervals 397
Exercise 398
15.2 Some Examples 398
Exercises 404
15.3 Confidence Intervals in the Presence of Nuisance Parameters 407
Exercise 410
15.4 Confidence Regions—Approximate Confidence Intervals 410
Exercises 412
15.5 Tolerance Intervals 413

Chapter 16 The General Linear Hypothesis 416

16.1 Introduction of the Model 416
16.2 Least Square Estimators—Normal Equations 418
16.3 Canonical Reduction of the Linear Model—Estimation of σ^2 424
Exercises 428
16.4 Testing Hypotheses About $\eta = E(\mathbf{Y})$ 429
Exercises 433
16.5 Derivation of the Distribution of the $\mathcal{F}$ Statistic 433
Exercises 436

Chapter 17 Analysis of Variance 440

17.1 One-way Layout (or One-way Classification) with the Same Number of Observations Per Cell 440
Exercise 446
17.2 Two-way Layout (Classification) with One Observation Per Cell 446
Exercises 451

17.3 Two-way Layout (Classification) with K (≥ 2) Observations Per Cell 452
Exercises 457
17.4 A Multicomparison method 458
Exercises 462

Chapter 18 The Multivariate Normal Distribution 463

18.1 Introduction 463
Exercises 466
18.2 Some Properties of Multivariate Normal Distributions 467
Exercise 469
18.3 Estimation of μ and Σ and a Test of Independence 469
Exercises 475

Chapter 19 Quadratic Forms 476

19.1 Introduction 476
19.2 Some Theorems on Quadratic Forms 477
Exercises 483

Chapter 20 Nonparametric Inference 485

20.1 Nonparametric Estimation 485
20.2 Nonparametric Estimation of a p.d.f. 487
Exercise 490
20.3 Some Nonparametric Tests 490
20.4 More About Nonparametric Tests: Rank Tests 493
Exercises 496
20.5 Sign Test 496
20.6 Relative Asymptotic Efficiency of Tests 498

Appendix I Topics from Vector and Matrix Algebra 499

I.1 Basic Definitions in Vector Spaces 499
I.2 Some Theorems on Vector Spaces 501
I.3 Basic Definitions About Matrices 502
I.4 Some Theorems About Matrices and Quadratic Forms 504

Appendix II Noncentral t-, χ^2-, and F-Distributions 508

II.1 Noncentral t-Distribution 508

II.2	Noncentral x^2-Distribution 508
II.3	Noncentral F-Distribution 509

Appendix III **Tables** 511

1. The Cumulative Binomial Distribution 511
2. The Cumulative Poisson Distribution 520
3. The Normal Distribution 523
4. Critical Values for Student's t-Distribution 526
5. Critical Values for the Chi-Square Distribution 529
6. Critical Values for the F-Distribution 532
7. Table of Selected Discrete and Continuous Distributions and Some of Their Characteristics 542

Some Notation and Abbreviations 545

Answers to Selected Exercises 547

Index 561

Preface to the Second Edition

This is the second edition of a book published for the first time in 1973 by Addison-Wesley Publishing Company, Inc., under the title *A First Course in Mathematical Statistics*. The first edition has been out of print for a number of years now, although its reprint in Taiwan is still available. That issue, however, is meant for circulation only in Taiwan.

The first issue of the book was very well received from an academic viewpoint. I have had the pleasure of hearing colleagues telling me that the book filled an existing gap between a plethora of textbooks of lower mathematical level and others of considerably higher level. A substantial number of colleagues, holding senior academic appointments in North America and elsewhere, have acknowledged to me that they made their entrance into the wonderful world of probability and statistics through my book. I have also heard of the book as being in a class of its own, and also as forming a collector's item, after it went out of print. Finally, throughout the years, I have received numerous inquiries as to the possibility of having the book reprinted. It is in response to these comments and inquiries that I have decided to prepare a second edition of the book.

This second edition preserves the unique character of the first issue of the book, whereas some adjustments are affected. The changes in this issue consist in correcting some rather minor factual errors and a considerable number of misprints, either kindly brought to my attention by users of the book or located by my students and myself. Also, the reissuing of the book has provided me with an excellent opportunity to incorporate certain rearrangements of the material.

One change occurring throughout the book is the grouping of exercises of each chapter in clusters added at the end of sections. Associating exercises with material discussed in sections clearly makes their assignment easier. In the process of doing this, a handful of exercises were omitted, as being too complicated for the level of the book, and a few new ones were inserted. In

Chapters 1 through 8, some of the materials were pulled out to form separate sections. These sections have also been marked by an asterisk (*) to indicate the fact that their omission does not jeopardize the flow of presentation and understanding of the remaining material.

Specifically, in Chapter 1, the concepts of a field and of a σ-field, and basic results on them, have been grouped together in Section 1.2*. They are still readily available for those who wish to employ them to add elegance and rigor in the discussion, but their inclusion is not indispensable. In Chapter 2, the number of sections has been doubled from three to six. This was done by discussing independence and product probability spaces in separate sections. Also, the solution of the problem of the probability of matching is isolated in a section by itself. The section on the problem of the probability of matching and the section on product probability spaces are also marked by an asterisk for the reason explained above. In Chapter 3, the discussion of random variables as measurable functions and related results is carried out in a separate section, Section 3.5*. In Chapter 4, two new sections have been created by discussing separately marginal and conditional distribution functions and probability density functions, and also by presenting, in Section 4.4*, the proofs of two statements, Statements 1 and 2, formulated in Section 4.1; this last section is also marked by an asterisk. In Chapter 5, the discussion of covariance and correlation coefficient is carried out in a separate section; some additional material is also presented for the purpose of further clarifying the interpretation of correlation coefficient. Also, the justification of relation (2) in Chapter 2 is done in a section by itself, Section 5.6*. In Chapter 6, the number of sections has been expanded from three to five by discussing in separate sections characteristic functions for the one-dimensional and the multidimensional case, and also by isolating in a section by itself definitions and results on moment-generating functions and factorial moment generating functions. In Chapter 7, the number of sections has been doubled from two to four by presenting the proof of Lemma 2, stated in Section 7.1, and related results in a separate section; also, by grouping together in a section marked by an asterisk definitions and results on independence. Finally, in Chapter 8, a new theorem, Theorem 10, especially useful in estimation, has been added in Section 8.5. Furthermore, the proof of Pólya's lemma and an alternative proof of the Weak Law of Large Numbers, based on truncation, are carried out in a separate section, Section 8.6*, thus increasing the number of sections from five to six.

In the remaining chapters, no changes were deemed necessary, except that in Chapter 13, the proof of Theorem 2 in Section 13.3 has been facilitated by the formulation and proof in the same section of two lemmas, Lemma 1 and Lemma 2. Also, in Chapter 14, the proof of Theorem 1 in Section 14.1 has been somewhat simplified by the formulation and proof of Lemma 1 in the same section.

Finally, a table of some commonly met distributions, along with their means, variances and other characteristics, has been added. The value of such a table for reference purposes is obvious, and needs no elaboration.

This book contains enough material for a year course in probability and statistics at the advanced undergraduate level, or for first-year graduate students not having been exposed before to a serious course on the subject matter. Some of the material can actually be omitted without disrupting the continuity of presentation. This includes the sections marked by asterisks, perhaps, Sections 13.4–13.6 in Chapter 13, and all of Chapter 14. The instructor can also be selective regarding Chapters 11 and 18. As for Chapter 19, it has been included in the book for completeness only.

The book can also be used independently for a one-semester (or even one quarter) course in probability alone. In such a case, one would strive to cover the material in Chapters 1 through 10 with the exclusion, perhaps, of the sections marked by an asterisk. One may also be selective in covering the material in Chapter 9.

In either case, presentation of results involving characteristic functions may be perfunctory only, with emphasis placed on moment-generating functions. One should mention, however, why characteristic functions are introduced in the first place, and therefore what one may be missing by not utilizing this valuable tool.

In closing, it is to be mentioned that this author is fully aware of the fact that the audience for a book of this level has diminished rather than increased since the time of its first edition. He is also cognizant of the trend of having recipes of probability and statistical results parading in textbooks, depriving the reader of the challenge of thinking and reasoning instead delegating the "thinking" to a computer. It is hoped that there is still room for a book of the nature and scope of the one at hand. Indeed, the trend and practices just described should make the availability of a textbook such as this one exceedingly useful if not imperative.

G. G. Roussas
Davis, California
May 1996

Preface to the First Edition

This book is designed for a first-year course in mathematical statistics at the undergraduate level, as well as for first-year graduate students in statistics—or graduate students, in general—with no prior knowledge of statistics. A typical three-semester course in calculus and some familiarity with linear algebra should suffice for the understanding of most of the mathematical aspects of this book. Some advanced calculus—perhaps taken concurrently—would be helpful for the complete appreciation of some fine points.

There are basically two streams of textbooks on mathematical statistics that are currently on the market. One category is the advanced level texts which demonstrate the statistical theories in their full generality and mathematical rigor; for that purpose, they require a high level, mathematical background of the reader (for example, measure theory, real and complex analysis). The other category consists of intermediate level texts, where the concepts are demonstrated in terms of intuitive reasoning, and results are often stated without proofs or with partial proofs that fail to satisfy an inquisitive mind. Thus, readers with a modest background in mathematics and a strong motivation to understand statistical concepts are left somewhere in between. The advanced texts are inaccessible to them, whereas the intermediate texts deliver much less than they hope to learn in a course of mathematical statistics. The present book attempts to bridge the gap between the two categories, so that students without a sophisticated mathematical background can assimilate a fairly broad spectrum of the theorems and results from mathematical statistics. This has been made possible by developing the fundamentals of modern probability theory and the accompanying mathematical ideas at the beginning of this book so as to prepare the reader for an understanding of the material presented in the later chapters.

This book consists of two parts, although it is not formally so divided. Part 1 (Chapters 1–10) deals with probability and distribution theory, whereas Part

2 (Chapters 11–20) is devoted to statistical inference. More precisely, in Part 1 the concepts of a field and σ-field, and also the definition of a random variable as a measurable function, are introduced. This allows us to state and prove fundamental results in their full generality that would otherwise be presented vaguely using statements such as "it may be shown that . . . ," "it can be proved that . . . ," etc. This we consider to be one of the distinctive characteristics of this part. Other important features are as follows: a detailed and systematic discussion of the most useful distributions along with figures and various approximations for several of them; the establishment of several moment and probability inequalities; the systematic employment of characteristic functions—rather than moment generating functions—with all the well-known advantages of the former over the latter; an extensive chapter on limit theorems, including all common modes of convergence and their relationship; a *complete* statement and proof of the Central Limit Theorem (in its classical form); statements of the Laws of Large Numbers and several proofs of the Weak Law of Large Numbers, and further useful limit theorems; and also an extensive chapter on transformations of random variables with numerous illustrative examples discussed in detail.

The second part of the book opens with an extensive chapter on sufficiency. The concept of sufficiency is usually treated only in conjunction with estimation and testing hypotheses problems. In our opinion, this does not do justice to such an important concept as that of sufficiency. Next, the point estimation problem is taken up and is discussed in great detail and as large a generality as is allowed by the level of this book. Special attention is given to estimators derived by the principles of unbiasedness, uniform minimum variance and the maximum likelihood and minimax principles. An abundance of examples is also found in this chapter. The following chapter is devoted to testing hypotheses problems. Here, along with the examples (most of them numerical) and the illustrative figures, the reader finds a discussion of families of probability density functions which have the monotone likelihood ratio property and, in particular, a discussion of exponential families. These latter topics are available only in more advanced texts. Other features are a complete formulation and treatment of the general Linear Hypothesis and the discussion of the Analysis of Variance as an application of it. In many textbooks of about the same level of sophistication as the present book, the above two topics are approached either separately or in the reverse order from the one used here, which is pedagogically unsound, although historically logical. Finally, there are special chapters on sequential procedures, confidence regions—tolerance intervals, the Multivariate Normal distribution, quadratic forms, and nonparametric inference.

A few of the proofs of theorems and some exercises have been drawn from recent publications in journals.

For the convenience of the reader, the book also includes an appendix summarizing all necessary results from vector and matrix algebra.

There are more than 120 examples and applications discussed in detail in the text. Also, there are more than 530 exercises, appearing at the end of the chapters, which are of both theoretical and practical importance.

The careful selection of the material, the inclusion of a large variety of topics, the abundance of examples, and the existence of a host of exercises of both theoretical and applied nature will, we hope, satisfy people of both theoretical and applied inclinations. All the application-oriented reader has to do is to skip some fine points of some of the proofs (or some of the proofs altogether!) when studying the book. On the other hand, the careful handling of these same fine points should offer some satisfaction to the more mathematically inclined readers.

The material of this book has been presented several times to classes of the composition mentioned earlier; that is, classes consisting of relatively mathematically immature, eager, and adventurous sophomores, as well as juniors and seniors, and statistically unsophisticated graduate students. These classes met three hours a week over the academic year, and most of the material was covered in the order in which it is presented with the occasional exception of Chapters 14 and 20, Section 5 of Chapter 5, and Section 3 of Chapter 9. We feel that there is enough material in this book for a three-quarter session if the classes meet three or even four hours a week.

At various stages and times during the organization of this book several students and colleagues helped improve it by their comments. In connection with this, special thanks are due to G. K. Bhattacharyya. His meticulous reading of the manuscripts resulted in many comments and suggestions that helped improve the quality of the text. Also thanks go to B. Lind, K. G. Mehrotra, A. Agresti, and a host of others, too many to be mentioned here. Of course, the responsibility in this book lies with this author alone for all omissions and errors which may still be found.

As the teaching of statistics becomes more widespread and its level of sophistication and mathematical rigor (even among those with limited mathematical training but yet wishing to know "why" and "how") more demanding, we hope that this book will fill a gap and satisfy an existing need.

G. G. R.
Madison, Wisconsin
November 1972

Chapter 1

Basic Concepts of Set Theory

1.1 Some Definitions and Notation

A *set S* is a (well defined) collection of distinct objects which we denote by s. The fact that s is *a member of S*, *an element of S*, or that it *belongs to S* is expressed by writing $s \in S$. The negation of the statement is expressed by writing $s \notin S$. We say that S' *is a subset of S*, or that S' *is contained in S*, and write $S' \subseteq S$, if for every $s \in S'$, we have $s \in S$. S' is said to be a *proper subset of S*, and we write $S' \subset S$, if $S' \subseteq S$ and there exists $s \in S$ such that $s \notin S'$. Sets are denoted by capital letters, while lower case letters are used for elements of sets.

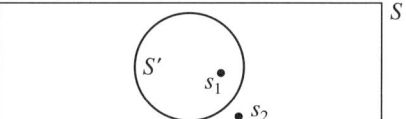

Figure 1.1 $S' \subseteq S$; in fact, $S' \subset S$, since $s_2 \in S$, but $s_2 \notin S'$.

These concepts can be illustrated pictorially by a drawing called a *Venn diagram* (Fig. 1.1). From now on a *basic*, or *universal* set, or *space* (which may be different from situation to situation), to be denoted by S, will be considered and all other sets in question will be subsets of S.

1.1.1 Set Operations

1. The *complement* (with respect to S) of the set A, denoted by A^c, is defined by $A^c = \{s \in S; s \notin A\}$. (See Fig. 1.2.)

Figure 1.2 A^c is the shaded region.

2. The *union* of the sets A_j, $j = 1, 2, \ldots, n$, to be denoted by

$$A_1 \cup A_2 \cup \cdots \cup A_n \quad \text{or} \quad \bigcup_{j=1}^{n} A_j,$$

is defined by

$$\bigcup_{j=1}^{n} A_j = \{s \in S; \, s \in A_j \text{ for } \textit{at least} \text{ one } j = 1, 2, \ldots, n\}.$$

For $n = 2$, this is pictorially illustrated in Fig. 1.3. The definition extends to an infinite number of sets. Thus for denumerably many sets, one has

$$\bigcup_{j=1}^{\infty} A_j = \{s \in S; \, s \in A_j \text{ for } \textit{at least} \text{ one } j = 1, 2, \ldots\}.$$

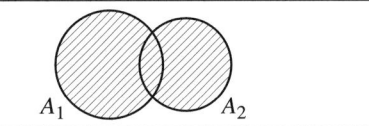

Figure 1.3 $A_1 \cup A_2$ is the shaded region.

3. The *intersection* of the sets A_j, $j = 1, 2, \ldots, n$, to be denoted by

$$A_1 \cap A_2 \cap \cdots \cap A_n \quad \text{or} \quad \bigcap_{j=1}^{n} A_j,$$

is defined by

$$\bigcap_{j=1}^{n} A_j = \{s \in S; \, s \in A_j \text{ for } \textit{all } j = 1, 2, \ldots, n\}.$$

For $n = 2$, this is pictorially illustrated in Fig. 1.4. This definition extends to an infinite number of sets. Thus for denumerably many sets, one has

$$\bigcap_{j=1}^{\infty} A_j = \{s \in S; \, s \in A_j \text{ for } \textit{all } j = 1, 2, \ldots\}.$$

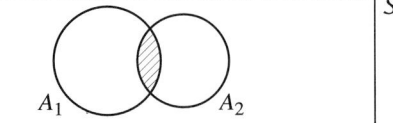

Figure 1.4 $A_1 \cap A_2$ is the shaded region.

4. The *difference* $A_1 - A_2$ is defined by
$$A_1 - A_2 = \{s \in S; s \in A_1, s \notin A_2\}.$$

Symmetrically,
$$A_2 - A_1 = \{s \in S; s \in A_2, s \notin A_1\}.$$

Note that $A_1 - A_2 = A_1 \cap A_2^c$, $A_2 - A_1 = A_2 \cap A_1^c$, and that, in general, $A_1 - A_2 \neq A_2 - A_1$. (See Fig. 1.5.)

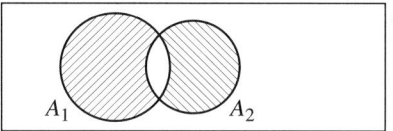

Figure 1.5 $A_1 - A_2$ is ////.
$A_2 - A_1$ is \\\\.

5. The *symmetric difference* $A_1 \triangle A_2$ is defined by
$$A_1 \triangle A_2 = (A_1 - A_2) \cup (A_2 - A_1).$$

Note that
$$A_1 \triangle A_2 = (A_1 \cup A_2) - (A_1 \cap A_2).$$

Pictorially, this is shown in Fig. 1.6. It is worthwhile to observe that operations (4) and (5) can be expressed in terms of operations (1), (2), and (3).

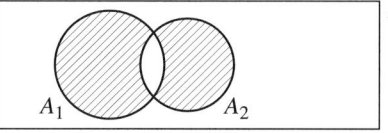

Figure 1.6 $A_1 \triangle A_2$ is the shaded area.

1.1.2 Further Definitions and Notation

A set which contains no elements is called the *empty set* and is denoted by $\emptyset$. Two sets A_1, A_2 are said to be *disjoint* if $A_1 \cap A_2 = \emptyset$. Two sets A_1, A_2 are said to be *equal*, and we write $A_1 = A_2$, if both $A_1 \subseteq A_2$ and $A_2 \subseteq A_1$. The sets A_j, $j = 1, 2, \ldots$ are said to be *pairwise* or *mutually disjoint* if $A_i \cap A_j = \emptyset$ for all $i \neq j$ (Fig. 1.7). In such a case, it is customary to write
$$A_1 + A_2, \; A_1 + \cdots + A_n = \sum_{j=1}^{n} A_j \quad \text{and} \quad A_1 + A_2 + \cdots = \sum_{j=1}^{\infty} A_j$$
instead of $A_1 \cup A_2$, $\bigcup_{j=1}^{n} A_j$, and $\bigcup_{j=1}^{\infty} A_j$, respectively. We will write $\bigcup_j A_j$, $\sum_j A_j$, $\bigcap_j A_j$ (or $\cup_j A_j$, $\Sigma_j A_j$, $\cap_j A_j$ where this is expedient for typgraphical reasons),

where we do not wish to specify the range of j, which will usually be either the (finite) set $\{1, 2, \ldots, n\}$, or the (infinite) set $\{1, 2, \ldots\}$.

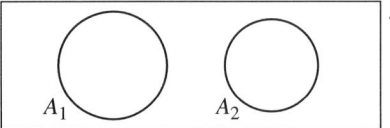

Figure 1.7 A_1 and A_2 are disjoint; that is, $A_1 \cap A_2 = \emptyset$. Also $A_1 \cup A_2 = A_1 + A_2$ for the same reason.

1.1.3 Properties of the Operations on Sets

1. $S^c = \emptyset$, $\emptyset^c = S$, $(A^c)^c = A$.
2. $S \cup A = S$, $\emptyset \cup A = A$, $A \cup A^c = S$, $A \cup A = A$.
3. $S \cap A = A$, $\emptyset \cap A = \emptyset$, $A \cap A^c = \emptyset$, $A \cap A = A$.

The previous statements are all obvious as is the following: $\emptyset \subseteq A$ for every subset A of S. Also

4. $\left. \begin{array}{l} A_1 \cup (A_2 \cup A_3) = (A_1 \cup A_2) \cup A_3 \\ A_1 \cap (A_2 \cap A_3) = (A_1 \cap A_2) \cap A_3 \end{array} \right\}$ (Associative laws)

5. $\left. \begin{array}{l} A_1 \cup A_2 = A_2 \cup A_1 \\ A_1 \cap A_2 = A_2 \cap A_1 \end{array} \right\}$ (Commutative laws)

6. $\left. \begin{array}{l} A \cap (\cup_j A_j) = \cup_j (A \cap A_j) \\ A \cup (\cap_j A_j) = \cap_j (A \cup A_j) \end{array} \right\}$ (Distributive laws)

are easily seen to be true.

The following identity is a useful tool in writing a union of sets as a sum of disjoint sets.

An identity:
$$\bigcup_j A_j = A_1 + A_1^c \cap A_2 + A_1^c \cap A_2^c \cap A_3 + \cdots.$$

There are two more important properties of the operation on sets which relate complementation to union and intersection. They are known as **De Morgan's laws**:

i) $\left(\bigcup_j A_j \right)^c = \bigcap_j A_j^c$,

ii) $\left(\bigcap_j A_j \right)^c = \bigcup_j A_j^c$.

As an example of a set theoretic proof, we prove (i).

PROOF OF (i) We wish to establish

a) $(\cup_j A_j)^c \subseteq \cap_j A_j^c$ and b) $\cap_j A_j^c \subseteq (\cup_j A_j)^c$.

We will then, by definition, have verified the desired equality of the two sets.

a) Let $s \in (\bigcup_j A_j)^c$. Then $s \notin \bigcup_j A_j$, hence $s \notin A_j$ for any j. Thus $s \in A_j^c$ for every j and therefore $s \in \bigcap_j A_j^c$.

b) Let $s \in \bigcap_j A_j^c$. Then $s \in A_j^c$ for every j and hence $s \notin A_j$ for any j. Then $s \notin \bigcup_j A_j$ and therefore $s \in (\bigcup_j A_j)^c$.

The proof of (ii) is quite similar. ▲

This section is concluded with the following:

DEFINITION 1 The sequence $\{A_n\}$, $n = 1, 2, \ldots$, is said to be a *monotone sequence* of sets if:

i) $A_1 \subseteq A_2 \subseteq A_3 \subseteq \cdots$ (that is, A_n is *increasing*, to be denoted by $A_n \uparrow$), or

ii) $A_1 \supseteq A_2 \supseteq A_3 \supseteq \cdots$ (that is, A_n is *decreasing*, to be denoted by $A_n \downarrow$).

The *limit* of a monotone sequence is defined as follows:

i) If $A_n \uparrow$, then $\lim\limits_{n \to \infty} A_n = \bigcup\limits_{n=1}^{\infty} A_n$, and

ii) If $A_n \downarrow$, then $\lim\limits_{n \to \infty} A_n = \bigcap\limits_{n=1}^{\infty} A_n$.

More generally, for any sequence $\{A_n\}$, $n = 1, 2, \ldots$, we define

$$\underline{A} = \liminf_{n \to \infty} A_n = \bigcup_{n=1}^{\infty} \bigcap_{j=n}^{\infty} A_j,$$

and

$$\overline{A} = \limsup_{n \to \infty} A_n = \bigcap_{n=1}^{\infty} \bigcup_{j=n}^{\infty} A_j.$$

The sets $\underline{A}$ and $\overline{A}$ are called the *inferior limit* and *superior limit*, respectively, of the sequence $\{A_n\}$. The sequence $\{A_n\}$ has a *limit* if $\underline{A} = \overline{A}$.

Exercises

1.1.1 Let A_j, $j = 1, 2, 3$ be arbitrary subsets of S. Determine whether each of the following statements is correct or incorrect.

i) $(A_1 - A_2) \cup A_2 = A_2$;

ii) $(A_1 \cup A_2) - A_1 = A_2$;

iii) $(A_1 \cap A_2) \cap (A_1 - A_2) = \emptyset$;

iv) $(A_1 \cup A_2) \cap (A_2 \cup A_3) \cap (A_3 \cup A_1) = (A_1 \cap A_2) \cup (A_2 \cap A_3) \cup (A_3 \cap A_1)$.

1.1.2 Let $S = \{(x, y)' \in \mathbb{R}^2; -5 \leq x \leq 5, 0 \leq y \leq 5, x, y = \text{integers}\}$, where prime denotes transpose, and define the subsets $A_j, j = 1, \ldots, 7$ of S as follows:

$$A_1 = \left\{(x,y)' \in S; x = y\right\}; \qquad A_2 = \left\{(x,y)' \in S; x = -y\right\};$$

$$A_3 = \left\{(x,y)' \in S; x^2 = y^2\right\}; \qquad A_4 = \left\{(x,y)' \in S; x^2 \leq y^2\right\};$$

$$A_5 = \left\{(x,y)' \in S; x^2 + y^2 \leq 4\right\}; \qquad A_6 = \left\{(x,y)' \in S; x \leq y^2\right\};$$

$$A_7 = \left\{(x,y)' \in S; x^2 \geq y\right\}.$$

List the members of the sets just defined.

1.1.3 Refer to Exercise 1.1.2 and show that:

i) $A_1 \cap \left(\bigcup_{j=2}^{7} A_j\right) = \bigcup_{j=2}^{7} (A_1 \cap A_j);$

ii) $A_1 \cup \left(\bigcap_{j=2}^{7} A_j\right) = \bigcap_{j=2}^{7} (A_1 \cup A_j);$

iii) $\left(\bigcup_{j=1}^{7} A_j\right)^c = \bigcap_{j=1}^{7} A_j^c;$

iv) $\left(\bigcap_{j=1}^{7} A_j\right)^c = \bigcup_{j=1}^{7} A_j^c$

by listing the members of each one of the eight sets appearing on either side of each one of the relations (i)–(iv).

1.1.4 Let A, B and C be subsets of S and suppose that $A \subseteq B$ and $B \subseteq C$. Then show that $A \subseteq C$; that is, the subset relationship is transitive. Verify it by taking $A = A_1$, $B = A_3$ and $C = A_4$, where A_1, A_3 and A_4 are defined in Exercise 1.1.2.

1.1.5 Establish the distributive laws stated on page 4.

1.1.6 In terms of the events A_1, A_2, A_3, and perhaps their complements, express each one of the following events:

i) B_i = "exactly i of the events A_1, A_2, A_3 occur," $i = 0, 1, 2, 3$;

ii) C = "all events A_1, A_2, A_3 occur";

iii) D = "none of the events A_1, A_2, A_3 occurs";

iv) E = "at most 2 of the events A_1, A_2, A_3 occur";

v) F = "at least 1 of the events A_1, A_2, A_3 occurs."

1.1.7 Establish the identity stated on page 4.

1.1.8 Give a detailed proof of the second identity in De Morgan's laws; that is, show that

$$\left(\bigcap_j A_j\right)^c = \bigcup_j A_j^c.$$

1.1.9 Refer to Definition 1 and show that

i) $\underline{A} = \{s \in S; s \text{ belongs to all but finitely many } A\text{'s}\}$;

ii) $\bar{A} = \{s \in S; s \text{ belongs to infinitely many } A\text{'s}\}$;

iii) $\underline{A} \subseteq \bar{A}$;

iv) If $\{A_n\}$ is a monotone sequence, then $\underline{A} = \bar{A} = \lim_{n \to \infty} A_n$.

1.1.10 Let $S = \mathbb{R}^2$ and define the subsets A_n, B_n, $n = 1, 2, \ldots$ of S as follows:

$$A_n = \left\{(x, y)' \in \mathbb{R}^2; 3 + \frac{1}{n} \leq x < 6 - \frac{2}{n}, 0 \leq y \leq 2 - \frac{1}{n^2}\right\},$$

$$B_n = \left\{(x, y)' \in \mathbb{R}^2; x^2 + y^2 \leq \frac{1}{n^3}\right\}.$$

Then show that $A_n \uparrow A$, $B_n \downarrow B$ and identify A and B.

1.1.11 Let $S = \mathbb{R}$ and define the subsets A_n, B_n, $n = 1, 2, \ldots$ of S as follows:

$$A_n = \left\{x \in \mathbb{R}; -5 + \frac{1}{n} < x < 20 - \frac{1}{n}\right\}, \quad B_n = \left\{x \in \mathbb{R}; 0 < x < 7 + \frac{3}{n}\right\}.$$

Then show that $A_n \uparrow$ and $B_n \downarrow$, so that $\lim_{n \to \infty} A_n = A$ and $\lim_{n \to \infty} B_n = B$ exist (by Exercise 1.1.9(iv)). Also identify the sets A and B.

1.1.12 Let A and B be subsets of S and for $n = 1, 2, \ldots$, define the sets A_n as follows: $A_{2n-1} = A$, $A_{2n} = B$. Then show that

$$\liminf_{n \to \infty} A_n = A \cap B, \quad \limsup_{n \to \infty} A_n = A \cup B.$$

1.2* Fields and σ-Fields

In this section, we introduce the concepts of a field and of a σ-field, present a number of examples, and derive some basic results.

DEFINITION 2 A class (set) of subsets of S is said to be a *field*, and is denoted by $\mathcal{F}$, if

($\mathcal{F}$1) $\mathcal{F}$ is a non-empty class.
($\mathcal{F}$2) $A \in \mathcal{F}$ implies that $A^c \in \mathcal{F}$ (that is, $\mathcal{F}$ is closed under complementation).
($\mathcal{F}$3) $A_1, A_2 \in \mathcal{F}$ implies that $A_1 \cup A_2 \in \mathcal{F}$ (that is, $\mathcal{F}$ is closed under pairwise unions).

1.2.1 Consequences of the Definition of a Field

1. $S, \emptyset \in \mathcal{F}$.
2. If $A_j \in \mathcal{F}, j = 1, 2, \ldots, n$, then $\bigcup_{j=1}^{n} A_j \in \mathcal{F}$, $\bigcap_{j=1}^{n} A_j \in \mathcal{F}$ for any *finite* n.

(That is, $\mathcal{F}$ is closed under finite unions and intersections. Notice, however, that $A_j \in \mathcal{F}, j = 1, 2, \ldots$ need not imply that their union or intersection is in $\mathcal{F}$; for a counterexample, see consequence 2 on page 10.)

PROOF OF (1) AND (2) (1) ($\mathcal{F}$1) implies that there exists $A \in \mathcal{F}$ and ($\mathcal{F}$2) implies that $A^c \in \mathcal{F}$. By ($\mathcal{F}$3), $A \cup A^c = S \in \mathcal{F}$. By ($\mathcal{F}$2), $S^c = \emptyset \in \mathcal{F}$.

(2) The proof will be by induction on n and by one of the De Morgan's laws. By ($\mathcal{F}$3), if $A_1, A_2 \in \mathcal{F}$, then $A_1 \cup A_2 \in \mathcal{F}$; hence the statement for unions is true for $n = 2$. (It is trivially true for $n = 1$.) Now assume the statement for unions is true for $n = k - 1$; that is, if

$$A_1, A_2, \ldots, A_{k-1} \in \mathcal{F}, \quad \text{then} \quad \bigcup_{j=1}^{k-1} A_j \in \mathcal{F}.$$

Consider $A_1, A_2, \ldots, A_k \in \mathcal{F}$. By the associative law for unions of sets,

$$\bigcup_{j=1}^{k} A_j = \left(\bigcup_{j=1}^{k-1} A_j \right) \cup A_k.$$

By the induction hypothesis, $\bigcup_{j=1}^{k-1} A_j \in \mathcal{F}$. Since $A_k \in \mathcal{F}$, ($\mathcal{F}$3) implies that

$$\left(\bigcup_{j=1}^{k-1} A_j \right) \cup A_k = \bigcup_{j=1}^{k} A_j \in \mathcal{F}$$

and by induction, the statement for unions is true for any finite n. By observing that

$$\bigcap_{j=1}^{n} A_j = \left(\bigcup_{j=1}^{n} A_j^c \right)^c,$$

---

* The reader is reminded that sections marked by an asterisk may be omitted without jeopardizing the understanding of the remaining material.

we see that ($\mathcal{F}2$) and the above statement for unions imply that if $A_1, \ldots, A_n \in \mathcal{F}$, then $\bigcap_{j=1}^{n} A_j \in \mathcal{F}$ for any finite n. ▲

1.2.2 Examples of Fields

1. $C_1 = \{\emptyset, S\}$ is a field (*trivial field*).
2. $C_2 = \{$all subsets of $S\}$ is a field (*discrete field*).
3. $C_3 = \{\emptyset, S, A, A^c\}$, for some $\emptyset \subset A \subset S$, is a field.
4. Let S be infinite (countably so or not) and let C_4 be the class of subsets of S which are finite, or whose complements are finite; that is, $C_4 = \{A \subset S; A$ or A^c is finite$\}$.

As an example, we shall verify that C_4 is a field.

PROOF THAT C_4 IS A FIELD

i) Since $S^c = \emptyset$ is finite, $S \in C_4$, so that C_4 is non-empty.
ii) Suppose that $A \in C_4$. Then A or A^c is finite. If A is finite, then $(A^c)^c = A$ is finite and hence $A^c \in C_4$ also. If A^c is finite, then $A^c \in C_4$.
iii) Suppose that $A_1, A_2 \in C_4$. Then A_1 or A_1^c is finite and A_2 or A_2^c is finite.

   a) Suppose that A_1, A_2 are both finite. Then $A_1 \cup A_2$ is finite, so that $A_1 \cup A_2 \in C_4$.
   b) Suppose that A_1^c, A_2 are finite. Then $(A_1 \cup A_2)^c = A_1^c \cap A_2^c$ is finite since A_1^c is. Hence $A_1 \cup A_2 \in C_4$.

The other two possibilities follow just as in (b). Hence ($\mathcal{F}1$), ($\mathcal{F}2$), ($\mathcal{F}3$) are satisfied. ▲

We now formulate and prove the following theorems about fields.

THEOREM 1 Let I be any non-empty index set (finite, or countably infinite, or uncountable), and let $\mathcal{F}_j, j \in I$ be fields of subsets of S. Define $\mathcal{F}$ by $\mathcal{F} = \bigcap_{j \in I} \mathcal{F}_j = \{A; A \in \mathcal{F}_j$ for all $j \in I\}$. Then $\mathcal{F}$ is a field.

PROOF

i) $S, \emptyset \in \mathcal{F}_j$ for every $j \in I$, so that $S, \emptyset \in \mathcal{F}$ and hence $\mathcal{F}$ is non-empty.
ii) If $A \in \mathcal{F}$, then $A \in \mathcal{F}_j$ for every $j \in I$. Thus $A^c \in \mathcal{F}_j$ for every $j \in I$, so that $A^c \in \mathcal{F}$.
iii) If $A_1, A_2 \in \mathcal{F}$, then $A_1, A_2 \in \mathcal{F}_j$ for every $j \in I$. Then $A_1 \cup A_2 \in \mathcal{F}_j$ for every $j \in I$, and hence $A_1 \cup A_2 \in \mathcal{F}$. ▲

THEOREM 2 Let C be an arbitrary class of subsets of S. Then there is a unique minimal field $\mathcal{F}$ containing C. (We say that $\mathcal{F}$ is *generated* by C and write $\mathcal{F} = \mathcal{F}(C)$.)

PROOF Clearly, C is contained in the discrete field. Next, let $\{\mathcal{F}_j, j \in I\}$ be the class of all fields containing C and define $\mathcal{F}(C)$ by

$$\mathcal{F}(C) = \bigcap_{j \in I} \mathcal{F}_j.$$

By Theorem 1, $\mathcal{F}(C)$ is a field containing C. It is obviously the smallest such field, since it is the intersection of all fields containing C, and is unique. Hence $\mathcal{F} = \mathcal{F}(C)$. ▲

DEFINITION 3 A class of subsets of S is said to be a *σ-field*, and is denoted by $\mathcal{A}$, if it is a field and furthermore ($\mathcal{F}3$) is replaced by ($\mathcal{A}3$): If $A_j \in \mathcal{A}, j = 1, 2, \ldots$, then $\bigcup_{j=1}^{\infty} A_j \in \mathcal{A}$ (that is, $\mathcal{A}$ is closed under denumerable unions).

1.2.3 Consequences of the Definition of a σ-Field

1. If $A_j \in \mathcal{A}, j = 1, 2, \ldots$, then $\bigcap_{j=1}^{\infty} A_j \in \mathcal{A}$ (that is, $\mathcal{A}$ is closed under denumerable intersections).

2. By definition, a σ-field is a field, but the converse is not true. In fact, in Example 4 on page 9, take $S = (-\infty, \infty)$, and define $A_j = \{\text{all integers in } [-j, j]\}$, $j = 1, 2, \ldots$. Then $\bigcup_{j=1}^{\infty} A_j$ is the set A, say, of all integers. Thus A is infinite and furthermore so is A^c. Hence $A \notin \mathcal{F}$, whereas $A_j \in \mathcal{F}$ for all j.

1.2.4 Examples of σ-Fields

1. $C_1 = \{\emptyset, S\}$ is a σ-field (*trivial σ-field*).
2. $C_2 = \{\text{all subsets of } S\}$ is a σ-field (*discrete σ-field*).
3. $C_3 = \{\emptyset, S, A, A^c\}$ for some $\emptyset \subset A \subset S$ is a σ-field.
4. Take S to be uncountable and define C_4 as follows:
$C_4 = \{\text{all subsets of } S \text{ which are countable or whose complements are countable}\}$.

As an example, we prove that C_4 is a σ-field.

PROOF

i) $S = \emptyset$ is countable, so C_4 is non-empty.

ii) If $A \in C_4$, then A or A^c is countable. If A is countable, then $(A^c)^c = A$ is countable, so that $A^c \in C_4$. If A^c is countable, by definition $A^c \in C_4$.

iii) The proof of this statement requires knowledge of the fact that a countable union of countable sets is countable. (For proof of this fact see page 36 in Tom M. Apostol's book *Mathematical Analysis*, published by Addison-Wesley, 1957.) Let $A_j, j = 1, 2, \ldots \in \mathcal{A}$. Then either each A_j is countable, or there exists some A_j for which A_j is not countable but A_j^c is. In the first case, we invoke the previously mentioned theorem on the countable union of countable sets. In the second case, we note that

$$\left(\bigcup_{j=1}^{\infty} A_j \right)^c = \bigcap_{j=1}^{\infty} A_j^c,$$

which is countable, since it is the intersection of sets, one of which is countable. ▲

We now introduce some useful theorems about σ-fields.

THEOREM 3 Let I be as in Theorem 1, and let $\mathcal{A}_j$, $j \in I$, be σ-fields. Define $\mathcal{A}$ by $\mathcal{A} = \bigcap_{j \in I} \mathcal{A}_j = \{A;\ A \in \mathcal{A}_j \text{ for all } j \in I\}$. Then $\mathcal{A}$ is a σ-field.

PROOF

i) $S, \emptyset \in \mathcal{A}_j$ for every $j \in I$ and hence they belong in $\mathcal{A}$.

ii) If $A \in \mathcal{A}$, then $A \in \mathcal{A}_j$ for every $j \in I$, so that $A^c \in \mathcal{A}_j$ for every $j \in I$. Thus $A^c \in \mathcal{A}$.

iii) If $A_1, A_2, \ldots, \in \mathcal{A}$, then $A_1, A_2, \ldots \in \mathcal{A}_j$ for every $j \in I$ and hence $\bigcup_{j=1}^{\infty} A_j \in \mathcal{A}_j$; for every $j \in I$; thus $\bigcup_{j=1}^{\infty} A_j \in \mathcal{A}$. ▲

THEOREM 4 Let C be an arbitrary class of subsets of S. Then there is a unique minimal σ-field $\mathcal{A}$ containing C. (We say that $\mathcal{A}$ is the σ-field *generated* by C and write $\mathcal{A} = \sigma(C)$.)

PROOF Clearly, C is contained in the discrete σ-field. Define

$$\sigma(C) = \bigcap \{\text{all σ-fields containing } C\}.$$

By Theorem 3, $\sigma(C)$ is a σ-field which obviously contains C. Uniqueness and minimality again follow from the definition of $\sigma(C)$. Hence $\mathcal{A} = \sigma(C)$. ▲

REMARK 1 For later use, we note that if $\mathcal{A}$ is a σ-field and $A \in \mathcal{A}$, then $\mathcal{A}_A = \{C;\ C = B \cap A \text{ for some } B \in \mathcal{A}\}$ is a σ-field, where complements of sets are formed with respect to A, which now plays the role of the entire space. This is easily seen to be true by the distributive property of intersection over union (see also Exercise 1.2.5).

In all that follows, if S is countable (that is, finite or denumerably infinite), we will always take $\mathcal{A}$ to be the discrete σ-field. However, if S is uncountable, then for certain technical reasons, we take the σ-field to be "smaller" than the discrete one. In both cases, the pair $(S, \mathcal{A})$ is called a *measurable space*.

1.2.5 Special Cases of Measurable Spaces

1. Let S be $\mathbb{R}$ (the set of real numbers, or otherwise known as the real line) and define C_0 as follows:

$$C_0 = \{\text{all intervals in } \mathbb{R}\} = \left\{\begin{array}{l} (-\infty, x),\ (-\infty, x],\ (x, \infty),\ [x, \infty),\ (x, y), \\ (x, y],\ [x, y),\ [x, y];\ x, y \in \mathbb{R},\ x < y \end{array}\right\}.$$

By Theorem 4, there is a σ-field $\mathcal{A} = \sigma(C_0)$; we denote this σ-field by $\mathcal{B}$ and call it the *Borel σ-field* (over the real line). The pair $(\mathbb{R}, \mathcal{B})$ is called the *Borel real line*.

THEOREM 5 Each one of the following classes generates the Borel σ-field.

$$C_1 = \{(x, y]; x, y \in \mathbb{R}, x < y\},$$
$$C_2 = \{[x, y); x, y \in \mathbb{R}, x < y\},$$
$$C_3 = \{[x, y]; x, y \in \mathbb{R}, x < y\},$$
$$C_4 = \{(x, y); x, y \in \mathbb{R}, x < y\},$$
$$C_5 = \{(x, \infty); x \in \mathbb{R}\},$$
$$C_6 = \{[x, \infty); x \in \mathbb{R}\},$$
$$C_7 = \{(-\infty, x); x \in \mathbb{R}\},$$
$$C_8 = \{(-\infty, x]; x \in \mathbb{R}\}.$$

Also the classes $C'_j, j = 1, \ldots, 8$ generate the Borel σ-field, where for $j = 1, \ldots, 8$, C'_j is defined the same way as C_j is except that x, y are restricted to the rational numbers.

PROOF Clearly, if C, C' are two classes of subsets of S such that $C \subseteq C'$, then $\sigma(C) \subseteq \sigma(C')$. Thus, in order to prove the theorem, it suffices to prove that $\mathcal{B} \subseteq \sigma(C_j), \mathcal{B} \subseteq \sigma(C'_j), j = 1, 2, \ldots, 8$, and in order to prove this, it suffices to show that $C_0 \subseteq \sigma(C_j), C_0 \subseteq \sigma(C'_j), j = 1, 2, \ldots, 8$. As an example, we show that $C_0 \subseteq \sigma(C_7)$. Consider $x_n \downarrow x$. Then $(-\infty, x_n) \in \sigma(C_7)$ and hence $\bigcap_{n=1}^{\infty}(-\infty, x_n) \in \sigma(C_7)$. But

$$\bigcap_{n=1}^{\infty}(-\infty, x_n) = (-\infty, x].$$

Thus $(-\infty, x] \in \sigma(C_7)$ for every $x \in \mathbb{R}$. Since

$$(x, \infty) = (-\infty, x]^c, \quad [x, \infty) = (-\infty, x)^c,$$

it also follows that $(x, \infty), [x, \infty) \in \sigma(C_7)$. Next,

$$(x, y) = (-\infty, y) - (-\infty, x] = (-\infty, y) \cap (x, \infty) \in \sigma(C_7),$$
$$(x, y] = (-\infty, y] \cap (x, \infty) \in \sigma(C_7), \quad [x, y) = (-\infty, y) \cap [x, \infty) \in \sigma(C_7),$$
$$[x, y] = (-\infty, y] \cap [x, \infty) \in \sigma(C_7).$$

Thus $C_0 \subseteq \sigma(C_7)$. In the case of $C'_j, j = 1, 2, \ldots, 8$, consider monotone sequences of rational numbers convergent to given irrationals x, y. ▲

2. Let $S = \mathbb{R} \times \mathbb{R} = \mathbb{R}^2$ and define C_0 as follows:

$$C_0 = \{\text{all rectangles in } \mathbb{R}^2\} = \{(-\infty, x) \times (-\infty, x'), (-\infty, x) \times (-\infty, x'],$$
$$(-\infty, x] \times (-\infty, x'), (-\infty, x] \times (-\infty, x'],$$
$$(x, \infty) \times (x', \infty), \ldots, [x, \infty) \times [x', \infty), \ldots,$$
$$(x, y) \times (x', y'), \ldots, [x, y] \times [x', y'],$$
$$x, y, x', y' \in \mathbb{R}, x < y, x' < y'\}.$$

The σ-field generated by C_0 is denoted by $\mathcal{B}^2$ and is the *two-dimensional Borel σ-field*. A theorem similar to Theorem 5 holds here too.

3. Let $S = \mathbb{R} \times \mathbb{R} \times \cdots \times \mathbb{R} = \mathbb{R}^k$ (k copies of $\mathbb{R}$) and define C_0 in a way similar to that in (2) above. The σ-field generated by C_0 is denoted by $\mathcal{B}^k$ and is the *k-dimensional Borel σ-field*. A theorem similar to Theorem 5 holds here too.

Exercises

1.2.1 Verify that the classes defined in Examples 1, 2 and 3 on page 9 are fields.

1.2.2 Show that in the definition of a field (Definition 2), property ($\mathcal{F}3$) can be replaced by ($\mathcal{F}3'$) which states that if $A_1, A_2 \in \mathcal{F}$, then $A_1 \cap A_2 \in \mathcal{F}$.

1.2.3 Show that in Definition 3, property ($\mathcal{A}3$) can be replaced by ($\mathcal{A}3'$), which states that if

$$A_j \in \mathcal{A}, \quad j = 1, 2, \ldots \quad \text{then} \quad \bigcap_{j=1}^{\infty} A_j \in \mathcal{A}.$$

1.2.4 Refer to Example 4 on σ-fields on page 10 and explain why S was taken to be uncountable.

1.2.5 Give a formal proof of the fact that the class $\mathcal{A}_A$ defined in Remark 1 is a σ-field.

1.2.6 Refer to Definition 1 and show that all three sets $\underline{A}$, $\overline{A}$ and $\lim_{n \to \infty} A_n$, whenever it exists, belong to $\mathcal{A}$ provided $A_n, n \geq 1$, belong to $\mathcal{A}$.

1.2.7 Let $S = \{1, 2, 3, 4\}$ and define the class C of subsets of S as follows:

$$C = \{\emptyset, \{1\}, \{2\}, \{3\}, \{4\}, \{1, 2\}, \{1, 3\}, \{1, 4\}, \{2, 3\}, \{2, 4\},$$
$$\{1, 2, 3\}, \{1, 3, 4\}, \{2, 3, 4\}, S\}.$$

Determine whether or not C is a field.

1.2.8 Complete the proof of the remaining parts in Theorem 5.

Chapter 2

Some Probabilistic Concepts and Results

2.1 Probability Functions and Some Basic Properties and Results

Intuitively by an *experiment* one pictures a procedure being carried out under a certain set of conditions whereby the procedure can be repeated any number of times under the same set of conditions, and upon completion of the procedure certain results are observed. An experiment is a *deterministic experiment* if, given the conditions under which the experiment is carried out, the outcome is completely determined. If, for example, a container of pure water is brought to a temperature of 100°C and 760 mmHg of atmospheric pressure the outcome is that the water will boil. Also, a certificate of deposit of $1,000 at the annual rate of 5% will yield $1,050 after one year, and $$(1.05)^n \times 1,000$ after n years when the (constant) interest rate is compounded. An experiment for which the outcome cannot be determined, except that it is known to be one of a set of possible outcomes, is called a *random experiment*. Only random experiments will be considered in this book. Examples of random experiments are tossing a coin, rolling a die, drawing a card from a standard deck of playing cards, recording the number of telephone calls which arrive at a telephone exchange within a specified period of time, counting the number of defective items produced by a certain manufacturing process within a certain period of time, recording the heights of individuals in a certain class, etc. The set of all possible outcomes of a random experiment is called a *sample space* and is denoted by S. The elements s of S are called *sample points*. Certain subsets of S are called *events*. Events of the form $\{s\}$ are called *simple events*, while an event containing at least two sample points is called a *composite event*. S and $\emptyset$ are always events, and are called the *sure* or *certain event* and the *impossible event*, respectively. The class of all events has got to be sufficiently rich in order to be meaningful. Accordingly, we require that, if A is an event, then so is its complement A^c. Also, if A_j, $j = 1, 2, \ldots$ are events, then so is their union $\bigcup_j A_j$.

2.1 Probability Functions and Some Basic Properties and Results

(In the terminology of Section 1.2, we require that the events associated with a sample space form a *σ-field of subsets* in that space.) It follows then that $\cap_j A_j$ is also an event, and so is $A_1 - A_2$, etc. If the random experiment results in s and $s \in A$, we say that the event A *occurs* or *happens*. The $\cup_j A_j$ occurs if at least one of the A_j occurs, the $\cap_j A_j$ occurs if all A_j occur, $A_1 - A_2$ occurs if A_1 occurs but A_2 does not, etc.

The next basic quantity to be introduced here is that of a probability function (or of a *probability measure*).

DEFINITION 1 A *probability function* denoted by P is a (set) function which assigns to each event A a number denoted by $P(A)$, called the *probability of A*, and satisfies the following requirements:

(P1) P is *non-negative*; that is, $P(A) \geq 0$, for every event A.

(P2) P is *normed*; that is, $P(S) = 1$.

(P3) P is *σ-additive*; that is, for every collection of pairwise (or mutually) disjoint events A_j, $j = 1, 2, \ldots$, we have $P(\Sigma_j A_j) = \Sigma_j P(A_j)$.

This is the axiomatic (Kolmogorov) definition of probability. The triple (S, class of events, P) (or (S, $\mathcal{A}$, P)) is known as a *probability space*.

REMARK 1 If S is finite, then every subset of S is an event (that is, $\mathcal{A}$ is taken to be the discrete σ-field). In such a case, there are only finitely many events and hence, in particular, finitely many pairwise disjoint events. Then (P3) is reduced to:

(P3′) P is *finitely additive*; that is, for every collection of pairwise disjoint events, A_j, $j = 1, 2, \ldots, n$, we have

$$P\left(\sum_{j=1}^{n} A_j\right) = \sum_{j=1}^{n}(A_j).$$

Actually, in such a case it is sufficient to assume that (P3′) holds for any two disjoint events; (P3′) follows then from this assumption by induction.

2.1.1 Consequences of Definition 1

(C1) $P(\varnothing) = 0$. In fact, $S = S + \varnothing + \cdots$,

so that

$$P(S) = P(S + \varnothing + \cdots) = P(S) + P(\varnothing) + \cdots,$$

or

$$1 = 1 + P(\varnothing) + \cdots \quad \text{and} \quad P(\varnothing) = 0,$$

since $P(\varnothing) \geq 0$. (So $P(\varnothing) = 0$. Any event, possibly $\neq \varnothing$, with probability 0 is called a *null* event.)

(C2) P is *finitely additive*; that is for any event A_j, $j = 1, 2, \ldots, n$ such that $A_i \cap A_j = \varnothing$, $i \neq j$,

$$P\left(\sum_{j=1}^{n} A_j\right) = \sum_{j=1}^{n} P(A_j).$$

Indeed, for $A_j = 0, j \geq n+1, P\left(\sum_{j=1}^{n} A_j\right) = P\left(\sum_{j=1}^{\infty} A_j\right) = \sum_{j=1}^{\infty} P(A_j) = \sum_{j=1}^{n} P(A_j)$.

(C3) For every event A, $P(A^c) = 1 - P(A)$. In fact, since $A + A^c = S$,

$$P(A + A^c) = P(S), \quad \text{or} \quad P(A) + P(A^c) = 1,$$

so that $P(A^c) = 1 - P(A)$.

(C4) P is a *non-decreasing function*; that is $A_1 \subseteq A_2$ implies $P(A_1) \leq P(A_2)$. In fact,

$$A_2 = A_1 + (A_2 - A_1),$$

hence

$$P(A_2) = P(A_1) + P(A_2 - A_1),$$

and therefore $P(A_2) \geq P(A_1)$.

REMARK 2 If $A_1 \subseteq A_2$, then $P(A_2 - A_1) = P(A_2) - P(A_1)$, but *this is not true, in general*.

(C5) $0 \leq P(A) \leq 1$ for every event A. This follows from (C1), (P2), and (C4).

(C6) For any events A_1, A_2, $P(A_1 \cup A_2) = P(A_1) + P(A_2) - P(A_1 \cap A_2)$. In fact,

$$A_1 \cup A_2 = A_1 + (A_2 - A_1 \cap A_2).$$

Hence

$$P(A_1 \cup A_2) = P(A_1) + P(A_2 - A_1 \cap A_2)$$
$$= P(A_1) + P(A_2) - P(A_1 \cap A_2),$$

since $A_1 \cap A_2 \subseteq A_2$ implies

$$P(A_2 - A_1 \cap A_2) = P(A_2) - P(A_1 \cap A_2).$$

(C7) P is *subadditive*; that is,

$$P\left(\bigcup_{j=1}^{\infty} A_j\right) \leq \sum_{j=1}^{\infty} P(A_j)$$

and also

$$P\left(\bigcup_{j=1}^{n} A_j\right) \leq \sum_{j=1}^{n} P(A_j).$$

This follows from the identities

$$\bigcup_{j=1}^{\infty} A_j = A_1 + \left(A_1^c \cap A_2\right) + \cdots + \left(A_1^c \cap \cdots \cap A_{n-1}^c \cap A_n\right) + \cdots,$$

$$\bigcup_{j=1}^{n} A_j = A_1 + \left(A_1^c \cap A_2\right) + \cdots + \left(A_1^c \cap \cdots \cap A_{n-1}^c \cap A_n\right),$$

(P3) and (C2), respectively, and (C4).

A special case of a probability space is the following: Let $S = \{s_1, s_2, \ldots, s_n\}$, let the class of events be the class of all subsets of S, and define P as $P(\{s_j\}) = 1/n$, $j = 1, 2, \ldots, n$. With this definition, P clearly satisfies (P1)–(P3′) and this is the *classical* definition of probability. Such a probability function is called a *uniform probability function*. This definition is adequate as long as S is finite and the simple events $\{s_j\}$, $j = 1, 2, \ldots, n$, may be assumed to be "equally likely," but it breaks down if either of these two conditions is not satisfied. However, this classical definition together with the following *relative frequency (or statistical)* definition of probability served as a motivation for arriving at the axioms (P1)–(P3) in the Kolmogorov definition of probability. The relative frequency definition of probability is this: Let S be any sample space, finite or not, supplied with a class of events A. A random experiment associated with the sample space S is carried out n times. Let $n(A)$ be the number of times that the event A occurs. If, as $n \to \infty$, $\lim[n(A)/n]$ exists, it is called the probability of A, and is denoted by $P(A)$. Clearly, this definition satisfies (P1), (P2) and (P3′).

Neither the classical definition nor the relative frequency definition of probability is adequate for a deep study of probability theory. The relative frequency definition of probability provides, however, an intuitively satisfactory interpretation of the concept of probability.

We now state and prove some general theorems about probability functions.

THEOREM 1 (Additive Theorem) For any finite number of events, we have

$$P\left(\bigcup_{j=1}^{n} A_j\right) = \sum_{j=1}^{n} P(A_j) - \sum_{1 \le j_1 < j_2 \le n} P\left(A_{j_1} \cap A_{j_2}\right) \\ + \sum_{1 \le j_1 < j_2 < j_3 \le n} P\left(A_{j_1} \cap A_{j_2} \cap A_{j_3}\right) \\ - \cdots + (-1)^{n+1} P\left(A_1 \cap A_2 \cap \cdots \cap A_n\right).$$

PROOF (By induction on n). For $n = 1$, the statement is trivial, and we have proven the case $n = 2$ as consequence (C6) of the definition of probability functions. Now assume the result to be true for $n = k$, and prove it for $n = k + 1$.

We have

$$P\left(\bigcup_{j=1}^{k+1} A_j\right) = P\left(\left(\bigcup_{j=1}^{k} A_j\right) \cup A_{k+1}\right)$$

$$= P\left(\bigcup_{j=1}^{k} A_j\right) + P(A_{k+1}) - P\left(\left(\bigcup_{j=1}^{k} A_j\right) \cap A_{k+1}\right)$$

$$= \left[\sum_{j=1}^{k} P(A_j) - \sum_{1 \leq j_1 < j_2 \leq k} P(A_{j_1} \cap A_{j_2})\right.$$

$$+ \sum_{1 \leq j_1 < j_2 < j_3 \leq k} P(A_{j_1} \cap A_{j_2} \cap A_{j_3}) - \cdots$$

$$\left. + (-1)^{k+1} P(A_1 \cap A_2 \cap \cdots \cap A_k)\right] + P(A_{k+1}) - P\left(\bigcup_{j=1}^{k}(A_j \cap A_{k+1})\right)$$

$$= \sum_{j=1}^{k+1} P(A_j) - \sum_{1 \leq j_1 < j_2 \leq k} P(A_{j_1} \cap A_{j_2})$$

$$+ \sum_{1 \leq j_1 < j_2 < j_3 \leq k} P(A_{j_1} \cap A_{j_2} \cap A_{j_3}) - \cdots$$

$$+ (-1)^{k+1} P(A_1 \cap A_2 \cdots \cap A_k) - P\left(\bigcup_{j=1}^{k}(A_j \cap A_{k+1})\right). \tag{1}$$

But

$$P\left(\bigcup_{j=1}^{k}(A_j \cap A_{k+1})\right) = \sum_{j=1}^{k} P(A_j \cap A_{k+1}) - \sum_{1 \leq j_1 < j_2 \leq k} P(A_{j_1} \cap A_{j_2} \cap A_{k+1})$$

$$+ \sum_{1 \leq j_1 < j_2 < j_3 \leq k} P(A_{j_1} \cap A_{j_2} \cap A_{j_3} \cap A_{k+1}) - \cdots$$

$$+ (-1)^{k} \sum_{1 \leq j_1 < j_2 \cdots j_{k-1} \leq k} P(A_{j_1} \cap \cdots \cap A_{j_{k-1}} \cap A_{k+1})$$

$$+ (-1)^{k+1} P(A_1 \cap \cdots \cap A_k \cap A_{k+1}).$$

Replacing this in (1), we get

$$P\left(\bigcup_{j=1}^{k+1} A_j\right) = \sum_{j=1}^{k+1} P(A_j) - \left[\sum_{1 \leq j_1 < j_2 \leq k} P(A_{j_1} \cap A_{j_2}) + \sum_{j=1}^{k} P(A_j \cap A_{k+1})\right]$$

$$+ \left[\sum_{1 \leq j_1 < j_2 < j_3 \leq k} P(A_{j_1} \cap A_{j_2} \cap A_{j_3}) + \sum_{1 \leq j_1 < j_2 \leq k} P(A_{j_1} \cap A_{j_2} \cap A_{k+1})\right]$$

$$- \cdots + (-1)^{k+1} \left[P(A_1 \cap \cdots \cap A_k)\right.$$

$$\left. + \sum_{1 \leq j_1 < j_2 < \cdots < j_{k-1} \leq k} P(A_{j_1} \cap \cdots \cap A_{j_{k-1}} \cap A_{k+1})\right]$$

$$+(-1)^{k+2} P(A_1 \cap \cdots \cap A_k \cap A_{k+1})$$

$$= \sum_{j=1}^{k+1} P(A_j) - \sum_{1 \le j_1 < j_2 \le k+1} P(A_{j_1} \cap A_{j_2})$$

$$+ \sum_{1 \le j_1 < j_2 < j_3 \le k+1} P(A_{j_1} \cap A_{j_2} \cap A_{j_3}) - \cdots$$

$$+(-1)^{k+2} P(A_1 \cap \cdots \cap A_{k+1}). \quad \blacktriangle$$

THEOREM 2 Let $\{A_n\}$ be a sequence of events such that, as $n \to \infty$, $A_n \uparrow$ or $A_n \downarrow$. Then,

$$P\left(\lim_{n \to \infty} A_n\right) = \lim_{n \to \infty} P(A_n).$$

PROOF Let us first assume that $A_n \uparrow$. Then

$$\lim_{n \to \infty} A_n = \bigcup_{j=1}^{\infty} A_j.$$

We recall that

$$\bigcup_{j=1}^{\infty} A_j = A_1 + (A_1^c \cap A_2) + (A_1^c \cap A_2^c \cap A_3) + \cdots$$

$$= A_1 + (A_2 - A_1) + (A_3 - A_2) + \cdots,$$

by the assumption that $A_n \uparrow$. Hence

$$P\left(\lim_{n \to \infty} A_n\right) = P\left(\bigcup_{j=1}^{\infty} A_j\right) = P(A_1) + P(A_2 - A_1)$$

$$+ P(A_3 - A_2) + \cdots + P(A_n - A_{n-1}) + \cdots$$

$$= \lim_{n \to \infty}\left[P(A_1) + P(A_2 - A_1) + \cdots + P(A_n - A_{n-1})\right]$$

$$= \lim_{n \to \infty}\left[P(A_1) + P(A_2) - P(A_1)\right.$$

$$\left. + P(A_3) - P(A_2) + \cdots + P(A_n) - P(A_{n-1})\right]$$

$$= \lim_{n \to \infty} P(A_n).$$

Thus

$$P\left(\lim_{n \to \infty} A_n\right) = \lim_{n \to \infty} P(A_n).$$

Now let $A_n \downarrow$. Then $A_n^c \uparrow$, so that

$$\lim_{n \to \infty} A_n^c = \bigcup_{j=1}^{\infty} A_j^c.$$

Hence

$$P\left(\lim_{n\to\infty} A_n^c\right) = P\left(\bigcup_{j=1}^{\infty} A_j^c\right) = \lim_{n\to\infty} P\left(A_n^c\right),$$

or equivalently,

$$P\left[\left(\bigcap_{j=1}^{\infty} A_j\right)^c\right] = \lim_{n\to\infty}\left[1 - P(A_n)\right], \text{ or } 1 - P\left(\bigcap_{j=1}^{\infty} A_j\right) = 1 - \lim_{n\to\infty} P(A_n).$$

Thus

$$\lim_{n\to\infty} P(A_n) = P\left(\bigcap_{j=1}^{\infty} A_j\right) = P\left(\lim_{n\to\infty} A_n\right),$$

and the theorem is established. ▲

This theorem will prove very useful in many parts of this book.

Exercises

2.1.1 If the events A_j, $j = 1, 2, 3$ are such that $A_1 \subset A_2 \subset A_3$ and $P(A_1) = \frac{1}{4}$, $P(A_2) = \frac{5}{12}$, $P(A_3) = \frac{7}{12}$, compute the probability of the following events:

$$A_1^c \cap A_2, \ A_1^c \cap A_3, \ A_2^c \cap A_3, \ A_1 \cap A_2^c \cap A_3^c, \ A_1^c \cap A_2^c \cap A_3^c.$$

2.1.2 If two fair dice are rolled once, what is the probability that the total number of spots shown is

i) Equal to 5?
ii) Divisible by 3?

2.1.3 Twenty balls numbered from 1 to 20 are mixed in an urn and two balls are drawn successively and without replacement. If x_1 and x_2 are the numbers written on the first and second ball drawn, respectively, what is the probability that:

i) $x_1 + x_2 = 8$?
ii) $x_1 + x_2 \leq 5$?

2.1.4 Let $S = \{x \text{ integer}; 1 \leq x \leq 200\}$ and define the events A, B, and C by:

$$A = \{x \in S;\ x \text{ is divisible by } 7\},$$
$$B = \{x \in S;\ x = 3n + 10 \text{ for some positive integer } n\},$$
$$C = \{x \in S;\ x^2 + 1 \leq 375\}.$$

Compute $P(A)$, $P(B)$, $P(C)$, where P is the equally likely probability function on the events of S.

2.1.5 Let S be the number of all outcomes when flipping a fair coin four times and let P be the uniform probability function on the events of S. Define the events A, B as follows:

$$A = \{s \in S;\ s \text{ contains more } T\text{'s than } H\text{'s}\},$$
$$B = \{s \in S;\ \text{any } T \text{ in } s \text{ precedes every } H \text{ in } s\}.$$

Compute the probabilities $P(A)$, $P(B)$.

2.1.6 Suppose that the events A_j, $j = 1, 2, \ldots$ are such that

$$\sum_{j=1}^{\infty} P(A_j) < \infty.$$

Use Definition 1 in Chapter 1 and Theorem 2 in this chapter in order to show that $P(\overline{A}) = 0$.

2.1.7 Consider the events A_j, $j = 1, 2, \ldots$ and use Definition 1 in Chapter 1 and Theorem 2 herein in order to show that

$$P(\underline{A}) \leq \liminf_{n \to \infty} P(A_n) \leq \limsup_{n \to \infty} P(A_n) \leq P(\overline{A}).$$

2.2 Conditional Probability

In this section, we shall introduce the concepts of conditional probability and stochastic independence. Before the formal definition of conditional probability is given, we shall attempt to provide some intuitive motivation for it. To this end, consider a balanced die and suppose that the sides bearing the numbers 1, 4 and 6 are painted red, whereas the remaining three sides are painted black. The die is rolled once and we are asked for the probability that the upward side is the one bearing the number 6. Assuming the uniform probability function, the answer is, clearly, $\frac{1}{6}$. Next, suppose that the die is rolled once as before and all that we can observe is the color of the upward side but not the number on it (for example, we may be observing the die from a considerable distance, so that the color is visible but not the numbers on the sides). The same question as above is asked, namely, what is the probability that the number on the uppermost side is 6. Again, by assuming the uniform probability function, the answer now is $\frac{1}{3}$. This latter probability is called the conditional probability of the number 6 turning up, given the information that the uppermost side was painted red. Letting B stand for the event that number 6 appears and A for the event that the uppermost side is red, the above-mentioned conditional probability is denoted by $P(B|A)$, and we observe that this is equal to the quotient $P(A \cap B)/P(A)$. Or suppose that, for the purposes of a certain study, we observe two-children families in a certain locality, and record the gender of the children. A sample space for this experiment is the following: $S = \{bb, bg, gb, gg\}$, where b stands for boy and g for girl, and bg, for example, indicates that the boy is older than the girl. Suppose further (although this is not exactly correct) that: $P(\{bb\}) = P(\{bg\}) = P(\{gb\}) = P(\{gg\}) = \frac{1}{4}$, and define the events A and B as follows: A = "children of one gender" = $\{bb, gg\}$, B = "at least one boy" = $\{bb, bg, gb\}$. Then $P(A|B) = P(A \cap B)/P(B) = \frac{1}{3}$.

From these and other examples, one is led to the following definition of conditional probability.

DEFINITION 2 Let A be an event such that $P(A) > 0$. Then the conditional probability, given A, is the (set) function denoted by $P(\cdot|A)$ and defined for every event B as follows:

$$P(B|A) = \frac{P(A \cap B)}{P(A)}.$$

$P(B|A)$ is called the *conditional probability* of B, given A.

The set function $P(\cdot|A)$ is actually a probability function. To see this, it suffices to prove the $P(\cdot|A)$ satisfies (P1)–(P3). We have: $P(B|A) \geq 0$ for every event B, clearly. Next,

$$P(S|A) = \frac{P(S \cap A)}{P(A)} = \frac{P(A)}{P(A)} = 1,$$

and if A_j, $j = 1, 2, \ldots$, are events such that $A_i \cap A_j = \emptyset$, $i \neq j$, we have

$$P\left(\sum_{j=1}^{\infty} A_j \Big| A\right) = \frac{P\left[\left(\sum_{j=1}^{\infty} A_j\right) \cap A\right]}{P(A)} = \frac{P\left[\sum_{j=1}^{\infty}(A_j \cap A)\right]}{P(A)}$$

$$= \frac{1}{P(A)} \sum_{j=1}^{\infty} P(A_j \cap A) = \sum_{j=1}^{\infty} \frac{P(A_j \cap A)}{P(A)} = \sum_{j=1}^{\infty} P(A_j|A).$$

The conditional probability can be used in expressing the probability of the intersection of a finite number of events.

THEOREM 3 (Multiplicative Theorem) Let A_j, $j = 1, 2, \ldots, n$, be events such that

$$P\left(\bigcap_{j=1}^{n-1} A_j\right) > 0.$$

Then

$$P\left(\bigcap_{j=1}^{n} A_j\right) = P(A_n | A_1 \cap A_2 \cap \cdots \cap A_{n-1})$$
$$\times P(A_{n-1} | A_1 \cap \cdots \cap A_{n-2}) \cdots P(A_2 | A_1) P(A_1).$$

(The proof of this theorem is left as an exercise; see Exercise 2.2.4.)

REMARK 3 The value of the above formula lies in the fact that, in general, it is easier to calculate the conditional probabilities on the right-hand side. This point is illustrated by the following simple example.

EXAMPLE 1 An urn contains 10 identical balls (except for color) of which five are black, three are red and two are white. Four balls are drawn without replacement. Find the probability that the first ball is black, the second red, the third white and the fourth black.

Let A_1 be the event that the first ball is black, A_2 be the event that the second ball is red, A_3 be the event that the third ball is white and A_4 be the event that the fourth ball is black. Then

$$P(A_1 \cap A_2 \cap A_3 \cap A_4)$$
$$= P(A_4 | A_1 \cap A_2 \cap A_3) P(A_3 | A_1 \cap A_2) P(A_2 | A_1) P(A_1),$$

and by using the uniform probability function, we have

$$P(A_1) = \frac{5}{10}, \quad P(A_2 | A_1) = \frac{3}{9}, \quad P(A_3 | A_1 \cap A_2) = \frac{2}{8},$$
$$P(A_4 | A_1 \cap A_2 \cap A_3) = \frac{4}{7}.$$

Thus the required probability is equal to $\frac{1}{42} \simeq 0.0238$.

Now let A_j, $j = 1, 2, \ldots$, be events such that $A_i \cap A_j = \emptyset$, $i \neq j$, and $\Sigma_j A_j = S$. Such a collection of events is called a *partition* of S. The partition is *finite* or (denumerably) *infinite*, accordingly, as the events A_j are finitely or (denumerably) infinitely many. For any event, we clearly have:

$$B = \sum_j (B \cap A_j).$$

Hence

$$P(B) = \sum_j (B \cap A_j) = \sum_j P(B | A_j) P(A_j),$$

provided $P(A_j) > 0$, for all j. Thus we have the following theorem.

THEOREM 4 (Total Probability Theorem) Let $\{A_j, j = 1, 2, \ldots\}$ be a partition of S with $P(A_j) > 0$, all j. Then for $B \in \mathcal{A}$, we have $P(B) = \Sigma_j P(B|A_j) P(A_j)$.

This formula gives a way of evaluating $P(B)$ in terms of $P(B|A_j)$ and $P(A_j)$, $j = 1, 2, \ldots$. Under the condition that $P(B) > 0$, the above formula can be "reversed" to provide an expression for $P(A_j|B)$, $j = 1, 2, \ldots$. In fact,

$$P(A_j|B) = \frac{P(A_j \cap B)}{P(B)} = \frac{P(B|A_j)P(A_j)}{P(B)} = \frac{P(B|A_j)P(A_j)}{\sum_i P(B|A_i)P(A_i)}.$$

Thus

THEOREM 5 (Bayes Formula) If $\{A_j, j = 1, 2, \ldots\}$ is a partition of S and $P(A_j) > 0$, $j = 1, 2, \ldots$, and if $P(B) > 0$, then

$$P(A_j|B) = \frac{P(B|A_j)P(A_j)}{\sum_i P(B|A_i)P(A_i)}.$$

REMARK 4 It is important that one checks to be sure that the collection $\{A_j, j \geq 1\}$ forms a partition of S, as only then are the above theorems true.

The following simple example serves as an illustration of Theorems 4 and 5.

EXAMPLE 2 A multiple choice test question lists five alternative answers, of which only one is correct. If a student has done the homework, then he/she is certain to identify the correct answer; otherwise he/she chooses an answer at random. Let p denote the probability of the event A that the student does the homework and let B be the event that he/she answers the question correctly. Find the expression of the conditional probability $P(A|B)$ in terms of p.

By noting that A and A^c form a partition of the appropriate sample space, an application of Theorems 4 and 5 gives

$$P(A|B) = \frac{P(B|A)P(A)}{P(B|A)P(A) + P(B|A^c)P(A^c)} = \frac{1 \cdot p}{1 \cdot p + \frac{1}{5}(1-p)} = \frac{5p}{4p+1}.$$

Furthermore, it is easily seen that $P(A|B) = P(A)$ if and only if $p = 0$ or 1.

For example, for $p = 0.7, 0.5, 0.3$, we find, respectively, that $P(A|B)$ is approximately equal to: 0.92, 0.83 and 0.68.

Of course, there is no reason to restrict ourselves to one partition of S only. We may consider, for example, two partitions $\{A_i, i = 1, 2, \ldots\}$ $\{B_j, j = 1, 2, \ldots\}$. Then, clearly,

$$A_i = \sum_j (A_i \cap B_j), \quad i = 1, 2, \ldots,$$

$$B_j = \sum_i (B_j \cap A_i), \quad j = 1, 2, \ldots,$$

and

$$\{A_i \cap B_j, i = 1, 2, \ldots, j = 1, 2, \ldots\}$$

is a partition of S. In fact,
$$(A_i \cap B_j) \cap (A_{i'} \cap B_{j'}) = \emptyset \quad \text{if} \quad (i,j) \neq (i',j')$$
and
$$\sum_{i,j}(A_i \cap B_j) = \sum_i \sum_j (A_i \cap B_j) = \sum_i A_i = S.$$

The expression $P(A_i \cap B_j)$ is called the *joint probability* of A_i and B_j. On the other hand, from
$$A_i = \sum_j (A_i \cap B_j) \quad \text{and} \quad B_j = \sum_i (A_i \cap B_j),$$
we get
$$P(A_i) = \sum_j P(A_i \cap B_j) = \sum_j P(A_i|B_j)P(B_j),$$
provided $P(B_j) > 0$, $j = 1, 2, \ldots$, and
$$P(B_j) = \sum_i P(A_i \cap B_j) = \sum_i P(B_j|A_i)P(A_i),$$
provided $P(A_i) > 0$, $i = 1, 2, \ldots$. The probabilities $P(A_i)$, $P(B_j)$ are called *marginal probabilities*. We have analogous expressions for the case of more than two partitions of S.

Exercises

2.2.1 If $P(A|B) > P(A)$, then show that $P(B|A) > P(B)$ $(P(A)P(B) > 0)$.

2.2.2 Show that:

i) $P(A^c|B) = 1 - P(A|B)$;
ii) $P(A \cup B|C) = P(A|C) + P(B|C) - P(A \cap B|C)$.

Also show, by means of counterexamples, that the following equations need not be true:

iii) $P(A|B^c) = 1 - P(A|B)$;
iv) $P(C|A + B) = P(C|A) + P(C|B)$.

2.2.3 If $A \cap B = \emptyset$ and $P(A + B) > 0$, express the probabilities $P(A|A + B)$ and $P(B|A + B)$ in terms of $P(A)$ and $P(B)$.

2.2.4 Use induction to prove Theorem 3.

2.2.5 Suppose that a multiple choice test lists n alternative answers of which only one is correct. Let p, A and B be defined as in Example 2 and find $P_n(A|B)$

in terms of n and p. Next show that if p is fixed but different from 0 and 1, then $P_n(A|B)$ increases as n increases. Does this result seem reasonable?

2.2.6 If A_j, $j = 1, 2, 3$ are any events in S, show that $\{A_1, A_1^c \cap A_2, A_1^c \cap A_2^c \cap A_3, (A_1 \cup A_2 \cup A_3)^c\}$ is a partition of S.

2.2.7 Let $\{A_j, j = 1, \ldots, 5\}$ be a partition of S and suppose that $P(A_j) = j/15$ and $P(A|A_j) = (5 - j)/15$, $j = 1, \ldots, 5$. Compute the probabilities $P(A_j|A)$, $j = 1, \ldots, 5$.

2.2.8 A girl's club has on its membership rolls the names of 50 girls with the following descriptions:

>20 blondes, 15 with blue eyes and 5 with brown eyes;
>25 brunettes, 5 with blue eyes and 20 with brown eyes;
>5 redheads, 1 with blue eyes and 4 with green eyes.

If one arranges a blind date with a club member, what is the probability that:

i) The girl is blonde?

ii) The girl is blonde, if it was only revealed that she has blue eyes?

2.2.9 Suppose that the probability that both of a pair of twins are boys is 0.03 and that the probability that they are both girls is 0.26. Given that the probability of a child being a boy is 0.52, what is the probability that:

i) The second twin is a boy, given that the first is a boy?

ii) The second twin is a girl, given that the first is a girl?

2.2.10 Three machines I, II and III manufacture 30%, 30% and 40%, respectively, of the total output of certain items. Of them, 4%, 3% and 2%, respectively, are defective. One item is drawn at random, tested and found to be defective. What is the probability that the item was manufactured by each one of the machines I, II and III?

2.2.11 A shipment of 20 TV tubes contains 16 good tubes and 4 defective tubes. Three tubes are chosen at random and tested successively. What is the probability that:

i) The third tube is good, if the first two were found to be good?

ii) The third tube is defective, if one of the other two was found to be good and the other one was found to be defective?

2.2.12 Suppose that a test for diagnosing a certain heart disease is 95% accurate when applied to both those who have the disease and those who do not. If it is known that 5 of 1,000 in a certain population have the disease in question, compute the probability that a patient actually has the disease if the test indicates that he does. (Interpret the answer by intuitive reasoning.)

2.2.13 Consider two urns U_j, $j = 1, 2$, such that urn U_j contains m_j white balls and n_j black balls. A ball is drawn at random from each one of the two urns and is placed into a third urn. Then a ball is drawn at random from the third urn. Compute the probability that the ball is black.

2.2.14 Consider the urns of Exercise 2.2.13. A balanced die is rolled and if an even number appears, a ball, chosen at random from U_1, is transferred to urn U_2. If an odd number appears, a ball, chosen at random from urn U_2, is transferred to urn U_1. What is the probability that, after the above experiment is performed twice, the number of white balls in the urn U_2 remains the same?

2.2.15 Consider three urns U_j, $j = 1, 2, 3$ such that urn U_j contains m_j white balls and n_j black balls. A ball, chosen at random, is transferred from urn U_1 to urn U_2 (color unnoticed), and then a ball, chosen at random, is transferred from urn U_2 to urn U_3 (color unnoticed). Finally, a ball is drawn at random from urn U_3. What is the probability that the ball is white?

2.2.16 Consider the urns of Exercise 2.2.15. One urn is chosen at random and one ball is drawn from it also at random. If the ball drawn was white, what is the probability that the urn chosen was urn U_1 or U_2?

2.2.17 Consider six urns U_j, $j = 1, \ldots, 6$ such that urn U_j contains m_j (≥ 2) white balls and n_j (≥ 2) black balls. A balanced die is tossed once and if the number j appears on the die, two balls are selected at random from urn U_j. Compute the probability that one ball is white and one ball is black.

2.2.18 Consider k urns U_j, $j = 1, \ldots, k$ each of which contain m white balls and n black balls. A ball is drawn at random from urn U_1 and is placed in urn U_2. Then a ball is drawn at random from urn U_2 and is placed in urn U_3 etc. Finally, a ball is chosen at random from urn U_{k-1} and is placed in urn U_k. A ball is then drawn at random from urn U_k. Compute the probability that this last ball is black.

2.3 Independence

For any events A, B with $P(A) > 0$, we defined $P(B|A) = P(A \cap B)/P(A)$. Now $P(B|A)$ may be $>P(B)$, $<P(B)$, or $= P(B)$. As an illustration, consider an urn containing 10 balls, seven of which are red, the remaining three being black. Except for color, the balls are identical. Suppose that two balls are drawn successively and without replacement. Then (assuming throughout the uniform probability function) the conditional probability that the second ball is red, given that the first ball was red, is $\frac{6}{9}$, whereas the conditional probability that the second ball is red, given that the first was black, is $\frac{7}{9}$. Without any knowledge regarding the first ball, the probability that the second ball is red is $\frac{7}{10}$. On the other hand, if the balls are drawn with replacement, the probability that the second ball is red, given that the first ball was red, is $\frac{7}{10}$. This probability is the same even if the first ball was black. In other words, knowledge of the event which occurred in the first drawing provides no additional information in

calculating the probability of the event that the second ball is red. Events like these are said to be independent.

As another example, revisit the two-children families example considered earlier, and define the events A and B as follows: A = "children of both genders," B = "older child is a boy." Then $P(A) = P(B) = P(B|A) = \frac{1}{2}$. Again knowledge of the event A provides no additional information in calculating the probability of the event B. Thus A and B are independent.

More generally, let A, B be events with $P(A) > 0$. Then if $P(B|A) = P(B)$, we say that the even B is (statistically or stochastically or in the probability sense) *independent* of the event A. If $P(B)$ is also > 0, then it is easily seen that A is also independent of B. In fact,

$$P(A|B) = \frac{P(A \cap B)}{P(B)} = \frac{P(B|A)P(A)}{P(B)} = \frac{P(B)P(A)}{P(B)} = P(A).$$

That is, if $P(A)$, $P(B) > 0$, and one of the events is independent of the other, then this second event is also independent of the first. Thus, independence is a symmetric relation, and we may simply say that A and B are independent. In this case $P(A \cap B) = P(A)P(B)$ and we may take *this* relationship as the definition of independence of A and B. That is,

DEFINITION 3 The events A, B are said to be (*statistically* or *stochastically* or in the *probability sense*) *independent* if $P(A \cap B) = P(A)P(B)$.

Notice that this relationship is true even if one or both of $P(A)$, $P(B) = 0$.

As was pointed out in connection with the examples discussed above, independence of two events simply means that knowledge of the occurrence of one of them helps in no way in re-evaluating the probability that the other event happens. This is true for any two independent events A and B, as follows from the equation $P(A|B) = P(A)$, provided $P(B) > 0$, or $P(B|A) = P(B)$, provided $P(A) > 0$. Events which are intuitively independent arise, for example, in connection with the descriptive experiments of successively drawing balls with replacement from the same urn with always the same content, or drawing cards with replacement from the same deck of playing cards, or repeatedly tossing the same or different coins, etc.

What actually happens in practice is to consider events which are independent in the intuitive sense, and then define the probability function P appropriately to reflect this independence.

The definition of independence generalizes to any finite number of events. Thus:

DEFINITION 4 The events A_j, $j = 1, 2, \ldots, n$ are said to be (*mutually or completely*) *independent* if the following relationships hold:

$$P(A_{j_1} \cap \cdots \cap A_{j_k}) = P(A_{j_1}) \cdots P(A_{j_k})$$

for any $k = 2, \ldots, n$ and $j_1, \ldots, j_k = 1, 2, \ldots, n$ such that $1 \leq j_1 < j_2 < \cdots < j_k \leq n$. These events are said to be *pairwise independent* if $P(A_i \cap A_j) = P(A_i)P(A_j)$ for all $i \neq j$.

It follows that, if the events $A_j, j = 1, 2, \ldots, n$ are mutually independent, then they are pairwise independent. The converse need not be true, as the example below illustrates. Also there are

$$\binom{n}{2} + \binom{n}{3} + \cdots + \binom{n}{n} = 2^n - \binom{n}{1} - \binom{n}{0} = 2^n - n - 1$$

relationships characterizing the independence of $A_j, j = 1, \ldots, n$ and they are *all* necessary. For example, for $n = 3$ we will have:

$$P(A_1 \cap A_2 \cap A_3) = P(A_1)P(A_2)P(A_3),$$
$$P(A_1 \cap A_2) = P(A_1)P(A_2),$$
$$P(A_1 \cap A_3) = P(A_1)P(A_3),$$
$$P(A_2 \cap A_3) = P(A_2)P(A_3).$$

That these four relations are necessary for the characterization of independence of A_1, A_2, A_3 is illustrated by the following examples:

Let $S = \{1, 2, 3, 4\}$, $P(\{1\}) = \cdots = P(\{4\}) = \frac{1}{4}$, and set $A_1 = \{1, 2\}$, $A_2 = \{1, 3\}$, $A_3 = \{1, 4\}$. Then

$$A_1 \cap A_2 = A_1 \cap A_3 = A_2 \cap A_3 = \{1\}, \quad \text{and} \quad A_1 \cap A_2 \cap A_3 = \{1\}.$$

Thus

$$P(A_1 \cap A_2) = P(A_1 \cap A_3) = P(A_2 \cap A_3) = P(A_1 \cap A_2 \cap A_3) = \frac{1}{4}.$$

Next,

$$P(A_1 \cap A_2) = \frac{1}{4} = \frac{1}{2} \cdot \frac{1}{2} = P(A_1)P(A_2),$$
$$P(A_1 \cap A_3) = \frac{1}{4} = \frac{1}{2} \cdot \frac{1}{2} = P(A_1)P(A_3),$$
$$P(A_1 \cap A_3) = \frac{1}{4} = \frac{1}{2} \cdot \frac{1}{2} = P(A_2)P(A_3),$$

but

$$P(A_1 \cap A_2 \cap A_3) = \frac{1}{4} \neq \frac{1}{2} \cdot \frac{1}{2} \cdot \frac{1}{2} = P(A_1)P(A_2)P(A_3).$$

Now let $S = \{1, 2, 3, 4, 5\}$, and define P as follows:

$$P(\{1\}) = \frac{1}{8}, \quad P(\{2\}) = P(\{3\}) = P(\{4\}) = \frac{3}{16}, \quad P(\{5\}) = \frac{5}{16}.$$

Let
$$A_1 = \{1, 2, 3\}, \quad A_2 = \{1, 2, 4\}, \quad A_3 = \{1, 3, 4\}.$$
Then
$$A_1 \cap A_2 = \{1, 2\}, \quad A_1 \cap A_2 \cap A_3 = \{1\}.$$
Thus
$$P(A_1 \cap A_2 \cap A_3) = \frac{1}{8} = \frac{1}{2} \cdot \frac{1}{2} \cdot \frac{1}{2} = P(A_1)P(A_2)P(A_3),$$
but
$$P(A_1 \cap A_2) = \frac{5}{16} \neq \frac{1}{2} \cdot \frac{1}{2} = P(A_1)P(A_2).$$

The following result, regarding independence of events, is often used by many authors without any reference to it. It is the theorem below.

THEOREM 6 If the events $A_1, \ldots, A_n$ are independent, so are the events $A'_1, \ldots, A'_n$, where A'_j is either A_j or A_j^c, $j = 1, \ldots, n$.

PROOF The proof is done by (a double) induction. For $n = 2$, we have to show that $P(A'_1 \cap A'_2) = P(A'_1)P(A'_2)$. Indeed, let $A'_1 = A_1$ and $A'_2 = A_2^c$. Then $P(A'_1 \cap A'_2) = P(A_1 \cap A_2^c) = P[A_1 \cap (S - A_2)] = P(A_1 - A_1 \cap A_2) = P(A_1) - P(A_1 \cap A_2) = P(A_1) - P(A_1)P(A_2) = P(A_1)[1 - P(A_2)] = P(A_1)P(A_2^c) = P(A'_1)P(A'_2)$. Similarly if $A'_1 = A_1^c$ and $A'_2 = A_2$. For $A'_1 = A_1^c$ and $A'_2 = A_2^c$, $P(A'_1 \cap A'_2) = P(A_1^c \cap A_2^c) = P[(S - A_1) \cap A_2^c] = P(A_2^c - A_1 \cap A_2^c) = P(A_2^c) - P(A_1 \cap A_2^c) = P(A_2^c) - P(A_1)P(A_2^c) = P(A_2^c)[1 - P(A_1)] = P(A_2^c)P(A_1^c) = P(A'_1)P(A'_2)$.

Next, assume the assertion to be true for k events and show it to be true for $k + 1$ events. That is, we suppose that $P(A'_1 \cap \cdots \cap A'_k) = P(A'_1) \cdots P(A'_k)$, and we shall show that $P(A'_1 \cap \cdots \cap A'_{k+1}) = P(A'_1) \cdots P(A'_{k+1})$. First, assume that $A'_{k+1} = A_{k+1}$, and we have to show that

$$P(A'_1 \cap \cdots \cap A'_{k+1}) = P(A'_1 \cap \cdots \cap A'_k \cap A_{k+1})$$
$$= P(A'_1) \cdots P(A'_k)P(A_{k+1}).$$

This relationship is established also by induction as follows: If $A'_1 = A_1^c$ and $A'_i = A_i$, $i = 2, \ldots, k$, then

$$P(A_1^c \cap A_2 \cap \cdots \cap A_k \cap A_{k+1}) = P[(S - A_1) \cap A_2 \cap \cdots \cap A_k \cap A_{k+1}]$$
$$= P(A_2 \cap \cdots \cap A_k \cap A_{k+1} - A_1 \cap A_2 \cap \cdots \cap A_k \cap A_{k+1})$$
$$= P(A_2 \cap \cdots \cap A_k \cap A_{k+1}) - P(A_1 \cap A_2 \cap \cdots \cap A_k \cap A_{k+1})$$
$$= P(A_2) \cdots P(A_k)P(A_{k+1}) - P(A_1)P(A_2) \cdots P(A_k)P(A_{k+1})$$
$$= P(A_2) \cdots P(A_k)P(A_{k+1})[1 - P(A_1)] = P(A_1^c)P(A_2) \cdots P(A_k)P(A_{k+1}).$$

This is, clearly, true if A_1^c is replaced by any other A_i^c, $i = 2, \ldots, k$. Now, for $\ell < k$, assume that

$$P\left(A_1^c \cap \cdots \cap A_\ell^c \cap A_{\ell+1} \cap \cdots \cap A_{k+1}\right)$$
$$= P\left(A_1^c\right) \cdots P\left(A_\ell^c\right) P\left(A_{\ell+1}\right) \cdots P\left(A_{k+1}\right)$$

and show that

$$P\left(A_1^c \cap \cdots \cap A_\ell^c \cap A_{\ell+1}^c \cap A_{\ell+2} \cap \cdots \cap A_{k+1}\right)$$
$$= P\left(A_1^c\right) \cdots P\left(A_\ell^c\right) P\left(A_{\ell+1}^c\right) P\left(A_{\ell+2}\right) \cdots P\left(A_{k+1}\right).$$

Indeed,

$$P\left(A_1^c \cap \cdots \cap A_\ell^c \cap A_{\ell+1}^c \cap A_{\ell+2} \cap \cdots \cap A_{k+1}\right)$$
$$= P\left[A_1^c \cap \cdots \cap A_\ell^c \cap (S - A_{\ell+1}) \cap A_{\ell+2} \cap \cdots \cap A_{k+1}\right]$$
$$= P\big(A_1^c \cap \cdots \cap A_\ell^c \cap A_{\ell+2} \cap \cdots \cap A_{k+1}$$
$$\quad - A_1^c \cap \cdots \cap A_\ell^c \cap A_{\ell+1} \cap A_{\ell+2} \cap \cdots \cap A_{k+1}\big)$$
$$= P\left(A_1^c \cap \cdots \cap A_\ell^c \cap A_{\ell+2} \cap \cdots \cap A_{k+1}\right)$$
$$\quad - P\left(A_1^c \cap \cdots \cap A_\ell^c \cap A_{\ell+1} \cap A_{\ell+2} \cap \cdots \cap A_{k+1}\right)$$
$$= P\left(A_1^c\right) \cdots P\left(A_\ell^c\right) P\left(A_{\ell+2}\right) \cdots P\left(A_{k+1}\right)$$
$$\quad - P\left(A_1^c\right) \cdots P\left(A_\ell^c\right) P\left(A_{\ell+1}\right) P\left(A_{\ell+2}\right) \cdots P\left(A_{k+1}\right)$$

(by the induction hypothesis)

$$= P\left(A_1^c\right) \cdots P\left(A_\ell^c\right) P\left(A_{\ell+2}\right) \cdots P\left(A_{k+1}\right)\left[1 - P\left(A_{\ell+1}\right)\right]$$
$$= P\left(A_1^c\right) \cdots P\left(A_\ell^c\right) P\left(A_{\ell+2}\right) \cdots P\left(A_{k+1}\right) P\left(A_{\ell+1}^c\right)$$
$$= P\left(A_1^c\right) \cdots P\left(A_\ell^c\right) P\left(A_{\ell+1}^c\right) P\left(A_{\ell+2}\right) \cdots P\left(A_{k+1}\right),$$

as was to be seen. It is also, clearly, true that the same result holds if the ℓ A_i''s which are A_i^c are chosen in any one of the $\binom{k}{\ell}$ possible ways of choosing ℓ out of k. Thus, we have shown that

$$P\left(A_1' \cap \cdots \cap A_k' \cap A_{k+1}\right) = P\left(A_1'\right) \cdots P\left(A_k'\right) P\left(A_{k+1}\right).$$

Finally, under the assumption that

$$P\left(A_1' \cap \cdots \cap A_k'\right) = P\left(A_1'\right) \cdots P\left(A_k'\right),$$

take $A_{k+1}' = A_{k+1}^c$, and show that

$$P\left(A_1' \cap \cdots \cap A_k' \cap A_{k+1}^c\right) = P\left(A_1'\right) \cdots P\left(A_k'\right) P\left(A_{k+1}^c\right).$$

In fact,

$$P(A_1' \cap \cdots \cap A_k' \cap A_{k+1}^c) = P[(A_1' \cap \cdots \cap A_k' \cap (S - A_{k+1}))]$$
$$= P(A_1' \cap \cdots \cap A_k' - A_1' \cap \cdots \cap A_k' \cap A_{k+1})$$
$$= P(A_1' \cap \cdots \cap A_k') - P(A_1' \cap \cdots \cap A_k' \cap A_{k+1})$$
$$= P(A_1') \cdots P(A_k') - P(A_1') \cdots P(A_k')P(A_{k+1})$$

(by the induction hypothesis and what was last proved)

$$= P(A_1') \cdots P(A_k')[1 - P(A_{k+1})]$$
$$= P(A_1') \cdots P(A_k')P(A_{k+1}^c).$$

This completes the proof of the theorem. ▲

Now, for $j = 1, 2$, let $\mathcal{E}_j$ be an experiment having the sample space S_j. One may look at the pair $(\mathcal{E}_1, \mathcal{E}_2)$ of experiments, and then the question arises as to what is the appropriate sample space for this *composite* or *compound experiment*, also denoted by $\mathcal{E}_1 \times \mathcal{E}_2$. If S stands for this sample space, then, clearly, $S = S_1 \times S_2 = \{(s_1, s_2); s_1 \in S_1, s_2 \in S_2\}$. The corresponding events are, of course, subsets of S. The notion of independence also carries over to experiments. Thus, we say that the experiments $\mathcal{E}_1$ and $\mathcal{E}_2$ are *independent* if $P(B_1 \cap B_2) = P(B_1)P(B_2)$ for all events B_1 associated with $\mathcal{E}_1$ alone, and all events B_2 associated $\mathcal{E}_2$ alone.

What actually happens in practice is to start out with two experiments $\mathcal{E}_1$, $\mathcal{E}_2$ which are intuitively independent, such as the descriptive experiments (also mentioned above) of successively drawing balls with replacement from the same urn with always the same content, or drawing cards with replacement from the same deck of playing cards, or repeatedly tossing the same or different coins etc., and have the corresponding probability spaces (S_1, class of events, P_1) and (S_2, class of events, P_2), and then define the probability function P, in terms of P_1 and P_2, on the class of events in the space $S_1 \times S_2$ so that it reflects the intuitive independence.

The above definitions generalize in a straightforward manner to any finite number of experiments. Thus, if $\mathcal{E}_j$, $j = 1, 2, \ldots, n$, are n experiments with corresponding sample spaces S_j and probability functions P_j on the respective classes of events, then the compound experiment

$$(\mathcal{E}_1, \mathcal{E}_2, \ldots, \mathcal{E}_n) = \mathcal{E}_1 \times \mathcal{E}_2 \times \cdots \times \mathcal{E}_n$$

has sample space S, where

$$S = S_1 \times \cdots \times S_n = \{(s_1, \cdots, s_n); s_j \in S_j, j = 1, 2, \ldots, n\}.$$

The class of events are subsets of S, and the experiments are said to be *independent* if for all events B_j associated with experiment $\mathcal{E}_j$ alone, $j = 1, 2, \ldots, n$, it holds

$$P(B_1 \cap \cdots \cap B_n) = P(B_1) \cdots P(B_2).$$

Again, the probability function P is defined, in terms of P_j, $j = 1, 2, \ldots, n$, on the class of events in S so that to reflect the intuitive independence of the experiments $\mathcal{E}_j$, $j = 1, 2, \ldots, n$.

In closing this section, we mention that events and experiments which are not independent are said to be *dependent*.

Exercises

2.3.1 If A and B are disjoint events, then show that A and B are independent if and only if at least one of $P(A)$, $P(B)$ is zero.

2.3.2 Show that if the event A is independent of itself, then $P(A) = 0$ or 1.

2.3.3 If A, B are independent, A, C are independent and $B \cap C = \emptyset$, then A, $B + C$ are independent. Show, by means of a counterexample, that the conclusion need not be true if $B \cap C \neq \emptyset$.

2.3.4 For each $j = 1, \ldots, n$, suppose that the events $A_1, \ldots, A_m$, B_j are independent and that $B_i \cap B_j = \emptyset$, $i \neq j$. Then show that the events $A_1, \ldots, A_m$, $\Sigma_{j=1}^{n} B_j$ are independent.

2.3.5 If A_j, $j = 1, \ldots, n$ are independent events, show that

$$P\left(\bigcup_{j=1}^{n} A_j\right) = 1 - \prod_{j=1}^{n} P(A_j^c).$$

2.3.6 Jim takes the written and road driver's license tests repeatedly until he passes them. Given that the probability that he passes the written test is 0.9 and the road test is 0.6 and that tests are independent of each other, what is the probability that he will pass both tests on his nth attempt? (Assume that the road test cannot be taken unless he passes the written test, and that once he passes the written test he does not have to take it again, no matter whether he passes or fails his next road test. Also, the written and the road tests are considered distinct attempts.)

2.3.7 The probability that a missile fired against a target is not intercepted by an antimissile missile is $\frac{2}{3}$. Given that the missile has not been intercepted, the probability of a successful hit is $\frac{3}{4}$. If four missiles are fired independently, what is the probability that:

i) All will successfully hit the target?

ii) At least one will do so?

How many missiles should be fired, so that:

iii) At least one is not intercepted with probability ≥ 0.95?

iv) At least one successfully hits its target with probability ≥ 0.99?

2.3.8 Two fair dice are rolled repeatedly and independently. The first time a total of 10 appears, player A wins, while the first time that a total of 6 appears, player B wins, and the game is terminated. Compute the probabilities that:

i) The game terminates on the nth throw and player A wins;

ii) The same for player B;

iii) Player A wins;

iv) Player B wins;

v) Does the game terminate ever?

2.3.9 Electric current is transmitted from point A to point B provided at least one of the circuits #1 through #n below is closed. If the circuits close independently of each other and with respective probabilities p_i, $i = 1, \ldots, n$, determine the probability that:

i) Exactly one circuit is closed;

ii) At least one circuit is closed;

iii) Exactly m circuits are closed for $0 \leq m \leq n$;

iv) At least m circuits are closed with m as in part (iii);

v) What do parts (i)–(iv) become for $p_1 = \cdots = p_n = p$, say?

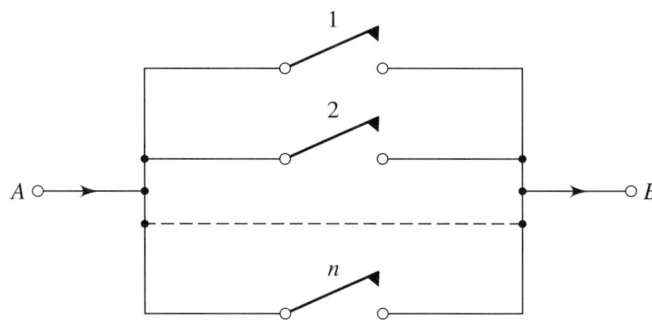

2.4 Combinatorial Results

In this section, we will restrict ourselves to finite sample spaces and uniform probability functions. Some combinatorial results will be needed and we proceed to derive them here. Also examples illustrating the theorems of previous sections will be presented.

2.4 Combinatorial Results

The following theorem, known as the *Fundamental Principle of Counting*, forms the backbone of the results in this section.

THEOREM 7 Let a task T be completed by carrying out all of the subtasks $T_j, j = 1, 2, \ldots, k$, and let it be possible to perform the subtask T_j in n_j (different) ways, $j = 1, 2, \ldots, k$. Then the total number of ways the task T may be performed is given by $\prod_{j=1}^{k} n_j$.

PROOF The assertion is true for $k = 2$, since by combining each one of the n_1 ways of performing subtask T_1 with each one of the n_2 ways of performing subtask T_2, we obtain $n_1 n_2$ as the total number of ways of performing task T. Next, assume the result to be true for $k = m$ and establish it for $k = m + 1$. The reasoning is the same as in the step just completed, since by combining each one of the $\prod_{j=1}^{m} n_j$ ways of performing the first m subtasks with each one of n_{m+1} ways of performing substask T_{m+1}, we obtain $\left(\prod_{j=1}^{m} n_j\right) \times n_{m+1} = \prod_{j=1}^{m+1} n_j$ for the total number of ways of completing task T. ▲

The following examples serve as an illustration to Theorem 7.

EXAMPLE 3
i) A man has five suits, three pairs of shoes and two hats. Then the number of different ways he can attire himself is $5 \cdot 3 \cdot 2 = 30$.

ii) Consider the set $S = \{1, \ldots, N\}$ and suppose that we are interested in finding the number of its subsets. In forming a subset, we consider for each element whether to include it or not. Then the required number is equal to the following product of N factors $2 \cdots 2 = 2^N$.

iii) Let $n_j = n(S_j)$ be the number of points of the sample space $S_j, j = 1, 2, \ldots, k$. Then the sample space $S = S_1 \times \cdots \times S_k$ has $n(S) = n_1 \cdots n_k$ sample points. Or, if n_j is the number of outcomes of the experiment $\mathcal{E}_j, j = 1, 2, \ldots, k$, then the number of outcomes of the compound experiment $\mathcal{E}_1 \times \ldots \times \mathcal{E}_k$ is $n_1 \ldots n_k$.

In the following, we shall consider the problems of selecting balls from an urn and also placing balls into cells which serve as general models of many interesting real life problems. The main results will be formulated as theorems and their proofs will be applications of the Fundamental Principle of Counting.

Consider an urn which contains n numbered (distinct, but otherwise identical) balls. If k balls are drawn from the urn, we say that a *sample of size k* was drawn. The sample is *ordered* if the order in which the balls are drawn is taken into consideration and *unordered* otherwise. Then we have the following result.

THEOREM 8
i) The number of *ordered* samples of size k is $n(n-1) \cdots (n-k+1) = P_{n,k}$ (*permutations* of k objects out of n, and in particular, if $k = n$, $P_{n,n} = 1 \cdot 2 \cdots n = n!$), provided the sampling is done *without replacement*; and is equal to n^k if the sampling is done *with replacement*.

ii) The number of *unordered* samples of size k is

$$\frac{P_{n,k}}{k!} = C_{n,k} = \binom{n}{k} = \frac{n!}{k!(n-k)!}$$

if the sampling is done *without replacement*; and is equal to

$$N(n, k) = \binom{n+k-1}{k}$$

if the sampling is done *with replacement*. [See also Theorem 9(iii).]

PROOF

i) The first part follows from Theorem 6 by taking $n_j = (n - j + 1)$, $j = 1, \ldots, k$, and the second part follows from the same theorem by taking $n_j = n$, $j = 1, \ldots, k$.

ii) For the first part, we have that, if order counts, this number is $P_{n,k}$. Since for every sample of size k one can form $k!$ ordered samples of the same size, if x is the required number, then $P_{n,k} = xk!$. Hence the desired result.

The proof of the second part may be carried out by an appropriate induction method. However, we choose to present the following short alternative proof which is due to S. W. Golomb and appeared in the *American Mathematical Monthly*, 75, 1968, p. 530. For clarity, consider the n balls to be cards numbered from 1 to n and adjoin $k - 1$ extra cards numbered from $n + 1$ to $n + k - 1$ and bearing the respective instructions: "repeat lowest numbered card," "repeat 2nd lowest numbered card," ..., "repeat $(k - 1)$st lowest numbered card." Then a sample of size k without replacement from this enlarged $(n + k - 1)$-card deck corresponds uniquely to a sample of size k from the original deck with replacement. (That is, take k out of $n + k - 1$, without replacement so that there will be at least one out of $1, 2, \ldots, n$, and then apply the instructions.) Thus, by the first part, the required number is

$$\binom{n+k-1}{k} = N(n, k),$$

as was to be seen. ▲

For the sake of illustration of Theorem 8, let us consider the following examples.

EXAMPLE 4 (i) Form all possible three digit numbers by using the numbers 1, 2, 3, 4, 5.
(ii) Find the number of all subsets of the set $S = \{1, \ldots, N\}$.

In part (i), clearly, the order in which the numbers are selected is relevant. Then the required number is $P_{5,3} = 5 \cdot 4 \cdot 3 = 60$ without repetitions, and $5^3 = 125$ with repetitions.

In part (ii) the order is, clearly, irrelevant and the required number is $\binom{N}{0} + \binom{N}{1} + \cdots + \binom{N}{N} = 2^N$, as already found in Example 3.

EXAMPLE 5 An urn contains 8 balls numbered 1 to 8. Four balls are drawn. What is the probability that the smallest number is 3?

Assuming the uniform probability function, the required probabilities are as follows for the respective four possible sampling cases:

Order does not count/replacements not allowed: $\dfrac{\binom{5}{3}}{\binom{8}{4}} = \dfrac{1}{7} \approx 0.14;$

Order does not count/replacements allowed: $\dfrac{\binom{6+3-1}{3}}{\binom{8+4-1}{4}} = \dfrac{28}{165} \approx 0.17;$

Order counts/replacements not allowed: $\dfrac{(5 \cdot 4 \cdot 3)4}{8 \cdot 7 \cdot 6 \cdot 5} = \dfrac{1}{7} \approx 0.14;$

Order counts/replacements allowed: $\dfrac{\binom{4}{1} \cdot 5^3 + \binom{4}{2} \cdot 5^2 + \binom{4}{3} \cdot 5 + \binom{4}{4}}{84} = \dfrac{671}{4{,}056} \approx 0.16.$

EXAMPLE 6 What is the probability that a poker hand will have exactly one pair?

A poker hand is a 5-subset of the set of 52 cards in a full deck, so there are

$$\binom{52}{5} = N = 2{,}598{,}960$$

different poker hands. We thus let S be a set with N elements and assign the uniform probability measure to S. A poker hand with one pair has two cards of the same face value and three cards whose faces are all different among themselves and from that of the pair. We arrive at a unique poker hand with one pair by completing the following tasks in order:

a) Choose the face value of the pair from the 13 available face values. This can be done in $\binom{13}{1} = 13$ ways.

b) Choose two cards with the face value selected in (a). This can be done in $\binom{4}{2} = 6$ ways.

c) Choose the three face values for the other three cards in the hand. Since there are 12 face values to choose from, this can be done in $\binom{12}{3} = 220$ ways.

d) Choose one card (from the four at hand) of each face value chosen in (c). This can be done in $4 \cdot 4 \cdot 4 = 4^3 = 64$ ways.

Then, by Theorem 6, there are $13 \cdot 6 \cdot 220 \cdot 64 = 1{,}098{,}240$ poker hands with one pair. Hence, by assuming the uniform probability measure, the required probability is equal to

$$\frac{1{,}098{,}240}{2{,}598{,}960} \approx 0.42.$$

THEOREM 9
i) The number of ways in which n *distinct* balls can be distributed into k *distinct* cells is k^n.

ii) The number of ways that n *distinct* balls can be distributed into k *distinct* cells so that the jth cell contains n_j balls ($n_j \geq 0$, $j = 1, \ldots, k$, $\Sigma_{j=1}^{k} n_j = n$) is

$$\frac{n!}{n_1! n_2! \cdots n_k!} = \binom{n}{n_1, n_2, \ldots, n_k}.$$

iii) The number of ways that n *indistinguishable* balls can be distributed into k *distinct* cells is

$$\binom{k+n-1}{n}.$$

Furthermore, if $n \geq k$ and no cell is to be empty, this number becomes

$$\binom{n-1}{k-1}.$$

PROOF

i) Obvious, since there are k places to put each of the n balls.

ii) This problem is equivalent to partitioning the n balls into k groups, where the jth group contains exactly n_j balls with n_j as above. This can be done in the following number of ways:

$$\binom{n}{n_1}\binom{n-n_1}{n_2}\cdots\binom{n-n_1-\cdots-n_{k-1}}{n_k} = \frac{n!}{n_1! n_2! \cdots n_k!}.$$

iii) We represent the k cells by the k spaces between $k+1$ vertical bars and the n balls by n stars. By fixing the two extreme bars, we are left with $k + n - 1$ bars and stars which we may consider as $k + n - 1$ spaces to be filled in by a bar or a star. Then the problem is that of selecting n spaces for the n stars which can be done in $\binom{k+n-1}{n}$ ways. As for the second part, we now have the condition that there should not be two adjacent bars. The n stars create $n - 1$ spaces and by selecting $k - 1$ of them in $\binom{n-1}{k-1}$ ways to place the $k - 1$ bars, the result follows. ▲

2.4 Combinatorial Results

REMARK 5

i) The numbers n_j, $j = 1, \ldots, k$ in the second part of the theorem are called *occupancy* numbers.

ii) The answer to (ii) is also the answer to the following different question: Consider n numbered balls such that n_j are identical among themselves and distinct from all others, $n_j \geq 0$, $j = 1, \ldots, k$, $\sum_{j=1}^{k} n_j = n$. Then the number of different permutations is

$$\binom{n}{n_1, n_2, \ldots, n_k}.$$

Now consider the following examples for the purpose of illustrating the theorem.

EXAMPLE 7 Find the probability that, in dealing a bridge hand, each player receives one ace.

The number of possible bridge hands is

$$N = \binom{52}{13, 13, 13, 13} = \frac{52!}{(13!)^4}.$$

Our sample space S is a set with N elements and assign the uniform probability measure. Next, the number of sample points for which each player, North, South, East and West, has one ace can be found as follows:

a) Deal the four aces, one to each player. This can be done in

$$\binom{4}{1, 1, 1, 1} = \frac{4!}{1! \, 1! \, 1! \, 1!} = 4! \text{ ways.}$$

b) Deal the remaining 48 cards, 12 to each player. This can be done in

$$\binom{48}{12, 12, 12, 12} = \frac{48!}{(12!)^4} \text{ ways.}$$

Thus the required number is $4!48!/(12!)^4$ and the desired probability is $4!48!(13!)^4/[(12!)^4(52!)]$. Furthermore, it can be seen that this probability lies between 0.10 and 0.11.

EXAMPLE 8 The eleven letters of the word MISSISSIPPI are scrambled and then arranged in some order.

i) What is the probability that the four I's are consecutive letters in the resulting arrangement?

There are eight possible positions for the first I and the remaining seven letters can be arranged in $\binom{7}{1, 4, 2}$ distinct ways. Thus the required probability is

$$\frac{8\binom{7}{1,\,4,\,2}}{\binom{11}{1,\,4,\,4,\,2}} = \frac{4}{165} \approx 0.02.$$

ii) What is the conditional probability that the four I's are consecutive (event A), given B, where B is the event that the arrangement starts with M and ends with S?

Since there are only six positions for the first I, we clearly have

$$P(A|B) = \frac{6\binom{5}{2}}{\binom{9}{4,\,3,\,2}} = \frac{1}{21} \approx 0.05.$$

iii) What is the conditional probability of A, as defined above, given C, where C is the event that the arrangement ends with four consecutive S's?

Since there are only four positions for the first I, it is clear that

$$P(A|C) = \frac{4\binom{3}{2}}{\binom{7}{1,\,2,\,4}} = \frac{4}{35} \approx 0.11.$$

Exercises

2.4.1 A combination lock can be unlocked by switching it to the left and stopping at digit a, then switching it to the right and stopping at digit b and, finally, switching it to the left and stopping at digit c. If the distinct digits a, b and c are chosen from among the numbers $0, 1, \ldots, 9$, what is the number of possible combinations?

2.4.2 How many distinct groups of n symbols in a row can be formed, if each symbol is either a dot or a dash?

2.4.3 How many different three-digit numbers can be formed by using the numbers $0, 1, \ldots, 9$?

2.4.4 Telephone numbers consist of seven digits, three of which are grouped together, and the remaining four are also grouped together. How many numbers can be formed if:

i) No restrictions are imposed?

ii) If the first three numbers are required to be 752?

2.4.5 A certain state uses five symbols for automobile license plates such that the first two are letters and the last three numbers. How many license plates can be made, if:

i) All letters and numbers may be used?

ii) No two letters may be the same?

2.4.6 Suppose that the letters C, E, F, F, I and O are written on six chips and placed into an urn. Then the six chips are mixed and drawn one by one without replacement. What is the probability that the word "OFFICE" is formed?

2.4.7 The 24 volumes of the *Encyclopaedia Britannica* are arranged on a shelf. What is the probability that:

i) All 24 volumes appear in ascending order?

ii) All 24 volumes appear in ascending order, given that volumes 14 and 15 appeared in ascending order and that volumes 1–13 precede volume 14?

2.4.8 If n countries exchange ambassadors, how many ambassadors are involved?

2.4.9 From among n eligible draftees, m men are to be drafted so that all possible combinations are equally likely to be chosen. What is the probability that a specified man is not drafted?

2.4.10 Show that

$$\frac{\binom{n+1}{m+1}}{\binom{n}{m}} = \frac{n+1}{m+1}.$$

2.4.11 Consider five line segments of length 1, 3, 5, 7 and 9 and choose three of them at random. What is the probability that a triangle can be formed by using these three chosen line segments?

2.4.12 From 10 positive and 6 negative numbers, 3 numbers are chosen at random and without repetitions. What is the probability that their product is a negative number?

2.4.13 In how many ways can a committee of $2n + 1$ people be seated along one side of a table, if the chairman must sit in the middle?

2.4.14 Each of the $2n$ members of a committee flips a fair coin in deciding whether or not to attend a meeting of the committee; a committee member attends the meeting if an H appears. What is the probability that a majority will show up in the meeting?

2.4.15 If the probability that a coin falls H is p $(0 < p < 1)$, what is the probability that two people obtain the same number of H's, if each one of them tosses the coin independently n times?

2.4.16

i) Six fair dice are tossed once. What is the probability that all six faces appear?

ii) Seven fair dice are tossed once. What is the probability that every face appears at least once?

2.4.17 A shipment of 2,000 light bulbs contains 200 defective items and 1,800 good items. Five hundred bulbs are chosen at random, are tested and the entire shipment is rejected if more than 25 bulbs from among those tested are found to be defective. What is the probability that the shipment will be accepted?

2.4.18 Show that

$$\binom{M}{m} = \binom{M-1}{m} + \binom{M-1}{m-1},$$

where N, m are positive integers and $m < M$.

2.4.19 Show that

$$\sum_{x=0}^{r} \binom{m}{x}\binom{n}{r-x} = \binom{m+n}{r},$$

where

$$\binom{k}{x} = 0 \quad \text{if} \quad x > k.$$

2.4.20 Show that

i) $\sum_{j=0}^{n} \binom{n}{j} = 2^n;$ ii) $\sum_{j=0}^{n} (-1)^j \binom{n}{j} = 0.$

2.4.21 A student is given a test consisting of 30 questions. For each question there are supplied 5 different answers (of which only one is correct). The student is required to answer correctly at least 25 questions in order to pass the test. If he knows the right answers to the first 20 questions and chooses an answer to the remaining questions at random and independently of each other, what is the probability that he will pass the test?

2.4.22 A student committee of 12 people is to be formed from among 100 freshmen (60 male + 40 female), 80 sophomores (50 male + 30 female), 70 juniors (46 male + 24 female), and 40 seniors (28 male + 12 female). Find the total number of different committees which can be formed under each one of the following requirements:

i) No restrictions are imposed on the formation of the committee;

ii) Seven students are male and five female;

iii) The committee contains the same number of students from each class;

iv) The committee contains two male students and one female student from each class;

v) The committee chairman is required to be a senior;

vi) The committee chairman is required to be both a senior and male;

vii) The chairman, the secretary and the treasurer of the committee are all required to belong to different classes.

2.4.23 Refer to Exercise 2.4.22 and suppose that the committee is formed by choosing its members at random. Compute the probability that the committee to be chosen satisfies each one of the requirements (i)–(vii).

2.4.24 A fair die is rolled independently until all faces appear at least once. What is the probability that this happens on the 20th throw?

2.4.25 Twenty letters addressed to 20 different addresses are placed at random into the 20 envelopes. What is the probability that:

i) All 20 letters go into the right envelopes?

ii) Exactly 19 letters go into the right envelopes?

iii) Exactly 17 letters go into the right envelopes?

2.4.26 Suppose that each one of the 365 days of a year is equally likely to be the birthday of each one of a given group of 73 people. What is the probability that:

i) Forty people have the same birthdays and the other 33 also have the same birthday (which is different from that of the previous group)?

ii) If a year is divided into five 73-day specified intervals, what is the probability that the birthday of: 17 people falls into the first such interval, 23 into the second, 15 into the third, 10 into the fourth and 8 into the fifth interval?

2.4.27 Suppose that each one of n sticks is broken into one long and one short part. Two parts are chosen at random. What is the probability that:

i) One part is long and one is short?

ii) Both parts are either long or short?

The $2n$ parts are arranged at random into n pairs from which new sticks are formed. Find the probability that:

iii) The parts are joined in the original order;

iv) All long parts are paired with short parts.

2.4.28 Derive the third part of Theorem 9 from Theorem 8(ii).

2.4.29 Three cards are drawn at random and with replacement from a standard deck of 52 playing cards. Compute the probabilities $P(A_j)$, $j = 1, \ldots, 5$, where the events A_j, $j = 1, \ldots, 5$ are defined as follows:

$$A_1 = \{s \in S;\ \text{all 3 cards in } s \text{ are black}\},$$
$$A_2 = \{s \in S;\ \text{at least 2 cards in } s \text{ are red}\},$$
$$A_3 = \{s \in S;\ \text{exactly 1 card in } s \text{ is an ace}\},$$
$$A_4 = \{s \in S;\ \text{the first card in } s \text{ is a diamond,}$$
$$\text{the second is a heart and the third is a club}\},$$
$$A_5 = \{s \in S;\ \text{1 card in } s \text{ is a diamond, 1 is a heart and 1 is a club}\}.$$

2.4.30 Refer to Exercise 2.4.29 and compute the probabilities $P(A_j)$, $j = 1, \ldots, 5$ when the cards are drawn at random but without replacement.

2.4.31 Consider hands of 5 cards from a standard deck of 52 playing cards. Find the number of all 5-card hands which satisfy one of the following requirements:

i) Exactly three cards are of one color;

ii) Three cards are of three suits and the other two of the remaining suit;

iii) At least two of the cards are aces;

iv) Two cards are aces, one is a king, one is a queen and one is a jack;

v) All five cards are of the same suit.

2.4.32 An urn contains n_R red balls, n_B black balls and n_W white balls. r balls are chosen at random and with replacement. Find the probability that:

i) All r balls are red;

ii) At least one ball is red;

iii) r_1 balls are red, r_2 balls are black and r_3 balls are white ($r_1 + r_2 + r_3 = r$);

iv) There are balls of all three colors.

2.4.33 Refer to Exercise 2.4.32 and discuss the questions (i)–(iii) for $r = 3$ and $r_1 = r_2 = r_3\ (= 1)$, if the balls are drawn at random but without replacement.

2.4.34 Suppose that all 13-card hands are equally likely when a standard deck of 52 playing cards is dealt to 4 people. Compute the probabilities $P(A_j)$, $j = 1, \ldots, 8$, where the events A_j, $j = 1, \ldots, 8$ are defined as follows:

$$A_1 = \{s \in S;\ s \text{ consists of 1 color cards}\},$$
$$A_2 = \{s \in S;\ s \text{ consists only of diamonds}\},$$
$$A_3 = \{s \in S;\ s \text{ consists of 5 diamonds, 3 hearts, 2 clubs and 3 spades}\},$$
$$A_4 = \{s \in S;\ s \text{ consists of cards of exactly 2 suits}\},$$
$$A_5 = \{s \in S;\ s \text{ contains at least 2 aces}\},$$
$$A_6 = \{s \in S;\ s \text{ does not contain aces, tens and jacks}\},$$

$$A_7 = \{s \in S; \ s \text{ consists of 3 aces, 2 kings and exactly 7 red cards}\},$$
$$A_8 = \{s \in S; \ s \text{ consists of cards of all different denominations}\}.$$

2.4.35 Refer to Exercise 2.4.34 and for $j = 0, 1, \ldots, 4$, define the events A_j and also A as follows:

$$A_j = \{s \in S; \ s \text{ contains exactly } j \text{ tens}\},$$
$$A = \{s \in S; \ s \text{ contains exactly 7 red cards}\}.$$

For $j = 0, 1, \ldots, 4$, compute the probabilities $P(A_j)$, $P(A_j|A)$ and also $P(A)$; compare the numbers $P(A_j)$, $P(A_j|A)$.

2.4.36 Let S be the set of all n^3 3-letter words of a language and let P be the equally likely probability function on the events of S. Define the events A, B and C as follows:

$$A = \{s \in S; \ s \text{ begins with a specific letter}\},$$
$$B = \{s \in S; \ s \text{ has the specified letter (mentioned in the definition of } A\text{)}$$
$$\text{in the middle entry}\},$$
$$C = \{s \in S; \ s \text{ has exactly two of its letters the same}\}.$$

Then show that:

i) $P(A \cap B) = P(A)P(B)$;
ii) $P(A \cap C) = P(A)P(C)$;
iii) $P(B \cap C) = P(B)P(C)$;
iv) $P(A \cap B \cap C) \neq P(A)P(B)P(C)$.

Thus the events A, B, C are pairwise independent but not mutually independent.

2.5* Product Probability Spaces

The concepts discussed in Section 2.3 can be stated precisely by utilizing more technical language. Thus, if we consider the experiments $\mathcal{E}_1$ and $\mathcal{E}_2$ with respective probability spaces $(S_1, \mathcal{A}_1, P_1)$ and $(S_2, \mathcal{A}_2, P_2)$, then the compound experiment $(\mathcal{E}_1, \mathcal{E}_2) = \mathcal{E}_1 \times \mathcal{E}_2$ has sample space $S = S_1 \times S_2$ as defined earlier. The appropriate σ-field $\mathcal{A}$ of events in S is defined as follows: First define the class C by:

$$C = \{A_1 \times A_2; \ A_1 \in \mathcal{A}_1, \ A_2 \in \mathcal{A}_2\},$$
$$\text{where } A_1 \times A_2 = \{(s_1, s_2); \ s_1 \in A_1, \ s_2 \in A_2\}.$$

Then $\mathcal{A}$ is taken to be the σ-field generated by C (see Theorem 4 in Chapter 1). Next, define on C the set function P by $P(A_1 \times A_2) = P_1(A_1)P_2(A_2)$. It can be shown that P determines uniquely a probability measure on $\mathcal{A}$ (by means of the so-called Carathéodory extension theorem). This probability measure is usually denoted by $P_1 \times P_2$ and is called the *product probability measure* (with factors P_1 and P_2), and the probability space $(S, \mathcal{A}, P)$ is called the *product probability space* (with factors $(S_j, \mathcal{A}_j, P_j), j = 1, 2$). It is to be noted that events which refer to $\mathcal{E}_1$ alone are of the form $B_1 = A_1 \times S_2$, $A_1 \in \mathcal{A}_1$, and those referring to $\mathcal{E}_2$ alone are of the form $B_2 = S_1 \times A_2$, $A_2 \in \mathcal{A}_2$. The experiments $\mathcal{E}_1$ and $\mathcal{E}_2$ are then said to be *independent* if $P(B_1 \cap B_2) = P(B_1)P(B_2)$ for all events B_1 and B_2 as defined above.

For n experiments $\mathcal{E}_j, j = 1, 2, \ldots, n$ with corresponding probability spaces $(S_j, \mathcal{A}_j, P_j)$, the compound experiment $(\mathcal{E}_1, \ldots, \mathcal{E}_n) = \mathcal{E}_1 \times \cdots \times \mathcal{E}_n$ has probability space $(S, \mathcal{A}, P)$, where

$$S = S_1 \times \cdots \times S_n = \left\{ (s_1, \ldots, s_n);\ s_j \in S_j,\ j = 1, 2, \ldots, n \right\},$$

$\mathcal{A}$ is the σ-field generated by the class C, where

$$C = \left\{ A_1 \times \cdots \times A_n;\ A_j \in \mathcal{A}_j,\ j = 1, 2, \ldots, n \right\},$$

and P is the unique probability measure defined on $\mathcal{A}$ through the relationships

$$P(A_1 \times \cdots \times A_n) = P(A_1) \cdots P(A_n),\ A_j \in \mathcal{A}_j,\ j = 1, 2, \ldots, n.$$

The probability measure P is usually denoted by $P_1 \times \cdots \times P_n$ and is called the *product probability measure* (with factors $P_j, j = 1, 2, \ldots, n$), and the probability space $(S, \mathcal{A}, P)$ is called the *product probability space* (with factors $(S_j, \mathcal{A}_j, P_j), j = 1, 2, \ldots, n$). Then the experiments $\mathcal{E}_j, j = 1, 2, \ldots, n$ are said to be *independent* if $P(B_1 \cap \cdots \cap B_2) = P(B_1) \cdots P(B_2)$, where B_j is defined by

$$B_j = S_1 \times \cdots \times S_{j-1} \times A_j \times S_{j+1} \times \cdots \times S_n,\ j = 1, 2, \ldots, n.$$

The definition of independent events carries over to σ-fields as follows. Let $\mathcal{A}_1, \mathcal{A}_2$ be two sub-σ-fields of $\mathcal{A}$. We say that $\mathcal{A}_1, \mathcal{A}_2$ are *independent* if $P(A_1 \cap A_2) = P(A_1)P(A_2)$ for any $A_1 \in \mathcal{A}_1, A_2 \in \mathcal{A}_2$. More generally, the σ-fields $\mathcal{A}_j, j = 1, 2, \ldots, n$ (sub-σ-fields of $\mathcal{A}$) are said to be *independent* if

$$P\left(\bigcap_{j=1}^{n} A_j \right) = \prod_{j=1}^{n} P(A_j) \quad \text{for any} \quad A_j \in \mathcal{A}_j,\ j = 1, 2, \ldots, n.$$

Of course, σ-fields which are not independent are said to be *dependent*.

At this point, notice that the factor σ-fields $\mathcal{A}_j, j = 1, 2, \ldots, n$ may be considered as sub-σ-fields of the product σ-field $\mathcal{A}$ by identifying A_j with B_j, where the B_j's are defined above. Then independence of the experiments $\mathcal{E}_j$, $j = 1, 2, \ldots, n$ amounts to independence of the corresponding σ-fields $\mathcal{A}_j$, $j = 1, 2, \ldots, n$ (looked upon as sub-σ-fields of the product σ-field $\mathcal{A}$).

Exercises

2.5.1 Form the Cartesian products $A \times B$, $A \times C$, $B \times C$, $A \times B \times C$, where $A = \{\text{stop, go}\}$, $B = \{\text{good, defective}\}$, $C = \{(1, 1), (1, 2), (2, 2)\}$.

2.5.2 Show that $A \times B = \varnothing$ if and only if at least one of the sets A, B is $\varnothing$.

2.5.3 If $A \subseteq B$, show that $A \times C \subseteq B \times C$ for any set C.

2.5.4 Show that

i) $(A \times B)^c = (A \times B^c) + (A^c \times B) + (A^c \times B^c)$;
ii) $(A \times B) \cap (C \times D) = (A \cap C) \times (B \cap D)$;
iii) $(A \times B) \cup (C \times D) = (A \cup C) \times (B \cup D) - [(A \cap C^c) \times (B^c \cap D) + (A^c \cap C) \times (B \cap D^c)]$.

2.6* The Probability of Matchings

In this section, an important result, Theorem 10, is established providing an expression for the probability of occurrence of exactly m events out of possible M events. The theorem is then illustrated by means of two interesting examples. For this purpose, some additional notation is needed which we proceed to introduce. Consider M events A_j, $j = 1, 2, \ldots, M$ and set

$$S_0 = 1,$$
$$S_1 = \sum_{j=1}^{M} P(A_j),$$
$$S_2 = \sum_{1 \leq j_1 < j_2 \leq M} P(A_{j_1} \cap A_{j_2}),$$
$$\vdots$$
$$S_r = \sum_{1 \leq j_1 < j_2 < \cdots < j_r \leq M} P(A_{j_1} \cap A_{j_2} \cap \cdots \cap A_{j_r}),$$
$$\vdots$$
$$S_M = P(A_1 \cap A_2 \cap \cdots \cap A_M).$$

Let also

$$\left.\begin{array}{l} B_m = \text{exactly} \\ C_m = \text{at least} \\ D_m = \text{at most} \end{array}\right\} m \text{ of the events } A_j, j = 1, 2, \ldots, M \text{ occur.}$$

Then we have

THEOREM 10   With the notation introduced above

$$P(B_m) = S_m - \binom{m+1}{m}S_{m+1} + \binom{m+2}{m}S_{m+2} - \cdots + (-1)^{M-m}\binom{M}{m}S_M \qquad (2)$$

which for $m = 0$ is

$$P(B_0) = S_0 - S_1 + S_2 - \cdots + (-1)^M S_M, \qquad (3)$$

and

$$P(C_m) = P(B_m) + P(B_{m+1}) + \cdots + P(B_M), \qquad (4)$$

and

$$P(D_m) = P(B_0) + P(B_1) + \cdots + P(B_m). \qquad (5)$$

For the proof of this theorem, all that one has to establish is (2), since (4) and (5) follow from it. This will be done in Section 5.6 of Chapter 5. For a proof where S is discrete the reader is referred to the book *An Introduction to Probability Theory and Its Applications*, Vol. I, 3rd ed., 1968, by W. Feller, pp. 99–100.

The following examples illustrate the above theorem.

EXAMPLE 9   The matching problem (*case of sampling without replacement*). Suppose that we have M urns, numbered 1 to M. Let M balls numbered 1 to M be inserted randomly in the urns, with one ball in each urn. If a ball is placed into the urn bearing the same number as the ball, a *match* is said to have occurred.

i) Show the probability of at least one match is

$$1 - \frac{1}{2!} + \frac{1}{3!} - \cdots + (-1)^{M+1}\frac{1}{M!} \approx 1 - e^{-1} \approx 0.63$$

for large M, and

ii) exactly m matches will occur, for $m = 0, 1, 2, \ldots, M$ is

$$\frac{1}{m!}\left(1 - 1 + \frac{1}{2!} - \frac{1}{3!} + \cdots + (-1)^{M-m}\frac{1}{(M-m)!}\right)$$

$$= \frac{1}{m!}\sum_{k=0}^{M-m}(-1)^k\frac{1}{k!} \approx \frac{1}{m!}e^{-1} \quad \text{for} \quad M - m \text{ large.}$$

DISCUSSION   To describe the distribution of the balls among the urns, write an M-tuple $(z_1, z_2, \ldots, z_M)$ whose jth component represents the number of the ball inserted in the jth urn. For $k = 1, 2, \ldots, M$, the event A_k that a match will occur in the kth urn may be written $A_k = \{(z_1, \ldots, z_M)' \in \mathbb{R}^M; z_j \text{ integer}, 1 \leq z_j$

2.6* The Probability of Matchings

$\leq M, j = 1, \ldots, M, z_k = k\}$. It is clear that for any integer $r = 1, 2, \ldots, M$ and any r unequal integers $k_1, k_2, \ldots, k_r,$ from 1 to M,

$$P\left(A_{k_1} \cap A_{k_2} \cap \cdots \cap A_{k_r}\right) = \frac{(M-r)!}{M!}.$$

It then follows that S_r is given by

$$S_r = \binom{M}{r} \frac{(M-r)!}{M!} = \frac{1}{r!}.$$

This implies the desired results.

EXAMPLE 10 Coupon collecting (*case of sampling with replacement*). Suppose that a manufacturer gives away in packages of his product certain items (which we take to be coupons), each bearing one of the integers 1 to M, in such a way that each of the M items is equally likely to be found in any package purchased. If n packages are bought, show that the probability that exactly m of the integers, 1 to M, will not be obtained is equal to

$$\binom{M}{m} \sum_{k=0}^{M-m} (-1)^k \binom{M-m}{k} \left(1 - \frac{m+k}{M}\right)^n.$$

Many variations and applications of the above problem are described in the literature, one of which is the following. If n distinguishable balls are distributed among M urns, numbered 1 to M, what is the probability that there will be exactly m urns in which no ball was placed (that is, exactly m urns remain empty after the n balls have been distributed)?

DISCUSSION To describe the coupons found in the n packages purchased, we write an n-tuple $(z_1, z_2, \cdots, z_n)$, whose jth component z_j represents the number of the coupon found in the jth package purchased. We now define the events $A_1, A_2, \ldots, A_M$. For $k = 1, 2, \ldots, M$, A_k is the event that the number k will *not* appear in the sample, that is,

$$A_k = \left\{(z_1, \cdots, z_n)' \in \mathbb{R}^n; z_j \text{ integer}, 1 \leq z_j \leq M, z_j \neq k, j = 1, 2, \ldots, n\right\}.$$

It is easy to see that we have the following results:

$$P(A_k) = \left(\frac{M-1}{M}\right)^n = \left(1 - \frac{1}{M}\right)^n, \quad k = 1, 2, \ldots, M,$$

$$P(A_{k_1} \cap A_{k_2}) = \left(\frac{M-2}{M}\right)^n = \left(1 - \frac{2}{M}\right)^n, \quad \begin{aligned} k_1 &= 1, 2, \ldots, n \\ k_2 &= k_1 + 1, \ldots, n \end{aligned}$$

and, in general,

$$P(A_{k_1} \cap A_{k_2} \cap \cdots \cap A_{k_r}) = \left(1 - \frac{r}{M}\right)^n, \quad \begin{aligned} k_1 &= 1, 2, \ldots, n \\ k_2 &= k_1 + 1, \ldots, n \\ &\vdots \\ k_r &= k_{r-1} + 1, \ldots, n. \end{aligned}$$

Thus the quantities S_r are given by

$$S_r = \binom{M}{r}\left(1 - \frac{r}{M}\right)^n, \quad r = 0, 1, \ldots, M. \tag{6}$$

Let B_m be the event that exactly m of the integers 1 to M will not be found in the sample. Clearly, B_m is the event that exactly m of the events $A_1, \ldots, A_M$ will occur. By relations (2) and (6), we have

$$\begin{aligned}P(B_m) &= \sum_{r=m}^{M} (-1)^{r-m} \binom{r}{m}\binom{M}{r}\left(1 - \frac{r}{M}\right)^n \\ &= \binom{M}{m} \sum_{k=0}^{M-m} (-1)^k \binom{M-m}{k}\left(1 - \frac{m+k}{M}\right)^n, \end{aligned} \tag{7}$$

by setting $r - m = k$ and using the identity

$$\binom{m+k}{m}\binom{M}{m+k} = \binom{M}{m}\binom{M-m}{k}. \tag{8}$$

This is the desired result.

This section is concluded with the following important result stated as a theorem.

THEOREM 11 Let A and B be two disjoint events. Then in a series of independent trials, show that:

$$P(A \text{ occurs before } B \text{ occurs}) = \frac{P(A)}{P(A) + P(B)}.$$

PROOF For $i = 1, 2, \ldots$, define the events A_i and B_i as follows:

$A_i = $ "A occurs on the ith trial," $B_i = $ "B occurs on the ith trial."

Then, clearly, required the event is the sum of the events

$$A_1, \; A_1^c \cap B_1^c \cap A_2, \; A_1^c \cap B_1^c \cap A_2^c \cap B_2^c \cap A_3, \ldots,$$
$$A_1^c \cap B_1^c \cap \cdots \cap A_n^c \cap B_n^c \cap A_{n+1}, \ldots$$

and therefore

2.6* The Probability of Matchings

$P(A \text{ occurs before } B \text{ occurs})$

$$= P\Big[A_1 + \big(A_1^c \cap B_1^c \cap A_2\big) + \big(A_1^c \cap B_1^c \cap A_2^c \cap B_2^c \cap A_3\big)$$
$$+ \cdots + \big(A_1^c \cap B_1^c \cap \cdots \cap A_n^c \cap B_n^c \cap A_{n+1}\big) + \cdots \Big]$$

$$= P(A_1) + P\big(A_1^c \cap B_1^c \cap A_2\big) + P\big(A_1^c \cap B_1^c \cap A_2^c \cap B_2^c \cap A_3\big)$$
$$+ \cdots + P\big(A_1^c \cap B_1^c \cap \cdots \cap A_n^c \cap B_n^c \cap A_{n+1}\big) + \cdots$$

$$= P(A_1) + P\big(A_1^c \cap B_1^c\big)P(A_2) + P\big(A_1^c \cap B_1^c\big)P\big(A_2^c \cap B_2^c\big)P(A_3)$$
$$+ \cdots + P\big(A_1^c \cap B_1^c\big) \cdots P\big(A_n^c \cap B_n^c\big)P(A_{n+1}) + \cdots \quad \text{(by Theorem 6)}$$

$$= P(A) + P\big(A^c \cap B^c\big)P(A) + P^2\big(A^c \cap B^c\big)P(A)$$
$$+ \cdots + P^n\big(A^c \cap B^c\big)P(A) \cdots$$

$$= P(A)\Big[1 + P\big(A^c \cap B^c\big) + P^2\big(A^c \cap B^c\big) + \cdots + P^n\big(A^c \cap B^c\big) + \cdots \Big]$$

$$= P(A)\frac{1}{1 - P\big(A^c \cap B^c\big)}.$$

But

$$P\big(A^c \cap B^c\big) = P\Big[\big(A \cup B\big)^c\Big] = 1 - P(A \cup B)$$
$$= 1 - P(A + B) = 1 - P(A) - P(B),$$

so that

$$1 - P\big(A^c \cap B^c\big) = P(A) + P(B).$$

Therefore

$$P(A \text{ occurs before } B \text{ occurs}) = \frac{P(A)}{P(A) + P(B)},$$

as asserted. ▲

It is possible to interpret B as a catastrophic event, and A as an event consisting of taking certain precautionary and protective actions upon the energizing of a signaling device. Then the significance of the above probability becomes apparent. As a concrete illustration, consider the following simple example (see also Exercise 2.6.3).

EXAMPLE 11 In repeated (independent) draws with replacement from a standard deck of 52 playing cards, calculate the probability that an ace occurs before a picture.

$$\text{Let } A = \text{``an ace occurs,''} \quad B = \text{``a picture occurs.''}$$

Then $P(A) = \frac{4}{52} = \frac{1}{13}$ and $P(B) = \frac{12}{52} = \frac{4}{13}$, so that $P(A$ occurs before B occurs$) = \frac{1}{13} / \frac{4}{13} = \frac{1}{4}$.

Exercises

2.6.1 Show that

$$\binom{m+k}{m}\binom{M}{m+k} = \binom{M}{m}\binom{M-m}{k},$$

as asserted in relation (8).

2.6.2 Verify the transition in (7) and that the resulting expression is indeed the desired result.

2.6.3 Consider the following game of chance. Two fair dice are rolled repeatedly and independently. If the sum of the outcomes is either 7 or 11, the player wins immediately, while if the sum is either 2 or 3 or 12, the player loses immediately. If the sum is either 4 or 5 or 6 or 8 or 9 or 10, the player continues rolling the dice until either the same sum appears before a sum of 7 appears in which case he wins, or until a sum of 7 appears before the original sum appears in which case the player loses. It is assumed that the game terminates the first time the player wins or loses. What is the probability of winning?

Chapter 3

On Random Variables and Their Distributions

3.1 Some General Concepts

Given a probability space (S, class of events, P), the main objective of probability theory is that of calculating probabilities of events which may be of importance to us. Such calculations are facilitated by a transformation of the sample space S, which may be quite an abstract set, into a subset of the real line $\mathbb{R}$ with which we are already familiar. This is, actually, achieved by the introduction of the concept of a random variable. A *random variable* (r.v.) is a function (in the usual sense of the word), which assigns to each sample point $s \in S$ a real number, the value of the r.v. at s. We require that an r.v. be a well-behaving function. This is satisfied by stipulating that r.v.'s are measurable functions. For the precise definition of this concept and related results, the interested reader is referred to Section 3.5 below. Most functions as just defined, which occur in practice are, indeed, r.v.'s, and we leave the matter to rest here. The notation $X(S)$ will be used for the set of values of the r.v. X, the range of X.

Random variables are denoted by the last letters of the alphabet X, Y, Z, etc., with or without subscripts. For a subset B of $\mathbb{R}$, we usually denote by $(X \in B)$ the following event in S: $(X \in B) = \{s \in S; X(s) \in B\}$ for simplicity. In particular, $(X = x) = \{s \in S; X(s) = x\}$. The *probability distribution function* (or just the *distribution*) of an r.v. X is usually denoted by P_X and is a probability function defined on subsets of $\mathbb{R}$ as follows: $P_X(B) = P(X \in B)$. An r.v. X is said to be of the *discrete type* (or just *discrete*) if there are countable (that is, finitely many or denumerably infinite) many points in $\mathbb{R}$, $x_1, x_2, \ldots$, such that $P_X(\{x_j\}) > 0, j \geq 1$, and $\Sigma_j P_X(\{x_j\})(= \Sigma_j P(X = x_j)) = 1$. Then the function f_X defined on the entire $\mathbb{R}$ by the relationships:

$$f_X(x) = P_X(\{x_j\})(= P(X = x_j)) \quad \text{for } x = x_j,$$

and $f_X(x) = 0$ otherwise has the properties:

$$f_X(x) \geq 0 \quad \text{for all } x, \text{ and} \quad \sum_j f_X(x_j) = 1.$$

Furthermore, it is clear that

$$P(X \in B) = \sum_{x_j \in B} f_X(x_j).$$

Thus, instead of striving to calculate the probability of the event $\{s \in S; X(s) \in B\}$, all we have to do is to sum up the values of $f_X(x_j)$ for all those x_j's which lie in B; this assumes, of course, that the function f_X is known. The function f_X is called the *probability density function* (p.d.f.) of X. The distribution of a discrete r.v. will also be referred to as a *discrete distribution*. In the following section, we will see some discrete r.v.'s (distributions) often occurring in practice. They are the Binomial, Poisson, Hypergeometric, Negative Binomial, and the (discrete) Uniform distributions.

Next, suppose that X is an r.v. which takes values in a (finite or infinite but proper) interval I in $\mathbb{R}$ with the following qualification: $P(X = x) = 0$ for every single x in I. Such an r.v. is called an r.v. of the *continuous type* (or just a *continuous* r.v.). Also, it often happens for such an r.v. to have a function f_X satisfying the properties $f_X(x) \geq 0$ for all $x \in I$, and $P(X \in J) = \int_J f_X(x)dx$ for any sub-interval J of I. Such a function is called the *probability density function* (p.d.f.) of X in analogy with the discrete case. It is to be noted, however, that here $f_X(x)$ does *not* represent the $P(X = x)$! A continuous r.v. X with a p.d.f. f_X is called *absolutely continuous* to differentiate it from those continuous r.v.'s which do not have a p.d.f. In this book, however, we are not going to concern ourselves with non-absolutely continuous r.v.'s. Accordingly, the term "continuous" r.v. will be used instead of "absolutely continuous" r.v. Thus, the r.v.'s to be considered will be either discrete or continuous (= absolutely continuous). Roughly speaking, the idea that $P(X = x) = 0$ for all x for a continuous r.v. may be interpreted that X takes on "too many" values for each one of them to occur with positive probability. The fact that $P(X = x)$ also follows formally by the fact that $P(X = x) = \int_x^x f_X(y)dy$, and this is 0. Other interpretations are also possible. It is true, nevertheless, that X takes values in as small a neighborhood of x as we please with positive probability. The distribution of a continuous r.v. is also referred to as a *continuous distribution*. In Section 3.3, we will discuss some continuous r.v.'s (distributions) which occur often in practice. They are the Normal, Gamma, Chi-square, Negative Exponential, Uniform, Beta, Cauchy, and Lognormal distributions. Reference will also be made to t and F r.v.'s (distributions).

Often one is given a function f and is asked whether f is a p.d.f. (of some r.v.). All one has to do is to check whether f is non-negative for all values of its argument, and whether the sum or integral of its values (over the appropriate set) is equal to 1.

When (a well-behaving) function X is defined on a sample space S and takes values in the plane or the three-dimensional space or, more generally, in the k-dimensional space $\mathbb{R}^k$, it is called a *k-dimensional random vector* (r. vector) and is denoted by $\mathbf{X}$. Thus, an r.v. is a one-dimensional r. vector. The distribution of $\mathbf{X}$, $P_\mathbf{X}$, is defined as in the one-dimensional case by simply replacing B with subsets of $\mathbb{R}^k$. The r. vector $\mathbf{X}$ is discrete if $P(\mathbf{X} = \mathbf{x}_j) > 0$, $j = 1, 2, \ldots$ with $\Sigma_j P(\mathbf{X} = \mathbf{x}_j) = 1$, and the function $f_\mathbf{X}(\mathbf{x}) = P(\mathbf{X} = \mathbf{x}_j)$ for $\mathbf{x} = \mathbf{x}_j$, and $f_\mathbf{X}(\mathbf{x}) = 0$ otherwise is the p.d.f. of $\mathbf{X}$. Once again, $P(\mathbf{X} \in B) = \Sigma_{\mathbf{x}_j \in B} f_\mathbf{X}(\mathbf{x}_j)$ for B subsets of $\mathbb{R}^k$. The r. vector $\mathbf{X}$ is (absolutely) *continuous* if $P(\mathbf{X} = \mathbf{x}) = 0$ for all $\mathbf{x} \in I$, but there is a function $f_\mathbf{X}$ defined on $\mathbb{R}^k$ such that:

$$f_\mathbf{X}(\mathbf{x}) \geq 0 \quad \text{for all} \quad \mathbf{x} \in \mathbb{R}^k, \quad \text{and} \quad P(\mathbf{X} \in J) = \int_J f_\mathbf{X}(\mathbf{x}) d\mathbf{x}$$

for any sub-rectangle J of I. The function $f_\mathbf{X}$ is the p.d.f. of $\mathbf{X}$. The distribution of a k-dimensional r. vector is also referred to as a *k-dimensional discrete* or (absolutely) *continuous* distribution, respectively, for a discrete or (absolutely) continuous r. vector. In Sections 3.2 and 3.3, we will discuss two representative multidimensional distributions; namely, the Multinomial (discrete) distribution, and the (continuous) Bivariate Normal distribution.

We will write f rather than $f_\mathbf{X}$ when no confusion is possible. Again, when one is presented with a function f and is asked whether f is a p.d.f. (of some r. vector), all one has to check is non-negativity of f, and that the sum of its values or its integral (over the appropriate space) is equal to 1.

3.2 Discrete Random Variables (and Random Vectors)

3.2.1 Binomial

The Binomial distribution is associated with a *Binomial experiment*; that is, an experiment which results in two possible outcomes, one usually termed as a "success," S, and the other called a "failure," F. The respective probabilities are p and q. It is to be noted, however, that the experiment does not really have to result in two outcomes only. Once some of the possible outcomes are called a "failure," any experiment can be reduced to a Binomial experiment. Here

$$X(S) = \{0, 1, 2, \ldots, n\}, \quad P(X = x) = f(x) = \binom{n}{x} p^x q^{n-x},$$

where $0 < p < 1$, $q = 1 - p$, and $x = 0, 1, 2, \ldots, n$. That this is in fact a p.d.f. follows from the fact that $f(x) \geq 0$ and

$$\sum_{x=0}^{n} f(x) = \sum_{x=0}^{n} \binom{n}{x} p^x q^{n-x} = (p+q)^n = 1^n = 1.$$

The appropriate S here is:

$$S = \{S, F\} \times \cdots \times \{S, F\} \quad (n \text{ copies}).$$

In particular, for $n = 1$, we have the *Bernoulli* or *Point Binomial* r.v. The r.v. X may be interpreted as representing the number of S's ("successes") in the compound experiment $\mathcal{E} \times \cdots \times \mathcal{E}$ (n copies), where $\mathcal{E}$ is the experiment resulting in the sample space $\{S, F\}$ and the n experiments are independent (or, as we say, the n trials are independent). $f(x)$ is the probability that exactly x S's occur. In fact, $f(x) = P(X = x) = P$ (of all n sequences of S's and F's with exactly x S's). The probability of one such a sequence is $p^x q^{n-x}$ by the independence of the trials and this also does not depend on the particular sequence we are considering. Since there are $\binom{n}{x}$ such sequences, the result follows.

The distribution of X is called the *Binomial distribution* and the quantities n and p are called the *parameters of the Binomial distribution*. We denote the Binomial distribution by $B(n, p)$. Often the notation $X \sim B(n, p)$ will be used to denote the fact that the r.v. X is distributed as $B(n, p)$. Graphs of the p.d.f. of the $B(n, p)$ distribution for selected values of n and p are given in Figs. 3.1 and 3.2.

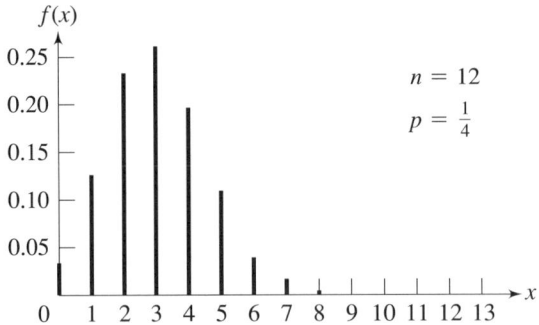

Figure 3.1 Graph of the p.d.f. of the Binomial distribution for $n = 12$, $p = \frac{1}{4}$.

$$f(0) = 0.0317 \qquad f(7) = 0.0115$$
$$f(1) = 0.1267 \qquad f(8) = 0.0024$$
$$f(2) = 0.2323 \qquad f(9) = 0.0004$$
$$f(3) = 0.2581 \qquad f(10) = 0.0000$$
$$f(4) = 0.1936 \qquad f(11) = 0.0000$$
$$f(5) = 0.1032 \qquad f(12) = 0.0000$$
$$f(6) = 0.0401$$

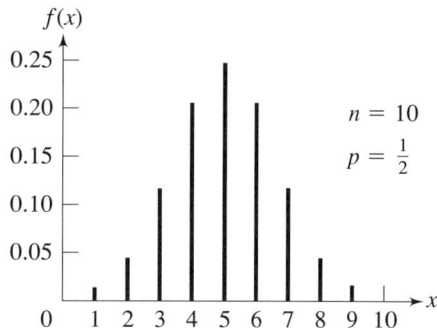

Figure 3.2 Graph of the p.d.f. of the Binomial distribution for $n = 10$, $p = \frac{1}{2}$.

$$f(0) = 0.0010 \qquad f(6) = 0.2051$$
$$f(1) = 0.0097 \qquad f(7) = 0.1172$$
$$f(2) = 0.0440 \qquad f(8) = 0.0440$$
$$f(3) = 0.1172 \qquad f(9) = 0.0097$$
$$f(4) = 0.2051 \qquad f(10) = 0.0010$$
$$f(5) = 0.2460$$

3.2.2 Poisson

$$X(S) = \{0, 1, 2, \ldots\}, \qquad P(X = x) = f(x) = e^{-\lambda}\frac{\lambda^x}{x!},$$

$x = 0, 1, 2, \ldots$; $\lambda > 0$. f is, in fact, a p.d.f., since $f(x) \geq 0$ and

$$\sum_{x=0}^{\infty} f(x) = e^{-\lambda} \sum_{x=0}^{\infty} \frac{\lambda^x}{x!} = e^{-\lambda} e^{\lambda} = 1.$$

The distribution of X is called the *Poisson distribution* and is denoted by $P(\lambda)$. λ is called the *parameter of the distribution*. Often the notation $X \sim P(\lambda)$ will be used to denote the fact that the r.v. X is distributed as $P(\lambda)$. The Poisson distribution is appropriate for predicting the number of phone calls arriving at a given telephone exchange within a certain period of time, the number of particles emitted by a radioactive source within a certain period of time, etc. The reader who is interested in the applications of the Poisson distribution should see W. Feller, *An Introduction to Probability Theory*, Vol. I, 3rd ed., 1968, Chapter 6, pages 156–164, for further examples.

In Theorem 1 in Section 3.4, it is shown that the Poisson distribution may be taken as the limit of Binomial distributions. Roughly speaking, suppose that $X \sim B(n, p)$, where n is large and p is small. Then $P(X = x) = \binom{n}{x} p^x (1-p)^{n-x} \approx e^{-np} \frac{(np)^x}{x!}$, $x \geq 0$. For the graph of the p.d.f. of the $P(\lambda)$ distribution for $\lambda = 5$ see Fig. 3.3.

A visualization of such an approximation may be conceived by stipulating that certain events occur in a time interval $[0,t]$ in the following manner: events occurring in nonoverlapping subintervals are independent; the probability that one event occurs in a small interval is approximately proportional to its length; and two or more events occur in such an interval with probability approximately 0. Then dividing $[0,t]$ into a large number n of small intervals of length t/n, we have that the probability that exactly x events occur in $[0,t]$ is approximately $\binom{n}{x} \left(\frac{\lambda t}{n}\right)^x \left(1 - \frac{\lambda t}{n}\right)^{n-x}$, where λ is the factor of proportionality. Setting $p_n = \frac{\lambda t}{n}$, we have $np_n = \lambda t$ and Theorem 1 in Section 3.4 gives that $\binom{n}{x} \left(\frac{\lambda t}{n}\right)^x \left(1 - \frac{\lambda t}{n}\right)^{n-x} \approx e^{-\lambda t} \frac{(\lambda t)^x}{x!}$. Thus Binomial probabilities are approximated by Poisson probabilities.

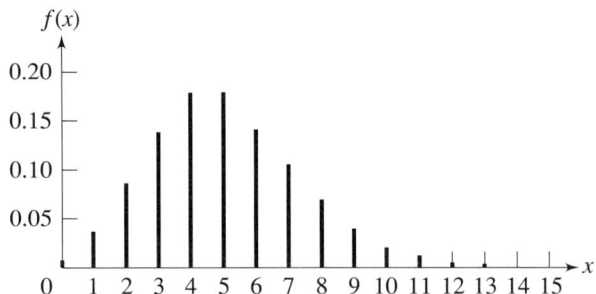

Figure 3.3 Graph of the p.d.f. of the Poisson distribution with $\lambda = 5$.

$f(0) = 0.0067$ $f(9) = 0.0363$

$f(1) = 0.0337$ $f(10) = 0.0181$

$f(2) = 0.0843$ $f(11) = 0.0082$

$f(3) = 0.1403$ $f(12) = 0.0035$

$f(4) = 0.1755$ $f(13) = 0.0013$

$f(5) = 0.1755$ $f(14) = 0.0005$

$f(6) = 0.1462$ $f(15) = 0.0001$

$f(7) = 0.1044$

$f(8) = 0.0653$ $f(n)$ is negligible for $n \geq 16$.

3.2.3 Hypergeometric

$$X(S) = \{0, 1, 2, \ldots, r\}, \quad f(x) = \frac{\binom{m}{x}\binom{n}{r-x}}{\binom{m+n}{r}},$$

where $\binom{m}{x} = 0$, by definition, for $x > m$. f is a p.d.f., since $f(x) \geq 0$ and

$$\sum_{x=0}^{r} f(x) = \frac{1}{\binom{m+n}{r}} \sum_{x=0}^{r} \binom{m}{x}\binom{n}{r-x} = \frac{1}{\binom{m+n}{r}} \binom{m+n}{r} = 1.$$

The distribution of X is called the *Hypergeometric distribution* and arises in situations like the following. From an urn containing m red balls and n black balls, r balls are drawn at random *without replacement*. Then X represents the number of red balls among the r balls selected, and $f(x)$ is the probability that this number is exactly x. Here $S = \{$all r-sequences of R's and B's$\}$, where R stands for a red ball and B stands for a black ball. The urn/balls model just described is a generic model for situations often occurring in practice. For instance, the urn and the balls may be replaced by a box containing certain items manufactured by a certain process over a specified period of time, out of which m are defective and n meet set specifications.

3.2.4 Negative Binomial

$$X(S) = \{0, 1, 2, \ldots\}, \quad f(x) = p^r \binom{r+x-1}{x} q^x,$$

$0 < p < 1$, $q = 1 - p$, $x = 0, 1, 2, \ldots$. f is, in fact, a p.d.f. since $f(x) \geq 0$ and

$$\sum_{x=0}^{\infty} f(x) = p^r \sum_{x=0}^{\infty} \binom{r+x-1}{x} q^x = \frac{p^r}{(1-q)^r} = \frac{p^r}{p^r} = 1.$$

This follows by the Binomial theorem, according to which

$$\frac{1}{(1-x)^n} = \sum_{j=0}^{\infty} \binom{n+j-1}{j} x^j, \quad |x| < 1.$$

The distribution of X is called the *Negative Binomial* distribution. This distribution occurs in situations which have as a model the following. A Binomial experiment $\mathcal{E}$, with sample space $\{S, F\}$, is repeated independently until exactly r S's appear and then it is terminated. Then the r.v. X represents the number of times beyond r that the experiment is required to be carried out, and $f(x)$ is the probability that this number of times is equal to x. In fact, here $S =$

{all $(r + x)$-sequences of S's and F's such that the rth S is at the end of the sequence}, $x = 0, 1, \ldots$ and $f(x) = P(X = x) = P[$all $(r + x)$-sequences as above for a specified $x]$. The probability of one such sequence is $p^{r-1}q^x p$ by the independence assumption, and hence

$$f(x) = \binom{r+x-1}{x} p^{r-1} q^x p = p^r \binom{r+x-1}{x} q^x.$$

The above interpretation also justifies the name of the distribution. For $r = 1$, we get the *Geometric* (or *Pascal*) *distribution*, namely $f(x) = pq^x, x = 0, 1, 2, \ldots$.

3.2.5 Discrete Uniform

$$X(S) = \{0, 1, \ldots, n-1\}, \quad f(x) = \frac{1}{n}, \quad x = 0, 1, \ldots, n-1.$$

This is the uniform probability measure. (See Fig. 3.4.)

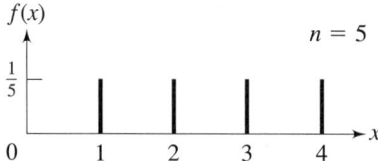

Figure 3.4 Graph of the p.d.f. of a Discrete Uniform distribution.

3.2.6 Multinomial

Here

$$\mathbf{X}(S) = \left\{ \mathbf{x} = (x_1, \ldots, x_k)'; \ x_j \geq 0, j = 1, 2, \ldots, k, \sum_{j=1}^{k} x_j = n \right\},$$

$$f(\mathbf{x}) = \frac{n!}{x_1! x_2! \cdots x_k!} p_1^{x_1} p_2^{x_2} \cdots p_k^{x_k}, \quad p_j > 0, j = 1, 2, \ldots, k, \sum_{j=1}^{k} p_j = 1.$$

That f is, in fact, a p.d.f. follows from the fact that

$$\sum_{\mathbf{X}} f(\mathbf{x}) = \sum_{x_1, \ldots, x_k} \frac{n!}{x_1! \cdots x_k!} p_1^{x_1} \cdots p_k^{x_k} = (p_1 + \cdots + p_k)^n = 1^n = 1,$$

where the summation extends over all x_j's such that $x_j \geq 0$, $j = 1, 2, \ldots, k$, $\sum_{j=1}^{k} x_j = n$. The distribution of $\mathbf{X}$ is also called the *Multinomial distribution* and $n, p_1, \ldots, p_k$ are called the *parameters* of the distribution. This distribution occurs in situations like the following. A Multinomial experiment $\mathcal{E}$ with k possible outcomes O_j, $j = 1, 2, \ldots, k$, and hence with sample space $S = \{$all n-sequences of O_j's$\}$, is carried out n independent times. The probability of the O_j's occurring is p_j, $j = 1, 2, \ldots k$ with $p_j > 0$ and $\sum_{j=1}^{k} p_j = 1$. Then $\mathbf{X}$ is the random vector whose jth component X_j represents the number of times x_j the outcome O_j occurs, $j = 1, 2, \ldots, k$. By setting $\mathbf{x} = (x_1, \ldots, x_k)'$, then f is the probability that the outcome O_j occurs exactly x_j times. In fact $f(\mathbf{x}) = P(\mathbf{X} = \mathbf{x})$ = P("all n-sequences which contain exactly $x_j O_j$'s, $j = 1, 2, \ldots, k$). The probability of each one of these sequences is $p_1^{x_1} \cdots p_k^{x_k}$ by independence, and since there are $n!/(x_1! \cdots x_k!)$ such sequences, the result follows.

The fact that the r. vector $\mathbf{X}$ has the Multinomial distribution with parameters n and $p_1, \ldots, p_k$ may be denoted thus: $\mathbf{X} \sim M(n; p_1, \ldots, p_k)$.

REMARK 1 When the tables given in the appendices are not directly usable because the underlying parameters are not included there, we often resort to *linear interpolation*. As an illustration, suppose $X \sim B(25, 0.3)$ and we wish to calculate $P(X = 10)$. The value $p = 0.3$ is not included in the Binomial Tables in Appendix III. However, $\frac{4}{16} = 0.25 < 0.3 < 0.3125 = \frac{5}{16}$ and the probabilities $P(X = 10)$, for $p = \frac{4}{16}$ and $p = \frac{5}{16}$ are, respectively, 0.9703 and 0.8756. Therefore linear interpolation produces the value:

$$0.9703 - (0.9703 - 0.8756) \times \frac{0.3 - 0.25}{0.3125 - 0.25} = 0.8945.$$

Likewise for other discrete distributions. The same principle also applies appropriately to continuous distributions.

REMARK 2 In discrete distributions, we are often faced with calculations of the form $\sum_{x=1}^{\infty} x\theta^x$. Under appropriate conditions, we may apply the following approach:

$$\sum_{x=1}^{\infty} x\theta^x = \theta \sum_{x=1}^{\infty} x\theta^{x-1} = \theta \sum_{x=1}^{\infty} \frac{d}{d\theta}\theta^{x-1} = \theta \frac{d}{d\theta} \sum_{x=1}^{\infty} \theta^{x-1} = \theta \frac{d}{d\theta}\left(\frac{1}{1-\theta}\right) = \frac{\theta}{(1-\theta)^2}.$$

Similarly for the expression $\sum_{x=2}^{\infty} x(x-1)\theta^{x-2}$.

Exercises

3.2.1 A fair coin is tossed independently four times, and let X be the r.v. defined on the usual sample space S for this experiment as follows:

$$X(s) = \text{the number of } H\text{'s in } s.$$

i) What is the set of values of X?
ii) What is the distribution of X?
iii) What is the partition of S induced by X?

3.2.2 It has been observed that 12.5% of the applicants fail in a certain screening test. If X stands for the number of those out of 25 applicants who fail to pass the test, what is the probability that:

i) $X \geq 1$?
ii) $X \leq 20$?
iii) $5 \leq X \leq 20$?

3.2.3 A manufacturing process produces certain articles such that the probability of each article being defective is p. What is the minimum number, n, of articles to be produced, so that at least one of them is defective with probability at least 0.95? Take $p = 0.05$.

3.2.4 If the r.v. X is distributed as $B(n, p)$ with $p > \frac{1}{2}$, the Binomial Tables in Appendix III cannot be used directly. In such a case, show that:

i) $P(X = x) = P(Y = n - x)$, where $Y \sim B(n, q)$, $x = 0, 1, \ldots, n$, and $q = 1 - p$;
ii) Also, for any integers a, b with $0 \leq a < b \leq n$, one has: $P(a \leq X \leq b) = P(n - b \leq Y \leq n - a)$, where Y is as in part (i).

3.2.5 Let X be a Poisson distributed r.v. with parameter λ. Given that $P(X = 0) = 0.1$, compute the probability that $X > 5$.

3.2.6 Refer to Exercise 3.2.5 and suppose that $P(X = 1) = P(X = 2)$. What is the probability that $X < 10$? If $P(X = 1) = 0.1$ and $P(X = 2) = 0.2$, calculate the probability that $X = 0$.

3.2.7 It has been observed that the number of particles emitted by a radioactive substance which reach a given portion of space during time t follows closely the Poisson distribution with parameter λ. Calculate the probability that:

i) No particles reach the portion of space under consideration during time t;
ii) Exactly 120 particles do so;
iii) At least 50 particles do so;
iv) Give the numerical values in (i)–(iii) if $\lambda = 100$.

3.2.8 The phone calls arriving at a given telephone exchange within one minute follow the Poisson distribution with parameter $\lambda = 10$. What is the probability that in a given minute:

i) No calls arrive?
ii) Exactly 10 calls arrive?
iii) At least 10 calls arrive?

3.2.9 (*Truncation of a Poisson r.v.*) Let the r.v. X be distributed as Poisson with parameter λ and define the r.v. Y as follows:

$$Y = X \text{ if } X \geq k \text{ (a given positive integer) and } Y = 0 \text{ otherwise.}$$

Find:

i) $P(Y = y)$, $y = k, k+1, \ldots$;

ii) $P(Y = 0)$.

3.2.10 A university dormitory system houses 1,600 students, of whom 1,200 are undergraduates and the remaining are graduate students. From the combined list of their names, 25 names are chosen at random. If X stands for the r.v. denoting the number of graduate students among the 25 chosen, what is the probability that $X \geq 10$?

3.2.11 (*Multiple Hypergeometric distribution*) For $j = 1, \ldots, k$, consider an urn containing n_j balls with the number j written on them. n balls are drawn at random and without replacement, and let X_j be the r.v. denoting the number of balls among the n ones with the number j written on them. Then show that the joint distribution of $X_j, j = 1, \ldots, k$ is given by

$$P(X_j = x_j, j = 1, \ldots, k) = \frac{\prod_{j=1}^{k}\binom{n_j}{x_j}}{\binom{n_1 + \cdots + n_k}{n}},$$

$$0 \leq x_j \leq n_j, j = 1, \ldots, k, \sum_{j=1}^{k} x_j = n.$$

3.2.12 Refer to the manufacturing process of Exercise 3.2.3 and let Y be the r.v. denoting the minimum number of articles to be manufactured until the first two defective articles appear.

i) Show that the distribution of Y is given by

$$P(Y = y) = p^2(y-1)(1-p)^{y-2}, \quad y = 2, 3, \ldots;$$

ii) Calculate the probability $P(Y \geq 100)$ for $p = 0.05$.

3.2.13 Show that the function $f(x) = (\frac{1}{2})^x I_A(x)$, where $A = \{1, 2, \ldots\}$, is a p.d.f.

3.2.14 For what value of c is the function f defined below a p.d.f.?

$$f(x) = c\alpha^x I_A(x), \quad \text{where} \quad A = \{0, 1, 2, \ldots\} \quad (0 < \alpha < 1).$$

3.2.15 Suppose that the r.v. X takes on the values $0, 1, \ldots$ with the following probabilities:

$$f(j) = P(X = j) = \frac{c}{3^j}, \quad j = 0, 1, \ldots;$$

i) Determine the constant c.

Compute the following probabilities:

ii) $P(X \geq 10)$;

iii) $P(X \in A)$, where $A = \{j; j = 2k + 1, k = 0, 1, \ldots\}$;

iv) $P(X \in B)$, where $B = \{j; j = 3k + 1, k = 0, 1, \ldots\}$.

3.2.16 There are four distinct types of human blood denoted by O, A, B and AB. Suppose that these types occur with the following frequencies: 0.45, 0.40, 0.10, 0.05, respectively. If 20 people are chosen at random, what is the probability that:

i) All 20 people have blood of the same type?

ii) Nine people have blood type O, eight of type A, two of type B and one of type AB?

3.2.17 A balanced die is tossed (independently) 21 times and let X_j be the number of times the number j appears, $j = 1, \ldots, 6$.

i) What is the joint p.d.f. of the X's?

ii) Compute the probability that $X_1 = 6, X_2 = 5, X_3 = 4, X_4 = 3, X_5 = 2, X_6 = 1$.

3.2.18 Suppose that three coins are tossed (independently) n times and define the r.v.'s $X_j, j = 0, 1, 2, 3$ as follows:

$$X_j = \text{the number of times } j \; H\text{'s appear}.$$

Determine the joint p.d.f. of the X_j's.

3.2.19 Let X be an r.v. distributed as $P(\lambda)$, and set $E = \{0, 2, \ldots\}$ and $O = \{1, 3, \ldots\}$. Then:

i) In terms of λ, calculate the probabilities: $P(X \in E)$ and $P(X \in O)$;

ii) Find the numerical values of the probabilities in part (i) for $\lambda = 5$. (*Hint*: If $S_E = \sum_{k \in E} \frac{\lambda^k}{k!}$ and $S_O = \sum_{k \in O} \frac{\lambda^k}{k!}$, notice that $S_E + S_O = e^\lambda$, and $S_E - S_O = e^{-\lambda}$.)

3.2.20 The following recursive formulas may be used for calculating Binomial, Poisson and Hypergeometric probabilities. To this effect, show that:

i) If $X \sim B(n, p)$, then $f(x+1) = \frac{p}{q} \frac{n-x}{x+1} f(x), \; x = 0, 1, \ldots, n-1$;

ii) If $X \sim P(\lambda)$, then $f(x+1) = \frac{\lambda}{x+1} f(x), \; x = 0, 1, \ldots$;

iii) If X has the Hypergeometric distribution, then

$$f(x+1) = \frac{(m-x)(r-x)}{(n-r+x+1)(x+1)} f(x), \quad x = 0, 1, \ldots, \min\{m, r\}.$$

3.2.21

i) Suppose the r.v.'s $X_1, \ldots, X_k$ have the Multinomial distribution, and let j be a fixed number from the set $\{1, \ldots, k\}$. Then show that X_j is distributed as $B(n, p_j)$;

ii) If m is an integer such that $2 \le m \le k-1$ and $j_1, \ldots, j_m$ are m distinct integers from the set $\{1, \ldots, k\}$, show that the r.v.'s $X_{j_1}, \ldots, X_{j_m}$ have Multinomial distributions with parameters n and $p_{j_1}, \cdots p_{j_m}$.

3.2.22 (*Polya's urn scheme*) Consider an urn containing b black balls and r red balls. One ball is drawn at random, is replaced and c balls of the same color as the one drawn are placed into the urn. Suppose that this experiment is repeated independently n times and let X be the r.v. denoting the number of black balls drawn. Then show that the p.d.f. of X is given by

$$P(X = x) = \binom{n}{x} \frac{b(b+c)(b+2c) \cdots [b+(x-1)c] \times r(r+c) \cdots [r+(n-x-1)c]}{(b+r)(b+r+c) \times (b+r+2c) \cdots [b+r+(m-1)c]}.$$

(This distribution can be used for a rough description of the spread of contagious diseases. For more about this and also for a certain approximation to the above distribution, the reader is referred to the book *An Introduction to Probability Theory and Its Applications*, Vol. 1, 3rd ed., 1968, by W. Feller, pp. 120–121 and p. 142.)

3.3 Continuous Random Variables (and Random Vectors)

3.3.1 Normal (or Gaussian)

$$X(S) = \mathbb{R}, \; f(x) = \frac{1}{\sqrt{2\pi}\sigma} \exp\left[-\frac{(x-\mu)^2}{2\sigma^2}\right], \quad x \in \mathbb{R}.$$

We say that X is distributed as normal (μ, σ^2), denoted by $N(\mu, \sigma^2)$, where μ, σ^2 are called the *parameters of the distribution of X* which is also called the *Normal distribution* (μ = mean, $\mu \in \mathbb{R}$, σ^2 = variance, $\sigma > 0$). For $\mu = 0$, $\sigma = 1$, we get what is known as the *Standard Normal distribution*, denoted by $N(0, 1)$. Clearly $f(x) > 0$; that $I = \int_{-\infty}^{\infty} f(x)dx = 1$ is proved by showing that $I^2 = 1$. In fact,

$$I^2 = \left[\int_{-\infty}^{\infty} f(x)dx\right]^2 = \int_{-\infty}^{\infty} f(x)dx \int_{-\infty}^{\infty} f(y)dy$$

$$= \frac{1}{2\pi\sigma} \int_{-\infty}^{\infty} \exp\left[-\frac{(x-\mu)^2}{2\sigma^2}\right]dx \cdot \frac{1}{\sigma}\int_{-\infty}^{\infty} \exp\left[-\frac{(y-\mu)^2}{2\sigma^2}\right]dy$$

$$= \frac{1}{2\pi} \cdot \frac{1}{\sigma}\int_{-\infty}^{\infty} e^{-z^2/2}\sigma dz \cdot \frac{1}{\sigma}\int_{-\infty}^{\infty} e^{-v^2/2}\sigma dv,$$

upon letting $(x-\mu)/\sigma = z$, so that $z \in (-\infty, \infty)$, and $(y-\mu)/\sigma = v$, so that $v \in (-\infty, \infty)$. Thus

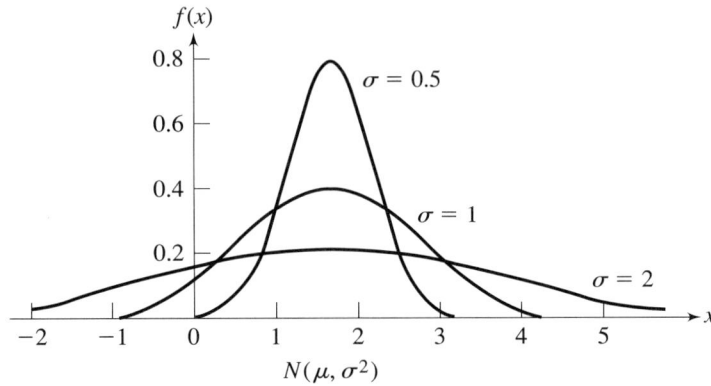

Figure 3.5 Graph of the p.d.f. of the Normal distribution with $\mu = 1.5$ and several values of σ.

$$I^2 = \frac{1}{2\pi} \int_{-\infty}^{\infty} \int_{-\infty}^{\infty} e^{-(z^2+v^2)/2} \, dz \, dv = \frac{1}{2\pi} \int_0^{\infty} \int_0^{2\pi} e^{-r^2/2} r \, dr \, d\theta$$

by the standard transformation to polar coordinates. Or

$$I^2 = \frac{1}{2\pi} \int_0^{\infty} e^{-r^2/2} r \, dr \int_0^{2\pi} d\theta = \int_0^{\infty} e^{-r^2/2} r \, dr = -e^{-r^2/2}\Big|_0^{\infty} = 1;$$

that is, $I^2 = 1$ and hence $I = 1$, since $f(x) > 0$.

It is easily seen that $f(x)$ is symmetric about $x = \mu$, that is, $f(\mu - x) = f(\mu + x)$ and that $f(x)$ attains its *maximum* at $x = \mu$ which is equal to $1/(\sqrt{2\pi}\sigma)$. From the fact that

$$\max_{x \in \mathbb{R}} f(x) = \frac{1}{\sqrt{2\pi}\sigma}$$

and the fact that

$$\int_{-\infty}^{\infty} f(x) dx = 1,$$

we conclude that the larger σ is, the more spread-out $f(x)$ is and vice versa. It is also clear that $f(x) \to 0$ as $x \to \pm\infty$. Taking all these facts into consideration, we are led to Fig. 3.5.

The Normal distribution is a good approximation to the distribution of grades, heights or weights of a (large) group of individuals, lifetimes of various manufactured items, the diameters of hail hitting the ground during a storm, the force required to punctuate a cardboard, errors in numerous measurements, etc. However, the main significance of it derives from the Central Limit Theorem to be discussed in Chapter 8, Section 8.3.

3.3.2 Gamma

$$X(S) = \mathbb{R} \left(\text{actually } X(S) = (0, \infty) \right)$$

Here

$$f(x) = \begin{cases} \dfrac{1}{\Gamma(\alpha)\beta^\alpha} x^{\alpha-1} e^{-x/\beta}, & x > 0 \\ 0, & x \leq 0 \end{cases} \quad \alpha > 0, \ \beta > 0,$$

where $\Gamma(\alpha) = \int_0^\infty y^{\alpha-1} e^{-y} dy$ (which exists and is finite for $\alpha > 0$). (This integral is known as the *Gamma function*.) The distribution of X is also called the *Gamma distribution* and α, β are called the *parameters of the distribution*. Clearly, $f(x) \geq 0$ and that $\int_{-\infty}^\infty f(x)dx = 1$ is seen as follows.

$$\int_{-\infty}^\infty f(x)dx = \frac{1}{\Gamma(\alpha)\beta^\alpha} \int_0^\infty x^{\alpha-1} e^{-x/\beta} dx = \frac{1}{\Gamma(\alpha)\beta^\alpha} \int_0^\infty y^{\alpha-1} e^{-y} \beta^\alpha dy,$$

upon letting $x/\beta = y$, $x = \beta y$, $dx = \beta dy$, $y \in (0, \infty)$; that is,

$$\int_{-\infty}^\infty f(x)dx = \frac{1}{\Gamma(\alpha)} \int_0^\infty y^{\alpha-1} e^{-y} dy = \frac{1}{\Gamma(\alpha)} \cdot \Gamma(\alpha) = 1.$$

REMARK 3 One easily sees, by integrating by parts, that

$$\Gamma(\alpha) = (\alpha - 1)\Gamma(\alpha - 1),$$

and if α is an integer, then

$$\Gamma(\alpha) = (\alpha - 1)(\alpha - 2) \cdots \Gamma(1),$$

where

$$\Gamma(1) = \int_0^\infty e^{-y} dy = 1; \quad \text{that is,} \quad \Gamma(\alpha) = (\alpha - 1)!$$

We often use this notation even if α is not an integer, that is, we write

$$\Gamma(\alpha) = (\alpha - 1)! = \int_0^\infty y^{\alpha-1} e^{-y} dy \quad \text{for} \quad \alpha > 0.$$

For later use, we show that

$$\Gamma\left(\frac{1}{2}\right) = \left(-\frac{1}{2}\right)! = \sqrt{\pi}.$$

We have

$$\Gamma\left(\frac{1}{2}\right) = \int_0^\infty y^{-(1/2)} e^{-y} dy.$$

By setting

$$y^{(1/2)} = \frac{t}{\sqrt{2}}, \quad \text{so that} \quad y = \frac{t^2}{2}, \quad dy = t\, dt, \quad t \in (0, \infty).$$

we get

$$\Gamma\left(\frac{1}{2}\right) = \sqrt{2}\int_0^\infty \frac{1}{t} e^{-t^2/2} t\, dt = \sqrt{2}\int_0^\infty e^{-t^2/2}\, dt = \sqrt{\pi};$$

that is,

$$\Gamma\left(\frac{1}{2}\right) = \left(-\frac{1}{2}\right)! = \sqrt{\pi}.$$

From this we also get that

$$\Gamma\left(\frac{3}{2}\right) = \frac{1}{2}\Gamma\left(\frac{1}{2}\right) = \frac{\sqrt{\pi}}{2}, \text{ etc.}$$

Graphs of the p.d.f. of the Gamma distribution for selected values of α and β are given in Figs. 3.6 and 3.7.

The Gamma distribution, and in particular its special case the Negative Exponential distribution, discussed below, serve as satisfactory models for

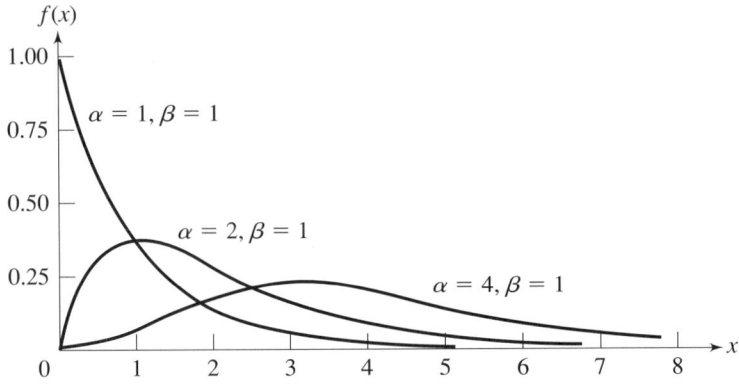

Figure 3.6 Graphs of the p.d.f. of the Gamma distribution for several values of α, β.

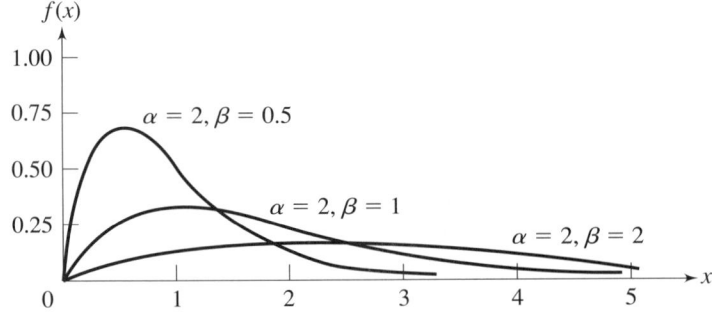

Figure 3.7 Graphs of the p.d.f. of the Gamma distribution for several values of α, β.

describing lifetimes of various manufactured items, among other things, as well as in statistics.

For specific choices of the parameters α and β in the Gamma distribution, we obtain the Chi-square and the Negative Exponential distributions given below.

3.3.3 Chi-square

For $\alpha = r/2$, $r \geq 1$, integer, $\beta = 2$, we get what is known as the *Chi-square distribution*, that is,

$$f(x) = \begin{cases} \dfrac{1}{\Gamma\left(\frac{1}{2}r\right)2^{r/2}} x^{(r/2)-1} e^{-x/2}, & x > 0 \\ 0, & x \leq 0 \end{cases} \quad r > 0, \text{ integer}.$$

The distribution with this p.d.f. is denoted by χ_r^2 and r is called the number of *degrees of freedom (d.f.) of the distribution*. The Chi-square distribution occurs often in statistics, as will be seen in subsequent chapters.

3.3.4 Negative Exponential

For $\alpha = 1$, $\beta = 1/\lambda$, we get

$$f(x) = \begin{cases} \lambda e^{-\lambda x}, & x > 0 \\ 0, & x \leq 0 \end{cases} \quad \lambda > 0,$$

which is known as the *Negative Exponential distribution*. The Negative Exponential distribution occurs frequently in statistics and, in particular, in waiting-time problems. More specifically, if X is an r.v. denoting the waiting time between successive occurrences of events following a Poisson distribution, then X has the Negative Exponential distribution. To see this, suppose that events occur according to the Poisson distribution $P(\lambda)$; for example, particles emitted by a radioactive source with the average of λ particles per time unit. Furthermore, we suppose that we have just observed such a particle, and let X be the r.v. denoting the waiting time until the next particle occurs. We shall show that X has the Negative Exponential distribution with parameter λ. To this end, it is mentioned here that the distribution function F of an r.v., to be studied in the next chapter, is defined by $F(x) = P(X \leq x)$, $x \in \mathbb{R}$, and if X is a continuous r.v., then $\frac{dF(x)}{dx} = f(x)$. Thus, it suffices to determine F here. Since $X \geq 0$, it follows that $F(x) = 0$, $x \leq 0$. So let $x > 0$ be the the waiting time for the emission of the next item. Then $F(x) = P(X \leq x) = 1 - P(X > x)$. Since λ is the average number of emitted particles per time unit, their average number during time x will be λx. Then $P(X > x) = e^{-\lambda x} \frac{(\lambda x)^0}{0!} = e^{-\lambda x}$, since no particles are emitted in $(0, x]$. That is, $F(x) = 1 - e^{-\lambda x}$, so that $f(x) = \lambda e^{-\lambda x}$. To summarize: $f(x) = 0$ for $x \leq 0$, and $f(x) = \lambda e^{-\lambda x}$ for $x > 0$, so that X is distributed as asserted.

Consonant with previous notation, we may use the notation $X \sim \Gamma(\alpha, \beta)$ or $X \sim NE(\lambda)$, or $X \sim \chi_r^2$ in order to denote the fact that X is distributed as Gamma with parameters α and β, or Negative Exponental with parameter λ, or Chi-square with r degrees of freedom, respectively.

3.3.5 Uniform U(α, β) or Rectangular R(α, β)

$$X(S) = \mathbb{R} \left(\text{actually } X(S) = [\alpha, \beta]\right) \text{ and}$$

$$f(x) = \begin{cases} 1/(\beta - \alpha), & \alpha \leq x \leq \beta \\ 0, & \text{otherwise} \end{cases} \quad \alpha < \beta.$$

Clearly,

$$f(x) \geq 0, \quad \int_{-\infty}^{\infty} f(x)dx = \frac{1}{\beta - \alpha} \int_{\alpha}^{\beta} dx = 1.$$

The distribution of X is also called *Uniform* or *Rectangular* (α, β), and α and β are the *parameters of the distribution*. The interpretation of this distribution is that subintervals of $[\alpha, \beta]$, of the same length, are assigned the same probability of being observed regardless of their location. (See Fig. 3.8.)

The fact that the r.v. X has the Uniform distribution with parameters α and β may be denoted by $X \sim U(\alpha, \beta)$.

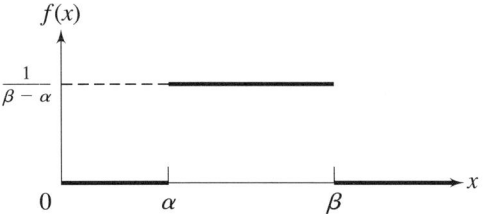

Figure 3.8 Graph of the p.d.f. of the $U(\alpha, \beta)$ distribution.

3.3.6 Beta

$$X(S) = \mathbb{R} \left(\text{actually } X(S) = (0, 1)\right) \text{ and}$$

$$f(x) = \begin{cases} \dfrac{\Gamma(\alpha + \beta)}{\Gamma(\alpha)\Gamma(\beta)} x^{\alpha-1}(1-x)^{\beta-1}, & 0 < x < 1 \\ 0 \text{ elsewhere}, & \alpha > 0, \beta > 0. \end{cases}$$

Clearly, $f(x) \geq 0$. That $\int_{-\infty}^{\infty} f(x)dx = 1$ is seen as follows.

$$\Gamma(\alpha)\Gamma(\beta) = \left(\int_0^\infty x^{\alpha-1}e^{-x}dx\right)\left(\int_0^\infty y^{\beta-1}e^{-y}dy\right)$$

$$= \int_0^\infty \int_0^\infty x^{\alpha-1}y^{\beta-1}e^{-(x+y)}dx\,dy$$

which, upon setting $u = x/(x + y)$, so that

$$x = \frac{uy}{1-u}, \quad dx = \frac{y\,du}{(1-u)^2}, \quad u \in (0, 1) \quad \text{and} \quad x+y = \frac{y}{1-u},$$

becomes

$$= \int_0^\infty \int_0^1 \frac{u^{\alpha-1}}{(1-u)^{\alpha-1}} y^{\alpha-1} y^{\beta-1} e^{-y/(1-u)} y \frac{du}{(1-u)^2} dy$$

$$= \int_0^\infty \int_0^1 \frac{u^{\alpha-1}}{(1-u)^{\alpha+1}} y^{\alpha+\beta-1} e^{-y/(1-u)} du\,dy.$$

Let $y/(1 - u) = v$, so that $y = v(1 - u)$, $dy = (1 - u)dv$, $v \in (0, \infty)$. Then the integral is

$$= \int_0^\infty \int_0^1 u^{\alpha-1}(1-u)^{\beta-1} v^{\alpha+\beta-1} e^{-v} du\,dv$$

$$= \int_0^\infty v^{\alpha+\beta-1} e^{-v} dv \int_0^1 u^{\alpha-1}(1-u)^{\beta-1} du$$

$$= \Gamma(\alpha+\beta)\int_0^1 u^{\alpha-1}(1-u)^{\beta-1} du;$$

that is,

$$\Gamma(\alpha)\Gamma(\beta) = \Gamma(\alpha+\beta)\int_0^1 x^{\alpha-1}(1-x)^{\beta-1} dx$$

and hence

$$\int_{-\infty}^\infty f(x)dx = \frac{\Gamma(\alpha+\beta)}{\Gamma(\alpha)\Gamma(\beta)}\int_0^1 x^{\alpha-1}(1-x)^{\beta-1} dx = 1.$$

Graphs of the p.d.f. of the Beta distribution for selected values of α and β are given in Fig. 3.9.

REMARK 4 For $\alpha = \beta = 1$, we get the $U(0, 1)$, since $\Gamma(1) = 1$ and $\Gamma(2) = 1$. The distribution of X is also called the *Beta distribution* and occurs rather often in statistics. α, β are called the *parameters of the distribution* and the function defined by $\int_0^1 x^{\alpha-1}(1-x)^{\beta-1} dx$ for $\alpha, \beta > 0$ is called the *Beta function*.

Again the fact that X has the Beta distribution with parameters α and β may be expressed by writing $X \sim B(\alpha, \beta)$.

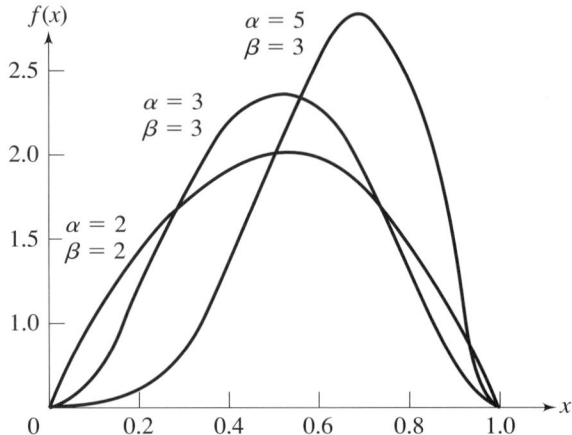

Figure 3.9 Graphs of the p.d.f. of the Beta distribution for several values of α, β.

3.3.7 Cauchy

Here

$$X(S) = \mathbb{R} \quad \text{and} \quad f(x) = \frac{\sigma}{\pi} \cdot \frac{1}{\sigma^2 + (x-\mu)^2}, \quad x \in \mathbb{R}, \ \mu \in \mathbb{R}, \ \sigma > 0.$$

Clearly, $f(x) > 0$ and

$$\int_{-\infty}^{\infty} f(x)dx = \frac{\sigma}{\pi} \int_{-\infty}^{\infty} \frac{1}{\sigma^2 + (x-\mu)^2} dx = \frac{1}{\sigma\pi} \int_{-\infty}^{\infty} \frac{1}{1+\left[(x-\mu)/\sigma\right]^2} dx$$

$$= \frac{1}{\pi} \int_{-\infty}^{\infty} \frac{dy}{1+y^2} = \frac{1}{\pi} \arctan y \Big|_{-\infty}^{\infty} = 1,$$

upon letting

$$y = \frac{x-\mu}{\sigma}, \quad \text{so that} \quad \frac{dx}{\sigma} = dy.$$

The distribution of X is also called the *Cauchy distribution* and μ, σ are called the *parameters of the distribution* (see Fig. 3.10). We may write $X \sim$ Cauchy(μ, σ^2) to express the fact that X has the Cauchy distribution with parameters μ and σ^2.

(The p.d.f. of the Cauchy distribution looks much the same as the Normal p.d.f., except that the tails of the former are heavier.)

3.3.8 Lognormal

Here $X(S) = \mathbb{R}$ (actually $X(S) = (0, \infty)$) and

3.3 Continuous Random Variables (and Random Vectors)

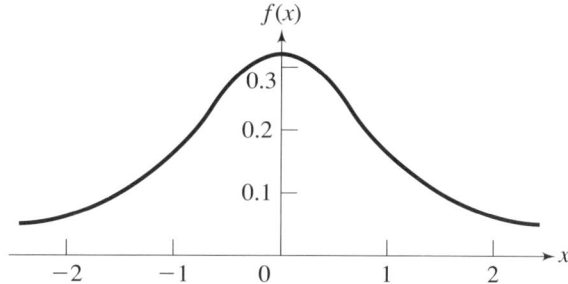

Figure 3.10 Graph of the p.d.f. of the Cauchy distribution with $\mu = 0$, $\sigma = 1$.

$$f(x) = \begin{cases} \dfrac{1}{x\beta\sqrt{2\pi}} \exp\left[-\dfrac{(\log x - \log \alpha)^2}{2\beta^2}\right], & x > 0 \\ 0, & x \leq 0 \end{cases} \text{ where } \alpha, \beta > 0.$$

Now $f(x) \geq 0$ and

$$\int_{-\infty}^{\infty} f(x)\,dx = \frac{1}{\beta\sqrt{2\pi}} \int_0^{\infty} \frac{1}{x} \exp\left[-\frac{(\log x - \log \alpha)^2}{2\beta^2}\right] dx$$

which, letting $x = e^y$, so that $\log x = y$, $dx = e^y dy$, $y \in (-\infty, \infty)$, becomes

$$= \frac{1}{\beta\sqrt{2\pi}} \int_{-\infty}^{\infty} \frac{1}{e^y} \exp\left[-\frac{(y - \log \alpha)^2}{2\beta^2}\right] e^y \, dy.$$

But this is the integral of an $N(\log \alpha, \beta^2)$ density and hence is equal to 1; that is, if X is lognormally distributed, then $Y = \log X$ is normally distributed with parameters $\log \alpha$ and β^2. The distribution of X is called *Lognormal* and α, β are called the *parameters of the distribution* (see Fig. 3.11). The notation $X \sim$ Lognormal(α, β) may be used to express the fact that X has the Lognormal distribution with parameters α and β.

(For the many applications of the Lognormal distribution, the reader is referred to the book *The Lognormal Distribution* by J. Aitchison and J. A. C. Brown, Cambridge University Press, New York, 1957.)

3.3.9 t \
3.3.10 F These distributions occur very often in Statistics (interval estimation, testing hypotheses, analysis of variance, etc.) and their densities will be presented later (see Chapter 9, Section 9.2).

We close this section with an example of a continuous random vector.

3.3.11 Bivariate Normal

Here $\mathbf{X}(S) = \mathbb{R}^2$ (that is, $\mathbf{X}$ is a 2-dimensional random vector) with

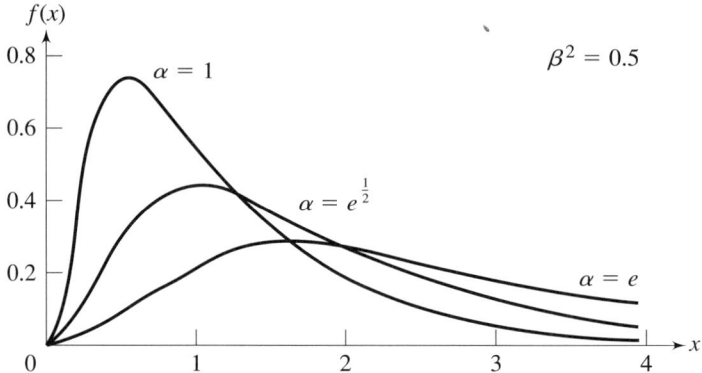

Figure 3.11 Graphs of the p.d.f. of the Lognormal distribution for several values of α, β.

$$f(x_1, x_2) = \frac{1}{2\pi \sigma_1 \sigma_2 \sqrt{1-\rho^2}} e^{-q/2},$$

where $x_1, x_2 \in \mathbb{R}$; $\sigma_1, \sigma_2 > 0$; $-1 < \rho < 1$ and

$$q = \frac{1}{1-\rho^2}\left[\left(\frac{x_1-\mu_1}{\sigma_1}\right)^2 - 2\rho\left(\frac{x_1-\mu_1}{\sigma_1}\right)\left(\frac{x_2-\mu_2}{\sigma_2}\right) + \left(\frac{x_2-\mu_2}{\sigma_2}\right)^2\right]$$

with $\mu_1, \mu_2 \in \mathbb{R}$. The distribution of **X** is also called the *Bivariate Normal distribution* and the quantities $\mu_1, \mu_2, \sigma_1, \sigma_2, \rho$ are called the *parameters of the distribution*. (See Fig. 3.12.)

Clearly, $f(x_1, x_2) > 0$. That $\iint_{\mathbb{R}^2} f(x_1, x_2)dx_1 dx_2 = 1$ is seen as follows:

$$(1-\rho^2)q = \left[\left(\frac{x_1-\mu_1}{\sigma_1}\right)^2 - 2\rho\left(\frac{x_1-\mu_1}{\sigma_1}\right)\left(\frac{x_2-\mu_2}{\sigma_2}\right) + \left(\frac{x_2-\mu_2}{\sigma_2}\right)^2\right]$$

$$= \left[\left(\frac{x_2-\mu_2}{\sigma_2}\right) - \rho\left(\frac{x_1-\mu_1}{\sigma_1}\right)\right]^2 + (1-\rho^2)\left(\frac{x_1-\mu_1}{\sigma_1}\right)^2.$$

Furthermore,

$$\left(\frac{x_2-\mu_2}{\sigma_2}\right) - \rho\left(\frac{x_1-\mu_1}{\sigma_1}\right) = \frac{x_2-\mu_2}{\sigma_2} - \frac{1}{\sigma_2} \cdot \rho\sigma_2 \cdot \frac{x_1-\mu_1}{\sigma_1}$$

$$= \frac{1}{\sigma_2}\left[x_2 - \left(\mu_2 + \rho\sigma_2\frac{x_1-\mu_1}{\sigma_1}\right)\right] = \frac{1}{\sigma_2}(x_2 - b),$$

where

$$b = \mu_2 + \frac{\rho\sigma_2}{\sigma_1}(x_1-\mu_1).$$

3.3 Continuous Random Variables (and Random Vectors)

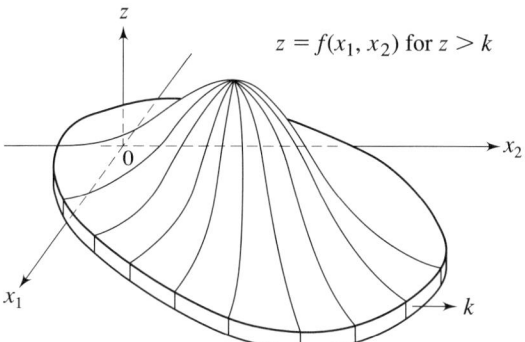

Figure 3.12 Graph of the p.d.f. of the Bivariate Normal distribution.

Thus

$$(1-\rho^2)q = \left(\frac{x_2-b}{\sigma_2}\right)^2 + (1-\rho^2)\left(\frac{x_1-\mu_1}{\sigma_1}\right)^2$$

and hence

$$\int_{-\infty}^{\infty} f(x_1, x_2)dx_2 = \frac{1}{\sqrt{2\pi}\sigma_1}\exp\left[-\frac{(x_1-\mu_1)^2}{2\sigma_1^2}\right]$$

$$\times \int_{-\infty}^{\infty}\frac{1}{\sqrt{2\pi}\sigma_2\sqrt{1-\rho^2}}\exp\left[-\frac{(x_2-b)^2}{2\sigma_2^2(1-\rho^2)}\right]dx_2$$

$$= \frac{1}{\sqrt{2\pi}\sigma_1}\exp\left[-\frac{(x_1-\mu_1)^2}{2\sigma_1^2}\right]\cdot 1,$$

since the integral above is that of an $N(b, \sigma_2^2(1-\rho^2))$ density. Since the first factor is the density of an $N(\mu_1, \sigma_1^2)$ random variable, integrating with respect to x_1, we get

$$\int_{-\infty}^{\infty}\int_{-\infty}^{\infty} f(x_1, x_2)dx_1\,dx_2 = 1.$$

REMARK 5 From the above derivations, it follows that, if $f(x_1, x_2)$ is Bivariate Normal, then

$$f_1(x_1) = \int_{-\infty}^{\infty} f(x_1, x_2)dx_2 \quad \text{is} \quad N(\mu_1, \sigma_1^2),$$

and similarly,

$$f_2(x_2) = \int_{-\infty}^{\infty} f(x_1, x_2)dx_1 \quad \text{is} \quad N(\mu_2, \sigma_2^2).$$

As will be seen in Chapter 4, the p.d.f.'s f_1 and f_2 above are called marginal p.d.f.'s of f.

The notation $\mathbf{X} \sim N(\mu_1, \mu_2, \sigma_1^2, \sigma_2^2, \rho)$ may be used to express the fact that $\mathbf{X}$ has the Bivariate Normal distribution with parameters $\mu_1, \mu_2, \sigma_1^2, \sigma_2^2, \rho$. Then $X_1 \sim N(\mu_1, \sigma_1^2)$ and $X_2 \sim N(\mu_2, \sigma_2^2)$.

Exercises

3.3.1 Let f be the p.d.f. of the $N(\mu, \sigma^2)$ distribution and show that:

i) f is symmetric about μ;

ii) $\max_{x \in \mathbb{R}} f(x) = \dfrac{1}{\sqrt{2\pi}\sigma}$.

3.3.2 Let X be distributed as $N(\mu, \sigma^2)$, and for $a < b$, let $p = P(a < X < b)$. Then use the symmetry of the p.d.f. f in order to show that:

i) For $0 \leq a < b$, $p = \Phi(b) - \Phi(a)$;
ii) For $a \leq 0 < b$, $p = \Phi(b) + \Phi(-a) - 1$;
iii) For $a \leq b < 0$, $p = \Phi(-a) - \Phi(-b)$;
iv) For $c > 0$, $P(-c < X < c) = 2\Phi(c) - 1$.

(See Normal Tables in Appendix III for the definition of Φ.)

3.3.3 If $X \sim N(0, 1)$, use the Normal Tables in Appendix III in order to show that:

i) $P(-1 < X < 1) = 0.68269$;
ii) $P(-2 < X < 2) = 0.9545$;
iii) $P(-3 < X < 3) = 0.9973$.

3.3.4 Let X be a χ_r^2. In Table 5, Appendix III, the values $\gamma = P(X \leq x)$ are given for r ranging from 1 to 45, and for selected values of γ. From the entries of the table, observe that, for a fixed γ, the values of x increase along with the number of degrees of freedom r. Select some values of γ and record the corresponding values of x for a set of increasing values of r.

3.3.5 Let X be an r.v. distributed as χ_{10}^2. Use Table 5 in Appendix III in order to determine the numbers a and b for which the following are true:

i) $P(X < a) = P(X > b)$;
ii) $P(a < X < b) = 0.90$.

3.3.6 Consider certain events which in every time interval $[t_1, t_2]$ ($0 < t_1 < t_2$) occur according to the Poisson distribution $P(\lambda(t_2 - t_1))$. Let T be the r.v. denoting the time which lapses between two consecutive such events. Show that the distribution of T is Negative Exponential with parameter λ by computing the probability that $T > t$.

3.3.7 Let X be an r.v. denoting the life length of a TV tube and suppose that its p.d.f. f is given by:
$$f(x) = \lambda e^{-\lambda x} I_{(0, \infty)}(x).$$
Compute the following probabilities:

i) $P(j < X \leq j + 1)$, $j = 0, 1, \ldots$;
ii) $P(X > t)$ for some $t > 0$;
iii) $P(X > s + t | X > s)$ for some $s, t > 0$;
iv) Compare the probabilities in parts (ii) and (iii) and conclude that the Negative Exponential distribution is "memoryless";
v) If it is known that $P(X > s) = \alpha$, express the parameter λ in terms of α and s.

3.3.8 Suppose that the life expectancy X of each member of a certain group of people is an r.v. having the Negative Exponential distribution with parameter $\lambda = 1/50$ (years). For an individual from the group in question, compute the probability that:

i) He will survive to retire at 65;
ii) He will live to be at least 70 years old, given that he just celebrated his 40th birthday;
iii) For what value of c, $P(X > c) = \frac{1}{2}$?

3.3.9 Let X be an r.v. distributed as $U(-\alpha, \alpha)$. Determine the values of the parameter α for which the following are true:

i) $P(-1 < X < 2) = 0.75$;
ii) $P(|X| < 1) = P(|X| > 2)$.

3.3.10 Refer to the Beta distribution and set:
$$B(\alpha, \beta) = \int_0^1 x^{\alpha-1}(1-x)^{\beta-1} dx.$$
Then show that $B(\alpha, \beta) = B(\beta, \alpha)$.

3.3.11 Establish the following identity:
$$n\binom{n-1}{m-1}\int_0^p x^{m-1}(1-x)^{n-m} dx = \frac{n!}{(m-1)(n-m)!}\int_0^p x^{m-1}(1-x)^{n-m} dx$$
$$= \sum_{j=m}^{n}\binom{n}{j}p^j(1-p)^{n-j}.$$

3.3.12 Let X be an r.v. with p.d.f given by $f(x) = 1/[\pi(1 + x^2)]$. Calculate the probability that $X^2 \leq c$.

3.3.13 Show that the following functions are p.d.f.'s:

i) $f(x) = xe^{-x^2/2}I_{(0,\infty)}(x)$ (Raleigh distribution);

ii) $f(x) = \sqrt{2/\pi}\, x^2 e^{-x^2/2}I_{(0,\infty)}(x)$ (Maxwell's distribution);

iii) $f(x) = \frac{1}{2}e^{-|x-\upsilon|}$ (Double Exponential);

iv) $f(x) = \left(\frac{a}{c}\right)\left(\frac{c}{x}\right)^{\alpha+1}I_A(x)$, $A = (c, \infty)$, $\alpha, c > 0$ (Pareto distribution).

3.3.14 Show that the following functions are p.d.f.'s:

i) $f(x) = \cos x\, I_{(0,\pi/2)}(x)$;

ii) $f(x) = xe^{-x}I_{(0,\infty)}(x)$.

3.3.15 For what values of the constant c are the following functions p.d.f.'s?

i) $f(x) = \begin{cases} ce^{-6x}, & x > 0 \\ -cx, & -1 < x \leq 0; \\ 0, & x \leq -1 \end{cases}$

ii) $f(x) = cx^2 e^{-x^3}I_{(0,\infty)}(x)$.

3.3.16 Let X be an r.v. with p.d.f. given by 3.3.15(ii). Compute the probability that $X > x$.

3.3.17 Let X be the r.v. denoting the life length of a certain electronic device expressed in hours, and suppose that its p.d.f. f is given by:

$$f(x) = \frac{c}{x^n} I_{[1,000,\, 3,000]}(x).$$

i) Determine the constant c in terms of n;

ii) Calculate the probability that the life span of one electronic device of the type just described is at least 2,000 hours.

3.3.18 Refer to Exercise 3.3.15(ii) and compute the probability that X exceeds $s + t$, given that $X > s$. Compare the answer with that of Exercise 3.3.7(iii).

3.3.19 Consider the function $f(x) = \alpha\beta x^{\beta-1}e^{-x^\beta}$, $x > 0$ ($\alpha, \beta > 0$), and:

i) Show that it is a p.d.f. (called the Weibull p.d.f. with parameters α and β);

ii) Observe that the Negative Exponential p.d.f. is a special case of a Weibull p.d.f., and specify the values of the parameters for which this happens;

iii) For $\alpha = 1$ and $\beta = \frac{1}{2}$, $\beta = 1$ and $\beta = 2$, draw the respective graphs of the p.d.f.'s involved.

(*Note:* The Weibull distribution is employed for describing the lifetime of living organisms or of mechanical systems.)

3.3.20 Let X and Y be r.v.'s having the joint p.d.f. f given by:

$$f(x, y) = c(25 - x^2 - y^2)I_{(0,5)}(x^2 + y^2).$$

Determine the constant c and compute the probability that $0 < X^2 + Y^2 < 4$.

3.3.21 Let X and Y be r.v.'s whose joint p.d.f. f is given by $f(x, y) = cxyI_{(0,2)\times(0,5)}(x, y)$. Determine the constant c and compute the following probabilities:

i) $P(\frac{1}{2} < X < 1, 0 < Y < 3)$;
ii) $P(X < 2, 2 < Y < 4)$;
iii) $P(1 < X < 2, Y > 5)$;
iv) $P(X > Y)$.

3.3.22 Verify that the following function is a p.d.f.:

$$f(x, y) = \frac{1}{4\pi}(\cos y)I_A(x, y), \quad A = (-\pi, \pi] \times \left(-\frac{\pi}{2}, \frac{\pi}{2}\right].$$

3.3.23 (*A mixed distribution*) Show that the following function is a p.d.f.

$$f(x) = \begin{cases} \frac{1}{4}e^x, & x \leq 0 \\ \frac{1}{8}, & 0 < x < 2 \\ \left(\frac{1}{2}\right)^x, & x = 2, 3, \ldots \\ 0, & \text{otherwise.} \end{cases}$$

3.4 The Poisson Distribution as an Approximation to the Binomial Distribution and the Binomial Distribution as an Approximation to the Hypergeometric Distribution

In this section, we first establish rigorously the assertion made earlier that the Poisson distribution may be obtained as the limit of Binomial distributions. To this end, consider a sequence of Binomial distributions, so that the nth distribution has probability p_n of a success, and we assume that as $n \to \infty$, $p_n \to 0$ and that

$$\lambda_n = np_n \to \lambda,$$

for some $\lambda > 0$. Then the following theorem is true.

THEOREM 1 With the above notation and assumptions, we have

$$\binom{n}{x} p_n^x q_n^{n-x} \xrightarrow[n \to \infty]{} e^{-\lambda} \frac{\lambda^x}{x!} \quad \text{for each fixed } x = 0, 1, 2, \ldots.$$

80   3   On Random Variables and Their Distributions

PROOF   We have

$$\binom{x}{n} p_n^x q_n^{n-x} = \frac{n(n-1)\cdots(n-x+1)}{x!} p_n^x q_n^{n-x}$$

$$= \frac{n(n-1)\cdots(n-x+1)}{x!} \left(\frac{\lambda_n}{n}\right)^x \left(1-\frac{\lambda_n}{n}\right)^{n-x}$$

$$= \frac{n(n-1)\cdots(n-x+1)}{n^x} \cdot \frac{1}{x!} \lambda_n^x \cdot \left(1-\frac{\lambda_n}{n}\right)^n \cdot \frac{1}{\left(1-\frac{\lambda_n}{n}\right)^x}$$

$$= 1\cdot\left(1-\frac{1}{n}\right)\cdots\left(1-\frac{x-1}{n}\right)\frac{\lambda_n^x}{x!}$$

$$\times \left(1-\frac{\lambda_n}{n}\right)^n \cdot \frac{1}{\left(1-\frac{\lambda_n}{n}\right)^x} \xrightarrow[n\to\infty]{} \frac{\lambda^x}{x!} e^{-\lambda},$$

since, if $\lambda_n \to \lambda$, then

$$\left(1-\frac{\lambda_n}{n}\right)^n \to e^{-\lambda}.$$

This is merely a generalization of the better-known fact that

$$\left(1-\frac{\lambda}{n}\right)^n \to e^{-\lambda}. \quad \blacktriangle$$

REMARK 6   The meaning of the theorem is the following: If n is large the probabilities $\binom{x}{n} p^x q^{n-x}$ are not easily calculated. Then we can approximate them by $e^{-\lambda}(\lambda^x/x!)$, provided p is small, where we replace λ be np. This is true for all $x = 0, 1, 2, \ldots, n$.

We also meet with difficulty in calculating probabilities in the Hypergeometric distribution

$$\binom{m}{x}\binom{n}{r-x} \bigg/ \binom{m+n}{r}$$

if m, n are large. What we do is approximate the Hypergeometric distribution by an appropriate Binomial distribution, and then, if need be, we can go one step further in approximating the Binomial by the appropriate Poisson distribution according to Theorem 1. Thus we have

THEOREM 2   Let $m, n \to \infty$ and suppose that $m/(m+n) = p_{m,n} \to p$, $0 < p < 1$. Then

3.4 The Poisson Distribution

$$\frac{\binom{m}{x}\binom{n}{r-x}}{\binom{m+n}{r}} \to \binom{r}{x} p^x q^{r-x}, \qquad x = 0, 1, 2, \ldots, r.$$

PROOF We have

$$\frac{\binom{m}{x}\binom{n}{r-x}}{\binom{m+n}{r}} = \frac{m!n!(m+n-r)!}{(m-x)![n-(r-x)]!(m+n)!} \cdot \frac{r!}{x!(r-x)!}$$

$$= \binom{r}{x}\frac{(m-x+1)\cdots m(n-r+x+1)\cdots n}{(m+n-r+1)\cdots (m+n)}$$

$$= \binom{r}{x}\frac{m(m-1)\cdots [m-(x-1)] \cdot n(n-1)\cdots [n-(r-x-1)]}{(m+n)\cdots [(m+n)-(r-1)]}.$$

Both numerator and denominator have r factors. Dividing through by $(m+n)$, we get

$$\frac{\binom{m}{x}\binom{n}{r-x}}{\binom{m+n}{r}} = \binom{r}{x}\left(\frac{m}{m+n}\right)\left(\frac{m}{m+n} - \frac{1}{m+n}\right)\cdots\left(\frac{m}{m+n} - \frac{x-1}{m+n}\right)$$

$$\times \left(\frac{n}{m+n}\right)\left(\frac{n}{m+n} - \frac{1}{m+n}\right)\cdots\left(\frac{n}{m+n} - \frac{r-x-1}{m+n}\right)$$

$$\times \left[1\cdot\left(1 - \frac{1}{m+n}\right)\cdots\left(1 - \frac{r-1}{m+n}\right)\right]^{-1}$$

$$\xrightarrow[m,n\to\infty]{} \binom{r}{x} p^x q^{r-x},$$

since

$$\frac{m}{m+n} \to p \quad \text{and hence} \quad \frac{n}{m+n} \to 1-p = q. \quad \blacktriangle$$

REMARK 7 The meaning of the theorem is that if m, n are large, we can approximate the probabilities

$$\frac{\binom{m}{x}\binom{n}{r-x}}{\binom{m+n}{r}} \text{ by } \binom{r}{x}p^x q^{r-x}$$

by setting $p = m/(m + n)$. This is true for all $x = 0, 1, 2, \ldots, r$. It is to be observed that $\binom{r}{x}\left(\frac{m}{m+n}\right)^x\left(1-\frac{m}{m+n}\right)^{r-x}$ is the exact probability of having exactly x successes in r trials when sampling is done *with* replacement, so that the probability of a success, $\frac{m}{m+n}$, remains constant. The Hypergeometric distribution is appropriate when sampling is done *without* replacement. If, however, m and n are large ($m, n \to \infty$) and $\frac{m}{n}$ remains approximately constant $\left(\frac{m}{n} \to c = p/q\right)$, then the probabilities of having exactly x successes in r (independent) trials are approximately equal under both sampling schemes.

Exercises

3.4.1 For the following values of n, p and $\lambda = np$, draw graphs of $B(n, p)$ and $P(\lambda)$ on the same coordinate axes:

i) $n = 10, p = \frac{4}{16}$, so that $\lambda = 2.5$;
ii) $n = 16, p = \frac{2}{16}$, so that $\lambda = 2$;
iii) $n = 20, p = \frac{2}{16}$, so that $\lambda = 2.5$;
iv) $n = 24, p = \frac{1}{16}$, so that $\lambda = 1.5$;
v) $n = 24, p = \frac{2}{16}$, so that $\lambda = 3$.

3.4.2 Refer to Exercise 3.2.2 and suppose that the number of applicants is equal to 100. Compute the probabilities (i)–(iii) by using the Poisson approximation to Binomial (Theorem 1).

3.4.3 Refer to Exercise 3.2.10 and use Theorem 2 in order to obtain an approximate value of the required probability.

3.5* Random Variables as Measurable Functions and Related Results

In this section, random variables and random vectors are introduced as special cases of measurable functions. Certain results related to σ-fields are also derived. Consider the probability space $(S, \mathcal{A}, P)$ and let $\mathcal{T}$ be a space and X be a function defined on S into $\mathcal{T}$, that is, $X: S \to \mathcal{T}$. For $T \subseteq \mathcal{T}$, define *the inverse image of T*, under X, denoted by $X^{-1}(T)$, as follows:

$$X^{-1}(T) = \{s \in S;\ X(s) \in T\}.$$

This set is also denoted by $[X \in T]$ or $(X \in T)$. Then the following properties are immediate consequences of the definition (and the fact X is a function):

$$X^{-1}\left(\bigcup_j T_j\right) = \bigcup_j X^{-1}(T_j). \tag{1}$$

If $T_1 \cap T_2 = \emptyset$, then $X^{-1}(T_1) \cap X^{-1}(T_2) = \emptyset.$ $\quad(2)$

Hence by (1) and (2) we have

$$X^{-1}\left(\sum_j T_j\right) = \sum_j X^{-1}(T_j). \tag{3}$$

Also
$$X^{-1}\left(\bigcap_j T_j\right) = \bigcap_j X^{-1}(T_j), \tag{4}$$

$$X^{-1}(T^c) = [X^{-1}(T)]^c, \tag{5}$$

$$X^{-1}(\mathcal{T}) = S, \tag{6}$$

$$X^{-1}(\emptyset) = \emptyset. \tag{7}$$

Let now $\mathcal{D}$ be a σ-field of subsets of $\mathcal{T}$ and define the class $X^{-1}(\mathcal{D})$ of subsets of S as follows:

$$X^{-1}(\mathcal{D}) = \{A \subseteq S;\ A = X^{-1}(T) \text{ for some } T \in \mathcal{D}\}.$$

By means of (1), (5), (6) above, we immediately have

THEOREM 3   The class $X^{-1}(\mathcal{D})$ is a σ-field of subsets of S.

The above theorem is the reason we require measurability in our definition of a random variable. It guarantees that the probability distribution function of a random vector, to be defined below, is well defined.

If $X^{-1}(\mathcal{D}) \subseteq \mathcal{A}$, then we say that X is $(\mathcal{A}, \mathcal{D})$-*measurable*, or just measurable if there is no confusion possible. If $(\mathcal{T}, \mathcal{D}) = (\mathbb{R}, \mathcal{B})$ and X is $(\mathcal{A}, \mathcal{B})$-measurable, we say that X is a *random variable* (r.v.). More generally, if $(\mathcal{T}, \mathcal{D}) = (\mathbb{R}^k, \mathcal{B}^k)$, where $\mathbb{R}^k = \mathbb{R} \times \mathbb{R} \times \cdots \times \mathbb{R}$ (k copies of $\mathbb{R}$), and X is $(\mathcal{A}, \mathcal{B}^k)$-measurable, we say that X is a *k-dimensional random vector* (r. vector). In this latter case, we shall write **X** if $k \geq 1$, and just X if $k = 1$. A random variable is a one-dimensional random vector.

On the basis of the properties (1)–(7) of X^{-1}, the following is immediate.

THEOREM 4   Define the class C^* of subsets of $\mathcal{T}$ as follows: $C^* = \{T \subseteq \mathcal{T}; X^{-1}(T) = A$ for some $A \in \mathcal{A}\}$. Then C^* is a σ-field.

COROLLARY   Let $\mathcal{D} = \sigma(C)$, where C is a class of subsets of $\mathcal{T}$. Then X is $(\mathcal{A}, \mathcal{D})$-measurable if and only if $X^{-1}(C) \subseteq \mathcal{A}$. In particular, X is a random variable if and only if

$X^{-1}(C_o)$, or $X^{-1}(C_j)$, or $X^{-1}(C'_j) \subseteq \mathcal{A}$, $j = 1, 2, \ldots, 8$, and similarly for the case of k-dimensional random vectors. The classes C_o, C_j, C'_j, $j = 1, \ldots, 8$ are defined in Theorem 5 and the paragraph before it in Chapter 1.

PROOF The σ-field C^* of Theorem 4 has the property that $C^* \supseteq C$. Then $C^* \supseteq \mathcal{D} = \sigma(C)$ and hence $X^{-1}(C^*) \supseteq X^{-1}(\mathcal{D})$. But $X^{-1}(C^*) \subseteq \mathcal{A}$. Thus $X^{-1}(\mathcal{D}) \subseteq \mathcal{A}$. The converse is a direct consequence of the definition of $(\mathcal{A}, \mathcal{D})$-measurability. ▲

Now, by means of an r. vector $\mathbf{X} : (S, \mathcal{A}, P) \to (\mathbb{R}^k, \mathcal{B}^k)$, define on $\mathcal{B}^k$ the set function $P_\mathbf{X}$ as follows:

$$P_\mathbf{X}(B) = P[\mathbf{X}^{-1}(B)] = P(\mathbf{X} \in B) = P(\{s \in S; \mathbf{X}(s) \in B\}). \tag{8}$$

By the Corollary to Theorem 4, the sets $\mathbf{X}^{-1}(B)$ in S are actually events due to the assumption that $\mathbf{X}$ is an r. vector. Therefore $P_\mathbf{X}$ is well defined by (8); i.e., $P[\mathbf{X}^{-1}(B)]$ makes sense, is well defined. It is now shown that $P_\mathbf{X}$ is a probability measure on $\mathcal{B}^k$. In fact, $P_\mathbf{X}(B) \geq 0$, $B \in \mathcal{B}^k$, since P is a probability measure. Next, $P_\mathbf{X}(\mathbb{R}^k) = P[\mathbf{X}^{-1}(\mathbb{R}^k)] = P(S) = 1$, and finally,

$$P_\mathbf{X}\left(\sum_{j=1}^\infty B_j\right) = P\left[\mathbf{X}^{-1}\left(\sum_{j=1}^\infty B_j\right)\right] = P\left[\sum_{j=1}^\infty \mathbf{X}^{-1}(B_j)\right] = \sum_{j=1}^\infty P[\mathbf{X}^{-1}(B_j)] = \sum_{j=1}^\infty P_\mathbf{X}(B_j).$$

The probability measure $P_\mathbf{X}$ is called the *probability distribution function* (or just the *distribution*) of $\mathbf{X}$.

Exercises

3.5.1 Consider the sample space S supplied with the σ-field of events $\mathcal{A}$. For an event A, the *indicator* I_A of A is defined by: $I_A(s) = 1$ if $s \in A$ and $I_A(s) = 0$ if $s \in A^c$.

i) Show that I_A is r.v. for any $A \in \mathcal{A}$.

ii) What is the partition of S induced by I_A?

iii) What is the σ-field induced by I_A?

3.5.2 Write out the proof of Theorem 1 by using (1), (5) and (6).

3.5.3 Write out the proof of Theorem 2.

Chapter 4

Distribution Functions, Probability Densities, and Their Relationship

4.1 The Cumulative Distribution Function (c.d.f. or d.f.) of a Random Vector—Basic Properties of the d.f. of a Random Variable

The distribution of a k-dimensional r. vector $\mathbf{X}$ has been defined through the relationship: $P_\mathbf{X}(B) = P(\mathbf{X} \in B)$, where B is a subset of $\mathbb{R}^k$. In particular, one may choose B to be an "interval" in $\mathbb{R}^k$; i.e., $B = \{\mathbf{y} \in \mathbb{R}^k; \mathbf{y} \leq \mathbf{x}\}$ in the sense that, if $\mathbf{x} = (x_1, \ldots, x_k)'$ and $\mathbf{y} = (y_1, \ldots, y_k)$, then $y_j \leq x_j, j = 1, \ldots, k$. For such a choice of B, $P_\mathbf{X}(B)$ is denoted by $F_\mathbf{X}(\mathbf{x})$ and is called the *cumulative distribution function* of $\mathbf{X}$ (evaluated at $\mathbf{x}$), or just the *distribution function* (d.f.) of $\mathbf{X}$. We omit the subscript $\mathbf{X}$ if no confusion is possible. Thus, the d.f. F of a r. vector $\mathbf{X}$ is an ordinary point function defined on $\mathbb{R}^k$ (and taking values in $[0, 1]$). Now we restrict our attention to the case $k = 1$ and prove the following basic properties of the d.f. of an r.v.

THEOREM 1 The distribution function F of a random variable X satisfies the following properties:

i) $0 \leq F(x) \leq 1, x \in \mathbb{R}$.

ii) F is nondecreasing.

iii) F is continuous from the right.

iv) $F(x) \to 0$ as $x \to -\infty$, $F(x) \to 1$, as $x \to +\infty$.

We express this by writing $F(-\infty) = 0$, $F(+\infty) = 1$.

PROOF In the course of this proof, we set Q for the distribution P_X of X, for the sake of simplicity. We have then:

i) Obvious.

ii) This means that $x_1 < x_2$ implies $F(x_1) \leq F(x_2)$. In fact,

86   **4 Distribution Functions, Probability Densities, and Their Relationship**

$$x_1 < x_2 \text{ implies } (-\infty, x_1] \subset (-\infty, x_2]$$

and hence

$$Q(-\infty, x_1] \leq Q(-\infty, x_2]; \text{ equivalently, } F(x_1) \leq F(x_2).$$

iii) This means that, if $x_n \downarrow x$, then $F(x_n) \downarrow F(x)$. In fact,

$$x_n \downarrow x \text{ implies } (-\infty, x_n] \downarrow (-\infty, x]$$

and hence

$$Q(-\infty, x_n] \to Q(-\infty, x]$$

by Theorem 2, Chapter 2; equivalently, $F(x_n) \downarrow F(x)$.

iv) Let $x_n \to -\infty$. We may assume that $x_n \downarrow -\infty$ (see also Exercise 4.1.6). Then

$$(-\infty, x_n] \downarrow \varnothing, \text{ so that } Q(-\infty, x_n] \downarrow Q(\varnothing) = 0$$

by Theorem 2, Chapter 2. Equivalently, $F(x_n) \to 0$. Similarly, if $x_n \to +\infty$. We may assume $x_n \uparrow \infty$. Then

$$(-\infty, x_n] \uparrow \mathbb{R} \text{ and hence } Q(-\infty, x_n] \uparrow Q(\mathbb{R}) = 1; \text{ equivalently, } F(x_n) \to 1. \ \blacktriangle$$

Graphs of d.f.'s of several distributions are given in Fig. 4.1.

REMARK 1

i) $F(x)$ can be used to find probabilities of the form $P(a < X \leq b)$; that is

$$P(a < X \leq b) = F(b) - F(a).$$

In fact,

$$(a < X \leq b) = (-\infty < X \leq b) - (-\infty < X \leq a)$$

and

$$(-\infty < X \leq a) \subseteq (-\infty < X \leq b).$$

Thus

$$P(a < X \leq b) = P(-\infty < X \leq b) - P(-\infty < X \leq a) = F(b) - F(a).$$

ii) The limit from the left of $F(x)$ at x, denoted by $F(x-)$, is defined as follows:

$$F(x-) = \lim_{n \to \infty} F(x_n) \text{ with } x_n \uparrow x.$$

This limit always exists, since $F(x_n)\uparrow$, but need not be equal to $F(x+)(=$limit from the right$) = F(x)$. The quantities $F(x)$ and $F(x-)$ are used to express the probability $P(X = a)$; that is, $P(X = a) = F(a) - F(a-)$. In fact, let $x_n \uparrow a$ and set $A = (X = a)$, $A_n = (x_n < X \leq a)$. Then, clearly, $A_n \downarrow A$ and hence by Theorem 2, Chapter 2,

$$P(A_n) \downarrow P(A), \text{ or } \lim_{n \to \infty} P(x_n < X \leq a) = P(X = a),$$

or

$$\lim_{n \to \infty} [F(a) - F(x_n)] = P(X = a),$$

or
$$F(a) - \lim_{n\to\infty} F(x_n) = P(X = a),$$
or
$$F(a) - F(a-) = P(X = a).$$

It is known that a nondecreasing function (such as F) may have discontinuities which can only be jumps. Then $F(a) - F(a-)$ is the length of the jump of F at a. Of course, if F is continuous then $F(x) = F(x-)$ and hence $P(X = x) = 0$ for all x.

iii) If X is discrete, its d.f. is a "step" function, the value of it at x being defined by
$$F(x) = \sum_{x_j \leq x} f(x_j) \quad \text{and} \quad f(x_j) = F(x_j) - F(x_{j-1}),$$
where it is assumed that $x_1 < x_2 < \cdots$.

iv) If X is of the continuous type, its d.f. F is continuous. Furthermore,
$$\frac{dF(x)}{dx} = f(x)$$

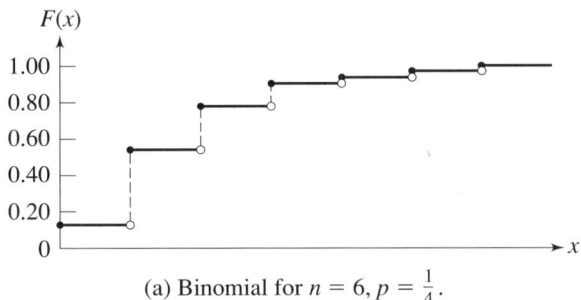

(a) Binomial for $n = 6$, $p = \frac{1}{4}$.

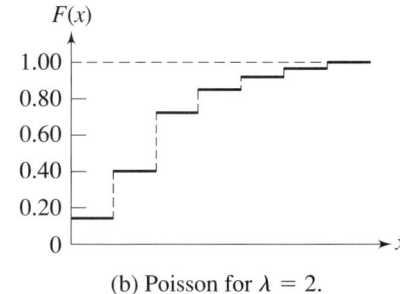

(b) Poisson for $\lambda = 2$.

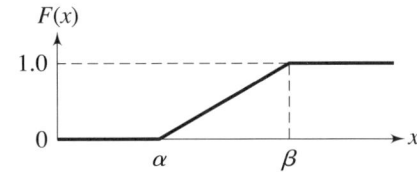

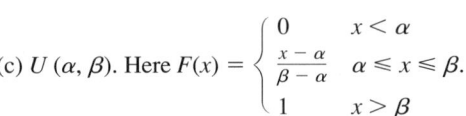

(c) $U(\alpha, \beta)$. Here $F(x) = \begin{cases} 0 & x < \alpha \\ \frac{x - \alpha}{\beta - \alpha} & \alpha \leq x \leq \beta. \\ 1 & x > \beta \end{cases}$

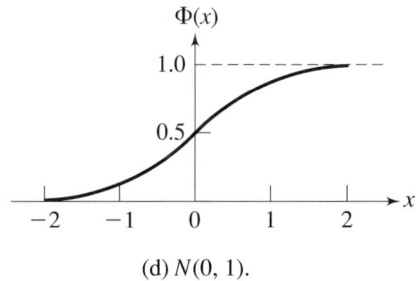

(d) $N(0, 1)$.

Figure 4.1 Examples of graphs of c.d.f.'s.

at continuity points of f, as is well known from calculus. Through the relations

$$F(x) = \int_{-\infty}^{x} f(t)dt \quad \text{and} \quad \frac{dF(x)}{dx} = f(x),$$

we see that if f is continuous, f determines F ($f \Rightarrow F$) and F determines f ($F \Rightarrow f$); that is, $F \Leftrightarrow f$. Two important applications of this are the following two theorems.

Often one has to deal with functions of an r. vector itself. In such cases, the resulting entities have got to be r. vectors, since we operate in a probability framework. The following statement is to this effect. Its precise formulation and justification is given as Theorem 7 on page 104.

STATEMENT 1 Let $\mathbf{X}$ be a k-dimensional r. vector defined on the sample space S, and let g be a (well-behaving) function defined on $\mathbb{R}^k$ and taking values in $\mathbb{R}^m$. Then $g(\mathbf{X})$ is defined on the underlying sample space S, takes values in $\mathbb{R}^m$, and is an r. vector. (That is, well-behaving functions of r. vectors are r. vectors.) In particular, $g(\mathbf{X})$ is an r. vector if g is continuous.

Now a k-dimensional r. vector $\mathbf{X}$ may be represented in terms of its coordinates; i.e., $\mathbf{X} = (X_1, \ldots, X_k)'$, where X_j, $j = 1, \ldots, k$ are real-valued functions defined on S. The question then arises as to how $\mathbf{X}$ and X_j, $j = 1, \ldots, k$ are related from the point of view of being r. vectors. The answer is provided by the following statement, whereas the precise statement and justification are given as Theorem 8 below.

STATEMENT 2 Let $\mathbf{X}$ and X_j, $j = 1, \ldots, k$ be functions defined on the sample space S and taking values in $\mathbb{R}^k$ and $\mathbb{R}$, respectively, and let $\mathbf{X} = (X_1, \ldots, X_k)$. Then $\mathbf{X}$ is an r. vector if and only if X_j, $j = 1, \ldots, k$ are r.v.'s.

The following two theorems provide applications of Statement 1.

THEOREM 2 Let X be an $N(\mu, \sigma^2)$-distributed r.v. and set $Y = \frac{X-\mu}{\sigma}$. Then Y is an r.v. and its distribution is $N(0, 1)$.

PROOF In the first place, Y is an r.v. by Statement 1. Then it suffices to show that the d.f. of Y is Φ, where

$$\Phi(y) = \frac{1}{\sqrt{2\pi}} \int_{-\infty}^{y} e^{-t^2/2} dt.$$

We have

$$P(Y \le y) = P\left(\frac{X-\mu}{\sigma} \le y\right) = P(X \le y\sigma + \mu)$$

$$= \frac{1}{\sqrt{2\pi}\sigma} \int_{-\infty}^{y\sigma+\mu} \exp\left[-\frac{(t-\mu)^2}{2\sigma^2}\right] dt = \frac{1}{\sqrt{2\pi}} \int_{-\infty}^{y} e^{-u^2/2} du = \Phi(y),$$

where we let $u = (t - \mu)/\sigma$ in the transformation of the integral. ▲

REMARK 2 The transformation $\frac{X-\mu}{\sigma}$ of X is referred to as *normalization* of X.

THEOREM 3 (i) Let X be an $N(0, 1)$-distributed r.v. Then $Y = X^2$ is distributed as χ_1^2.
(ii) If X is a $N(\mu, \sigma^2)$-distributed r.v., then the r.v. $\left(\frac{X-\mu}{\sigma}\right)^2$ is distributed as χ_1^2.

PROOF (i) We will show that the p.d.f. of Y is that of a χ_1^2-distributed r.v. by deriving the d.f. of Y first and then differentiating it in order to obtain f_Y. To this end, let us observe first that Y is an r.v. on account of Statement 1. Next, for $y > 0$, we have

$$F_Y(y) = P(Y \leq y) = P\left(-\sqrt{y} \leq \frac{X-\mu}{\sigma} \leq \sqrt{y}\right)$$

$$= \frac{1}{\sqrt{2\pi}} \int_{-\sqrt{y}}^{\sqrt{y}} e^{-x^2/2} dx = 2 \cdot \frac{1}{\sqrt{2\pi}} \int_0^{\sqrt{y}} e^{-x^2/2} dx.$$

Let $x = \sqrt{t}$. Then $dx = dt/2\sqrt{t}$, $t \in (0, y]$ and

$$F_Y(y) = 2 \cdot \frac{1}{\sqrt{2\pi}} \int_0^y \frac{1}{2\sqrt{t}} e^{-t/2} dt.$$

Hence

$$\frac{dF_Y(y)}{dy} = \frac{1}{\sqrt{2\pi}} \frac{1}{\sqrt{y}} e^{-y/2} = \frac{1}{\sqrt{2\pi}} y^{(1/2)-1} e^{-y/2}.$$

Since $f_Y(y) = 0$ for $y \leq 0$ (because $F_Y(y) = 0$, $y \leq 0$), it follows that

$$f_Y(y) = \begin{cases} \dfrac{1}{\sqrt{\pi} 2^{1/2}} y^{(1/2)-1} e^{-y/2}, & y > 0 \\ 0, & y \leq 0, \end{cases}$$

and this is the p.d.f. of χ_1^2. (Observe that here we used the fact that $\Gamma(\frac{1}{2}) = \sqrt{\pi}$.) ▲

Exercises

4.1.1 Refer to Exercise 3.2.13, in Chapter 3, and determine the d.f.'s corresponding to the p.d.f.'s given there.

4.1.2 Refer to Exercise 3.2.14, in Chapter 3, and determine the d.f.'s corresponding to the p.d.f.'s given there.

4.1.3 Refer to Exercise 3.3.13, in Chapter 3, and determine the d.f.'s corresponding to the p.d.f.'s given there.

4.1.4 Refer to Exercise 3.3.14, in Chapter 3, and determine the d.f.'s corresponding to the p.d.f.'s given there.

4.1.5 Let X be an r.v. with d.f. F. Determine the d.f. of the following r.v.'s: $-X$, X^2, $aX + b$, $XI_{[a,b)}(X)$ when:

i) X is continuous and F is strictly increasing;

ii) X is discrete.

4.1.6 Refer to the proof of Theorem 1 (iv) and show that we may assume that $x_n \downarrow -\infty$ ($x_n \uparrow \infty$) instead of $x_n \to -\infty$ ($x_n \to \infty$).

4.1.7 Let f and F be the p.d.f. and the d.f., respectively, of an r.v. X. Then show that F is continuous, and $dF(x)/dx = f(x)$ at the continuity points x of f.

4.1.8

i) Show that the following function F is a d.f. (*Logistic distribution*) and derive the corresponding p.d.f., f.

$$F(x) = \frac{1}{1 + e^{-(\alpha x + \beta)}}, \quad x \in \mathbb{R}, \quad \alpha > 0, \quad \beta \in \mathbb{R};$$

ii) Show that $f(x) = \alpha F(x)[1 - F(x)]$.

4.1.9 Refer to Exercise 3.3.17 in Chapter 3 and determine the d.f. F corresponding to the p.d.f. f given there. Write out the expressions of F and f for $n = 2$ and $n = 3$.

4.1.10 If X is an r.v. distributed as $N(3, 0.25)$, use Table 3 in Appendix III in order to compute the following probabilities:

i) $P(X < -1)$;

ii) $P(X > 2.5)$;

iii) $P(-0.5 < X < 1.3)$.

4.1.11 The distribution of IQ's of the people in a given group is well approximated by the Normal distribution with $\mu = 105$ and $\sigma = 20$. What proportion of the individuals in the group in question has an IQ:

i) At least 150?

ii) At most 80?

iii) Between 95 and 125?

4.1.12 A certain manufacturing process produces light bulbs whose life length (in hours) is an r.v. X distributed as $N(2,000, 200^2)$. A light bulb is supposed to be defective if its lifetime is less than 1,800. If 25 light bulbs are tested, what is the probability that at most 15 of them are defective? (Use the required independence.)

4.1.13 A manufacturing process produces $\frac{1}{2}$-inch ball bearings, which are assumed to be satisfactory if their diameter lies in the interval 0.5 ± 0.0006 and defective otherwise. A day's production is examined, and it is found that the distribution of the actual diameters of the ball bearings is approximately normal with mean $\mu = 0.5007$ inch and $\sigma = 0.0005$ inch. Compute the proportion of defective ball bearings.

4.1.14 If X is an r.v. distributed as $N(\mu, \sigma^2)$, find the value of c (in terms of μ and σ) for which $P(X < c) = 2 - 9P(X > c)$.

4.1.15 Refer to the Weibull p.d.f., f, given in Exercise 3.3.19 in Chapter 3 and do the following:

i) Calculate the corresponding d.f. F and the *reliability* function $\mathbb{R}(x) = 1 - F(x)$;

ii) Also, calculate the *failure (or hazard) rate* $H(x) = \frac{f(x)}{\mathbb{R}(x)}$, and draw its graph for $\alpha = 1$ and $\beta = \frac{1}{2}, 1, 2$;

iii) For s and $t > 0$, calculate the probability $P(X > s + t | X > t)$ where X is an r.v. having the Weibull distribution;

iv) What do the quantities $F(x)$, $\mathbb{R}(x)$, $H(x)$ and the probability in part (iii) become in the special case of the Negative Exponential distribution?

4.2 The d.f. of a Random Vector and Its Properties—Marginal and Conditional d.f.'s and p.d.f.'s

For the case of a two-dimensional r. vector, a result analogous to Theorem 1 can be established. So consider the case that $k = 2$. We then have $\mathbf{X} = (X_1, X_2)'$ and the d.f. F (or $F_\mathbf{X}$ or F_{X_1, X_2}) of $\mathbf{X}$, or the *joint distribution function* of X_1, X_2, is $F(x_1, x_2) = P(X_1 \leq x_1, X_2 \leq x_2)$. Then the following theorem holds true.

With the above notation we have

THEOREM 4

i) $0 \leq F(x_1, x_2) \leq 1$, $x_1, x_2 \in \mathbb{R}$.

ii) The variation of F over rectangles with sides parallel to the axes, given in Fig. 4.2, is ≥ 0.

iii) F is continuous from the right with respect to each of the coordinates x_1, x_2, or both of them jointly.

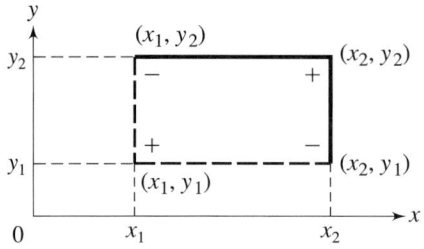

Figure 4.2 The variation V of F over the rectangle is:
$F(x_1, y_1) + F(x_2, y_2) - F(x_1, y_2) - F(x_2, y_1)$

iv) If both $x_1, x_2, \to \infty$, then $F(x_1, x_2) \to 1$, and if at least one of the $x_1, x_2 \to -\infty$, then $F(x_1, x_2) \to 0$. We express this by writing $F(\infty, \infty) = 1$, $F(-\infty, x_2) = F(x_1, -\infty) = F(-\infty, -\infty) = 0$, where $-\infty < x_1, x_2 < \infty$.

PROOF

i) Obvious.

ii) $V = P(x_1 < X_1 \leq x_2, y_1 < X_2 \leq y_2)$ and is hence, clearly, ≥ 0.

iii) Same as in Theorem 3. (If $\mathbf{x} = (x_1, x_2)'$, and $\mathbf{z}_n = (x_{1n}, x_{2n})'$, then $\mathbf{z}_n \downarrow \mathbf{x}$ means $x_{1n} \downarrow x_1, x_{2n} \downarrow x_2$).

iv) If $x_1, x_2 \uparrow \infty$, then $(-\infty, x_1] \times (-\infty, x_2] \uparrow R^2$, so that $F(x_1, x_2) \to P(S) = 1$. If at least one of x_1, x_2 goes ($\downarrow$) to $-\infty$, then $(-\infty, x_1] \times (-\infty, x_2] \downarrow \emptyset$, hence
$$F(x_1, x_2) \to P(\emptyset) = 0. \blacktriangle$$

REMARK 3 The function $F(x_1, \infty) = F_1(x_1)$ is the d.f. of the random variable X_1. In fact, $F(x_1, \infty) = F_1(x_1)$ is the d.f. of the random variable X_1. In fact,
$$F(x_1, \infty) = \lim_{x_n \uparrow \infty} P(X_1 \leq x_1, X_2 \leq x_n)$$
$$= P(X_1 \leq x_1, -\infty < X_2 < \infty) = P(X_1 \leq x_1) = F_1(x_1).$$

Similarly $F(\infty, x_2) = F_2(x_2)$ is the d.f. of the random variable X_2. F_1, F_2 are called *marginal d.f.'s*.

REMARK 4 It should be pointed out here that results like those discussed in parts (i)–(iv) in Remark 1 still hold true here (appropriately interpreted). In particular, part (iv) says that $F(x_1, x_2)$ has second order partial derivatives and
$$\frac{\partial^2}{\partial x_1 \partial x_2} F(x_1, x_2) = f(x_1, x_2)$$
at continuity points of f.

For $k > 2$, we have a theorem strictly analogous to Theorems 3 and 6 and also remarks such as Remark 1(i)–(iv) following Theorem 3. In particular, the analog of (iv) says that $F(x_1, \ldots, x_k)$ has kth order partial derivatives and

$$\frac{\partial^k}{\partial x_1 \partial x_2 \cdots \partial x_k} F(x_1, \ldots, x_k) = f(x_1, \ldots, x_k)$$

at continuity points of f, where F, or $F_\mathbf{X}$, or $F_{X_1,\ldots,X_k}$, is the d.f. of $\mathbf{X}$, or the *joint distribution function* of $X_1, \ldots, X_k$. As in the two-dimensional case,

$$F(\infty, \ldots, \infty, x_j, \infty, \ldots, \infty) = F_j(x_j)$$

is the d.f. of the random variable X_j, and if m x_j's are replaced by ∞ ($1 < m < k$), then the resulting function is the joint d.f. of the random variables corresponding to the remaining $(k - m)$ X_j's. All these d.f.'s are called *marginal distribution functions*.

In Statement 2, we have seen that if $\mathbf{X} = (X_1, \ldots, X_k)'$ is an r. vector, then X_j, $j = 1, 2, \ldots, k$ are r.v.'s and vice versa. Then the p.d.f. of $\mathbf{X}$, $f(\mathbf{x}) = f(x_1, \ldots, x_k)$, is also called the joint p.d.f. of the r.v.'s $X_1, \ldots, X_k$.

Consider first the case $k = 2$; that is, $\mathbf{X} = (X_1, X_2)'$, $f(\mathbf{x}) = f(x_1, x_2)$ and set

$$f_1(x_1) = \begin{cases} \sum_{x_2} f(x_1, x_2) \\ \int_{-\infty}^{\infty} f(x_1, x_2) dx_2 \end{cases}$$

$$f_2(x_2) = \begin{cases} \sum_{x_1} f(x_1, x_2) \\ \int_{-\infty}^{\infty} f(x_1, x_2) dx_1. \end{cases}$$

Then f_1, f_2 are p.d.f.'s. In fact, $f_1(x_1) \geq 0$ and

$$\sum_{x_1} f_1(x_1) = \sum_{x_1} \sum_{x_2} f(x_1, x_2) = 1,$$

or

$$\int_{-\infty}^{\infty} f_1(x_1) dx_1 = \int_{-\infty}^{\infty} \int_{-\infty}^{\infty} f(x_1, x_2) dx_1 dx_2 = 1.$$

Similarly we get the result for f_2. Furthermore, f_1 is the p.d.f. of X_1, and f_2 is the p.d.f. of X_2. In fact,

$$P(X_1 \in B) = \begin{cases} \sum_{x_1 \in B, x_2 \in \mathbb{R}} f(x_1, x_2) = \sum_{x_1 \in B} \sum_{x_2 \in \mathbb{R}} f(x_1, x_2) = \sum_{x_1 \in B} f_1(x_1) \\ \int_B \int_\mathbb{R} f(x_1, x_2) dx_1 dx_2 = \int_B \left[\int_\mathbb{R} f(x_1, x_2) dx_2 \right] dx_1 = \int_B f_1(x_1) dx_1. \end{cases}$$

Similarly f_2 is the p.d.f. of the r.v. X_2. We call f_1, f_2 the *marginal p.d.f.'s*. Now suppose $f_1(x_1) > 0$. Then define $f(x_2|x_1)$ as follows:

$$f(x_2|x_1) = \frac{f(x_1, x_2)}{f_1(x_1)}.$$

This is considered as a function of x_2, x_1 being an arbitrary, but fixed, value of X_1 ($f_1(x_1) > 0$). Then $f(\cdot|x_1)$ is a p.d.f. In fact, $f(x_2|x_1) \geq 0$ and

$$\sum_{x_2} f(x_2|x_1) = \frac{1}{f_1(x_1)} \sum_{x_2} f(x_1, x_2) = \frac{1}{f_1(x_1)} \cdot f_1(x_1) = 1,$$

$$\int_{-\infty}^{\infty} f(x_2|x_1) dx_2 = \frac{1}{f_1(x_1)} \int_{-\infty}^{\infty} f(x_1, x_2) dx_2 = \frac{1}{f_1(x_1)} \cdot f_1(x_1) = 1.$$

In a similar fashion, if $f_2(x_2) > 0$, we define $f(x_1|x_2)$ by:

$$f(x_1|x_2) = \frac{f(x_1, x_2)}{f_2(x_2)}$$

and show that $f(\cdot|x_2)$ is a p.d.f. Furthermore, if X_1, X_2 are both discrete, the $f(x_2|x_1)$ has the following interpretation:

$$f(x_2|x_1) = \frac{f(x_1, x_2)}{f_1(x_1)} = \frac{P(X_1 = x_1, X_2 = x_2)}{P(X_1 = x_1)} = P(X_2 = x_2 | X_1 = x_1).$$

Hence $P(X_2 \in B | X_1 = x_1) = \sum_{x_2 \in B} f(x_2|x_1)$. For this reason, we call $f(\cdot|x_2)$ the *conditional p.d.f. of* X_2, *given that* $X_1 = x_1$ (provided $f_1(x_1) > 0$). For a similar reason, we call $f(\cdot|x_2)$ the *conditional p.d.f. of* X_1, *given that* $X_2 = x_2$ (provided $f_2(x_2) > 0$). For the case that the p.d.f.'s f and f_2 are of the continuous type, the conditional p.d.f. $f(x_1|x_2)$ may be given an interpretation similar to the one given above. By assuming (without loss of generality) that $h_1, h_2 > 0$, one has

$$(1/h_1) P(x_1 < X_1 \leq x_1 + h_1 | x_2 < X_2 \leq x_2 + h_2)$$
$$= \frac{(1/h_1 h_2) P(x_1 < X_1 \leq x_1 + h_1, x_2 < X_2 \leq x_2 + h_2)}{(1/h_2) P(x_2 < X_2 \leq x_2 + h_2)}$$
$$= \frac{(1/h_1 h_2)[F(x_1, x_2) + F(x_1 + h_1, x_2 + h_2) - F(x_1, x_2 + h_2) - F(x_1 + h_1, x_2)]}{(1/h_2)[F_2(x_2 + h_2) - F_2(x_2)]}$$

where F is the joint d.f. of X_1, X_2 and F_2 is the d.f. of X_2. By letting $h_1, h_2 \to 0$ and assuming that $(x_1, x_2)'$ and x_2 are continuity points of f and f_2, respectively, the last expression on the right-hand side above tends to $f(x_1, x_2)/f_2(x_2)$ which was denoted by $f(x_1|x_2)$. Thus for small h_1, h_2, $h_1 f(x_1|x_2)$ is approximately equal to $P(x_1 < X_1 \leq x_1 + h_1 | x_2 < X_2 \leq x_2 + h_2)$, so that $h_1 f(x_1|x_2)$ is approximately the conditional probability that X_1 lies in a small neighborhood (of length h_1) of x_1, given that X_2 lies in a small neighborhood of x_2. A similar interpretation may be given to $f(x_2|x_1)$. We can also define the *conditional d.f. of* X_2, *given* $X_1 = x_1$, by means of

$$F(x_2|x_1) = \begin{cases} \sum_{x_2' \leq x_2} f(x_2'|x_1) \\ \int_{-\infty}^{x_2} f(x_2'|x_1)dx_2', \end{cases}$$

and similarly for $F(x_1|x_2)$.

The concepts introduced thus far generalize in a straightforward way for $k > 2$. Thus if $\mathbf{X} = (X_1, \ldots, X_k)'$ with p.d.f. $f(x_1, \ldots, x_k)$, then we have called $f(x_1, \ldots, x_k)$ the *joint p.d.f. of the r.v.'s* $X_1, X_2, \ldots, X_k$. If we sum (integrate) over n of the variables $x_1, \ldots, x_k$ keeping the remaining m fixed ($n + m = k$), the resulting function is the *joint p.d.f. of the r.v.'s corresponding to the remaining m variables*; that is,

$$f_{i_1, \ldots, i_m}(x_{i_1}, \ldots, x_{i_m}) = \begin{cases} \sum_{x_{j_1}, \ldots, x_{j_n}} f(x_1, \ldots, x_k) \\ \int_{-\infty}^{\infty} \cdots \int_{-\infty}^{\infty} f(x_1, \ldots, x_k) dx_{j_1} \cdots dx_{j_n}. \end{cases}$$

There are

$$\binom{k}{1} + \binom{k}{2} + \cdots + \binom{k}{k-1} = 2^k - 2$$

such p.d.f.'s which are also called *marginal p.d.f.'s*. Also if $x_{i_1}, \ldots, x_{i_m}$ are such that $f_{i_1, \ldots, i_m}(x_{i_1}, \ldots, x_{i_m}) > 0$, then the function (of $x_{j_1}, \ldots, x_{j_n}$) defined by

$$f(x_{j_1}, \ldots, x_{j_n}|x_{i_1}, \ldots, x_{i_m}) = \frac{f(x_1, \ldots, x_k)}{f_{i_1, \ldots, i_m}(x_{i_1}, \ldots, x_{i_m})}$$

is a p.d.f. called the *joint conditional p.d.f. of the r.v.'s* $X_{j_1}, \ldots, X_{j_n}$, given $X_{i_1} = x_{i_1}, \ldots, X_{i_m} = x_{i_m}$, or just given $X_{i_1}, \ldots, X_{i_m}$. Again there are $2^k - 2$ joint conditional p.d.f.'s involving all k r.v.'s $X_1, \ldots, X_k$. Conditional distribution functions are defined in a way similar to the one for $k = 2$. Thus

$$F(x_{j_1}, \ldots, x_{j_n}|x_{i_1}, \ldots, x_{i_m})$$
$$= \begin{cases} \sum_{(x_{j_1}', \ldots, x_{j_n}') \leq (x_{j_1}, \ldots, x_{j_n})} f(x_{j_1}', \ldots, x_{j_n}'|x_{i_1}, \ldots, x_{i_m}) \\ \int_{-\infty}^{x_{j_1}} \cdots \int_{-\infty}^{x_{j_n}} f(x_{j_1}', \ldots, x_{j_n}'|x_{i_1}, \ldots, x_{i_m}) dx_{j_1}' \ldots dx_{j_n}'. \end{cases}$$

We now present two examples of marginal and conditional p.d.f.'s, one taken from a discrete distribution and the other taken from a continuous distribution.

EXAMPLE 1 Let the r.v.'s $X_1, \ldots, X_k$ have the Multinomial distribution with parameters n and $p_1, \ldots, p_k$. Also, let m and n be integers such that $1 \leq m, n < k$ and $m + n = k$. Then in the notation employed above, we have:

i)
$$f_{i_1,\ldots,i_m}(x_{i_1},\ldots,x_{i_m}) = \frac{n!}{x_{i_1}!\cdots x_{i_m}!(n-r)!} p_{i_1}^{x_{i_1}} \cdots p_{i_m}^{x_{i_m}} q^{n-r},$$
$$q = 1-(p_{i_1}+\cdots+p_{i_m}),\ r = x_{i_1}+\cdots+x_{i_m};$$

that is, the r.v.'s $X_{i_1},\ldots,X_{i_m}$ and $Y = n - (X_{i_1}+\cdots+X_{i_m})$ have the Multinomial distribution with parameters n and $p_{i_1},\ldots,p_{i_m},q$.

ii)
$$f(x_{j_1},\ldots,x_{j_n}|x_{i_1},\ldots,x_{i_m}) = \frac{(n-r)!}{x_{j_1}!\cdots x_{j_n}!}\left(\frac{p_{j_1}}{q}\right)^{x_{j_1}}\cdots\left(\frac{p_{j_n}}{q}\right)^{x_{j_n}},$$
$$r = x_{i_1}+\cdots+x_{i_m};$$

that is, the (joint) conditional distribution of $X_{j_1},\ldots,X_{j_n}$ given $X_{i_1},\ldots,X_{i_m}$ is Multinomial with parameters $n-r$ and $p_{j_1}/q,\ldots,p_{j_n}/q$.

DISCUSSION

i) Clearly,
$$(X_{i_1}=x_{i_1},\ldots,X_{i_m}=x_{i_m}) \subseteq (X_{i_1}+\cdots+X_{i_m}=r) = (n-Y=r) = (Y=n-r),$$
so that
$$(X_{i_1}=x_{i_1},\ldots,X_{i_m}=x_{i_m}) = (X_{i_1}=x_{i_1},\ldots,X_{i_m}=x_{i_m},Y=n-r).$$

Denoting by O the outcome which is the grouping of all n outcomes distinct from those designated by $i_1,\ldots,i_m$, we have that the probability of O is q, and the number of its occurrences is Y. Thus, the r.v.'s $X_{i_1},\ldots,X_{i_m}$ and Y are distributed as asserted.

ii) We have

$$f(x_{j_1},\ldots,x_{j_n}|x_{i_1},\ldots,x_{i_m}) = \frac{f(x_{j_1},\ldots,x_{j_n},x_{i_1},\ldots,x_{i_m})}{f(x_{i_1},\ldots,x_{i_m})} = \frac{f(x_1,\ldots,x_k)}{f(x_{i_1},\ldots,x_{i_m})}$$

$$= \left(\frac{n!}{x_1!\cdots x_k!}p_1^{x_1}\cdots p_k^{x_k}\right) \Big/ \left(\frac{n!}{x_{i_1}!\cdots x_{i_m}!(n-r)!}p_{i_1}^{x_{i_1}}\cdots p_{i_m}^{x_{i_m}}q^{n-r}\right)$$

$$= \left(\frac{p_{i_1}^{x_{i_1}}\cdots p_{i_m}^{x_{i_m}}\cdot p_{j_1}^{x_{j_1}}\cdots p_{j_n}^{x_{j_n}}}{x_{i_1}!\cdots x_{i_m}!x_{j_1}!\cdots x_{j_n}!}\right) \Big/ \left(\frac{p_{i_1}^{x_{i_1}}\cdots p_{i_m}^{x_{i_m}}q^{x_{j_1}+\cdots+x_{j_n}}}{x_{i_1}!\cdots x_{i_m}!(n-r)!}\right)$$

$$\left(\text{since } n-r = n-(x_{i_1}+\cdots+x_{i_m}) = x_{j_1}+\cdots+x_{j_n}\right)$$

$$= \frac{(n-r)!}{x_{j_1}!\cdots x_{j_n}!}\left(\frac{p_{j_1}}{q}\right)^{x_{j_1}}\cdots\left(\frac{p_{j_n}}{q}\right)^{x_{j_n}},$$

as was to be seen.

EXAMPLE 2 Let the r.v.'s X_1 and X_2 have the Bivariate Normal distribution, and recall that their (joint) p.d.f. is given by:

$$f(x_1, x_2) = \frac{1}{2\pi\sigma_1\sigma_2\sqrt{1-\rho^2}}$$
$$\times \exp\left\{-\frac{1}{2(1-\rho^2)}\left[\left(\frac{x_1-\mu_1}{\sigma_1}\right)^2 - 2\rho\left(\frac{x_1-\mu_1}{\sigma_1}\right)\left(\frac{x_2-\mu_2}{\sigma_2}\right) + \left(\frac{x_2-\mu_2}{\sigma_2}\right)^2\right]\right\}.$$

We saw that the marginal p.d.f.'s f_1, f_2 are $N(\mu_1, \sigma_1^2)$, $N(\mu_2, \sigma_2^2)$, respectively; that is, X_1, X_2 are also normally distributed. Furthermore, in the process of proving that $f(x_1, x_2)$ is a p.d.f., we rewrote it as follows:

$$f(x_1, x_2) = \frac{1}{2\pi\sigma_1\sigma_2\sqrt{1-\rho^2}} \exp\left[-\frac{(x_1-\mu_1)^2}{2\sigma_1^2}\right] \cdot \exp\left[-\frac{(x_2-b)^2}{2\left(\sigma_2\sqrt{1-\rho^2}\right)^2}\right],$$

where
$$b = \mu_2 + \rho\frac{\sigma_2}{\sigma_1}(x_1-\mu_1).$$

Hence
$$f(x_2|x_1) = \frac{f(x_1, x_2)}{f_1(x_1)} = \frac{1}{\sqrt{2\pi}\sigma_2\sqrt{1-\rho^2}} \exp\left[-\frac{(x_2-b)^2}{2\left(\sigma_2\sqrt{1-\rho^2}\right)^2}\right]$$

which is the p.d.f. of an $N(b, \sigma_2^2(1-\rho^2))$ r.v. Similarly $f(x_1|x_2)$ is seen to be the p.d.f. of an $N(b', \sigma_1^2(1-\rho^2))$ r.v., where

$$b' = \mu_1 + \rho\frac{\sigma_1}{\sigma_2}(x_2-\mu_2).$$

Exercises

4.2.1 Refer to Exercise 3.2.17 in Chapter 3 and:

i) Find the marginal p.d.f.'s of the r.v.'s X_j, $j = 1, \cdots, 6$;
ii) Calculate the probability that $X_1 \geq 5$.

4.2.2 Refer to Exercise 3.2.18 in Chapter 3 and determine:

i) The marginal p.d.f. of each one of X_1, X_2, X_3;
ii) The conditional p.d.f. of X_1, X_2, given X_3; X_1, X_3, given X_2; X_2, X_3, given X_1;

iii) The conditional p.d.f. of X_1, given X_2, X_3; X_2, given X_3, X_1; X_3, given X_1, X_2.

If $n = 20$, compute the following probabilities:

iv) $P(3X_1 + X_2 \leq 5)$;

v) $P(X_1 < X_2 < X_3)$;

vi) $P(X_1 + X_2 = 10 | X_3 = 5)$;

vii) $P(3 \leq X_1 \leq 10 | X_2 = X_3)$;

viii) $P(X_1 < 3X_2 | X_1 > X_3)$.

4.2.3 Let X, Y be r.v.'s jointly distributed with p.d.f. f given by $f(x, y) = 2/c^2$ if $0 \leq x \leq y$, $0 \leq y \leq c$ and 0 otherwise.

i) Determine the constant c;

ii) Find the marginal p.d.f.'s of X and Y;

iii) Find the conditional p.d.f. of X, given Y, and the conditional p.d.f. of Y, given X;

iv) Calculate the probability that $X \leq 1$.

4.2.4 Let the r.v.'s X, Y be jointly distributed with p.d.f. f given by $f(x, y) = e^{-x-y} I_{(0,\infty) \times (0,\infty)}(x, y)$. Compute the following probabilities:

i) $P(X \leq x)$;

ii) $P(Y \leq y)$;

iii) $P(X < Y)$;

iv) $P(X + Y \leq 3)$.

4.2.5 If the joint p.d.f. f of the r.v.'s X_j, $j = 1, 2, 3$, is given by

$$f(x_1, x_2, x_3) = c^3 e^{-c(x_1 + x_2 + x_3)} I_A(x_1, x_2, x_3),$$

where

$$A = (0, \infty) \times (0, \infty) \times (0, \infty),$$

i) Determine the constant c;

ii) Find the marginal p.d.f. of each one of the r.v.'s X_j, $j = 1, 2, 3$;

iii) Find the conditional (joint) p.d.f. of X_1, X_2, given X_3, and the conditional p.d.f. of X_1, given X_2, X_3;

iv) Find the conditional d.f.'s corresponding to the conditional p.d.f.'s in (iii).

4.2.6 Consider the function given below:

$$f(x|y) = \begin{cases} \dfrac{y^x e^{-y}}{x!}, & x = 0, 1, \ldots; y \geq 0 \\ 0, & \text{otherwise.} \end{cases}$$

i) Show that for each fixed y, $f(\cdot|y)$ is a p.d.f., the conditional p.d.f. of an r.v. X, given that another r.v. Y equals y;

ii) If the marginal p.d.f. of Y is Negative Exponential with parameter $\lambda = 1$, what is the joint p.d.f. of X, Y?

iii) Show that the marginal p.d.f. of X is given by $f(x) = (\frac{1}{2})^{x+1} I_A(x)$, where $A = \{0, 1, 2, \dots\}$.

4.2.7 Let Y be an r.v. distributed as $P(\lambda)$ and suppose that the conditional distribution of the r.v. X, given $Y = n$, is $B(n, p)$. Determine the p.d.f. of X and the conditional p.d.f. of Y, given $X = x$.

4.2.8 Consider the function f defined as follows:

$$f(x_1, x_2) = \frac{1}{2\pi} \exp\left(-\frac{x_1^2 + x_2^2}{2}\right) + \frac{1}{4\pi e} x_1^3 x_2^3 I_{[-1,1] \times [-1,1]}(x_1, x_2)$$

and show that:

i) f is a non-Normal Bivariate p.d.f.

ii) Both marginal p.d.f.'s

$$f_1(x_1) = \int_{-\infty}^{\infty} f(x_1, x_2) dx_2$$

and

$$f_2(x_2) = \int_{-\infty}^{\infty} f(x_1, x_2) dx_1$$

are Normal p.d.f.'s.

4.3 Quantiles and Modes of a Distribution

Let X be an r.v. with d.f. F and consider a number p such that $0 < p < 1$. A *p*th *quantile* of the r.v. X, or of its d.f. F, is a number denoted by x_p and having the following property: $P(X \leq x_p) \geq p$ and $P(X \geq x_p) \geq 1 - p$. For $p = 0.25$ we get a *quartile* of X, or its d.f., and for $p = 0.5$ we get a *median* of X, or its d.f. For illustrative purposes, consider the following simple examples.

EXAMPLE 3 Let X be an r.v. distributed as $U(0, 1)$ and let $p = 0.10, 0.20, 0.30, 0.40, 0.50, 0.60, 0.70, 0.80$ and 0.90. Determine the respective $x_{0.10}, x_{0.20}, x_{0.30}, x_{0.40}, x_{0.50}, x_{0.60}, x_{0.70}, x_{0.80}$, and $x_{0.90}$.

Since for $0 \leq x \leq 1$, $F(x) = x$, we get: $x_{0.10} = 0.10$, $x_{0.20} = 0.20$, $x_{0.30} = 0.30$, $x_{0.40} = 0.40$, $x_{0.50} = 0.50$, $x_{0.60} = 0.60$, $x_{0.70} = 0.70$, $x_{0.80} = 0.80$, and $x_{0.90} = 0.90$.

EXAMPLE 4 Let X be an r.v. distributed as $N(0, 1)$ and let $p = 0.10, 0.20, 0.30, 0.40, 0.50, 0.60, 0.70, 0.80$ and 0.90. Determine the respective $x_{0.10}, x_{0.20}, x_{0.30}, x_{0.40}, x_{0.50}, x_{0.60}, x_{0.70}, x_{0.80}$, and $x_{0.90}$.

Typical cases:

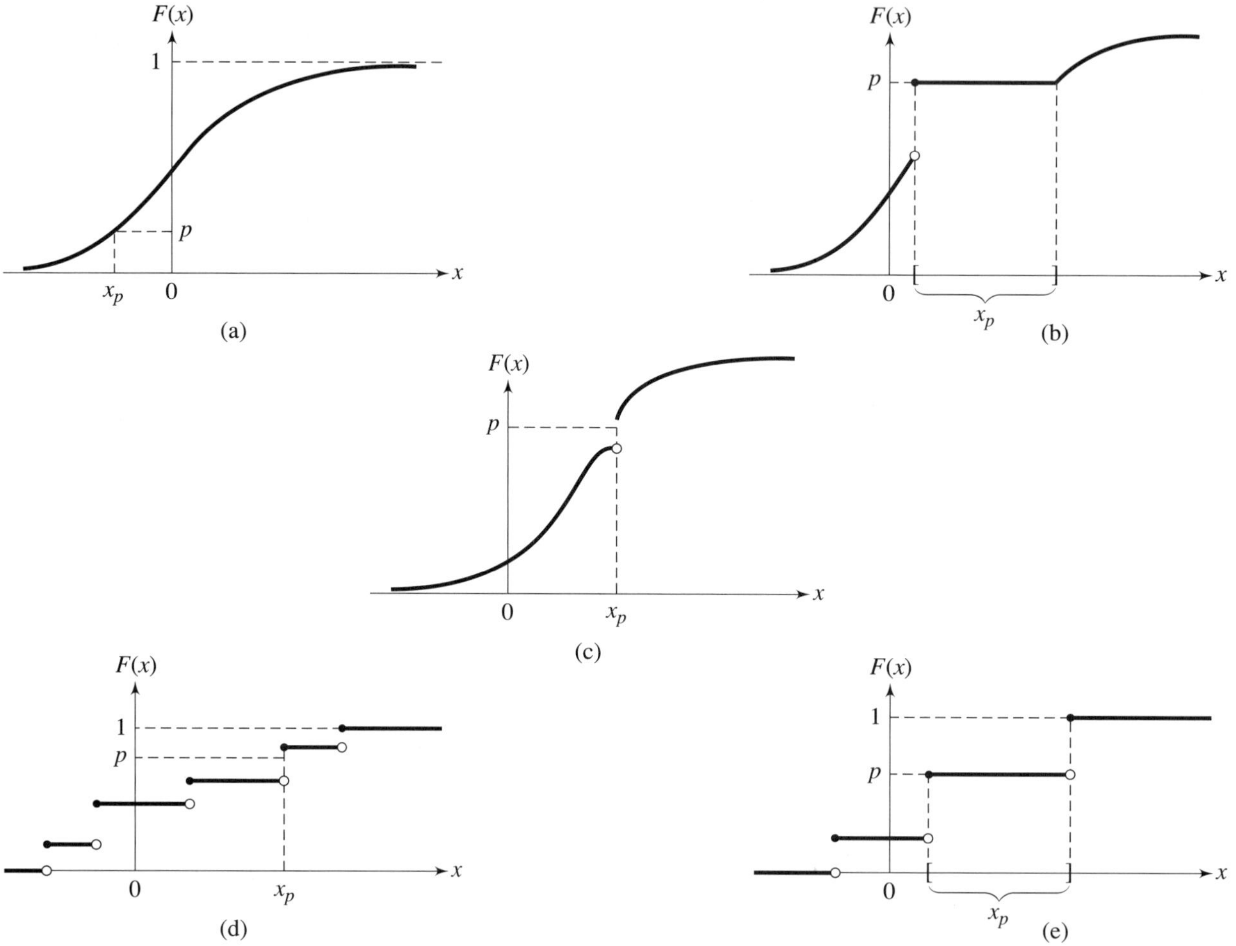

Figure 4.3 Observe that the figures demonstrate that, as defined, x_p need not be unique.

From the Normal Tables (Table 3 in Appendix III), by linear interpolation and symmetry, we find: $x_{0.10} = -1.282$, $x_{0.20} = -0.842$, $x_{0.30} = -0.524$, $x_{0.40} = -0.253$, $x_{0.50} = 0$, $x_{0.60} = 0.253$, $x_{0.70} = 0.524$, $x_{0.80} = 0.842$, and $x_{0.90} = 1.282$.

Knowledge of quantiles x_p for several values of p provides an indication as to how the unit probability mass is distributed over the real line. In Fig. 4.3 various cases are demonstrated for determining graphically the pth quantile of a d.f.

Let X be an r.v. with a p.d.f. f. Then *a mode* of f, if it exists, is any number which maximizes $f(x)$. In case f is a p.d.f. which is twice differentiable, a mode can be found by differentiation. This process breaks down in the discrete cases. The following theorems answer the question for two important discrete cases.

4.3 Quantiles and Modes of a Distribution

THEOREM 5 Let X be $B(n, p)$; that is,

$$f(x) = \binom{n}{x} p^x q^{n-x}, \quad 0 < p < 1, \quad q = 1 - p, \quad x = 0, 1, \ldots, n.$$

Consider the number $(n + 1)p$ and set $m = [(n + 1)p]$, where $[y]$ denotes the largest integer which is $\leq y$. Then if $(n + 1)p$ is not an integer, $f(x)$ has a unique mode at $x = m$. If $(n + 1)p$ is an integer, then $f(x)$ has two modes obtained for $x = m$ and $x = m - 1$.

PROOF For $x \geq 1$, we have

$$\frac{f(x)}{f(x-1)} = \frac{\binom{n}{x} p^x q^{n-x}}{\binom{n}{x-1} p^{x-1} q^{n-x+1}}$$

$$= \frac{\dfrac{n!}{x!(n-x)!} p^x q^{n-x}}{\dfrac{n!}{(x-1)!(n-x+1)!} p^{x-1} q^{n-x+1}} = \frac{n-x+1}{x} \cdot \frac{p}{q}.$$

That is,

$$\frac{f(x)}{f(x-1)} = \frac{n-x+1}{x} \cdot \frac{p}{q}.$$

Hence $f(x) > f(x - 1)$ (f is increasing) if and only if

$$(n - x + 1)p > x(1 - p), \quad \text{or} \quad np - xp + p > x - xp, \quad \text{or} \quad (n+1)p > x.$$

Thus if $(n + 1)p$ is *not* an integer, $f(x)$ keeps increasing for $x \leq m$ and then decreases so the maximum occurs at $x = m$. If $(n + 1)p$ is an integer, then the maximum occurs at $x = (n + 1)p$, where $f(x) = f(x - 1)$ (from above calculations). Thus

$$x = (n+1)p - 1$$

is a second point which gives the maximum value. ▲

THEOREM 6 Let X be $P(\lambda)$; that is,

$$f(x) = e^{-\lambda} \frac{\lambda^x}{x!}, \quad x = 0, 1, 2, \ldots, \quad \lambda > 0.$$

Then if λ is not an integer, $f(x)$ has a unique mode at $x = [\lambda]$. If λ is an integer, then $f(x)$ has two modes obtained for $x = \lambda$ and $x = \lambda - 1$.

PROOF For $x \geq 1$, we have

$$\frac{f(x)}{f(x-1)} = \frac{e^{-\lambda}(\lambda^x/x!)}{e^{-\lambda}[\lambda^{x-1}/(x-1)!]} = \frac{\lambda}{x}.$$

Hence $f(x) > f(x-1)$ if and only if $\lambda > x$. Thus if λ is *not* an integer, $f(x)$ keeps increasing for $x \leq [\lambda]$ and then decreases. Then the maximum of $f(x)$ occurs at $x = [\lambda]$. If λ is an integer, then the maximum occurs at $x = \lambda$. But in this case $f(x) = f(x-1)$ which implies that $x = \lambda - 1$ is a second point which gives the maximum value to the p.d.f. ▲

Exercises

4.3.1 Determine the pth quantile x_p for each one of the p.d.f.'s given in Exercises 3.2.13–15, 3.3.13–16 (Exercise 3.2.14 for $\alpha = \frac{1}{4}$) in Chapter 3 if $p = 0.75, 0.50$.

4.3.2 Let X be an r.v. with p.d.f. f symmetric about a constant c (that is, $f(c - x) = f(c + x)$ for all $x \in \mathbb{R}$). Then show that c is a median of f.

4.3.3 Draw four graphs—two each for $B(n, p)$ and $P(\lambda)$—which represent the possible occurrences for modes of the distributions $B(n, p)$ and $P(\lambda)$.

4.3.4 Consider the same p.d.f.'s mentioned in Exercise 4.3.1 from the point of view of a mode.

4.4* Justification of Statements 1 and 2

In this section, a rigorous justification of Statements 1 and 2 made in Section 4.1 will be presented. For this purpose, some preliminary concepts and results are needed and will be also discussed.

DEFINITION 1 A set G in $\mathbb{R}$ is called *open* if for every x in G there exists an open interval containing x and contained in G. Without loss of generality, such intervals may be taken to be centered at x.

It follows from this definition that an open interval is an open set, the entire real line $\mathbb{R}$ is an open set, and so is the empty set (in a vacuous manner).

LEMMA 1 Every open set in $\mathbb{R}$ is measurable.

PROOF Let G be an open set in $\mathbb{R}$, and for each $x \in G$, consider an open interval centered at x and contained in G. Clearly, the union over x, as x varies in G, of such intervals is equal to G. The same is true if we consider only those intervals corresponding to all rationals x in G. These intervals are countably many and each one of them is measurable; then so is their union. ▲

DEFINITION 2 A set G in $\mathbb{R}^m$, $m \geq 1$, is called *open* if for every x in G there exists an open cube in $\mathbb{R}^m$ containing x and contained in G; by the term open "cube" we mean the Cartesian product of m open intervals of equal length. Without loss of generality, such cubes may be taken to be centered at x.

LEMMA 2 Every open set in $\mathbb{R}^n$ is measurable.

PROOF It is analogous to that of Lemma 1. Indeed, let G be an open set in $\mathbb{R}^m$, and for each $x \in G$, consider an open cube centered at x and contained in G. The union over x, as x varies in G, of such cubes clearly is equal to G. The same is true if we restrict ourselves to x's in G whose m coordinates are rationals. Then the resulting cubes are countably many, and therefore their union is measurable, since so is each cube. ▲

DEFINITION 3 Recall that a function $g: S \subseteq \mathbb{R} \to \mathbb{R}$ is said to be *continuous at* $x_0 \in S$ if for every $\varepsilon > 0$ there exists a $\delta = \delta(\varepsilon, x_0) > 0$ such that $|x - x_0| < \varepsilon$ implies $|g(x) - g(x_0)| < \delta$. The function g is *continuous in* S if it is continuous for every $x \in S$.

It follows from the concept of continuity that $\varepsilon \to 0$ implies $\delta \to 0$.

LEMMA 3 Let $g: \mathbb{R} \to \mathbb{R}$ be continuous. Then g is measurable.

PROOF By Theorem 5 in Chapter 1 it suffices to show that $g^{-1}(G)$ are measurable sets for all open intevals G in $\mathbb{R}$. Set $B = g^{-1}(G)$. Thus if $B = \emptyset$, the assertion is valid, so let $B \neq \emptyset$ and let x_0 be an arbitrary point of B, so that $g(x_0) \in G$. Continuity of g at x_0 implies that for every $\varepsilon > 0$ there exists $\delta = \delta(\varepsilon, x_0) > 0$ such that $|x - x_0| < \varepsilon$ implies $|g(x) - g(x_0)| < \delta$. Equivalently, $x \in (x_0 - \varepsilon, x_0 + \varepsilon)$ implies $g(x) \in (g(x_0) - \delta, g(x_0) + \delta)$. Since $g(x_0) \in G$ and G is open, by choosing ε sufficiently small, we can make δ so small that $(g(x_0) - \delta, g(x_0) + \delta)$ is contained in G. Thus, for such a choice of ε and δ, $x \in (x_0 - \varepsilon, x_0 + \varepsilon)$ implies that $(g(x_0) - \delta, g(x_0) + \delta) \subset G$. But $B(= g^{-1}(G))$ is the set of *all* $x \in \mathbb{R}$ for which $g(x) \in G$. As all $x \in (x_0 - \varepsilon, x_0 + \varepsilon)$ have this property, it follows that $(x_0 - \varepsilon, x_0 + \varepsilon) \subset B$. Since x_0 is arbitrary in B, it follows that B is open. Then by Lemma 1, it is measurable. ▲

The concept of continuity generalizes, of course, to Euclidean spaces of higher dimensions, and then a result analogous to the one in Lemma 3 also holds true.

DEFINITION 4 A function $g: S \subseteq \mathbb{R}^k \to \mathbb{R}^m$ ($k, m \geq 1$) is said to be *continuous at* $\mathbf{x}_0 \in \mathbb{R}^k$ if for every $\varepsilon > 0$ there exists a $\delta = \delta(\varepsilon, \mathbf{x}_0) > 0$ such that $\|\mathbf{x} - \mathbf{x}_0\| < \varepsilon$ implies $\|g(\mathbf{x}) - g(\mathbf{x}_0)\| < \delta$. The function g is *continuous in* S if it is continuous for every $\mathbf{x} \in S$. Here $\|\mathbf{x}\|$ stands for the usual norm in $\mathbb{R}^k$; i.e., for $\mathbf{x} = (x_1, \ldots, x_k)'$, $\|\mathbf{x}\| = \left(\sum_{i=1}^k x_i^2\right)^{1/2}$, and similarly for the other quantities.

Once again, from the concept of continuity it follows that $\varepsilon \to 0$ implies $\delta \to 0$.

LEMMA 4 Let $g: \mathbb{R}^k \to \mathbb{R}^m$ be continuous. Then g is measurable.

PROOF The proof is similar to that of Lemma 3. The details are presented here for the sake of completeness. Once again, it suffices to show that $g^{-1}(G)$ are measurable sets for all open cubes G in $\mathbb{R}^m$. Set $B = g^{-1}(G)$. If $B = \emptyset$ the assertion is true, and therefore suppose that $B \neq \emptyset$ and let $\mathbf{x}_0$ be an arbitrary point of B. Continuity of g at $\mathbf{x}_0$ implies that for every $\varepsilon > 0$ there exists a $\delta = \delta(\varepsilon, \mathbf{x}_0) > 0$ such that $\|\mathbf{x} - \mathbf{x}_0\| < \varepsilon$ implies $\|g(\mathbf{x}) - g(\mathbf{x}_0)\| < \delta$; equivalently, $\mathbf{x} \in S(\mathbf{x}_0, \varepsilon)$ implies $g(\mathbf{x}) \in S(g(\mathbf{x}_0), \delta)$, where $S(c, r)$ stands for the open sphere with center $\mathbf{c}$ and radius r. Since $g(\mathbf{x}_0) \in G$ and G is open, we can choose ε so small that the corresponding δ is sufficiently small to imply that $g(\mathbf{x}) \in S(g(\mathbf{x}_0), \delta)$. Thus, for such a choice of ε and δ, $\mathbf{x} \in S(\mathbf{x}_0, \varepsilon)$ implies that $g(\mathbf{x}) \in S(g(\mathbf{x}_0), \delta)$. Since $B(= g^{-1}(G))$ is the set of *all* $\mathbf{x} \in \mathbb{R}^k$ for which $g(\mathbf{x}) \in G$, and $\mathbf{x} \in S(\mathbf{x}_0, \varepsilon)$ implies that $g(\mathbf{x}) \in S(g(\mathbf{x}_0), \delta)$, it follows that $S(\mathbf{x}_0, \varepsilon) \subset B$. At this point, observe that it is clear that there is a cube containing $\mathbf{x}_0$ and contained in $S(\mathbf{x}_0, \varepsilon)$; call it $C(\mathbf{x}_0, \varepsilon)$. Then $C(\mathbf{x}_0, \varepsilon) \subset B$, and therefore B is open. By Lemma 2, it is also measurable. ▲

We may now proceed with the justification of Statement 1.

THEOREM 7 Let $X : (S, \mathcal{A}) \to (\mathbb{R}^k, \mathcal{B}^k)$ be a random vector, and let $g : (\mathbb{R}^k, \mathcal{B}^k) \to (\mathbb{R}^m, \mathcal{B}^m)$ be measurable. Then $g(\mathbf{X}) : (S, \mathcal{A}) \to (\mathbb{R}^m, \mathcal{B}^m)$ and is a random vector. (That is, measurable functions of random vectors are random vectors.)

PROOF To prove that $[g(\mathbf{X})]^{-1}(B) \in \mathcal{A}$ if $B \in \mathcal{B}^m$, we have

$$[g(\mathbf{X})]^{-1}(B) = \mathbf{X}^{-1}[g^{-1}(B)] = \mathbf{X}^{-1}(B_1), \quad \text{where} \quad B_1 = g^{-1}(B) \in \mathcal{B}^k$$

by the measurability of g. Also, $\mathbf{X}^{-1}(B_1) \in \mathcal{A}$ since $\mathbf{X}$ is measurable. The proof is completed. ▲

To this theorem, we have the following

COROLLARY Let $\mathbf{X}$ be as above and g be continuous. Then $g(\mathbf{X})$ is a random vector. (That is, continuous functions of random vectors are random vectors.)

PROOF The continuity of g implies its measurability by Lemma 3, and therefore the theorem applies and gives the result. ▲

DEFINITION 5 For $j = 1, \ldots, k$, the *j*th *projection function* g_j is defined by: $g_j : \mathbb{R}^k \to \mathbb{R}$ and $g_j(\mathbf{x}) = g_j(x_1, \ldots, x_k) = x_j$.

It so happens that projection functions are continuous; that is,

LEMMA 5 The coordinate functions g_j, $j = 1, \ldots, k$, as defined above, are continuous.

PROOF For an arbitrary point $\mathbf{x}_0$ in $\mathbb{R}^K$, consider $\mathbf{x} \in \mathbb{R}^K$ such that $\|\mathbf{x} - \mathbf{x}_0\| < \varepsilon$ for some $\varepsilon > 0$. This is equivalent to $\|\mathbf{x} - \mathbf{x}_0\|^2 < \varepsilon^2$ or $\sum_{j=1}^{k}(x_j - x_{0j})^2 < \varepsilon^2$ which implies that $(x_j - x_{0j})^2 < \varepsilon^2$ for $j = 1, \ldots, k$, or $|x_j - x_{0j}| < \varepsilon$, $j = 1, \ldots, k$. This last expression is equivalent to $|g_j(\mathbf{x}) - g_j(\mathbf{x}_0)| < \varepsilon$, $j = 1, \ldots, k$. Thus the definition of continuity of g_j is satisfied here for $\delta = \varepsilon$. ▲

Now consider a k-dimensional function $\mathbf{X}$ defined on the sample space S. Then $\mathbf{X}$ may be written as $\mathbf{X} = (X_1, \ldots, X_k)'$, where X_j, $j = 1, \ldots, k$ are real-valued functions. The question then arises as to how $\mathbf{X}$ and X_j, $j = 1, \ldots, k$ are related from a measurability point of view. To this effect, we have the following result.

THEOREM 8 Let $\mathbf{X} = (X_1, \ldots, X_k)': (S, \mathcal{A}) \to (\mathbb{R}^k, \mathcal{B}^k)$. Then $\mathbf{X}$ is an r. vector if and only if $X_j, j = 1, \ldots, k$ are r.v.'s.

PROOF Suppose $\mathbf{X}$ is an r. vector and let $g_j, j = 1, \ldots, k$ be the coordinate functions defined on $\mathbb{R}^k$. Then g_j's are continuous by Lemma 5 and therefore measurable by Lemma 4. Then for each $j = 1, \ldots, k$, $g_j(\mathbf{X}) = g_j(X_1, \ldots, X_k) = X_j$ is measurable and hence an r.v.

Next, assume that $X_j, j = 1, \ldots, k$ are r.v.'s. To show that $\mathbf{X}$ is an r. vector, by special case 3 in Section 2 of Chapter 1, it suffices to show that $\mathbf{X}^{-1}(B) \in \mathcal{A}$ for each $B = (-\infty, x_1] \times \cdots \times (-\infty, x_k]$, $x_1, \ldots, x_k \in \mathbb{R}$. Indeed,

$$\mathbf{X}^{-1}(B) = (\mathbf{X} \in B) = \left(X_j \in (-\infty, x_j], j = 1, \ldots, k\right) = \bigcap_{j=1}^{k} X_j^{-1}\left((-\infty, x_j]\right) \in \mathcal{A}.$$

The proof is completed. ▲

Exercises

4.4.1 If X and Y are functions defined on the sample space S into the real line $\mathbb{R}$, show that:

$$\{s \in S;\ X(s) + Y(s) < x\} = \bigcup_{r \in Q}\left[\{s \in S;\ X(s) < r\} \cap \{s \in S;\ Y(s) < x - r\}\right],$$

where Q is the set of rationals in $\mathbb{R}$.

4.4.2 Use Exercise 4.4.1 in order to show that, if X and Y are r.v.'s, then so is the function $X + Y$.

4.4.3

i) If X is an r.v., then show that so is the function $-X$.

ii) Use part (i) and Exercise 4.4.2 to show that, if X and Y are r.v.'s, then so is the function $X - Y$.

4.4.4

i) If X is an r.v., then show that so is the function X^2.

ii) Use the identity: $XY = \frac{1}{2}(X + Y)^2 - \frac{1}{2}(X^2 + Y^2)$ in conjunction with part (i) and Exercises 4.4.2 and 4.4.3(ii) to show that, if X and Y are r.v.'s, then so is the function XY.

4.4.5

i) If X is an r.v., then show that so is the function $\frac{1}{X}$, provided $X \neq 0$.

ii) Use part (i) in conjunction with Exercise 4.4.4(ii) to show that, if X and Y are r.v.'s, then so is the function $\frac{X}{Y}$, provided $Y \neq 0$.

Chapter 5

Moments of Random Variables—Some Moment and Probability Inequalities

5.1 Moments of Random Variables

In the definitions to be given shortly, the following remark will prove useful.

REMARK 1 We say that the (infinite) series $\sum_{\mathbf{x}} h(\mathbf{x})$, where $\mathbf{x} = (x_1, \ldots, x_k)'$ varies over a discrete set in $\mathbb{R}^k$, $k \geq 1$, *converges absolutely* if $\sum_{\mathbf{x}} |h(\mathbf{x})| < \infty$. Also we say that the integral $\int_{-\infty}^{\infty} \cdots \int_{-\infty}^{\infty} h(x_1, \ldots, x_k) dx_1 \cdots dx_k$ converges absolutely if

$$\int_{-\infty}^{\infty} \cdots \int_{-\infty}^{\infty} |h(x_1, x_2, \ldots, x_k)| dx_1 dx_2 \cdots dx_k < \infty.$$

In what follows, when we write (infinite) series or integrals *it will always be assumed that they converge absolutely.* In this case, we say that the moments to be defined below *exist.*

Let $\mathbf{X} = (X_1, \ldots, X_k)'$ be an r. vector with p.d.f. f and consider the (measurable) function $g: \mathbb{R}^k \to \mathbb{R}$, so that $g(\mathbf{X}) = g(X_1, \ldots, X_k)$ is an r.v. Then we give the

DEFINITION 1 i) For $n = 1, 2, \ldots$, the *nth moment of $g(\mathbf{X})$* is denoted by $E[g(\mathbf{X})]^n$ and is defined by:

$$E[g(\mathbf{X})]^n = \begin{cases} \sum_{\mathbf{x}} [g(\mathbf{x})]^n f(\mathbf{x}), & \mathbf{x} = (x_1, \ldots, x_k)' \\ \int_{-\infty}^{\infty} \cdots \int_{-\infty}^{\infty} [g(x_1, \ldots, x_k)]^n f(x_1, \ldots, x_k) dx_1 \cdots dx_k. \end{cases}$$

For $n = 1$, we get

$$E[g(\mathbf{X})] = \begin{cases} \sum_{\mathbf{x}} g(\mathbf{x}) f(\mathbf{x}) \\ \int_{-\infty}^{\infty} \cdots \int_{-\infty}^{\infty} g(x_1, \ldots, x_k) f(x_1, \ldots, x_k) dx_1 \cdots dx_k \end{cases}$$

and call it the *mathematical expectation or mean value* or just *mean* of $g(\mathbf{X})$. Another notation for $E[g(\mathbf{X})]$ which is often used is $\mu_{g(\mathbf{X})}$, or $\mu[g(\mathbf{X})]$, or just μ, if no confusion is possible.

ii) For $r > 0$, the *rth absolute moment* of $g(\mathbf{X})$ is denoted by $E|g(\mathbf{X})|^r$ and is defined by:

$$E|g(\mathbf{X})|^r = \begin{cases} \sum_{\mathbf{x}} |g(\mathbf{x})|^r f(\mathbf{x}), & \mathbf{x} = (x_1, \ldots, x_k)' \\ \int_{-\infty}^{\infty} \cdots \int_{-\infty}^{\infty} |g(x_1, \ldots, x_k)|^r f(x_1, \ldots, x_k) dx_1 \cdots dx_k. \end{cases}$$

iii) For an arbitrary constant c, and n and r as above, the *nth moment* and *rth absolute moment of $g(\mathbf{X})$ about c* are denoted by $E[g(\mathbf{X}) - c]^n$, $E|g(\mathbf{X}) - c|^r$, respectively, and are defined as follows:

$$E[g(\mathbf{X}) - c]^n = \begin{cases} \sum_{\mathbf{x}} [g(\mathbf{x}) - c]^n f(\mathbf{x}), & \mathbf{x} = (x_1, \ldots, x_k)' \\ \int_{-\infty}^{\infty} \cdots \int_{-\infty}^{\infty} [g(x_1, \ldots, x_k) - c]^n f(x_1, \ldots, x_k) dx_1 \cdots dx_k, \end{cases}$$

and

$$E|g(\mathbf{X}) - c|^r = \begin{cases} \sum_{\mathbf{x}} |g(\mathbf{x}) - c|^r f(\mathbf{x}), & \mathbf{x} = (x_1, \ldots, x_k)' \\ \int_{-\infty}^{\infty} \cdots \int_{-\infty}^{\infty} |g(x_1, \ldots, x_k) - c|^r f(x_1, \ldots, x_k) dx_1 \cdots dx_k. \end{cases}$$

For $c = E[g(\mathbf{X})]$, the moments are called *central moments*. The 2nd central moment of $g(\mathbf{X})$, that is,

$$E\{g(\mathbf{X}) - E[g(\mathbf{X})]\}^2$$

$$= \begin{cases} \sum_{\mathbf{x}} [g(\mathbf{x}) - Eg(\mathbf{X})]^2 f(\mathbf{x}), & \mathbf{x} = (x_1, \ldots, x_k)' \\ \int_{-\infty}^{\infty} \cdots \int_{-\infty}^{\infty} [g(x_1, \ldots, x_k) - Eg(\mathbf{X})]^2 f(x_1, \ldots, x_k) dx_1 \cdots dx_k \end{cases}$$

is called the *variance* of $g(\mathbf{X})$ and is also denoted by $\sigma^2[g(\mathbf{X})]$, or $\sigma^2_{g(\mathbf{X})}$, or just σ^2, if no confusion is possible. The quantity $+\sqrt{\sigma^2[g(\mathbf{X})]} = \sigma[g(\mathbf{X})]$ is called the *standard deviation (s.d.) of $g(\mathbf{X})$* and is also denoted by $\sigma_{g(\mathbf{X})}$, or just σ, if no confusion is possible. The variance of an r.v. is referred to as the *moment of inertia* in Mechanics.

5.1.1 Important Special Cases

1. Let $g(X_1, \ldots, X_k) = X_1^{n_1} \cdots X_k^{n_k}$, where $n_j \geq 0$ are integers. Then $E(X_1^{n_1} \cdots X_k^{n_k})$ is called the $(n_1, \ldots, n_k)$-*joint moment* of $X_1, \ldots, X_k$. In particular, for $n_1 = \cdots = n_{j-1} = n_{j+1} = \cdots = n_k = 0$, $n_j = n$, we get

$$E(X_j^n) = \begin{cases} \sum_{\mathbf{x}} x_j^n f(\mathbf{x}) = \sum_{(x_1,\ldots,x_k)} x_j^n f(x_1,\ldots,x_k) \\ \int_{-\infty}^{\infty} \cdots \int_{-\infty}^{\infty} x_j^n f(x_1,\ldots,x_k) dx_1 \cdots dx_k \end{cases}$$

$$= \begin{cases} \sum_{x_j} x_j^n f_j(x_j) \\ \int_{-\infty}^{\infty} x_j^n f_j(x_j) dx_j \end{cases}$$

which is the *nth moment of the r.v.* X_j. Thus the *n*th moment of an r.v. X with p.d.f. f is

$$E(X^n) = \begin{cases} \sum_{x} x^n f(x) \\ \int_{-\infty}^{\infty} x^n f(x) dx. \end{cases}$$

For $n = 1$, we get

$$E(X) = \begin{cases} \sum_{x} x f(x) \\ \int_{-\infty}^{\infty} x f(x) dx \end{cases}$$

which is the *mathematical expectation* or *mean value* or just *mean* of X. This quantity is also denoted by μ_X or $\mu(X)$ or just μ when no confusion is possible.

The quantity μ_X can be interpreted as follows: It follows from the definition that if X is a discrete uniform r.v., then μ_X is just the arithmetic average of the possible outcomes of X. Also, if one recalls from physics or elementary calculus the definition of *center of gravity* and its physical interpretation as the point of balance of the distributed mass, the interpretation of μ_X as the *mean* or *expected* value of the random variable is the natural one, provided the probability distribution of X is interpreted as the unit mass distribution.

REMARK 2 In Definition 1, suppose X is a continuous r.v. Then $E[g(X)] = \int_{-\infty}^{\infty} g(x)f(x)dx$. On the other hand, from the last expression above, $E(X) = \int_{-\infty}^{\infty} xf(x)dx$. There seems to be a discrepancy between these two definitions. More specifically, in the definition of $E[g(X)]$, one would expect to use the p.d.f. of $g(X)$ rather than that of X. Actually, the definition of $E[g(X)]$, as given, is correct and its justification is roughly as follows: Consider $E[g(x)] = \int_{-\infty}^{\infty} g(x)f(x)dx$ and set $y = g(x)$. Suppose that g is differentiable and has an inverse g^{-1}, and that some further conditions are met. Then

$$\int_{-\infty}^{\infty} g(x)f(x)dx = \int_{-\infty}^{\infty} yf[g^{-1}(y)]\left|\frac{d}{dy}g^{-1}(y)\right|dy.$$

On the other hand, if f_Y is the p.d.f. of Y, then $f_Y(y) = f[g^{-1}(y)] |\frac{d}{dy} g^{-1}(y)|$. Therefore the last integral above is equal to $\int_{-\infty}^{\infty} y f_Y(y) dy$, which is consonant with the definition of $E(X) = \int_{-\infty}^{\infty} x f(x) dx$. (A justification of the above derivations is given in Theorem 2 of Chapter 9.)

2. For g as above, that is, $g(X_1, \ldots, X_k) = X_1^{n_1} \cdots X_k^{n_k}$ and $n_1 = \cdots = n_{j-1} = n_{j+1} = \cdots = n_k = 0$, $n_j = 1$, and $c = E(X_j)$, we get

$$E(X_j - EX_j)^n = \begin{cases} \sum_{\mathbf{x}} (x_j - EX_j)^n f(\mathbf{x}), & \mathbf{x} = (x_1, \ldots, x_k)' \\ \int_{-\infty}^{\infty} \cdots \int_{-\infty}^{\infty} (x_j - EX_j)^n f(x_1, \ldots, x_k) dx_1 \cdots dx_k \end{cases}$$

$$= \begin{cases} \sum_{x_j} (x_j - EX_j)^n f_j(x_j) \\ \int_{-\infty}^{\infty} (x_j - EX_j)^n f_j(x_j) dx_j \end{cases}$$

which is the *n*th *central moment of the r.v.* X_j (or *the nth moment of* X_j *about its mean*).

Thus the *n*th central moment of an r.v. X with p.d.f. f and mean μ is

$$E(X - EX)^n = E(X - \mu)^n = \begin{cases} \sum_x (x - EX)^n f(x) = \sum_x (x - \mu)^n f(x) \\ \int_{-\infty}^{\infty} (x - EX)^n f(x) dx = \int_{-\infty}^{\infty} (x - \mu)^n f(x) dx. \end{cases}$$

In particular, for $n = 2$ the 2nd central moment of X is denoted by σ_X^2 or $\sigma^2(X)$ or just σ^2 when no confusion is possible, and is called the *variance of* X. Its positive square root σ_X or $\sigma(X)$ or just σ is called the *standard deviation* (*s.d.*) *of* X.

As in the case of μ_X, σ_X^2 has a physical interpretation also. Its definition corresponds to that of the second moment, or moment of inertia. One recalls that a *large* moment of inertia means the mass of the body is spread widely about its center of gravity. Likewise a *large variance* corresponds to a probability distribution which is not well concentrated about its mean value.

3. For $g(X_1, \ldots, X_k) = (X_1 - EX_1)^{n_1} \cdots (X_k - EX_k)^{n_k}$, the quantity

$$E\left[(X_1 - EX_1)^{n_1} \cdots (X_k - EX_k)^{n_k}\right]$$

is the $(n_1, \ldots, n_k)$-*central joint moment of* $X_1, \ldots, X_k$ or the $(n_1, \ldots, n_k)$-*joint moment of* $X_1, \ldots, X_k$ *about their means*.

4. For $g(X_1, \ldots, X_k) = X_j(X_j - 1) \cdots (X_j - n + 1), j = 1, \ldots, k$, the quantity

$$E[X_j(X_j - 1) \cdots (X_j - n + 1)] = \begin{cases} \sum_{x_j} x_j(x_j - 1) \cdots (x_j - n + 1) f_j(x_j) \\ \int_{-\infty}^{\infty} x_j(x_j - 1) \cdots (x_j - n + 1) f_j(x_j) dx_j \end{cases}$$

is the *n*th factorial moment of the r.v. X_j. Thus the *n*th *factorial moment* of an r.v. X with p.d.f. f is

$$E[X(X-1)\cdots(X-n+1)] = \begin{cases} \sum_x x(x-1)\cdots(x-n+1)f(x) \\ \int_{-\infty}^{\infty} x(x-1)\cdots(x-n+1)f(x)dx. \end{cases}$$

5.1.2 Basic Properties of the Expectation of an R.V.

From the very definition of $E[g(\mathbf{X})]$, the following properties are immediate.

(E1) $E(c) = c$, where c is a constant.
(E2) $E[cg(\mathbf{X})] = cE[g(\mathbf{X})]$, and, in particular, $E(cX) = cE(X)$ if X is an r.v.
(E3) $E[g(\mathbf{X}) + d] = E[g(\mathbf{X})] + d$, where d is a constant. In particular, $E(X + d) = E(X) + d$ if X is an r.v.
(E4) Combining (E2) and (E3), we get $E[cg(\mathbf{X}) + d] = cE[g(\mathbf{X})] + d$, and, in particular, $E(cX + d) = cE(X) + d$ if X is an r.v.
(E4′) $E\left[\sum_{j=1}^n c_j g_j(\mathbf{X})\right] = \sum_{j=1}^n c_j E[g_j(\mathbf{X})]$.

In fact, for example, in the continuous case, we have

$$E\left[\sum_{j=1}^n c_j g_j(\mathbf{X})\right] = \int_{-\infty}^{\infty} \cdots \int_{-\infty}^{\infty} \left[\sum_{j=1}^n c_j g_j(x_1, \ldots, x_k)\right] f(x_1, \ldots, x_k) dx_1 \cdots dx_k$$

$$= \sum_{j=1}^n c_j \int_{-\infty}^{\infty} \cdots \int_{-\infty}^{\infty} g_j(x_1, \ldots, x_k) f(x_1, \ldots, x_k) dx_1 \cdots dx_k$$

$$= \sum_{j=1}^n c_j E[g_j(\mathbf{X})].$$

The discrete case follows similarly. In particular,

(E4″) $E\left(\sum_{j=1}^n c_j X_j\right) = \sum_{j=1}^n c_j E(X_j)$.
(E5) If $X \geq 0$, then $E(X) \geq 0$.

Consequently, by means of (E5) and (E4″), we get that

(E5′) If $X \geq Y$, then $E(X) \geq E(Y)$, where X and Y are r.v.'s (with finite expectations).
(E6) $|E[g(\mathbf{X})]| \leq E|g(\mathbf{X})|$.
(E7) If $E|X|^r < \infty$ for some $r > 0$, where X is an r.v., then $E|X|^{r'} < \infty$ for all $0 < r' < r$.

This is a consequence of the obvious inequality $|X|^{r'} \leq 1 + |X|^r$ and (E5′).

Furthermore, since of $n = 1, 2, \ldots$, we have $|X^n| = |X|^n$, by means of (E6), it follows that

(E7′) If $E(X^n)$ exists (that is, $E|X|^n < \infty$) for some $n = 2, 3, \ldots$, then $E(X^{n'})$ also exists for all $n' = 1, 2, \ldots$ with $n' < n$.

5.1.3 Basic Properties of the Variance of an R.V.

Regarding the variance, the following properties are easily established by means of the definition of the variance.

(V1) $\sigma^2(c) = 0$, where c is a constant.
(V2) $\sigma^2[cg(\mathbf{X})] = c^2\sigma^2[g(\mathbf{X})]$, and, in particular, $\sigma^2(cX) = c^2\sigma^2(X)$, if X is an r.v.
(V3) $\sigma^2[g(\mathbf{X}) + d] = \sigma^2[g(\mathbf{X})]$, where d is a constant. In particular, $\sigma^2(X + d) = \sigma^2(X)$, if X is an r.v.

In fact,

$$\sigma^2\big[g(\mathbf{X}) + d\big] = E\Big\{\big[g(\mathbf{X}) + d\big] - E\big[g(\mathbf{X}) + d\big]\Big\}^2$$
$$= E\big[g(\mathbf{X}) - Eg(\mathbf{X})\big]^2 = \sigma^2\big[g(\mathbf{X})\big].$$

(V4) Combining (V2) and (V3), we get $\sigma^2[cg(\mathbf{X}) + d] = c^2\sigma^2[g(\mathbf{X})]$, and, in particular, $\sigma^2(cX + d) = c^2\sigma^2(X)$, if X is an r.v.
(V5) $\sigma^2[g(\mathbf{X})] = E[g(\mathbf{X})]^2 - [Eg(\mathbf{X})]^2$, and, in particular,
(V5′) $\sigma^2(X) = E(X^2) - (EX)^2$, if X is an r.v.

In fact,

$$\sigma^2\big[g(\mathbf{X})\big] = E\big[g(\mathbf{X}) - Eg(\mathbf{X})\big]^2 = E\Big\{\big[g(\mathbf{X})\big]^2 - 2g(\mathbf{X})Eg(\mathbf{X}) + \big[Eg(\mathbf{X})\big]^2\Big\}$$
$$= E\big[g(\mathbf{X})\big]^2 - 2\big[Eg(\mathbf{X})\big]^2 + \big[Eg(\mathbf{X})\big]^2 = E\big[g(\mathbf{X})\big]^2 - \big[Eg(\mathbf{X})\big]^2,$$

the equality before the last one being true because of (E4′).

(V6) $\sigma^2(X) = E[X(X-1)] + EX - (EX)^2$, if X is an r.v., as is easily seen. This formula is especially useful in calculating the variance of a discrete r.v., as is seen below.

Exercises

5.1.1 Verify the details of properties $(E1)$–$(E7)$.

5.1.2 Verify the details of properties $(V1)$–$(V5)$.

5.1.3 For $r' < r$, show that $|X|^{r'} \leq 1 + |X|^r$ and conclude that if $E|X|^r < \infty$, then $E|X|^{r'}$ for all $0 < r' < r$.

5.1.4 Verify the equality $(E[g(X)]=)\int_{-\infty}^{\infty} g(x)f_X(x)dx = \int_{-\infty}^{\infty} yf_Y(y)dy$ for the case that $X \sim N(0, 1)$ and $Y = g(X) = X^2$.

5.1.5 For any event A, consider the r.v. $X = I_A$, the indicator of A defined by $I_A(s) = 1$ for $s \in A$ and $I_A(s) = 0$ for $s \in A^c$, and calculate EX^r, $r > 0$, and also $\sigma^2(X)$.

5.1.6 Let X be an r.v. such that

$$P(X = -c) = P(X = c) = \frac{1}{2}.$$

Calculate EX, $\sigma^2(X)$ and show that

$$P(|X - EX| \leq c) = \frac{\sigma^2(X)}{c^2}.$$

5.1.7 Let X be an r.v. with finite EX.

i) For any constant c, show that $E(X - c)^2 = E(X - EX)^2 + (EX - c)^2$;

ii) Use part (i) to conclude that $E(X - c)^2$ is minimum for $c = EX$.

5.1.8 Let X be an r.v. such that $EX^4 < \infty$. Then show that

i) $E(X - EX)^3 = EX^3 - 3(EX)(EX)^2 + 2(EX)^3$;

ii) $E(X - EX)^4 = EX^4 - 4(EX)(EX^3) + 6(EX)^2(EX^2) - 3(EX)^4$.

5.1.9 If $EX^4 < \infty$, show that:

$$E[X(X-1)] = EX^2 - EX;\ E[X(X-1)(X-2)] = EX^3 - 3EX^2 + 2EX;$$
$$E[X(X-1)(X-2)(X-3)] = EX^4 - 6EX^3 + 11EX^2 - 6EX.$$

(These relations provide a way of calculating EX^k, $k = 2, 3, 4$ by means of the factorial moments $E[X(X - 1)]$, $E[X(X - 1)(X - 2)]$, $E[X(X - 1)(X - 2)(X - 3)]$.)

5.1.10 Let X be the r.v. denoting the number of claims filed by a policyholder of an insurance company over a specified period of time. On the basis of an extensive study of the claim records, it may be assumed that the distribution of X is as follows:

x	0	1	2	3	4	5	6
$f(x)$	0.304	0.287	0.208	0.115	0.061	0.019	0.006

i) Calculate the EX and the $\sigma^2(X)$;

ii) What premium should the company charge in order to break even?

iii) What should be the premium charged if the company is to expect to come ahead by \$M for administrative expenses and profit?

5.1.11 A roulette wheel has 38 slots of which 18 are red, 18 black, and 2 green.

  i) Suppose a gambler is placing a bet of $M on red. What is the gambler's expected gain or loss and what is the standard deviation?

  ii) If the same bet of $M is placed on green and if $kM is the amount the gambler wins, calculate the expected gain or loss and the standard deviation.

  iii) For what value of k do the two expectations in parts (i) and (ii) coincide?

  iv) Does this value of k depend on M?

  v) How do the respective standard deviations compare?

5.1.12 Let X be an r.v. such that $P(X = j) = (\frac{1}{2})^j, j = 1, 2, \ldots$.

  i) Compute EX, $E[X(X-1)]$;

  ii) Use (i) in order to compute $\sigma^2(X)$.

5.1.13 If X is an r.v. distributed as $U(\alpha, \beta)$, show that

$$EX = \frac{\alpha + \beta}{2}, \quad \sigma^2(X) = \frac{(\alpha - \beta)^2}{12}.$$

5.1.14 Let the r.v. X be distributed as $U(\alpha, \beta)$. Calculate EX^n for any positive integer n.

5.1.15 Let X be an r.v. with p.d.f. f symmetric about a constant c (that is, $f(c - x) = f(c + x)$ for every x).

  i) Then if EX exists, show that $EX = c$;

  ii) If $c = 0$ and EX^{2n+1} exists, show that $EX^{2n+1} = 0$ (that is, those moments of X of odd order which exist are all equal to zero).

5.1.16 Refer to Exercise 3.3.13(iv) in Chapter 3 and find the EX for those α's for which this expectation exists, where X is an r.v. having the distribution in question.

5.1.17 Let X be an r.v. with p.d.f. given by

$$f(x) = \frac{|x|}{c^2} I_{(-c,c)}(x).$$

Compute EX^n for any positive integer n, $E|X^r|$, $r > 0$, $\sigma^2(X)$.

5.1.18 Let X be an r.v. with finite expectation and d.f. F.

  i) Show that

$$EX = \int_0^\infty [1 - F(x)] dx - \int_{-\infty}^0 F(x) dx;$$

ii) Use the interpretation of the definite integral as an area in order to give a geometric interpretation of EX.

5.1.19 Let X be an r.v. of the continuous type with finite EX and p.d.f. f.

i) If m is a median of f and c is any constant, show that

$$E|X-c| = E|X-m| + 2\int_m^c (c-x)f(x)dx;$$

ii) Utilize (i) in order to conclude that $E|X-c|$ is minimized for $c = m$. (Hint: Consider the two cases that $c \geq m$ and $c < m$, and in each one split the integral from $-\infty$ to c and c to ∞ in order to remove the absolute value. Then the fact that $\int_{-\infty}^m f(x)dx = \int_m^\infty f(x)dx = \frac{1}{2}$ and simple manipulations prove part (i). For part (ii), observe that $\int_m^c (c-x)f(x)dx \geq 0$ whether $c \geq m$ or $c < m$.)

5.1.20 If the r.v. X is distributed according to the *Weibull* distribution (see Exercise 4.1.15 in Chapter 4), then:

i) Show that $EX = \Gamma\left(1+\dfrac{1}{\beta}\right)\Big/\alpha^{1/\beta}$, $EX^2 = \Gamma\left(1+\dfrac{2}{\beta}\right)\Big/\alpha^{2/\beta}$, so that

$$\sigma^2(X) = \left[\Gamma\left(1+\frac{2}{\beta}\right) - \Gamma^2\left(1+\frac{1}{\beta}\right)\right]\Big/\sigma^{2/\beta},$$

where recall that the Gamma function Γ is defined by $\Gamma(\gamma) = \int_0^\infty t^{\gamma-1}e^{-t}dt$, $\gamma > 0$;

ii) Determine the numerical values of EX and $\sigma^2(X)$ for $\alpha = 1$ and $\beta = \frac{1}{2}$, $\beta = 1$ and $\beta = 2$.

5.2 Expectations and Variances of Some r.v.'s

5.2.1 Discrete Case

1. Let X be $B(n, p)$. Then $E(X) = np$, $\sigma^2(X) = npq$. In fact,

$$E(X) = \sum_{x=0}^n x\binom{n}{x}p^x q^{n-x} = \sum_{x=1}^n x\frac{n!}{x!(n-x)!}p^x q^{n-x}$$

$$= \sum_{x=1}^n \frac{n(n-1)!}{(x-1)!(n-x)!}p^x q^{n-x}$$

$$= np\sum_{x=1}^n \frac{(n-1)!}{(x-1)!\left[(n-1)-(x-1)\right]!}p^{x-1}q^{(n-1)-(x-1)}$$

$$= np\sum_{x=0}^{n-1} \frac{(n-1)!}{x!\left[(n-1)-x\right]!}p^x q^{(n-1)-x} = np(p+q)^{n-1} = np.$$

Next,
$$E[X(X-1)]$$
$$= \sum_{x=0}^{n} x(x-1)\frac{n!}{x!(n-x)!}p^x q^{n-x}$$
$$= \sum_{x=2}^{n} x(x-1)\frac{n(n-1)(n-2)!}{x(x-1)(x-2)![(n-2)-(x-2)]!}p^2 p^{x-2} q^{(n-2)-(x-2)}$$
$$= n(n-1)p^2 \sum_{x=2}^{n} \frac{(n-2)!}{(x-2)![(n-2)-(x-2)]!}p^{x-2} q^{(n-2)-(x-2)}$$
$$= n(n-1)p^2 \sum_{x=0}^{n-2} \frac{(n-2)!}{x![(n-2)-x]!}p^x q^{(n-2)-x}$$
$$= n(n-1)p^2 (p+q)^{n-2} = n(n-1)p^2.$$

That is,
$$E[X(X-1)] = n(n-1)p^2.$$

Hence, by (V6),
$$\sigma^2(X) = E[X(X-1)] + EX - (EX)^2 = n(n-1)p^2 + np - n^2 p^2$$
$$= n^2 p^2 - np^2 + np - n^2 p^2 = np(1-p) = npq.$$

2. Let X be $P(\lambda)$. Then $E(X) = \sigma^2(X) = \lambda$. In fact,
$$E(X) = \sum_{x=0}^{\infty} x e^{-\lambda} \frac{\lambda^x}{x!} = \sum_{x=1}^{\infty} x e^{-\lambda} \frac{\lambda^x}{x(x-1)!} = \lambda e^{-\lambda} \sum_{x=1}^{\infty} \frac{\lambda^{x-1}}{(x-1)!}$$
$$= \lambda e^{-\lambda} \sum_{x=0}^{\infty} \frac{\lambda^x}{x!} = \lambda e^{-\lambda} e^{\lambda} = \lambda.$$

Next,
$$E[X(X-1)] = \sum_{x=0}^{\infty} x(x-1) e^{-\lambda} \frac{\lambda^x}{x!}$$
$$= \sum_{x=2}^{\infty} x(x-1) e^{-\lambda} \frac{\lambda^x}{x(x-1)(x-2)!} = \lambda^2 e^{-\lambda} \sum_{x=0}^{\infty} \frac{\lambda^x}{x!} = \lambda^2.$$

Hence $EX^2 = \lambda^2 + \lambda$, so that, $\sigma^2(X) = \lambda^2 + \lambda - \lambda^2 = \lambda$.

REMARK 3 One can also prove that the nth factorial moment of X is λ^n; that is, $E[X(X-1) \cdots (X-n+1)] = \lambda^n$.

5.2.2 Continuous Case

1. Let X be $N(0, 1)$. Then

$$E(X^{2n+1}) = 0, \quad E(X^{2n}) = \frac{(2n)!}{2^n(n!)}, \quad n \geq 0.$$

In particular, then

$$E(X) = 0, \quad \sigma^2(X) = E(X^2) = \frac{2}{2 \cdot 1!} = 1.$$

In fact,

$$E(X^{2n+1}) = \frac{1}{\sqrt{2\pi}} \int_{-\infty}^{\infty} x^{2n+1} e^{-x^2/2} dx.$$

But

$$\int_{-\infty}^{\infty} x^{2n+1} e^{-x^2/2} dx = \int_{-\infty}^{0} x^{2n+1} e^{-x^2/2} dx + \int_{0}^{\infty} x^{2n+1} e^{-x^2/2} dx$$

$$= \int_{\infty}^{0} y^{2n+1} e^{-y^2/2} dy + \int_{0}^{\infty} x^{2n+1} e^{-x^2/2} dx$$

$$= -\int_{0}^{\infty} x^{2n+1} e^{-x^2/2} dx + \int_{0}^{\infty} x^{2n+1} e^{-x^2/2} dx = 0.$$

Thus $E(X^{2n+1}) = 0$. Next,

$$\int_{-\infty}^{\infty} x^{2n} e^{-x^2/2} dx = 2 \int_{0}^{\infty} x^{2n} e^{-x^2/2} dx,$$

as is easily seen, and

$$\int_{0}^{\infty} x^{2n} e^{-x^2/2} dx = -\int_{0}^{\infty} x^{2n-1} de^{-x^2/2}$$

$$= -x^{2n-1} e^{-x^2/2} \Big|_{0}^{\infty} + (2n-1) \int_{0}^{\infty} x^{2n-2} e^{-x^2/2} dx$$

$$= (2n-1) \int_{0}^{\infty} x^{2n-2} e^{-x^2/2} dx,$$

and if we set $m_{2n} = E(X^{2n})$, we get then

$$m_{2n} = (2n-1)m_{2n-2}, \text{ and similarly,}$$
$$m_{2n-2} = (2n-3)m_{2n-4}$$
$$\vdots$$
$$m_2 = 1 \cdot m_0$$
$$m_0 = 1 \left(\text{since } m_0 = E(X^0) = E(1) = 1 \right).$$

Multiplying them out, we obtain

$$m_{2n} = (2n-1)(2n-3)\cdots 1$$
$$= \frac{1\cdot 2\cdots(2n-3)(2n-2)(2n-1)(2n)}{2\cdots(2n-2)(2n)} = \frac{(2n)!}{(2\cdot 1)\cdots[2(n-1)](2\cdot n)}$$
$$= \frac{(2n)!}{2^n[1\cdots(n-1)n]} = \frac{(2n)!}{2^n(n!)}.$$

REMARK 4 Let now X be $N(\mu, \sigma^2)$. Then $(X - \mu)/\sigma$ is $N(0, 1)$. Hence

$$E\left(\frac{X-\mu}{\sigma}\right) = 0, \quad \sigma^2\left(\frac{X-\mu}{\sigma}\right) = 1.$$

But

$$E\left(\frac{X-\mu}{\sigma}\right) = \frac{1}{\sigma}E(X) - \frac{\mu}{\sigma}.$$

Hence

$$\frac{1}{\sigma}E(X) - \frac{\mu}{\sigma} = 0,$$

so that $E(X) = \mu$. Next,

$$\sigma^2\left(\frac{X-\mu}{\sigma}\right) = \frac{1}{\sigma^2}\sigma^2(X)$$

and then

$$\frac{1}{\sigma^2}\sigma^2(X) = 1,$$

so that $\sigma^2(X) = \sigma^2$.

2. Let X be *Gamma* with parameters α and β. Then $E(X) = \alpha\beta$ and $\sigma^2(X) = \alpha\beta^2$. In fact,

$$E(X) = \frac{1}{\Gamma(\alpha)\beta^\alpha}\int_0^\infty xx^{\alpha-1}e^{-x/\beta}dx = \frac{1}{\Gamma(\alpha)\beta^\alpha}\int_0^\infty x^\alpha e^{-x/\beta}dx$$
$$= \frac{-\beta}{\Gamma(\alpha)\beta^\alpha}\int_0^\infty x^\alpha de^{-x/\beta} = -\frac{\beta}{\Gamma(\alpha)\beta^\alpha}\left(x^\alpha e^{-x/\beta}\Big|_0^\infty - \alpha\int_0^\infty x^{\alpha-1}e^{-x/\beta}dx\right)$$
$$= \alpha\beta\frac{1}{\Gamma(\alpha)\beta^\alpha}\int_0^\infty x^{\alpha-1}e^{-x/\beta}dx = \alpha\beta.$$

Next,

$$E(X^2) = \frac{1}{\Gamma(\alpha)\beta^\alpha} \int_0^\infty x^{\alpha+1} e^{-x/\beta} dx = \beta^2 \alpha(\alpha+1)$$

and hence

$$\sigma^2(X) = \beta^2 \alpha(\alpha+1) - \alpha^2 \beta^2 = \alpha\beta^2(\alpha+1-\alpha) = \alpha\beta^2.$$

REMARK 5

i) If X is χ_r^2, that is, if $\alpha = r/2$, $\beta = 2$, we get $E(X) = r$, $\sigma^2(X) = 2r$.

ii) If X is *Negative Exponential*, that is, if $\alpha = 1$, $\beta = 1/\lambda$, we get $E(X) = 1/\lambda$, $\sigma^2(X) = 1/\lambda^2$.

3. Let X be *Cauchy*. Then $E(X^n)$ does *not* exist for any $n \geq 1$. For example, for $n = 1$, we get

$$I = \frac{\sigma}{\pi} \int_{-\infty}^\infty \frac{x\, dx}{\sigma^2 + (x-\mu)^2}.$$

For simplicity, we set $\mu = 0$, $\sigma = 1$ and we have

$$I = \frac{1}{\pi} \int_{-\infty}^\infty \frac{x\, dx}{1+x^2} = \frac{1}{\pi} \left(\frac{1}{2} \int_{-\infty}^\infty \frac{d(x^2)}{1+x^2} \right)$$

$$= \frac{1}{\pi} \frac{1}{2} \int_{-\infty}^\infty \frac{d(1+x^2)}{1+x^2} = \frac{1}{2\pi} \log(1+x^2) \Big|_{-\infty}^\infty$$

$$= \frac{1}{2\pi} (\infty - \infty),$$

which is an indeterminate form. Thus the Cauchy distribution is an example of a distribution without a mean.

REMARK 6 In somewhat advanced mathematics courses, one encounters sometimes the so-called *Cauchy Principal Value Integral*. This coincides with the improper Riemann integral when the latter exists, and it often exists even if the Riemann integral does not. It is an improper integral in which the limits are taken symmetrically. As an example, for $\sigma = 1$, $\mu = 0$, we have, in terms of the principal value integral,

$$I^* = \lim_{A \to \infty} \frac{1}{\pi} \int_{-A}^A \frac{x\, dx}{1+x^2} = \frac{1}{2\pi} \lim_{A \to \infty} \log(1+x^2) \Big|_{-A}^A$$

$$= \frac{1}{2\pi} \lim_{A \to \infty} \left[\log(1+A^2) - \log(1+A^2) \right] = 0.$$

Thus the mean of the Cauchy exists in terms of principal value, but not in the sense of our definition which requires absolute convergence of the improper Riemann integral involved.

Exercises

5.2.1 If X is an r.v. distributed as $B(n, p)$, calculate the kth factorial moment $E[X(X − 1) \cdots (X − k + 1)]$.

5.2.2 An honest coin is tossed independently n times and let X be the r.v. denoting the number of H's that occur.

i) Calculate $E(X/n)$, $\sigma^2(X/n)$;
ii) If $n = 100$, find a lower bound for the probability that the observed frequency X/n does not differ from 0.5 by more than 0.1;
iii) Determine the smallest value of n for which the probability that X/n does not differ from 0.5 by more 0.1 is at least 0.95;
iv) If $n = 50$ and $P(|(X/n) − 0.5| < c) \geq 0.9$, determine the constant c. (Hint: In (ii)–(iv), utilize Tchebichev's inequality.)

5.2.3 Refer to Exercise 3.2.16 in Chapter 3 and suppose that 100 people are chosen at random. Find the expected number of people with blood of each one of the four types and the variance about these numbers.

5.2.4 If X is an r.v. distributed as $P(\lambda)$, calculate the kth factorial moment $E[X(X − 1) \cdots (X − k + 1)]$.

5.2.5 Refer to Exercise 3.2.7 in Chapter 3 and find the expected number of particles to reach the portion of space under consideration there during time t and the variance about this number.

5.2.6 If X is an r.v. with a Hypergeometric distribution, use an approach similar to the one used in the Binomial example in order to show that

$$EX = \frac{mr}{m+n}, \quad \sigma^2(X) = \frac{mnr(m+n-r)}{(m+n)^2(m+n-1)}.$$

5.2.7 Let X be an r.v. distributed as Negative Binomial with parameters r and p.

i) By working as in the Binomial example, show that $EX = rq/p$, $\sigma^2(X) = rq/p^2$;
ii) Use (i) in order to show that $EX = q/p$ and $\sigma^2(X) = q/p^2$, if X has the Geometric distribution.

5.2.8 Let f be the Gamma density with parameters $\alpha = n$, $\beta = 1$. Then show that

$$\int_\lambda^\infty f(x)dx = \sum_{x=0}^{n-1} e^{-\lambda} \frac{\lambda^x}{x!}.$$

Conclude that in this case, one may utilize the Incomplete Gamma tables (see, for example, *Tables of the Incomplete Γ-Function*, Cambridge University Press, 1957, Karl Paerson, editor) in order to evaluate the d.f. of a Poisson distribution at the points $j = 1, 2, \ldots$.

5.2.9 Refer to Exercise 3.3.7 in Chapter 3 and suppose that each TV tube costs $7 and that it sells for $11. Suppose further that the manufacturer sells an item on money-back guarantee terms if the lifetime of the tube is less than c.

i) Express his expected gain (or loss) in terms of c and λ;

ii) For what value of c will he break even?

5.2.10 Refer to Exercise 4.1.12 in Chapter 4 and suppose that each bulb costs 30 cents and sells for 50 cents. Furthermore, suppose that a bulb is sold under the following terms: The entire amount is refunded if its lifetime is <1,000 and 50% of the amount is refunded if its lifetime is <2,000. Compute the expected gain (or loss) of the dealer.

5.2.11 If X is an r.v. having the Beta distribution with parameters α and β, then

i) Show that

$$EX^n = \frac{\Gamma(\alpha+\beta)\Gamma(\alpha+n)}{\Gamma(\alpha)\Gamma(\alpha+\beta+n)}, \quad n = 1, 2, \ldots;$$

ii) Use (i) in order to find EX and $\sigma^2(X)$.

5.2.12 Let X be an r.v. distributed as Cauchy with parameters μ and σ^2. Then show that $E|X| = \infty$.

5.2.13 If the r.v. X is distributed as Lognormal with parameters α and β^2, compute EX, $\sigma^2(X)$.

5.2.14 Suppose that the average monthly water consumption by the residents of a certain community follows the Lognormal distribution with $\mu = 10^4$ cubic feet and $\sigma = 10^3$ cubic feet monthly. Compute the proportion of the residents who consume more than 15×10^3 cubic feet monthly.

5.2.15 Let X be an r.v. with finite third moment and set $\mu = EX$, $\sigma^2 = \sigma^2(X)$. Define the (dimensionless quantity, pure number) γ_1 by

$$\gamma_1 = E\left(\frac{X-\mu}{\sigma}\right)^3.$$

γ_1 is called the *skewness* of the distribution of the r.v. X and is a measure of asymmetry of the distribution. If $\gamma_1 > 0$, the distribution is said to be *skewed to the right* and if $\gamma_1 < 0$, the distribution is said to be *skewed to the left*. Then show that:

i) If the p.d.f. of X is symmetric about μ, then $\gamma_1 = 0$;

ii) The Binomial distribution $B(n, p)$ is skewed to the right for $p < \frac{1}{2}$ and is skewed to the left for $p > \frac{1}{2}$;

iii) The Poisson distribution $P(\lambda)$ and the Negative Exponential distribution are always skewed to the right.

5.2.16 Let X be an r.v. with $EX^4 < \infty$ and define the (pure number) γ_2 by

$$\gamma_2 = E\left(\frac{X-\mu}{\sigma}\right)^4 - 3, \quad \text{where} \quad \mu = EX, \ \sigma^2 = \sigma^2(X).$$

γ_2 is called the *kurtosis* of the distribution of the r.v. X and is a measure of "peakedness" of this distribution, where the $N(0, 1)$ p.d.f. is a measure of reference. If $\gamma_2 > 0$, the distribution is called *leptokurtic* and if $\gamma_2 < 0$, the distribution is called *platykurtic*. Then show that:

i) $\gamma_2 < 0$ if X is distributed as $U(\alpha, \beta)$;

ii) $\gamma_2 > 0$ if X has the Double Exponential distribution (see Exercise 3.3.13(iii) in Chapter 3).

5.2.17 Let X be an r.v. taking on the values j with probability $p_j = P(X = j)$, $j = 0, 1, \ldots$. Set

$$G(t) = \sum_{j=0}^{\infty} p_j t^j, \quad -1 \leq t \leq 1.$$

The function G is called the *generating function of the sequence* $\{p_j\}, j \geq 0$.

i) Show that if $|EX| < \infty$, then $EX = d/dt\, G(t)|_{t=1}$;

ii) Also show that if $|E[X(X-1)\cdots(X-k+1)]| < \infty$, then

$$E[X(X-1)\cdots(X-k+1)] = \frac{d^k}{dt^k} G(t)\Big|_{t=1};$$

iii) Find the generating function of the sequences

$$\left\{\binom{n}{j} p^j q^{n-j}\right\}, \quad j \geq 0, \ 0 < p < 1, \ q = 1-p$$

and

$$\left\{e^{-\lambda}\frac{\lambda j}{j!}\right\}, \quad j \geq 0, \ \lambda > 0;$$

iv) Utilize (ii) and (iii) in order to calculate the kth factorial moments of X being $B(n, p)$ and X being $P(\lambda)$. Compare the results with those found in Exercises 5.2.1 and 5.2.4, respectively.

5.3 Conditional Moments of Random Variables

If, in the preceding definitions, the p.d.f. f of the r. vector $\mathbf{X}$ is replaced by a conditional p.d.f. $f(x_{j_1}, \ldots, x_{j_n} | x_{i_1}, \ldots, x_{i_m})$, the resulting moments are called *conditional moments*, and they are functions of $x_{i_1}, \ldots, x_{i_m}$.
Thus

$$E(X_2 | X_1 = x_1) = \begin{cases} \sum_{x_2} x_2 f(x_2 | x_1) \\ \int_{-\infty}^{\infty} x_2 f(x_2 | x_1) dx_2, \end{cases}$$

$$\sigma^2(X_2 | X_1 = x_1) = \begin{cases} \sum_{x_2} [x_2 - E(X_2 | X_1 = x_1)]^2 f(x_2 | x_1) \\ \int_{-\infty}^{\infty} [x_2 - E(X_2 | X_1 = x_1)]^2 f(x_2 | x_1) dx_2. \end{cases}$$

For example, if $(X_1, X_2)'$ has the Bivariate Normal distribution, then $f(x_2 | x_1)$ is the p.d.f. of an $N(b, \sigma_2^2(1 - \rho^2))$ r.v., where

$$b = \mu_2 + \frac{\rho \sigma_2}{\sigma_1}(x_1 - \mu_1).$$

Hence

$$E(X_2 | X_1 = x_1) = \mu_2 + \frac{\rho \sigma_2}{\sigma_1}(x_1 - \mu_1).$$

Similarly,

$$E(X_1 | X_2 = x_2) = \mu_1 + \frac{\rho \sigma_1}{\sigma_2}(x_2 - \mu_2).$$

Let X_1, X_2 be two r.v.'s with joint p.d.f $f(x_1, x_2)$. We just gave the definition of $E(X_2 | X_1 = x_1)$ for all x_1 for which $f(x_2 | x_1)$ is defined; that is, for all x_1 for which $f_{X_1}(x_1) > 0$. Then $E(X_2 | X_1 = x_1)$ is a function of x_1. Replacing x_1 by X_1 and writing $E(X_2 | X_1)$ instead of $E(X_2 | X_1 = x_1)$, we then have that $E(X_2 | X_1)$ is itself an r.v., and a function of X_1. Then we may talk about the $E[E(X_2 | X_1)]$. In connection with this, we have the following properties:

5.3.1 Some Basic Properties of the Conditional Expectation

(CE1) If $E(X_2)$ and $E(X_2|X_1)$ exist, then $E[E(X_2|X_1)] = E(X_2)$ (that is, the expectation of the conditional expectation of an r.v. is the same as the (unconditional) expectation of the r.v. in question).

It suffices to establish the property for the continuous case only, for the proof for the discrete case is quite analogous. We have

$$E[E(X_2|X_1)] = \int_{-\infty}^{\infty}\left[\int_{-\infty}^{\infty} x_2 f(x_2|x_1)dx_2\right]f_{X_1}(x_1)dx_1$$
$$= \int_{-\infty}^{\infty}\int_{-\infty}^{\infty} x_2 f(x_2|x_1)f_{X_1}(x_1)dx_2 dx_1$$
$$= \int_{-\infty}^{\infty}\int_{-\infty}^{\infty} x_2 f(x_1, x_2)dx_2 dx_1 = \int_{-\infty}^{\infty}\int_{-\infty}^{\infty} x_2 f(x_1, x_2)dx_1 dx_2$$
$$= \int_{-\infty}^{\infty} x_2 \left(\int_{-\infty}^{\infty} f(x_1, x_2)dx_1\right)dx_2 = \int_{-\infty}^{\infty} x_2 f_{X_2}(x_2)dx_2 = E(X_2).$$

REMARK 7 Note that here all interchanges of order of integration are legitimate because of the absolute convergence of the integrals involved.

(CE2) Let X_1, X_2 be two r.v.'s, $g(X_1)$ be a (measurable) function of X_1 and let that $E(X_2)$ exists. Then for all x_1 for which the conditional expectations below exist, we have

$$E[X_2 g(X_1)|X_1 = x_1] = g(x_1)E(X_2|X_1 = x_1)$$

or

$$E[X_2 g(X_1)|X_1] = g(X_1)E(X_2|X_1).$$

Again, restricting ourselves to the continuous case, we have

$$E[X_2 g(X_1)|X_1 = x_1] = \int_{-\infty}^{\infty} x_2 g(x_1) f(x_2|x_1)dx_2 = g(x_1)\int_{-\infty}^{\infty} x_2 f(x_2|x_1)dx_2$$
$$= g(x_1)E(X_2|X_1 = x_1).$$

In particular, by taking $X_2 = 1$, we get

(CE2′) For all x_1 for which the conditional expectations below exist, we have $E[g(X_1)|X_1 = x_1] = g(x_1)$ (or $E[g(X_1)|X_1] = g(X_1)$).

(CV) Provided the quantities which appear below exist, we have

$$\sigma^2[E(X_2|X_1)] \leq \sigma^2(X_2)$$

and the inequality is strict, unless X_2 is a function of X_1 (on a set of probability one).

Set $\mu = E(X_2)$, $\phi(X_1) = E(X_2|X_1)$.

Then

$$\sigma^2(X_2) = E(X_2 - \mu)^2 = E\{[X_2 - \phi(X_1)] + [\phi(X_1) - \mu]\}^2$$
$$= E[X_2 - \phi(X_1)]^2 + E[\phi(X_1) - \mu]^2 + 2E\{[X_2 - \phi(X_1)][\phi(X_1) - \mu]\}.$$

Next,

$$E\{[X_2 - \phi(X_1)][\phi(X_1) - \mu]\}$$
$$= E[X_2\phi(X_1)] - E[\phi^2(X_1)] - \mu E(X_2) + \mu E[\phi(X_1)]$$
$$= E\{E[X_2\phi(X_1)|X_1]\} - E[\phi^2(X_1)] - \mu E[E(X_2|X_1)]$$
$$+ \mu E[\phi(X_1)] \quad \text{(by (CE1))},$$

and this is equal to

$$E[\phi^2(X_1)] - E[\phi^2(X_1)] - \mu E[\phi(X_1)] + \mu E[\phi(X_1)] \quad \text{(by (CE2))},$$

which is 0. Therefore

$$\sigma^2(X_2) = E[X_2 - \phi(X_1)]^2 + E[\phi(X_1) - \mu]^2,$$

and since

$$E[X_2 - \phi(X_1)]^2 \geq 0,$$

we have

$$\sigma^2(X_2) \geq E[\phi(X_1) - \mu^2] = \sigma^2[E(X_2|X_1)].$$

The inequality is strict unless

$$E[X_2 - \phi(X_1)]^2 = 0.$$

But

$$E[X_2 - \phi(X_1)]^2 = \sigma^2[X_2 - \phi(X_1)], \quad \text{since} \quad E[X_2 - \phi(X_1)] = \mu - \mu = 0.$$

Thus $\sigma^2[X_2 - \phi(X_1)] = 0$ and therefore $X_2 = \phi(X_1)$ with probability one, by Remark 8, which follows.

Exercises

5.3.1 Establish properties (CEI) and (CE2) for the discrete case.

5.3.2 Let the r.v.'s X, Y be jointly distributed with p.d.f. given by

$$f(x, y) = \frac{2}{n(n+1)}$$

if $y = 1, \ldots, x$; $x = 1, \ldots, n$, and 0 otherwise. Compute the following quantities: $E(X|Y = y)$, $E(Y|X = x)$. (Hint: Recall that $\sum_{x=1}^{n} x = \frac{n(n+1)}{2}$, and $\sum_{x=1}^{n} x^2 = \frac{n(n+1)(2n+1)}{6}$.)

5.3.3 Let X, Y be r.v.'s with p.d.f. f given by $f(x, y) = (x + y)I_{(0,1)\times(0,1)}(x, y)$. Calculate the following quantities: EX, $\sigma^2(X)$, EY, $\sigma^2(Y)$, $E(X|Y = y)$, $\sigma^2(X|Y = y)$.

5.3.4 Let X, Y be r.v.'s with p.d.f. f given by $f(x, y) = \lambda^2 e^{-\lambda(x+y)}I_{(0,\infty)\times(0,\infty)}(x, y)$. Calculate the following quantities: EX, $\sigma^2(X)$, EY, $\sigma^2(Y)$, $E(X|Y = y)$, $\sigma^2(X|Y = y)$.

5.3.5 Let X be an r.v. with finite EX. Then for any r.v. Y, show that $E[E(X|Y)] = EX$. (Assume the existence of all p.d.f.'s needed.)

5.3.6 Consider the r.v.'s X, Y and let h, g be (measurable) functions on $\mathbb{R}$ into itself such that $E[h(X)g(Y)]$ and $Eg(X)$ exist. Then show that

$$E\big[h(X)g(Y)\big|X = x\big] = h(x)E\big[g(Y)\big|X = x\big].$$

5.4 Some Important Applications: Probability and Moment Inequalities

THEOREM 1 Let $\mathbf{X}$ be a k-dimensional r. vector and $g \geq 0$ be a real-valued (measurable) function defined on $\mathbb{R}^k$, so that $g(\mathbf{X})$ is an r.v., and let $c > 0$. Then

$$P\big[g(\mathbf{X}) \geq c\big] \leq \frac{E\big[g(\mathbf{X})\big]}{c}.$$

PROOF Assume $\mathbf{X}$ is continuous with p.d.f. f. Then

$$\begin{aligned}E\big[g(\mathbf{X})\big] &= \int_{-\infty}^{\infty} \cdots \int_{-\infty}^{\infty} g(x_1, \ldots, x_k) f(x_1, \ldots, x_k) dx_1 \cdots dx_k \\ &= \int_A g(x_1, \ldots, x_k) f(x_1, \ldots, x_k) dx_1 \cdots dx_k + \int_{A^c} g(x_1, \ldots, x_k) \\ &\quad \times f(x_1, \ldots, x_k) dx_1 \cdots dx_k,\end{aligned}$$

where $A = \{(x_1, \ldots, x_k)' \in \mathbb{R}^k; g(x_1, \ldots, x_k) \geq c\}$. Then

$$E[g(\mathbf{X})] \geq \int_A g(x_1, \ldots, x_k) f(x_1, \ldots, x_k) dx_1 \cdots dx_k$$
$$\geq c \int_A f(x_1, \ldots, x_k) dx_1 \cdots dx_k$$
$$= cP[g(\mathbf{X}) \in A] = cP[g(\mathbf{X}) \geq c].$$

Hence $P[g(\mathbf{X}) \geq c] \leq E[g(\mathbf{X})]/c$. The proof is completely analogous if $\mathbf{X}$ is of the discrete type; all one has to do is to replace integrals by summation signs. ▲

5.4.1 Special Cases of Theorem 1

1. Let X be an r.v. and take $g(X) = |X - \mu|^r$, $\mu = E(X)$, $r > 0$. Then

$$P[|X - \mu| \geq c] = P[|X - \mu|^r \geq c^r] \leq \frac{E|X - \mu|^r}{c^r}.$$

This is known as *Markov's inequality*.

2. In Markov's inequality replace r by 2 to obtain

$$P[|X - \mu| \geq c]\left(= P[|X - \mu|^2 \geq c^2]\right) \leq \frac{E|X - \mu|^2}{c^2} = \frac{\sigma^2(X)}{c^2} = \frac{\sigma^2}{c^2}.$$

This is known as *Tchebichev's inequality*. In particular, if $c = k\sigma$, then

$$P[|X - \mu| \geq k\sigma] \leq \frac{1}{k^2}; \quad \text{equivalently,} \quad P[|X - \mu| < k\sigma] \geq 1 - \frac{1}{k^2}.$$

REMARK 8 Let X be an r.v. with mean μ and variance $\sigma^2 = 0$. Then Tchebichev's inequality gives: $P[|X - \mu| \geq c] = 0$ for every $c > 0$. This result and Theorem 2, Chapter 2, imply then that $P(X = \mu) = 1$ (see also Exercise 5.4.6).

LEMMA 1 Let X and Y be r.v.'s such that

$$E(X) = E(Y) = 0, \quad \sigma^2(X) = \sigma^2(Y) = 1.$$

Then

$$E^2(XY) \leq 1 \quad \text{or, equivalently,} \quad -1 \leq E(XY) \leq 1,$$

and

$$E(XY) = 1 \quad \text{if any only if} \quad P(Y = X) = 1,$$
$$E(XY) = -1 \quad \text{if any only if} \quad P(Y = -X) = 1.$$

PROOF We have

$$0 \leq E(X - Y)^2 = E(X^2 - 2XY + Y^2)$$
$$= EX^2 - 2E(XY) + EY^2 = 2 - 2E(XY)$$

and

$$0 \le E(X+Y)^2 = E(X^2 + 2XY + Y^2)$$
$$= EX^2 + 2E(XY) + EY^2 = 2 + 2E(XY).$$

Hence $E(XY) \le 1$ and $-1 \le E(XY)$, so that $-1 \le E(XY) \le 1$. Now let $P(Y = X) = 1$. Then $E(XY) = EY^2 = 1$, and if $P(Y = -X) = 1$, then $E(XY) = -EY^2 = -1$. Conversely, let $E(XY) = 1$. Then

$$\sigma^2(X-Y) = E(X-Y)^2 - [E(X-Y)]^2 = E(X-Y)^2$$
$$= EX^2 - 2E(XY) + EY^2 = 1 - 2 + 1 = 0,$$

so that $P(X = Y) = 1$ by Remark 8; that is, $P(X = Y) = 1$. Finally, let $E(XY) = -1$. Then $\sigma^2(X+Y) = 2 + 2E(XY) = 2 - 2 = 0$, so that

$$P(X = -Y) = 1. \quad \blacktriangle$$

THEOREM 2 (Cauchy–Schwarz inequality) Let X and Y be two random variables with means μ_1, μ_2 and (positive) variances σ_1^2, σ_2^2, respectively. Then

$$E^2[(X-\mu_1)(Y-\mu_2)] \le \sigma_1^2 \sigma_2^2,$$

or, equivalently,

$$-\sigma_1\sigma_2 \le E[(X-\mu_1)(Y-\mu_2)] \le \sigma_1\sigma_2,$$

and

$$E[(X-\mu_1)(Y-\mu_2)] = \sigma_1\sigma_2$$

if and only if

$$P\left[Y = \mu_2 + \frac{\sigma_2}{\sigma_1}(X-\mu_1)\right] = 1$$

and

$$E[(X-\mu_1)(Y-\mu_2)] = -\sigma_1\sigma_2$$

if and only if

$$P\left[Y = \mu_2 - \frac{\sigma_2}{\sigma_1}(X-\mu_1)\right] = 1.$$

PROOF Set

$$X_1 = \frac{X-\mu_1}{\sigma_1}, \quad Y_1 = \frac{Y-\mu_2}{\sigma_2}.$$

Then X_1, Y_1 are as in the previous lemma, and hence

$$E^2(X_1Y_1) \leq 1$$

if and only if

$$-1 \leq E(X_1Y_1) \leq 1$$

becomes

$$\frac{E^2[(X-\mu_1)(Y-\mu_2)]}{\sigma_1^2\sigma_2^2} \leq 1$$

if and only if

$$-\sigma_1\sigma_2 \leq E[(X-\mu_1)(Y-\mu_2)] \leq \sigma_1\sigma_2.$$

The second half of the conclusion follows similarly and will be left as an exercise (see Exercise 5.4.6). ▲

REMARK 9 A more familiar form of the Cauchy–Schwarz inequality is $E^2(XY) \leq (EX^2)(EY^2)$. This is established as follows: Since the inequality is trivially true if either one of EX^2, EY^2 is ∞, suppose that they are both finite and set $Z = \lambda X - Y$, where λ is a real number. Then $0 \leq EZ^2 = (EX^2)\lambda^2 - 2[E(XY)]\lambda + EY^2$ for all λ, which happens if and only if $E^2(XY) - (EX^2)(EY^2) \leq 0$ (by the discriminant test for quadratic equations), or $E^2(XY) \leq (EX^2)(EY^2)$.

Exercises

5.4.1 Establish Theorem 1 for the discrete case.

5.4.2 Let g be a (measurable) function defined on $\mathbb{R}$ into $(0, \infty)$ and suppose that g is even (that is, $g(-x) = g(x)$) and nondecreasing for $x \geq 0$. Then for any r.v. X and any $\varepsilon > 0$, show that

$$P[g(X) \geq \varepsilon] \leq \frac{Eg(X)}{g(\varepsilon)}.$$

5.4.3 For an r.v. X with $EX = \mu$ and $\sigma^2(X) = \sigma^2$, both finite, use Tchebichev's inequality in order to find a lower bound for the probability $P(|X - \mu| \geq k\sigma)$. Compare the lower bounds for $k = 1, 2, 3$ with the respective probabilities when $X \sim N(\mu, \sigma^2)$.

5.4.4 Let X be an r.v. distributed as χ^2_{40}. Use Tchebichev's inequality in order to find a lower bound for the probability $P(|(X/n) - 1| \leq 0.5)$, and compare this bound with the exact value found from Table 3 in Appendix III.

5.4.5 Refer to Remark 8 and show that if X is an r.v. with $EX = \mu$ (finite) such that $P(|X - \mu| \geq c) = 0$ for every $c > 0$, then $P(X = \mu) = 1$.

5.4.6 Prove the second conclusion of Theorem 2.

5.4.7 For any r.v. X, use the Cauchy–Schwarz inequality in order to show that $E|X| \leq E^{1/2} X^2$.

5.5 Covariance, Correlation Coefficient and Its Interpretation

In this section, we introduce the concepts of covariance and correlation coefficient of two r.v.'s and provide an interpretation for the latter. To this end, for two r.v.'s X and Y with means μ_1, μ_2, the $(1, 1)$-joint central mean, that is, $E[(X - \mu_1)(Y - \mu_2)]$, is called the *covariance* of X, Y and is denoted by $\text{Cov}(X, Y)$. If σ_1, σ_2 are the standard deviations of X and Y, which are assumed to be positive, then the covariance of $(X - \mu_1)/\sigma_1$, $(Y - \mu_2)/\sigma_2$ is called the *correlation coefficient* of X, Y and is denoted by $\rho(X, Y)$ or $\rho_{X,Y}$ or ρ_{12} or just ρ if no confusion is possible; that is,

$$\rho = E\left[\left(\frac{X - \mu_1}{\sigma_1}\right)\left(\frac{Y - \mu_2}{\sigma_2}\right)\right] = \frac{E[(X - \mu_1)(Y - \mu_2)]}{\sigma_1 \sigma_2} = \frac{\text{Cov}(X, Y)}{\sigma_1 \sigma_2}$$

$$= \frac{E(XY) - \mu_1 \mu_2}{\sigma_1 \sigma_2}.$$

From the Cauchy–Schwarz inequality, we have that $\rho^2 \leq 1$; that is $-1 \leq \rho \leq 1$, and $\rho = 1$ if and only if

$$Y = \mu_2 + \frac{\sigma_2}{\sigma_1}(X - \mu_1)$$

with probability 1, and $\rho = -1$ if and only if

$$Y = \mu_2 - \frac{\sigma_2}{\sigma_1}(X - \mu_1)$$

with probability 1. So $\rho = \pm 1$ means X and Y are *linearly related*. From this stems the significance of ρ as a measure of *linear dependence* between X and Y. (See Fig. 5.1.) If $\rho = 0$, we say that X and Y are *uncorrelated*, while if $\rho = \pm 1$, we say that X and Y are *completely correlated* (positively if $\rho = 1$, negatively if $\rho = -1$).

For $-1 < \rho < 1, \rho \neq 0$, we say that X and Y are *correlated* (positively if $\rho > 0$, negatively if $\rho < 0$). Positive values of ρ may indicate that there is a tendency of large values of Y to correspond to large values of X and small values of Y to correspond to small values of X. Negative values of ρ may indicate that small

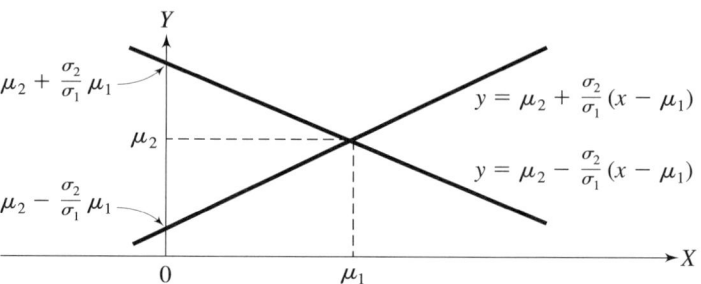

Figure 5.1

values of Y correspond to large values of X and large values of Y to small values of X. Values of ρ close to zero may also indicate that these tendencies are weak, while values of ρ close to ± 1 may indicate that the tendencies are strong.

The following elaboration sheds more light on the intuitive interpretation of the correlation coefficient $\rho(=\rho(X, Y))$ as a measure of co-linearity of the r.v.'s X and Y. To this end, for $\rho > 0$, consider the line $y = \mu_2 + \frac{\sigma_2}{\sigma_1}(x - \mu_1)$ in the xy-plane and let D be the distance of the (random) point (X, Y) from the above line. Recalling that the distance of the point (x_0, y_0) from the line $ax + by + c = 0$ is given by $|ax_0 + by_0 + c| \sqrt{a^2 + b^2}$, we have in the present case:

$$D = \left|X - \frac{\sigma_1}{\sigma_2}Y + \left(\frac{\sigma_1 \mu_2}{\sigma_2} - \mu_1\right)\right| \sqrt{1 + \frac{\sigma_1^2}{\sigma_2^2}},$$

since here $a = 1$, $b = -\frac{\sigma_1}{\sigma_2}$ and $c = \frac{\sigma_1 \mu_2}{\sigma_2} - \mu_1$. Thus,

$$D^2 = \left[X - \frac{\sigma_1}{\sigma_2}Y + \left(\frac{\sigma_1 \mu_2}{\sigma_2} - \mu_1\right)\right]^2 \bigg/ \left(1 + \frac{\sigma_1^2}{\sigma_2^2}\right),$$

and we wish to evaluate the expected squared distance of (X, Y) from the line $y = \mu_2 + \frac{\sigma_2}{\sigma_1}(x - \mu_1)$; that is, ED^2. Carrying out the calculations, we find

$$\left(\sigma_1^2 + \sigma_2^2\right)D^2 = \sigma_2^2 X^2 + \sigma_1^2 Y^2 - 2\sigma_1 \sigma_2 XY + 2\sigma_2\left(\sigma_1 \mu_2 - \sigma_2 \mu_1\right)X$$
$$- 2\sigma_1\left(\sigma_1 \mu_2 - \sigma_2 \mu_1\right)Y + \left(\sigma_1 \mu_2 - \sigma_2 \mu_1\right)^2. \tag{1}$$

Taking the expectations of both sides in (1) and recalling that

$$EX^2 = \sigma_1^2 + \mu_1^2, \ EY^2 = \sigma_2^2 + \mu_2^2 \quad \text{and} \quad E(XY) = \rho\sigma_1\sigma_2 + \mu_1\mu_2,$$

we obtain

$$ED^2 = \frac{2\sigma_1^2\sigma_2^2}{\sigma_1^2 + \sigma_2^2}(1-\rho) \quad (\rho > 0). \tag{2}$$

Working likewise for the case that $\rho < 0$, we get

$$ED^2 = \frac{2\sigma_1^2\sigma_2^2}{\sigma_1^2 + \sigma_2^2}(1+\rho) \quad (\rho < 0). \tag{3}$$

For $\rho = 1$ or $\rho = -1$, we already know that (X, Y) lies on the line $y = \mu_2 + \frac{\sigma_2}{\sigma_1}(x - \mu_1)$ or $y = \mu_2 - \frac{\sigma_2}{\sigma_1}(x - \mu_1)$, respectively (with probability 1). Therefore, regardless of the value of ρ, by the observation just made, relations (2) and (3) are summarized by the expression

$$ED^2 = \frac{2\sigma_1^2\sigma_2^2}{\sigma_1^2 + \sigma_2^2}(1-|\rho|). \tag{4}$$

At this point, exploiting the interpretation of an expectation as an average, relation (4) indicates the following: For $\rho > 0$, the pairs (X, Y) tend to be arranged along the line $y = \mu_2 + \frac{\sigma_2}{\sigma_1}(x - \mu_1)$. These points get closer and closer to this line as ρ gets closer to 1, and lie on the line for $\rho = 1$. For $\rho < 0$, the pairs (X, Y) tend to be arranged along the line $y = \mu_2 - \frac{\sigma_2}{\sigma_1}(x - \mu_1)$. These points get closer and closer to this line as ρ gets closer to -1, and lie on this line for $\rho = -1$. For $\rho = 0$, the expected distance is constantly equal to $2\sigma_1^2\sigma_2^2/(\sigma_1^2 + \sigma_2^2)$ from either one of the lines $y = \mu_2 + \frac{\sigma_2}{\sigma_1}(x - \mu_1)$ and $y = \mu_2 - \frac{\sigma_2}{\sigma_1}(x - \mu_1)$, which is equivalent to saying that the pairs (X, Y) may lie anywhere in the xy-plane. It is in this sense that ρ is a measure of co-linearity of the r.v.'s X and Y.

The preceding discussion is based on the paper "A Direct Development of the Correlation Coefficient" by Leo Katz, published in the *American Statistician*, Vol. 29 (1975), page 170. His approach is somewhat different and is outlined below. First, consider the r.v.'s X_1 and Y_1 as defined in the proof of Theorem 2; unlike the original r.v.'s X and Y, the "normalized" r.v.'s X_1 and Y_1 are dimensionless. Through the transformations $x_1 = \frac{x-\mu_1}{\sigma_1}$ and $y_1 = \frac{y-\mu_2}{\sigma_2}$, we move from the xy-plane to the x_1y_1-plane. In this latter plane, look at the point (X_1, Y_1) and seek the line $Ax_1 + By_1 + C = 0$ from which the expected squared distance of (X_1, Y_1) is minimum. That is, determine the coefficients A, B and C, so that ED_1^2 is minimum, where

$D_1 = |AX_1 + BY_1 + C_1| \sqrt{A^2 + B^2}$. Expanding D_1^2, taking expectations, and noticing that

$$EX_1 = EY_1 = 0, \quad EX_1^2 = EY_1^2 = 1, \quad \text{and} \quad E(X_1 Y_1) = \rho,$$

we obtain

$$ED_1^2 = 1 + \frac{2AB\rho}{A^2 + B^2} + \frac{C^2}{A^2 + B^2}. \tag{5}$$

Clearly, for ED_1^2 to be minimized it is necessary that $C = 0$. Then, by (5), the expression to be minimized is

$$ED_1^2 = 1 + \frac{2AB\rho}{A^2 + B^2}. \tag{6}$$

At this point, observe that

$$-(A+B)^2 = -(A^2 + B^2) - 2AB \le 0 \le (A^2 + B^2) - 2AB = (A-B)^2,$$

or equivalently,

$$-1 \le \frac{2AB}{A^2 + B^2} \le 1. \tag{7}$$

From (6) and (7), we conclude that:

If $\rho > 0$, ED_1^2 is minimized for $\frac{2AB}{A^2+B^2} = -1$ and the minimum is $1 - \rho$.

If $\rho < 0$, ED_1^2 is minimized for $\frac{2AB}{A^2+B^2} = 1$ and the minimum is $1 + \rho$.

Finally, if $\rho = 0$, the ED_1^2 is constantly equal to 1; there is no minimizing line (through the origin) $Ax_1 + By_1 = 0$. However, $\frac{2AB}{A^2+B^2} = -1$ if and only if $A = B$, and $\frac{2AB}{A^2+B^2} = 1$ if and only if $A = -B$. The corresponding lines are $y_1 = x_1$, the main diagonal, and $y_1 = -x_1$. Also observe that both minima of ED_1^2 (for $\rho > 0$ and $\rho < 0$), and its constant value 1 (for $\rho = 0$) are expressed by a single form, namely, $1 - |\rho|$.

To summarize: For $\rho > 0$, the ED_1^2 is minimized for the line $y_1 = x_1$; for $\rho < 0$, the ED_1^2 is minimized for the line $y_1 = -x_1$; for $\rho = 0$, $ED_1^2 = 1$, there is no minimizing line. From this point on, the interpretation of ρ as a measure of co-linearity (of X_1 and Y_1) is argued as above, with the lines $y = \mu_2 + \frac{\sigma_2}{\sigma_1}(x - \mu_1)$ and $y = \mu_2 - \frac{\sigma_2}{\sigma_1}(x - \mu_1)$ being replaced by the lines $y_1 = x_1$ and $y_1 = -x_1$, respectively.

Exercises

5.5.1 Let X be an r.v. taking on the values $-2, -1, 1, 2$ each with probability $\frac{1}{4}$. Set $Y = X^2$ and compute the following quantities: EX, $\sigma^2(X)$, EY, $\sigma^2(Y)$, $\rho(X, Y)$.

5.5.2 Go through the details required in establishing relations (2), (3) and (4).

5.5.3 Do likewise in establishing relation (5).

5.5.4 Refer to Exercise 5.3.2 (including the hint given there) and

i) Calculate the covariance and the correlation coefficient of the r.v.'s X and Y;

ii) Referring to relation (4), calculate the expected squared distance of (X, Y) from the appropriate line $y = \mu_2 + \frac{\sigma_2}{\sigma_1}(x - \mu_1)$ or $y = \mu_2 - \frac{\sigma_2}{\sigma_1}(x - \mu_1)$ (which one?);

iii) What is the minimum expected squared distance of (X_1, Y_1) from the appropriate line $y = x$ or $y = -x$ (which one?) where $X_1 = \frac{X - \mu_1}{\sigma_1}$ and $Y_1 = \frac{Y - \mu_2}{\sigma_2}$.

$$\left(\text{Hint: Recall that } \sum_{x=1}^{n} x^3 = \left[\frac{n(n+1)}{2} \right]^2 . \right)$$

5.5.5 Refer to Exercise 5.3.2 and calculate the covariance and the correlation coefficient of the r.v.'s X and Y.

5.5.6 Do the same in reference to Exercise 5.3.3.

5.5.7 Repeat the same in reference to Exercise 5.3.4.

5.5.8 Show that $\rho(aX + b, cY + d) = \text{sgn}(ac)\rho(X, Y)$, where a, b, c, d are constants and sgn x is 1 if $x > 0$ and is -1 if $x < 0$.

5.5.9 Let X and Y be r.v.'s representing temperatures in two localities, A and B, say, given in the Celsius scale, and let U and V be the respective temperatures in the Fahrenheit scale. Then it is known that U and X are related as follows: $U = \frac{9}{5}X + 32$, and likewise for V and Y. Fit this example in the model of Exercise 5.5.5, and conclude that the correlation coefficients of X, Y and U, V are identical, as one would expect.

5.5.10 Consider the jointly distributed r.v.'s X, Y with finite second moments and $\sigma^2(X) > 0$. Then show that the values $\hat{\alpha}$ and $\hat{\beta}$ for which $E[Y - (\alpha X + \beta)]^2$ is minimized are given by

$$\hat{\beta} = EY - \hat{\alpha}EX, \quad \hat{\alpha} = \frac{\sigma(Y)}{\sigma(X)}\rho(X, Y).$$

(The r.v. $\hat{Y} = \hat{\alpha} X + \hat{\beta}$ is called the *best linear predictor* or Y, given X.)

5.5.11 If the r.v.'s X_1 and X_2 have the Bivariate Normal distribution, show that the parameter ρ is, actually, the correlation coefficient of X_1 and X_2. (Hint: Observe that the exponent in the joint p.d.f. of X_1 and X_2 may be written as follows:

$$\frac{1}{2(1-\rho^2)}\left[\left(\frac{x_1-\mu_1}{\sigma_1}\right)^2 - 2\rho\left(\frac{x_1-\mu_1}{\sigma_1}\right)\left(\frac{x_2-\mu_2}{\sigma_2}\right) + \left(\frac{x_2-\mu_2}{\sigma_2}\right)^2\right]$$

$$= \frac{(x_1-\mu_1)^2}{2\sigma_1^2} + \frac{(x_2-b)^2}{2\left(\sigma_2\sqrt{1-\rho^2}\right)^2}, \quad \text{where } b = \mu_2 + \rho\frac{\sigma_2}{\sigma_1}(x_1-\mu_1).$$

This facilitates greatly the integration in calculating $E(X_1X_2)$.

5.5.12 If the r.v.'s X_1 and X_2 have jointly the Bivariate Normal distribution with parameters $\mu_1, \mu_2, \sigma_1^2, \sigma_2^2$ and ρ, calculate $E(c_1X_1 + c_2X_2)$ and $\sigma^2(c_1X_1 + c_2X_2)$ in terms of the parameters involved, where c_1 and c_2 are real constants.

5.5.13 For any two r.v.'s X and Y, set $U = X + Y$ and $V = X - Y$. Then

i) Show that $P(UV < 0) = P(|X| < |Y|)$;

ii) If $EX^2 = EY^2 < \infty$, then show that $E(UV) = 0$;

iii) If $EX^2, EY^2 < \infty$ and $\sigma^2(X) = \sigma^2(Y)$, then U and V are uncorrelated.

5.5.14 If the r.v.'s $X_i, i = 1, \ldots, m$ and $Y_j, j = 1, \ldots, n$ have finite second moments, show that

$$\text{Cov}\left(\sum_{i=1}^{m} X_i, \sum_{j=1}^{n} Y_j\right) = \sum_{i=1}^{m}\sum_{j=1}^{n} \text{Cov}(X_i, Y_j).$$

5.6* Justification of Relation (2) in Chapter 2

As a final application of the results of this chapter, we give a general proof of Theorem 9, Chapter 2. To do this we remind the reader of the definition of the concept of the *indicator function of a set A*.

Let A be an event in the sample space S. Then the *indicator function* of A, denoted by I_A, is a function on S defined as follows:

$$I_A(s) = \begin{cases} 1 & \text{if } s \in A \\ 0 & \text{if } s \in A^c. \end{cases}$$

The following are simple consequences of the indicator function:

$$I_{\cap_{j=1}^n A_j} = \prod_{j=1}^n I_{A_j} \tag{8}$$

$$I_{\sum_{j=1}^n A_j} = \sum_{j=1}^n I_{A_j}, \tag{9}$$

and, in particular,

$$I_{A^c} = 1 - I_A. \tag{10}$$

Clearly,

$$E(I_A) = P(A) \tag{11}$$

and for any $X_1, \ldots, X_r$, we have

$$(1 - X_1)(1 - X_2) \cdots (1 - X_r) = 1 - H_1 + H_2 - \cdots + (-1)^r H_r, \tag{12}$$

where H_j stands for the sum of the products $X_{i_1} \cdots X_{i_j}$, where the summation extends over all subsets $\{i_1, i_2, \ldots, i_j\}$ of the set $\{1, 2, \ldots, r\}$, $j = 1, \ldots, r$. Let α, β be such that: $0 < \alpha, \beta$ and $\alpha + \beta \leq r$. Then the following is true:

$$\sum_{J_\alpha} X_{i_1} \cdots X_{i_\alpha} H_\beta(J_\alpha) = \binom{\alpha + \beta}{\alpha} H_{\alpha + \beta}, \tag{13}$$

where $J_\alpha = \{i_1, \ldots, i_\alpha\}$ is the typical member of all subsets of size α of the set $\{1, 2, \ldots, r\}$, $H_\beta(J_\alpha)$ is the sum of the products $X_{j_1} \cdots X_{j_\beta}$, where the summation extends over all subsets of size β of the set $\{1, \ldots, r\} - J_\alpha$, and $\sum_{J_\alpha}$ is meant to extend over all subsets J_α of size α of the set $\{1, 2, \ldots, r\}$.

The justification of (13) is as follows: In forming $H_{\alpha+\beta}$, we select $(\alpha + \beta)$ X's from the available r X's in all possible ways, which is $\binom{r}{\alpha+\beta}$. On the other hand, for each choice of J_α, there are $\binom{r-\alpha}{\beta}$ ways of choosing β X's from the remaining $(r - \alpha)$ X's. Since there are $\binom{r}{\alpha}$ choices of J_α, we get $\binom{r}{\alpha}\binom{r-\alpha}{\beta}$ groups (products) of $(\alpha + \beta)$ X's out of r X's. The number of different groups of $(\alpha + \beta)$ X's out of r X's is $\binom{r}{\alpha+\beta}$. Thus among the $\binom{r}{\alpha}\binom{r-\alpha}{\beta}$ groups of $(\alpha+\beta)$ X's out of r X's, the number of distinct ones is given by

$$\frac{\binom{r}{\alpha}\binom{r-\alpha}{\beta}}{\binom{r}{\alpha+\beta}} = \frac{\frac{r!}{\alpha!(r-\alpha)!} \frac{(r-\alpha)!}{\beta!(r-\alpha-\beta)!}}{\frac{r!}{(\alpha+\beta)!(r-\alpha-\beta)!}}$$

$$= \frac{(\alpha+\beta)!}{\alpha!\beta!} = \binom{\alpha+\beta}{\alpha}.$$

This justifies (13).

Now clearly,

$$B_m = \sum_{J_m} A_{i_1} \cap \cdots \cap A_{i_m} \cap A^c_{i_{m+1}} \cap \cdots \cap A^c_{i_M},$$

where the summation extends over all choices of subsets $J_m = \{i_1, \ldots, i_m\}$ of the set $\{1, 2, \ldots, M\}$ and B_m is the one used in Theorem 9, Chapter 2. Hence

$$I_{B_m} = \sum_{J_m} I_{A_{i_1} \cap \cdots \cap A_{i_m} \cap A^c_{i_{m+1}} \cap \cdots \cap A^c_{i_M}}$$

$$= \sum_{J_m} I_{A_{i_1}} \cdots I_{A_{i_m}} (1 - I_{A_{i_{m+1}}}) \cdots (1 - I_{A_{i_M}}) \quad \text{(by (8), (9), (10))}$$

$$= \sum_{J_m} I_{A_{i_1}} \cdots I_{A_{i_m}} \left[1 - H_1(J_m) + H_2(J_m) - \cdots + (-1)^{M-m} H_{M-m}(J_m) \right]$$

$$\text{(by (12))}.$$

Since

$$\sum_{J_m} I_{A_{i_1}} \cdots I_{A_{i_m}} H_k(J_m) = \binom{m+k}{m} H_{m+k} \quad \text{(by (13))},$$

we have

$$I_{B_m} = H_m - \binom{m+1}{m} H_{m+1} + \binom{m+2}{m} H_{m+2} - \cdots + (-1)^{M-m} \binom{M}{m} H_M.$$

Taking expectations of both sides, we get (from (11) and the definition of S_r in Theorem 9, Chapter 2)

$$P(B_m) = S_m - \binom{m+1}{m} S_{m+1} + \binom{m+2}{m} S_{m+2} - \cdots + (-1)^{M-m} \binom{M}{m} S_M,$$

as was to be proved.

(For the proof just completed, also see pp. 80–85 in E. Parzen's book *Modern Probability Theory and Its Applications* published by Wiley, 1960.)

REMARK 10 In measure theory the quantity I_A is sometimes called the characteristic function of the set A and is usually denoted by χ_A. In probability theory the term characteristic function is reserved for a different concept and will be a major topic of the next chapter.

Chapter 6

Characteristic Functions, Moment Generating Functions and Related Theorems

6.1 Preliminaries

The main subject matter of this chapter is the introduction of the concept of the characteristic function of an r.v. and the discussion of its main properties. The characteristic function is a powerful mathematical tool, which is used profitably for probabilistic purposes, such as producing the moments of an r.v., recovering its distribution, establishing limit theorems, etc. To this end, recall that for $z \in \mathbb{R}$, $e^{iz} = \cos z + i \sin z$, $i = \sqrt{-1}$, and in what follows, i may be treated formally as a real number, subject to its usual properties: $i^2 = -1$, $i^3 = -i$, $i^4 = 1$, $i^5 = i$, etc.

The sequence of lemmas below will be used to justify the theorems which follow, as well as in other cases in subsequent chapters. A brief justification for some of them is also presented, and relevant references are given at the end of this section.

LEMMA A Let $g_1, g_2 : \{x_1, x_2, \ldots\} \to [0, \infty)$ be such that

$$g_1(x_j) \leq g_2(x_j), \quad j = 1, 2, \ldots,$$

and that $\sum_{x_j} g_2(x_j) < \infty$. Then $\sum_{x_j} g_1(x_j) < \infty$.

PROOF If the summations are finite, the result is immediate; if not, it follows by taking the limits of partial sums, which satisfy the inequality. ▲

LEMMA A′ Let $g_1, g_2 : \mathbb{R} \to [0, \infty)$ be such that $g_1(x) \leq g_2(x)$, $x \in \mathbb{R}$, and that $\int_a^b g_1(x)dx$ exists for every $a, b, \in \mathbb{R}$ with $a < b$, and that $\int_{-\infty}^{\infty} g_2(x)dx < \infty$. Then $\int_{-\infty}^{\infty} g_1(x)dx < \infty$.

PROOF Same as above replacing sums by integrals. ▲

LEMMA B Let $g:\{x_1, x_2, \ldots\} \to \mathbb{R}$ and $\sum_{x_j}|g(x_j)| < \infty$. Then $\sum_{x_j}g(x_j)$ also converges.

PROOF The result is immediate for finite sums, and it follows by taking the limits of partial sums, which satisfy the inequality. ▲

LEMMA B′ Let $g:\mathbb{R} \to \mathbb{R}$ be such that $\int_a^b g(x)dx$ exists for every $a, b, \in \mathbb{R}$ with $a < b$, and that $\int_{-\infty}^{\infty}|g(x)|dx < \infty$. Then $\int_{-\infty}^{\infty} g(x)dx$ also converges.

PROOF Same as above replacing sums by integrals. ▲

The following lemma provides conditions under which the operations of taking limits and expectations can be interchanged. In more advanced probability courses this result is known as the *Dominated Convergence Theorem*.

LEMMA C Let $\{X_n\}$, $n = 1, 2, \ldots$, be a sequence of r.v.'s, and let Y, X be r.v.'s such that $|X_n(s)| \leq Y(s)$, $s \in S$, $n = 1, 2, \ldots$ and $X_n(s) \to X(s)$ (on a set of s's of probability 1) and $E(Y) < \infty$. Then $E(X)$ exists and $E(X_n) \xrightarrow[n \to \infty]{} E(X)$, or equivalently,

$$\lim_{n \to \infty} E(X_n) = E\left(\lim_{n \to \infty} X_n\right).$$

REMARK 1 The index n can be replaced by a continuous variable.

The next lemma gives conditions under which the operations of differentiation and taking expectations commute.

LEMMA D For each $t \in T$ (where T is $\mathbb{R}$ or an appropriate subset of it, such as the interval $[a, b]$), let $X(\cdot; t)$ be an r.v. such that $(\partial/\partial_t)X(s; t)$ exists for each $s \in S$ and $t \in T$. Furthermore, suppose there exists an r.v. Y with $E(Y) < \infty$ and such that

$$\left|\frac{\partial}{\partial t}X(s; t)\right| \leq Y(s), \quad s \in S, \quad t \in T.$$

Then

$$\frac{d}{dt}E[X(\cdot; t)] = E\left[\frac{\partial}{\partial t}X(\cdot; t)\right], \quad \text{for all} \quad t \in T.$$

The proofs of the above lemmas can be found in any book on real variables theory, although the last two will be stated in terms of weighting functions rather than expectations; for example, see *Advanced Calculus*, Theorem 2, p. 285, Theorem 7, p. 292, by D. V. Widder, Prentice-Hall, 1947; *Real Analysis*, Theorem 7.1, p. 146, by E. J. McShane and T. A. Botts, Van Nostrand, 1959; *The Theory of Lebesgue Measure and Integration*, pp. 66–67, by S. Hartman and J. Mikusiński, Pergamon Press, 1961. Also *Mathematical Methods of Statistics*, pp. 45–46 and pp. 66–68, by H. Cramér, Princeton University Press, 1961.

6.2 Definitions and Basic Theorems—The One-Dimensional Case

Let X be an r.v. with p.d.f. f. Then the *characteristic function of X* (ch.f. of X), denoted by ϕ_X (or just ϕ when no confusion is possible) is a function defined on $\mathbb{R}$, taking complex values, in general, and defined as follows:

$$\phi_X(t) = E\left[e^{itX}\right] = \begin{cases} \sum_x e^{itx} f(x) = \sum_x \left[\cos(tx)f(x) + i\sin(tx)f(x)\right] \\ \int_{-\infty}^{\infty} e^{itx} f(x)dx = \int_{-\infty}^{\infty} \left[\cos(tx)f(x) + i\sin(tx)f(x)\right]dx \end{cases}$$

$$= \begin{cases} \sum_x \left[\cos(tx)f(x)\right] + i\sum_x \left[\sin(tx)f(x)\right] \\ \int_{-\infty}^{\infty} \cos(tx)f(x)dx + i\int_{-\infty}^{\infty} \sin(tx)f(x)dx. \end{cases}$$

By Lemmas A, A', B, B', $\phi_X(t)$ exists for all $t \in \mathbb{R}$. The ch.f. ϕ_X is also called the *Fourier transform* of f.

The following theorem summarizes the basic properties of a ch.f.

THEOREM 1 (Some properties of ch.f.'s)

 i) $\phi_X(0) = 1$.
 ii) $|\phi_X(t)| \leq 1$.
 iii) ϕ_X is continuous, and, in fact, uniformly continuous.
 iv) $\phi_{X+d}(t) = e^{itd}\phi_X(t)$, where d is a constant.
 v) $\phi_{cX}(t) = \phi_X(ct)$, where c is a constant.
 vi) $\phi_{cX+d}(t) = e^{itd}\phi_X(ct)$.
 vii) $\left.\dfrac{d^n}{dt^n}\phi_X(t)\right|_{t=0} = i^n E(X^n)$, $n = 1, 2, \ldots$, if $E|X^n| < \infty$.

PROOF

 i) $\phi_X(t) = Ee^{itX}$. Thus $\phi_X(0) = Ee^{i0X} = E(1) = 1$.
 ii) $|\phi_X(t)| = |Ee^{itX}| \leq E|e^{itX}| = E(1) = 1$, because $|e^{itX}| = 1$. (For the proof of the inequality, see Exercise 6.2.1.)
 iii) $\left|\phi_X(t+h) - \phi_X(t)\right| = \left|E\left[e^{i(t+h)X} - e^{itX}\right]\right|$
 $= \left|E\left[e^{itX}\left(e^{ihX} - 1\right)\right]\right| \leq E\left|e^{itX}\left(e^{ihX} - 1\right)\right|$
 $= E\left|e^{ihX} - 1\right|.$

Then

6.2 Definitions and Basic Theorems—The One-Dimensional Case

$$\lim_{h\to 0}|\phi_X(t+h) - \phi_X(t)| \leq \lim_{h\to 0} E|e^{ihX} - 1| = E\left[\lim_{h\to 0}|e^{ihX} - 1|\right] = 0,$$

provided we can interchange the order of lim and E, which here can be done by Lemma C. We observe that uniformity holds since the last expression on the right is independent of t.

iv) $\phi_{X+d}(t) = Ee^{it(X+d)} = E(e^{itX}e^{itd}) = e^{itd} Ee^{itX} = e^{itd}\phi_X(t)$.

v) $\phi_{cX}(t) = Ee^{it(cX)} = Ee^{i(ct)X} = \phi_X(ct)$.

vi) Follows trivially from (iv) and (v).

vii) $\dfrac{d^n}{dt^n}\phi_X(t) = \dfrac{d^n}{dt^n} Ee^{itX} = E\left(\dfrac{\partial^n}{\partial t^n} e^{itX}\right) = E(i^n X^n e^{itX}),$

provided we can interchange the order of differentiation and E. This can be done here, by Lemma D (applied successively n times to ϕ_X and its $n-1$ first derivatives), since $E|X^n| < \infty$ implies $E|X^k| < \infty$, $k = 1, \ldots, n$ (see Exercise 6.2.2). Thus

$$\left.\dfrac{d^n}{dt^n}\phi_X(t)\right|_{t=0} = i^n E(X^n). \blacktriangle$$

REMARK 2 From part (vii) of the theorem we have that $E(X^n) = (-i)^n \dfrac{d^n}{dt^n} \varphi_X(t)|_{t=0}$, so that the ch.f. produces the nth moment of the r.v.

REMARK 3 If X is an r.v. whose values are of the form $x = a + kh$, where a, h are constants, $h > 0$, and k runs through the integral values $0, 1, \ldots, n$ or $0, 1, \ldots,$ or $0, \pm 1, \ldots, \pm n$ or $0, \pm 1, \ldots,$ then the distribution of X is called a *lattice distribution*. For example, if X is distributed as $B(n, p)$, then its values are of the form $x = a + kh$ with $a = 0$, $h = 1$, and $k = 0, 1, \ldots, n$. If X is distributed as $P(\lambda)$, or it has the Negative Binomial distribution, then again its values are of the same form with $a = 0$, $h = 1$, and $k = 0, 1, \ldots$. If now ϕ is the ch.f. of X, it can be shown that the distribution of X is a lattice distribution if and only if $|\phi(t)| = 1$ for some $t \neq 0$. It can be readily seen that this is indeed the case in the cases mentioned above (for example, $\phi(t) = 1$ for $t = 2\pi$). It can also be shown that the distribution of X is a lattice distribution, if and only if the ch.f. ϕ is periodic with period 2π (that is, $\phi(t + 2\pi) = \phi(t)$, $t \in \mathbb{R}$).

In the following result, the ch.f. serves the purpose of recovering the distribution of an r.v. by way of its ch.f.

THEOREM 2 (Inversion formula) Let X be an r.v. with p.d.f. f and ch.f. ϕ. Then if X is of the discrete type, taking on the (distinct) values x_j, $j \geq 1$, one has

i) $f(x_j) = \lim\limits_{T\to\infty} \dfrac{1}{2T} \int_{-T}^{T} e^{-itx_j} \phi(t)\,dt,\ j \geq 1.$

If X is of the continuous type, then

ii) $f(x) = \lim_{h \to 0} \lim_{T \to \infty} \frac{1}{2\pi} \int_{-T}^{T} \frac{1-e^{-ith}}{ith} e^{-itx} \phi(t) dt$

and, in particular, if $\int_{-\infty}^{\infty} |\phi(t)| dt < \infty$, then ($f$ is bounded and continuous and)

ii') $f(x) = \frac{1}{2\pi} \int_{-\infty}^{\infty} e^{-itx} \phi(t) dt$.

PROOF (outline) i) The ch.f. ϕ is continuous, by Theorem 1(iii), and since so is e^{-itx_j}, it follows that the integral $\int_{-T}^{T} e^{-itx_j} \phi(t) dt$ exists for every $T(>0)$. We have then

$$\frac{1}{2T} \int_{-T}^{T} e^{-itx_j} \phi(t) dt = \frac{1}{2T} \int_{-T}^{T} \left[e^{-itx_j} \sum_k e^{itx_k} f(x_k) \right] dt$$

$$= \frac{1}{2T} \int_{-T}^{T} \left[\sum_k e^{it(x_k - x_j)} f(x_k) \right] dt$$

$$= \sum_k f(x_k) \frac{1}{2T} \int_{-T}^{T} e^{it(x_k - x_j)} dt$$

(the interchange of the integral and summations is valid here). That is,

$$\frac{1}{2T} \int_{-T}^{T} e^{-itx_j} \phi(t) dt = \sum_k f(x_k) \frac{1}{2T} \int_{-T}^{T} e^{it(x_k - x_j)} dt. \tag{1}$$

But

$$\int_{-T}^{T} e^{it(x_k - x_j)} dt = \int_{-T}^{T} \left[\cos t(x_k - x_j) + i \sin t(x_k - x_j) \right] dt$$

$$= \int_{-T}^{T} \cos t(x_k - x_j) dt + i \int_{-T}^{T} \sin t(x_k - x_j) dt$$

$$= \int_{-T}^{T} \cos t(x_k - x_j) dt, \text{ since } \sin z \text{ is an odd function. That is,}$$

$$\int_{-T}^{T} e^{it(x_k - x_j)} dt = \int_{-T}^{T} \cos t(x_k - x_j) dt. \tag{2}$$

If $x_k = x_j$, the above integral is equal to $2T$, whereas, for $x_k \neq x_j$,

$$\int_{-T}^{T} \cos t(x_k - x_j) dt = \frac{1}{x_k - x_j} \int_{-T}^{T} d \sin t(x_k - x_j)$$

$$= \frac{\sin T(x_k - x_j) - \sin[-T(x_k - x_j)]}{x_k - x_j}$$

$$= \frac{2 \sin T(x_k - x_j)}{x_k - x_j}.$$

Therefore,

$$\frac{1}{2T}\int_{-T}^{T} e^{it(x_k - x_j)} dt = \begin{cases} 1, & \text{if } x_k = x_j \\ \dfrac{\sin T(x_k - x_j)}{T(x_k - x_j)}, & \text{if } x_k \neq x_j. \end{cases} \quad (3)$$

But $\left|\frac{\sin T(x_k - x_j)}{x_k - x_j}\right| \leq \frac{1}{|x_k - x_j|}$, a constant independent of T, and therefore, for $x_k \neq x_j$, $\lim_{T \to \infty} \frac{\sin T(x_k - x_j)}{T(x_k - x_j)} = 0$, so that

$$\lim_{T \to \infty} \frac{1}{2T}\int_{-T}^{T} e^{it(x_k - x_j)} dt = \begin{cases} 1, & \text{if } x_k = x_j \\ 0, & \text{if } x_k \neq x_j. \end{cases} \quad (4)$$

By means of (4), relation (1) yields

$$\lim_{T \to \infty} \frac{1}{2T}\int_{-T}^{T} e^{-itx_j} \phi(t) dt = \lim_{T \to \infty} \sum_k f(x_k) \frac{1}{2T}\int_{-T}^{T} e^{it(x_k - x_j)} dt$$

$$= \sum_k f(x_k) \lim_{T \to \infty} \frac{1}{2T}\int_{-T}^{T} e^{it(x_k - x_j)} dt$$

(the interchange of the limit and summation is legitimate here)

$$= f(x_j) + \sum_{k \neq j} \lim_{T \to \infty} \frac{1}{2T}\int_{-T}^{T} e^{it(x_k - x_j)} dt = f(x_j),$$

as was to be seen

ii) (ii') Strictly speaking, (ii') follows from (ii). We are going to omit (ii) entirely and attempt to give a rough justification of (ii'). The assumption that $\int_{-\infty}^{\infty} |\phi(t)| dt < \infty$ implies that $\int_{-\infty}^{\infty} e^{-itx} \phi(t) dt$ exists, and is taken as follows for every arbitrary but fixed $x \in \mathbb{R}$:

$$\int_{-\infty}^{\infty} e^{-itx} \phi(t) dt = \lim_{(0<)T \to \infty} \int_{-T}^{T} e^{-itx} \phi(t) dt. \quad (5)$$

But

$$\int_{-T}^{T} e^{-itx} \phi(t) dt = \int_{-T}^{T} e^{-itx} \left[\int_{-\infty}^{\infty} e^{ity} f(y) dy\right] dt$$

$$= \int_{-T}^{T} \int_{-\infty}^{\infty} e^{it(y-x)} f(y) dy \, dt = \int_{-\infty}^{\infty} f(y) \left[\int_{-T}^{T} e^{it(y-x)} dt\right] dy, \quad (6)$$

where the interchange of the order of integration is legitimate here. Since the integral (with respect to y) is zero over a single point, we may assume in the sequel that $y \neq x$. Then

$$\int_{-T}^{T} e^{it(y-x)} dt = \frac{2 \sin T(y - x)}{y - x}, \quad (7)$$

as was seen in part (i). By means of (7), relation (6) yields

$$\int_{-\infty}^{\infty} e^{-itx}\phi(t)dt = 2\lim_{T\to\infty}\int_{-\infty}^{\infty} f(y)\frac{\sin T(y-x)}{y-x}dy. \quad (8)$$

Setting $T(y-x) = z$, expression (8) becomes

$$\int_{-\infty}^{\infty} e^{-itx}\phi(t)dt = 2\lim_{T\to\infty}\int_{-\infty}^{\infty} f\left(x+\frac{z}{T}\right)\frac{\sin z}{z}dz$$

$$= 2f(x)\pi = 2\pi f(x),$$

by taking the limit under the integral sign, and by using continuity of f and the fact that $\int_{-\infty}^{\infty}\frac{\sin z}{z}dz = \pi$. Solving for $f(x)$, we have

$$f(x) = \frac{1}{2\pi}\int_{-\infty}^{\infty} e^{-itx}\phi(t)dt,$$

as asserted. ▲

EXAMPLE 1 Let X be $B(n, p)$. In the next section, it will be seen that $\phi_X(t) = (pe^{it} + q)^n$. Let us apply (i) to this expression. First of all, we have

$$\frac{1}{2T}\int_{-T}^{T} e^{-itx}\phi(t)dt = \frac{1}{2T}\int_{-T}^{T}(pe^{it}+q)^n e^{-itx}dt$$

$$= \frac{1}{2T}\int_{-T}^{T}\left[\sum_{r=0}^{n}\binom{n}{r}(pe^{it})^r q^{n-r} e^{-itx}\right]dt$$

$$= \frac{1}{2T}\sum_{r=0}^{n}\binom{n}{r}p^r q^{n-r}\int_{-T}^{T} e^{i(r-x)t}dt$$

$$= \frac{1}{2T}\sum_{\substack{r=0\\r\neq x}}^{n}\binom{n}{r}p^r q^{n-r}\frac{1}{i(r-x)}\int_{-T}^{T} e^{i(r-x)t}i(r-x)dt$$

$$+ \frac{1}{2T}\binom{n}{x}p^x q^{n-x}\int_{-T}^{T}dt$$

$$= \sum_{\substack{r=0\\r\neq x}}^{n}\binom{n}{r}p^r q^{n-r}\frac{e^{i(r-x)T}-e^{-i(r-x)T}}{2Ti(r-x)} + \frac{1}{2T}\binom{n}{x}p^x q^{n-x}2T$$

$$= \sum_{\substack{r=0\\r\neq x}}^{n}\binom{n}{r}p^r q^{n-r}\frac{\sin(r-x)T}{(r-x)T} + \binom{n}{x}p^x q^{n-x}.$$

Taking the limit as $T \to \infty$, we get the desired result, namely

$$f(x) = \binom{n}{x}p^x q^{n-x}.$$

(One could also use (i') for calculating $f(x)$, since ϕ is, clearly, periodic with period 2π.)

EXAMPLE 2 For an example of the continuous type, let X be $N(0, 1)$. In the next section, we will see that $\phi_X(t) = e^{-t^2/2}$. Since $|\phi(t)| = e^{-t^2/2}$, we know that $\int_{-\infty}^{\infty} |\phi(t)| dt < \infty$, so that (ii') applies. Thus we have

$$f(x) = \frac{1}{2\pi} \int_{-\infty}^{\infty} e^{-itx} \phi(t) dt = \frac{1}{2\pi} \int_{-\infty}^{\infty} e^{-itx} e^{-t^2/2} dt$$

$$= \frac{1}{2\pi} \int_{-\infty}^{\infty} e^{-(1/2)(t^2 + 2itx)} dt = \frac{1}{2\pi} \int_{-\infty}^{\infty} e^{-(1/2)\left[t^2 + 2t(ix) + (ix)^2\right]} e^{(1/2)(ix)^2} dt$$

$$= \frac{e^{-(1/2)x^2}}{\sqrt{2\pi}} \int_{-\infty}^{\infty} \frac{1}{\sqrt{2\pi}} e^{-(1/2)(t+ix)^2} dt = \frac{e^{-(1/2)x^2}}{\sqrt{2\pi}} \int_{-\infty}^{\infty} \frac{1}{\sqrt{2\pi}} e^{-(1/2)u^2} du$$

$$= \frac{e^{-(1/2)x^2}}{\sqrt{2\pi}} \cdot 1 = \frac{1}{\sqrt{2\pi}} e^{-x^2/2},$$

as was to be shown.

THEOREM 3 (Uniqueness Theorem) There is a one-to-one correspondence between the characteristic function and the p.d.f. of a random variable.

PROOF The p.d.f. of an r.v. determines its ch.f. through the definition of the ch.f. The converse, which is the involved part of the theorem, follows from Theorem 2. ▲

Exercises

6.2.1 Show that for any r.v. X and every $t \in \mathbb{R}$, one has $|Ee^{itX}| \leq E|e^{itX}|(= 1)$. (Hint: If $z = a + ib$, $a, b \in \mathbb{R}$, recall that $|z| = \sqrt{a^2 + b^2}$. Also use Exercise 5.4.7 in Chapter 5 in order to conclude that $(EY)^2 \leq EY^2$ for any r.v. Y.)

6.2.2 Write out detailed proofs for parts (iii) and (vii) of Theorem 1 and justify the use of Lemmas C, D.

6.2.3 For any r.v. X with ch.f. ϕ_X, show that $\phi_{-X}(t) = \bar{\phi}_X(t)$, $t \in \mathbb{R}$, where the bar over ϕ_X denotes conjugate, that is, if $z = a + ib$, $a, b \in \mathbb{R}$, then $\bar{z} = a - ib$.

6.2.4 Show that the ch.f. ϕ_X of an r.v. X is real if and only if the p.d.f. f_X of X is symmetric about 0 (that is, $f_X(-x) = f_X(x)$, $x \in \mathbb{R}$). (Hint: If ϕ_X is real, then the conclusion is reached by means of the previous exercise and Theorem 2. If f_X is symmetric, show that $f_{-X}(x) = f_X(-x)$, $x \in \mathbb{R}$.)

6.2.5 Let X be an r.v. with p.d.f. f and ch.f. ϕ given by: $\phi(t) = 1 - |t|$ if $|t| \leq 1$ and $\phi(t) = 0$ if $|t| > 1$. Use the appropriate inversion formula to find f.

6.2.6 Consider the r.v. X with ch.f. $\phi(t) = e^{-|t|}$, $t \in \mathbb{R}$, and utilize Theorem 2(ii') in order to determine the p.d.f. of X.

6.3 The Characteristic Functions of Some Random Variables

In this section, the ch.f.'s of some distributions commonly occurring will be derived, both for illustrative purposes and for later use.

6.3.1 Discrete Case

1. Let X be $B(n, p)$. Then $\phi_X(t) = (pe^{it} + q)^n$. In fact,

$$\phi_X(t) = \sum_{x=0}^{n} e^{itx} \binom{n}{x} p^x q^{n-x} = \sum_{x=0}^{n} \binom{n}{x} (pe^{it})^x q^{n-x} = (pe^{it} + q)^n,$$

Hence

$$\frac{d}{dt}\phi_X(t)\bigg|_{t=0} = n(pe^{it} + q)^{n-1} ipe^{it}\bigg|_{t=0} = inp,$$

so that $E(X) = np$. Also,

$$\frac{d^2}{dt^2}\phi_X(t)\bigg|_{t=0} = inp \frac{d}{dt}\left[(pe^{it} + q)^{n-1} e^{it}\right]\bigg|_{t=0}$$

$$= inp\left[(n-1)(pe^{it} + q)^{n-2} ipe^{it}e^{it} + (pe^{it} + q)^{n-1} ie^{it}\right]\bigg|_{t=0}$$

$$= i^2 np\left[(n-1)p + 1\right] = -np\left[(n-1)p + 1\right] = -E(X^2),$$

so that

$$E(X^2) = np\left[(n-1)p + 1\right] \text{ and } \sigma^2(X) = E(X^2) - (EX)^2$$
$$= n^2 p^2 - np^2 + np - n^2 p^2 = np(1-p) = npq;$$

that is, $\sigma^2(X) = npq$.

2. Let X be $P(\lambda)$. Then $\phi_X(t) = e^{\lambda e^{it} - \lambda}$. In fact,

$$\phi_X(t) = \sum_{x=0}^{\infty} e^{itx} e^{-\lambda} \frac{\lambda^x}{x!} = e^{-\lambda} \sum_{x=0}^{\infty} \frac{(\lambda e^{it})^x}{x!} = e^{-\lambda} e^{\lambda e^{it}} = e^{\lambda e^{it} - \lambda}.$$

Hence

$$\frac{d}{dt}\phi_X(t)\bigg|_{t=0} = e^{\lambda e^{it}-\lambda}i\lambda e^{it}\bigg|_{t=0} = i\lambda,$$

so that $E(X) = \lambda$. Also,

$$\begin{aligned}\frac{d^2}{dt^2}\phi_X(t)\bigg|_{t=0} &= \frac{d}{dt}\left(i\lambda e^{-\lambda}e^{\lambda e^{it}+it}\right)\bigg|_{t=0}\\ &= i\lambda e^{-\lambda}\frac{d}{dt}e^{\lambda e^{it}+it}\bigg|_{t=0}\\ &= i\lambda e^{-\lambda}\cdot e^{\lambda e^{it}+it}\cdot\left(\lambda e^{it}\cdot i+i\right)\bigg|_{t=0}\\ &= i\lambda e^{-\lambda}\cdot e^{\lambda}\left(\lambda i+i\right)\\ &= i^2\lambda(\lambda+1) = -\lambda(\lambda+1) = -E(X^2),\end{aligned}$$

so that

$$\sigma^2(X) = E(X^2) - (EX)^2 = \lambda(\lambda+1) - \lambda^2 = \lambda;$$

that is, $\sigma^2(X) = \lambda$.

6.3.2 Continuous Case

1. Let X be $N(\mu, \sigma^2)$. Then $\phi_X(t) = e^{it\mu-(\sigma^2 t^2/2)}$, and, in particular, if X is $N(0, 1)$, then $\phi_X(t) = e^{-t^2/2}$. If X is $N(\mu, \sigma^2)$, then $(X - \mu)/\sigma$, is $N(0, 1)$. Thus

$$\phi_{(X-\mu)/\sigma}(t) = \phi_{(1/\sigma)X-(\mu/\sigma)}(t) = e^{-it\mu/\sigma}\phi_X(t/\sigma), \quad \text{and} \quad \phi_X(t/\sigma) = e^{it\mu/\sigma}\phi_{(X-\mu)/\sigma}(t).$$

So it suffices to find the ch.f. of an $N(0, 1)$ r.v. Y, say. Now

$$\begin{aligned}\phi_Y(t) &= \frac{1}{\sqrt{2\pi}}\int_{-\infty}^{\infty}e^{ity}e^{-y^2/2}dy = \frac{1}{\sqrt{2\pi}}\int_{-\infty}^{\infty}e^{-(y^2-2ity)/2}dy\\ &= \frac{1}{\sqrt{2\pi}}e^{-t^2/2}\int_{-\infty}^{\infty}e^{-(y-it)^2/2}dy = e^{-t^2/2}.\end{aligned}$$

Hence $\phi_X(t/\sigma) = e^{it\mu/\sigma}e^{-t^2/2}$ and replacing t/σ by t, we get, finally:

$$\phi_X(t) = \exp\left(it\mu - \frac{\sigma^2 t^2}{2}\right).$$

Hence

$$\left.\frac{d}{dt}\phi_X(t)\right|_{t=0} = \exp\left(it\mu - \frac{\sigma^2 t^2}{2}\right)(i\mu - \sigma^2 t)\bigg|_{t=0} = i\mu, \quad \text{so that} \quad E(X) = \mu.$$

$$\left.\frac{d^2}{dt^2}\phi_X(t)\right|_{t=0} = \exp\left(it\mu - \frac{\sigma^2 t^2}{2}\right)(i\mu - \sigma^2 t)^2 - \sigma^2 \exp\left(it\mu - \frac{\sigma^2 t^2}{2}\right)\bigg|_{t=0}$$

$$= i^2\mu^2 - \sigma^2 = i^2(\mu^2 + \sigma^2).$$

Then $E(X^2) = \mu^2 + \sigma^2$ and $\sigma^2(X) = \mu^2 + \sigma^2 - \mu^2 = \sigma^2$.

2. Let X be *Gamma* distributed with parameters α and β. Then $\phi_X(t) = (1 - i\beta t)^{-\alpha}$. In fact,

$$\phi_X(t) = \frac{1}{\Gamma(\alpha)\beta^\alpha}\int_0^\infty e^{itx} x^{\alpha-1} e^{-x/\beta} dx = \frac{1}{\Gamma(\alpha)\beta^\alpha}\int_0^\infty x^{\alpha-1} e^{-x(1-i\beta t)/\beta} dx.$$

Setting $x(1 - i\beta t) = y$, we get

$$x = \frac{y}{1 - i\beta t}, \quad dx = \frac{dy}{1 - i\beta t}, \quad y \in [0, \infty).$$

Hence the above expression becomes

$$\frac{1}{\Gamma(\alpha)\beta^\alpha}\int_0^\infty \frac{1}{(1-i\beta t)^{\alpha-1}} y^{\alpha-1} e^{-y/\beta} \frac{dy}{1-i\beta t}$$

$$= (1 - i\beta t)^{-\alpha} \frac{1}{\Gamma(\alpha)\beta^\alpha}\int_0^\infty y^{\alpha-1} e^{-y/\beta} dy = (1 - i\beta t)^{-\alpha}.$$

Therefore

$$\left.\frac{d}{dt}\phi_X(t)\right|_{t=0} = \frac{i\alpha\beta}{(1-i\beta t)^{\alpha+1}}\bigg|_{t=0} = i\alpha\beta,$$

so that $E(X) = \alpha\beta$, and

$$\left.\frac{d^2}{dt^2}\phi_X\right|_{t=0} = i^2 \frac{\alpha(\alpha+1)\beta^2}{(1-i\beta t)^{\alpha+2}}\bigg|_{t=0} = i^2\alpha(\alpha+1)\beta^2,$$

so that $E(X^2) = \alpha(\alpha + 1)\beta^2$. Thus $\sigma^2(X) = \alpha(\alpha + 1)\beta^2 - \alpha^2\beta^2 = \alpha\beta^2$.

For $\alpha = r/2$, $\beta = 2$, we get the corresponding quantities for χ_r^2, and for $\alpha = 1$, $\beta = 1/\lambda$, we get the corresponding quantities for the *Negative Exponential* distribution. So

$$\phi_X(t) = (1-2it)^{-r/2}, \quad \phi_X(t) = \left(1-\frac{it}{\lambda}\right)^{-1} = \frac{\lambda}{\lambda - it},$$

respectively.

3. Let X be *Cauchy* distributed with $\mu = 0$ and $\sigma = 1$. Then $\phi_X(t) = e^{-|t|}$. In fact,

$$\phi_X(t) = \int_{-\infty}^{\infty} e^{itx} \frac{1}{\pi} \frac{1}{1+x^2} dx = \frac{1}{\pi} \int_{-\infty}^{\infty} \frac{\cos(tx)}{1+x^2} dx$$
$$+ \frac{i}{\pi} \int_{-\infty}^{\infty} \frac{\sin(tx)}{1+x^2} dx = \frac{2}{\pi} \int_{0}^{\infty} \frac{\cos(tx)}{1+x^2} dx$$

because

$$\int_{-\infty}^{\infty} \frac{\sin(tx)}{1+x^2} dx = 0,$$

since $\sin(tx)$ is an odd function, and $\cos(tx)$ is an even function. Further, it can be shown by complex variables theory that

$$\int_{0}^{\infty} \frac{\cos(tx)}{1+x^2} dx = \frac{\pi}{2} e^{-|t|}.$$

Hence

$$\phi_X(t) = e^{-|t|}.$$

Now

$$\frac{d}{dt}\phi_X(t) = \frac{d}{dt} e^{-|t|}$$

does *not* exist for $t = 0$. This is consistent with the fact of nonexistence of $E(X)$, as has been seen in Chapter 5.

Exercises

6.3.1 Let X be an r.v. with p.d.f. f given in Exercise 3.2.13 of Chapter 3. Derive its ch.f. ϕ, and calculate EX, $E[X(X-1)]$, $\sigma^2(X)$, provided they are finite.

6.3.2 Let X be an r.v. with p.d.f. f given in Exercise 3.2.14 of Chapter 3. Derive its ch.f. ϕ, and calculate EX, $E[X(X-1)]$, $\sigma^2(X)$, provided they are finite.

6.3.3 Let X be an r.v. with p.d.f. f given by $f(x) = \lambda e^{-\lambda(x-\alpha)} I_{(\alpha,\infty)}(x)$. Find its ch.f. ϕ, and calculate EX, $\sigma^2(X)$, provided they are finite.

6.3.4 Let X be an r.v. distributed as Negative Binomial with parameters r and p.

i) Show that its ch.f., ϕ, is given by

$$\phi(t) = \frac{p^r}{\left(1 - qe^{it}\right)^r};$$

ii) By differentiating ϕ, show that $EX = rq/p$ and $\sigma^2(X) = rq/p^2$;

iii) Find the quantities mentioned in (i) and (ii) for the Geometric distribution.

6.3.5 Let X be an r.v. distributed as $U(\alpha, \beta)$.

i) Show that its ch.f., ϕ, is given by

$$\phi(t) = \frac{e^{it\beta} - e^{it\alpha}}{it(\beta - \alpha)};$$

ii) By differentiating ϕ, show that $EX = \frac{\alpha + \beta}{2}$ and $\sigma^2(X) = \frac{(\alpha - \beta)^2}{12}$.

6.3.6 Consider the r.v. X with p.d.f. f given in Exercise 3.3.14(ii) of Chapter 3, and by using the ch.f. of X, calculate EX^n, $n = 1, 2, \ldots$, provided they are finite.

6.4 Definitions and Basic Theorems—The Multidimensional Case

In this section, versions of Theorems 1, 2 and 3 are presented for the case that the r.v. X is replaced by a k-dimensional r. vector $\mathbf{X}$. Their interpretation, usefulness and usage is analogous to the ones given in the one-dimensional case. To this end, let now $\mathbf{X} = (X_1, \ldots, X_k)'$ be a random vector. Then the ch.f. of the r. vector $\mathbf{X}$, or the *joint ch.f.* of the r.v.'s $X_1, \ldots, X_k$, denoted by $\phi_{\mathbf{X}}$ or $\phi_{X_1, \ldots, X_k}$, is defined as follows:

$$\phi_{X_1, \ldots, X_k}(t_1, \ldots, t_k) = E\left[e^{it_1 X_1 + it_2 X_2 + \cdots + it_k X_k}\right], \quad t_j \in \mathbb{R},$$

$j = 1, 2, \ldots, k$. The ch.f. $\phi_{X_1, \ldots, X_k}$ always exists by an obvious generalization of Lemmas A, A' and B, B'. The joint ch.f. $\phi_{X_1, \ldots, X_k}$ satisfies properties analogous to properties (i)–(vii). That is, one has

THEOREM 1' (Some properties of ch.f.'s)

i') $\phi_{X_1, \ldots, X_k}(0, \ldots, 0) = 1$.

ii') $|\phi_{X_1, \ldots, X_k}(t_1, \ldots, t_k)| \leq 1$.

iii') $\phi_{X_1, \ldots, X_k}$ is uniformly continuous.

iv') $\phi_{X_1+d_1, \ldots, X_k+d_k}(t_1, \ldots, t_k) = e^{it_1 d_1 + \cdots + it_k d_k} \phi_{X_1, \ldots, X_k}(t_1, \ldots, t_k)$.

v') $\phi_{c_1 X_1, \ldots, c_k X_k}(t_1, \ldots, t_k) = \phi_{X_1, \ldots, X_k}(c_1 t_1, \ldots, c_k t_k)$.

vi') $\phi_{c_1 X_1 + d_1, \ldots, c_k X_k + d_k}(t_1, \ldots, t_k) = e^{it_1 d_1 + \cdots + it_k d_k} \phi_{X_1, \ldots, X_k}(c_1 t_1, \ldots, c_k t_k)$.

vii') If the absolute $(n_1, \ldots, n_k)$-joint moment, as well as all lower order joint moments of $X_1, \ldots, X_k$ are finite, then

$$\left. \frac{\partial^{n_1 + \cdots + n_k}}{\partial t_1^{n_1} \cdots \partial t_k^{n_k}} \phi_{X_1, \ldots, X_k}(t_1, \ldots, t_k) \right|_{t_1 = \cdots = t_k = 0} = i^{\sum_{j=1}^{k} n_j} E\left(X_1^{n_1} \cdots X_k^{n_k}\right),$$

and, in particular,

$$\left. \frac{\partial^n}{\partial t_j^n} \phi_{X_1, \ldots, X_k}(t_1, \ldots, t_k) \right|_{t_1 = \cdots = t_k = 0} = i^n E\left(X_j^n\right), \quad j = 1, 2, \ldots, k.$$

viii) If in the $\phi_{X_1, \ldots, X_k}(t_1, \ldots, t_k)$ we set $t_{j_1} = \cdots = t_{j_n} = 0$, then the resulting expression is the joint ch.f. of the r.v.'s $X_{i_1}, \ldots, X_{i_m}$, where the j's and the i's are different and $m + n = k$.

Multidimensional versions of Theorem 2 and Theorem 3 also hold true. We give their formulations below.

THEOREM 2' (Inversion formula) Let $\mathbf{X} = (X_1, \ldots, X_k)'$ be an r. vector with p.d.f. f and ch.f. ϕ. Then

i) $f_{X_1, \ldots, X_k}(x_1, \ldots, x_k) = \lim_{T \to \infty} \left(\frac{1}{2T}\right)^k \int_{-T}^{T} \cdots \int_{-T}^{T} e^{-it_1 x_1 - \cdots - it_k x_k}$

$\times \phi_{X_1, \ldots, X_k}(t_1, \ldots, t_k) dt_1 \cdots dt_k,$

if $\mathbf{X}$ is of the discrete type, and

ii) $f_{X_1, \ldots, X_k}(x_1, \ldots, x_k) = \lim_{h \to 0} \lim_{T \to \infty} \left(\frac{1}{2\pi}\right)^k \int_{-T}^{T} \cdots \int_{-T}^{T} \prod_{j=1}^{k} \left(\frac{1 - e^{-it_j h}}{it_j h}\right)$

$\times e^{-it_1 x_1 - \cdots - it_k x_k} \phi_{X_1, \ldots, X_k}(t_1, \ldots, t_k) dt_1 \cdots dt_k,$

if $\mathbf{X}$ is of the continuous type, with the analog of (ii') holding if the integral of $|\phi_{X_1, \ldots, X_k}(t_1, \ldots, t_k)|$ is finite.

THEOREM 3' (Uniqueness Theorem) There is a one-to-one correspondence between the ch.f. and the p.d.f. of an r. vector.

PROOFS The justification of Theorem 1' is entirely analogous to that given for Theorem 1, and so is the proof of Theorem 2'. As for Theorem 3', the fact that the p.d.f. of $\mathbf{X}$ determines its ch.f. follows from the definition of the ch.f. That the ch.f. of $\mathbf{X}$ determines its p.d.f. follows from Theorem 2'. ▲

6.4.1 The Ch.f. of the Multinomial Distribution

Let $\mathbf{X} = (X_1, \ldots, X_k)'$ be *Multinomially* distributed; that is,

$$P(X_1 = x_1, \ldots, X_k = x_k) = \frac{n!}{x_1! \cdots x_k!} p_1^{x_1} \cdots p_k^{x_k}.$$

Then

$$\phi_{X_1, \ldots, X_k}(t_1, \ldots, t_k) = \left(p_1 e^{it_1} + \cdots + p_k e^{it_k}\right)^n.$$

In fact,

$$\phi_{X_1, \ldots, X_k}(t_1, \ldots, t_k) = \sum_{x_1, \ldots, x_k} e^{it_1 x_1 + \cdots + it_k x_k} \frac{n!}{x_1! \cdots x_k!} \times p_1^{x_1} \cdots p_k^{x_k}$$

$$= \sum_{x_1, \ldots, x_k} \frac{n!}{x_1! \cdots x_k!} \left(p_1 e^{it_1}\right)^{x_1} \cdots \left(p_k e^{it_k}\right)^{x_k}$$

$$= \left(p_1 e^{it_1} + \cdots + p_k e^{it_k}\right)^n.$$

Hence

$$\left.\frac{\partial^k}{\partial t_1 \cdots \partial t_k} \phi_{X_1, \ldots, X_k}(t_1, \ldots, t_k)\right|_{t_1 = \cdots = t_k = 0}$$

$$= n(n-1) \cdots (n-k+1) i^k p_1 \cdots p_k \left(p_1 e^{it_1} + \cdots + p_k e^{it_k}\right)^{n-k}\bigg|_{t_1 = \cdots = t_k = 0} = i^k n(n-1) \cdots (n-k+1) p_1 p_2 \cdots p_k.$$

Hence

$$E(X_1 \cdots X_k) = n(n-1) \cdots (n-k+1) p_1 p_2 \cdots p_k.$$

Finally, the *ch.f.* of a (measurable) function $g(\mathbf{X})$ of the r. vector $\mathbf{X} = (X_1, \ldots, X_k)'$ is defined by:

$$\phi_{g(\mathbf{x})}(t) = E\left[e^{itg(\mathbf{X})}\right] = \begin{cases} \sum_{\mathbf{x}} e^{itg(\mathbf{x})} f(\mathbf{x}), & \mathbf{x} = (x_1, \ldots, x_k)' \\ \int_{-\infty}^{\infty} \cdots \int_{-\infty}^{\infty} e^{itg(x_1, \ldots, x_k)} f(x_1, \ldots, x_k) dx_1, \ldots, dx_k. \end{cases}$$

Exercise

6.4.1 (*Cramér–Wold*) Consider the r.v.'s X_j, $j = 1, \ldots, k$ and for $c_j \in \mathbb{R}$, $j = 1, \ldots, k$, set

$$Y_c = \sum_{j=1}^{k} c_j X_j.$$

Then

i) Show that $\phi_{Y_c}(t) = \phi_{X_1,\ldots,X_k}(c_1 t, \ldots, c_k t)$, $t \in \mathbb{R}$, and $\phi_{X_1,\ldots,X_k}(c_1, \ldots, c_k) = \phi_{Y_c}(1)$;

ii) Conclude that the distribution of the X's determines the distribution of Y_c for every $c_j \in \mathbb{R}$, $j = 1, \ldots, k$. Conversely, the distribution of the X's is determined by the distribution of Y_c for every $c_j \in \mathbb{R}$, $j = 1, \ldots, k$.

6.5 The Moment Generating Function and Factorial Moment Generating Function of a Random Variable

The ch.f. of an r.v. or an r. vector is a function defined on the entire real line and taking values in the complex plane. Those readers who are not well versed in matters related to complex-valued functions may feel uncomfortable in dealing with ch.f.'s. There is a partial remedy to this potential problem, and that is to replace a ch.f. by an entity which is called moment generating function. However, there is a price to be paid for this: namely, a moment generating function may exist (in the sense of being finite) only for $t = 0$. There are cases where it exists for t's lying in a proper subset of $\mathbb{R}$ (containing 0), and yet other cases, where the moment generating function exists for all real t. All three cases will be illustrated by examples below.

First, consider the case of an r.v. X. Then the *moment generating function* (m.g.f.) M_X (or just M when no confusion is possible) of a random variable X, which is also called the *Laplace transform* of f, is defined by $M_X(t) = E(e^{tX})$, $t \in \mathbb{R}$, if this expectation exists. For $t = 0$, $M_X(0)$ always exists and equals 1. However, it may fail to exist for $t \neq 0$. If $M_X(t)$ exists, then formally $\phi_X(t) = M_X(it)$ and therefore the m.g.f. satisfies most of the properties analogous to properties (i)–(vii) cited above in connection with the ch.f., under suitable conditions. In particular, property (vii) in Theorem 1 yields $\frac{d^n}{dt^n} M_X(t)\big|_{t=0} = E(X^n)$, provided Lemma D applies. In fact,

$$\frac{d^n}{dt^n} M_X(t)\bigg|_{t=0} = \frac{d^n}{dt^n}\left(Ee^{tX}\right)\bigg|_{t=0} = E\left(\frac{d^n}{dt^n} e^{tX}\right)\bigg|_{t=0}$$
$$= E\left(X^n e^{tX}\right)\bigg|_{t=0} = E(X^n).$$

This is the property from which the m.g.f. derives its name.

Here are some examples of m.g.f.'s. It is instructive to derive them in order to see how conditions are imposed on t in order for the m.g.f. to be finite. It so happens that part (vii) of Theorem 1, as it would be formulated for an m.g.f., is applicable in all these examples, although no justification will be supplied.

6.5.1 The M.G.F.'s of Some R.V.'s

1. If $X \sim B(n, p)$, then $M_X(t) = (pe^t + q)^n$, $t \in \mathbb{R}$. Indeed,

$$M_X(t) = \sum_{x=0}^{n} e^{tx} \binom{n}{x} p^x q^{n-x} = \sum_{x=0}^{n} \binom{n}{x} (pe^t)^x q^{n-x} = (pe^t + q)^n,$$

which, clearly, is finite for all $t \in \mathbb{R}$.
Then

$$\left. \frac{d}{dt} M_X(t) \right|_{t=0} = \left. \frac{d}{dt} (pe^t + q)^n \right|_{t=0} = \left. n(pe^t + q)^{n-1} pe^t \right|_{t=0} = np = E(X),$$

and

$$\left. \frac{d^2}{dt^2} M_X(t) \right|_{t=0} = \left. \frac{d}{dt} \left[np(pe^t + q)^{n-1} e^t \right] \right|_{t=0}$$

$$= \left. np \left[(n-1)(pe^t + q)^{n-2} pe^t e^t + (pe^t + q)^{n-1} e^t \right] \right|_{t=0}$$

$$= n(n-1)p^2 + np = n^2 p^2 - np^2 + np = E(X^2),$$

so that $\sigma^2(X) = n^2 p^2 - np^2 + np - n^2 p^2 = np(1-p) = npq$.

2. If $X \sim P(\lambda)$, then $M_X(t) = e^{\lambda e^t - \lambda}$, $t \in \mathbb{R}$. In fact,

$$M_X(t) = \sum_{x=0}^{\infty} e^{tx} e^{-\lambda} \frac{\lambda^x}{x!} = e^{-\lambda} \sum_{x=0}^{\infty} \frac{(\lambda e^t)^x}{x!} = e^{-\lambda} e^{\lambda e^t} = e^{\lambda e^t - \lambda}.$$

Then

$$\left. \frac{d}{dt} M_X(t) \right|_{t=0} = \left. \frac{d}{dt} e^{\lambda e^t - \lambda} \right|_{t=0} = \left. \lambda e^t e^{\lambda e^t - \lambda} \right|_{t=0} = \lambda = E(X),$$

and

$$\left. \frac{d^2}{dt^2} M_X(t) \right|_{t=0} = \left. \frac{d}{dt} \left(\lambda e^t e^{\lambda e^t - \lambda} \right) \right|_{t=0} = \left. \lambda \left(e^t e^{\lambda e^t - \lambda} + e^t e^{\lambda e^t - \lambda} \lambda e^t \right) \right|_{t=0}$$

$$= \lambda(1 + \lambda) = E(X^2), \quad \text{so that } \sigma^2(X) = \lambda + \lambda^2 - \lambda^2 = \lambda.$$

3. If $X \sim N(\mu, \sigma^2)$, then $M_X(t) = e^{t\mu + \frac{\sigma^2 t^2}{2}}$, $t \in \mathbb{R}$, and, in particular, if $X \sim N(0, 1)$, then $M_X(t) = e^{t^2/2}$, $t \in \mathbb{R}$. By the property for m.g.f. analogous to property (vi) in Theorem 1,

$$M_{\frac{X-\mu}{\sigma}}(t) = e^{-\mu t/\sigma} M_X\left(\frac{t}{\sigma}\right), \quad \text{so that } M_X\left(\frac{t}{\sigma}\right) = e^{\mu t/\sigma} M_{\frac{X-\mu}{\sigma}}(t)$$

for all $t \in \mathbb{R}$. Therefore

$$M_X\left(\frac{t}{\sigma}\right) = e^{\frac{\mu t}{\sigma}} e^{t^2/2} = e^{\frac{\mu t}{\sigma} + \frac{t^2}{2}}.$$

Replacing t by σt, we get, finally, $M_X(t) = e^{\mu t + \frac{\sigma^2 t^2}{2}}$. Then

$$\frac{d}{dt} M_X(t) \bigg|_{t=0} = \frac{d}{dt} e^{\mu t + \frac{\sigma^2 t^2}{2}} \bigg|_{t=0} = (\mu + \sigma^2 t) e^{\mu t + \frac{\sigma^2 t^2}{2}} \bigg|_{t=0} = \mu = E(X),$$

and

$$\frac{d^2}{dt^2} M_X(t) \bigg|_{t=0} = \frac{d}{dt} \left[(\mu + \sigma^2 t) e^{\mu t + \frac{\sigma^2 t^2}{2}} \right]_{t=0}$$

$$= \left[\sigma^2 e^{\mu t + \frac{\sigma^2 t^2}{2}} + (\mu + \sigma^2 t)^2 e^{\mu t + \frac{\sigma^2 t^2}{2}} \right]_{t=0} = \sigma^2 + \mu^2$$

$$= E(X^2), \quad \text{so that } \sigma^2(X) = \sigma^2 + \mu^2 - \mu^2 = \sigma^2.$$

4. If X is distributed as Gamma with parameters α and β, then $M_X(t) = (1 - \beta t)^{-\alpha}$, $t < 1/\beta$. Indeed,

$$M_X(t) = \frac{1}{\Gamma(\alpha)\beta^\alpha} \int_0^\infty e^{tx} x^{\alpha-1} e^{-x/\beta} dx = \frac{1}{\Gamma(\alpha)\beta^\alpha} \int_0^\infty x^{\alpha-1} e^{-x(1-\beta t)/\beta} dx.$$

Then by setting $x(1 - \beta t) = y$, so that $x = \frac{y}{1-\beta t}$, $dx = \frac{dy}{1-\beta t}$, and $y \in [0, \infty)$, the above expression is equal to

$$\frac{1}{(1-\beta t)^\alpha} \cdot \frac{1}{\Gamma(\alpha)\beta^\alpha} \int_0^\infty y^{\alpha-1} e^{-y/\beta} dy = \frac{1}{(1-\beta t)^\alpha},$$

provided $1 - \beta t > 0$, or equivalently, $t < 1/\beta$. Then

$$\frac{d}{dt} M_X(t) \bigg|_{t=0} = \frac{d}{dt} (1 - \beta t)^{-\alpha} \bigg|_{t=0} = \alpha\beta = E(X),$$

and

$$\left.\frac{d^2}{dt^2}M_X(t)\right|_{t=0} = \left.\alpha\beta\frac{d}{dt}(1-\beta t)^{-\alpha-1}\right|_{t=0} = \left.\alpha(\alpha+1)\beta^2(1-\beta t)^{-\alpha-2}\right|_{t=0}$$
$$= \alpha(\alpha+1)\beta^2 = (EX^2), \text{ so that } \sigma^2(X) = \alpha\beta^2.$$

In particular, for $\alpha = \frac{r}{2}$ and $\beta = 2$, we get the m.g.f. of the χ_r^2, and its mean and variance; namely,

$$M_X(t) = (1-2t)^{-r/2}, \ t < \frac{1}{2}, \ E(X) = r, \ \sigma^2(X) = 2r.$$

For $\alpha = 1$ and $\beta = \frac{1}{\lambda}$, we obtain the m.g.f. of the Negative Exponential distribution, and its mean and variance; namely

$$M_X(t) = \frac{\lambda}{\lambda - t}, \ t < \lambda, \ EX = \frac{1}{\lambda}, \ \sigma^2(X) = \frac{1}{\lambda^2}.$$

5. Let X have the Cauchy distribution with parameters μ and σ, and without loss of generality, let $\mu = 0$, $\sigma = 1$. Then the $M_X(t)$ exists *only* for $t = 0$. In fact,

$$M_X(t) = E(e^{tX}) = \int_{-\infty}^{\infty} e^{tx} \frac{1}{\pi} \frac{1}{1+x^2} dx$$
$$> \frac{1}{\pi}\int_0^{\infty} e^{tx}\frac{1}{1+x^2}dx > \frac{1}{\pi}\int_0^{\infty}(tx)\frac{1}{1+x^2}dx$$

if $t > 0$, since $e^z > z$, for $z > 0$, and this equals

$$\frac{t}{2\pi}\int_0^{\infty}\frac{2x\,dx}{1+x^2} = \frac{t}{2\pi}\int_1^{\infty}\frac{du}{u} = \frac{t}{2\pi}\left(\lim_{x\to\infty}\log u\right).$$

Thus for $t > 0$, $M_X(t)$ obviously is equal to ∞. If $t < 0$, by using the limits $-\infty$, 0 in the integral, we again reach the conclusion that $M_X(t) = \infty$ (see Exercise 6.5.9).

REMARK 4 The examples just discussed exhibit all three cases regarding the existence or nonexistence of an m.g.f. In Examples 1 and 3, the m.g.f.'s exist for all $t \in \mathbb{R}$; in Examples 2 and 4, the m.g.f.'s exist for proper subsets of $\mathbb{R}$; and in Example 5, the m.g.f. exists only for $t = 0$.

For an r.v. X, we also define what is known as its factorial moment generating function. More precisely, the *factorial m.g.f.* η_X (or just η when no confusion is possible) of an r.v. X is defined by:

$$\eta_X(t) = E(t^X), \ t \in \mathbb{R}, \text{ if } E(t^X) \text{ exists}.$$

This function is sometimes referred to as the *Mellin or Mellin–Stieltjes transform of f*. Clearly, $\eta_X(t) = M_X(\log t)$ for $t > 0$.

Formally, the *n*th factorial moment of an r.v. X is taken from its factorial m.g.f. by differentiation as follows:

6.5 The Moment Generating Function

$$\frac{d^n}{dt^n}\eta_X(t)\bigg|_{t=1} = E[X(X-1)\cdots(X-n+1)].$$

In fact,

$$\frac{d^n}{dt^n}\eta_X(t) = \frac{d^n}{dt^n}E(t^X) = E\left(\frac{\partial^n}{\partial t^n}t^X\right) = E[X(X-1)\cdots(X-n+1)t^{X-n}],$$

provided Lemma D applies, so that the interchange of the order of differentiation and expectation is valid. Hence

$$\frac{d^n}{dt^n}\eta_X(t)\bigg|_{t=1} = E[X(X-1)\cdots(X-n+1)]. \tag{9}$$

REMARK 5 The factorial m.g.f. derives its name from the property just established. As has already been seen in the first two examples in Section 2 of Chapter 5, factorial moments are especially valuable in calculating the variance of discrete r.v.'s. Indeed, since

$$\sigma^2(X) = E(X^2) - (EX)^2, \quad \text{and} \quad E(X^2) = E[X(X-1)] + E(X),$$

we get

$$\sigma^2(X) = E[X(X-1)] + E(X) - (EX)^2;$$

that is, an expression of the variance of X in terms of derivatives of its factorial m.g.f. up to order two.

Below we derive some factorial m.g.f.'s. Property (9) (for $n = 2$) is valid in all these examples, although no specific justification will be provided.

6.5.2 The Factorial M.G.F.'s of some R.V.'s

1. If $X \sim B(n, p)$, then $\eta_X(t) = (pt + q)^n$, $t \in \mathbb{R}$. In fact,

$$\eta_X(t) = \sum_{x=0}^{n} t^x \binom{n}{x} p^x q^{n-x} = \sum_{x=0}^{n} \binom{n}{x}(pt)^x q^{n-x} = (pt+q)^n.$$

Then

$$\frac{d^2}{dt^2}\eta_X(t)\bigg|_{t=1} = n(n-1)p^2(pt+q)^{n-2} = n(n-1)p^2,$$

so that $\sigma^2(X) = n(n-1)p^2 + np - n^2p^2 = npq$.

2. If $X \sim P(\lambda)$, then $\eta_X(t) = e^{\lambda t - \lambda}$, $t \in \mathbb{R}$. In fact,

$$\eta_X(t) = \sum_{x=0}^{\infty} t^x e^{-\lambda}\frac{\lambda^x}{x!} = e^{-\lambda}\sum_{x=0}^{\infty}\frac{(\lambda t)^x}{x!} = e^{-\lambda}e^{\lambda t} = e^{\lambda t - \lambda}, \; t \in \mathbb{R}.$$

Hence

$$\left.\frac{d^2}{dt^2}\eta_X(t)\right|_{t=1} = \lambda^2 e^{\lambda t - \lambda}\bigg|_{t=1} = \lambda^2, \quad \text{so that } \sigma^2(X) = \lambda^2 + \lambda - \lambda^2 = \lambda.$$

The m.g.f. of an r. vector $\mathbf{X}$ or the *joint* m.g.f. of the r.v.'s $X_1, \ldots, X_k$, denoted by $M_\mathbf{X}$ or $M_{X_1, \ldots, X_k}$, is defined by:

$$M_{X_1, \ldots, X_k}(t_1, \ldots, t_k) = E\left(e^{t_1 X_1 + \cdots + t_k X_k}\right), \quad t_j \in \mathbb{R}, \quad j = 1, 2, \ldots, k,$$

for those t_j's in $\mathbb{R}$ for which this expectation exists. If $M_{X_1, \ldots, X_k}(t_1, \ldots, t_k)$ exists, then formally $\phi_{X_1, \ldots, X_k}(t_1, \ldots, t_k) = M_{X_1, \ldots, X_k}(it_1, \ldots, it_k)$ and properties analogous to (i')–(vii'), (viii) in Theorem 1' hold true under suitable conditions. In particular,

$$\left.\frac{\partial^{n_1 + \cdots + n_k}}{\partial t_1^{n_1} \cdots \partial t_k^{n_k}} M_{X_1, \ldots, X_k}(t_1, \ldots, t_k)\right|_{t_1 = \cdots = t_k = 0} = E\left(X_1^{n_1} \cdots X_k^{n_k}\right), \quad (10)$$

where $n_1, \ldots, n_k$ are non-negative integers.

Below, we present two examples of m.g.f.'s of r. vectors.

6.5.3 The M.G.F.'s of Some R. Vectors

1. If the r.v.'s $X_1, \ldots, X_k$ have jointly the Multinomial distribution with parameters n and $p_1, \ldots, p_k$, then

$$M_{X_1, \ldots, X_k}(t_1, \ldots, t_k) = \left(p_1 e^{t_1} + \cdots + p_k e^{t_k}\right)^n, \quad t_j \in \mathbb{R}, \quad j = 1, \ldots, k.$$

In fact,

$$M_{X_1, \ldots, X_k}(t_1, \ldots, t_k) = E e^{t_1 X_1 + \cdots + t_k X_k} = \sum e^{t_1 X_1 + \cdots + t_k X_k} \frac{n!}{x_1! \cdots x_k!} p_1^{x_1} \cdots p_k^{x_k}$$

$$= \sum \frac{n!}{x_1! \cdots x_k!} \left(p_1 e^{t_1}\right)^{x_1} \cdots \left(p_k e^{t_k}\right)^{x_k}$$

$$= \left(p_1 e^{t_1} + \cdots + p_k e^{t_k}\right)^n,$$

where the summation is over all integers $x_1, \ldots, x_k \geq 0$ with $x_1 + \cdots + x_k = n$. Clearly, the above derivations hold true for all $t_j \in \mathbb{R}$, $j = 1, \ldots, k$.

2. If the r.v.'s X_1 and X_2 have the Bivariate Normal distribution with parameters $\mu_1, \mu_2, \sigma_1^2, \sigma_2^2$ and ρ, then their joint m.g.f. is

$$M_{X_1, X_2}(t_1, t_2) = \exp\left[\mu_1 t_1 + \mu_2 t_2 + \frac{1}{2}\left(\sigma_1^2 t_1^2 + 2\rho\sigma_1\sigma_2 t_1 t_2 + \sigma_2^2 t_2^2\right)\right], \quad t_1, t_2 \in \mathbb{R}. \quad (11)$$

An analytical derivation of this formula is possible, but we prefer to use the matrix approach, which is more elegant and compact. Recall that the joint p.d.f. of X_1 and X_2 is given by

$$f(x_1, x_2)$$
$$= \frac{1}{2\pi\sigma_1\sigma_2\sqrt{1-\rho^2}} \exp\left\{-\frac{1}{2}\left[\left(\frac{x_1-\mu_1}{\sigma_1}\right)^2 - 2\rho\left(\frac{x_1-\mu_1}{\sigma_1}\right)\left(\frac{x_2-\mu_2}{\sigma_2}\right) + \left(\frac{x_2-\mu_2}{\sigma_2}\right)^2\right]\right\}.$$

Set $\mathbf{x} = (x_1\ x_2)'$, $\boldsymbol{\mu} = (\mu_1\ \mu_2)'$, and

$$\boldsymbol{\Sigma} = \begin{pmatrix} \sigma_1^2 & \rho\sigma_1\sigma_2 \\ \rho\sigma_1\sigma_2 & \sigma_2^2 \end{pmatrix}.$$

Then the determinant of $\boldsymbol{\Sigma}$, $|\boldsymbol{\Sigma}|$, is $|\boldsymbol{\Sigma}| = \sigma_1^2\sigma_2^2(1-\rho^2)$, and the inverse, $\boldsymbol{\Sigma}^{-1}$, is

$$\boldsymbol{\Sigma}^{-1} = \frac{1}{|\boldsymbol{\Sigma}|}\begin{pmatrix} \sigma_2^2 & -\rho\sigma_1\sigma_2 \\ -\rho\sigma_1\sigma_2 & \sigma_1^2 \end{pmatrix}.$$

Therefore

$$(\mathbf{x}-\boldsymbol{\mu})'\boldsymbol{\Sigma}^{-1}(\mathbf{x}-\boldsymbol{\mu}) = \frac{1}{|\boldsymbol{\Sigma}|}(x_1-\mu_1\ x_2-\mu_2)\begin{pmatrix} \sigma_2^2 & -\rho\sigma_1\sigma_2 \\ -\rho\sigma_1\sigma_2 & \sigma_1^2 \end{pmatrix}\begin{pmatrix} x_1-\mu_1 \\ x_2-\mu_2 \end{pmatrix}$$

$$= \frac{1}{\sigma_1^2\sigma_2^2(1-\rho^2)}\left[\sigma_2^2(x_1-\mu_1)^2 - 2\rho\sigma_1\sigma_2(x_1-\mu_1)(x_2-\mu_2) + \sigma_1^2(x_2-\mu_2)^2\right]$$

$$= \frac{1}{1-\rho^2}\left[\left(\frac{x_1-\mu_1}{\sigma_1}\right)^2 - 2\rho\left(\frac{x_1-\mu_1}{\sigma_1}\right)\left(\frac{x_2-\mu_2}{\sigma_2}\right) + \left(\frac{x_2-\mu_2}{\sigma_2}\right)^2\right].$$

Therefore the p.d.f. is written as follows in matrix notation:

$$f(\mathbf{x}) = \frac{1}{2\pi|\boldsymbol{\Sigma}|^{1/2}}\exp\left[-\frac{1}{2}(\mathbf{x}-\boldsymbol{\mu})'\boldsymbol{\Sigma}^{-1}(\mathbf{x}-\boldsymbol{\mu})\right].$$

In this form, $\boldsymbol{\mu}$ is the *mean vector* of $\mathbf{X} = (X_1\ X_2)'$, and $\boldsymbol{\Sigma}$ is the *covariance matrix* of $\mathbf{X}$.

Next, for $\mathbf{t} = (t_1\ t_2)'$, we have

$$M_{\mathbf{X}}(\mathbf{t}) = Ee^{\mathbf{t}'\mathbf{X}} = \int_{R^2} \exp(\mathbf{t}'\mathbf{x})f(\mathbf{x})d\mathbf{x}$$

$$= \frac{1}{2\pi|\boldsymbol{\Sigma}|^{1/2}}\int_{R^2} \exp\left[\mathbf{t}'\mathbf{x} - \frac{1}{2}(\mathbf{x}-\boldsymbol{\mu})'\boldsymbol{\Sigma}^{-1}(\mathbf{x}-\boldsymbol{\mu})\right]d\mathbf{x}. \quad (11)$$

The exponent may be written as follows:

$$\left(\boldsymbol{\mu}'\mathbf{t}+\frac{1}{2}\mathbf{t}'\boldsymbol{\Sigma}\mathbf{t}\right)-\frac{1}{2}\left[2\boldsymbol{\mu}'\mathbf{t}+\mathbf{t}'\boldsymbol{\Sigma}\mathbf{t}-2\mathbf{t}'\mathbf{x}+(\mathbf{x}-\boldsymbol{\mu})'\boldsymbol{\Sigma}^{-1}(\mathbf{x}-\boldsymbol{\mu})\right]. \quad (13)$$

Focus on the quantity in the bracket, carry out the multiplication, and observe that $\boldsymbol{\Sigma}' = \boldsymbol{\Sigma}$, $(\boldsymbol{\Sigma}^{-1})' = \boldsymbol{\Sigma}^{-1}$, $\mathbf{x}'\mathbf{t} = \mathbf{t}'\mathbf{x}$, $\boldsymbol{\mu}'\mathbf{t} = \mathbf{t}'\boldsymbol{\mu}$, and $\mathbf{x}'\boldsymbol{\Sigma}^{-1}\boldsymbol{\mu} = \boldsymbol{\mu}'\boldsymbol{\Sigma}^{-1}\mathbf{x}$, to obtain

$$2\boldsymbol{\mu}'\mathbf{t}+\mathbf{t}'\boldsymbol{\Sigma}\mathbf{t}-2\mathbf{t}'\mathbf{x}+(\mathbf{x}-\boldsymbol{\mu})'\boldsymbol{\Sigma}^{-1}(\mathbf{x}-\boldsymbol{\mu})=\left[\mathbf{x}-(\boldsymbol{\mu}+\boldsymbol{\Sigma}\mathbf{t})\right]'\boldsymbol{\Sigma}^{-1}\left[\mathbf{x}-(\boldsymbol{\mu}+\boldsymbol{\Sigma}\mathbf{t})\right]. \quad (14)$$

By means of (13) and (14), the m.g.f. in (12) becomes

$$M_{\mathbf{X}}(\mathbf{t}) = \exp\left(\boldsymbol{\mu}'\mathbf{t}+\frac{1}{2}\mathbf{t}'\boldsymbol{\Sigma}\mathbf{t}\right)\frac{1}{2\pi|\boldsymbol{\Sigma}|^{1/2}}\int_{\mathbb{R}^2}\exp\left[-\frac{1}{2}(\mathbf{x}-(\boldsymbol{\mu}+\boldsymbol{\Sigma}\mathbf{t}))'\boldsymbol{\Sigma}^{-1}(\mathbf{x}-(\boldsymbol{\mu}+\boldsymbol{\Sigma}\mathbf{t}))\right]d\mathbf{x}.$$

However, the second factor above is equal to 1, since it is the integral of a Bivariate Normal distribution with mean vector $\boldsymbol{\mu}+\boldsymbol{\Sigma}\mathbf{t}$ and covariance matrix $\boldsymbol{\Sigma}$. Thus

$$M_{\mathbf{X}}(\mathbf{t}) = \exp\left(\boldsymbol{\mu}'\mathbf{t}+\frac{1}{2}\mathbf{t}'\boldsymbol{\Sigma}\mathbf{t}\right). \quad (15)$$

Observing that

$$\mathbf{t}'\boldsymbol{\Sigma}\mathbf{t} = (t_1\ t_2)\begin{pmatrix}\sigma_1^2 & \rho\sigma_1\sigma_2 \\ \rho\sigma_1\sigma_2 & \sigma_2^2\end{pmatrix}\begin{pmatrix}t_1 \\ t_2\end{pmatrix} = \sigma_1^2 t_1^2 + 2\rho\sigma_1\sigma_2 t_1 t_2 + \sigma_2^2 t_2^2,$$

it follows that the m.g.f. is, indeed, given by (11).

Exercises

6.5.1 Derive the m.g.f. of the r.v. X which denotes the number of spots that turn up when a balanced die is rolled.

6.5.2 Let X be an r.v. with p.d.f. f given in Exercise 3.2.13 of Chapter 3. Derive its m.g.f. and factorial m.g.f., $M(t)$ and $\eta(t)$, respectively, for those t's for which they exist. Then calculate EX, $E[X(X-1)]$ and $\sigma^2(X)$, provided they are finite.

6.5.3 Let X be an r.v. with p.d.f. f given in Exercise 3.2.14 of Chapter 3. Derive its m.g.f. and factorial m.g.f., $M(t)$ and $\eta(t)$, respectively, for those t's for which they exist. Then calculate EX, $E[X(X-1)]$ and $\sigma^2(X)$, provided they are finite.

6.5.4 Let X be an r.v. with p.d.f. f given by $f(x) = \lambda e^{-\lambda(x-\alpha)} I_{(\alpha,\infty)}(x)$. Find its m.g.f. $M(t)$ for those t's for which it exists. Then calculate EX and $\sigma^2(X)$, provided they are finite.

6.5.5 Let X be an r.v. distributed as $B(n, p)$. Use its factorial m.g.f. in order to calculate its kth factorial moment. Compare with Exercise 5.2.1 in Chapter 5.

6.5.6 Let X be an r.v. distributed as $P(\lambda)$. Use its factorial m.g.f. in order to calculate its kth factorial moment. Compare with Exercise 5.2.4 in Chapter 5.

6.5.7 Let X be an r.v. distributed as Negative Binomial with parameters r and p.

i) Show that its m.g.f and factorial m.g.f., $M(t)$ and $\eta(t)$, respectively, are given by

$$M_X(t) = \frac{p^r}{(1-qe^t)^r}, \quad t < -\log q, \quad \eta_X(t) = \frac{p^r}{(1-qt)^r}, \quad |t| < \frac{1}{q};$$

ii) By differentiation, show that $EX = rq/p$ and $\sigma^2(X) = rq/p^2$;

iii) Find the quantities mentioned in parts (i) and (ii) for the Geometric distribution.

6.5.8 Let X be an r.v. distributed as $U(\alpha, \beta)$.

i) Show that its m.g.f., M, is given by

$$M(t) = \frac{e^{t\beta} - e^{t\alpha}}{t(\beta - \alpha)};$$

ii) By differentiation, show that $EX = \frac{\alpha+\beta}{2}$ and $\sigma^2(X) = \frac{(\alpha-\beta)^2}{12}$.

6.5.9 Refer to Example 3 in the Continuous case and show that $M_X(t) = \infty$ for $t < 0$ as asserted there.

6.5.10 Let X be an r.v. with m.g.f. M given by $M(t) = e^{\alpha t + \beta t^2}$, $t \in \mathbb{R}$ ($\alpha \in \mathbb{R}$, $\beta > 0$). Find the ch.f. of X and identify its p.d.f. Also use the ch.f. of X in order to calculate EX^4.

6.5.11 For an r.v. X, define the function γ by $\gamma(t) = E(1+t)^X$ for those t's for which $E(1+t)^X$ is finite. Then, if the nth factorial moment of X is finite, show that

$$\left(d^{r\alpha}/dt^n\right)\gamma(t)\Big|_{t=0} = E\left[X(X-1)\cdots(X-n+1)\right].$$

6.5.12 Refer to the previous exercise and let X be $P(\lambda)$. Derive $\gamma(t)$ and use it in order to show that the nth factorial moment of X is λ^n.

6.5.13 Let X be an r.v. with m.g.f. M and set $K(t) = \log M(t)$ for those t's for which $M(t)$ exists. Furthermore, suppose that $EX = \mu$ and $\sigma^2(X) = \sigma^2$ are both finite. Then show that

$$\frac{d}{dt}K(t)\bigg|_{t=0} = \mu \quad \text{and} \quad \frac{d^2}{dt^2}K(t)\bigg|_{t=0} = \sigma^2.$$

(The function K just defined is called the *cumulant generating function* of X.)

6.5.14 Let X be an r.v. such that EX^n is finite for all $n = 1, 2, \ldots$. Use the expansion

$$e^x = \sum_{n=0}^{\infty} \frac{x^n}{n!}$$

in order to show that, under appropriate conditions, one has that the m.g.f. of X is given by

$$M(t) = \sum_{n=0}^{\infty} (EX^n) \frac{t^n}{n!}.$$

6.5.15 If X is an r.v. such that $EX^n = n!$, then use the previous exercise in order to find the m.g.f. $M(t)$ of X for those t's for which it exists. Also find the ch.f. of X and from this, deduce the distribution of X.

6.5.16 Let X be an r.v. such that

$$EX^{2k} = \frac{(2k)!}{2^k k!}, \quad EX^{2k+1} = 0,$$

$k = 0, 1, \ldots$. Find the m.g.f. of X and also its ch.f. Then deduce the distribution of X.

6.5.17 Let X_1, X_2 be two r.v.'s with m.f.g. given by

$$M(t_1, t_2) = \left[\frac{1}{3}\left(e^{t_1+t_2} + 1\right) + \frac{1}{6}\left(e^{t_1} + e^{t_2}\right)\right]^2, \quad t_1, t_2 \in \mathbb{R}.$$

Calculate EX_1, $\sigma^2(X_1)$ and $\text{Cov}(X_1, X_2)$, provided they are finite.

6.5.18 Refer to Exercise 4.2.5. in Chapter 4 and find the joint m.g.f. $M(t_1, t_2, t_3)$ of the r.v.'s X_1, X_2, X_3 for those t_1, t_2, t_3 for which it exists. Also find their joint ch.f. and use it in order to calculate $E(X_1 X_2 X_3)$, provided the assumptions of Theorem 1' (vii') are met.

6.5.19 Refer to the previous exercise and derive the m.g.f. $M(t)$ of the r.v. $g(X_1, X_2, X_3) = X_1 + X_2 + X_3$ for those t's for which it exists. From this, deduce the distribution of g.

6.5.20 Let X_1, X_2 be two r.v.'s with m.g.f. M and set $K(t_1, t_2) = \log M(t_1, t_2)$ for those t_1, t_2 for which $M(t_1, t_2)$ exists. Furthermore, suppose that expectations, variances, and covariances of these r.v.'s are all finite. Then show that for $j = 1, 2$,

$$\left.\frac{\partial}{\partial t_j} K(t_1, t_2)\right|_{t_1=t_2=0} = EX_j, \quad \left.\frac{\partial^2}{\partial t_j^2} K(t_1, t_2)\right|_{t_1=t_2=0} = \sigma^2(X_j),$$

$$\left.\frac{\partial^2}{\partial t_1 \partial t_2} K(t_1, t_2)\right|_{t_1=t_2=0} = \text{Cov}(X_1, X_2).$$

6.5.21 Suppose the r.v.'s $X_1, \ldots, X_k$ have the Multinomial distribution with parameters n and $p_1, \ldots, p_k$, and let i, j, be arbitrary but fixed, $1 \leq i < j \leq k$. Consider the r.v.'s X_i, X_j, and set $X = n - X_i - X_j$, so that these r.v.'s have the Multinomial distribution with parameters n and p_i, p_j, p, where $p = 1 - p_i - p_j$.

i) Write out the joint m.g.f. of X_i, X_j, X, and by differentiation, determine the $E(X_i X_j)$;

ii) Calculate the covariance of X_i, X_j, $\text{Cov}(X_i, X_j)$, and show that it is negative. Note: Two r.v.'s whose covariance is nonpositive are said to be *negatively quadrant dependent*. Accordingly, any two r.v.'s X_i and X_j out of the Multinomially distributed r.v.'s $X_1, \ldots, X_k$ are negatively quadrant dependent.

6.5.22 If the r.v.'s X_1 and X_2 have the Bivariate Normal distribution with parameters $\mu_1, \mu_2, \sigma_1^2, \sigma_2^2$ and ρ, show that $\text{Cov}(X_1, X_2) \geq 0$ if $\rho \geq 0$, and $\text{Cov}(X_1, X_2) < 0$ if $\rho < 0$. Note: Two r.v.'s whose covariance is nonnegative are said to be *positively quadrant dependent*. Accordingly, any two r.v.'s X_1 and X_2 which have the Bivariate Normal distribution are positively quadrant dependent if $\rho \geq 0$; they are *negatively quadrant dependent* if $\rho < 0$.

6.5.23 Verify the validity of relation (13).

6.5.24

i) If the r.v.'s X_1 and X_2 have the Bivariate Normal distribution with parameters $\mu_1, \mu_2, \sigma_1^2, \sigma_2^2$ and ρ, use their joint m.g.f. given by (11) and property (10) in order to determine $E(X_1 X_2)$;

ii) Show that ρ is, indeed, the correlation coefficient of X_1 and X_2.

6.5.25 Both parts of Exercise 6.4.1 hold true if the ch.f.'s involved are replaced by m.g.f.'s, provided, of course, that these m.g.f.'s exist.

i) Use Exercise 6.4.1 for $k = 2$ and formulated in terms of m.g.f.'s in order to show that the r.v.'s X_1 and X_2 have a Bivariate Normal distribution if and only if for every $c_1, c_2 \in \mathbb{R}$, $Y_c = c_1 X_1 + c_2 X_2$ is normally distributed;

ii) In either case, show that $c_1 X_1 + c_2 X_2 + c_3$ is also normally distributed for any $c_3 \in \mathbb{R}$.

Chapter 7

Stochastic Independence with Some Applications

7.1 Stochastic Independence: Criteria of Independence

Let S be a sample space, consider a class of events associated with this space, and let P be a probability function defined on the class of events. In Chapter 2 (Section 2.3), the concept of independence of events was defined and was heavily used there, as well as in subsequent chapters. Independence carries over to r.v.'s also, and is the most basic assumption made in this book. Independence of r.v.'s, in essence, reduces to that of events, as will be seen below. In this section, the not-so-rigorous definition of independence of r.v.'s is presented, and two criteria of independence are also discussed. A third criterion of independence, and several applications, based primarily on independence, are discussed in subsequent sections. A rigorous treatment of some results is presented in Section 7.4.

DEFINITION 1 The r.v.'s X_j, $j = 1, \ldots, k$ are said to be *independent* if, for sets $B_j \subseteq \mathbb{R}$, $j = 1, \ldots, k$, it holds

$$P(X_j \in B_j, j = 1, \ldots, k) = \prod_{j=1}^{k} P(X_j \in B_j).$$

The r.v.'s X_j, $j = 1, 2, \ldots$ are said to be *independent* if every finite subcollection of them is a collection of independent r.v.'s. Non-independent r.v.'s are said to be *dependent*. (See also Definition 3 in Section 7.4, and the comment following it.)

REMARK 1 (i) The sets B_j, $j = 1, \ldots, k$ may not be chosen entirely arbitrarily, but there is plenty of leeway in their choice. For example, taking $B_j = (-\infty, x_j]$, $x_j \in \mathbb{R}$, $j = 1, \ldots, k$ would be sufficient. (See Lemma 3 in Section 7.4.)

(ii) Definition 1 (as well as Definition 3 in Section 7.4) also applies to m-dimensional r. vectors when $\mathbb{R}$ (and $\mathcal{B}$ in Definition 3) is replaced by $\mathbb{R}^m$ ($\mathcal{B}^m$).

THEOREM 1 (Factorization Theorem) The r.v.'s $X_j, j = 1, \ldots, k$ are independent if and only if any one of the following two (equivalent) conditions holds:

i) $F_{X_1, \ldots, X_k}(x_1, \ldots, x_k) = \prod_{j=1}^{k} F_{X_j}(x_j)$, for all $x_j \in \mathbb{R}, j = 1, \ldots, k$.

ii) $f_{X_1, \ldots, X_k}(x_1, \ldots, x_k) = \prod_{j=1}^{k} f_{X_j}(x_j)$, for all $x_j \in \mathbb{R}, j = 1, \ldots, k$.

PROOF

i) If $X_j, j = 1, \cdots, k$ are independent, then

$$P(X_j \in B_j, j = 1, \ldots, k) = \prod_{j=1}^{k} P(X_j \in B_j), \; B_j \subseteq \mathbb{R}, j = 1, \ldots, k.$$

In particular, this is true for $B_j = (-\infty, x_j], x_j \in \mathbb{R}, j = 1, \ldots, k$ which gives

$$F_{X_1, \ldots, X_k}(x_1, \ldots, x_k) = \prod_{j=1}^{k} F_{X_j}(x_j).$$

The proof of the converse is a deep probability result, and will, of course, be omitted. Some relevant comments will be made in Section 7.4, Lemma 3.

ii) For the discrete case, we set $B_j = \{x_j\}$, where x_j is in the range of $X_j, j = 1, \ldots, k$. Then if $X_j, j = 1, \ldots, k$ are independent, we get

$$P(X_1 = x_1, \ldots, X_k = x_k) = \prod_{j=1}^{k} P(X_j = x_j),$$

or

$$f_{X_1, \ldots, X_k}(x_1, \ldots, x_k) = \prod_{j=1}^{k} f_{X_j}(x_j).$$

Let now

$$f_{X_1, \ldots, X_k}(x_1, \ldots, x_k) = \prod_{j=1}^{k} f_{X_j}(x_j).$$

Then for any sets $B_j = (-\infty, y_j], y_j \in \mathbb{R}, j = 1, \ldots, k$, we get

$$\sum_{B_1 \times \cdots \times B_k} f_{X_1, \ldots, X_k}(x_1, \ldots, x_k) = \sum_{B_1 \times \cdots \times B_k} f_{X_1}(x_1) \cdots f_{X_k}(x_k)$$

$$= \prod_{j=1}^{k} \left[\sum_{B_j} f_{X_j}(x_j) \right],$$

or

$$F_{X_1, \ldots, X_k}(y_1, \ldots, y_k) = \prod_{j=1}^{k} F_{X_j}(y_j).$$

Therefore $X_j, j = 1, \ldots, k$ are independent by (i). For the continuous case, we have: Let

$$f_{X_1, \ldots, X_k}(x_1, \ldots, x_k) = \prod_{j=1}^{k} f_{X_j}(x_j)$$

and let

$$C_j = (-\infty, y_j], \quad y_j \in \mathbb{R}. \quad j = 1, \ldots, k.$$

Then integrating both sides of this last relationship over the set $C_1 \times \cdots \times C_k$, we get

$$F_{X_1, \ldots, X_k}(y_1, \ldots, y_k) = \prod_{j=1}^{k} F_{X_j}(y_j),$$

so that $X_j, j = 1, \ldots, k$ are independent by (i). Next, assume that

$$F_{X_1, \ldots, X_k}(x_1, \ldots, x_k) = \prod_{j=1}^{k} F_{X_j}(x_j)$$

(that is, the X_j's are independent). Then differentiating both sides, we get

$$f_{X_1, \ldots, X_k}(x_1, \ldots, x_k) = \prod_{j=1}^{k} f_{X_j}(x_j). \quad \blacktriangle$$

REMARK 2 It is noted that this step also is justifiable (by means of calculus) for the continuity points of the p.d.f. only.

Consider independent r.v.'s and suppose that g_j is a function of the jth r.v. alone. Then it seems intuitively clear that the r.v.'s $g_j(X_j), j = 1, \ldots, k$ ought to be independent. This is, actually, true and is the content of the following

LEMMA 1 For $j = 1, \ldots, k$, let the r.v.'s X_j be independent and consider (measurable) functions $g_j: \mathbb{R} \to \mathbb{R}$, so that $g_j(X_j), j = 1, \ldots, k$ are r.v.'s. Then the r.v.'s $g_j(X_j)$, $j = 1, \ldots, k$ are also independent. The same conclusion holds if the r.v.'s are replaced by m-dimensional r. vectors, and the functions $g_j, j = 1, \ldots, k$ are defined on $\mathbb{R}^m$ into $\mathbb{R}$. (That is, functions of independent r.v.'s (r. vectors) are independent r.v.'s.)

PROOF See Section 7.4. $\blacktriangle$

Independence of r.v.'s also has the following consequence stated as a lemma. Both this lemma, as well as Lemma 1, are needed in the proof of Theorem 1' below.

LEMMA 2 Consider the r.v.'s $X_j, j = 1, \ldots, k$ and let $g_j: \mathbb{R} \to \mathbb{R}$ be (measurable) functions, so that $g_j(X_j), j = 1, \ldots, k$ are r.v.'s. Then, if the r.v.'s $X_j, j = 1, \ldots, k$ are independent, we have

$$E\left[\prod_{j=1}^{k} g_j(X_j)\right] = \prod_{j=1}^{k} E[g_j(X_j)],$$

provided the expectations considered exist. The same conclusion holds if the g_j's are complex-valued.

PROOF See Section 7.2. $\blacktriangle$

REMARK 3 The converse of the above statement need not be true as will be seen later by examples.

THEOREM 1' (Factorization Theorem) The r.v.'s $X_j, j = 1, \ldots, k$ are independent if and only if:

$$\phi_{X_1, \ldots, X_k}(t_1, \ldots, t_k) = \prod_{j=1}^{k} \phi_{X_j}(t_j), \quad \text{for all} \quad t_j \in \mathbb{R}, j = 1, \ldots, k.$$

7.1 Stochastic Independence: Criteria of Independence

PROOF If $X_1, j = 1, \ldots, k$ are independent, then by Theorem 1(ii),

$$f_{X_1, \ldots, X_k}(x_1, \ldots, x_k) = \prod_{j=1}^{k} f_{X_j}(x_j).$$

Hence

$$\phi_{X_1, \ldots, X_k}(x_1, \ldots, x_k) = E\left(\exp\left(i\sum_{j=1}^{k} t_j X_j\right)\right) = E\left(\prod_{j=1}^{k} e^{it_j X_j}\right) = \prod_{j=1}^{k} E e^{it_j X_j}$$

by Lemmas 1 and 2, and this is $\prod_{j=1}^{k} \phi_{X_j}(t_j)$. Let us assume now that

$$\phi_{X_1, \ldots, X_k}(t_1, \ldots, t_k) = \prod_{j=1}^{k} \phi_{X_j}(t_j).$$

For the discrete case, we have (see Theorem 2(i) in Chapter 6)

$$f_{X_j}(x_j) = \lim_{T \to \infty} \frac{1}{2T} \int_{-T}^{T} e^{-it_j x_j} \phi_{X_j}(t_j) dt_j, \quad j = 1, \ldots, k,$$

and for the multidimensional case, we have (see Theorem 2'(i) in Chapter 6)

$$f_{X_1, \ldots, X_k}(x_1, \ldots, x_k) = \lim_{T \to \infty} \left(\frac{1}{2T}\right)^k \int_{-T}^{T} \cdots \int_{-T}^{T} \exp\left(-i\sum_{j=1}^{k} t_j x_j\right)$$

$$\times \phi_{X_1, \ldots, X_k}(t_1, \ldots, t_k) dt_1 \cdots dt_k$$

$$= \lim_{T \to \infty} \left(\frac{1}{2T}\right)^k \int_{-T}^{T} \cdots \int_{-T}^{T} \exp\left(-i\sum_{j=1}^{k} t_j x_j\right) \prod_{j=1}^{k} \phi_{X_j}(t_j)(dt_1 \cdots dt_k)$$

$$= \prod_{j=1}^{k} \left[\lim_{T \to \infty} \frac{1}{2T} \int_{-T}^{T} e^{-it_j x_j} \phi_{X_j}(t_j) dt_j\right] = \prod_{j=1}^{k} f_{X_j(x_j)}.$$

That is, $X_j, j = 1, \ldots, k$ are independent by Theorem 1(ii). For the continuous case, we have

$$f_{X_j}(x_j) = \lim_{h \to 0} \lim_{T \to \infty} \frac{1}{2\pi} \int_{-T}^{T} \frac{1-e^{-it_j h}}{it_j h} e^{-it_j h} \phi_{X_j}(t_j) dt_j, \quad j = 1, \ldots, k,$$

and for the multidimensional case, we have (see Theorem 2'(ii) in Chapter 6)

$$f_{X_1, \ldots, X_k}(x_1, \ldots, x_k) = \lim_{h \to 0} \lim_{T \to \infty} \left(\frac{1}{2\pi}\right)^k \int_{-T}^{T} \cdots \int_{-T}^{T} \prod_{j=1}^{k} \left(\frac{1-e^{-it_j h}}{it_j h} e^{-it_j x_j}\right)$$

$$\times \phi_{X_1, \ldots, X_k}(t_1, \ldots, t_k) dt_1 \cdots dt_k$$

$$= \lim_{h \to 0} \lim_{T \to \infty} \left(\frac{1}{2\pi}\right)^k \int_{-T}^{T} \cdots \int_{-T}^{T} \prod_{j=1}^{k} \left[\frac{1-e^{-it_j h}}{it_j h} e^{-it_j x_j} \phi_{X_j}(t_j)\right]$$

$$\times dt_1 \cdots dt_k$$

$$= \prod_{j=1}^{k} \left[\lim_{h \to 0} \lim_{T \to \infty} \frac{1}{2\pi} \int_{-T}^{T} \frac{1-e^{-it_j h}}{it_j h} e^{-it_j x_j} \phi_{X_j}(t_j) dt_j\right]$$

$$= \prod_{j=1}^{k} f_{X_j(x_j)},$$

which again establishes independence of X_j, $j = 1, \ldots, k$ by Theorem 1(ii). ▲

REMARK 4 A version of this theorem involving m.g.f.'s can be formulated, if the m.g.f.'s exist.

COROLLARY Let X_1, X_2 have the Bivariate Normal distribution. Then X_1, X_2 are independent if and only if they are uncorrelated.

PROOF We have seen that (see Bivariate Normal in Section 3.3 of Chapter 3)

$$f_{X_1, X_2}(x_1, x_2) = \frac{1}{2\pi\sigma_1\sigma_2\sqrt{1-\rho^2}} e^{-q/2},$$

where

$$q = \frac{1}{1-\rho^2} \left[\left(\frac{x_1 - \mu_1}{\sigma_1} \right)^2 - 2\rho \left(\frac{x_1 - \mu_2}{\sigma_1} \right) \left(\frac{x_2 - \mu_2}{\sigma_2} \right) - \left(\frac{x_2 - \mu_2}{\sigma_2} \right)^2 \right],$$

and

$$f_{X_1}(x_1) = \frac{1}{\sqrt{2\pi}\sigma_1} \exp\left[-\frac{(x_1 - \mu_1)^2}{2\sigma_1^2} \right], \quad f_{X_2}(x_2) = \frac{1}{\sqrt{2\pi}\sigma_2} \exp\left[-\frac{(x_2 - \mu_2)^2}{2\sigma_2^2} \right].$$

Thus, if X_1, X_2 are uncorrelated, so that $\rho = 0$, then

$$f_{X_1, X_2}(x_1, x_2) = f_{X_1}(x_1) \cdot f_{X_2}(x_2),$$

that is, X_1, X_2 are independent. The converse is always true by Corollary 1 in Section 7.2. ▲

Exercises

7.1.1 Let X_j, $j = 1, \ldots, n$ be i.i.d. r.v.'s with p.d.f. f and d.f. F. Set

$$X_{(1)} = \min(X_1, \ldots, X_n), \quad X_n = \max(X_1, \ldots, X_n);$$

that is,

$$X_{(1)}(s) = \min[X_1(s), \ldots, X_n(s)], \quad X_{(n)}(s) = \max[X_1(s), \ldots, X_n(s)].$$

Then express the d.f. and p.d.f. of $X_{(1)}$, $X_{(n)}$ in terms of f and F.

7.1.2 Let the r.v.'s X_1, X_2 have p.d.f. f given by $f(x_1, x_2) = I_{(0,1) \times (0,1)}(x_1, x_2)$.

i) Show that X_1, X_2 are independent and identify their common distribution;
ii) Find the following probabilities: $P(X_1 + X_2 < \frac{1}{3})$, $P(X_1^2 + X_2^2 < \frac{1}{4})$, $P(X_1 X_2 > \frac{1}{2})$.

7.1.3 Let X_1, X_2 be two r.v.'s with p.d.f. f given by $f(x_1, x_2) = g(x_1)h(x_2)$.

i) Derive the p.d.f. of X_1 and X_2 and show that X_1, X_2 are independent;

ii) Calculate the probability $P(X_1 > X_2)$ if $g = h$ and h is of the continuous type.

7.1.4 Let X_1, X_2, X_3 be r.v.'s with p.d.f. f given by $f(x_1, x_2, x_3) = 8x_1 x_2 x_3 \, I_A(x_1, x_2, x_3)$, where $A = (0, 1) \times (0, 1) \times (0, 1)$.

i) Show that these r.v.'s are independent;

ii) Calculate the probability $P(X_1 < X_2 < X_3)$.

7.1.5 Let X_1, X_2 be two r.v.'s with p.d.f f given by $f(x_1, x_2) = cI_A(x_1, x_2)$, where $A = \{(x_1, x_2)' \in \mathbb{R}^2; \; x_1^2 + x_2^2 \leq 9\}$.

i) Determine the constant c;

ii) Show that X_1, X_2 are dependent.

7.1.6 Let the r.v.'s X_1, X_2, X_3 be jointly distributed with p.d.f. f given by

$$f(x_1, x_2, x_3) = \frac{1}{4} I_A(x_1, x_2, x_3),$$

where

$$A = \{(1, 0, 0), (0, 1, 0), (0, 0, 1), (1, 1, 1)\}.$$

Then show that

i) X_i, X_j, $i \neq j$, are independent;

ii) X_1, X_2, X_3 are dependent.

7.1.7 Refer to Exercise 4.2.5 in Chapter 4 and show that the r.v.'s X_1, X_2, X_3 are independent. Utilize this result in order to find the p.d.f. of $X_1 + X_2$ and $X_1 + X_2 + X_3$.

7.1.8 Let X_j, $j = 1, \ldots, n$ be i.i.d. r.v.'s with p.d.f. f and let B be a (Borel) set in $\mathbb{R}$.

i) In terms of f, express the probability that at least k of the X's lie in B for some fixed k with $1 \leq k \leq n$;

ii) Simplify this expression if f is the Negative Exponential p.d.f. with parameter λ and $B = (1/\lambda, \infty)$;

iii) Find a numerical answer for $n = 10$, $k = 5$, $\lambda = \frac{1}{2}$.

7.1.9 Let X_1, X_2 be two independent r.v.'s and let $g: \mathbb{R} \to \mathbb{R}$ be measurable. Let also $Eg(X_2)$ be finite. Then show that $E[g(X_2) \mid X_1 = x_1] = Eg(X_2)$.

7.1.10 If X_j, $j = 1, \ldots, n$ are i.i.d. r.v.'s with ch.f. ϕ and sample mean $\overline{X}$, express the ch.f. of $\overline{X}$ in terms of ϕ.

7.1.11 For two i.i.d. r.v.'s X_1, X_2, show that $\phi_{X_1 - X_2}(t) = |\phi_{X_1}(t)|^2$, $t \in \mathbb{R}$. (Hint: Use Exercise 6.2.3 in Chapter 6.)

7.1.12 Let X_1, X_2 be two r.v.'s with joint and marginal ch.f.'s ϕ_{X_1,X_2}, ϕ_{X_1} and ϕ_{X_2}. Then X_1, X_2 are independent if and only if

$$\phi_{X_1,X_2}(t_1, t_2) = \phi_{X_1}(t_1)\phi_{X_2}(t_2), \qquad t_1, t_2 \in \mathbb{R}.$$

By an example, show that

$$\phi_{X_1,X_2}(t, t) = \phi_{X_1}(t)\phi_{X_2}(t), \qquad t \in \mathbb{R},$$

does not imply independence of X_1, X_2.

7.2 Proof of Lemma 2 and Related Results

We now proceed with the proof of Lemma 2.

PROOF OF LEMMA 2 Suppose that the r.v.'s involved are continuous, so that we use integrals. Replace integrals by sums in the discrete case. Thus,

$$E\left[\prod_{j=1}^{k} g_j(X_j)\right] = \int_{-\infty}^{\infty} \cdots \int_{-\infty}^{\infty} g_1(x_1) \cdots g_k(x_k)$$
$$\times f_{X_1, \ldots, X_k}(x_1, \ldots, x_k) dx_1 \cdots dx_k$$
$$= \int_{-\infty}^{\infty} \cdots \int_{-\infty}^{\infty} g_1(x_1) \cdots g_k(x_k) f_{X_1}(x_1) \cdots f_{X_k}(x_k) dx_1 \cdots dx_k$$
(by independence)
$$= \left[\int_{-\infty}^{\infty} g_1(x_1) f_{X_1}(x_1) dx_1\right] \cdots \left[\int_{-\infty}^{\infty} g_k(x_k) f_{X_k}(x_k) dx_k\right]$$
$$= E[g_1(X_1)] \cdots E[g_k(X_k)].$$

Now suppose that the g_j's are complex-valued, and for simplicity, set $g_j(X_j) = Y_j = Y_{j1} + Y_{j2}$, $j = 1, \ldots, k$. For $k = 2$,

$$E(Y_1 Y_2) = E[(Y_{11} + iY_{12})(Y_{21} + iY_{22})]$$
$$= E(Y_{11}Y_{21} - Y_{12}Y_{22}) + iE(Y_{11}Y_{22} + Y_{12}Y_{21})$$
$$= [E(Y_{11}Y_{21}) - E(Y_{12}Y_{22})] + i[E(Y_{11}Y_{22}) + E(Y_{12}Y_{21})]$$
$$= [(EY_{11})(EY_{21}) - (EY_{12})(EY_{22})] + i[(EY_{11})(EY_{22}) - (EY_{12})(EY_{21})]$$
$$= (EY_{11} + iEY_{12})(EY_{21} + iEY_{22}) = (EY_1)(EY_2).$$

Next, assume the result to be true for $k = m$ and establish it for $k = m + 1$. Indeed,

$$E(Y_1 \cdots Y_{m+1}) = E[(Y_1 \cdots Y_m)Y_{m+1}]$$
$$= E(Y_1 \cdots Y_m)(EY_{m+1}) \quad \text{(by the part just established)}$$
$$= (EY_1) \cdots (EY_m)(EY_{m+1}) \quad \text{(by the induction hypothesis)}. \blacktriangle$$

COROLLARY 1 The covariance of an r.v. X and of any other r.v. which is equal to a constant c (with probability 1) is equal to 0; that is, $\text{Cov}(X, c) = 0$.

PROOF $\text{Cov}(X, c) = E(cX) - (Ec)(EX) = cEX - cEX = 0$. $\blacktriangle$

COROLLARY 2 If the r.v.'s X_1 and X_2 are independent, then they have covariance equal to 0, provided their second moments are finite. In particular, if their variances are also positive, then they are uncorrelated.

PROOF In fact,
$$\text{Cov}(X_1, X_2) = E(X_1 X_2) - (EX_1)(EX_2)$$
$$= (EX_1)(EX_2) - (EX_1)(EX_2) = 0,$$
by independence and Lemma 2.

The second assertion follows since $\rho(X, Y) = \text{Cov}(X, Y)/\sigma(X)\sigma(Y)$. $\blacktriangle$

REMARK 5 The converse of the above corollary need not be true. Thus uncorrelated r.v.'s in general are not independent. (See, however, the corollary to Theorem 1 after the proof of part (iii).)

COROLLARY 3
i) For any k r.v.'s $X_j, j = 1, \ldots, k$ with finite second moments and variances $\sigma_j^2 = \sigma^2(X_j)$, and any constants $c_j, j = 1, \ldots, k$, it holds:

$$\sigma^2\left(\sum_{j=1}^k c_j X_j\right) = \sum_{j=1}^k c_j^2 \sigma_j^2 + \sum_{1 \leq i \neq j \leq k} c_i c_j \text{Cov}(X_i, X_j)$$
$$= \sum_{j=1}^k c_j^2 \sigma_j^2 + 2 \sum_{1 \leq i < j \leq k} c_i c_j \text{Cov}(X_i, X_j).$$

ii) If also $\sigma_j > 0, j = 1, \ldots, k$, and $\rho_{ij} = \rho(X_i, X_j), i \neq j$, then:

$$\sigma^2\left(\sum_{j=1}^k c_j X_j\right) = \sum_{j=1}^k c_j^2 \sigma_j^2 + \sum_{1 \leq i \neq j \leq k} c_i c_j \sigma_i \sigma_j \rho_{ij}$$
$$= \sum_{j=1}^k c_j^2 \sigma_j^2 + 2 \sum_{1 \leq i < j \leq k} c_i c_j \sigma_i \sigma_j \rho_{ij}.$$

In particular, if the r.v.'s are independent or have pairwise covariances 0 (are pairwise uncorrelated), then:

iii) $\sigma^2(\sum_{j=1}^k c_j X_j) = \sum_{j=1}^k c_j^2 \sigma_j^2$, and

iii') $\sigma^2(\sum_{j=1}^k X_j) = \sum_{j=1}^k \sigma_j^2$ (Bienaymé equality).

PROOF

i) Indeed,

$$\sigma^2\left(\sum_{j=1}^k c_j X_j\right) = E\left[\sum_{j=1}^k c_j X_j - E\left(\sum_{j=1}^k c_j X_j\right)\right]^2$$

$$= E\left[\sum_{j=1}^k c_j (X_j - EX_j)\right]^2$$

$$= E\left[\sum_{j=1}^k c_j^2 (X_j - EX_j)^2 + \sum_{i \neq j} c_i c_j (X_i - EX_i)(X_j - EX_j)\right]$$

$$= \sum_{j=1}^k c_j^2 \sigma_j^2 + \sum_{1 \leq i \neq j \leq k} c_i c_j \text{Cov}(X_i, X_j)$$

$$= \sum_{j=1}^k c_j^2 \sigma_j^2 + 2 \sum_{1 \leq i < j \leq k} c_i c_j \text{Cov}(X_i, X_j)$$

$$\left(\text{since } \text{Cov}(X_i, X_j) = \text{Cov}(X_j, X_i)\right).$$

This establishes part (i). Part (ii) follows by the fact that $\text{Cov}(X_i, X_j) = \sigma_i \sigma_j \rho_{ij} = \sigma_j \sigma_i \rho_{ji}$.

iii) Here $\text{Cov}(X_i, X_j) = 0$, $i \neq j$, either because of independence and Corollary 2, or $\rho_{ij} = 0$, in case $\sigma_j > 0$, $j = 1, \ldots, k$. Then the assertion follows from either part (i) or part (ii), respectively.

iii') Follows from part (iii) for $c_1 = \cdots = c_k = 1$. ▲

Exercises

7.2.1 For any k r.v.'s X_j, $j = 1, \ldots, k$ for which $E(X_j) = \mu$ (finite) $j = 1, \ldots, k$, show that

$$\sum_{j=1}^k (X_j - \mu)^2 = \sum_{j=1}^k (X_j - \overline{X})^2 + k(\overline{X} - \mu)^2 = kS^2 + k(\overline{X} - \mu)^2,$$

where

$$\overline{X} = \frac{1}{k} \sum_{j=1}^k X_j \quad \text{and} \quad S^2 = \frac{1}{k} \sum_{j=1}^k (X_j - \overline{X})^2.$$

7.2.2 Refer to Exercise 4.2.5 in Chapter 4 and find the $E(X_1 X_2)$, $E(X_1 X_2 X_3)$, $\sigma^2(X_1 + X_2)$, $\sigma^2(X_1 + X_2 + X_3)$ without integration.

7.2.3 Let X_j, $j = 1, \ldots, n$ be independent r.v.'s with finite moments of third order. Then show that

$$E\left[\sum_{j=1}^n (X_j - EX_j)\right]^3 = \sum_{j=1}^n E(X_j - EX_j)^3.$$

7.2.4 Let X_j, $j = 1, \ldots, n$ be i.i.d. r.v.'s with mean μ and variance σ^2, both finite.

i) In terms of α, c and σ, find the smallest value of n for which the probability that $\bar{X}$ (the sample mean of the X's) and μ differ in absolute value at most by c is at least α;

ii) Give a numerical answer if $\alpha = 0.90$, $c = 0.1$ and $\sigma = 2$.

7.2.5 Let X_1, X_2 be two r.v.'s taking on the values $-1, 0, 1$ with the following respective probabilities:

$$f(-1, 1) = \alpha, \quad f(-1, 0) = \beta, \quad f(-1, -1) = \alpha$$
$$f(0, -1) = \beta, \quad f(0, 0) = 0, \quad f(0, 1) = \beta \quad ; \quad \alpha, \beta > 0, \quad \alpha + \beta = \frac{1}{4}.$$
$$f(1, -1) = \alpha, \quad f(1, 0) = \beta, \quad f(1, 1) = \alpha.$$

Then show that:

i) $\text{Cov}(X_1, X_2) = 0$, so that $\rho = 0$;

ii) X_1, X_2 are dependent.

7.3 Some Consequences of Independence

The basic assumption throughout this section is that the r.v.'s involved are independent. Then ch.f.'s are used very effectively in deriving certain "classic" theorems, as will be seen below. The m.g.f.'s, when they exist, can be used in the same way as the ch.f.'s. However, we will restrict ourselves to the case of ch.f.'s alone. In all cases, the conclusions of the theorems will be reached by way of Theorem 3 in Chapter 6, without explicitly mentioning it.

THEOREM 2 Let X_j be $B(n_j, p)$, $j = 1, \ldots, k$ and independent. Then

$$X = \sum_{j=1}^{k} X_j \text{ is } B(n, p), \text{ where } n = \sum_{j=1}^{k} n_j.$$

(That is, the sum of independent Binomially distributed r.v.'s with the *same* parameter p and possibly distinct n_j's is also Binomially distributed.)

PROOF It suffices to prove that the ch.f. of X is that of a $B(n, p)$ r.v., where n is as above. For simplicity, writing $\Sigma_j X_j$ instead of $\Sigma_{j=1}^{k} X_j$, when this last expression appears as a subscript here and thereafter, we have

$$\phi_X(t) = \phi_{\Sigma_j X_j}(t) = \prod_{j=1}^{k} \phi_{X_j}(t) = \prod_{j=1}^{k} (pe^{it} + q)^{n_j} = (pe^{it} + q)^n$$

which is the ch.f. of a $B(n, p)$ r.v., as we desired to prove. ▲

THEOREM 3 Let X_j be $P(\lambda_j)$, $j = 1, \ldots, k$ and independent. Then

$$X = \sum_{j=1}^{k} X_j \text{ is } P(\lambda), \text{ where } \lambda = \sum_{j=1}^{k} \lambda_j.$$

(That is, the sum of independent Poisson distributed r.v.'s is also Poisson distributed.)

PROOF   We have

$$\phi_X(t) = \phi_{\Sigma_j X_j}(t) = \prod_{j=1}^{k} \phi_{X_j}(t) = \prod_{j=1}^{k} \exp(\lambda_j e^{it} - \lambda_j)$$

$$= \exp\left(e^{it}\sum_{j=1}^{k}\lambda_j - \sum_{j=1}^{k}\lambda_j\right) = \exp(\lambda e^{it} - \lambda)$$

which is the ch.f. of a $P(\lambda)$ r.v. ▲

THEOREM 4   Let X_j, be $N(\mu_j, \sigma_j^2)$, $j = 1, \ldots, k$ and independent. Then

  i) $X = \sum_{j=1}^{k} X_j$ is $N(\mu, \sigma^2)$, where $\mu = \sum_{j=1}^{k}\mu_j$, $\sigma^2 = \sum_{j=1}^{k}\sigma_j^2$, and, more generally,
  ii) $X = \sum_{j=1}^{k} c_j X_j$ is $N(\mu, \sigma^2)$, where $\mu = \sum_{j=1}^{k} c_j \mu_j$, $\sigma^2 = \sum_{j=1}^{k} c_j^2 \sigma_j^2$.

(That is, the sum of independent Normally distributed r.v.'s is Normally distributed.)

PROOF   (ii) We have

$$\phi_X(t) = \phi_{\Sigma_j c_j X_j}(t) = \prod_{j=1}^{k}\phi_{X_j}(c_j t) = \prod_{j=1}^{k}\left[\exp\left(ic_j t\mu_j - \frac{\sigma_j^2 c_j^2 t^2}{2}\right)\right]$$

$$= \exp\left(it\mu - \frac{\sigma^2 t^2}{2}\right)$$

with μ and σ^2 as in (ii) above. Hence X is $N(\mu, \sigma^2)$. (i) Follows from (ii) by setting $c_1 = c_2 = \cdots = c_k = 1$. ▲

Now let X_j, $j = 1, \ldots, k$ be any k independent r.v.'s with

$$E(X_j) = \mu, \quad \sigma^2(X_j) = \sigma^2, \quad j = 1, \ldots, k.$$

Set

$$\overline{X} = \frac{1}{k}\sum_{j=1}^{k} X_j.$$

By assuming that the X's are normal, we get

COROLLARY   Let X_j be $N(\mu, \sigma^2)$, $j = 1, \ldots, k$ and independent. Then $\overline{X}$ is $N(\mu, \sigma^2/k)$, or equivalently, $[\sqrt{k}(\overline{X} - \mu)]/\sigma$ is $N(0, 1)$.

PROOF   In (ii) of Theorem 4, we set

$$c_1 = \cdots = c_k = \frac{1}{k}, \quad \mu_1 = \cdots = \mu_k = \mu, \quad \text{and} \quad \sigma_1^2 = \cdots = \sigma_k^2 = \sigma^2$$

and get the first conclusion. The second follows from the first by the use of Theorem 4, Chapter 4, since

7.3 Some Consequences of Independence

$$\frac{\sqrt{k}(\overline{X} - \mu)}{\sigma} = \frac{(\overline{X} - \mu)}{\sqrt{\sigma^2/k}}. \quad \blacktriangle$$

THEOREM 5 Let X_j be $\chi^2_{r_j}$, $j = 1, \ldots, k$ and independent. Then

$$X = \sum_{j=1}^{k} X_j \text{ is } \chi^2_r, \text{ where } r = \sum_{j=1}^{k} r_j.$$

PROOF We have

$$\phi_X(t) = \phi_{\Sigma_j X_j}(t) = \prod_{j=1}^{k} \phi_{X_j}(t) = \prod_{j=1}^{k}(1 - 2it)^{-r_j/2} = (1 - 2it)^{-r/2}$$

which is the ch.f. of a χ^2_r r.v. $\blacktriangle$

COROLLARY 1 Let X_j be $N(\mu_j, \sigma_j^2)$, $j = 1, \ldots, k$ and independent. Then

$$X = \sum_{j=1}^{k} \left(\frac{X_j - \mu_j}{\sigma_j}\right)^2 \text{ is } \chi^2_k.$$

PROOF By Lemma 1,

$$\left(\frac{X_j - \mu_j}{\sigma_j}\right)^2, \quad j = 1, \ldots, k$$

are independent, and by Theorem 3, Chapter 4,

$$\left(\frac{X_j - \mu_j}{\sigma_j}\right)^2 \text{ are } \chi^2_1, \quad j = 1, \ldots, k.$$

Thus Theorem 3 applies and gives the result. $\blacktriangle$

Now let X_j, $j = 1, \ldots, k$ be any k r.v.'s such that $E(X_j) = \mu$, $j = 1, \ldots, k$. Then the following useful identity is easily established:

$$\sum_{j=1}^{k}(X_j - \mu)^2 = \sum_{j=1}^{k}(X_j - \overline{X})^2 + k(\overline{X} - \mu)^2 = kS^2 + k(\overline{X} - \mu)^2,$$

where

$$S^2 = \frac{1}{k}\sum_{j=1}^{k}(X_j - \overline{X})^2.$$

If, in particular, X_j, $j = 1, \ldots, k$ are $N(\mu, \sigma^2)$ and independent, then it will be shown that $\overline{X}$ and S^2 are independent. (For this, see Theorem 6, Chapter 9.)

COROLLARY 2 Let X_j, be $N(\mu, \sigma^2)$, $j = 1, \ldots, k$ and independent. Then kS^2/σ^2 is χ^2_{k-1}.

PROOF We have

$$\sum_{j=1}^{k}\left(\frac{X_j - \mu}{\sigma}\right)^2 = \left[\frac{\sqrt{k}(\overline{X} - \mu)}{\sigma}\right]^2 + \frac{kS^2}{\sigma^2}.$$

Now

$$\sum_{j=1}^{k}\left(\frac{X_j - \mu}{\sigma}\right)^2 \text{ is } \chi_k^2$$

by Corollary 1 above, and

$$\left[\frac{\sqrt{k}(\overline{X} - \mu)}{\sigma}\right]^2 \text{ is } \chi_1^2,$$

by Theorem 3, Chapter 4. Then taking ch.f.'s of both sides of the last identity above, we get $(1 - 2it)^{-k/2} = (1 - 2it)^{-1/2} \phi_{kS^2/\sigma^2}(t)$.

Hence $\phi_{kS^2/\sigma^2}(t) = (1 - 2it)^{-(k-1)/2}$ which is the ch.f. of a χ_{k-1}^2 r.v. ▲

REMARK 6 It thus follows that,

$$E\left(\frac{kS^2}{\sigma^2}\right) = k - 1, \text{ and } \sigma^2\left(\frac{kS^2}{\sigma^2}\right) = 2(k-1),$$

or

$$ES^2 = \frac{k-1}{k}\sigma^2, \text{ and } \sigma^2(S^2) = \frac{2(k-1)}{k^2}\sigma^4.$$

The following result demonstrates that the sum of independent r.v.'s having a certain distribution need not have a distribution of the same kind, as was the case in Theorems 2–5 above.

THEOREM 6 Let $X_j, j = 1, \ldots, k$ be independent r.v.'s having the Cauchy distribution with parameters $\mu = 0$ and $\sigma = 1$. Then $X = \sum_{j=1}^{k} X_j$ is kY, where Y is Cauchy with $\mu = 0$, $\sigma = 1$, and hence, $X/k = \overline{X}$ is Cauchy with $\mu = 0$, $\sigma = 1$.

PROOF We have $\phi_X(t) = \phi_{\Sigma X_j}(t) = [\phi_{X_1}(t)]^k = (e^{-|t|})^k = e^{-k|t|}$, which is the ch.f. of kY, where Y is Cauchy with $\mu = 0$, $\sigma = 1$. The second statement is immediate. ▲

Exercises

7.3.1 For $j = 1, \ldots, n$, let X_j be independent r.v.'s distributed as $P(\lambda_j)$, and set

$$T = \sum_{j=1}^{n} X_j, \quad \lambda = \sum_{j=1}^{n} \lambda_j.$$

Then show that

i) The conditional p.d.f. of X_j, given $T = t$, is $B(t, \lambda_j/\lambda)$, $j = 1, \ldots, n$;

ii) The conditional joint p.d.f. of $X_j, j = 1, \ldots, n$, given $T = t$, is the Multinomial p.d.f. with parameters t and $p_j = \lambda_j/\lambda$, $j = 1, \ldots, n$.

7.3.2 If the independent r.v.'s $X_j, j = 1, \ldots, r$ have the Geometric distribution with parameter p, show that the r.v. $X = X_1 + \cdots + X_r$ has the Negative Binomial distribution with parameters r and p.

7.3.3 The life of a certain part in a new automobile is an r.v. X whose p.d.f. is Negative Exponential with parameter $\lambda = 0.005$ day.

i) Find the expected life of the part in question;

ii) If the automobile comes supplied with a spare part, whose life is an r.v. Y distributed as X and independent of it, find the p.d.f. of the combined life of the part and its spare;

iii) What is the probability that $X + Y \geq 500$ days?

7.3.4 Let X_1, X_2 be independent r.v.'s distributed as $B(n_1, p_1)$ and $B(n_2, p_2)$, respectively. Determine the distribution of the r.v.'s $X_1 + X_2$, $X_1 - X_2$ and $X_1 - X_2 + n_2$.

7.3.5 Let X_1, X_2 be independent r.v.'s distributed as $N(\mu_1, \sigma_1^2)$, and $N(\mu_2, \sigma_2^2)$, respectively. Calculate the probability $P(X_1 - X_2 > 0)$ as a function of μ_1, μ_2 and σ_1, σ_2. (For example, X_1 may represent the tensile strength (measured in p.s.i.) of a steel cable and X_2 may represent the strains applied on this cable. Then $P(X_1 - X_2 > 0)$ is the probability that the cable does not break.)

7.3.6 Let $X_i, i = 1, \ldots, m$ and $Y_j, j = 1, \ldots, n$ be independent r.v.'s such that the X's are distributed as $N(\mu_1, \sigma_1^2)$ and the Y's are distributed as $N(\mu_2, \sigma_2^2)$. Then

i) Calculate the probability $P(\bar{X} > \bar{Y})$ as a function of m, n, μ_1, μ_2 and σ_1, σ_2;

ii) Give the numerical value of this probability for $m = 10$, $n = 15$, $\mu_1 = \mu_2$ and $\sigma_1^2 = \sigma_2^2 = 6$.

7.3.7 Let X_1 and X_2 be independent r.v.'s distributed as $\chi_{r_1}^2$ and $\chi_{r_2}^2$, respectively, and for any two constants c_1 and c_2, set $X = c_1 X_1 + c_2 X_2$. Under what conditions on c_1 and c_2 is the r.v. X distributed as χ_r^2? Also, specify r.

7.3.8 Let $X_j, j = 1, \ldots, n$ be independent r.v.'s distributed as $N(\mu, \sigma^2)$ and set

$$X = \sum_{j=1}^{n} \alpha_j X_j, \quad Y = \sum_{j=1}^{n} \beta_j X_j,$$

where the α's and β's are constants. Then

i) Find the p.d.f.'s of the r.v.'s X, Y;

ii) Under what conditions on the α's and β's are the r.v.'s X and Y independent?

7.4* Independence of Classes of Events and Related Results

In this section, we give an alternative definition of independence of r.v.'s, which allows us to present a proof of Lemma 1. An additional result, Lemma 3, is also stated, which provides a parsimonious way of checking independence of r.v.'s.

To start with, consider the probability space $(S, \mathcal{A}, P)$ and recall that k events $A_1, \ldots, A_k$ are said to be independent if for all $2 \leq m \leq k$ and all $1 \leq i_1 < \cdots < i_m \leq k$, it holds that $P(A_{i_1} \cap \cdots \cap A_{i_m}) = P(A_{i_1}) \cdots P(A_{i_m})$. This definition is extended to any subclasses of $C_j, j = 1, \ldots, k$, as follows:

DEFINITION 2 We say that $C_j, j = 1, \ldots, k$ are (stochastically) *independent* (or *independent in the probability sense*, or *statistically independent*) if for every $A_j \in C_j, j = 1, \ldots, k$, the events $A_1, \ldots, A_k$ are independent.

It is an immediate consequence of this definition that subclasses of independent classes are independent. The next step is to carry over the definition of independence to r.v.'s. To this end, let X be a random variable. Then we have seen (Theorem 1, Chapter 3) that $X^{-1}(\mathcal{B})$ is a σ-field, sub-σ-field of $\mathcal{A}$, the σ-field *induced* by X. Thus, if we consider the r.v.'s $X_j, j = 1, \ldots, k$, we will have the σ-fields induced by them which we denote by $\mathcal{A}_j = X_j^{-1}(\mathcal{B})$, $j = 1, \ldots, k$.

DEFINITION 3 We say that the r.v.'s $X_j, j = 1, \ldots, k$ are *independent* (in any one of the modes mentioned in the previous definition) if the σ-fields induced by them are independent.

From the very definition of $X_j^{-1}(\mathcal{B})$, for every $A_j \in X_j^{-1}(\mathcal{B})$ there exists $B_j \in \mathcal{B}$ such that $A_j = X_j^{-1}(B_j), j = 1, \ldots, k$. The converse is also obviously true; that is, $X_j^{-1}(B_j) \in X_j^{-1}(\mathcal{B})$, for every $B_j \in \mathcal{B}$, $j = 1, \ldots, k$. On the basis of these observations, the previous definition is equivalent to Definition 1. Actually, Definition 3 can be weakened considerably, as explained in Lemma 3 below.

According to the following statement, in order to establish independence of the r.v.'s $X_j, j = 1, \ldots, k$, it suffices to establish independence of the (much "smaller") classes $C_j, j = 1, \ldots, k$, where $C_j = X_j^{-1}(\{(-\infty, x], x \in \mathbb{R}\})$. More precisely,

LEMMA 3 Let

$$\mathcal{A}_j = X_j^{-1}(\mathcal{B}) \quad \text{and} \quad C_j = X_j^{-1}\left(\{(-\infty, x]; x \in \mathbb{R}\}\right), j = 1, \ldots, k.$$

Then if C_j are independent, so are $\mathcal{A}_j, j = 1, \ldots, k$.

PROOF By Definition 3, independence of the r.v.'s $X_j, j = 1, \ldots, k$ means independence of the σ-fields. That independence of those σ-fields is implied by independence of the classes $C_j, j = 1, \ldots, k$, is an involved result in probability theory and it cannot be discussed here. ▲

We may now proceed with the proof of Lemma 1.

PROOF OF LEMMA 1 In the first place, if X is an r.v. and $\mathcal{A}_X = X^{-1}(\mathcal{B})$, and if $g(X)$ is a measurable function of X and $\mathcal{A}_{g(X)} = [g(X)]^{-1}(\mathcal{B})$, then $\mathcal{A}_{g(X)} \subseteq \mathcal{A}_X$. In fact, let $A \in \mathcal{A}_{g(X)}$. Then there exists $B \in \mathcal{B}$ such that $A = [g(X)]^{-1}(B)$. But

$$A = \left[g(X)\right]^{-1}(B) = X^{-1}\left[g^{-1}(B)\right] = X^{-1}(B'),$$

where $B' = g^{-1}(B)$ and by the measurability of g, $B' \in \mathcal{B}$. It follows that $X^{-1}(B') \in \mathcal{A}_X$ and thus, $A \in \mathcal{A}_X$. Let now $\mathcal{A}_j = X_j^{-1}(\mathcal{B})$ and

$$\mathcal{A}_j^* = [g(X_j)]^{-1}(\mathcal{B}), \quad j = 1, \ldots, k.$$

Then

$$\mathcal{A}_j^* \subseteq \mathcal{A}_j, \quad j = 1, \ldots, k,$$

and since $\mathcal{A}_j, j = 1, \cdots, k$, are independent, so are $\mathcal{A}_j^*, j = 1, \ldots, k$. ▲

Exercise

7.4.1 Consider the probability space $(S, \mathcal{A}, P)$ and let A_1, A_2 be events. Set $X_1 = I_{A_1}$, $X_2 = I_{A_2}$ and show that X_1, X_2 are independent if and only if A_1, A_2 are independent. Generalize it for the case of n events $A_j, j = 1, \ldots, n$.

Chapter 8

Basic Limit Theorems

8.1 Some Modes of Convergence

Let $\{X_n\}, n = 1, 2, \ldots$ be a sequence of random variables and let X be a random variable defined on the sample space S supplied with a class of events $\mathcal{A}$ and a probability function P (that is, the sequence of the r.v.'s and the r.v. X are defined on the probability space $(S, \mathcal{A}, P)$). For such a sequence of r.v.'s four kinds of convergence are defined, and some comments are provided as to their nature. An illustrative example is also discussed.

DEFINITION 1
i) We say that $\{X_n\}$ *converges almost surely* (a.s.), or *with probability one*, to X as $n \to \infty$, and we write $X_n \xrightarrow[n\to\infty]{a.s.} X$, or $X_n \xrightarrow[n\to\infty]{} X$ with probability 1, or $P[X_n \xrightarrow[n\to\infty]{} X] = 1$, if $X_n(s) \xrightarrow[n\to\infty]{} X(s)$ for all $s \in S$ except possibly for a subset N of S such that $P(N) = 0$.

Thus $X_n \xrightarrow[n\to\infty]{a.s.} X$ means that for every $\varepsilon > 0$ and for every $s \in N^c$ there exists $N(\varepsilon, s) > 0$ such that

$$|X_n(s) - X(s)| < \varepsilon$$

for all $n \geq N(\varepsilon, s)$. This type of convergence is also known as *strong convergence*.

ii) We say that $\{X_n\}$ *converges in probability* to X as $n \to \infty$, and we write $X_n \xrightarrow[n\to\infty]{P} X$, if for every $\varepsilon > 0$, $P[|X_n - X| > \varepsilon] \xrightarrow[n\to\infty]{} 0$.

Thus $X_n \xrightarrow[n\to\infty]{P} X$ means that: For every $\varepsilon, \delta > 0$ there exists $N(\varepsilon, \delta) > 0$ such that $P[|X_n - X| > \varepsilon] < \delta$ for all $n \geq N(\varepsilon, \delta)$.

REMARK 1 Since $P[|X_n - X| > \varepsilon] + P[|X_n - X| \leq \varepsilon] = 1$, then $X_n \xrightarrow[n\to\infty]{P} X$ is equivalent to: $P[|X_n - X| \leq \varepsilon] \xrightarrow[n\to\infty]{} 1$. Also if $P[|X_n - X| > \varepsilon] \xrightarrow[n\to\infty]{} 0$ for every $\varepsilon > 0$, then clearly $P[|X_n - X| \geq \varepsilon] \xrightarrow[n\to\infty]{} 0$.

Let now $F_n = F_{X_n}$, $F = F_X$. Then

iii) We say that $\{X_n\}$ *converges in distribution* to X as $n \to \infty$, and we write $X_n \xrightarrow[n \to \infty]{d} X$, if $F_n(x) \xrightarrow[n \to \infty]{} F(x)$ for all $x \in \mathbb{R}$ for which F is continuous. Thus $X_n \xrightarrow[n \to \infty]{d} X$ means that: For every $\varepsilon > 0$ and every x for which F is continuous there exists $N(\varepsilon, x)$ such that $|F_n(x) - F(x)| < \varepsilon$ for all $n \geq N(\varepsilon, x)$. This type of convergence is also known as *weak* convergence.

REMARK 2 If F_n have p.d.f.'s f_n, then $X_n \xrightarrow[n \to \infty]{d} X$ does *not* necessarily imply the convergence of $f_n(x)$ to a p.d.f., as the following example illustrates.

EXAMPLE 1

For $n = 1, 2, \ldots$, consider the p.d.f.'s defined by

$$f_n(x) = \begin{cases} \frac{1}{2}, & \text{if } x = 1 - (1/n) \text{ or } x = 1 + (1/n) \\ 0, & \text{otherwise.} \end{cases}$$

Then, clearly, $f_n(x) \xrightarrow[n \to \infty]{} f(x) = 0$ for all $x \in \mathbb{R}$ and $f(x)$ is not a p.d.f.

Next, the d.f. F_n corresponding to f_n is given by

$$F_n(x) = \begin{cases} 0, & \text{if } x < 1 - (1/n) \\ \frac{1}{2}, & \text{if } 1 - (1/n) \leq x < 1 + (1/n) \\ 1, & \text{if } x \geq 1 + (1/n). \end{cases}$$

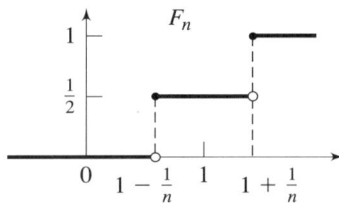

Figure 8.1

One sees that $F_n(x) \xrightarrow[n \to \infty]{} F(x)$ for all $x \neq 1$, where $F(x)$ is defined by

$$F(x) = \begin{cases} 0, & \text{if } x < 1 \\ 1, & \text{if } x \geq 1, \end{cases}$$

which is a d.f.

Under further conditions on f_n, f, it may be the case, however, that f_n converges to a p.d.f. f.

We now assume that $E|X_n|^2 < \infty$, $n = 1, 2, \ldots$, Then:

iv) We say that $\{X_n\}$ converges to X in *quadratic mean* (q.m.) as $n \to \infty$, and we write $X_n \xrightarrow[n \to \infty]{q.m.} X$, if $E|X_n - X|^2 \xrightarrow[n \to \infty]{} 0$.

Thus $X_n \xrightarrow[n \to \infty]{q.m.} X$ means that: For every $\varepsilon > 0$, there exists $N(\varepsilon) > 0$ such that $E|X_n - X|^2 < \varepsilon$ for all $n \geq N(\varepsilon)$.

REMARK 3 Almost sure convergence is the familiar pointwise convergence of the sequence of numbers $\{X_n(s)\}$ for every s outside of an event N of probability zero (a null event). Convergence in distribution is also a pointwise convergence of the sequence of numbers $\{F_n(x)\}$ for every x for which F is continuous. Convergence in probability, however, is of a different nature. By setting $A_n = \{s \in S; |X_n(s) - X(s)| > \varepsilon\}$ for an arbitrary but fixed $\varepsilon > 0$, we have that $X_n \xrightarrow[n \to \infty]{P} X$, if $P(A_n) \xrightarrow[n \to \infty]{} 0$. So the sequence of numbers $\{P(A_n)\}$ tends to 0, as $n \to \infty$, but the events A_n themselves keep wandering around the sample space S. Finally, convergence in quadratic mean simply signifies that the averages $E|X_n - X|^2$ converge to 0 as $n \to \infty$.

Exercises

8.1.1 For $n = 1, \ldots, n$, let X_n be independent r.v.'s such that

$$P(X_n = 1) = p_n, \quad P(X_n = 0) = 1 - p_n.$$

Under what conditions on the p_n's does $X_n \xrightarrow[n \to \infty]{P} 0$?

8.1.2 For $n = 1, 2, \ldots$, let X_n be an r.v. with d.f. F_n given by $F_n(x) = 0$ if $x < n$ and $F_n(x) = 1$ if $x \geq n$. Then show that $F_n(x) \xrightarrow[n \to \infty]{} 0$ for every $x \in \mathbb{R}$. Thus a convergent sequence of d.f.'s need not converge to a d.f.

8.1.3 Let $X_j, j = 1, \ldots, n$, be i.i.d. r.v.'s such that $EX_j = \mu$, $\sigma^2(X_j) = \sigma^2$, both finite. Show that $E(\bar{X}_n - \mu)^2 \xrightarrow[n \to \infty]{} 0$.

8.1.4 For $n = 1, 2, \ldots$, let X_n, Y_n be r.v.'s such that $E(X_n - Y_n)^2 \xrightarrow[n \to \infty]{} 0$ and suppose that $E(X_n - X)^2 \xrightarrow[n \to \infty]{} 0$ for some r.v. X. Then show that $Y_n \xrightarrow[n \to \infty]{q.m.} X$.

8.1.5 Let $X_j, j = 1, \ldots, n$ be independent r.v.'s distributed as $U(0, 1)$, and set $Y_n = \min(X_1, \ldots, X_n)$, $Z_n = \max(X_1, \ldots, X_n)$, $U_n = nY_n$, $V_n = n(1 - Z_n)$. Then show that, as $n \to \infty$, one has

i) $Y_n \xrightarrow{P} 0$;

ii) $Z_n \xrightarrow{P} 1$;

iii) $U_n \xrightarrow{d} U$;

iv) $V_n \xrightarrow{d} V$, where U and V have the negative exponential distribution with parameter $\lambda = 1$.

8.2 Relationships Among the Various Modes of Convergence

The following theorem states the relationships which exist among the various modes of convergence.

8.2 Relationships Among the Various Modes of Convergence

THEOREM 1
i) $X_n \xrightarrow[n\to\infty]{\text{a.s.}} X$ implies $X_n \xrightarrow[n\to\infty]{P} X$.
ii) $X_n \xrightarrow[n\to\infty]{\text{q.m.}} X$ implies $X_n \xrightarrow[n\to\infty]{P} X$.
iii) $X_n \xrightarrow[n\to\infty]{P} X$ implies $X_n \xrightarrow[n\to\infty]{d} X$. The converse is also true if X is *degenerate*; that is, $P[X = c] = 1$ for some constant c. In terms of a diagram this is

$$\text{a.s. conv.} \Rightarrow \text{conv. in prob.} \Rightarrow \text{conv. in dist.}$$
$$\Uparrow$$
$$\text{conv. in q.m.}$$

PROOF

i) Let A be the subset of S on which $X_n \xrightarrow[n\to\infty]{} X$. Then it is not hard to see (see Exercise 8.2.4) that

$$A = \bigcap_{k=1}^{\infty} \bigcup_{n=1}^{\infty} \bigcap_{r=1}^{\infty} \left(|X_{n+r} - X| < \frac{1}{k} \right),$$

so that the set A^c for which $X_n \not\xrightarrow[n\to\infty]{} X$ is given by

$$A^c = \bigcup_{k=1}^{\infty} \bigcap_{n=1}^{\infty} \bigcup_{r=1}^{\infty} \left(|X_{n+r} - X| \geq \frac{1}{k} \right).$$

The sets A, A^c as well as those appearing in the remaining of this discussion are all events, and hence we can take their probabilities. By setting

$$B_k = \bigcap_{n=1}^{\infty} \bigcup_{r=1}^{\infty} \left(|X_{n+r} - X| \geq \frac{1}{k} \right),$$

we have $B_k \uparrow A^c$, as $k \to \infty$, so that $P(B_k) \to P(A^c)$, by Theorem 2, Chapter 2. Thus if $X_n \xrightarrow[n\to\infty]{\text{a.s.}} X$, then $P(A^c) = 0$, and therefore $P(B_k) = 0, k \geq 1$. Next, it is clear that for every fixed k, and as $n \to \infty$, $C_n \downarrow B_k$, where

$$C_n = \bigcup_{r=1}^{\infty} \left(|X_{n+r} - X| \geq \frac{1}{k} \right).$$

Hence $P(C_n) \downarrow P(B_k) = 0$ by Theorem 2, Chapter 2, again. To summarize, if $X_n \xrightarrow[n\to\infty]{\text{a.s.}} X$, which is equivalent to saying that $P(A^c) = 0$, one has that $P(C_n) \xrightarrow[n\to\infty]{} 0$. But for any fixed positive integer m,

$$\left(|X_{n+m} - X| \geq \frac{1}{k} \right) \subseteq \bigcup_{r=1}^{\infty} \left(|X_{n+r} - X| \geq \frac{1}{k} \right),$$

so that

$$P\left(|X_{n+m} - X| \geq \frac{1}{k} \right) \leq P\left[\bigcup_{r=1}^{\infty} \left(|X_{n+r} - X| \geq \frac{1}{k} \right) \right] = P(C_n) \xrightarrow[n\to\infty]{} 0$$

for every $k \geq 1$. However, this is equivalent to saying that $X_n \xrightarrow[n\to\infty]{P} X$, as was to be seen.

ii) By special case 1 (applied with $r = 2$) of Theorem 1, we have

$$P[|X_n - X| > \varepsilon] \leq \frac{E|X_n - X|^2}{\varepsilon^2}.$$

Thus, if $X_n \xrightarrow[n\to\infty]{q.m.} X$, then $E|X_n - X|^2 \xrightarrow[n\to\infty]{} 0$ implies $P[|X_n - X| > \varepsilon] \xrightarrow[n\to\infty]{} 0$ for every $\varepsilon > 0$, or equivalently, $X_n \xrightarrow[n\to\infty]{P} X$.

iii) Let $x \in \mathbb{R}$ be a continuity point of F and let $\varepsilon > 0$ be given. Then we have

$$[X \leq x - \varepsilon] = [X_n \leq x, X \leq x - \varepsilon] + [X_n > x, X \leq x - \varepsilon]$$
$$\subseteq [X_n \leq x] + [X_n > x, X \leq x - \varepsilon]$$
$$\subseteq [X_n \leq x] \cup [|X_n - X| \geq \varepsilon],$$

since

$$[X_n > x, X \leq x - \varepsilon] = [X_n > x, -X \geq -x + \varepsilon]$$
$$\subseteq [X_n - X \geq \varepsilon] \subseteq [|X_n - X| \geq \varepsilon].$$

So

$$[X \leq x - \varepsilon] \subseteq [X_n \leq x] \cup [|X_n - X| \geq \varepsilon]$$

implies

$$P[X \leq x - \varepsilon] \leq P[X_n \leq x] + P[|X_n - X| \geq \varepsilon],$$

or

$$F(x - \varepsilon) \leq F_n(x) + P[|X_n - X| \geq \varepsilon].$$

Thus, if $X_n \xrightarrow[n\to\infty]{P} X$, then we have by taking limits

$$F(x - \varepsilon) \leq \liminf_{n\to\infty} F_n(x). \tag{1}$$

In a similar manner one can show that

$$\limsup_{n\to\infty} F_n(x) \leq F(x + \varepsilon). \tag{2}$$

But (1) and (2) imply $F(x - \varepsilon) \leq \liminf_{n\to\infty} F_n(x) \leq \limsup_{n\to\infty} F_n(x) \leq F(x + \varepsilon)$. Letting $\varepsilon \to 0$, we get (by the fact that x is a continuity point of F) that

$$F(x) \leq \liminf_{n\to\infty} F_n(x) \leq \limsup_{n\to\infty} F_n(x) \leq F(x).$$

Hence $\lim_{n\to\infty} F_n(x)$ exists and equals $F(x)$. Assume now that $P[X = c] = 1$. Then

$$F(x) = \begin{cases} 0, & x < c \\ 1, & x \geq c \end{cases}$$

and our assumption is that $F_n(x) \xrightarrow[n\to\infty]{} F(x)$, $x \neq c$. We must show that $X_n \xrightarrow[n\to\infty]{P} c$. We have

$$P[|X_n - c| \leq \varepsilon] = P[-\varepsilon \leq X_n - c \leq \varepsilon]$$
$$= P[c - \varepsilon \leq X_n \leq c + \varepsilon]$$
$$= P[X_n \leq c + \varepsilon] - P[X_n < c - \varepsilon]$$
$$\geq P[X_n \leq c + \varepsilon] - P[X_n \leq c - \varepsilon]$$
$$= F_n(c + \varepsilon) - F_n(c - \varepsilon).$$

Since $c - \varepsilon$, $c + \varepsilon$ are continuity points of F, we get
$$\lim_{n\to\infty} P[|X_n - c| \leq \varepsilon] \geq F(c + \varepsilon) - F(c - \varepsilon) = 1 - 0 = 1.$$
Thus
$$P[|X_n - c| \leq \varepsilon] \xrightarrow[n\to\infty]{} 1. \quad \blacktriangle$$

REMARK 4 It is shown by the following example that the converse in (i) is not true.

EXAMPLE 2 Let $S = (0, 1]$, and let P be the probability function which assigns to subintervals of $(0, 1]$ as measures of their length. (This is known as the Lebesgue measure over $(0, 1]$.) Define the sequence $X_1, X_2, \ldots$ of r.v.'s as follows: For each $k = 1, 2, \ldots$, divide $(0, 1]$ into 2^{k-1} subintervals of equal length. These intervals are then given by
$$\left(\frac{j-1}{2^{k-1}}, \frac{j}{2^{k-1}}\right], \quad j = 1, 2, \ldots, 2^{k-1}.$$

For each $k = 1, 2, \ldots$, we define a group of 2^{k-1} r.v.'s, whose subscripts range from 2^{k-1} to $2^k - 1$, in the following way: There are $(2^k - 1) - (2^{k-1} - 1) = 2^{k-1}$ r.v.'s within this group. We define the jth r.v. in this group to be equal to 1 for
$$s \in \left(\frac{j-1}{2^{k-1}}, \frac{j}{2^{k-1}}\right] \quad \text{and} \quad 0, \text{ otherwise.}$$

We assert that the so constructed sequence $X_1, X_2, \ldots$ of r.v.'s converges to 0 in probability, while it converges nowhere pointwise, not even for a single $s \in (0, 1]$. In fact, by Theorem 1(ii), it suffices to show that $X_n \xrightarrow[n\to\infty]{q.m.} 0$; that is, $EX_n^2 \xrightarrow[n\to\infty]{} 0$. For any $n \geq 1$, we have that X_n is the indicator of an interval
$$\left(\frac{j-1}{2^{k-1}}, \frac{j}{2^{k-1}}\right]$$
for some k and j as above. Hence $EX_n^2 = 1/2^{k-1}$. It is also clear that for $m > n$, $EX_m^2 \leq 1/2^{k-1}$. Since for every $\varepsilon > 0$, $1/2^{k-1} < \varepsilon$ for all sufficiently large k, the proof that $EX_n^2 \xrightarrow[n\to\infty]{} 0$ is complete.

The example just discussed shows that $X_n \xrightarrow[n\to\infty]{P} X$ need not imply that $X_n \xrightarrow[n\to\infty]{a.s.} X$, and also that $X_n \xrightarrow[n\to\infty]{q.m.} X$ need not imply $X_n \xrightarrow[n\to\infty]{a.s.} X$. That $X_n \xrightarrow[n\to\infty]{a.s.} X$ need not imply that $X_n \xrightarrow[n\to\infty]{q.m.} X$ is seen by the following example.

EXAMPLE 3 Let S and P be as in Example 2, and for $n \geq 1$, let X_n be defined by $X_n = \sqrt{n} I_{(0,1/n]}$. Then, clearly, $X_n \xrightarrow[n\to\infty]{} 0$ but $EX_n^2 = n(1/n) = 1$, so that $X_n \xrightarrow[n\to\infty]{q.m.} 0$.

REMARK 5 In (ii), if $P[X = c] = 1$, then: $X_n \xrightarrow[n\to\infty]{q.m.} X$ if and only if
$$E(X_n) \xrightarrow[n\to\infty]{} c, \quad \sigma^2(X_n) \xrightarrow[n\to\infty]{} 0.$$

In fact,

$$E(X_n - c)^2 = E[(X_n - EX_n) + (EX_n - c)]^2$$
$$= E(X_n - EX_n)^2 + (EX_n - c)^2$$
$$= \sigma^2(X_n) + (EX_n - c)^2.$$

Hence $E(X_n - c)^2 \xrightarrow[n \to \infty]{} 0$ if and only if $\sigma^2(X_n) \xrightarrow[n \to \infty]{} 0$ and $EX_n \xrightarrow[n \to \infty]{} c$.

REMARK 6 The following example shows that the converse of (iii) is not true.

EXAMPLE 4 Let $S = \{1, 2, 3, 4\}$, and on the subsets of S, let P be the discrete uniform function. Define the following r.v.'s:

$$X_n(1) = X_n(2) = 1, \; X_n(3) = X_n(4) = 0, \; n = 1, 2, \ldots,$$

and

$$X(1) = X(2) = 0, \quad X(3) = X(4) = 1.$$

Then

$$|X_n(s) - X(s)| = 1 \quad \text{for all} \quad s \in S.$$

Hence X_n does *not* converge in probability to X, as $n \to \infty$. Now,

$$F_{X_n}(x) = \begin{cases} 0, & x < 0 \\ \tfrac{1}{2}, & 0 \le x < 1 \\ 1, & x \ge 1 \end{cases}, \quad F_X(x) = \begin{cases} 0, & x < 0 \\ \tfrac{1}{2}, & 0 \le x < 1 \\ 1, & x \ge 1 \end{cases}$$

so that $F_{X_n}(x) = F_X(x)$ for all $x \in \mathbb{R}$. Thus, trivially, $F_{X_n}(x) \xrightarrow[n \to \infty]{} F_X(x)$ for all continuity points of F_X; that is, $X_n \xrightarrow[n \to \infty]{d} X$, but X_n does not converge in probability to X.

Very often one is confronted with the problem of proving convergence in distribution. The following theorem replaces this problem with that of proving convergence of ch.f.'s, which is much easier to deal with.

THEOREM 2 (P. Lévy's Continuity Theorem) Let $\{F_n\}$ be a sequence of d.f.'s, and let F be a d.f. Let ϕ_n be the ch.f. corresponding to F_n and ϕ be the ch.f. corresponding to F. Then,

i) If $F_n(x) \xrightarrow[n \to \infty]{} F(x)$ for all continuity points x of F, then $\phi_n(t) \xrightarrow[n \to \infty]{} \phi(t)$, for every $t \in \mathbb{R}$.

ii) If $\phi_n(t)$ converges, as $n \to \infty$, and $t \in \mathbb{R}$, to a function $g(t)$ which is continuous at $t = 0$, then g is a ch.f., and if F is the corresponding d.f., then $F_n(x) \xrightarrow[n \to \infty]{} F(x)$, for all continuity points x of F.

PROOF Omitted.

REMARK 7 The assumption made in the second part of the theorem above according to which the function g is continuous at $t = 0$ is essential. In fact, let X_n be an r.v. distributed as $N(0, n)$, so that its ch.f. is given by $\phi_n(t) = e^{-t^2 n/2}$. Then

$\phi_n(t) \xrightarrow[n\to\infty]{} g(t)$, where $g(t) = 0$, if $t \neq 0$, and $g(0) = 1$, so that g is not continuous at 0. The conclusion in (ii) does not hold here because

$$F_{X_n}(x) = P(X_n \leq x) = P\left(\frac{X_n}{\sqrt{n}} \leq \frac{x}{\sqrt{n}}\right) = \Phi\left(\frac{x}{\sqrt{n}}\right) \xrightarrow[n\to\infty]{} \frac{1}{2}$$

for every $x \in \mathbb{R}$ and $F(x) = \frac{1}{2}$, $x \in \mathbb{R}$, is not a d.f. of an r.v.

Exercises

8.2.1 (Rényi) Let $S = [0, 1)$ and let P be the probability function on subsets of S, which assigns probability to intervals equal to their lengths. For $n = 1, 2, \ldots$, define the r.v.'s X_n as follows:

$$X_{N^2+j}(s) = \begin{cases} N, & \text{if } \frac{j}{2N+1} \leq s < \frac{j+1}{2N+1} \\ 0, & \text{otherwise,} \end{cases}$$

$j = 0, 1, \ldots, 2N$, $N = 1, 2, \ldots$. Then show that

i) $X_n \xrightarrow[n\to\infty]{P} 0$;

ii) $X_n(s) \not\xrightarrow[n\to\infty]{} 0$ for any $s \in [0, 1)$;

iii) $X_n^2(s) \xrightarrow[n\to\infty]{} 0$, $s \in (0, 1)$;

iv) $EX_n \not\xrightarrow[n\to\infty]{} 0$.

8.2.2 For $n = 1, 2, \ldots$, let X_n be r.v.'s distributed as $B(n, p_n)$, where $np_n = \lambda_n \xrightarrow[n\to\infty]{} \lambda(>0)$. Then, by using ch.f.'s, show that $X_n \xrightarrow[n\to\infty]{d} X$, where X is an r.v. distributed as $P(\lambda)$.

8.2.3 For $n = 1, 2, \ldots$, let X_n be r.v.'s having the negative binomial distribution with p_n and r_n such that $p_n \xrightarrow[n\to\infty]{} 1$, $r_n \xrightarrow[n\to\infty]{} \infty$, so that $r_n(1 - p_n) = \lambda_n \xrightarrow[n\to\infty]{} \lambda(>0)$. Show that $X_n \xrightarrow[n\to\infty]{d} X$, where X is an r.v. distributed as $P(\lambda)$. (Use ch.f.'s.)

8.2.4 If the i.i.d. r.v.'s $X_j, j = 1, \ldots, n$ have a Cauchy distribution, show that there is no finite constant c for which $\bar{X}_n \xrightarrow[n\to\infty]{P} c$. (Use ch.f.'s.)

8.2.5 In reference to the proof of Theorem 1, show that the set A of convergence of $\{X_n\}$ to X is, indeed, expressed by $A = \bigcap_{k=1}^{\infty} \bigcup_{n=1}^{\infty} \bigcap_{r=1}^{\infty} (|X_{n+r} - X| < \frac{1}{k})$.

8.3 The Central Limit Theorem

We are now ready to formulate and prove the celebrated Central Limit Theorem (CLT) in its simplest form.

THEOREM 3 (Central Limit Theorem) Let $X_1, \ldots, X_n$ be i.i.d. r.v.'s with mean μ (finite) and (finite and positive) variance σ^2. Let

$$\overline{X}_n = \frac{1}{n}\sum_{j=1}^{n} X_j, \quad G_n(x) = P\left[\frac{\sqrt{n}(\overline{X}_n - \mu)}{\sigma} \leq x\right], \quad \text{and} \quad \Phi(x) = \frac{1}{\sqrt{2\pi}}\int_{-\infty}^{x} e^{-t^2/2}dt.$$

Then $G_n(x) \xrightarrow[n \to \infty]{} \Phi(x)$ for every x in $\mathbb{R}$.

REMARK 8

i) We often express (loosely) the CLT by writing

$$\frac{\sqrt{n}(\overline{X}_n - \mu)}{\sigma} \approx N(0, 1), \quad \text{or} \quad \frac{S_n - E(S_n)}{\sigma(S_n)} \approx N(0, 1),$$

for large n, where

$$S_n = \sum_{j=1}^{n} X_j, \quad \text{since} \quad \frac{\sqrt{n}(\overline{X}_n - \mu)}{\sigma} = \frac{S_n - E(S_n)}{\sigma(S_n)}.$$

ii) In part (i), the notation S_n was used to denote the sum of the r.v.'s $X_1, \ldots, X_n$. This is a generally accepted notation, and we are going to adhere to it here. It should be pointed out, however, that the same or similar symbols have been employed elsewhere to denote different quantities (see, for example, Corollaries 1 and 2 in Chapter 7, or Theorem 9 and Corollary to Theorem 8 in this chapter). This point should be kept in mind throughout.

iii) In the proof of Theorem 3 and elsewhere, the "little o" notation will be employed as a convenient notation for the remainder in Taylor series expansions. A relevant comment would then be in order. To this end, let $\{a_n\}, \{b_n\}, n = 1, 2, \ldots$ be two sequences of numbers. We say that $\{a_n\}$ is $o(b_n)$ (little o of b_n) and we write $a_n = o(b_n)$, if $a_n/b_n \xrightarrow[n \to \infty]{} 0$. For example, if $a_n = n$ and $b_n = n^2$, then $a_n = o(b_n)$, since $n/n^2 = 1/n \xrightarrow[n \to \infty]{} 0$. Clearly, if $a_n = o(b_n)$, then $a_n = b_n o(1)$. Therefore $o(b_n) = b_n o(1)$.

iv) We recall the following fact which was also employed in the proof of Theorem 3, Chapter 3. Namely, if $a_n \xrightarrow[n \to \infty]{} a$, then

$$\left(1 + \frac{a_n}{n}\right)^n \xrightarrow[n \to \infty]{} e^a.$$

PROOF OF THEOREM 3 We may now begin the proof. Let g_n be the ch.f. of G_n and ϕ be the ch.f. of Φ; that is, $\phi(t) = e^{-t^2/2}$, $t \in \mathbb{R}$. Then, by Theorem 2, it suffices to prove that $g_n(t) \xrightarrow[n \to \infty]{} \phi(t), t \in \mathbb{R}$. This will imply that $G_n(x) \to \Phi(x)$, $x \in \mathbb{R}$. We have

$$\frac{\sqrt{n}(\overline{X}_n - \mu)}{\sigma} = \frac{n\overline{X}_n - n\mu}{\sigma\sqrt{n}} = \frac{1}{\sqrt{n}}\sum_{j=1}^{n}\frac{X_j - \mu}{\sigma}$$

$$= \frac{1}{\sqrt{n}}\sum_{j=1}^{n} Z_j,$$

where $Z_j = (X_j - \mu)/\sigma$, $j = 1, \ldots, n$ are i.i.d. with $E(Z_j) = 0$, $\sigma^2(Z_j) = E(Z_j^2) = 1$. Hence, for simplicity, writing $\Sigma_j Z_j$ instead of $\Sigma_{j=1}^n Z_j$, when this last expression appears as a subscript, we have

$$g_n(t) = g_{(1/\sqrt{n})\Sigma_j Z_j}(t) = g_{\Sigma_j Z_j}\left(\frac{t}{\sqrt{n}}\right) = \left[g_{Z_1}\left(\frac{1}{\sqrt{n}}\right)\right]^n.$$

Now consider the Taylor expansion of g_{Z_1} around zero up to the second order term. Then

$$g_{Z_1}\left(\frac{t}{\sqrt{n}}\right) = g_{Z_1}(0) + \frac{t}{\sqrt{n}} g'_{Z_1}(0) + \frac{1}{2!}\left(\frac{t}{\sqrt{n}}\right)^2 g''_{Z_1}(0) + o\left(\frac{t^2}{n}\right).$$

Since

$$g_{Z_1}(0) = 1, \quad g'_{Z_1}(0) = iE(Z_1) = 0, \quad g''_{Z_1}(0) = i^2 E(Z_1^2) = -1,$$

we get

$$g_{Z_1}\left(\frac{t}{\sqrt{n}}\right) = 1 - \frac{t^2}{2n} + o\left(\frac{t^2}{n}\right) = 1 - \frac{t^2}{2n} + \frac{t^2}{n}o(1) = 1 - \frac{t^2}{2n}[1 - o(1)].$$

Thus

$$g_n(t) = \left\{1 - \frac{t^2}{2n}[1 - o(1)]\right\}^n.$$

Taking limits as $n \to \infty$ we have, $g_n(t) \xrightarrow[n \to \infty]{} e^{-t^2/2}$, which is the ch.f. of Φ. ▲

The theorem just established has the following corollary, which along with the theorem itself provides the justification for many approximations.

COROLLARY The convergence $G_n(x) \xrightarrow[n \to \infty]{} \Phi(x)$ is *uniform* in $x \in \mathbb{R}$.
(That is, for every $x \in \mathbb{R}$ and every $\varepsilon > 0$ there exists $N(\varepsilon) > 0$ independent of x, such that $|G_n(x) - \Phi(x)| < \varepsilon$ for all $n \geq N(\varepsilon)$ and all $x \in \mathbb{R}$ simultaneously.)

PROOF It is an immediate consequence of Lemma 1 in Section 8.6*. ▲

The following examples are presented for the purpose of illustrating the theorem and its corollary.

8.3.1 Applications

1. If X_j, $j = 1, \ldots, n$ are i.i.d. with $E(X_j) = \mu$, $\sigma^2(X_j) = \sigma^2$, the CLT is used to give an approximation to $P[a < S_n \leq b]$, $-\infty < a < b < +\infty$. We have:

$$P[a < S_n \leq b] = P\left[\frac{a - E(S_n)}{\sigma(S_n)} < \frac{S_n - E(S_n)}{\sigma(S_n)} \leq \frac{b - E(S_n)}{\sigma(S_n)}\right]$$

$$= P\left[\frac{a - n\mu}{\sigma\sqrt{n}} < \frac{S_n - E(S_n)}{\sigma(S_n)} \leq \frac{b - n\mu}{\sigma\sqrt{n}}\right]$$

$$= P\left[\frac{S_n - E(S_n)}{\sigma(S_n)} \leq \frac{b - n\mu}{\sigma\sqrt{n}}\right] - P\left[\frac{S_n - E(S_n)}{\sigma(S_n)} \leq \frac{a - n\mu}{\sigma\sqrt{n}}\right]$$

$$\approx \Phi(b^*) - \Phi(a^*),$$

where

$$a^* = \frac{a - n\mu}{\sigma\sqrt{n}}, \quad b^* = \frac{b - n\mu}{\sigma\sqrt{n}}.$$

(Here is where the corollary is utilized. The points a^* and b^* do depend on n, and therefore move along $\mathbb{R}$ as $n \to \infty$. The above approximation would not be valid if the convergence was not uniform in $x \in \mathbb{R}$.) That is, $P(a < S_n \leq b) \approx \Phi(b^*) - \Phi(a^*)$.

2. *Normal approximation to the Binomial.* This is the same problem as above, where now X_j, $j = 1, \ldots, n$, are independently distributed as $B(1, p)$. We have $\mu = p$, $\sigma = \sqrt{pq}$. Thus:

$$P(a < S_n \leq b) \approx \Phi(b^*) - \Phi(a^*),$$

where

$$a^* = \frac{a - np}{\sqrt{npq}}, \quad b^* = \frac{b - np}{\sqrt{npq}},$$

REMARK 9 It is seen that the approximation is fairly good provided n and p are such that $npq \geq 20$. For a given n, the approximation is best for $p = \frac{1}{2}$ and deteriorates as p moves away from $\frac{1}{2}$. Some numerical examples will shed some light on these points. Also, the Normal approximation to the Binomial distribution presented above can be improved, if in the expressions of a^* and b^* we replace a and b by $a + 0.5$ and $b + 0.5$, respectively. This is called the *continuity correction*. In the following we give an explanation of the continuity correction. To start with, let

$$f_n(r) = \binom{n}{r} p^r q^{n-r}, \quad \text{and let} \quad \phi_n(x) = \frac{1}{\sqrt{2\pi npq}} e^{-x^2/2},$$

where

$$x = \frac{r - np}{\sqrt{npq}}.$$

Then it can be shown that $f_n(r)/\phi_n(x) \xrightarrow[n \to \infty]{} 1$ and this convergence is uniform for all x's in a finite interval $[a, b]$. (This is the De Moivre theorem.) Thus for large n, we have, in particular, that $f_n(r)$ is close to $\phi_n(x)$. That is, the probability $\binom{n}{r}p^r q^{n-r}$ is approximately equal to the value

$$\frac{1}{\sqrt{2\pi npq}} \exp\left[-\frac{(r-np)^2}{2npq}\right]$$

of the normal density with mean np and variance npq for sufficiently large n. Note that this asymptotic relationship of the p.d.f.'s is not implied, in general, by the convergence of the distribution functions in the CLT.

To give an idea of how the correction term $\frac{1}{2}$ comes in, we refer to Fig. 8.2 drawn for $n = 10$, $p = 0.2$.

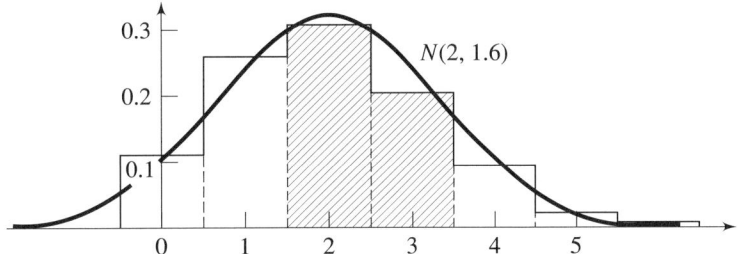

Figure 8.2

Now

$$P(1 < S_n \leq 3) = P(2 \leq S_n \leq 3) = f_n(2) + f_n(3)$$
$$= \text{shaded area},$$

while the approximation without correction is the area bounded by the normal curve, the horizontal axis, and the abscissas 1 and 3. Clearly, the correction, given by the area bounded by the normal curve, the horizontal axis and the abscissas 1.5 and 3.5, is closer to the exact area.

To summarize, under the conditions of the CLT, and for discrete r.v.'s,

$$P(a < S_n \leq b) \approx \Phi\left(\frac{b - n\mu}{\sigma\sqrt{n}}\right) - \Phi\left(\frac{a - n\mu}{\sigma\sqrt{n}}\right) \quad \text{without continuity correction,}$$

and

$$P(a < S_n \leq b) \approx \Phi\left(\frac{b + 0.5 - n\mu}{\sigma\sqrt{n}}\right) - \Phi\left(\frac{a + 0.5 - n\mu}{\sigma\sqrt{n}}\right)$$

with continuity correction.

In particular, for integer-valued r.v.'s and probabilities of the form $P(a \leq S_n \leq b_n)$, we first rewrite the expression as follows:

$$P(a \le S_n \le b_n) = P((a-1) < S_n \le b_n), \qquad (3)$$

and then apply the above approximations in order to obtain:

$$P(a \le S_n \le b) \approx \Phi(b^*) - \Phi(a^*) \quad \text{without continuity correction,}$$

where

$$a^* = \frac{a-1-n\mu}{\sigma\sqrt{n}}, \quad b^* = \frac{b-n\mu}{\sigma\sqrt{n}}, \qquad (4)$$

and

$$P(a \le S_n \le b) \approx \Phi(b') - \Phi(a') \quad \text{with continuity correction,}$$

where

$$a' = \frac{a-0.5-n\mu}{\sigma\sqrt{n}}, \quad b' = \frac{b+0.5-n\mu}{\sigma\sqrt{n}}. \qquad (5)$$

These expressions of a^*, b^* and a', b' in (4) and (5) will be used in calculating probabilities of the form (3) in the numerical examples below.

EXAMPLE 5 (Numerical) For $n = 100$ and $p_1 = \frac{1}{2}$, $p_2 = \frac{5}{16}$, find $P(45 \le S_n \le 55)$.

i) For $p_1 = \frac{1}{2}$: *Exact value*: 0.7288

Normal approximation without correction:

$$a^* = \frac{44 - 100 \cdot \frac{1}{2}}{\sqrt{100 \cdot \frac{1}{2} \cdot \frac{1}{2}}} = \frac{6}{5} = -1.2,$$

$$b^* = \frac{55 - 100 \cdot \frac{1}{2}}{\sqrt{100 \cdot \frac{1}{2} \cdot \frac{1}{2}}} = \frac{5}{5} = 1.$$

Thus

$$\Phi(b^*) - \Phi(a^*) = \Phi(1) - \Phi(-1.2) = \Phi(1) + \Phi(1.2) - 1$$
$$= 0.841345 + 0.884930 - 1 = 0.7263.$$

Normal approximation with correction:

$$a' = \frac{45 - 0.5 - 100 \cdot \frac{1}{2}}{\sqrt{100 \cdot \frac{1}{2} \cdot \frac{1}{2}}} = \frac{5.5}{5} = -1.1$$

$$b' = \frac{55 + 0.5 - 100 \cdot \frac{1}{2}}{\sqrt{100 \cdot \frac{1}{2} \cdot \frac{1}{2}}} = \frac{5.5}{5} = -1.1.$$

Thus
$$\Phi(b') - \Phi(a') = \Phi(1.1) - \Phi(-1.1) = 2\Phi(1.1) - 1 = 2 \times 0.864334 - 1 = 0.7286.$$

Error without correction: $0.7288 - 0.7263 = 0.0025$.
Error with correction: $\quad 0.7288 - 0.7286 = 0.0002$.

ii) For $p_2 = \frac{5}{16}$, working as above, we get:

Exact value: 0.0000.

$$a^* = 2.75, \quad b^* = 4.15, \quad \text{so that } \Phi(b^*) - \Phi(a^*) = 0.0030.$$
$$a' = 2.86, \quad b' = 5.23, \quad \text{so that } \Phi(b') - \Phi(a') = 0.0021.$$

Then:

Error without correction: 0.0030.
Error with correction: $\quad$ 0.0021.

3. *Normal approximation to Poisson.* This is the same problem as in (1), where now $X_j, j = 1, \ldots, n$ are independent $P(\lambda)$. We have $\mu = \lambda$, $\sigma = \sqrt{\lambda}$. Thus

$$P(a < S_n \leq b) \approx \Phi\left(\frac{b - n\lambda}{\sqrt{n\lambda}}\right) - \Phi\left(\frac{a - n\lambda}{\sqrt{n\lambda}}\right) \quad \text{without continuity correction,}$$

and

$$P(a < S_n \leq b) \approx \Phi\left(\frac{b + 0.5 - n\lambda}{\sqrt{n\lambda}}\right) - \Phi\left(\frac{a + 0.5 - n\lambda}{\sqrt{n\lambda}}\right)$$

with continuity correction.

Probabilities of the form $P(a \leq S_n \leq b)$ are approximated as follows:

$$P(a \leq S_n \leq b) \approx \Phi(b^*) - \Phi(a^*) \quad \text{without continuity correction,}$$

where

$$a^* = \frac{a - 1 - n\lambda}{\sqrt{n\lambda}}, \quad b^* = \frac{b - n\lambda}{\sqrt{n\lambda}},$$

and

$$P(a \leq S_n \leq b) \approx \Phi(b') - \Phi(a') \quad \text{with continuity correction,}$$

where

$$a' = \frac{a - 0.5 - n\lambda}{\sqrt{n\lambda}}, \quad b' = \frac{b + 0.5 - n\lambda}{\sqrt{n\lambda}}.$$

EXAMPLE 6 (Numerical) For $n\lambda = 16$, find $P(12 \leq S_n \leq 21)$. We have:

Exact value: 0.7838.

Normal approximation without correction:

$$a^* = \frac{11-16}{\sqrt{16}} = -\frac{5}{4} = -1.25, \quad b^* = \frac{21-16}{\sqrt{16}} = \frac{5}{4} = 1.25,$$

so that $\Phi(b^*) - \Phi(a^*) = \Phi(1.25) - \Phi(-1.25) = 2\Phi(1.25) - 1 = 2 \times 0.894350 - 1 = 0.7887.$

Normal approximation with correction:

$$a' = -1.125, \quad b' = 1.375, \quad \text{so that } \Phi(b') - \Phi(a') = 0.7851.$$

Error without correction: 0.0049.
Error with correction: 0.0013.

Exercises

8.3.1 Refer to Exercise 4.1.12 of Chapter 4 and suppose that another manufacturing process produces light bulbs whose mean life is claimed to be 10% higher than the mean life of the bulbs produced by the process described in the exercise cited above. How many bulbs manufactured by the new process must be examined, so as to establish the claim of their superiority with probability 0.95?

8.3.2 A fair die is tossed independently 1,200 times. Find the approximate probability that the number of aces X is such that $180 \leq X \leq 220$. (Use the CLT.)

8.3.3 Fifty balanced dice are tossed once and let X be the sum of the upturned spots. Find the approximate probability that $150 \leq X \leq 200$. (Use the CLT.)

8.3.4 Let $X_j, j = 1, \ldots, 100$ be independent r.v.'s distributed as $B(1, p)$. Find the exact and approximate value for the probability $P(\sum_{j=1}^{100} X_j = 50)$. (For the latter, use the CLT.)

8.3.5 One thousand cards are drawn with replacement from a standard deck of 52 playing cards, and let X be the total number of aces drawn. Find the approximate probability that $65 \leq X \leq 90$. (Use the CLT.)

8.3.6 A Binomial experiment with probability p of a success is repeated 1,000 times and let X be the number of successes. For $p = \frac{1}{2}$ and $p = \frac{1}{4}$, find the exact and approximate values of probability $P(1,000p - 50 \leq X \leq 1,000p + 50)$. (For the latter, use the CLT.)

8.3.7 From a large collection of bolts which is known to contain 3% defective bolts, 1,000 are chosen at random. If X is the number of the defective bolts among those chosen, what is probability that this number does not exceed 5% of 1,000? (Use the CLT.)

8.3.8 Suppose that 53% of the voters favor a certain legislative proposal. How many voters must be sampled, so that the observed relative frequency of those favoring the proposal will not differ from the assumed frequency by more than 2% with probability 0.99? (Use the CLT.)

8.3.9 In playing a game, you win or lose $1 with probability $\frac{1}{2}$. If you play the game independently 1,000 times, what is the probability that your fortune (that is, the total amount you won or lost) is at least $10? (Use the CLT.)

8.3.10 A certain manufacturing process produces vacuum tubes whose lifetimes in hours are independently distributed r.v.'s with Negative Exponential distribution with mean 1,500 hours. What is the probability that the total life of 50 tubes will exceed 75,000 hours? (Use the CLT.)

8.3.11 Let $X_j, j = 1, \ldots, n$ be i.i.d. r.v.'s such that $EX_j = \mu$ finite and $\sigma^2(X_j) = \sigma^2 = 4$. If $n = 100$, determine the constant c so that $P(|\bar{X}_n - \mu| \leq c) = 0.90$. (Use the CLT.)

8.3.12 Let $X_j, j = 1, \ldots, n$ be i.i.d. r.v.'s with $EX_j = \mu$ finite and $\sigma^2(X_j) = \sigma^2 \in (0, \infty)$.

i) Show that the smallest value of the sample size n for which $P(|\bar{X}_n - \mu| \leq k\sigma) \geq p$ is given by $n = \left[\frac{1}{k}\Phi^{-1}\left(\frac{1+p}{2}\right)\right]^2$ if this number is an integer, and n is the integer part of this number increased by 1, otherwise. (Use the CLT.);

ii) By using Tchebichev's inequality, show that the above value of n is given by $n = \frac{1}{k^2(1-p)}$ if this number is an integer, and n is the integer part of this number increased by 1, otherwise;

iii) For $p = 0.95$ and $k = 0.05, 0.1, 0.25$, compare the respective values of n in parts (i) and (ii).

8.3.13 Refer to Exercise 4.1.13 in Chapter 4 and let $X_j, j = 1, \ldots, n$ be the diameters of n ball bearings. If $EX_j = 0.5$ inch and $\sigma = 0.0005$ inch, what is the minimum value of n for which $P(|\bar{X}_n - \mu| \leq 0.0001) = 0.099$? (Use Exercise 8.3.12.)

8.3.14 Let $X_j, j = 1, \ldots, n$, $Y_j = 1, \ldots, n$ be independent r.v.'s such that the X's are identically distributed with $EX_j = \mu_1$, $\sigma^2(X_j) = \sigma^2$, both finite, and the Y's are identically distributed with $EY_j = \mu_2$ finite and $\sigma^2(Y_j) = \sigma^2$. Show that:

i) $E(\bar{X}_n - \bar{Y}_n) = \mu_1 - \mu_2$, $\sigma^2(\bar{X}_n - \bar{Y}_n) = \frac{2\sigma^2}{n}$;

ii) $\frac{\sqrt{n}[(\bar{X}_n - \bar{Y}_n) - (\mu_1 - \mu_2)]}{\sigma\sqrt{2}}$ is asymptotically distributed as $N(0, 1)$.

8.3.15 Let $X_j, j = 1, \ldots, n$, $Y_j, j = 1, \ldots, n$ be i.i.d. r.v.'s from the same distribution with $EX_j = EY_j = \mu$ and $\sigma^2(X_j) = \sigma^2(Y_j) = \sigma^2$, both finite. Determine the sample size n so that $P(|\bar{X}_n - \bar{Y}_n| \leq 0.25\sigma) = 0.95$. (Use Exercise 8.3.12.)

8.3.16 An academic department in a university wishes to admit c first-year graduate students. From past experience it follows that, on the average, $100p\%$ of the students admitted will, actually, accept an admission offer ($0 < p$

< 1). It may be assumed that acceptance and rejection of admission offers by the various students are independent events.

i) How many students n must be admitted, so that the probability $P(|X - c| \leq d)$ is maximum, where X is the number of students actually accepting an admission, and d is a prescribed number?

ii) What is the value of n for $c = 20$, $d = 2$, and $p = 0.6$?

iii) What is the maximum value of the probability $P(|X - 20| \leq 2)$ for $p = 0.6$?

Hint: For part (i), use the CLT (with continuity correction) in order to find the approximate value to $P(|X - c| \leq d)$. Then draw the picture of the normal curve, and conclude that the probability is maximized when n is close to c/p. For part (iii), there will be two successive values of n suggesting themselves as optimal values of n. Calculate the respective probabilities, and choose that value of n which gives the larger probability.)

8.4 Laws of Large Numbers

This section concerns itself with certain limit theorems which are known as *laws of large numbers* (*LLN*). We distinguish two categories of LLN: the strong LLN (SLLN) in which the convergence involved is strong (a.s.), and the weak LLN (WLLN), where the convergence involved is convergence in probability.

THEOREM 4 (SLLN) If $X_j, j = 1, \ldots, n$ are i.i.d. r.v.'s with (finite) mean μ, then

$$\overline{X}_n = \frac{X_1 + \cdots + X_n}{n} \xrightarrow[n \to \infty]{\text{a.s.}} \mu.$$

The converse is also true, that is, if $\overline{X}_n \xrightarrow[n \to \infty]{\text{a.s.}}$ to some finite constant μ, then $E(X_j)$ is finite and equal to μ.

PROOF Omitted; it is presented in a higher level probability course. ▲

Of course, $\overline{X}_n \xrightarrow[n \to \infty]{\text{a.s.}} \mu$ implies $\overline{X}_n \xrightarrow[n \to \infty]{P} \mu$. The latter are the weak LLN; that is,

THEOREM 5 (WLLN) If $X_j, j = 1, \ldots, n$, are i.i.d. r.v.'s with (finite) mean μ, then

$$\overline{X}_n = \frac{X_1 + \cdots + X_n}{n} \xrightarrow[n \to \infty]{P} \mu.$$

PROOF

i) The proof is a straightforward application of Tchebichev's inequality under the unnecessary assumption that the r.v.'s also have a finite variance σ^2. Then $E\overline{X}_n = \mu$, $\sigma^2(\overline{X}_n) = \sigma^2/n$, so that, for every $\varepsilon > 0$,

$$P\left[|\overline{X}_n - \mu| \geq \varepsilon\right] \leq \frac{1}{\varepsilon^2} \frac{\sigma^2}{n^2} \to 0 \quad \text{as} \quad n \to \infty.$$

ii) This proof is based on ch.f.'s (m.g.f.'s could also be used if they exist). By Theorems 1(iii) (the converse case) and 2(ii) of this chapter, in order to prove that $\overline{X}_n \xrightarrow[n\to\infty]{P} \mu$, it suffices to prove that

$$\phi_{\overline{X}_n}(t) \xrightarrow[n\to\infty]{} \phi_\mu(t) = e^{it\mu}, \quad \text{for} \quad t \in \mathbb{R}.$$

For simplicity, writing ΣX_j instead of $\Sigma_{j=1}^n X_j$, when this last expression appears as a subscript, we have

$$\phi_{\overline{X}_n}(t) = \phi_{1/n\Sigma X_j}(t) = \phi_{\Sigma X_j}\left(\frac{t}{n}\right) = \left[\phi_{X_1}\left(\frac{t}{n}\right)\right]^n$$

$$= \left[1 + \frac{t}{n}i\mu + o\left(\frac{t}{n}\right)\right]^n$$

$$= \left[1 + \frac{t}{n}i\mu + \frac{t}{n}o(1)\right]^n$$

$$= \left[1 + \frac{t}{n}[i\mu + o(1)]\right]^n \xrightarrow[n\to\infty]{} e^{i\mu t}. \blacktriangle$$

REMARK 10 An alternative proof of the WLLN, without the use of ch.f.'s, is presented in Lemma 1 in Section 8.6*. The underlying idea there is that of truncation, as will be seen.

Both laws of LLN hold in all concrete cases which we have studied except for the Cauchy case, where $E(X_j)$ does not exist. For example, in the Binomial case, we have:

If $X_j, j = 1, \ldots, n$ are independent and distributed as $B(1, p)$, then

$$\overline{X}_n = \frac{X_1 + \cdots + X_n}{n} \xrightarrow[n\to\infty]{} p \quad \text{a.s.}$$

and also in probability.

For the Poisson case we have:

If $X_j, j = 1, \ldots, n$ are independent and distributed as $P(\lambda)$, then:

$$\overline{X}_n = \frac{X_1 + \cdots + X_n}{n} \xrightarrow[n\to\infty]{} \lambda \quad \text{a.s.}$$

and also in probability.

8.4.1 An Application of SLLN and WLLN

Let $X_j, j = 1, \ldots, n$ be i.i.d. with d.f. F. The *sample* or *empirical* d.f. is denoted by F_n and is defined as follows:

$$\text{For } x \in \mathbb{R}, \quad F_n(x) = \frac{1}{n}\left[\text{the number of } X_1, \ldots, X_n \leq x\right].$$

F_n is a step function which is a d.f. for a fixed set of values of $X_1, \ldots, X_n$. It is also an r.v. as a function of the r.v.'s $X_1, \ldots, X_n$, for each x. Let

$$Y_j(x) = Y_j = \begin{cases} 1, & X_j \leq x \\ 0, & X_j > x, \end{cases} \quad j = 1, \ldots, n.$$

Then, clearly,

$$F_n(x) = \frac{1}{n} \sum_{j=1}^{n} Y_j.$$

On the other hand, Y_j, $j = 1, \ldots, n$ are independent since the X's are, and Y_j is $B(1, p)$, where

$$p = P(Y_j = 1) = P(X_j \leq x) = F(x).$$

Hence

$$E\left(\sum_{j=1}^{n} Y_j\right) = np = nF(x), \quad \sigma^2\left(\sum_{j=1}^{n} Y_j\right) = npq = nF(x)[1 - F(x)].$$

It follows that

$$E[F_n(x)] = \frac{1}{n} nF(x) = F(x).$$

So for each $x \in \mathbb{R}$, we get by the LLN

$$F_n(x) \xrightarrow[n \to \infty]{\text{a.s.}} F(x), \quad F_n(x) \xrightarrow[n \to \infty]{P} F(x).$$

Actually, more is true. Namely,

THEOREM 6 (Glivenko–Cantelli Lemma) With the above notation, we have

$$P\left[\sup\{|F_n(x) - F(x)|; x \in \mathbb{R}\} \xrightarrow[n \to \infty]{} 0\right] = 1$$

(that is, $F_n(x) \xrightarrow[n \to \infty]{\text{a.s.}} F(x)$ uniformly in $x \in \mathbb{R}$).

PROOF Omitted.

Exercises

8.4.1 Let X_j, $j = 1, \ldots, n$ be i.i.d. r.v.'s and suppose that EX_j^k is finite for a given positive integer k. Set

$$\overline{X}_n^{(k)} = \frac{1}{n} \sum_{j=1}^{n} X_j^k$$

for the kth sample moment of the distribution of the X's and show that $\overline{X}_n^{(k)} \xrightarrow[n \to \infty]{P} EX_1^k$.

8.4.2 Let X_j, $j = 1, \ldots, n$ be i.i.d. r.v.'s with p.d.f. given in Exercise 3.2.14 of Chapter 3 and show that the WLLN holds. (Calculate the expectation by means of the ch.f.)

8.4.3 Let X_j, $j = 1, \ldots, n$ be r.v.'s which need be neither independent nor identically distributed. Suppose that $EX_j = \mu_j$, $\sigma^2(X_j) = \sigma_j^2$, all finite, and set

$$\bar{\mu}_n = \frac{1}{n}\sum_{j=1}^{n}\mu_j.$$

Then a generalized version of the WLLN states that

$$\bar{X}_n - \bar{\mu}_n \xrightarrow[n\to\infty]{P} 0.$$

Show that if the X's are pairwise uncorrelated and $\sigma_j^2 \leq M(<\infty)$, $j \geq 1$, then the generalized version of the WLLN holds.

8.4.4 Let X_j, $j = 1, \ldots, n$ be pairwise uncorrelated r.v.'s such that

$$P(X_j = -\alpha^j) = P(X_j = \alpha^j) = \frac{1}{2}.$$

Show that for all α's such that $0 < \alpha \leq 1$, the generalized WLLN holds.

8.4.5 Decide whether the generalized WLLN holds for independent r.v.'s such that the jth r.v. has the Negative Exponential distribution with parameter $\lambda_j = 2^{j/2}$.

8.4.6 For $n = 1, 2, \ldots$, let X_n be independent r.v.'s such that X_n is distributed as χ_n^2. Show that the generalized WLLN holds.

8.4.7 For $n = 1, 2, \cdots$, let X_n be independent r.v.'s such that X_n is distributed as $P(\lambda_n)$. If

$$\frac{1}{n}\sum_{j=1}^{n}\lambda_j \xrightarrow[n\to\infty]{} 0,$$

show that the generalized WLLN holds.

8.5 Further Limit Theorems

In this section, we present some further limit theorems which will be used occasionally in the following chapters.

THEOREM 7 i) Let X_n, $n \geq 1$, and X be r.v.'s, and let $g: \mathbb{R} \to \mathbb{R}$ be continuous, so that $g(X_n)$, $n \geq 1$, and $g(X)$ are r.v.'s. Then $X_n \xrightarrow[n\to\infty]{\text{a.s.}} X$ implies $g(X_n) \xrightarrow[n\to\infty]{\text{a.s.}} g(X)$.

ii) More generally, if for $j = 1, \ldots, k$, $X_n^{(j)}$, $n \geq 1$, and X_j are r.v.'s, and $g: \mathbb{R}^k \to \mathbb{R}$ is continuous, so that $g(X_n^{(1)}, \ldots, X_n^{(k)})$ and $g(X_1, \ldots, X_k)$ are r.v.'s, then

$$X_n^{(j)} \xrightarrow[n\to\infty]{\text{a.s.}} X_j,$$

$j = 1, \ldots, k$ imply $g\left(X_n^{(1)}, \ldots, X_n^{(k)}\right) \xrightarrow[n\to\infty]{\text{a.s.}} g(X_1, \ldots, X_k)$.

PROOF Follows immediately from the definition of the a.s. convergence and the continuity of g. ▲

A similar result holds true when a.s. convergence is replaced by convergence in probability, but a justification is needed.

THEOREM 7′
i) Let $X_n, n \geq 1$, X and g be as in Theorem 7(i), and suppose that $X_n \xrightarrow[n \to \infty]{P} X$. Then $g(X_n) \xrightarrow[n \to \infty]{P} g(X)$.

ii) More generally, let again $X_n^{(j)}$, X_j and g be as in Theorem 7(ii), and suppose that $X_n^{(j)} \xrightarrow[n \to \infty]{P} X_j, j = 1, \ldots, k$. Then $g(X_n^{(1)}, \ldots, X_n^{(k)}) \xrightarrow[n \to \infty]{P} g(X_1, \ldots, X_k)$.

PROOF

i) We have $P(X \in \mathbb{R}) = 1$, and if $M_n \uparrow \infty (M_n > 0)$, then $P(X \in [-M_n, M_n]) \xrightarrow[n \to \infty]{} 1$. Thus there exist n_0 sufficiently large such that

$$P\bigl(\bigl[X \in (-\infty, -M_{n_0})\bigr] + \bigl[X \in (M_{n_0}, \infty)\bigr]\bigr) = P(|X| > M_{n_0}) < \varepsilon/2 \ (M_{n_0} > 1).$$

Define $M = M_{n_0}$; we then have

$$P(|X| > M) < \varepsilon/2.$$

g being continuous in $\mathbb{R}$, is uniformly continuous in $[-2M, 2M]$. Thus for every $\varepsilon > 0$, there exists $\delta(\varepsilon, M) = \delta(\varepsilon) \ (<1)$ such that $|g(x') - g(x'')| < \varepsilon$ for all $x', x'' \in [-2M, 2M]$ with $|x' - x''| < \delta(\varepsilon)$. From $X_n \xrightarrow[n \to \infty]{P} X$ we have that there exists $N(\varepsilon) > 0$ such that

$$P\bigl[|X_n - X| \geq \delta(\varepsilon)\bigr] < \varepsilon/2, \quad n \geq N(\varepsilon).$$

Set

$$A_1 = \bigl[|X| \leq M\bigr], \quad A_2(n) = \bigl[|X_n - X| < \delta(\varepsilon)\bigr],$$

and

$$A_3(n) = \bigl[|g(X_n) - g(X)| < \varepsilon\bigr] \quad (\text{for } n \geq N(\varepsilon)).$$

Then it is easily seen that on $A_1 \cap A_2(n)$, we have $-2M < X < 2M, -2M < X_n < 2M$, and hence

$$A_1 \cap A_2(n) \subseteq A_3(n),$$

which implies that

$$A_3^c(n) \subseteq A_1^c \cup A_2^c(n).$$

Hence

$$P\bigl[A_3^c(n)\bigr] \leq P(A_1^c) + P\bigl[A_2^c(n)\bigr] \leq \varepsilon/2 + \varepsilon/2 = \varepsilon$$
$$(\text{for } n \geq N(\varepsilon)).$$

That is, for $n \geq N(\varepsilon)$,

$$P\bigl[|g(X_n) - g(X)| \geq \varepsilon\bigr] < \varepsilon.$$

The proof is completed. (See also Exercise 8.6.1.)

ii) It is carried out along the same lines as the proof of part (i). (See also Exercises 8.5.3 and 8.6.2.) ▲

The following corollary to Theorem 7' is of wide applicability.

COROLLARY If $X_n \xrightarrow[n \to \infty]{P} X$, $Y_n \xrightarrow[n \to \infty]{P} Y$, then

  i) $X_n + Y_n \xrightarrow[n \to \infty]{P} X + Y$.
  ii) $aX_n + bY_n \xrightarrow[n \to \infty]{P} aX + bY$ (a, b constants).
  iii) $X_n Y_n \xrightarrow[n \to \infty]{P} XY$.
  iv) $X_n/Y_n \xrightarrow[n \to \infty]{P} X/Y$, provided $P(Y_n \neq 0) = P(Y \neq 0) = 1$.

PROOF In suffices to take g as follows and apply the second part of the theorem:

  i) $g(x, y) = x + y$,
  ii) $g(x, y) = ax + by$,
  iii) $g(x, y) = xy$,
  iv) $g(x, y) = x/y$, $y \neq 0$. ▲

The following is in itself a very useful theorem.

THEOREM 8 If $X_n \xrightarrow[n \to \infty]{d} X$ and $Y_n \xrightarrow[n \to \infty]{P} c$, constant, then

  i) $X_n + Y_n \xrightarrow[n \to \infty]{d} X + c$,
  ii) $X_n Y_n \xrightarrow[n \to \infty]{d} cX$,
  iii) $X_n/Y_n \xrightarrow[n \to \infty]{d} X/c$, provided $P(Y_n \neq 0) = 1$, $c \neq 0$.

Equivalently,

  i) $P(X_n + Y_n \leq z) = F_{X_n + Y_n}(z) \xrightarrow[n \to \infty]{} F_{X+c}(z)$
  $= P(X + c \leq z) = P(X \leq z - c) = F_X(z - c)$;

  ii) $P(X_n Y_n \leq z) = F_{X_n Y_n}(z) \xrightarrow[n \to \infty]{} F_{cX}(z)$
  $= P(cX \leq z) = \begin{cases} P\left(X \leq \dfrac{z}{c}\right) = F_X\left(\dfrac{z}{c}\right), & c > 0 \\ P\left(X \geq \dfrac{z}{c}\right) = 1 - F_X\left(\dfrac{z}{c}-\right), & c < 0; \end{cases}$

  iii) $P\left(\dfrac{X_n}{Y_n} \leq z\right) = F_{X_n/Y_n}(z) \xrightarrow[n \to \infty]{} F_{X/c}(z)$
  $= P\left(\dfrac{X}{c} \leq z\right) = \begin{cases} P(X \leq cz) = F_X(cz), & c > 0 \\ P(X \geq cz) = 1 - F_X(cz-), & c < 0, \end{cases}$

provided $P(Y_n \neq 0) = 1$.

REMARK 11 Of course, $F_X(z/c-) = F_X(z/c)$ and $F_X(cz-) = F_X(cz)$, if F is continuous.

PROOF As an illustration of how the proof of this theorem is carried out, we proceed to establish (iii) under the (unnecessary) additional assumption that F_X is continuous and for the case that $c > 0$. The case where $c < 0$ is treated similarly.

We first notice that $Y_n \xrightarrow[n\to\infty]{P} c\ (>0)$ implies that $P(Y_n > 0) \xrightarrow[n\to\infty]{} 1$. In fact, $Y_n \xrightarrow[n\to\infty]{P} c$ is equivalent to $P(|Y_n - c| \leq \varepsilon) \xrightarrow[n\to\infty]{} 1$ for every $\varepsilon > 0$, or $P(c - \varepsilon \leq Y_n \leq c + \varepsilon) \xrightarrow[n\to\infty]{} 1$. Thus, if we choose $\varepsilon < c$, we obtain the result. Next, since $P(Y_n \neq 0) = 1$, we may divide by Y_n except perhaps on a null set. Outside this null set, we have then

$$P\left(\frac{X_n}{Y_n} \leq z\right) = P\left[\left(\frac{X_n}{Y_n} \leq z\right) \cap (Y_n > 0)\right] + P\left[\left(\frac{X_n}{Y_n} \leq z\right) \cap (Y_n < 0)\right]$$

$$\leq P\left[\left(\frac{X_n}{Y_n} \leq z\right) \cap (Y_n > 0)\right] + P(Y_n < 0).$$

In the following, we will be interested in the limit of the above probabilities as $n \to \infty$. Since $P(Y_n < 0) \to 0$, we assume that $Y_n > 0$. We have then

$$\left(\frac{X_n}{Y_n} \leq z\right) = \left(\frac{X_n}{Y_n} \leq z\right) \cap (|Y_n - c| \geq \varepsilon) + \left(\frac{X_n}{Y_n} \leq z\right) \cap (|Y_n - c| < \varepsilon)$$

$$\subseteq (|Y_n - c| \geq \varepsilon) \cup (X_n \leq zY_n) \cap (|Y_n - c| < \varepsilon).$$

But $|Y_n - c| < \varepsilon$ is equivalent to $c - \varepsilon < Y_n < c + \varepsilon$. Therefore

$$(X_n \leq zY_n) \cap (|Y_n - c| < \varepsilon) \subseteq [X_n \leq z(c + \varepsilon)], \quad \text{if } z \geq 0,$$

and

$$(X_n \leq zY_n) \cap (|Y_n - c| < \varepsilon) \subseteq [X_n \leq z(c - \varepsilon)], \quad \text{if } z < 0.$$

That is, for every $z \in \mathbb{R}$,

$$(X_n \leq zY_n) \cap (|Y_n - c| < \varepsilon) \subseteq [X_n \leq z(c \pm \varepsilon)]$$

and hence

$$\left(\frac{X_n}{Y_n} \leq z\right) \subseteq (|Y_n - c| \geq \varepsilon) \cup [X_n \leq z(c \pm \varepsilon)], \quad z \in \mathbb{R}.$$

Thus

$$P\left(\frac{X_n}{Y_n} \leq z\right) \leq P(|Y_n - c| \geq \varepsilon) + P[X_n \leq z(c \pm \varepsilon)], \quad z \in \mathbb{R}.$$

Letting $n \to \infty$ and taking into consideration the fact that $P(|Y_n - c| \geq \varepsilon) \to 0$ and $P[X_n \leq z(c \pm \varepsilon)] \to F_X[z(c \pm \varepsilon)]$, we obtain

$$\limsup_{n\to\infty} P\left(\frac{X_n}{Y_n} \leq z\right) \leq F_X[z(c \pm \varepsilon)], \quad z \in \mathbb{R}.$$

Since, as $\varepsilon \to 0$, $F_X[z(c \pm \varepsilon)] \to F_X(zc)$, we have

$$\limsup_{n \to \infty} P\left(\frac{X_n}{Y_n} \leq z\right) \leq F_X(zc), \quad z \in \mathbb{R}. \tag{6}$$

Next,

$$[X_n \leq z(c \pm \varepsilon)] = [X_n \leq z(c \pm \varepsilon)] \cap (|Y_n - c| \geq \varepsilon) + [X_n \leq z(c \pm \varepsilon)]$$
$$\cap (|Y_n - c| < \varepsilon) \subseteq (|Y_n - c| \geq \varepsilon)$$
$$\cup [X_n \leq z(c \pm \varepsilon)] \cap (|Y_n - c| < \varepsilon).$$

By choosing $\varepsilon < c$, we have that $|Y_n - c| < \varepsilon$ is equivalent to $0 < c - \varepsilon < Y_n < c + \varepsilon$ and hence

$$[X_n \leq z(c - \varepsilon)] \cap (|Y_n - c| < \varepsilon) \subseteq \left(\frac{X_n}{Y_n} \leq z\right), \quad \text{if} \quad z \geq 0,$$

and

$$[X_n \leq z(c + \varepsilon)] \cap (|Y_n - c| < \varepsilon) \subseteq \left(\frac{X_n}{Y_n} \leq z\right), \quad \text{if} \quad z < 0.$$

That is, for every $z \in \mathbb{R}$,

$$[X_n \leq z(c \pm \varepsilon)] \cap (|Y_n - c| < \varepsilon) \subseteq \left(\frac{X_n}{Y_n} \leq z\right)$$

and hence

$$[X_n \leq z(c \pm \varepsilon)] \cap (|Y_n - c| \geq \varepsilon) \cup \left(\frac{X_n}{Y_n} \leq z\right), \quad z \in \mathbb{R}.$$

Thus

$$P[X_n \leq z(c \pm \varepsilon)] \leq P(|Y_n - c| \geq \varepsilon) + P\left(\frac{X_n}{Y_n} \leq z\right).$$

Letting $n \to \infty$ and taking into consideration the fact that $P(|Y_n - c| \geq \varepsilon) \to 0$ and $P[X_n \leq z(c \pm \varepsilon)] \to F_X[z(c \pm \varepsilon)]$, we obtain

$$F_X[z(c \pm \varepsilon)] \leq \liminf_{n \to \infty} P\left(\frac{X_n}{Y_n} \leq z\right), \quad z \in \mathbb{R}.$$

Since, as $\varepsilon \to 0$, $F_X[z(c \pm \varepsilon)] \to F_X(zc)$, we have

$$F_X(zc) \leq \liminf_{n \to \infty} P\left(\frac{X_n}{Y_n} \leq z\right), \quad z \in \mathbb{R}. \tag{7}$$

Relations (6) and (7) imply that $\lim_{n \to \infty} P(X_n/Y_n \leq z)$ exists and is equal to

$$F_X(zc) = P(X \leq zc) = P\left(\frac{X}{c} \leq z\right) = F_{X/c}(z).$$

Thus

$$P\left(\frac{X_n}{Y_n} \leq z\right) = F_{X_n/Y_n}(z) \xrightarrow[n\to\infty]{} F_{X/c}(z), \quad z \in \mathbb{R},$$

as was to be seen. ▲

REMARK 12 Theorem 8 is known as Slutsky's theorem.

Now, if X_j, $j = 1, \ldots, n$, are i.i.d. r.v.'s, we have seen that the sample variance

$$S_n^2 = \frac{1}{n}\sum_{j=1}^n (X_j - \overline{X}_n)^2 = \frac{1}{n}\sum_{j=1}^n X_j^2 - \overline{X}_n^2.$$

Next, the r.v.'s X_j^2, $j = 1, \ldots, n$ are i.i.d., since the X's are, and

$$E(X_j^2) = \sigma^2(X_j) + (EX_j)^2 = \sigma^2 + \mu^2, \quad \text{if} \quad \mu = E(X_j), \quad \sigma^2 = \sigma^2(X_j)$$

(which are assumed to exist). Therefore the SLLN and WLLN give the result that

$$\frac{1}{n}\sum_{j=1}^n X_j^2 \xrightarrow[n\to\infty]{} \sigma^2 + \mu^2 \quad \text{a.s.}$$

and also in probability. On the other hand, $\overline{X}_n \xrightarrow[n\to\infty]{} \mu$ a.s. and also in probability, and hence $\overline{X}_n^2 \xrightarrow[n\to\infty]{} \mu^2$ a.s. and also in probability (by Theorems 7(i) and 7'(i)). Thus

$$\frac{1}{n}\sum_{j=1}^n X_j^2 - \overline{X}_n^2 \to \sigma^2 + \mu^2 - \mu^2 = \sigma^2 \quad \text{a.s.}$$

and also in probability (by the same theorems just referred to). So we have proved the following theorem.

THEOREM 9 Let X_j, $j = 1, \ldots, n$, be i.i.d. r.v.'s with $E(X_j) = \mu$, $\sigma^2(X_j) = \sigma^2$, $j = 1, \ldots, n$. Then $S_n^2 \xrightarrow[n\to\infty]{} \sigma^2$ a.s. and also in probability.

REMARK 13 Of course,

$$S_n^2 \xrightarrow[n\to\infty]{P} \sigma^2 \quad \text{implies} \quad \frac{n}{n-1}\frac{S_n^2}{\sigma^2} \xrightarrow[n\to\infty]{P} 1,$$

since $n/(n-1) \xrightarrow[n\to\infty]{} 1$.

COROLLARY TO THEOREM 8 If $X_1, \ldots, X_n$ are i.i.d. r.v.'s with mean μ and (positive) variance σ^2, then

$$\frac{\sqrt{n-1}(\overline{X}_n - \mu)}{S_n} \xrightarrow[n\to\infty]{d} N(0,1) \quad \text{and also} \quad \frac{\sqrt{n}(\overline{X}_n - \mu)}{S_n} \xrightarrow[n\to\infty]{d} N(0,1).$$

PROOF In fact,

$$\frac{\sqrt{n}(\overline{X}_n - \mu)}{\sigma} \xrightarrow[n\to\infty]{d} N(0,1),$$

by Theorem 3, and
$$\frac{\sqrt{n}}{\sqrt{n-1}} \frac{S_n}{\sigma} \xrightarrow[n\to\infty]{P} 1,$$
by Remark 13. Hence the quotient of these r.v.'s which is
$$\frac{\sqrt{n-1}(\overline{X}_n - \mu)}{S_n}$$
converges in distribution to $N(0, 1)$ as $n \to \infty$, by Theorem 9. ▲

The following result is based on theorems established above and it is of significant importance.

THEOREM 10 For $n = 1, 2, \ldots$, let X_n and X be r.v.'s, let $g: \mathbb{R} \to \mathbb{R}$ be differentiable, and let its derivative $g'(x)$ be continuous at a point d. Finally, let c_n be constants such that $0 \neq c_n \to \infty$, and let $c_n(X_n - d) \xrightarrow{d} X$ as $n \to \infty$. Then $c_n[g(X_n) - g(d)] \xrightarrow{d} g'(d)X$ as $n \to \infty$.

PROOF In this proof, all limits are taken as $n \to \infty$. By assumption, $c_n(X_n - d) \xrightarrow{d} X$ and $c_n^{-1} \to 0$. Then, by Theorem 8(ii), $X_n - d \xrightarrow{d} 0$, or equivalently, $X_n - d \xrightarrow{P} 0$, and hence, by Theorem 7'(i),
$$|X_n - d| \xrightarrow{P} 0. \qquad (8)$$
Next, expand $g(X_n)$ around d according to Taylor's formula in order to obtain
$$g(X_n) = g(d) + (X_n - d)g'(X_n^*),$$
where X_n^* is an r.v. lying between d and X_n. Hence
$$c_n[g(X_n) - g(d)] = c_n(X_n - d)g'(X_n^*). \qquad (9)$$
However, $|X_n^* - d| \leq |X_n - d| \xrightarrow{P} 0$ by (8), so that $X_n^* \xrightarrow{P} d$, and therefore, by Theorem 7'(i) again,
$$g(X_n^*) \xrightarrow{P} g(d). \qquad (10)$$
By assumption, convergence (10) and Theorem 8(ii), we have $c_n(X_n - d)g'(X_n^*) \xrightarrow{d} g'(d)X$. This result and relation (9) complete the proof of the theorem. ▲

COROLLARY Let the r.v.'s $X_1, \ldots, X_n$ be i.i.d. with mean $\mu \in \mathbb{R}$ and variance $\sigma^2 \in (0, \infty)$, and let $g: \mathbb{R} \to \mathbb{R}$ be differentiable with derivative continuous at μ. Then, as $n \to \infty$,
$$\sqrt{n}[g(\overline{X}_n) - g(\mu)] \xrightarrow{d} N\left(0, [\sigma g'(\mu)]^2\right).$$

PROOF By the CLT, $\sqrt{n}(\overline{X}_n - \mu) \xrightarrow{d} X \sim N(0, \sigma^2)$, so that the theorem applies and gives

$$\sqrt{n}\left[g(\overline{X}_n) - g(\mu)\right] \xrightarrow{d} g'(\mu)X \sim N\left(0, \left[\sigma g'(\mu)\right]^2\right). \quad \blacktriangle$$

APPLICATION If the r.v.'s X_j, $j = 1, \ldots, n$ in the corollary are distributed as $B(1, p)$, then, as $n \to \infty$,

$$\sqrt{n}\left[\overline{X}_n(1 - \overline{X}_n) - pq\right] \xrightarrow{d} N\left(0, pq(1 - 2p)^2\right).$$

Here $\mu = p$, $\sigma^2 = pq$, and $g(x) = x(1 - x)$, so that $g'(x) = 1 - 2x$. The result follows.

Exercises

8.5.1 Use Theorem 8(ii) in order to show that if the CLT holds, then so does the WLLN.

8.5.2 Refer to the proof of Theorem 7′(i) and show that on the set $A_1 \cap A_2(n)$, we actually have $-2M < X < 2M$.

8.5.3 Carry out the proof of Theorem 7′(ii). (Use the usual Euclidean distance in $\mathbb{R}^k$.)

8.6* Pólya's Lemma and Alternative Proof of the WLLN

The following lemma is an analytical result of interest in its own right. It was used in the corollary to Theorem 3 to conclude uniform convergence.

LEMMA 1 (Pólya). Let F and $\{F_n\}$ be d.f.'s such that $F_n(x) \xrightarrow[n \to \infty]{} F(x)$, $x \in \mathbb{R}$, and let F be continuous. Then the convergence is uniform in $x \in \mathbb{R}$. That is, for every $\varepsilon > 0$ there exists $N(\varepsilon) > 0$ such that $n \geq N(\varepsilon)$ implies that $|F_n(x) - F(x)| < \varepsilon$ for every $x \in \mathbb{R}$.

PROOF Since $F(x) \to 0$ as $x \to -\infty$, and $F(x) \to 1$, as $x \to \infty$, there exists an interval $[\alpha, \beta]$ such that

$$F(\alpha) < \varepsilon/2, \quad F(\beta) > 1 - \varepsilon/2. \tag{11}$$

The continuity of F implies its uniform continuity in $[\alpha, \beta]$. Then there is a finite partition $\alpha = x_1 < x_2 < \cdots < x_r = \beta$ of $[\alpha, \beta]$ such that

$$F(x_{j+1}) - F(x_j) < \varepsilon/2, \quad j = 1, \ldots, r-1. \tag{12}$$

Next, $F_n(x_j) \xrightarrow[n \to \infty]{} F(x_j)$ implies that there exists $N_j(\varepsilon) > 0$ such that for all $n \geq N_j(\varepsilon)$,

$$\left|F_n(x_j) - F(x_j)\right| < \varepsilon/2, \quad j = 1, \ldots, r.$$

By taking

$$n \geq N(\varepsilon) = \max\left(N_1(\varepsilon), \ldots, N_r(\varepsilon)\right),$$

we then have that

$$|F_n(x_j) - F(x_j)| < \varepsilon/2, \quad j = 1, \ldots, r. \tag{13}$$

Let $x_0 = -\infty$, $x_{r+1} = \infty$. Then by the fact that $F(-\infty) = 0$ and $F(\infty) = 1$, relation (11) implies that

$$F(x_1) - F(x_0) < \varepsilon/2, \quad F(x_{r+1}) - F(x_r) < \varepsilon/2. \tag{14}$$

Thus, by means of (12) and (14), we have that

$$|F(x_{j+1}) - F(x_j)| < \varepsilon/2, \quad j = 0, 1, \ldots, r. \tag{15}$$

Also (13) trivially holds for $j = 0$ and $j = r + 1$; that is, we have

$$|F_n(x_j) - F(x_j)| < \varepsilon/2, \quad j = 0, 1, \ldots, r+1. \tag{16}$$

Next, let x be any real number. Then $x_j \leq x < x_{j+1}$ for some $j = 0, 1, \ldots, r$. By (15) and (16) and for $n \geq N(\varepsilon)$, we have the following string of inequalities:

$$F(x_j) - \varepsilon/2 < F_n(x_j) \leq F_n(x) \leq F_n(x_{j+1}) < F(x_{j+1}) + \varepsilon/2$$
$$< F(x_j) + \varepsilon \leq F(x) + \varepsilon \leq F(x_{j+1}) + \varepsilon.$$

Hence

$$0 \leq F(x) + \varepsilon - F_n(x) \leq F(x_{j+1}) + \varepsilon - F(x_j) + \varepsilon/2 < 2\varepsilon$$

and therefore $|F_n(x) - F(x)| < \varepsilon$. Thus for $n \geq N(\varepsilon)$, we have

$$|F_n(x) - F(x)| < \varepsilon \quad \text{for every} \quad x \in \mathbb{R}. \tag{17}$$

Relation (17) concludes the proof of the lemma. ▲

Below, a proof of the WLLN (Theorem 5) is presented without using ch.f.'s. The basic idea is that of suitably truncating the r.v.'s involved, and is due to Khintchine; it was also used by Markov.

ALTERNATIVE PROOF OF THEOREM 5 We proceed as follows: For any $\delta > 0$, we define

$$Y_j(n) = Y_j = \begin{cases} X_j, & \text{if } |X_j| \leq \delta \cdot n \\ 0, & \text{if } |X_j| > \delta \cdot n \end{cases}$$

and

$$Z_j(n) = Z_j = \begin{cases} 0, & \text{if } |X_j| \leq \delta \cdot n \\ X_j, & \text{if } |X_j| > \delta \cdot n, \end{cases} \quad j = 1, \ldots, n.$$

Then, clearly, $X_j = Y_j + Z_j$, $j = 1, \ldots, n$. Let us restrict ourselves to the continuous case and let f be the (common) p.d.f. of the X's. Then,

$$\sigma^2(Y_j) = \sigma^2(Y_1)$$
$$= E(Y_1^2) - (EY_1)^2 \leq E(Y_1^2)$$
$$= E\{X_1^2 \cdot I_{[|X_1| \leq \delta \cdot n]}(X_1)\}$$
$$= \int_{-\infty}^{\infty} x^2 I_{[|x| \leq \delta \cdot n]}(x) f(x) dx$$
$$= \int_{-\delta \cdot n}^{\delta \cdot n} x^2 f(x) dx \leq \delta \cdot n \int_{-\delta \cdot n}^{\delta \cdot n} |x| f(x) dx \leq \delta \cdot n \int_{-\infty}^{\infty} |x| f(x) dx$$
$$= \delta \cdot n E|X_1|;$$

that is,
$$\sigma^2(Y_j) \leq \delta \cdot n \cdot E|X_1|. \qquad (18)$$

Next,
$$E(Y_j) = E(Y_1) = E\{X_1 I_{[|X_1| \leq \delta \cdot n]}(X_1)\}$$
$$= \int_{-\infty}^{\infty} x I_{[|x| \leq \delta \cdot n]}(x) f(x) dx.$$

Now,
$$\left| x I_{[|x| \leq \delta \cdot n]}(x) f(x) \right| < |x| f(x), \quad x I_{[|x| \leq \delta \cdot n]}(x) f(x) \xrightarrow[n \to \infty]{} x f(x),$$

and
$$\int_{-\infty}^{\infty} |x| f(x) dx < \infty.$$

Therefore
$$\int_{-\infty}^{\infty} x I_{[|x| \leq \delta \cdot n]}(x) f(x) dx \xrightarrow[n \to \infty]{} \int_{-\infty}^{\infty} x f(x) dx = \mu$$

by Lemma C of Chapter 6; that is,
$$E(Y_j) \xrightarrow[n \to \infty]{} \mu. \qquad (19)$$

$$P\left[\left|\frac{1}{n}\sum_{j=1}^{n} Y_j - EY_j\right| \geq \varepsilon\right] = P\left[\left|\sum_{j=1}^{n} Y_j - E\left(\sum_{j=1}^{n} Y_j\right)\right| \geq n\varepsilon\right]$$
$$\leq \frac{1}{n^2 \varepsilon^2} \sigma^2\left(\sum_{j=1}^{n} Y_j\right)$$
$$= \frac{n \sigma^2(Y_1)}{n^2 \varepsilon^2}$$
$$\leq \frac{n \delta \cdot n \cdot E|X_1|}{n^2 \varepsilon^2}$$
$$= \frac{\delta}{\varepsilon^2} E|X_1|$$

by (18); that is,

$$P\left[\left|\frac{1}{n}\sum_{j=1}^{n}Y_j - EY_1\right| \geq \varepsilon\right] \leq \frac{\delta}{\varepsilon^2}E|X_1|. \quad (20)$$

Thus,

$$P\left[\left|\frac{1}{n}\sum_{j=1}^{n}Y_j - \mu\right| \geq 2\varepsilon\right] = P\left[\left|\left(\frac{1}{n}\sum_{j=1}^{n}Y_1 - E(Y_j)\right) + (E(Y_1) - \mu)\right| \geq 2\varepsilon\right]$$

$$\leq P\left[\left|\frac{1}{n}\sum_{j=1}^{n}Y_j - EY_1\right| + |EY_1 - \mu| \geq 2\varepsilon\right]$$

$$\leq P\left[\left|\frac{1}{n}\sum_{j=1}^{n}Y_j - EY_1\right| \geq \varepsilon\right] + P[|EY_1 - \mu| \geq \varepsilon]$$

$$\leq \frac{\delta}{\varepsilon^2}E|X_1|$$

for n sufficiently large, by (19) and (20); that is,

$$P\left[\left|\frac{1}{n}\sum_{j=1}^{n}Y_j - \mu\right| \geq 2\varepsilon\right] \leq \frac{\delta}{\varepsilon^2}E|X_1| \quad (21)$$

for n large enough. Next,

$$P(Z_j \neq 0) = P(|Z_j| > \delta \cdot n)$$
$$= P(|X_j| > \delta \cdot n)$$
$$= \int_{-\infty}^{-\delta \cdot n} f(x)dx + \int_{\delta \cdot n}^{\infty} f(x)dx$$
$$= \int_{(|x|>\delta \cdot n)} f(x)dx$$
$$= \int_{(|x|>\int in>1)} f(x)dx$$
$$< \int_{(|x|>\delta \cdot n)} \frac{|x|}{\delta \cdot n}f(x)dx$$
$$= \frac{1}{\delta \cdot n}\int_{(|x|>\delta \cdot n)}|x|f(x)dx$$
$$< \frac{1}{\delta \cdot n}\delta^2$$
$$= \frac{\delta}{n}, \quad \text{since } \int_{(|x|>\delta \cdot n)}|x|f(x)dx < \delta^2$$

for n sufficiently large. So $P(Z_j \neq 0) \leq \delta/n$ and hence

$$P\left[\sum_{j=1}^{n}Z_j \neq 0\right] \leq nP(Z_j \neq 0) \leq \delta \quad (22)$$

for n sufficiently large. Thus,

$$P\left[\left|\frac{1}{n}\sum_{j=1}^{n}X_j - \mu\right| \geq 4\varepsilon\right] = P\left[\left|\frac{1}{n}\sum_{j=1}^{n}Y_j + \frac{1}{n}\sum_{j=1}^{n}Z_j - \mu\right| \geq 4\varepsilon\right]$$

$$\leq P\left[\left|\frac{1}{n}\sum_{j=1}^{n}Y_j - \mu\right| + \left|\frac{1}{n}\sum_{j=1}^{n}Z_j\right| \geq 4\varepsilon\right]$$

$$\leq P\left[\left|\frac{1}{n}\sum_{j=1}^{n}Y_j - \mu\right| \geq 2\varepsilon\right] + P\left[\left|\frac{1}{n}\sum_{j=1}^{n}Z_j\right| \geq 2\varepsilon\right]$$

$$\leq P\left[\left|\frac{1}{n}\sum_{j=1}^{n}Y_j - \mu\right| \geq 2\varepsilon\right] + P\left[\sum_{j=1}^{n}Z_j \neq 0\right]$$

$$\leq \frac{\delta}{\varepsilon^2}E|X_1| + \delta$$

for n sufficiently large, by (21), (22).

Replacing δ by ε^3, for example, we get

$$P\left[\left|\frac{1}{n}\sum_{j=1}^{n}X_j - \mu\right| \geq 4\varepsilon\right] \leq \varepsilon E|X_1| + \varepsilon^3$$

for n sufficiently large. Since this is true for every $\varepsilon > 0$, the result follows. ▲

This section is concluded with a result relating convergence in probability and a.s. convergence. More precisely, in Remark 3, it was stated that $X_n \xrightarrow[n\to\infty]{P} X$ does *not* necessarily imply that $X_n \xrightarrow[n\to\infty]{\text{a.s.}} X$. However, the following is always true.

THEOREM 11 If $X_n \xrightarrow[n\to\infty]{P} X$, then there is a subsequence $\{n_k\}$ of $\{n\}$ (that is, $n_k \uparrow \infty$, $k \to \infty$) such that $X_{n_k} \xrightarrow[n\to\infty]{\text{a.s.}} X$.

PROOF Omitted.

As an application of Theorem 11, refer to Example 2 and consider the subsequence of r.v.'s $\{X_{2^k-1}\}$, where

$$X_{2^k-1} = I_{\left(\frac{2^{k-1}-1}{2^{k-1}}, 1\right]}.$$

Then for $\varepsilon > 0$ and large enough k, so that $1/2^{k-1} < \varepsilon$, we have

$$P(|X_{2^k-1}| > \varepsilon) = P(X_{2^k-1} = 1) = \frac{1}{2^{k-1}} < \varepsilon.$$

Hence the subsequence $\{X_{2^k-1}\}$ of $\{X_n\}$ converges to 0 in probability.

Exercises

8.6.1 Use Theorem 11 in order to prove Theorem 7′(i).

8.6.2 Do likewise in order to establish part (ii) of Theorem 7′.

Chapter 9

Transformations of Random Variables and Random Vectors

9.1 The Univariate Case

The problem we are concerned with in this section in its simplest form is the following:

Let X be an r.v. and let h be a (measurable) function on $\mathbb{R}$ into $\mathbb{R}$, so that $Y = h(X)$ is an r.v. Given the distribution of X, we want to determine the distribution of Y. Let P_X, P_Y be the distributions of X and Y, respectively. That is, $P_X(B) = P(X \in B)$, $P_Y(B) = P(Y \in B)$, B (Borel) subset of $\mathbb{R}$. Now $(Y \in B) = [h(X) \in B] = (X \in A)$, where $A = h^{-1}(B) = \{x \in \mathbb{R}; h(x) \in B\}$. Therefore $P_Y(B) = P(Y \in B) = P(X \in A) = P_X(A)$. Thus we have the following theorem.

THEOREM 1 Let X be an r.v. and let $h: \mathbb{R} \to \mathbb{R}$ be a (measurable) function, so that $Y = h(X)$ is an r.v. Then the distribution P_Y of the r.v. Y is determined by the distribution P_X of the r.v. X as follows: for any (Borel) subset B of $\mathbb{R}$, $P_Y(B) = P_X(A)$, where $A = h^{-1}(B)$.

9.1.1 Application 1: Transformations of Discrete Random Variables

Let X be a discrete r.v. taking the values $x_j, j = 1, 2, \ldots$, and let $Y = h(X)$. Then Y is also a discrete r.v. taking the values $y_j, j = 1, 2, \ldots$. We wish to determine $f_Y(y_j) = P(Y = y_j), j = 1, 2, \ldots$. By taking $B = \{y_j\}$, we have

$$A = \{x_i; h(x_i) = y_j\},$$

and hence

$$f_Y(y_j) = P(Y = y_j) = P_Y(\{y_j\}) = P_X(A) = \sum_{x_i \in A} f_X(x_i),$$

where
$$f_X(x_i) = P(X = x_i).$$

EXAMPLE 1 Let X take on the values $-n, \ldots, -1, 1, \ldots, n$ each with probability $1/2n$, and let $Y = X^2$. Then Y takes on the values $1, 4, \ldots, n^2$ with probability found as follows: If $B = \{r^2\}$, $r = \pm 1, \ldots, \pm n$, then
$$A = h^{-1}(B) = (x^2 = r^2) = (x = -r \ \text{or} \ x = r)$$
$$= (x = r) + (x = -r) = \{-r\} + \{r\}.$$

Thus
$$P_Y(B) = P_X(A) = P_X(\{-r\}) + P_X(\{r\}) = \frac{1}{2n} + \frac{1}{2n} = \frac{1}{n}.$$

That is,
$$P(Y = r^2) = 1/n, \quad r = 1, \ldots, n.$$

EXAMPLE 2 Let X be $P(\lambda)$ and let $Y = h(X) = X^2 + 2X - 3$. Then Y takes on the values
$$\{y = x^2 + 2x - 3; \ x = 0, 1, \ldots\} = \{-3, 0, 5, 12, \ldots\}.$$

From
$$x^2 + 2x - 3 = y,$$

we get
$$x^2 + 2x - (y + 3) = 0, \quad \text{so that} \quad x = -1 \pm \sqrt{y + 4}.$$

Hence $x = -1 + \sqrt{y + 4}$, the root $-1 - \sqrt{y + 4}$ being rejected, since it is negative. Thus, if $B = \{y\}$, then
$$A = h^{-1}(B) = \{-1 + \sqrt{y + 4}\},$$

and
$$P_Y(B) = P(Y = y) = P_X(A) = \frac{e^{-\lambda} \cdot \lambda^{-1 + \sqrt{y+4}}}{(-1 + \sqrt{y+4})!}.$$

For example, for $y = 12$, we have $P(Y = 12) = e^{-\lambda}\lambda^3/3!$.

It is a fact, proved in advanced probability courses, that the distribution P_X of an r.v. X is uniquely determined by its d.f. X. The same is true for r. vectors. (A first indication that such a result is feasible is provided by Lemma 3 in Chapter 7.) Thus, in determining the distribution P_Y of the r.v. Y above, it suffices to determine its d.f., F_Y. This is easily done if the transformation h is one-to-one from S onto T and monotone (increasing or decreasing), where S is the set of values of X for which f_X is positive and T is the image of S, under h: that is, the set to which S is transformed by h. By "one-to-one" it is meant that for each $y \in T$, there is only one $x \in S$ such that $h(x) = y$. Then the inverse transformation, h^{-1}, exists and, of course, $h^{-1}[h(x)] = x$. For such a transformation, we have

$$F_Y(y) = P(Y \leq y) = P[h(X) \leq y]$$
$$= P\{h^{-1}[h(X)] \leq h^{-1}(y)\}$$
$$= P(X \leq x) = F_X(x),$$

where $x = h^{-1}(y)$ and h is increasing. In the case where h is decreasing, we have

$$F_Y(y) = P[h(X) \leq y] = P\{h^{-1}[h(X)] \geq h^{-1}(y)\}$$
$$= P[X \geq h^{-1}(y)] = P(X \geq x)$$
$$= 1 - P(X < x) = 1 - F_X(x-),$$

where $F_X(x-)$ is the limit from the left of F_X at x; $F_X(x-) = \lim F_X(y)$, $y \uparrow x$.

REMARK 1 Figure 9.1 points out why the direction of the inequality is reversed when h^{-1} is applied if h in monotone *decreasing*.

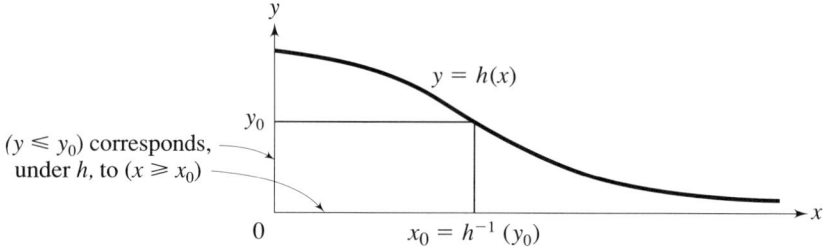

Figure 9.1

Thus we have the following corollary to Theorem 1.

COROLLARY Let $h: S \to T$ be one-to-one and monotone. Then $F_Y(y) = F_X(x)$ if h is increasing, and $F_Y(y) = 1 - F_X(x-)$ if h is decreasing, where $x = h^{-1}(y)$ in either case.

REMARK 2 Of course, it is possible that the d.f. F_Y of Y can be expressed in terms of the d.f. F_X of X even though h does not satisfy the requirements of the corollary above. Here is an example of such a case.

EXAMPLE 3 Let $Y = h(X) = X^2$. Then for $y \geq 0$,

$$F_Y(y) = P(Y \leq y) = P[h(X) \leq y] = P(X^2 \leq y) = P\left(-\sqrt{y} \leq X \leq \sqrt{y}\right)$$
$$= P\left(X \leq \sqrt{y}\right) - P\left(X < -\sqrt{y}\right) = F_X\left(\sqrt{y}\right) - F_X\left(-\sqrt{y}-\right);$$

that is,

$$F_Y(y) = F_X\left(\sqrt{y}\right) - F_X\left(-\sqrt{y}-\right)$$

for $y \geq 0$ and, of course, it is zero for $y < 0$.

We will now focus attention on the case that X has a p.d.f. and we will determine the p.d.f. of $Y = h(X)$, under appropriate conditions.

One way of going about this problem would be to find the d.f. F_Y of the r.v. Y by Theorem 1 (take $B = (-\infty, y]$, $y \in \mathbb{R}$), and then determine the p.d.f. f_Y of Y, provided it exists, by differentiating (for the continuous case) F_Y at continuity points of f_Y. The following example illustrates the procedure.

EXAMPLE 4 In Example 3, assume that X is $N(0, 1)$, so that

$$f_X(x) = \frac{1}{\sqrt{2\pi}} e^{-x^2/2}.$$

Then, if $Y = X^2$, we know that

$$F_Y(y) = F_X(\sqrt{y}) - F_X(-\sqrt{y}), \quad y \geq 0.$$

Next,

$$\frac{d}{dy} F_X(\sqrt{y}) = f_X(\sqrt{y}) \frac{d}{dy} \sqrt{y} = \frac{1}{2\sqrt{y}} f_X(\sqrt{y}) = \frac{1}{2\sqrt{2\pi}\sqrt{y}} e^{-y/2},$$

and

$$\frac{d}{dy} F_X(-\sqrt{y}) = -\frac{1}{2\sqrt{y}} f_X(-\sqrt{y}) = -\frac{1}{2\sqrt{2\pi}\sqrt{y}} e^{-y/2},$$

so that

$$\frac{d}{dy} F_Y(y) = f_Y(y) = \frac{1}{\sqrt{y}} \frac{1}{\sqrt{2\pi}} e^{-y/2} = \frac{1}{\Gamma(\frac{1}{2}) 2^{\frac{1}{2}}} y^{\frac{1}{2}-1} e^{-y/2} \left(\sqrt{\pi} = \Gamma\left(\tfrac{1}{2}\right)\right),$$

$y \geq 0$ and zero otherwise. We recognize it as being the p.d.f. of a χ_1^2 distributed r.v. which agrees with Theorem 3, Chapter 4.

Another approach to the same problem is the following. Let X be an r.v. whose p.d.f. f_X is continuous on the set S of positivity of f_X. Let $y = h(x)$ be a (measurable) transformation defined on $\mathbb{R}$ into $\mathbb{R}$ which is one-to-one on the set S onto the set T (the image of S under h). Then the inverse transformation $x = h^{-1}(y)$ exists for $y \in T$. It is further assumed that h^{-1} is differentiable and its derivative is continuous and different from zero on T. Set $Y = h(X)$, so that Y is an r.v. Under the above assumptions, the p.d.f. f_Y of Y is given by the following expression:

$$f_Y(y) = \begin{cases} f_X[h^{-1}(y)] \left| \frac{d}{dy} h^{-1}(y) \right|, & y \in T \\ 0, & \text{otherwise.} \end{cases}$$

For a sketch of the proof, let $B = [c, d]$ be any interval in T and set $A = h^{-1}(B)$. Then A is an interval in S and

$$P(Y \in B) = P[h(X) \in B] = P(X \in B) = \int_B f_X(x) dx.$$

Under the assumptions made, the theory of changing the variable in the integral on the right-hand side above applies (see for example, T. M. Apostol,

Mathematical Analysis, Addison-Wesley, 1957, pp. 216 and 270–271) and gives

$$\int_A f_X(x)dx = \int_B f_X[h^{-1}(y)]\left|\frac{d}{dy}h^{-1}(y)\right|dy.$$

That is, for any interval B in T,

$$P(Y \in B) = \int_B f_X[h^{-1}(y)]\left|\frac{d}{dy}h^{-1}(y)\right|dy.$$

Since for (measurable) subsets B of T^c, $P(Y \in B) = P[X \in h^{-1}(B)] \leq P(X \in S^c) = 0$, it follows from the definition of the p.d.f. of an r.v. that f_Y has the expression given above. Thus we have the following theorem.

THEOREM 2 Let the r.v. X have a continuous p.d.f. f_X on the set S on which it is positive, and let $y = h(x)$ be a (measurable) transformation defined on $\mathbb{R}$ into $\mathbb{R}$, so that $Y = h(X)$ is an r.v. Suppose that h is one-to-one on S onto T (the image of S under h), so that the inverse transformation $x = h^{-1}(y)$ exists for $y \in T$. It is further assumed that h^{-1} is differentiable and its derivative is continuous and $\neq 0$ on T. Then the p.d.f. f_Y of Y is given by

$$f_Y(y) = \begin{cases} f_X[h^{-1}(y)]\left|\frac{d}{dy}h^{-1}(y)\right|, & y \in T \\ 0, & \text{otherwise.} \end{cases}$$

EXAMPLE 5 Let X be $N(\mu, \sigma^2)$ and let $y = h(x) = ax + b$, where $a, b \in \mathbb{R}$, $a \neq 0$, are constants, so that $Y = aX + b$. We wish to determine the p.d.f. of the r.v. Y.

Here the transformation $h: \mathbb{R} \to \mathbb{R}$, clearly, satisfies the conditions of Theorem 2. We have

$$h^{-1}(y) = \frac{1}{a}(y - b) \quad \text{and} \quad \frac{d}{dy}h^{-1}(y) = \frac{1}{a}.$$

Therefore,

$$f_Y(y) = \frac{1}{\sqrt{2\pi}\sigma}\exp\left[-\frac{\left(\frac{y-b}{a} - M\right)}{2\sigma^2}\right]\cdot\frac{1}{|a|}$$

$$= \frac{1}{\sqrt{2\pi}|a|\sigma}\exp\left[\frac{-[y - (a\mu + b)]^2}{2a^2\sigma^2}\right]$$

which is the p.d.f. of a normally distributed r.v. with mean $a\mu + b$ and variance $a^2\sigma^2$. Thus, if X is $N(\mu, \sigma^2)$, then $aX + b$ is $N(a\mu + b, a^2\sigma^2)$.

Now it may happen that the transformation h satisfies all the requirements of Theorem 2 except that it is not one-to-one from S onto T. Instead, the following might happen: There is a (finite) partition of S, which we denote by

$\{S_j, j = 1, \ldots, r\}$, and there are r subsets of T, which we denote by $T_j, j = 1, \ldots, r$, (note that $\cup_{j=1}^{r} T_j = T$, but the T_j's need not be disjoint) such that $h: S_j \to T_j$, $j = 1, \ldots, r$ is one-to-one. Then by an argument similar to the one used in proving Theorem 2, we can establish the following theorem.

THEOREM 3 Let the r.v. X have a continuous p.d.f. f_X on the set S on which it is positive, and let $y = h(x)$ be a (measurable) transformation defined on $\mathbb{R}$ into $\mathbb{R}$, so that $Y = h(X)$ is an r.v. Suppose that there is a partition $\{S_j, j = 1, \ldots, r\}$ of S and subsets $T_j, j = 1, \ldots, r$ of T (the image of S under h), which need not be distinct or disjoint, such that $\cup_{j=1}^{r} T_j = T$ and that h defined on each one of S_j onto T_j, $j = 1, \ldots, r$, is one-to-one. Let h_j be the restriction of the transformation h to S_j and let h_j^{-1} be its inverse, $j = 1, \ldots, r$. Assume that h_j^{-1} is differentiable and its derivative is continuous and $\neq 0$ on $T_j, j = 1, \ldots, r$. Then the p.d.f. f_Y of Y is given by

$$f_Y(y) = \begin{cases} \sum_{j=1}^{r} \delta_j(y) f_{Y_j}(y), & y \in T \\ 0, & \text{otherwise,} \end{cases}$$

where for $j = 1, \ldots, r$,

$$f_{Y_j}(y) = f_X\left[h_j^{-1}(y)\right] \left|\frac{d}{dy} h_j^{-1}(y)\right|, \quad y \in T_j,$$

and $\delta_j(y) = 1$ if $y \in T_j$ and $\delta_j(y) = 0$ otherwise.

This result simply says that for each one of the r pairs of regions (S_j, T_j), $j = 1, \ldots, r$, we work as we did in Theorem 2 in order to find

$$f_{Y_j}(y) = f_X\left[h_j^{-1}(y)\right] \left|\frac{d}{dy} h_j^{-1}(y)\right|;$$

then if a y in T belongs to k of the regions $T_j, j = 1, \ldots, r$ ($0 \leq k \leq r$), we find $f_Y(y)$ by summing up the corresponding $f_{Y_j}(y)$'s. The following example will serve to illustrate the point.

EXAMPLE 6 Consider the r.v. X and let $Y = h(X) = X^2$. We want to determine the p.d.f. f_Y of the r.v. Y. Here the conditions of Theorem 3 are clearly satisfied with

$$S_1 = (-\infty, 0], \quad S_2 = (0, \infty), \quad T_1 = [0, \infty), \quad T_2 = (0, \infty)$$

by assuming that $f_X(x) > 0$ for every $x \in \mathbb{R}$. Next,

$$h_1^{-1}(y) = -\sqrt{y}, \quad h_2^{-1}(y) = \sqrt{y},$$

so that

$$\frac{d}{dy} h_1^{-1}(y) = -\frac{1}{2\sqrt{y}}, \quad \frac{d}{dy} h_2^{-1}(y) = \frac{1}{2\sqrt{y}}, \quad y > 0.$$

Therefore,

$$f_{Y_1}(y) = f_X(-\sqrt{y}) \frac{1}{2\sqrt{y}}, \qquad f_{Y_2}(y) = f_X(\sqrt{y}) \frac{1}{2\sqrt{y}},$$

and for $y > 0$, we then get

$$f_Y(y) = \frac{1}{2\sqrt{y}} \left[f_X(\sqrt{y}) + f_X(-\sqrt{y}) \right],$$

provided $\pm\sqrt{y}$ are continuity points of f_X. In particular, if X is $N(0, 1)$, we arrive at the conclusion that $f_Y(y)$ is the p.d.f. of a χ_1^2 r.v., as we also saw in Example 4 in a different way.

Exercises

9.1.1 Let X be an r.v. with p.d.f. f given in Exercise 3.2.14 of Chapter 3 and determine the p.d.f. of the r.v. $Y = X^3$.

9.1.2 Let X be an r.v. with p.d.f. of the continuous type and set $Y = \sum_{j=1}^{n} c_j I_{B_j}(X)$, where B_j, $j = 1, \ldots, n$, are pairwise disjoint (Borel) sets and c_j, $j = 1, \ldots, n$, are constants.

  i) Express the p.d.f. of Y in terms of that of X, and notice that Y is a discrete r.v. whereas X is an r.v. of the continuous type;

  ii) If $n = 3$, X is $N(99, 5)$ and $B_1 = (95, 105)$, $B_2 = (92, 95) + (105, 107)$, $B_3 = (-\infty, 92] + [107, \infty)$, determine the distribution of the r.v. Y defined above;

  iii) If X is interpreted as a specified measurement taken on each item of a product made by a certain manufacturing process and c_j, $j = 1, 2, 3$ are the profit (in dollars) realized by selling one item under the condition that $X \in B_j$, $j = 1, 2, 3$, respectively, find the expected profit from the sale of one item.

9.1.3 Let X, Y be r.v.'s representing the temperature of a certain object in degrees Celsius and Fahrenheit, respectively. Then it is known that $Y = \frac{9}{5}X + 32$. If X is distributed as $N(\mu, \sigma^2)$, determine the p.d.f. of Y, first by determining its d.f., and secondly directly.

9.1.4 If the r.v. X is distributed as Negative Exponential with parameter λ, find the p.d.f. of each one of the r.v.'s Y, Z, where $Y = e^X$, $Z = \log X$, first by determining their d.f.'s, and secondly directly.

9.1.5 If the r.v. X is distributed as $U(\alpha, \beta)$:

  i) Derive the p.d.f.'s of the following r.v.'s: $aX + b$ $(a > 0)$, $1/(X + 1)$, $X^2 + 1$, e^X, $\log X$ (for $\alpha > 0$), first by determining their d.f.'s, and secondly directly;

  ii) What do the p.d.f.'s in part (i) become for $\alpha = 0$ and $\beta = 1$?

  iii) For $\alpha = 0$ and $\beta = 1$, let $Y = \log X$ and suppose that the r.v.'s Y_j, $j = 1, \ldots, n$, are independent and distributed as the r.v. Y. Use the ch.f. approach to determine the p.d.f. of $-\sum_{j=1}^{n} Y_j$.

9.1.6 If the r.v. X is distributed as $U(-\frac{1}{2}\pi, \frac{1}{2}\pi)$, show that the r.v. $Y = \tan X$ is distributed as Cauchy. Also find the distribution of the r.v. $Z = \sin X$.

9.1.7 If the r.v. X has the Gamma distribution with parameters α, β, and $Y = 2X/\beta$, show that $Y \sim \chi^2_{2\alpha}$, provided 2α is an integer.

9.1.8 If X is an r.v. distributed as χ^2_r, set $Y = X/(1 + X)$ and determine the p.d.f. of Y.

9.1.9 If the r.v. X is distributed as Cauchy with $\mu = 0$ and $\sigma = 1$, show that the r.v. $Y = \tan^{-1} X$ is distributed as $U(-\frac{1}{2}\pi, \frac{1}{2}\pi)$.

9.1.10 Let X be an r.v. with p.d.f. f given by

$$f(x) = \frac{1}{\sqrt{2\pi}} x^{-2} e^{-1/(2x^2)}, \quad x \in \mathbb{R}$$

and show that the r.v. $Y = 1/X$ is distributed as $N(0, 1)$.

9.1.11 Suppose that the velocity X of a molecule of mass m is an r.v. with p.d.f. f given in Exercise 3.3.13(ii) of Chapter 3. Derive the distribution of the r.v. $Y = \frac{1}{2}mX^2$ (which is the kinetic energy of the molecule).

9.1.12 If the r.v. X is distributed as $N(\mu, \sigma^2)$, show, by means of a transformation, that the r.v. $Y = [(X - \mu)/\sigma]^2$ is distributed as χ^2_1.

9.2 The Multivariate Case

What has been discussed in the previous section carries over to the multidimensional case with the appropriate modifications.

THEOREM 1' Let $\mathbf{X} = (X_1, \ldots, X_k)'$ be a k-dimensional r. vector and let $h: \mathbb{R}^k \to \mathbb{R}^m$ be a (measurable) function, so that $\mathbf{Y} = h(\mathbf{X})$ is an r. vector. Then the distribution $P_\mathbf{Y}$ of the r. vector $\mathbf{Y}$ is determined by the distribution $P_\mathbf{X}$ of the r. vector $\mathbf{X}$ as follows: For any (Borel) subset B of $\mathbb{R}^m$, $P_\mathbf{Y}(B) = P_\mathbf{X}(A)$, where $A = h^{-1}(B)$.

The proof of this theorem is carried out in exactly the same way as that of Theorem 1. As in the univariate case, the distribution $P_\mathbf{Y}$ of the r. vector $\mathbf{Y}$ is uniquely determined by its d.f. $F_\mathbf{Y}$.

EXAMPLE 7 Let X_1, X_2 be independent r.v.'s distributed as $U(\alpha, \beta)$. We wish to determine the d.f. of the r.v. $Y = X_1 + X_2$. We have

$$F_Y(y) = P(X_1 + X_2 \leq y) = \iint_{\{x_1 + x_2 \leq y\}} f_{X_1, X_2}(x_1, x_2) dx_1 dx_2.$$

From Fig. 9.2, we see that for $y \leq 2\alpha$, $F_Y(y) = 0$. For

$$2\alpha < y \leq 2\beta, \quad F_Y(y) = \frac{1}{(\beta - \alpha)^2} \cdot A,$$

where A is the area of that part of the square lying to the left of the line $x_1 + x_2 = y$. Since for $y \leq \alpha + \beta$, $A = (y - 2\alpha)^2/2$, we get

$$F_Y(y) = \frac{(y-2\alpha)^2}{2(\beta-\alpha)^2} \quad \text{for} \quad 2\alpha < y \leq \alpha + \beta.$$

For $\alpha + \beta < y \leq 2\beta$, we have

$$F_Y(y) = \frac{1}{(\beta-\alpha)^2}\left[(\beta-\alpha)^2 - \frac{(2\beta-y)^2}{2}\right] = 1 - \frac{(2\beta-y)^2}{2(\beta-\alpha)^2}.$$

Thus we have:

$$F_Y(y) = \begin{cases} 0, & y \leq 2\alpha \\ \dfrac{(y-2\alpha)^2}{2(\beta-\alpha)^2}, & 2\alpha < y \leq \alpha + \beta \\ 1 - \dfrac{(2\beta-y)^2}{2(\beta-\alpha)^2}, & \alpha + \beta < y \leq 2\beta \\ 1, & y > 2\beta. \end{cases}$$

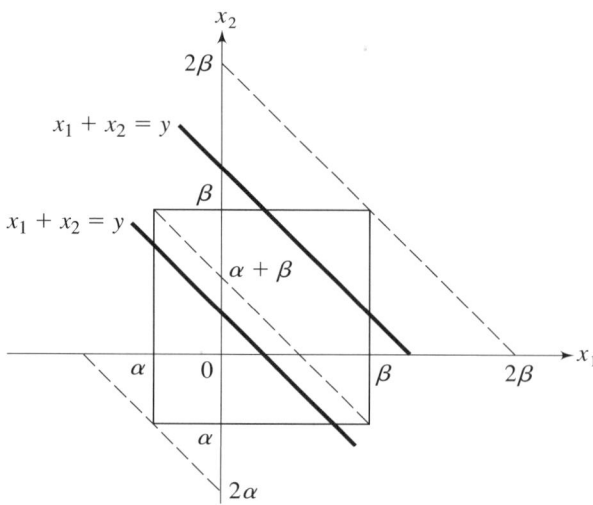

Figure 9.2

REMARK 3 The d.f. of $X_1 + X_2$ for any two independent r.v.'s (not necessarily $U(\alpha, \beta)$ distributed) is called the *convolution* of the d.f.'s of X_1, X_2 and is denoted by $F_{X_1+X_2} = F_{X_1} * F_{X_2}$. We also write $f_{X_1+X_2} = f_{X_1} * f_{X_2}$ for the corresponding p.d.f.'s. These concepts generalize to any (finite) number of r.v.'s.

EXAMPLE 8 Let X_1 be $B(n_1, p)$, X_2 be $B(n_2, p)$ and independent. Let $Y_1 = X_1 + X_2$ and $Y_2 = X_2$. We want to find the joint p.d.f. of Y_1, Y_2 and also the marginal p.d.f. of Y_1, and the conditional p.d.f. of Y_2, given $Y_1 = y_1$.

$$f_{Y_1,Y_2}(y_1, y_2) = P(Y_1 = y_1, Y_2 = y_2) = P(X_1 = y_1 - y_2, X_2 = y_2),$$

since $X_1 = Y_1 - Y_2$ and $X_2 = Y_2$. Furthermore, by independence, this is equal to

$$P(X_1 = y_1 - y_2)P(X_2 = y_2)$$
$$= \binom{n_1}{y_1 - y_2} p^{y_1 - y_2} q^{n_1 - (y_1 - y_2)} \cdot \binom{n_2}{y_2} p^{y_2} q^{n_2 - y_2}$$
$$= \binom{n_1}{y_1 - y_2}\binom{n_2}{y_2} p^{y_1} q^{(n_1 + n_2) - y_1};$$

that is

$$f_{Y_1,Y_2}(y_1, y_2) = \binom{n_1}{y_1 - y_2}\binom{n_2}{y_2} p^{y_1} q^{(n_1 + n_2) - y_1},$$

$$\begin{cases} 0 \leq y_1 \leq n_1 + n_2 \\ u = \max(0, y_1 - n_1) \leq y_2 \leq \min(y_1, n_2) = v. \end{cases}$$

Thus

$$f_{Y_1}(y_1) = P(Y_1 = y_1) = \sum_{y_2 = u}^{v} f_{Y_1,Y_2}(y_1, y_2) = p^{y_1} q^{(n_1 + n_2) - y_1} \sum_{y_2 = u}^{v} \binom{n_1}{y_1 - y_2}\binom{n_2}{y_2}.$$

Next, for the four possible values of the pair, (u, v), we have

$$\sum_{y_2 = 0}^{y_1} \binom{n_1}{y_1 - y_2}\binom{n_2}{y_2} = \sum_{y_2 = 0}^{n_2} \binom{n_1}{y_1 - y_2}\binom{n_2}{y_2} = \sum_{y_2 = y_1 - n_1}^{y_1} \binom{n_1}{y_1 - y_2}\binom{n_2}{y_2}$$
$$= \sum_{y_2 = y_1 - n_1}^{n_2} \binom{n_1}{y_1 - y_2}\binom{n_2}{y_2} = \binom{n_1 + n_2}{y_1};$$

that is, $Y_1 = X_1 + X_2$ is $B(n_1 + n_2, p)$. (Observe that this agrees with Theorem 2, Chapter 7.)

Finally, with y_1 and y_2 as above, it follows that

$$P(Y_2 = y_2 | Y_1 = y_1) = \frac{\binom{n_1}{y_1 - y_2}\binom{n_2}{y_2}}{\binom{n_1 + n_2}{y_1}},$$

the hypergeometric p.d.f., independent, of p!.

We next have two theorems analogous to Theorems 2 and 3 in Section 1. That is,

THEOREM 2′ Let the k-dimensional r. vector $\mathbf{X}$ have continuous p.d.f. $f_\mathbf{X}$ on the set S on which it is positive, and let

$$\mathbf{y} = h(\mathbf{x}) = \big(h_1(\mathbf{x}), \ldots, h_k(\mathbf{x})\big)'$$

be a (measurable) transformation defined on $\mathbb{R}^k$ into $\mathbb{R}^k$, so that $\mathbf{Y} = h(\mathbf{X})$ is a k-dimensional r. vector. Suppose that h is one-to-one on S onto T (the image of S under h), so that the inverse transformation

$$\mathbf{x} = h^{-1}(\mathbf{y}) = \big(g_1(\mathbf{y}), \ldots, g_k(\mathbf{y})\big)' \quad \text{exists for} \quad \mathbf{y} \in T.$$

It is further assumed that the partial derivatives

$$g_{ji}(\mathbf{y}) = \frac{\partial}{\partial y_i} g_j(y_1, \ldots, y_k), \quad i, j = 1, \ldots, k$$

exist and are continuous on T. Then the p.d.f. $f_\mathbf{Y}$ of $\mathbf{Y}$ is given by

$$f_\mathbf{Y}(\mathbf{y}) = \begin{cases} f_\mathbf{X}\big[h^{-1}(\mathbf{y})\big]|J| = f_\mathbf{X}\big[g_1(\mathbf{y}), \ldots, g_k(\mathbf{y})\big]|J|, & \mathbf{y} \in T \\ 0, & \text{otherwise,} \end{cases}$$

where the Jacobian J is a function of $\mathbf{y}$ and is defined as follows

$$J = \begin{vmatrix} g_{11} & g_{12} & \cdots & g_{1k} \\ g_{21} & g_{22} & \cdots & g_{2k} \\ \vdots & \vdots & & \vdots \\ g_{k1} & g_{k2} & \cdots & g_{kk} \end{vmatrix}$$

and is assumed to be $\neq 0$ on T.

REMARK 4 In Theorem 2′, the transformation h transforms the k-dimensional r. vector $\mathbf{X}$ to the k-dimensional r. vector $\mathbf{Y}$. In many applications, however, the dimensionality m of $\mathbf{Y}$ is less than k. Then in order to determine the p.d.f. of $\mathbf{Y}$, we work as follows. Let $\mathbf{y} = (h_1(\mathbf{x}), \ldots, h_m(\mathbf{x}))'$ and choose another $k - m$ transformations defined on $\mathbb{R}^k$ into $\mathbb{R}$, h_{m+j}, $j = 1, \ldots, k - m$, say, so that they are of the simplest possible form and such that the transformation

$$h = \big(h_1, \ldots, h_m, h_{m+1}, \ldots, h_k\big)'$$

satisfies the assumptions of Theorem 2′. Set $\mathbf{Z} = (Y_1, \ldots, Y_m, Y_{m+1}, \ldots, Y_k)'$, where $\mathbf{Y} = (Y_1, \ldots, Y_m)'$ and $Y_{m+j} = h_{m+j}(\mathbf{X})$, $j = 1, \ldots, k - m$. Then by applying Theorem 2′, we obtain the p.d.f. $f_\mathbf{Z}$ of $\mathbf{Z}$ and then integrating out the last $k - m$ arguments y_{m+j}, $j = 1, \ldots, k - m$, we have the p.d.f. of $\mathbf{Y}$.

A number of examples will be presented to illustrate the application of Theorem 2′ as well as of the preceding remark.

EXAMPLE 9 Let X_1, X_2 be i.i.d. r.v.'s distributed as $U(\alpha, \beta)$. Set $Y_1 = X_1 + X_2$ and find the p.d.f. of Y_1.

We have

$$f_{X_1,X_2}(x_1, x_2) = \begin{cases} \dfrac{1}{(\beta-\alpha)^2}, & \alpha < x_1, x_2 < \beta \\ 0, & \text{otherwise.} \end{cases}$$

Consider the transformation

$$h: \begin{cases} y_1 = x_1 + x_2 \\ y_2 = x_2 \end{cases}, \quad \alpha < x_1, x_2 < \beta; \quad \text{then} \begin{cases} Y_1 = X_1 + X_2 \\ Y_2 = X_2. \end{cases}$$

From h, we get

$$\begin{cases} x_1 = y_1 - y_2 \\ x_2 = y_2. \end{cases} \quad \text{Then} \quad J = \begin{vmatrix} 1 & -1 \\ 0 & 1 \end{vmatrix} = 1$$

and also $\alpha < y_2 < \beta$. Since $y_1 - y_2 = x_1$, $\alpha < x_1 < \beta$, we have $\alpha < y_1 - y_2 < \beta$. Thus the limits of y_1, y_2 are specified by $\alpha < y_2 < \beta$, $\alpha < y_1 - y_2 < \beta$. (See Figs. 9.3 and 9.4.)

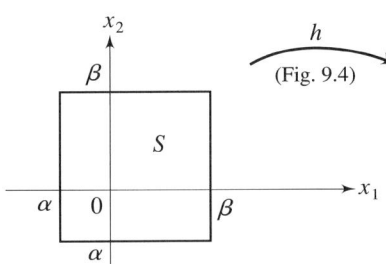

Figure 9.3 $S = \{(x_1, x_2)'; f_{X_1, X_2}(x_1, x_2) > 0\}$

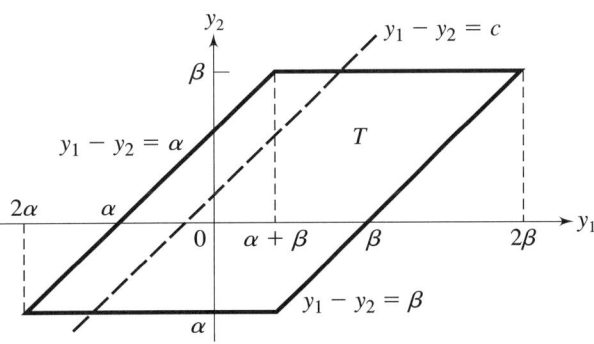

Figure 9.4 T = image of S under the transformation h.

Thus we get

$$f_{Y_1,Y_2}(y_1,y_2) = \begin{cases} \dfrac{1}{(\beta-\alpha)^2}, & 2\alpha < y_1 < 2\beta, \ \alpha < y_2 < \beta, \ \alpha < y_1 - y_2 < \beta \\ 0, & \text{otherwise.} \end{cases}$$

Therefore

$$f_{Y_1}(y_1) = \begin{cases} \dfrac{1}{(\beta-\alpha)^2}\int_{\alpha}^{y_1-\alpha} dy_2 = \dfrac{y_1 - 2\alpha}{(\beta-\alpha)^2}, & \text{for } 2\alpha < y_1 \leq \alpha+\beta \\ \dfrac{1}{(\beta-\alpha)^2}\int_{y_1-\beta}^{\beta} dy_2 = \dfrac{2\beta - y_1}{(\beta-\alpha)^2}, & \text{for } \alpha+\beta < y_1 < 2\beta \\ 0, & \text{otherwise.} \end{cases}$$

The graph of f_{Y_1} is given in Fig. 9.5.

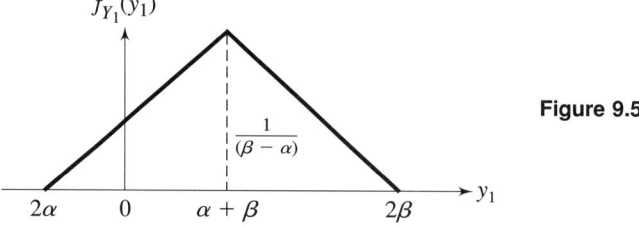

Figure 9.5

REMARK 5 This density is known as the *triangular p.d.f.*

EXAMPLE 10 Let X_1, X_2 be i.i.d. r.v.'s from $U(1, \beta)$. Set $Y_1 = X_1 X_2$ and find the p.d.f. of Y_1.

Consider the transformation

$$h: \begin{cases} y_1 = x_1 x_2 \\ y_2 = x_2 \end{cases}; \quad \text{then} \quad \begin{cases} Y_1 = X_1 X_2 \\ Y_2 = X_2. \end{cases}$$

From h, we get

$$\begin{cases} x_1 = \dfrac{y_1}{y_2} \\ x_2 = y_2 \end{cases} \quad \text{and} \quad J = \begin{vmatrix} \dfrac{1}{y_2} & -\dfrac{y_1}{y_2^2} \\ 0 & 1 \end{vmatrix} = \dfrac{1}{y_2}.$$

Now

$$S = \left\{(x_1, x_2)'; f_{X_1,X_2}(x_1, x_2) > 0\right\}$$

is transformed by h onto

$$T = \left\{(y_1, y_2)'; 1 < \dfrac{y_1}{y_2} < \beta, \ 1 < y_2 < \beta\right\}.$$

(See Fig. 9.6.) Thus, since

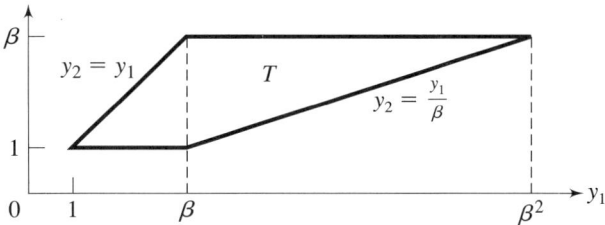

Figure 9.6

$$f_{Y_1,Y_2}(y_1,y_2) = \begin{cases} \dfrac{1}{(\beta-1)^2} \dfrac{1}{2}, & (y_1,y_2)' \in T \\ 0, & \text{otherwise,} \end{cases}$$

we have

$$f_{Y_1}(y_1) = \begin{cases} \dfrac{1}{(\beta-1)^2} \displaystyle\int_1^{y_1} \dfrac{dy_2}{y_2} = \dfrac{1}{(\beta-1)^2}\log y_1, & 1 < y_1 < \beta \\ \dfrac{1}{(\beta-1)^2} \displaystyle\int_{y_1/\beta}^{\beta} \dfrac{dy_2}{y_2} = \dfrac{1}{(\beta-1)^2}(2\log\beta - \log y_1), & \beta \le y_1 < \beta^2; \end{cases}$$

that is

$$f_{Y_1}(y_1) = \begin{cases} \dfrac{1}{(\beta-1)^2}\log y_1, & 1 < y_1 < \beta \\ \dfrac{1}{(\beta-1)^2}(2\log\beta - \log y_1), & \beta \le y_1 < \beta^2 \\ 0, & \text{otherwise.} \end{cases}$$

EXAMPLE 11 Let X_1, X_2 be i.i.d. r.v.'s from $N(0, 1)$. Show that the p.d.f. of the r.v. $Y_1 = X_1/X_2$ is Cauchy with $\mu = 0$, $\sigma = 1$; that is,

$$f_{Y_1}(y_1) = \dfrac{1}{\pi} \cdot \dfrac{1}{1+y_1^2}, \quad y_1 \in \mathbb{R}.$$

We have

$Y_1 = X_1/X_2$. Let $Y_2 = X_2$ and consider the transformation

$$h: \begin{cases} y_1 = x_1/x_2, & x_2 \neq 0 \\ y_2 = x_2; \end{cases} \quad \text{then} \quad \begin{cases} x_1 = y_1 y_2 \\ x_2 = y_2 \end{cases}$$

and

$$J = \begin{vmatrix} y_2 & y_1 \\ 0 & 1 \end{vmatrix} = y_2, \quad \text{so that} \quad |J| = |y_2|.$$

Since $-\infty < x_1, x_2 < \infty$ implies $-\infty < y_1, y_2 < \infty$, we have

$$f_{Y_1,Y_2}(y_1, y_2) = f_{X_1,X_2}(y_1 y_2, y_2) \cdot |y_2| = \frac{1}{2\pi} \exp\left(-\frac{y_1^2 y_2^2 + y_2^2}{2}\right)|y_2|$$

and therefore

$$f_{Y_1}(y_1) = \frac{1}{2\pi} \int_{-\infty}^{\infty} \exp\left(-\frac{y_1^2 y_2^2 + y_2^2}{2}\right)|y_2| dy_2 = \frac{1}{\pi}\int_0^\infty \exp\left[-\frac{(y_1^2+1)y_2^2}{2}\right] y_2 dy_2.$$

Set

$$\frac{(y_1^2+1)}{2} y_2^2 = t, \quad \text{so that} \quad y_2^2 = \frac{2t}{y_1^2+1}$$

and

$$2 y_2 dy_2 = \frac{2dt}{y_1^2+1}, \quad \text{or} \quad y_2 dy_2 = \frac{dt}{y_1^2+1}, \quad t \in [0, \infty).$$

Thus we continue as follows:

$$\frac{1}{\pi}\int_0^\infty e^{-t} \frac{dt}{y_1^2+1} = \frac{1}{\pi} \cdot \frac{1}{y_1^2+1} \int_0^\infty e^{-t} dt = \frac{1}{\pi} \cdot \frac{1}{y_1^2+1},$$

since

$$\int_0^\infty e^{-t} dt = 1;$$

that is,

$$f_{Y_1}(y_1) = \frac{1}{\pi} \cdot \frac{1}{y_1^2+1}.$$

EXAMPLE 12 Let X_1, X_2 be independent r.v.'s distributed as Gamma with parameters $(\alpha, 2)$ and $(\beta, 2)$, respectively. Set $Y_1 = X_1/(X_1 + X_2)$ and prove that Y_1 is distributed as Beta with parameters α, β.

We set $Y_2 = X_1 + X_2$ and consider the transformation:

$$h: \begin{cases} y_1 = \dfrac{x_1}{x_1+x_2}, & x_1, x_2 > 0; \\ y_2 = x_1 + x_2 \end{cases} \quad \text{then} \quad \begin{cases} x_1 = y_1 y_2 \\ x_2 = y_2 - y_1 y_2. \end{cases}$$

Hence
$$J = \begin{vmatrix} y_2 & y_1 \\ -y_2 & 1-y_1 \end{vmatrix} = y_2 - y_1y_2 + y_1y_2 = y_2 \text{ and } |J| = y_2.$$

Next,
$$f_{X_1,X_2}(x_1,x_2) = \begin{cases} \dfrac{1}{\Gamma(\alpha)\Gamma(\beta)2^\alpha 2^\beta} x_1^{\alpha-1} x_2^{\beta-1} \exp\left(-\dfrac{x_1+x_2}{2}\right), & x_1,x_2 > 0, \\ 0, & \text{otherwise}, \quad \alpha, \beta > 0. \end{cases}$$

From the transformation, it follows that for $x_1 = 0$, $y_1 = 0$ and for $x_1 \to \infty$,
$$y_1 = \frac{x_1}{x_1+x_2} = \frac{1}{1+(x_2/x_1)} \to 1.$$

Thus $0 < y_1 < 1$ and, clearly, $0 < y_2 < \infty$. Therefore, for $0 < y_1 < 1$, $0 < y_2 < \infty$, we get
$$f_{Y_1,Y_2}(y_1,y_2) = \frac{1}{\Gamma(\alpha)\Gamma(\beta)2^{\alpha+\beta}} y_1^{\alpha-1} y_2^{\alpha-1} y_2^{\beta-1} (1-y_1)^{\beta-1} \exp\left(-\frac{y_2}{2}\right) y_2$$
$$= \frac{1}{\Gamma(\alpha)\Gamma(\beta)2^{\alpha+\beta}} y_1^{\alpha-1} (1-y_1)^{\beta-1} y_2^{\alpha+\beta-1} e^{-y_2/2}.$$

Hence
$$f_{Y_1}(y_1) = \frac{1}{\Gamma(\alpha)\Gamma(\beta)2^{\alpha+\beta}} y_1^{\alpha-1}(1-y_1)^{\beta-1}$$
$$\times \int_0^\infty y_2^{\alpha+\beta-1} e^{-y_2/2} dy_2.$$

But
$$\int_0^\infty y_2^{\alpha+\beta-1} e^{-y_2/2} dy_2 = 2^{\alpha+\beta} \int_0^\infty t^{\alpha+\beta-1} e^{-t} dt = 2^{\alpha+\beta} \Gamma(\alpha+\beta).$$

Therefore
$$f_{Y_1}(y_1) = \begin{cases} \dfrac{\Gamma(\alpha+\beta)}{\Gamma(\alpha)\Gamma(\beta)} y_1^{\alpha-1}(1-y_1)^{\beta-1}, & 0 < y_1 < 1 \\ 0, & \text{otherwise}. \end{cases}$$

EXAMPLE 13 Let X_1, X_2, X_3 be i.i.d. r.v.'s with density
$$f(x) = \begin{cases} e^{-x}, & x > 0 \\ 0, & x \leq 0. \end{cases}$$

Set
$$Y_1 = \frac{X_1}{X_1+X_2}, \quad Y_2 = \frac{X_1+X_2}{X_1+X_2+X_3}, \quad Y_3 = X_1+X_2+X_3$$

and prove that Y_1 is $U(0, 1)$, Y_3 is distributed as Gamma with $\alpha = 3$, $\beta = 1$, and Y_1, Y_2, Y_3 are independent.

Consider the transformation

$$h: \begin{cases} y_1 = \dfrac{x_1}{x_1 + x_2} \\ y_2 = \dfrac{x_1 + x_2}{x_1 + x_2 + x_3}, \quad x_1, x_2, x_3 > 0; \quad \text{then} \quad \begin{cases} x_1 = y_1 y_2 y_3 \\ x_2 = -y_1 y_2 y_3 + y_2 y_3 \\ x_3 = -y_2 y_3 + y_3 \end{cases} \\ y_3 = x_1 + x_2 + x_3 \end{cases}$$

and

$$J = \begin{vmatrix} y_2 y_3 & y_1 y_3 & y_1 y_2 \\ -y_2 y_3 & -y_1 y_3 + y_3 & -y_1 y_2 + y_2 \\ 0 & -y_3 & -y_2 + 1 \end{vmatrix} = y_2 y_3^2.$$

Now from the transformation, it follows that $x_1, x_2, x_3 \in (0, \infty)$ implies that

$$y_1 \in (0, 1), \quad y_2 \in (0, 1), \quad y_3 \in (0, \infty).$$

Thus

$$f_{Y_1, Y_2, Y_3}(y_1, y_2, y_3) = \begin{cases} y_2 y_3^2 e^{-y_3}, & 0 < y_1 < 1, \quad 0 < y_2 < 1, \quad 0 < y_3 < \infty \\ 0, & \text{otherwise.} \end{cases}$$

Hence

$$f_{Y_1}(y_1) = \int_0^\infty \int_0^1 y_2 y_3^2 e^{-y_3} dy_2 dy_3 = 1, \quad 0 < y_1 < 1,$$

$$f_{Y_2}(y_2) = \int_0^\infty \int_0^1 y_2 y_3^2 e^{-y_3} dy_1 dy_3 = y_2 \int_0^\infty y_3^2 e^{-y_3} dy_3$$
$$= 2y_2, \quad 0 < y_2 < 1$$

and

$$f_{Y_3}(y_3) = \int_0^1 \int_0^1 y_2 y_3^2 e^{-y_3} dy_1 dy_2 = y_3^2 e^{-y_3} \int_0^1 y_2 dy_2$$
$$= \frac{1}{2} y_3^2 e^{-y_3}, \quad 0 < y_3 < \infty.$$

Since

$$f_{Y_1, Y_2, Y_3}(y_1, y_2, y_3) = f_{Y_1}(y_1) f_{Y_2}(y_2) f_{Y_3}(y_3),$$

the independence of Y_1, Y_2, Y_3 is established. The functional forms of f_{Y_1}, f_{Y_3} verify the rest.

9.2.1 Application 2: The *t* and *F* Distributions

The density of the t distribution with r degrees of freedom (t_r). Let the independent r.v.'s X and Y be distributed as N(0, 1) and χ_r^2, respectively, and set $T = X/\sqrt{Y/r}$. The r.v. *T* is said to have the (Student's) *t-distribution with r degrees of freedom (d.f.)* and is often denoted by t_r. We want to find its p.d.f. We have:

$$f_X(x) = \frac{1}{\sqrt{2\pi}} e^{-(1/2)x^2}, \quad x \in \mathbb{R},$$

$$f_Y(y) = \begin{cases} \dfrac{1}{\Gamma(\frac{1}{2}r)2^{(1/2)r}} y^{(r/2)-1} e^{-y/2}, & y > 0 \\ 0, & y \leq 0. \end{cases}$$

Set $U = Y$ and consider the transformation

$$h: \begin{cases} t = \dfrac{x}{\sqrt{y/r}}; \\ u = y \end{cases} \quad \text{then} \quad \begin{cases} x = \dfrac{1}{\sqrt{r}} t\sqrt{u} \\ y = u \end{cases}$$

and

$$J = \begin{vmatrix} \dfrac{\sqrt{u}}{\sqrt{r}} & \dfrac{t}{2\sqrt{u}\sqrt{r}} \\ 0 & 1 \end{vmatrix} = \dfrac{\sqrt{u}}{\sqrt{r}}.$$

Then for $t \in \mathbb{R}$, $u > 0$, we get

$$f_{T,U}(t,u) = \frac{1}{\sqrt{2\pi}} e^{-t^2 u/(2r)} \cdot \frac{1}{\Gamma(r/2)2^{r/2}} u^{(r/2)-1} e^{-u/2} \cdot \frac{\sqrt{u}}{\sqrt{r}}$$

$$= \frac{1}{\sqrt{2\pi r}\,\Gamma(r/2)2^{r/2}} u^{(1/2)(r+1)-1} \exp\left[-\frac{u}{2}\left(1 + \frac{t^2}{r}\right)\right].$$

Hence

$$f_T(t) = \int_0^\infty \frac{1}{\sqrt{2\pi r}\,\Gamma(r/2)2^{r/2}} u^{(1/2)(r+1)-1} \exp\left[-\frac{u}{2}\left(1 + \frac{t^2}{r}\right)\right] du.$$

We set

$$\frac{u}{2}\left(1 + \frac{t^2}{r}\right) = z, \quad \text{so that} \quad u = 2z\left(1 + \frac{t^2}{r}\right)^{-1}, \quad du = 2\left(1 + \frac{t^2}{r}\right)^{-1} dz,$$

and $z \in [0, \infty)$. Therefore we continue as follows:

$$f_T(t) = \int_0^\infty \frac{1}{\sqrt{2\pi r}\,\Gamma(r/2)2^{r/2}} \left[\frac{2z}{1+(t^2/r)}\right]^{(1/2)(r+1)-1} e^{-z} \frac{2}{1+(t^2/r)} dz$$

$$= \frac{1}{\sqrt{2\pi r}\,\Gamma(r/2)2^{r/2}} \frac{2^{(1/2)(r+1)}}{\left[1+(t^2/r)\right]^{(1/2)(r+1)}} \int_0^\infty z^{(1/2)(r+1)-1} e^{-z} dz$$

$$= \frac{1}{\sqrt{\pi r}\,\Gamma(r/2)} \frac{1}{\left[1+(t^2/r)\right]^{(1/2)(r+1)}} \Gamma\!\left[\tfrac{1}{2}(r+1)\right];$$

that is

$$f_T(t) = \frac{\Gamma\!\left[\tfrac{1}{2}(r+1)\right]}{\sqrt{\pi r}\,\Gamma(r/2)} \frac{1}{\left[1+(t^2/r)\right]^{(1/2)(r+1)}}, \quad t \in \mathbb{R}.$$

The probabilities $P(T \le t)$ for selected values of t and r are given in tables (the t-tables). (For the graph of f_T, see Fig. 9.7.)

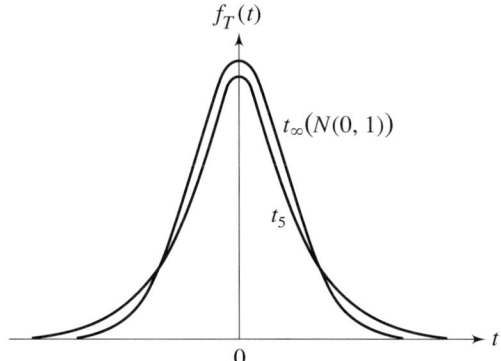

Figure 9.7

The density of the F distribution with r_1, r_2 d.f. (F_{r_1,r_2}). Let the independent r.v.'s X and Y be distributed as $\chi^2_{r_1}$ and $\chi^2_{r_2}$, respectively, and set $F = (X/r_1)/(Y/r_2)$. The r.v. F is said to have the *F distribution with r_1, r_2 degrees of freedom* (d.f.) and is often denoted by F_{r_1,r_2}.

We want to find its p.d.f. We have:

$$f_X(x) = \begin{cases} \dfrac{1}{\Gamma(\tfrac{1}{2}r_1)2^{r_1/2}} x^{(r_1/2)-1} e^{-x/2}, & x > 0 \\ 0, & x \le 0, \end{cases}$$

$$f_Y(y) = \begin{cases} \dfrac{1}{\Gamma\left(\frac{1}{2}r_2\right)2^{r_2/2}} y^{(r_2/2)-1} e^{-y/2}, & y > 0 \\ 0, & y \leq 0. \end{cases}$$

We set $Z = Y$, and consider the transformation

$$h: \begin{cases} f = \dfrac{x/r_1}{y/r_2}; \\ z = y \end{cases} \text{ then } \begin{cases} x = \dfrac{r_1}{r_2} fz \\ y = z \end{cases}$$

and

$$J = \begin{vmatrix} \dfrac{r_1}{r_2}z & \dfrac{r_1}{r_2}f \\ 0 & 1 \end{vmatrix} = \dfrac{r_1}{r_2}z, \text{ so that } |J| = \dfrac{r_1}{r_2}z.$$

For $f, z > 0$, we get:

$$f_{F,Z}(f,z) = \dfrac{1}{\Gamma\left(\frac{1}{2}r_1\right)\Gamma\left(\frac{1}{2}r_2\right)2^{(1/2)(r_1+r_2)}} \left(\dfrac{r_1}{r_2}\right)^{(r_1/2)-1} f^{(r_1/2)-1} z^{(r_1/2)-1} z^{(r_2/2)-1}$$

$$\times \exp\left(-\dfrac{r_1}{2r_2}\right) fz \, e^{-z/2} \dfrac{r_1}{r_2} z$$

$$= \dfrac{(r_1/r_2)^{r_1/2} f^{(r_1/2)-1}}{\Gamma\left(\frac{1}{2}r_1\right)\Gamma\left(\frac{1}{2}r_2\right)2^{(1/2)(r_1+r_2)}} z^{(1/2)(r_1+r_2)-1} \exp\left[-\dfrac{z}{2}\left(\dfrac{r_1}{r_2}f+1\right)\right].$$

Therefore

$$f_F(f) = \int_0^\infty f_{F,Z}(f,z) dz$$

$$= \dfrac{(r_1/r_2)^{r_1/2} f^{(r_1/2)-1}}{\Gamma\left(\frac{1}{2}r_1\right)\Gamma\left(\frac{1}{2}r_2\right)2^{(1/2)(r_1+r_2)}} \int_0^\infty z^{(1/2)(r_1+r_2)-1} \exp\left[-\dfrac{z}{2}\left(\dfrac{r_1}{r_2}f+1\right)\right] dz.$$

Set

$$\dfrac{z}{2}\left(\dfrac{r_1}{r_2}f+1\right) = t, \text{ so that } z = 2t\left(\dfrac{r_1}{r_2}f+1\right)^{-1},$$

$$dz = 2\left(\dfrac{r_1}{r_2}f+1\right)^{-1} dt, \quad t \in [0, \infty).$$

Thus continuing, we have

$$f_F(f) = \frac{(r_1/r_2)^{r_1/2} f^{(r_1/2)-1}}{\Gamma(\frac{1}{2}r_1)\Gamma(\frac{1}{2}r_2) 2^{(1/2)(r_1+r_2)}} 2^{(1/2)(r_1+r_2)-1} \left(\frac{r_1}{r_2} f + 1\right)^{-(1/2)(r_1+r_2)+1}$$

$$\times 2\left(\frac{r_1}{r_2} f + 1\right)^{-1} \int_0^\infty t^{(1/2)(r_1+r_2)-1} e^{-t} dt$$

$$= \frac{\Gamma[\frac{1}{2}(r_1+r_2)](r_1/r_2)^{r_1/2}}{\Gamma(\frac{1}{2}r_1)\Gamma(\frac{1}{2}r_2)} \cdot \frac{f^{(r_1/2)-1}}{[1+(r_1/r_2)f]^{(1/2)(r_1+r_2)}}.$$

Therefore

$$f_F(f) = \begin{cases} \dfrac{\Gamma[\frac{1}{2}(r_1+r_2)](r_1/r_2)^{r_1/2}}{\Gamma(\frac{1}{2}r_1)\Gamma(\frac{1}{2}r_2)} \cdot \dfrac{f^{(r_1/2)-1}}{[1+(r_1/r_2)f]^{(1/2)(r_1+r_2)}}, & \text{for } f > 0 \\ 0, & \text{for } f \leq 0. \end{cases}$$

The probabilities $P(F \leq f)$ for selected values of f and r_1, r_2 are given by tables (the F-tables). (For the graph of f_F, see Fig. 9.8.)

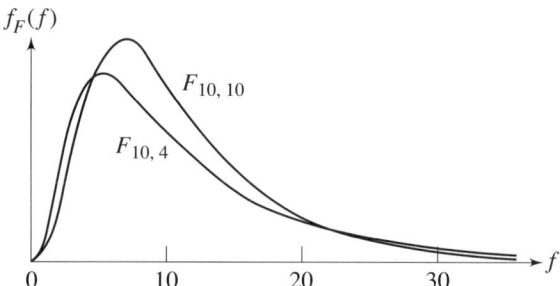

Figure 9.8

REMARK 6

i) If F is distributed as F_{r_1, r_2}, then, clearly, $1/F$ is distributed as F_{r_2, r_1}.

ii) If X is $N(0, 1)$, Y is χ_r^2 and X, Y are independent, so that $T = X/\sqrt{Y/r}$ is distributed as t_r, the n T^2 is distributed as $F_{1,r}$, since X^2 is χ_1^2.

We consider the multidimensional version of Theorem 3.

THEOREM 3' Let the k-dimensional r. vector $\mathbf{X}$ have continuous p.d.f. $f_\mathbf{X}$ on the set S on which it is positive, and let $\mathbf{y} = h(\mathbf{x}) = (h_1(\mathbf{x}), \ldots, h_k(\mathbf{x}))'$ be a (measurable) transformation defined on $\mathbb{R}^k$ into $\mathbb{R}^k$, so that $\mathbf{Y} = h(\mathbf{X})$ is a k-dimensional r. vector. Suppose that there is a partition $\{S_j, j = 1, \ldots, r\}$ of S and subsets T_j, $j = 1, \ldots, r$ of T (the image of S under h), which need not be distinct or disjoint, such that $\cup_{j=1}^r T_j = T$ and that h defined on each one of S_j onto T_j,

$j = 1, \ldots, r$ is one-to-one. Let h_j be the restriction of the transformation h to S_j and let $h_j^{-1}(\mathbf{y}) = (g_{j1}(\mathbf{y}), \ldots, g_{jk}(\mathbf{y}))'$ be its inverse, $j = 1, \ldots, r$. Assume that the partial derivatives $g_{jil}(\mathbf{y}) = (\partial/\partial y_l)g_{ji}(y_1, \cdots, y_k)$, $i, l = 1, \ldots, k$, $j = 1, \ldots, r$ exist and for each j, g_{jil}, $i, l = 1, \ldots, k$ are continuous, $j = 1, \ldots, r$. Then the p.d.f. $f_\mathbf{Y}$ of $\mathbf{Y}$ is given by

$$f_\mathbf{Y}(\mathbf{y}) = \begin{cases} \sum_{j=1}^{r} \delta_j(\mathbf{y}) f_{\mathbf{Y}_j}(\mathbf{y}), & \mathbf{y} \in T \\ 0, & \text{otherwise,} \end{cases}$$

where for $j = 1, \ldots, r$, $f_{\mathbf{Y}_j}(\mathbf{y}) = f_\mathbf{X}[h_j^{-1}(\mathbf{y})]|J_j|$, $\mathbf{y} \in T_j$, $\delta_j(\mathbf{y}) = 1$ if $\mathbf{y} \in T$ and $\delta_j(\mathbf{y}) = 0$ otherwise, and the Jacobians J_j which are functions of $\mathbf{y}$ are defined by

$$J_j = \begin{vmatrix} g_{j11} & g_{j12} & \cdots & g_{j1k} \\ g_{j21} & g_{j22} & \cdots & g_{j2k} \\ \vdots & \vdots & \vdots & \vdots \\ g_{jk1} & g_{jk2} & \cdots & g_{jkk} \end{vmatrix},$$

and are assumed to be $\neq 0$ on T_j, $j = 1, \ldots, r$.

In the next chapter (Chapter 10) on order statistics we will have the opportunity of applying Theorem 3'.

Exercises

9.2.1 Let X_1, X_2 be independent r.v.'s taking on the values $1, \ldots, 6$ with probability $f(x) = \frac{1}{6}$, $x = 1, \ldots, 6$. Derive the distribution of the r.v. $X_1 + X_2$.

9.2.2 Let X_1, X_2 be r.v.'s with joint p.d.f. f given by

$$f(x_1, x_2) = \frac{1}{\pi} I_A(x_1, x_2),$$

where

$$A = \left\{ (x_1, x_2)' \in \mathbb{R}^2; x_1^2 + x_2^2 \leq 1 \right\}.$$

Set $Z^2 = X_1^2 + X_2^2$ and derive the p.d.f. of the r.v. Z^2. (Hint: Use polar coordinates.)

9.2.3 Let X_1, X_2 be independent r.v.'s distributed as $N(0, 1)$. Then:

i) Find the p.d.f. of the r.v.'s $X_1 + X_2$ and $X_1 - X_2$;
ii) Calculate the probability $P(X_1 - X_2 < 0, X_1 + X_2 > 0)$.

9.2.4 Let X_1, X_2 be independent r.v.'s distributed as Negative Exponential with parameter $\lambda = 1$. Then:

i) Derive the p.d.f.'s of the following r.v.'s:
$$X_1 + X_2, \quad X_1 - X_2, \quad \text{and} \quad X_1/X_2;$$
ii) Show that $X_1 + X_2$ and X_1/X_2 are independent.

9.2.5 Let X_1, X_2 be independent r.v.'s distributed as $U(\alpha, \alpha + 1)$. Then:

i) Derive the p.d.f.'s of the r.v.'s $X_1 + X_2$ and $X_1 - X_2$;

ii) Determine whether these r.v.'s are independent or not.

9.2.6 Let the independent r.v.'s X_1, X_2 have p.d.f. f given by
$$f(x) = \frac{1}{x^2} I_{(1,\infty)}(x).$$
Determine the distribution of the r.v. $X = X_1/X_2$.

9.2.7 Let X be an r.v. distributed as t_r.

i) For $r = 1$, show that the p.d.f. of X becomes a Cauchy p.d.f.;

ii) Also show that the r.v. $Y = \dfrac{1}{1+(X^2/r)}$ is distributed as Beta.

9.2.8 If the r.v. X is distributed as F_{r_1, r_2}, then:

i) Find its expectation and variance;

ii) If $r_1 = r_2$, show that its median is equal to 1;

iii) The p.d.f. of $Y = \dfrac{1}{1+(r_1/r_2)X}$ is Beta;

iv) The p.d.f. of $r_1 X$ converges to that of $\chi^2_{r_1}$ as $r_2 \to \infty$.

9.2.9 Let X_1, X_2 be independent r.v.'s distributed as $\chi^2_{r_1}$ and $\chi^2_{r_2}$, respectively, and set $X = X_1 + X_2$, $Y = X_1/X_2$. Then show that:

i) The r.v. X is distributed as $\chi^2_{r_1+r_2}$ (as anticipated);

ii) The r.v. Y is distributed as $\frac{r_1}{r_2}Z$, where Z has the F_{r_1, r_2} distribution;

iii) The r.v.'s X and Y are independent.

9.2.10 Let X_1, X_2 be independent r.v.'s distributed as $N(0, \sigma^2)$. Then show that:

i) The r.v. $X_1^2 + X_2^2$ has the Negative Exponential distribution with parameter $\lambda = 1/2\sigma^2$;

ii) The r.v. X_1/X_2 has the Cauchy distribution with $\mu = 0$ and $\sigma = 1$;

iii) The r.v.'s $X_1^2 + X_2^2$ and X_1/X_2 are independent.

9.2.11 Let X_r be an r.v. distributed as t_r. Then show that:

$$EX_r = 0, \quad r \geq 2; \quad \sigma^2(X_r) = \frac{r}{r-2}, \quad r \geq 3.$$

9.2.12 Let X_{r_1,r_2} be an r.v. distributed as F_{r_1,r_2}. Then show that:

$$EX_{r_1,r_2} = \frac{r_2}{r_2-2}, \quad r_2 \geq 3; \quad \sigma^2(X_{r_1,r_2}) = \frac{2r_2^2(r_1+r_2-2)}{r_1(r_2-2)^2(r_2-4)}, \quad r_2 \geq 5.$$

9.2.13 Let X_r be an r.v. distributed as t_r, and let f_r be its p.d.f. Then show that:

$$f_r(x) \xrightarrow[r \to \infty]{} \frac{1}{\sqrt{2\pi}} \exp\left(-\frac{x^2}{2}\right), \quad x \in \mathbb{R}.$$

9.2.14 Let X_{r_1} and X_{r_1,r_2} be r.v.'s distributed as $\chi_{r_1}^2$ and F_{r_1,r_2}, respectively, and, for $\alpha \in (0,1)$, let $\chi_{r_1;\alpha}^2$ and $F_{r_1,r_2;\alpha}$ be defined by: $P(X_{r_1} \geq \chi_{r_1;\alpha}^2) = \alpha$, $P(X_{r_1,r_2} \geq F_{r_1,r_2;\alpha}) = \alpha$. Then show that:

$$F_{r_1,r_2;\alpha} \to \frac{1}{r_1}\chi_{r_1;\alpha}^2 \quad \text{as} \quad r_2 \to \infty.$$

9.3 Linear Transformations of Random Vectors

In this section we will restrict ourselves to a special and important class of transformations, the *linear transformations*. We first introduce some needed notation and terminology.

9.3.1 Preliminaries

A transformation $h: \mathbb{R}^k \to \mathbb{R}^k$ which transforms the variables $x_1, \ldots, x_k$ to the variables $y_1, \ldots, y_k$ in the following manner:

$$y_i = \sum_{j=1}^{k} c_{ij}x_j, \quad c_{ij} \text{ real constants}, \quad i,j = 1,2,\ldots,k \qquad (1)$$

is called a *linear transformation*. Let **C** be the $k \times k$ matrix whose elements are c_{ij}. That is, $\mathbf{C} = (c_{ij})$, and let $\Delta = |\mathbf{C}|$ be the determinant of **C**. If $\Delta \neq 0$, we can uniquely solve for the x's in (1) and get

$$x_i = \sum_{j=1}^{k} d_{ij}y_j, \quad d_{ij} \text{ real constants}, \quad i,j = 1,\ldots,k. \qquad (2)$$

Let $\mathbf{D} = (d_{ij})$ and $\Delta^* = |\mathbf{D}|$. Then, as is known from linear algebra (see also Appendix 1), $\Delta^* = 1/\Delta$. If, furthermore, the linear transformation above is such that the column vectors $(c_{1j}, c_{2j}, \ldots, c_{kj})'$, $j = 1, \ldots, k$ are *orthogonal*, that is

and
$$\left.\begin{array}{l}\sum_{i=1}^{k} c_{ij}c_{ij'} = 0 \quad \text{for} \quad j \neq j' \\ \sum_{i=1}^{k} c_{ij}^{2} = 1, \quad j = 1, \ldots, k,\end{array}\right\} \quad (3)$$

then the linear transformation is called *orthogonal*. The orthogonality relations (3) are equivalent to orthogonality of the row vectors $(c_{i1}, \ldots, c_{ik})'$ $i = 1, \ldots, k$. That is,

and
$$\left.\begin{array}{l}\sum_{j=1}^{k} c_{ij}c_{i'j} = 0 \quad \text{for} \quad i \neq i' \\ \sum_{j=1}^{k} c_{ij}^{2} = 1, \quad i = 1, \ldots, k.\end{array}\right\} \quad (4)$$

It is known from linear algebra that $|\Delta| = 1$ for an orthogonal transformation. Also in the case of an orthogonal transformation, we have $d_{ij} = c_{ji}$, $i, j = 1, \ldots, k$, so that

$$x_i = \sum_{j=1}^{k} c_{ji} y_j, \quad i = 1, \ldots, k.$$

This is seen as follows:

$$\sum_{j=1}^{k} c_{ji} y_j = \sum_{j=1}^{k} c_{ji} \left(\sum_{l=1}^{k} c_{jl} x_l \right) = \sum_{j=1}^{k} \sum_{l=1}^{k} c_{ji} c_{jl} x_l = \sum_{l=1}^{k} x_l \left(\sum_{j=1}^{k} c_{ji} c_{jl} \right) = x_i$$

by means of (3). Thus, for an orthogonal transformation, if

$$y_i = \sum_{j=1}^{k} c_{ij} x_j, \quad \text{then} \quad x_i = \sum_{j=1}^{k} c_{ji} y_j, \quad i = 1, \ldots, k.$$

According to what has been seen so far, the Jacobian of the transformation (1) is $J = \Delta^* = 1/\Delta$, and for the case that the transformation is orthogonal, we have $J = \pm 1$, so that $|J| = 1$. These results are now applied as follows:

Consider the r. vector $\mathbf{X} = (X_1, \ldots, X_k)'$ with p.d.f. $f_\mathbf{X}$ and let S be the subset of $\mathbb{R}^k$ over which $f_\mathbf{X} > 0$. Set

$$Y_i = \sum_{j=1}^{k} c_{ij} X_j, \quad i = 1, \ldots, k,$$

where we assume $\Delta = |(c_{ij})| \neq 0$. Then the p.d.f. of the r. vector $\mathbf{Y} = (Y_1, \ldots, Y_k)'$ is given by

$$f_\mathbf{Y}(y_1, \ldots, y_k) = \begin{cases} f_\mathbf{X}\left(\sum_{j=1}^{k} d_{1j} y_j, \ldots, \sum_{j=1}^{k} d_{kj} y_j\right) \cdot \frac{1}{|\Delta|}, & (y_1, \ldots, y_k)' \in T \\ 0, & \text{otherwise,} \end{cases}$$

where T is the image of S under the transformation in question. In particular, if the transformation is orthogonal,

$$f_\mathbf{Y}(y_1,\ldots,y_k) = \begin{cases} f_\mathbf{X}\left(\sum_{j=1}^k c_{j1}y_j, \ldots, \sum_{j=1}^k c_{jk}y_j\right), & (y_1,\ldots,y_k)' \in T \\ 0, & \text{otherwise.} \end{cases}$$

Another consequence of orthogonality of the transformation is that

$$\sum_{i=1}^k Y_i^2 = \sum_{i=1}^k X_i^2.$$

In fact,

$$\sum_{i=1}^k Y_i^2 = \sum_{i=1}^k \left(\sum_{j=1}^k c_{ij} X_j\right)^2 = \sum_{i=1}^k \left(\sum_{j=1}^k c_{ij} X_j\right)\left(\sum_{l=1}^k c_{il} X_l\right)$$
$$= \sum_{i=1}^k \sum_{j=1}^k \sum_{l=1}^k c_{ij} c_{il} X_j X_l = \sum_{j=1}^k \sum_{l=1}^k X_j X_l \left(\sum_{i=1}^k c_{ij} c_{il}\right)$$
$$= \sum_{i=1}^k X_i^2$$

because

$$\sum_{i=1}^k c_{ij} c_{il} = 1 \quad \text{for} \quad j = l \quad \text{and} \quad 0 \quad \text{for} \quad j \neq l.$$

We formulate these results as a theorem.

THEOREM 4 Consider the r. vector $\mathbf{X} = (X_1, \ldots, X_k)'$ with p.d.f. $f_\mathbf{X}$ which is > 0 on $S \subseteq \mathbb{R}^k$. Set

$$Y_i = \sum_{j=1}^k c_{ij} X_j, \quad i = 1, \ldots, k,$$

where $|(c_{ij})| = \Delta \neq 0$. Then

$$X_i = \sum_{j=1}^k d_{ij} Y_j, \quad i = 1, \ldots, k,$$

and the p.d.f. of the r. vector $\mathbf{Y} = (Y_1, \ldots, Y_k)'$ is

$$f_\mathbf{Y}(y_1,\ldots,y_k) = \begin{cases} f_\mathbf{X}\left(\sum_{j=1}^k d_{1j}y_j, \ldots, \sum_{j=1}^k c_{kj}y_j\right) \cdot \frac{1}{|\Delta|}, & (y_1,\ldots,y_k)' \in T \\ 0, & \text{otherwise.} \end{cases}$$

where T is the image of S under the given transformation. If, in particular, the transformation is orthogonal, then

$$f_\mathbf{Y}(y_1,\ldots,y_k) = \begin{cases} f_\mathbf{X}\left(\sum_{j=1}^k c_{j1}y_j, \ldots, \sum_{j=1}^k c_{jk}y_j\right), & (y_1,\ldots,y_k)' \in T \\ 0, & \text{otherwise.} \end{cases}$$

Furthermore, in the case of orthogonality, we also have

$$\sum_{j=1}^{k} X_j^2 = \sum_{j=1}^{k} Y_j^2.$$

The following theorem is an application of Theorem 4 to the normal case.

THEOREM 5 Let the r.v.'s X_i be $N(\mu_i, \sigma^2)$, $i = 1, \ldots, k$, and independent. Consider the orthogonal transformation

$$Y_i = \sum_{j=1}^{k} c_{ij} X_j, \quad i = 1, \ldots, k.$$

Then the r.v.'s $Y_1, \ldots, Y_k$ are also independent, normally distributed with common variance σ^2 and means given by

$$E(Y_i) = \sum_{j=1}^{k} c_{ij} \mu_j, \quad i = 1, \ldots, k.$$

PROOF With $\mathbf{X} = (X_1, \ldots, X_k)'$ and $\mathbf{Y} = (Y_1, \ldots, Y_k)'$, we have

$$f_{\mathbf{X}}(x_1, \ldots, x_k) = \left(\frac{1}{\sqrt{2\pi}\sigma}\right)^k \exp\left[-\frac{1}{2\sigma^2} \sum_{i=1}^{k} (x_i - \mu_i)^2\right],$$

and hence

$$f_{\mathbf{Y}}(y_1, \ldots, y_k) = \left(\frac{1}{\sqrt{2\pi}\sigma}\right)^k \exp\left[-\frac{1}{2\sigma^2} \sum_{i=1}^{k} \left(\sum_{j=1}^{k} c_{ji} y_j - \mu_i\right)^2\right].$$

Now

$$\sum_{i=1}^{k} \left(\sum_{j=1}^{k} c_{ji} y_j - \mu_i\right)^2 = \sum_{i=1}^{k} \left[\left(\sum_{j=1}^{k} c_{ji} y_j\right)^2 + \mu_i^2 - 2\mu_i \sum_{j=1}^{k} c_{ji} y_j\right]$$

$$= \sum_{i=1}^{k} \left(\sum_{j=1}^{k} \sum_{l=1}^{k} c_{ji} c_{li} y_j y_l + \mu_i^2 - 2\mu_i \sum_{j=1}^{k} c_{ji} y_j\right)$$

$$= \sum_{j=1}^{k} \sum_{l=1}^{k} y_j y_l \sum_{i=1}^{k} c_{ji} c_{li} + \sum_{i=1}^{k} \mu_i^2 - 2 \sum_{j=1}^{k} \sum_{i=1}^{k} \mu_i c_{ji} y_j$$

$$= \sum_{j=1}^{k} y_j^2 - 2 \sum_{j=1}^{k} \sum_{i=1}^{k} c_{ji} \mu_i y_j + \sum_{i=1}^{k} \mu_i^2$$

and this is equal to

$$\sum_{j=1}^{k} \left(y_j - \sum_{i=1}^{k} c_{ji} \mu_i\right)^2,$$

since expanding this last expression we get:

$$\sum_{j=1}^{k}\left(y_j^2 + \sum_{i=1}^{k}\sum_{l=1}^{k}c_{ji}c_{jl}\mu_i\mu_l - 2\sum_{i=1}^{k}c_{ji}\mu_i y_j\right)$$

$$= \sum_{j=1}^{k}y_j^2 - 2\sum_{j=1}^{k}\sum_{i=1}^{k}\mu_i c_{ji}y_j + \sum_{i=1}^{k}\sum_{l=1}^{k}\mu_i\mu_l\sum_{j=1}^{k}c_{ji}c_{jl}$$

$$= \sum_{j=1}^{k}y_j^2 - 2\sum_{j=1}^{k}\sum_{i=1}^{k}\mu_i c_{ji}y_j + \sum_{i=1}^{k}\mu_i^2,$$

as was to be seen. ▲

As a further application of Theorems 4 and 5, we consider the following result. Let $Z_1, \ldots, Z_k$ be independent $N(0, 1)$, and set

$$\begin{cases} Y_1 = \dfrac{1}{\sqrt{k}}Z_1 + \dfrac{1}{\sqrt{k}}Z_2 + \cdots + \dfrac{1}{\sqrt{k}}Z_k \\ Y_2 = \dfrac{1}{\sqrt{2\cdot 1}}Z_1 - \dfrac{1}{\sqrt{2\cdot 1}}Z_2 \\ Y_3 = \dfrac{1}{\sqrt{3\cdot 2}}Z_1 + \dfrac{1}{\sqrt{3\cdot 2}}Z_2 - \dfrac{2}{\sqrt{3\cdot 2}}Z_3 \\ \vdots \\ Y_k = \dfrac{1}{\sqrt{k(k-1)}}Z_1 + \cdots + \dfrac{1}{\sqrt{k(k-1)}}Z_{k-1} - \dfrac{k-1}{\sqrt{k(k-1)}}Z_k. \end{cases}$$

We thus have

$$c_{1j} = \frac{1}{\sqrt{k}}, j = 1, \ldots, k, \quad \text{and for} \quad i = 2, \ldots, k$$

$$c_{ij} = \frac{1}{\sqrt{i(i-1)}}, \quad \text{for} \quad j = 1, \ldots, i-1, \text{ and}$$

$$c_{ii} = -\frac{i-1}{\sqrt{i(i-1)}}.$$

Hence

$$\sum_{j=1}^{k}c_{1j}^2 = \frac{k}{k} = 1, \quad \text{and for} \quad i = 2, \ldots, k,$$

$$\sum_{j=1}^{k}c_{ij}^2 = \sum_{j=1}^{i}c_{ij}^2 = (i-1)\cdot\frac{1}{i(i-1)} + \frac{(i-1)^2}{i(i-1)}$$

$$= \frac{1}{i} + \frac{i-1}{i} = 1,$$

while for $i = 2, \ldots, k$, we get

$$\sum_{j=1}^{k}c_{1j}c_{ij} = \frac{1}{\sqrt{k}}\sum_{j=1}^{k}c_{ij} = \frac{1}{\sqrt{k}}\sum_{j=1}^{i}c_{ij} = \frac{1}{\sqrt{k}}\left(\frac{i-1}{\sqrt{i(i-1)}} - \frac{i-1}{\sqrt{i(i-1)}}\right) = 0,$$

and for $i, l = 2, \ldots, k (i \neq l)$, we have
$$\sum_{j=1}^{k} c_{ij}c_{lj} = \sum_{j=1}^{i} c_{ij}c_{lj} \quad \text{if} \quad i < l,$$
and
$$\sum_{j=1}^{l} c_{ij}c_{lj} \quad \text{if} \quad i > l.$$
For $i < l$, this is
$$\frac{1}{\sqrt{i(i-1)l(l-1)}}\left[(i-1)-(i-1)\right] = 0,$$
and for $i > l$, this is
$$\frac{1}{\sqrt{i(i-1)l(l-1)}}\left[(l-1)-(l-1)\right] = 0.$$

Thus the transformation is orthogonal. It follows, by Theorem 5, that $Y_1, \ldots, Y_k$ are independent, $N(0, 1)$, and that
$$\sum_{i=1}^{k} Y_i^2 = \sum_{i=1}^{k} Z_i^2 \quad \text{by Theorem 4.}$$
Thus
$$\sum_{i=2}^{k} Y_i^2 = \sum_{i=1}^{k} Y_i^2 - Y_1^2 = \sum_{i=1}^{k} Z_i^2 - \left(\sqrt{k}\overline{Z}\right)^2$$
$$= \sum_{i=1}^{k} Z_i^2 - k\overline{Z}^2 = \sum_{i=1}^{k} \left(Z_i - \overline{Z}\right)^2.$$

Since Y_1 is independent of $\sum_{i=2}^{k} Y_i^2$, we conclude that $\overline{Z}$ is independent of $\sum_{i=1}^{k}(Z_i - \overline{Z})^2$. Thus we have the following theorem.

THEOREM 6 Let $X_1, \ldots, X_k$ be independent r.v.'s distributed as $N(\mu, \sigma^2)$. Then $\overline{X}$ and S^2 are independent.

PROOF Set $Z_j = (X_j - \mu)/\sigma$, $j = 1, \ldots, k$. Then the Z's are as above, and hence
$$\overline{Z} = \frac{1}{\sigma}\left(\overline{X} - \mu\right) \quad \text{and} \quad \sum_{j=1}^{k}\left(Z_j - \overline{Z}\right)^2 = \frac{1}{\sigma^2}\sum_{j=1}^{k}\left(X_j - \overline{X}\right)^2$$
are independent. Hence $\overline{X}$ and S^2 are independent. ▲

Exercises

9.3.1 For $i = 1, 2, 3$, let X_i be independent r.v.'s distributed as $N(\mu_i, \sigma^2)$, and set:
$$Y_1 = -\frac{1}{\sqrt{2}}X_1 + \frac{1}{\sqrt{2}}X_2, \quad Y_2 = -\frac{1}{\sqrt{3}}X_1 - \frac{1}{\sqrt{3}}X_2 + \frac{1}{\sqrt{3}}X_3,$$
$$Y_3 = \frac{1}{\sqrt{6}}X_1 + \frac{1}{\sqrt{6}}X_2 + \frac{2}{\sqrt{6}}X_3.$$

Then:

i) Show that the r.v.'s Y_1, Y_2, Y_3 are also independent normally distributed with variance σ^2, and specify their respective means.
(Hint: Verify that the transformation is orthogonal, and then use Theorem 5);

ii) If $\mu_1 = \mu_2 = \mu_3 = 0$, use a conclusion in Theorem 4 in order to show that $Y_1^2 + Y_2^2 + Y_3^2 \sim \sigma^2 \chi_3^2$.

9.3.2 If the pair of r.v.'s (X, Y) has the Bivariate Normal distribution with parameters $\mu_1, \mu_2, \sigma_1^2, \sigma_2^2, \rho$, that is, $(X, Y) \sim N(\mu_1, \mu_2, \sigma_1^2, \sigma_2^2, \rho)$, then show that $\left(\frac{X-\mu_1}{\sigma_1}, \frac{Y-\mu_2}{\sigma_2}\right)' \sim N(0, 0, 1, 1, \rho)$, and vice versa.

9.3.3 If $(X, Y)' \sim N(0, 0, 1, 1, \rho)$, and c, d are constants with $cd \neq 0$, then show that $(cX, dY) \sim N(0, 0, c^2, d^2, \rho_0)$, where $\rho_0 = \rho$ if $cd > 0$, and $\rho_0 = -\rho$ if $cd < 0$.

9.3.4 If $(X, Y)' \sim N(0, 0, 1, 1, \rho)$, show that $X + Y \sim N(0, 2(1-\rho))$, $X - Y \sim N(0, 2(1-\rho))$, and $X+Y, X-Y$ are independent.

9.3.5 If $(X, Y)' \sim N(\mu_1, \mu_2, \sigma_1^2, \sigma_2^2, \rho)$, and $U = \frac{X-\mu_1}{\sigma_1}$, $V = \frac{Y-\mu_2}{\sigma_2}$, then:

i) Determine the distribution of the r.v.'s $U+V$, $U-V$, and show that these r.v.'s are independent;

ii) In particular, for $\sigma_1^2 = \sigma_2^2 = \sigma^2$, say, specify the distributions of the r.v.'s $X+Y, X-Y$, and show that r.v.'s are independent.

9.3.6 Let $(X, Y)' \sim N(0, 0, \sigma_1^2, \sigma_2^2, \rho)$. Then:

i) $(X+Y, X-Y)' \sim N(0, 0, \tau_1^2, \tau_2^2, \rho_0)$, where
$$\tau_1^2 = \sigma_1^2 + \sigma_2^2 + 2\rho\sigma_1\sigma_2, \quad \tau_2^2 = \sigma_1^2 + \sigma_2^2 - 2\rho\sigma_1\sigma_2, \text{ and } \rho_0 = \left(\sigma_1^2 - \sigma_2^2\right)/\tau_1\tau_2;$$

ii) $X + Y \sim N(0, \tau_1^2)$ and $X - Y \sim N(0, \tau_2^2)$;

iii) The r.v.'s $X+Y$ and $X-Y$ are independent if and only if $\sigma_1 = \sigma_2$. (Compare with the latter part of Exercise 9.3.5.)

9.3.7 Let $(X, Y)' \sim N(\mu_1, \mu_2, \sigma_1^2, \sigma_2^2, \rho)$, and let c, d be constants with $cd \neq 0$. Then:

i) $(cX, dY)' \sim N(c\mu_1, d\mu_2, c^2\sigma_1^2, d^2\sigma_2^2, \pm\rho)$, with $+\rho$ if $cd > 0$, and $-\rho$ if $cd < 0$;

ii) $(cX + dY, cX - dY)' \sim N(c\mu_1 + d\mu_2, c\mu_1 - d\mu_2, \tau_1^2, \tau_2^2, \rho_0)$, where
$$\tau_1^2 = c^2\sigma_1^2 + d^2\sigma_2^2 + 2pcd\sigma_1\sigma_2, \quad \tau_2^2 = c^2\sigma_1^2 + d^2\sigma_2^2 - 2\rho cd\sigma_1\sigma_2,$$
$$\text{and } \rho_0 = \frac{c^2\sigma_1^2 - d^2\sigma_2^2}{\tau_1\tau_2};$$

iii) The r.v.'s $cX + dY$ and $cX - dY$ are independent if and only if $\frac{c}{d} = \pm\frac{\sigma_2}{\sigma_1}$;

iv) The r.v.'s in part (iii) are distributed as $N(c\mu_1 + d\mu_2, \tau_1^2)$, and $N(c\mu_1 - d\mu_2, \tau_2^2)$, respectively.

9.3.8 Refer to Exercise 9.3.7 and:

i) Provide an expression for the probability $P(cX + dY > \lambda)$;

ii) Give the numerical value of the probability in part (i) for $c = 2$, $d = 3$, $\lambda = 15$, $\mu_1 = 3.5$, $\mu_2 = 1.5$, $\sigma_1 = 1$, $\sigma_2 = 0.9$, and $\rho = -0.5$.

9.3.9 For $j = 1, \ldots, n$, let $(X_j, Y_j)'$ be independent r. vectors with distribution $N(\mu_1, \mu_2, \sigma_1^2, \sigma_2^2, \rho)$. Then:

i) Determine the distribution of the r.v. $\bar{X} - \bar{Y}$;

ii) What does this distribution become for $\mu_1 = \mu_2$ and $\sigma_1^2 = \sigma_2^2 = \sigma^2$, say?

9.4 The Probability Integral Transform

In this short section, we derive two main results. According to the first result, if X is any r.v. with continuous d.f. F, and if $Y = F(X)$, then surprisingly enough $Y \sim U(0, 1)$. The name of this section is derived from the transformation $Y = F(X)$, since F is represented by the integral of a p.d.f. (in the absolutely continuous case). Next, in several instances a statement has been made to the effect that X is an r.v. with d.f. F. The question then arises as to whether such an r.v. can actually be constructed. The second result resolves this question as follows: Let F be any d.f. and let $Y \sim U(0, 1)$. Set $X = F^{-1}(Y)$. Then $X \sim F$.

The proof presented is an adaptation of the discussion in the Note "The Probability Integral Transformation: A Simple Proof" by E. F. Schuster, published in *Mathematics Magazine*, Vol. 49 (1976) No. 5, pages 242–243.

THEOREM 7 Let X be an r.v. with continuous d.f. F, and define the r.v. Y by $Y = F(X)$. Then the distribution of Y is $U(0, 1)$.

PROOF Let G be the d.f. of Y. We will show that $G(y) = y$, $0 < y < 1$; $G(0) = 0$; $G(1) = 1$. Indeed, let $y \in (0, 1)$. Since $F(x) \to 0$ as $x \to -\infty$, there exists a such that $(0 \leq) F(a) < y$; and since $F(x) \to 1$ as $x \to \infty$, there exists $\varepsilon > 0$ such that $y + \varepsilon < 1$ and $F(y) < F(y + \varepsilon) (\leq 1)$. Set $F(a) = c$, $y + \varepsilon = b$, and $F(b) = d$. Then the function F is continuous in the closed interval $[a, b]$ and all y of the form $y + \frac{\varepsilon}{n}$ ($n \geq 2$ integer) lie in (c, d). Therefore, by the Intermediate Value Theorem (see, for example, Theorem 3(ii) on page 95 in *Calculus and Analytic Geometry*, 3rd edition (1966), by George B. Thomas, Addison-Wesley Publishing Company, Inc., Reading, Massachusetts) there exist x_0 and x_n ($n \geq 2$) in (a, b) such that $F(x_0) = y$ and $F(x_n) = y + \frac{\varepsilon}{n}$. Then

$$(X \le x_0) \subseteq [F(X) \le F(x_0)] \quad \text{(since } F \text{ is nondecreasing)}$$
$$= [F(X) \le y] \quad \text{(since } F(x_0) = y\text{)}$$
$$\subseteq \left[F(X) < y + \frac{\varepsilon}{n}\right]$$
$$= [F(X) < F(x_n)] \quad \left(\text{since } F(x_n) = y + \frac{\varepsilon}{n}\right)$$
$$\subseteq (X < x_n) \quad \text{(by the fact that } F \text{ is nondecreasing and by contradiction)}.$$

That is $(X \le x_0) \subseteq [F(X) \le y] \subseteq (X \le x_n)$. Hence

$$P(X \le x_0) \le P[F(X) \le y] \le P(X \le x_n),$$

or $y = F(x_0) \le G(y) \le F(x_n) = y + \frac{\varepsilon}{n}.$

Letting $n \to \infty$, we obtain $G(y) = y$. Next, G is right-continuous, being a d.f. Thus, as $y \downarrow 0$, $G(0) = \lim G(y) = \lim y = 0$. Finally, as $y \uparrow 1$, $G(1-) = \lim G(y) = \lim y = 1$, so that $G(1) = 1$. The proof is completed. ▲

For the formulation and proof of the second result, we need some notation and a preliminary result. To this end, let X be an r.v. with d.f. F. Set $y = F(x)$ and define F^{-1} as follows:

$$F^{-1}(y) = \inf\{x \in \mathbb{R}; F(x) \ge y\}. \tag{5}$$

From this definition it is then clear that when F is strictly increasing, for each $x \in \mathbb{R}$, there is exactly one $y \in (0, 1)$ such that $F(x) = y$. It is also clear that, if F is continuous, then the above definition becomes as follows:

$$F^{-1}(y) = \inf\{x \in \mathbb{R}; F(x) = y\}. \tag{6}$$

(See also Figs. 9.9, 9.10 and 9.11.)

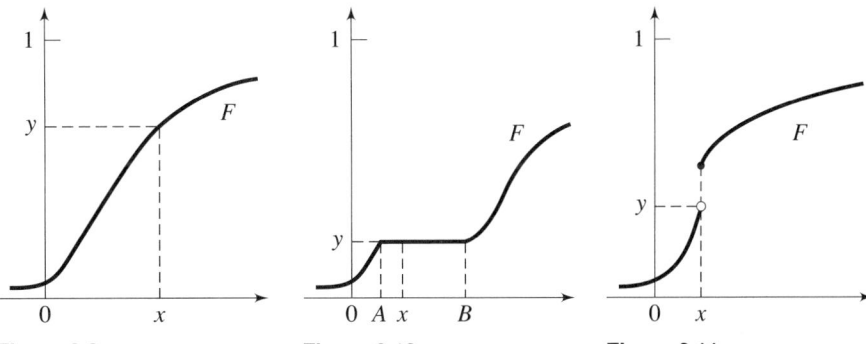

Figure 9.9   Figure 9.10   Figure 9.11

We now establish the result to be employed.

LEMMA 1 Let F^{-1} be defined by (5). Then $F^{-1}(y) \leq t$ if and only if $y \leq F(t)$.

PROOF We have $F^{-1}(y) = \inf\{x \in \mathbb{R}; F(x) \geq y\}$. Therefore there exists $x_n \in \{x \in \mathbb{R}; F(x_n) \geq y\}$ such that $x_n \downarrow F^{-1}(y)$. Hence $F(x_n) \to F[F^{-1}(y)]$, by the right continuity of F, and

$$F[F^{-1}(y)] \geq y. \tag{7}$$

Now assume that $F^{-1}(y) \leq t$. Then $F[F^{-1}(y)] \leq F(t)$, since F is nondecreasing. Combining this result with (7), we obtain $y \leq F(t)$.

Next assume, that $y \leq F(t)$. This means that t belongs to the set $\{x \in \mathbb{R}, F(x) \geq y\}$ and hence $F^{-1}(y) \leq t$. The proof of the lemma is completed. ▲

By means of the above lemma, we may now establish the following result.

THEOREM 8 Let Y be an r.v. distributed as $U(0, 1)$, and let F be a d.f. Define the r.v. X by $X = F^{-1}(Y)$, where F^{-1} is defined by (5). Then the d.f. of X is F.

PROOF We have

$$P(X \leq x) = P[F^{-1}(Y) \leq x] = P[Y \leq F(x)] = F(x),$$

where the last step follows from the fact that Y is distributed as $U(0, 1)$ and the one before it by Lemma 1. ▲

REMARK 7 As has already been stated, the theorem just proved provides a specific way in which one can construct an r.v. X whose d.f. is a given d.f. F.

Exercise

9.4.1 Let $X_j, j = 1, \ldots, n$ be independent r.v.'s such that X_j has continuous and strictly increasing d.f. F_j. Set $Y_j = F_j(X_j)$ and show that the r.v.

$$X = -2 \sum_{j=1}^{n} \log(1 - Y_j)$$

is distributed as χ^2_{2n}.

Chapter 10

Order Statistics and Related Theorems

In this chapter we introduce the concept of order statistics and also derive various distributions. The results obtained here will be used in the second part of this book for statistical inference purposes.

10.1 Order Statistics and Related Distributions

Let $X_1, X_2, \ldots, X_n$ be i.i.d. r.v.'s with d.f. F. The *jth order statistic* of $X_1, X_2, \ldots, X_n$ is denoted by $X_{(j)}$, or Y_j, for easier writing, and is defined as follows:

$$Y_j = j\text{th smallest of the } X_1, X_2, \ldots, X_n, j = 1, \ldots, n;$$

(that is, for each $s \in S$, look at $X_1(s), X_2(s), \ldots, X_n(s)$, and then $Y_j(s)$ is defined to be the jth smallest among the numbers $X_1(s), X_2(s), \ldots, X_n(s), j = 1, 2, \ldots, n$). It follows that $Y_1 \leq Y_2 \leq \cdots \leq Y_n$, and, in general, the Y's are not independent.

We assume now that the X's are of the continuous type with p.d.f. f such that $f(x) > 0$, $(-\infty \leq)a < x < b(\leq \infty)$ and zero otherwise. One of the problems we are concerned with is that of finding the joint p.d.f. of the Y's. By means of Theorem 3', Chapter 9, it will be established that:

THEOREM 1 If $X_1, \ldots, X_n$ are i.i.d. r.v.'s with p.d.f. f which is positive for $a < x < b$ and 0 otherwise, then the joint p.d.f. of the order statistics $Y_1, \ldots, Y_n$ is given by:

$$g(y_1, \ldots, y_n) = \begin{cases} n! f(y_1) \cdots f(y_n), & a < y_1 < y_2 < \cdots < y_n < b \\ 0, & \text{otherwise.} \end{cases}$$

PROOF The proof is carried out explicitly for $n = 3$, but it is easily seen, with the proper change in notation, to be valid in the general case as well. In the first place, since for $i \neq j$,

$$P(X_i = X_j) = \iint_{(x_i = x_j)} f(x_i)f(x_j) dx_i dx_j = \int_a^b \int_{x_j}^{x_j} f(x_i)f(x_j) dx_i dx_j = 0,$$

and therefore $P(X_i = X_j = X_k) = 0$ for $i \neq j \neq k$, we may assume that the joint p.d.f., $f(\cdot, \cdot, \cdot)$, of X_1, X_2, X_3 is zero if at least two of the arguments x_1, x_2, x_3 are equal. Thus we have

$$f(x_1, x_2, x_3) = \begin{cases} f(x_1)f(x_2)f(x_3), & a < x_1 \neq x_2 \neq x_3 < b \\ 0, & \text{otherwise.} \end{cases}$$

Thus $f(x_1, x_2, x_3)$ is positive on the set S, where

$$S = \left\{ (x_1, x_2, x_3)' \in \mathbb{R}^3; a < x_i < b, i = 1, 2, 3, x_1, x_2, x_3 \text{ all different} \right\}.$$

Let $S_{ijk} \subset S$ be defined by

$$S_{ijk} = \left\{ (x_1, x_2, x_3)'; a < x_i < x_j < x_k < b \right\}, \quad i, j, k = 1, 2, 3, \quad i \neq j \neq k.$$

Then we have

$$S = S_{123} + S_{132} + S_{213} + S_{231} + S_{312} + S_{321}.$$

Now on each one of the S_{ijk}'s there exists a one-to-one transformation from the x's to the y's defined as follows:

$$\begin{aligned} S_{123} &: y_1 = x_1, \quad y_2 = x_2, \quad y_3 = x_3 \\ S_{132} &: y_1 = x_1, \quad y_2 = x_3, \quad y_3 = x_2 \\ S_{213} &: y_1 = x_2, \quad y_2 = x_1, \quad y_3 = x_3 \\ S_{231} &: y_1 = x_2, \quad y_2 = x_3, \quad y_3 = x_1 \\ S_{312} &: y_1 = x_3, \quad y_2 = x_1, \quad y_3 = x_2 \\ S_{321} &: y_1 = x_3, \quad y_2 = x_2, \quad y_3 = x_1. \end{aligned}$$

Solving for the x's, we have then:

$$\begin{aligned} S_{123} &: x_1 = y_1, \quad x_2 = y_2, \quad x_3 = y_3 \\ S_{132} &: x_1 = y_1, \quad x_2 = y_3, \quad x_3 = y_2 \\ S_{213} &: x_1 = y_2, \quad x_2 = y_1, \quad x_3 = y_3 \\ S_{231} &: x_1 = y_3, \quad x_2 = y_1, \quad x_3 = y_2 \\ S_{312} &: x_1 = y_2, \quad x_2 = y_3, \quad x_3 = y_1 \\ S_{321} &: x_1 = y_3, \quad x_2 = y_2, \quad x_3 = y_1. \end{aligned}$$

The Jacobians are thus given by:

$$S_{123}: J_{123} = \begin{vmatrix} 1 & 0 & 0 \\ 0 & 1 & 0 \\ 0 & 0 & 1 \end{vmatrix} = 1 \qquad S_{231}: J_{231} = \begin{vmatrix} 0 & 0 & 1 \\ 1 & 0 & 0 \\ 0 & 1 & 0 \end{vmatrix} = 1$$

$$S_{132}: J_{132} = \begin{vmatrix} 1 & 0 & 0 \\ 0 & 0 & 1 \\ 0 & 1 & 0 \end{vmatrix} = -1; \qquad S_{312}: J_{312} = \begin{vmatrix} 0 & 1 & 0 \\ 0 & 0 & 1 \\ 1 & 0 & 0 \end{vmatrix} = 1$$

$$S_{213}: J_{213} = \begin{vmatrix} 0 & 1 & 0 \\ 1 & 0 & 0 \\ 0 & 0 & 1 \end{vmatrix} = -1 \qquad S_{321}: J_{321} = \begin{vmatrix} 0 & 0 & 1 \\ 0 & 1 & 0 \\ 1 & 0 & 0 \end{vmatrix} = -1.$$

Hence $|J_{123}| = \cdots = |J_{321}| = 1$, and Theorem 3', Chapter 9, gives

$$g(y_1, y_2, y_3) = \begin{cases} f(y_1)f(y_2)f(y_3) + f(y_1)f(y_3)f(y_2) + f(y_2)f(y_1)f(y_3) \\ + f(y_3)f(y_1)f(y_2) + f(y_2)f(y_3)f(y_1) + f(y_3)f(y_2)f(y_1), \\ \qquad a < y_1 < y_2 < y_3 < b \\ 0, \quad \text{otherwise}. \end{cases}$$

This is,

$$g(y_1, y_2, y_3) = \begin{cases} 3! f(y_1) f(y_2) f(y_3), & a < y_1 < y_2 < y_3 < b \\ 0, & \text{otherwise}. \end{cases} \quad \blacktriangle$$

Notice that the proof in the general case is exactly the same. One has $n!$ regions forming S, one for each permutation of the integers 1 through n. From the definition of a determinant and the fact that each row and column contains exactly one 1 and the rest all 0, it follows that the $n!$ Jacobians are either 1 or -1 and the remaining part of the proof is identical to the one just given except one adds up $n!$ like terms instead of 3!.

EXAMPLE 1 Let $X_1, \ldots, X_n$ be i.i.d. r.v.'s distributed as $N(\mu, \sigma^2)$. Then the joint p.d.f. of the order statistics $Y_1, \ldots, Y_n$ is given by

$$g(y_1, \ldots, y_n) = n! \left(\frac{1}{\sqrt{2\pi}\sigma} \right)^n \exp\left[-\frac{1}{2\sigma^2} \sum_{j=1}^{n} (y_j - \mu)^2 \right],$$

if $-\infty < y_1 < \cdots < y_n < \infty$ and zero otherwise.

EXAMPLE 2 Let $X_1, \ldots, X_n$ be i.i.d. r.v.'s distributed as $U(\alpha, \beta)$. Then the joint p.d.f. of the order statistics $Y_1, \ldots, Y_n$ is given by

$$g(y_1, \ldots, y_n) = \frac{n!}{(\beta - \alpha)^n},$$

if $\alpha < y_1 < \cdots < y_n < \beta$ and zero otherwise.

Another interesting problem is that of finding the marginal p.d.f. of each $Y_j, j = 1, \ldots, n$, as well as the joint p.d.f. of any number of the Y_j's. As a partial answer to this problem, we have the following theorem.

THEOREM 2 Let $X_1, \ldots, X_n$ be i.i.d. r.v.'s with d.f. F and p.d.f. f which is positive and continuous for $(-\infty \leq) a < x < b (\leq \infty)$ and zero otherwise, and let $Y_1, \ldots, Y_n$ be the order statistics. Then the p.d.f. g_j of $Y_j, j = 1, 2, \ldots, n$, is given by:

i) $g_j(y_j) = \begin{cases} \dfrac{n!}{(j-1)!(n-j)!} [F(y_j)]^{j-1} [1 - F(y_j)]^{n-j} f(y_j), & a < y_j < b \\ 0, & \text{otherwise.} \end{cases}$

In particular,

i') $g_1(y_1) = \begin{cases} n[1 - F(y_1)]^{n-1} f(y_1), & a < y_1 < b \\ 0, & \text{otherwise} \end{cases}$

and

i'') $g_n(y_n) = \begin{cases} n[F(y_n)]^{n-1} f(y_n), & a < y_n < b \\ 0, & \text{otherwise.} \end{cases}$

The joint p.d.f. g_{ij} of any Y_i, Y_j with $1 \leq i < j \leq n$, is given by:

ii) $g_{ij}(y_i, y_j) = \begin{cases} \dfrac{n!}{(i-1)!(j-i-1)!(n-j)!} [F(y_i)]^{i-1} [F(y_j) - F(y_i)]^{j-i-1} \\ \times [1 - F(y_j)]^{n-j} \cdot f(y_i) f(y_j), & a < y_i < y_j < b \\ 0, & \text{otherwise.} \end{cases}$

In particular,

ii') $g_{1n}(y_1, y_n) = \begin{cases} n(n-1)[F(y_n) - F(y_1)]^{n-2} f(y_1) f(y_n), & a < y_1 < y_n < b \\ 0, & \text{otherwise.} \end{cases}$

PROOF From Theorem 1, we have that $g(y_1, \ldots, y_n) = n! f(y_1) \cdots f(y_n)$ for $a < y_1 < \cdots < y_n < b$ and equals 0 otherwise. Since f is positive in (a, b), it follows that F is strictly increasing in (a, b) and therefore F^{-1} exists in this interval. Hence if $u = F(y)$, $y \in (a, b)$, then $y = F^{-1}(u)$, $u \in (0, 1)$ and

$$\frac{dy}{du} = \frac{1}{f[F^{-1}(u)]}, \quad u \in (0, 1).$$

Therefore by setting $U_j = F(Y_j)$, $j = 1, \ldots, n$, one has that the joint p.d.f. h of the U's is given by

$$h(u_1, \ldots, u_n) = n! f\left[F^{-1}(u_1)\right] \cdots f\left[F^{-1}(u_n)\right] \frac{1}{f\left[F^{-1}(u_1)\right] \cdots f\left[F^{-1}(u_n)\right]}$$

for $0 < u_1 < \cdots < u_n < 1$ and equals 0 otherwise; that is, $h(u_1, \ldots, u_n) = n!$ for $0 < u_1 < \cdots < u_n < 1$ and equals 0 otherwise. Hence for $u_j \in (0, 1)$,

$$h(u_j) = n! \int_0^{u_j} \cdots \int_0^{u_2} \int_{u_j}^1 \cdots \int_{u_{n-1}}^1 du_n \cdots du_{j+1} \, du_1 \cdots du_{j-1}.$$

The first $n - j$ integrations with respect to the variables $u_n, \ldots, u_{j+1}$ yield $[1/(n-j)!](1 - u_j)^{n-j}$ and the last $j - 1$ integrations with respect to the variables $u_1, \ldots, u_{j-1}$ yield $[1/(j-1)!] u_j^{j-1}$. Thus

$$h(u_j) = \frac{n!}{(j-1)!(n-j)!} u_j^{j-1} (1 - u_j)^{n-j}$$

for $u_j \in (0, 1)$ and equals 0 otherwise. Finally, using once again the transformation $U_j = F(Y_j)$, we obtain

$$g(y_j) = \frac{n!}{(j-1)!(n-j)!} \left[F(y_j)\right]^{j-1} \left[1 - F(y_j)\right]^{n-j} f(y_j)$$

for $y_j \in (a, b)$ and 0 otherwise. This completes the proof of (i).

Of course, (i′) and (i″) follow from (i) by setting $j = 1$ and $j = n$, respectively. An alternative and easier way of establishing (i′) and (i″) is the following:

$$G_n(y_n) = P(Y_n \leq y_n) = P\left(\text{all } X_j\text{'s} \leq y_n\right) = F^n(y_n).$$

Thus $g_n(y_n) = n[F(y_n)]^{n-1} f(y_n)$. Similarly,

$$1 - G_1(y_1) = P(Y_1 > y_1) = P\left(\text{all } X_j\text{'s} > y_1\right) = \left[1 - F(y_1)\right]^n.$$

Then

$$-g_1(y_1) = n\left[1 - F(y_1)\right]^{n-1} \left[-f(y_1)\right], \quad \text{or} \quad g_1(y_1) = n\left[1 - F(y_1)\right]^n f(y_1).$$

The proof of (ii) is similar to that of (i), and in fact the same method can be used to find the joint p.d.f. of any number of Y_j's (see also Exercise 10.1.19). ▲

EXAMPLE 3 Refer to Example 2. Then

$$F(x) = \begin{cases} 0, & x \leq \alpha \\ \dfrac{x - \alpha}{\beta - \alpha}, & a < x < \beta \\ 1, & x \geq \beta, \end{cases}$$

and therefore

$$g_j(y_j) = \begin{cases} \dfrac{n!}{(j-1)!(n-j)!}\left(\dfrac{y_j-\alpha}{\beta-\alpha}\right)^{j-1}\left(\dfrac{\beta-y_j}{\beta-\alpha}\right)^{n-j}\dfrac{1}{\beta-\alpha}, & \alpha < y_j < \beta \\ 0, & \text{otherwise} \end{cases}$$

$$= \begin{cases} \dfrac{n!}{(j-1)!(n-j)!(\beta-\alpha)^n}(y_j-\alpha)^{j-1}(\beta-y_j)^{n-j}, & \alpha < y_j < \beta \\ 0, & \text{otherwise.} \end{cases}$$

Thus

$$g_1(y_1) = \begin{cases} n\left(\dfrac{\beta-y_1}{\beta-\alpha}\right)^{n-1}\dfrac{1}{\beta-\alpha} = \dfrac{n}{(\beta-\alpha)^n}(\beta-y_1)^{n-1}, & \alpha < y_1 < \beta \\ 0, & \text{otherwise,} \end{cases}$$

$$g_n(y_n) = \begin{cases} n\left(\dfrac{y_n-\alpha}{\beta-\alpha}\right)^{n-1}\dfrac{1}{\beta-\alpha} = \dfrac{n}{(\beta-\alpha)^n}(y_n-\alpha)^{n-1}, & \alpha < y_n < \beta \\ 0, & \text{otherwise,} \end{cases}$$

$$g_{1n}(y_1, y_n) = \begin{cases} n(n-1)\left(\dfrac{y_n-y_1}{\beta-\alpha}\right)^{n-2}\dfrac{1}{(\beta-\alpha)^2} = \dfrac{n(n-1)}{(\beta-\alpha)^n}(y_n-y_1)^{n-2}, \\ \qquad\qquad\qquad\qquad\qquad\qquad\qquad\qquad \alpha < y_1 < y_n < \beta \\ 0, \qquad\qquad\qquad\qquad\qquad\qquad\qquad\qquad \text{otherwise.} \end{cases}$$

In particular, for $\alpha = 0$, $\beta = 1$, these formulas simplify as follows:

$$g_j(y_j) = \begin{cases} \dfrac{n!}{(j-1)!(n-j)!}y_j^{j-1}(1-y_j)^{n-j}, & 0 < y_j < 1 \\ 0, & \text{otherwise.} \end{cases}$$

Since $\Gamma(m) = (m-1)!$, this becomes

$$g_j(y_j) = \begin{cases} \dfrac{\Gamma(n+1)}{\Gamma(j)\Gamma(n-j+1)}y_j^{j-1}(1-y_j)^{n-j}, & 0 < y_j < 1 \\ 0, & \text{otherwise,} \end{cases}$$

which is the density of a Beta distribution with parameters $\alpha = j$, $\beta = n - j + 1$.
Likewise

$$g_1(y_1) = \begin{cases} n(1-y_1)^{n-1}, & 0 < y_1 < 1 \\ 0, & \text{otherwise,} \end{cases}$$

$$g_n(y_n) = \begin{cases} ny_n^{n-1}, & 0 < y_n < 1 \\ 0, & \text{otherwise} \end{cases}$$

and

$$g_{1n}(y_1, y_n) = \begin{cases} n(n-1)(y_n - y_1)^{n-2}, & 0 < y_1 < y_n < 1 \\ 0, & \text{otherwise.} \end{cases}$$

The r.v. $Y = Y_n - Y_1$ is called the (*sample*) *range* and is of some interest in statistical applications. The distribution of Y is found as follows. Consider the transformation

$$\begin{cases} y = y_n - y_1 \\ z = y_1. \end{cases} \quad \text{Then} \quad \begin{cases} y_1 = z \\ y_n = y + z \end{cases} \quad \text{and hence} \quad |J| = 1.$$

Therefore

$$f_{Y,Z}(y, z) = g_{1n}(z, y+z)$$

$$= n(n-1)\bigl[F(y+z) - F(z)\bigr]^{n-2} f(z)f(y+z), \begin{cases} 0 < y < b-a \\ a < z < b-y \end{cases}$$

and zero otherwise. Integrating with respect to z, one obtains

$$f_Y(y) = \begin{cases} n(n-1)\int_a^{b-y} \bigl[F(y+z) - F(z)\bigr]^{n-2} f(z)f(y+z)\,dz, & 0 < y < b-a \\ 0, & \text{otherwise.} \end{cases}$$

In particular, if X is an r.v. distributed as $U(0, 1)$, then

$$f_Y(y) = n(n-1)\int_0^{1-y} y^{n-2}\,dz = n(n-1)y^{n-2}(1-y), \quad 0 < y < 1;$$

that is

$$f_Y(y) = \begin{cases} n(n-1)y^{n-2}(1-y), & 0 < y < 1 \\ 0, & \text{otherwise.} \end{cases}$$

Let now U be χ_r^2 and independent of the sample range Y. Set

$$Z = \frac{Y}{\sqrt{U/r}}.$$

We are interested in deriving the distribution of the r.v. Z. To this end, we consider the transformation

$$\begin{cases} z = \dfrac{y}{\sqrt{u/r}} \\ w = u/r. \end{cases} \text{Then} \quad \begin{cases} u = rw \\ y = z\sqrt{w} \end{cases} \text{and hence} \quad |J| = r\sqrt{w}.$$

Therefore

$$f_{Z,W}(z, w) = f_Y\!\left(z\sqrt{w}\right) f_U(rw) r\sqrt{w},$$

if $0 < z, w < \infty$ and zero otherwise. Integrating out w, we get

$$f_Z(z) = \int_0^\infty f_Y\!\left(z\sqrt{w}\right) f_U(rw) r\sqrt{w}\, dw,$$

if $0 < z < \infty$ and zero otherwise.

Now if the r.v.'s $X_1, \ldots, X_n$ are i.i.d. $N(0, 1)$ and Y is as above, then the r.v. Z is called the *Studentized range*. Its density is given by $f_Z(z)$ above and the values of the points z_α for which $P(Z > z_\alpha) = \alpha$ are given by tables for selected values of α. (See, for example, Donald B. Owen's *Handbook of Statistical Tables*, pp. 144–149, published by Addison-Wesley.)

Exercises

Throughout these exercises, X_j, $j = 1, \ldots, n$, are i.i.d. r.v.'s and $Y_j = X_{(j)}$ is the jth order statistic of the X's. The r.v.'s X_j, $j = 1, \ldots, n$ may represent various physical quantities such as the breaking strength of certain steel bars, the crushing strength of bricks, the weight of certain objects, the life of certain items such as light bulbs, vacuum tubes, etc. From these examples, it is then clear that the distribution of the Y's and, in particular, of Y_1, Y_n as well as $Y_n - Y_1$, are quantities of great interest.

10.1.1 Let X_j, $j = 1, \ldots, n$ be i.i.d. r.v.'s with d.f. and p.d.f. F and f, respectively, and let m be the median of F. Use Theorem 2(i″) in order to calculate the probability that all X_j's exceed m; also calculate the probability $P(Y_n \leq m)$.

10.1.2 Let X_1, X_2, X_3 be independent r.v.'s with p.d.f. f given by:

$$f(x) = e^{-(x-\theta)} I_{(\theta, \infty)}(x_1).$$

Use Theorem 2(i″) in order to determine the constant $c = c(\theta)$ for which $P(\theta < Y_3 < c) = 0.90$.

10.1.3 If the independent r.v.'s $X_1, \ldots, X_n$ are distributed as $U(\alpha, \beta)$, then:

i) Calculate the probability that all X's are greater than $(\alpha + \beta)/2$;

ii) In particular, for $\alpha = 0$, $\beta = 1$, and $n = 2$, derive the p.d.f. of the r.v. Y_2/Y_1.

10.1.4 Let X_j, $j = 1, \ldots, n$ be independent r.v.'s distributed as $U(\alpha, \beta)$. Then:

i) Use the p.d.f. derived in Example 3 in order to show that

$$EY_j = \frac{(\beta-\alpha)j}{n+1} + \alpha \quad \text{and} \quad \sigma^2(Y_j) = \frac{(\beta-\alpha)^2 j(n-j+1)}{(n+1)^2(n+2)};$$

ii) Derive EY_1, $\sigma^2(Y_1)$, and EY_n, $\sigma^2(Y_n)$ from part (i);

iii) What do the quantities in parts (i) and (ii) become for $\alpha = 0$ and $\beta = 1$? (Hint: In part (i), use the appropriate Beta p.d.f.'s to facilitate the integrations.)

10.1.5

i) Refer to the p.d.f. g_{ij} derived in Theorem 2(ii), and show that, if $X_1, \ldots, X_n$ are independent with distribution $U(\alpha, \beta)$, then:

$$g_{1n}(y_1, y_n) = \frac{n(n-1)}{(\beta-\alpha)^n}(y_n - y_1)^{n-2}, \quad \alpha < y_1 < y_n < \beta;$$

ii) Set $Y = Y_n - Y_1$, and show that the p.d.f. of Y is given by:

$$f_Y(y) = \frac{n(n-1)}{(\beta-\alpha)^n}(\beta-\alpha-y)y^{n-2}, \quad 0 < y < \beta-\alpha;$$

iii) For a and b with $0 < a < b \leq \beta - \alpha$, show that:

$$P(a < Y < b) = \frac{n}{(\beta-\alpha)^n}\left[(\beta-\alpha-b)b^{n-1} - (\beta-\alpha-a)a^{n-1}\right] + \frac{b^n - a^n}{(\beta-\alpha)^n};$$

iv) What do the quantities in parts (i)–(iii) become for $\alpha = 0$ and $\beta = 1$?

10.1.6 Let $X_1, \ldots, X_n$ be independent r.v.'s distributed as $U(0, 1)$, and let $1 \leq i < j \leq n$.

i) Refer to the p.d.f. g_{ij} derived in Theorem 2(ii), and show that, in the present case:

$$g_{ij}(y_i, y_j) = \frac{n!}{(i-1)!(j-i-1)!(n-j)!} y_i^{i-1}(y_j - y_i)^{j-i-1}(1-y_j)^{n-j},$$

$$0 < y_i < y_j < 1;$$

ii) Integrating by parts repeatedly, show that:

$$\int_0^z y^i(z-y)^{j-i-1}\,dy = \frac{i!(j-i-1)!}{j!}z^j, \quad \int_0^1 z^{j+1}(1-z)^{n-j}\,dz = \frac{(j+1)!(n-j)!}{(n+2)!};$$

iii) Use parts (i) and (ii) to show that

$$E(Y_iY_j) = \frac{i(j+1)}{(n+1)(n+2)};$$

iv) By means of part (iii) and Exercise 1.4(i), show that:

$$\text{Cov}(Y_i, Y_j) = \frac{i(n-j+1)}{(n+1)^2(n+2)} \quad \text{and} \quad \rho(Y_i, Y_j) = \frac{i(n-j+1)}{[i(n-i+1)j(n-j+1)]^{1/2}};$$

v) From part (iv), derive $\text{Cov}(Y_1, Y_n)$ and $\rho(Y_1, Y_n)$.

10.1.7 Let the independent r.v.'s $X_1, \ldots, X_n$ be distributed as $U(0, 1)$, and let $1 < j < n$. Use the relevant results in Example 3, and Exercise 1.6(i) in order to derive:

i) The conditional p.d.f. of Y_j, given Y_1;

ii) The conditional p.d.f. of Y_j, given Y_n;

iii) The conditional p.d.f. of Y_n, given Y_1; and the conditional p.d.f. of Y_1, given Y_n.

10.1.8 Let the independent r.v.'s $X_1, \ldots, X_n$ be distributed as $U(0, 1)$, and define the r.v.'s Z_j, $j = 1, \ldots, n$ as follows: $Z_1 = Y_1$, $Z_j = Y_j - Y_{j-1}$, $j = 2, \ldots, n$. Then use the result in Theorem 1 in order to show that the r.v.'s Z_j, $j = 1, \ldots, n$ are uniformly distributed over the set

$$\left\{(z_1, \ldots, z_n)' \in \mathbb{R}^n; \ z_j \geq 0, j = 1, \ldots, n \text{ and } \sum_{j=1}^{n} z_j \leq 1\right\}.$$

(For $n = 2$, this set is a triangle in $\mathbb{R}^2$.)

10.1.9

i) Let the independent r.v.'s $X_1, \ldots, X_n$ have the Negative Exponential distribution with parameter λ. Then show that Y_1 has the same distribution with parameter $n\lambda$;

ii) Let F be the (common) d.f. of the independent r.v.'s $X_1, \ldots, X_n$, and suppose that their first order statistic Y_1 has the Negative Exponential distribution with parameter $n\lambda$. Then F is the Negative Exponential d.f. with parameter λ.

10.1.10 Let the independent r.v.'s $X_1, \ldots, X_n$ have the Negative Exponential distribution with parameter λ. Then:

i) Use Theorem 2(i') in order to show that the p.d.f. of Y_n is given by:

$$g_n(y_n) = n\lambda e^{-\lambda y_n}\left(1 - e^{\lambda y_n}\right)^{n-1}, \quad y_n > 0;$$

ii) Let Y be the (sample) range; that is, $Y = Y_n - Y_1$, and then use Theorem 2 (ii') in order to show that the p.d.f. of Y is given by:

$$f_Y(y) = (n-1)\lambda e^{-\lambda y}\left(1 - e^{-\lambda y}\right)^{n-2}, \quad y > 0;$$

iii) Calculate the probability $P(a < Y < b)$ for $0 < a < b$;

iv) For $a = 1/\lambda$, $b = 2/\lambda$, and $n = 10$, find a numerical value for the probability in part (iii).

10.1.11 Let the independent r.v.'s $X_1, \ldots, X_n$ have the Negative Exponential distribution with parameter λ, and let $1 < j < n$. Use Theorem 2(ii) and Exercises 1.9(i) and 1.10(i) in order to determine:

i) The conditional p.d.f. of Y_j, given Y_1;

ii) The conditional p.d.f. of Y_j, given Y_n;

iii) The conditional p.d.f. of Y_n, given Y_1; and the conditional p.d.f. of Y_1, given Y_n.

10.1.12 Let the independent r.v.'s $X_1, \ldots, X_n$ have the Negative Exponential distribution with parameter λ, and set: $Z_1 = Y_1$, $Z_j = Y_j - Y_{j-1}$, $j = 2, \ldots, n$. Then:

i) For $j = 1, \ldots, n$, show that Z_j has the Negative Exponential distribution with parameter $(n - j + 1)\lambda$, and that these r.v.'s are independent;

ii) From the definition of the Z_j's, it follows that $Y_j = Z_1 + \cdots + Z_j$, $j = 1, \ldots, n$. Use this expression and part (i) in order to conclude that:

$$EY_j = \frac{1}{\lambda}\left(\frac{1}{n} + \frac{1}{n-1} + \cdots + \frac{1}{n-j+1}\right);$$

iii) Use part (i) in order to show that, for $c > 0$:

$$P\left[\min_{i \neq j} |X_i - X_j| \geq c\right] = \exp\left[-\lambda n(n-1)c/2\right].$$

10.1.13 Refer to Exercise 10.1.12 and show that:

i) $\sigma^2(Y_j) = \sum_{i=1}^{j} \sigma_i^2$, where $\sigma_i^2 = [\lambda(n - i + 1)]^{-2}$, $i = 1, \ldots, n$;

ii) For $1 \leq i \leq j < n$, $\text{Cov}(Y_i, Y_j) = \sum_{k=1}^{i} \sigma_k^2$;

iii) From parts (i) and (ii), conclude that:

$$\sigma^2\left(\sum_{j=1}^{n} c_j Y_j\right) = \sum_{j=1}^{n} \sigma_j^2 \left(\sum_{i=j}^{n} c_i\right)^2, \quad c_i \in \mathbb{R} \text{ constants};$$

iv) Also utilize parts (i) and (ii) in order to show that:

$$\text{Cov}\left(\sum_{i=1}^{n} c_i Y_i, \sum_{j=1}^{n} d_j Y_j\right) = \sum_{i=1}^{n} \sigma_i^2 \left(\sum_{j=i}^{n} c_j d_j + 2 \sum_{1 \leq k < l \leq n} c_k d_l\right),$$

where $c_j \in \mathbb{R}$, $d_j \in \mathbb{R}$ constants.

Let X_j, $j = 1, \ldots, n$ be i.i.d. r.v.'s. Then the *sample median* S_M is defined as follows:

$$S_M = \begin{cases} Y_{\frac{n+1}{2}} & \text{if } n \text{ is odd} \\ \frac{1}{2}\left(Y_{\frac{n}{2}} + Y_{\frac{n}{2}}\right) & \text{if } n \text{ is even.} \end{cases} \quad (*)$$

10.1.14 If $X_j, j = 1, \ldots, n$ are i.i.d. r.v.'s, and n is odd, determine the p.d.f. of S_M when the underlying distribution is:

i) $U(\alpha, \beta)$;

ii) Negative Exponential with parameter λ.

10.1.15 If the r.v.'s $X_j, j = 1, \ldots, n$ are independently distributed as $N(\mu, \sigma^2)$, show that the p.d.f. of S_M is symmetric about μ, where S_M is defined by (*). Without integration, conclude that $ES_M = \mu$.

10.1.16 For n odd, let the independent r.v.'s $X_j, j = 1, \ldots, n$ have p.d.f. f with median m. Then determine the p.d.f. of S_M, and also calculate the probability $P(S_M > m)$ in each one of the following cases:

i) $f(x) = 2xI_{(0,1)}(x)$;

ii) $f(x) = 2(2 - x)I_{(1,2)}(x)$;

iii) $f(x) = 2(1 - x)I_{(0,1)}(x)$;

iv) What do parts (i)–(iii) become for $n = 3$?

10.1.17 Refer to Exercise 10.1.12 and derive the p.d.f. of S_M, where S_M is defined by (*).

10.1.18 Let $X_j, j = 1, \ldots, 6$ be i.i.d. r.v.'s with p.d.f. f given by $f(x) = \frac{1}{6}$, $x = 1, \ldots, 6$. Find the p.d.f.'s of Y_1 and Y_6. Also, observe that $P(Y_1 = y) = P(Y_6 = 7 - y), y = 1, \ldots, 6$.

10.1.19 Carry out the proof of part (ii) of Theorem 2.

10.2 Further Distribution Theory: Probability of Coverage of a Population Quantile

It has been shown in Theorem 7, Chapter 9, that if X is an r.v. with continuous d.f. F, then the r.v. $Y = F(X)$ is $U(0, 1)$. This result in conjunction with Theorem 1 of the present chapter gives the following theorem.

THEOREM 3 Let $X_1, \ldots, X_n$ be i.i.d. r.v.'s with continuous d.f. F and let $Z_j = F(Y_j)$, where Y_j, $j = 1, 2, \ldots, n$ are the order statistics. Then $Z_1, \ldots, Z_n$ are order statistics from $U(0, 1)$, and hence their joint p.d.f., h is given by:

$$h(z_1, \ldots, z_n) = \begin{cases} n!, & 0 < z_1 < \cdots < z_n < 1 \\ 0, & \text{otherwise.} \end{cases}$$

PROOF Set $W_j = F(X_j), j = 1, 2, \ldots, n$. Then the W_j's are independent, since the X_j's are, and also distributed as $U(0, 1)$, by Theorem 7, Chapter 9. Because F is nondecreasing, to each ordering of the X_j's, $X_{(1)} \leq X_{(2)} \leq \cdots \leq X_{(n)}$ there corresponds the ordering $F(X_{(1)}) \leq F(X_{(2)}) \leq \cdots \leq F(X_{(n)})$ of the $F(X_j)$'s, and conversely. Therefore $W_{(j)} = F(Y_j), j = 1, 2, \ldots, n$. That the joint p.d.f. of the Z_j's is the one given above follows from Theorem 1 of this chapter. ▲

The distributions of Z_j, Z_1, Z_n and (Z_1, Z_n) are given in Example 3 of this chapter.

Let now X be an r.v. with d.f. F. Consider a number p, $0 < p < 1$. Then in Chapter 4, a pth quantile, x_p, of F was defined to be a number with the following properties:

i) $P(X \leq x_p) \geq p$ and
ii) $P(X \geq x_p) \geq 1 - p$.

Now we would like to establish a certain theorem to be used in a subsequent chapter.

THEOREM 4 Let $X_1, \ldots, X_n$ be i.i.d. r.v.'s with continuous d.f. F and let $Y_1, \ldots, Y_n$ be the order statistics. For p, $0 < p < 1$, let x_p be the (unique by assumption) pth quantile. Then we have

$$P(Y_i \leq x_p \leq Y_j) = \sum_{k=i}^{j-1} \binom{n}{k} p^k q^{n-k}, \quad \text{where} \quad q = 1 - p.$$

PROOF Define the r.v.'s $W_j, j = 1, 2, \ldots, n$ as follows:

$$W_j = \begin{cases} 1, & X_j \leq x_p \\ 0, & X_j > x_p, \end{cases} \quad j = 1, 2, \ldots, n.$$

Then $W_1, \ldots, W_n$ are i.i.d. r.v.'s distributed as $B(1, p)$, since

$$P(W_1 = 1) = P(X_1 \leq x_p) = F(x_p) = p.$$

Therefore

$$P(\text{at least } i \text{ of } X_1, \ldots, X_n \leq x_p) = \sum_{k=i}^{n} \binom{n}{k} p^k q^{n-k};$$

or equivalently,

$$P(Y_i < x_p) = P(Y_i \leq x_p) = \sum_{k=i}^{n} \binom{n}{k} p^k q^{n-k}.$$

Next, for $1 \leq i < j \leq n$, we get

$$P(Y_i \leq x_p) = P(Y_i \leq x_p, Y_j \geq x_p) + P(Y_i \leq x_p, Y_j < x_p)$$
$$= P(Y_i \leq x_p \leq Y_j) + P(Y_j < x_p),$$

since
$$(Y_j < x_p) \subseteq (Y_i \leq x_p).$$

Therefore
$$P(Y_i \leq x_p \leq Y_j) = P(Y_i \leq x_p) - P(Y_j < x_p).$$

By means of (1), this gives
$$P(Y_i \leq x_p \leq Y_j) = \sum_{k=i}^{n} \binom{n}{k} p^k q^{n-k} - \sum_{k=j}^{n} \binom{n}{k} p^k q^{n-k}$$
$$= \sum_{k=i}^{j-1} \binom{n}{k} p^k q^{n-k}. \; \blacktriangle$$

Exercise

10.2.1 Let $X_j, j = 1, \ldots, n$ be i.i.d. r.v.'s with continuous d.f. F. Use Theorem 3 and the relevant part of Example 3 in order to determine the distribution of the r.v. $F(Y_1)$ and find its expectation.

Chapter 11

Sufficiency and Related Theorems

Let X be an r.v. with p.d.f. $f(\cdot; \theta)$ of known functional form but depending upon an unknown r-dimensional constant vector $\theta = (\theta_1, \ldots, \theta_r)'$ which is called a *parameter*. We let Ω stand for the set of all possible values of θ and call it the *parameter space*. So $\Omega \subseteq \mathbb{R}^r, r \geq 1$. By $\mathcal{F}$ we denote the family of all p.d.f.'s we get by letting θ vary over Ω; that is, $\mathcal{F} = \{f(\cdot; \theta); \theta \in \Omega\}$.

Let $X_1, \ldots, X_n$ be a *random sample of size n* from $f(\cdot; \theta)$, that is, n independent r.v.'s distributed as X above. One of the basic problems of statistics is that of making inferences about the parameter θ (such as estimating θ, testing hypotheses about θ, etc.) on the basis of the observed values $x_1, \ldots, x_n$, *the data*, of the r.v.'s $X_1, \ldots, X_n$. In doing so, the concept of sufficiency plays a fundamental role in allowing us to often substantially condense the data without ever losing any information carried by them about the parameter θ.

In most of the textbooks, the concept of sufficiency is treated exclusively in conjunction with estimation and testing hypotheses problems. We propose, however, to treat it in a separate chapter and gather together here all relevant results which will be needed in the sequel. In the same chapter, we also introduce and discuss other concepts such as: completeness, unbiasedness and minimum variance unbiasedness.

For $j = 1, \ldots, m$, let T_j be (measurable) functions defined on $\mathbb{R}^n$ into $\mathbb{R}$ and not depending on θ or any other unknown quantities, and set $\mathbf{T} = (T_1, \ldots, T_m)'$. Then

$$\mathbf{T}(X_1, \ldots, X_n) = \left(T_1(X_1, \ldots, X_n), \ldots, T_m(X_1, \ldots, X_n)\right)'$$

is called an m-dimensional *statistic*. We shall write $T(X_1, \ldots, X_n)$ rather than $\mathbf{T}(X_1, \ldots, X_n)$ if $m = 1$. Likewise, we shall write θ rather than $\boldsymbol{\theta}$ when $r = 1$. Also, we shall often write $\mathbf{T}$ instead of $\mathbf{T}(X_1, \ldots, X_n)$, by slightly abusing the notation.

11.1 Sufficiency: Definition and Some Basic Results

The basic notation and terminology introduced so far is enough to allow us to proceed with the main part of the present chapter.

Let us consider first some illustrative examples of families of p.d.f.'s.

EXAMPLE 1 Let $\mathbf{X} = (X_1, \ldots, X_r)'$ have the Multinomial distribution. Then by setting $\theta_j = p_j, j = 1, \ldots, r$, we have

$$\boldsymbol{\theta} = (\theta_1, \ldots, \theta_r)', \quad \Omega = \left\{ (\theta_1, \ldots, \theta_r)' \in \mathbb{R}^r; \theta_j > 0, j = 1, \ldots, r \text{ and } \sum_{j=1}^{r} \theta_j = 1 \right\}$$

and

$$f(\mathbf{x}; \boldsymbol{\theta}) = \frac{n!}{x_1! \cdots x_r!} \theta_1^{x_1} \cdots \theta_r^{x_r} I_A(\mathbf{x}) = \frac{n!}{\prod_{j=1}^{r-1} x_j! (n - x_1 - \cdots - x_{r-1})!}$$

$$\times \theta_1^{x_1} \cdots \theta_{r-1}^{x_{r-1}} (1 - \theta_1 - \cdots - \theta_{r-1})^{n - \sum_{j=1}^{r-1} x_j} I_A(\mathbf{x}),$$

$$A = \left\{ \mathbf{x} = (x_1, \ldots, x_r)' \in \mathbb{R}^r; x_j \geq 0, j = 1, \ldots, r, \sum_{j=1}^{r} x_j = n \right\}.$$

For example, for $r = 3$, Ω is that part of the plane through the points (1, 0, 0), (0, 1, 0) and (0, 0, 1) which lies in the first quadrant, whereas for $r = 2$, the distribution of $\mathbf{X} = (X_1, X_2)'$ is completely determined by that of $X_1 = X$ which is distributed as $B(n, \theta_1) = B(n, \theta)$.

EXAMPLE 2 Let X be $U(\alpha, \beta)$. By setting $\theta_1 = \alpha, \theta_2 = \beta$, we have $\boldsymbol{\theta} = (\theta_1, \theta_2)', \Omega = \{(\theta_1, \theta_2)' \in \mathbb{R}^2; \theta_1, \theta_2 \in \mathbb{R}, \theta_1 < \theta_2\}$ (that is, the part of the plane above the main diagonal) and

$$f(x; \boldsymbol{\theta}) = \frac{1}{\theta_2 - \theta_1} I_A(x), \quad A = [\theta_1, \theta_2].$$

If α is known and we put $\beta = \theta$, then $\Omega = (\alpha, \infty)$ and

$$f(x; \theta) = \frac{1}{\theta - \alpha} I_A(x), \quad A = [\alpha, \theta].$$

Similarly, if β is known and $\alpha = \theta$.

EXAMPLE 3 Let X be $N(\mu, \sigma^2)$. Then by setting $\theta_1 = \mu, \theta_2 = \sigma^2$, we have $\boldsymbol{\theta} = (\theta_1, \theta_2)'$,

$$\Omega = \left\{ (\theta_1, \theta_2)' \in \mathbb{R}^2; \theta_1 \in \mathbb{R}, \theta_2 > 0 \right\}$$

(that is, the part of the plane above the horizontal axis) and

$$f(x;\boldsymbol{\theta}) = \frac{1}{\sqrt{2\pi\theta_2}} \exp\left[-\frac{(x-\theta_1)^2}{2\theta_2}\right].$$

If σ is known and we set $\mu = \theta$, then $\Omega = \mathbb{R}$ and

$$f(x;\theta) = \frac{1}{\sqrt{2\pi}\sigma} \exp\left[-\frac{(x-\theta)^2}{2\sigma^2}\right].$$

Similarly if μ is known and $\sigma^2 = \theta$.

EXAMPLE 4 Let $\mathbf{X} = (X_1, X_2)'$ have the Bivariate Normal distribution. Setting $\theta_1 = \mu_1$, $\theta_2 = \mu_2$, $\theta_3 = \sigma_1^2$, $\theta_4 = \sigma_2^2$, $\theta_5 = \rho$, we have then $\boldsymbol{\theta} = (\theta_1, \ldots, \theta_5)'$ and

$$\Omega = \left\{ (\theta_1, \ldots, \theta_5)' \in \mathbb{R}^5;\ \theta_1,\ \theta_2 \in \mathbb{R},\ \theta_3,\ \theta_4 \in (0, \infty),\ \theta_5 \in (-1, 1) \right\}$$

and

$$f(\mathbf{x};\boldsymbol{\theta}) = \frac{1}{2\pi\sigma_1\sigma_2\sqrt{1-\rho^2}} e^{-q/2},$$

where

$$q = \frac{1}{1-\rho^2}\left[\left(\frac{x_1-\mu_1}{\sigma_1}\right)^2 - 2\rho\left(\frac{x_1-\mu_1}{\sigma_1}\right)\left(\frac{x_2-\mu_2}{\sigma_2}\right) + \left(\frac{x_2-\mu_2}{\sigma_2}\right)^2\right],$$

$$\mathbf{x} = (x_1, x_2)'.$$

Before the formal definition of sufficiency is given, an example will be presented to illustrate the underlying motivation.

EXAMPLE 5 Let $X_1, \ldots, X_n$ be i.i.d. r.v.'s from $B(1, \theta)$; that is,

$$f_{X_j}(x_j;\theta) = \theta^{x_j}(1-\theta)^{1-x_j} I_A(x_j),\quad j = 1, \ldots, n,$$

where $A = \{0, 1\}$, $\theta \in \Omega = (0, 1)$. Set $T = \sum_{j=1}^n X_j$. Then T is $B(n, \theta)$, so that

$$f_T(t;\theta) = \binom{n}{t}\theta^t(1-\theta)^{n-t} I_B(t),$$

where $B = \{0, 1, \ldots, n\}$. We suppose that the Binomial experiment in question is performed and that the observed values of X_j are x_j, $j = 1, \ldots, n$. Then the problem is to make some kind of inference about θ on the basis of x_j, $j = 1, \ldots, n$. As usual, we label as a success the outcome 1. Then the following question arises: Can we say more about θ if we know how many successes occurred and where rather than merely how many successes occurred? The answer to this question will be provided by the following argument. Given that the number of successes is t, that is, given that $T = t$, $t = 0, 1, \ldots, n$, find the probability of each one of the $\binom{n}{t}$ different ways in which the t successes can occur. Then, if there are values of θ for which particular occurrences of the t successes can happen with higher probability than others, we will say that knowledge of the positions where the t successes occurred is more informative about θ than simply knowledge of the total number of successes t. If, on the other hand, all possible outcomes, given the total number of successes t, have the same probability of occurrence, then clearly the positions where the t successes occurred are entirely irrelevant and the total number of successes t provides all possible information about θ. In the present case, we have

$$P_\theta(X_1 = x_1, \ldots, X_n = x_n \mid T = t) = \frac{P_\theta(X_1 = x_1, \ldots, X_n = x_n, T = t)}{P_\theta(T = t)}$$

$$= \frac{P_\theta(X_1 = x_1, \ldots, X_n = x_n)}{P_\theta(T = t)}$$

if $x_1 + \cdots + x_n = t$

and zero otherwise, and this is equal to

$$\frac{\theta^{x_1}(1-\theta)^{1-x_1} \cdots \theta^{x_n}(1-\theta)^{1-x_n}}{\binom{n}{t}\theta^t(1-\theta)^{n-t}} = \frac{\theta^t(1-\theta)^{n-t}}{\binom{n}{t}\theta^t(1-\theta)^{n-t}} = \frac{1}{\binom{n}{t}}$$

if $x_1 + \cdots + x_n = t$ and zero otherwise. Thus, we found that for all $x_1, \ldots, x_n$ such that $x_j = 0$ or 1, $j = 1, \ldots, n$ and

$$\sum_{j=1}^n x_j = t, \quad P_\theta(X_1 = x_1, \ldots, X_n = x_n \mid T = t) = 1 \Big/ \binom{n}{t}$$

independent of θ, and therefore the total number of successes t alone provides all possible information about θ.

This example motivates the following definition of a sufficient statistic.

DEFINITION 1 Let $X_j, j = 1, \ldots, n$ be i.i.d. r.v.'s with p.d.f. $f(\cdot; \boldsymbol{\theta})$, $\boldsymbol{\theta} = (\theta_1, \ldots, \theta_r)' \in \Omega \subseteq \mathbb{R}^r$, and let $\mathbf{T} = (T_1, \ldots, T_m)'$, where

$$T_j = T_j(X_1, \ldots, X_n), \quad j = 1, \ldots, m$$

are statistics. We say that $\mathbf{T}$ is an m-dimensional *sufficient statistic* for the family $\mathcal{F} = \{f(\cdot; \boldsymbol{\theta}); \boldsymbol{\theta} \in \Omega\}$, or for the parameter $\boldsymbol{\theta}$, if the conditional distribution of $(X_1, \ldots, X_n)'$, given $\mathbf{T} = \mathbf{t}$, is independent of $\boldsymbol{\theta}$ for all values of $\mathbf{t}$ (actually, for almost all (a.a.)$\mathbf{t}$, that is, except perhaps for a set N in $\mathbb{R}^m$ of values of $\mathbf{t}$ such that $P_\theta(\mathbf{T} \in N) = 0$ for all $\boldsymbol{\theta} \in \Omega$, where P_θ denotes the probability function associated with the p.d.f. $f(\cdot; \boldsymbol{\theta})$).

REMARK 1 Thus, $\mathbf{T}$ being a sufficient statistic for $\boldsymbol{\theta}$ implies that every (measurable) set A in $\mathbb{R}^n$, $P_\theta[(X_1, \ldots, X_n)' \in A \mid \mathbf{T} = \mathbf{t}]$ is independent of $\boldsymbol{\theta}$ for a.a.

11.1 Sufficiency: Definition and Some Basic Results

t. Actually, more is true. Namely, if $\mathbf{T}^* = (T_1^*, \ldots, T_k^*)'$ is any k-dimensional statistic, then the conditional distribution of $\mathbf{T}^*$, given $\mathbf{T} = \mathbf{t}$, is independent of $\boldsymbol{\theta}$ for a.a. **t**. To see this, let B be any (measurable) set in $\mathbb{R}^k$ and let $A = \mathbf{T}^{*-1}(B)$. Then

$$P_{\boldsymbol{\theta}}(\mathbf{T}^* \in B | \mathbf{T} = \mathbf{t}) = P_{\boldsymbol{\theta}}\left[(X_1, \ldots, X_n)' \in A | \mathbf{T} = \mathbf{t}\right]$$

and this is independent of $\boldsymbol{\theta}$ for a.a. **t**.

We finally remark that $\mathbf{X} = (X_1, \ldots, X_n)'$ is always a sufficient statistic for $\boldsymbol{\theta}$.

Clearly, Definition 1 above does not seem appropriate for identifying a sufficient statistic. This can be done quite easily by means of the following theorem.

THEOREM 1 (Fisher–Neyman factorization theorem) Let $X_1, \ldots, X_n$ be i.i.d. r.v.'s with p.d.f. $f(\cdot; \boldsymbol{\theta})$, $\boldsymbol{\theta} = (\theta_1, \ldots, \theta_r)' \in \Omega \subseteq \mathbb{R}^r$. An m-dimensional statistic

$$\mathbf{T} = \mathbf{T}(X_1, \ldots, X_n) = (T_1(X_1, \ldots, X_n), \ldots, T_m(X_1, \cdots, X_n))'$$

is sufficient for $\boldsymbol{\theta}$ if and only if the joint p.d.f. of $X_1, \ldots, X_n$ factors as follows,

$$f(x_1, \ldots, x_n; \boldsymbol{\theta}) = g[\mathbf{T}(x_1, \ldots, x_n); \boldsymbol{\theta}]h(x_1, \ldots, x_n),$$

where g depends on $x_1, \ldots, x_n$ only through $\mathbf{T}$ and h is (entirely) independent of $\boldsymbol{\theta}$.

PROOF The proof is given separately for the discrete and the continuous case.

Discrete case: In the course of this proof, we are going to use the notation $\mathbf{T}(x_1, \ldots, x_n) = \mathbf{t}$. In connection with this, it should be pointed out at the outset that by doing so we restrict attention only to those $x_1, \cdots, x_n$ for which $\mathbf{T}(x_1, \ldots, x_n) = \mathbf{t}$.

Assume that the factorization holds, that is,

$$f(x_1, \ldots, x_n; \boldsymbol{\theta}) = g[\mathbf{T}(x_1, \ldots, x_n); \boldsymbol{\theta}]h(x_1, \ldots, x_n),$$

with g and h as described in the theorem. Clearly, it suffices to restrict attention to those **t**'s for which $P_{\boldsymbol{\theta}}(\mathbf{T} = \mathbf{t}) > 0$. Next,

$$P_{\boldsymbol{\theta}}(\mathbf{T} = \mathbf{t}) = P_{\boldsymbol{\theta}}[\mathbf{T}(X_1, \ldots, X_n) = \mathbf{t}] = \sum P_{\boldsymbol{\theta}}(X_1 = x_1', \ldots, X_n = x_n'),$$

where the summation extends over all $(x_1', \ldots, x_n')'$ for which $\mathbf{T}(x_1', \ldots, x_n') = \mathbf{t}$. Thus

$$P_{\boldsymbol{\theta}}(\mathbf{T} = \mathbf{t}) = \sum f(x_1'; \boldsymbol{\theta}) \cdots f(x_n'; \boldsymbol{\theta}) = \sum g(\mathbf{t}; \boldsymbol{\theta})h(x_1', \ldots, x_n')$$
$$= g(\mathbf{t}; \boldsymbol{\theta})\sum h(x_1', \ldots, x_n').$$

Hence

$$P_\theta(X_1 = x_1, \ldots, X_n = x_n | T = t)$$
$$= \frac{P_\theta(X_1 = x_1, \ldots, X_n = x_n, T = t)}{P_\theta(T = t)} = \frac{P_\theta(X_1 = x_1, \ldots, X_n = x_n)}{P_\theta(T = t)}$$
$$= \frac{g(t; \theta)h(x_1, \ldots, x_n)}{g(t; \theta)\sum h(x_1, \ldots, x_n)} = \frac{h(x_1, \ldots, x_n)}{\sum h(x_1, \ldots, x_n)}$$

and this is independent of θ.

Now, let $\mathbf{T}$ be sufficient for θ. Then $P_\theta(X_1 = x_1, \ldots, X_n = x_n | \mathbf{T} = \mathbf{t})$ is independent of θ; call it $k[x_1, \ldots, x_n, \mathbf{T}(x_1, \ldots, x_n)]$. Then

$$P_\theta(X_1 = x_1, \ldots, X_n = x_n | \mathbf{T} = \mathbf{t}) = \frac{P_\theta(X_1 = x_1, \ldots, X_n = x_n)}{P_\theta(\mathbf{T} = \mathbf{t})}$$
$$= k[x_1, \ldots, x_n, \mathbf{T}(x_1, \ldots, x_n)]$$

if and only if

$$f(x_1; \theta) \cdots f(x_n; \theta) = P_\theta(X_1 = x_1, \ldots, X_n = x_n)$$
$$= P_\theta(\mathbf{T} = \mathbf{t})k[x_1, \ldots, x_n, \mathbf{T}(x_1, \ldots, x_n)].$$

Setting

$$g[\mathbf{T}(x_1, \ldots, x_n); \theta] = P_\theta(\mathbf{T} = \mathbf{t}) \quad \text{and} \quad h(x_1, \ldots, x_n)$$
$$= k[x_1, \ldots, x_n, \mathbf{T}(x_1, \ldots, x_n)],$$

we get

$$f(x_1; \theta) \cdots f(x_n; \theta) = g[\mathbf{T}(x_1, \ldots, x_n); \theta]h(x_1, \ldots, x_n),$$

as was to be seen.

Continuous case: The proof in this case is carried out under some further regularity conditions (and is not as rigorous as that of the discrete case). It should be made clear, however, that the theorem is true as stated. A proof without the regularity conditions mentioned above involves deeper concepts of measure theory the knowledge of which is not assumed here. From Remark 1, it follows that $m \leq n$. Then set $T_j = T_j(X_1, \ldots, X_n), j = 1, \ldots, m$, and assume that there exist other $n - m$ statistics $T_j = T_j(X_1, \ldots, X_n), j = m + 1, \ldots, n$, such that the transformation

$$t_j = T_j(x_1, \ldots, x_n), \quad j = 1, \ldots, n,$$

is invertible, so that

$$x_j = x_j(\mathbf{t}, t_{m+1}, \ldots, t_n), \quad j = 1, \ldots, n, \quad \mathbf{t} = (t_1, \ldots, t_m)'.$$

It is also assumed that the partial derivatives of x_j with respect to t_i, $i, j = 1, \ldots, n$, exist and are continuous, and that the respective Jacobian J (which is independent of $\boldsymbol{\theta}$) is different from 0.

Let first
$$f(x_1; \boldsymbol{\theta}) \cdots f(x_n; \boldsymbol{\theta}) = g[\mathbf{T}(x_1, \ldots, x_n); \boldsymbol{\theta}] h(x_1, \ldots, x_n).$$

Then
$$f_{\mathbf{T}, T_{m+1}, \ldots, T_n}(\mathbf{t}, t_{m+1}, \ldots, t_n; \boldsymbol{\theta})$$
$$= g(\mathbf{t}; \boldsymbol{\theta}) h[x_1(\mathbf{t}, t_{m+1}, \ldots, t_n), \ldots, x_n(\mathbf{t}, t_{m+1}, \ldots, t_n)]|J|$$
$$= g(\mathbf{t}; \boldsymbol{\theta}) h^*(\mathbf{t}, t_{m+1}, \ldots, t_n),$$

where we set
$$h^*(\mathbf{t}, t_{m+1}, \ldots, t_n) = h[x_1(\mathbf{t}, t_{m+1}, \ldots, t_n), \ldots, x_n(\mathbf{t}, t_{m+1}, \ldots, t_n)]|J|.$$

Hence
$$f_{\mathbf{T}}(\mathbf{t}; \boldsymbol{\theta}) = \int_{-\infty}^{\infty} \cdots \int_{-\infty}^{\infty} g(\mathbf{t}; \boldsymbol{\theta}) h^*(\mathbf{t}, t_{m+1}, \ldots, t_n) dt_{m+1} \cdots dt_n = g(\mathbf{t}; \boldsymbol{\theta}) h^{**}(\mathbf{t}),$$

where
$$h^{**}(\mathbf{t}) = \int_{-\infty}^{\infty} \cdots \int_{-\infty}^{\infty} h^*(\mathbf{t}, t_{m+1}, \ldots, t_n) dt_{m+1} \cdots dt_n.$$

That is, $f_{\mathbf{T}}(\mathbf{t}; \boldsymbol{\theta}) = g(\mathbf{t}; \boldsymbol{\theta}) h^{**}(\mathbf{t})$ and hence
$$f(t_{m+1}, \ldots, t_n | \mathbf{t}; \boldsymbol{\theta}) = \frac{g(\mathbf{t}; \boldsymbol{\theta}) h^*(\mathbf{t}, t_{m+1}, \ldots, t_n)}{g(\mathbf{t}; \boldsymbol{\theta}) h^{**}(\mathbf{t})} = \frac{h^*(\mathbf{t}, t_{m+1}, \ldots, t_n)}{h^{**}(\mathbf{t})}$$

which is independent of $\boldsymbol{\theta}$. That is, the conditional distribution of $T_{m+1}, \ldots, T_n$, given $\mathbf{T} = \mathbf{t}$, is independent of $\boldsymbol{\theta}$. It follows that the conditional distribution of $\mathbf{T}, T_{m+1}, \ldots, T_n$, given $\mathbf{T} = \mathbf{t}$, is independent of $\boldsymbol{\theta}$. Since, by assumption, there is a one-to-one correspondence between $\mathbf{T}, T_{m+1}, \ldots, T_n$, and $X_1, \ldots, X_n$, it follows that the conditional distribution of $X_1, \ldots, X_n$, given $\mathbf{T} = \mathbf{t}$, is independent of $\boldsymbol{\theta}$.

Let now $\mathbf{T}$ be sufficient for $\boldsymbol{\theta}$. Then, by using the inverse transformation of the one used in the first part of this proof, one has
$$f(x_1, \ldots, x_n; \boldsymbol{\theta}) = f_{\mathbf{T}, T_{m+1}, \ldots, T_n}(\mathbf{t}, t_{m+1}, \ldots, t_n; \boldsymbol{\theta})|J^{-1}|$$
$$= f(t_{m+1}, \ldots, t_n | \mathbf{t}; \boldsymbol{\theta}) f_{\mathbf{T}}(\mathbf{t}; \boldsymbol{\theta}) |J^{-1}|.$$

But $f(t_{m+1}, \ldots, t_n | \mathbf{t}; \boldsymbol{\theta})$ is independent of $\boldsymbol{\theta}$ by Remark 1. So we may set
$$f(t_{m+1}, \ldots, t_n | \mathbf{t}; \boldsymbol{\theta})|J^{-1}| = h^*(t_{m+1}, \ldots, t_n; \mathbf{t}) = h(x_1, \ldots, x_n).$$

If we also set
$$f_{\mathbf{T}}(\mathbf{t}; \boldsymbol{\theta}) = g[\mathbf{T}(x_1, \ldots, x_n); \boldsymbol{\theta}],$$

we get

$$f(x_1, \ldots, x_n; \boldsymbol{\theta}) = g[\mathbf{T}(x_1, \ldots, x_n); \boldsymbol{\theta}]h(x_1, \ldots, x_n),$$

as was to be seen. ▲

COROLLARY Let $\phi: \mathbb{R}^m \to \mathbb{R}^m$ ((measurable and independent) of $\boldsymbol{\theta}$) be one-to-one, so that the inverse ϕ^{-1} exists. Then, if $\mathbf{T}$ is sufficient for $\boldsymbol{\theta}$, we have that $\tilde{\mathbf{T}} = \phi(\mathbf{T})$ is also sufficient for $\boldsymbol{\theta}$ and $\mathbf{T}$ is sufficient for $\tilde{\boldsymbol{\theta}} = \psi(\boldsymbol{\theta})$, where $\psi: \mathbb{R}^r \to \mathbb{R}^r$ is one-to-one (and measurable).

PROOF We have $\mathbf{T} = \phi^{-1}[\phi(\mathbf{T})] = \phi^{-1}(\tilde{\mathbf{T}})$. Thus

$$f(x_1, \ldots, x_n; \boldsymbol{\theta}) = g[\mathbf{T}(x_1, \ldots, x_n); \boldsymbol{\theta}]h(x_1, \ldots, x_n)$$
$$= g\{\phi^{-1}[\tilde{\mathbf{T}}(x_1, \ldots, x_n)]; \boldsymbol{\theta}\}h(x_1, \ldots, x_n)$$

which shows that $\tilde{\mathbf{T}}$ is sufficient for $\boldsymbol{\theta}$. Next,

$$\boldsymbol{\theta} = \psi^{-1}[\psi(\boldsymbol{\theta})] = \psi^{-1}(\tilde{\boldsymbol{\theta}}).$$

Hence

$$f(x_1, \ldots, x_n; \boldsymbol{\theta}) = g[\mathbf{T}(x_1, \ldots, x_n); \boldsymbol{\theta}]h(x_1, \ldots, x_n)$$

becomes

$$\tilde{f}(x_1, \ldots, x_n; \tilde{\boldsymbol{\theta}}) = \tilde{g}[\mathbf{T}(x_1, \ldots, x_n); \tilde{\boldsymbol{\theta}}]h(x_1, \ldots, x_n),$$

where we set

$$\tilde{f}(x_1, \ldots, x_n; \tilde{\boldsymbol{\theta}}) = f[x_1, \ldots, x_n; \psi^{-1}(\tilde{\boldsymbol{\theta}})]$$

and

$$\tilde{g}[\mathbf{T}(x_1, \ldots, x_n); \tilde{\boldsymbol{\theta}}] = g[\mathbf{T}(x_1, \ldots, x_n); \psi^{-1}(\tilde{\boldsymbol{\theta}})].$$

Thus, $\mathbf{T}$ is sufficient for the new parameter $\tilde{\boldsymbol{\theta}}$. ▲

We now give a number of examples of determining sufficient statistics by way of Theorem 1 in some interesting cases.

EXAMPLE 6 Refer to Example 1, where

$$f(\mathbf{x}; \boldsymbol{\theta}) = \frac{n!}{x_1! \cdots x_r!} \theta_1^{x_1} \cdots \theta_r^{x_r} I_A(\mathbf{x}).$$

Then, by Theorem 1, it follows that the statistic $(X_1, \ldots, X_r)'$ is sufficient for $\boldsymbol{\theta} = (\theta_1, \ldots, \theta_r)'$. Actually, by the fact that $\sum_{j=1}^r \theta_j = 1$ and $\sum_{j=1}^r x_j = n$, we also have

$$f(\mathbf{x}; \boldsymbol{\theta}) = \frac{n!}{\prod_{j=1}^{r-1} x_j! (n - x_1 - \cdots - x_{r-1})!}$$
$$\times \theta_1^{x_1} \cdots \theta_{r-1}^{x_{r-1}} (1 - \theta_1 - \cdots - \theta_{r-1})^{n - \sum_{j=1}^{r-1} x_j} I_A(\mathbf{x})$$

from which it follows that the statistic $(X_1, \ldots, X_{r-1})'$ is sufficient for $(\theta_1, \ldots, \theta_{r-1})'$. In particular, for $r = 2$, $X_1 = X$ is sufficient for $\theta_1 = \theta$.

EXAMPLE 7 Let $X_1, \ldots, X_n$ be i.i.d. r.v.'s from $U(\theta_1, \theta_2)$. Then by setting $\mathbf{x} = (x_1, \ldots, x_n)'$ and $\boldsymbol{\theta} = (\theta_1, \theta_2)'$, we get

$$f(\mathbf{x}; \boldsymbol{\theta}) = \frac{1}{(\theta_2 - \theta_1)^n} I_{[\theta_1, \infty)}(x_{(1)}) I_{(-\infty, \theta_2]}(x_{(n)})$$

$$= \frac{1}{(\theta_2 - \theta_1)^n} g_1[x_{(1)}, \boldsymbol{\theta}] g_2[x_{(n)}, \boldsymbol{\theta}],$$

where $g_1[x_{(1)}, \boldsymbol{\theta}] = I_{[\theta_1, \infty)}(x_{(1)})$, $g_2[x_{(n)}, \boldsymbol{\theta}] = I_{(-\infty, \theta_2]}(x_{(n)})$. It follows that $(X_{(1)}, X_{(n)})'$ is sufficient for $\boldsymbol{\theta}$. In particular, if $\theta_1 = \alpha$ is known and $\theta_2 = \theta$, it follows that $X_{(n)}$ is sufficient for $\boldsymbol{\theta}$. Similarly, if $\theta_2 = \beta$ is known and $\theta_1 = \theta$, $X_{(1)}$ is sufficient for $\boldsymbol{\theta}$.

EXAMPLE 8 Let $X_1, \ldots, X_n$ be i.i.d. r.v.'s from $N(\mu, \sigma^2)$. By setting $\mathbf{x} = (x_1, \ldots, x_n)'$, $\mu = \theta_1$, $\sigma^2 = \theta_2$ and $\boldsymbol{\theta} = (\theta_1, \theta_2)'$, we have

$$f(\mathbf{x}; \boldsymbol{\theta}) = \left(\frac{1}{\sqrt{2\pi\theta_2}}\right)^n \exp\left[-\frac{1}{2\theta_2}\sum_{j=1}^n (x_j - \theta_1)^2\right].$$

But

$$\sum_{j=1}^n (x_j - \theta_1)^2 = \sum_{j=1}^n [(x_j - \bar{x}) + (\bar{x} - \theta_1)]^2 = \sum_{j=1}^n (x_j - \bar{x})^2 + n(\bar{x} - \theta_1)^2,$$

so that

$$f(\mathbf{x}; \boldsymbol{\theta}) = \left(\frac{1}{\sqrt{2\pi\theta_2}}\right)^n \exp\left[-\frac{1}{2\theta_2}\sum_{j=1}^n (x_j - \bar{x})^2 - \frac{n}{2\theta_2}(\bar{x} - \theta_1)^2\right].$$

It follows that $(\bar{X}, \sum_{j=1}^n (X_j - \bar{X})^2)'$ is sufficient for $\boldsymbol{\theta}$. Since also

$$f(\mathbf{x}; \boldsymbol{\theta}) = \left(\frac{1}{\sqrt{2\pi\theta_2}}\right)^n \exp\left(-\frac{n\theta_1^2}{2\theta_2}\right) \exp\left(\frac{\theta_1}{\theta_2}\sum_{j=1}^n x_j - \frac{1}{2\theta_2}\sum_{j=1}^n x_j^2\right),$$

it follows that, if $\theta_2 = \sigma^2$ is known and $\theta_1 = \theta$, then $\sum_{j=1}^n X_j$ is sufficient for θ, whereas if $\theta_1 = \mu$ is known and $\theta_2 = \theta$, then $\sum_{j=1}^n (X_j - \mu)^2$ is sufficient for θ, as follows from the form of $f(\mathbf{x}; \boldsymbol{\theta})$ at the beginning of this example. By the corollary to Theorem 1, it also follows that $(\bar{X}, S^2)'$ is sufficient for $\boldsymbol{\theta}$, where

$$S^2 = \frac{1}{n}\sum_{j=1}^n (X_j - \bar{X})^2, \quad \text{and} \quad \frac{1}{n}\sum_{j=1}^n (X_j - \mu)^2$$

is sufficient for $\theta_2 = \theta$ if $\theta_1 = \mu$ is known.

REMARK 2 In the examples just discussed it so happens that the dimensionality of the sufficient statistic is the same as the dimensionality of the parameter. Or to put it differently, the number of the real-valued statistics which are jointly sufficient for the parameter $\boldsymbol{\theta}$ coincides with the number of independent coordinates of $\boldsymbol{\theta}$. However, this need not always be the case. For example, if $X_1, \ldots, X_n$ are i.i.d. r.v.'s from the Cauchy distribution with parameter $\boldsymbol{\theta} = (\mu, \sigma^2)'$, it can be shown that no sufficient statistic of smaller dimensionality other than the (sufficient) statistic $(X_1, \ldots, X_n)'$ exists.

If m is the smallest number for which $\mathbf{T} = (T_1, \ldots, T_m)'$, $T_j = T_j(X_1, \ldots, X_n)$, $j = 1, \ldots, m$, is a sufficient statistic for $\boldsymbol{\theta} = (\theta_1, \ldots, \theta_r)'$, then T is called a *minimal* sufficient statistic for $\boldsymbol{\theta}$.

REMARK 3 In Definition 1, suppose that $m = r$ and that the conditional distribution of $(X_1, \ldots, X_n)'$, given $T_j = t_j$, is independent of θ_j. In a situation like this, one may be tempted to declare that T_j is sufficient for θ_j. This outlook, however, is not in conformity with the definition of a sufficient statistic. The notion of sufficiency is connected with a family of p.d.f.'s $\mathcal{F} = \{f(\cdot; \boldsymbol{\theta}); \boldsymbol{\theta} \in \Omega\}$, and we may talk about T_j being sufficient for θ_j, if all other θ_i, $i \neq j$, are known; otherwise T_j is to be either sufficient for the above family $\mathcal{F}$ or not sufficient at all.

As an example, suppose that $X_1, \ldots, X_n$ are i.i.d. r.v.'s from $N(\theta_1, \theta_2)$. Then $(\overline{X}, S^2)'$ is sufficient for $(\theta_1, \theta_2)'$, where

$$S^2 = \frac{1}{n}\sum_{j=1}^{n}(X_j - \overline{X})^2.$$

Now consider the conditional p.d.f. of $(X_1, \ldots, X_{n-1})'$, given $\sum_{j=1}^{n} X_j = y_n$. By using the transformation

$$y_j = x_j, \ j = 1, \ldots, n-1, \ y_n = \sum_{j=1}^{n} x_j,$$

one sees that the above mentioned conditional p.d.f. is given by the quotient of the following p.d.f.'s:

$$\left(\frac{1}{\sqrt{2\pi\theta_2}}\right)^n \exp\left\{-\frac{1}{2\theta_2}\left[(y_1 - \theta_1)^2 + \cdots + (y_{n-1} - \theta_1)^2 \right.\right.$$
$$\left.\left. + (y_n - y_1 - \cdots - y_{n-1} - \theta_1)^2\right]\right\}$$

and

$$\frac{1}{\sqrt{2\pi n\theta_2}} \exp\left[-\frac{1}{2n\theta_2}(y_n - n\theta_1)^2\right].$$

This quotient is equal to

$$\frac{\sqrt{2\pi n\theta_2}}{\left(\sqrt{2\pi\theta_2}\right)^n} \exp\left\{\frac{1}{2n\theta_2}\left[(y_n - n\theta_1)^2 - n(y_1 - \theta_1)^2 - \cdots - n(y_{n-1} - \theta_1)^2\right.\right.$$
$$\left.\left. - n(y_n - y_1 - \cdots - y_{n-1} - \theta_1)^2\right]\right\}$$

and

$$(y_n - n\theta_1)^2 - n(y_1 - \theta_1)^2 - \cdots - n(y_{n-1} - \theta_1)^2 - n(y_n - y_1 - \cdots - y_{n-1} - \theta_1)^2$$
$$= y_n^2 - n\left[y_1^2 + \cdots + y_{n-1}^2 + (y_n - y_1 - \cdots - y_{n-1})^2\right],$$

independent of θ_1. Thus the conditional p.d.f. under consideration is independent of θ_1 but it *does* depend on θ_2. Thus $\sum_{j=1}^n X_j$, or equivalently, $\overline{X}$ is not sufficient for $(\theta_1, \theta_2)'$. The concept of $\overline{X}$ being sufficient for θ_1 is not valid unless θ_2 is known.

Exercises

11.1.1 In each one of the following cases write out the p.d.f. of the r.v. X and specify the parameter space Ω of the parameter involved.

i) X is distributed as Poisson;
ii) X is distributed as Negative Binomial;
iii) X is distributed as Gamma;
iv) X is distributed as Beta.

11.1.2 Let $X_1, \ldots, X_n$ be i.i.d. r.v.'s distributed as stated below. Then use Theorem 1 and its corollary in order to show that:

i) $\sum_{j=1}^n X_j$ or $\overline{X}$ is a sufficient statistic for θ, if the X's are distributed as Poisson;

ii) $\sum_{j=1}^n X_j$ or $\overline{X}$ is a sufficient statistic for θ, if the X's are distributed as Negative Binomial;

iii) $(\Pi_{j=1}^n X_j, \sum_{j=1}^n X_j)'$ or $(\Pi_{j=1}^n X_j, \overline{X})'$ is a sufficient statistic for $(\theta_1, \theta_2)' = (\alpha, \beta)'$ if the X's are distributed as Gamma. In particular, $\Pi_{j=1}^n X_j$ is a sufficient statistic for $\alpha = \theta$ if β is known, and $\sum_{j=1}^n X_j$ or $\overline{X}$ is a sufficient statistic for $\beta = \theta$ if α is known. In the latter case, take $\alpha = 1$ and conclude that $\sum_{j=1}^n X_j$ or $\overline{X}$ is a sufficient statistic for the parameter $\tilde{\theta} = 1/\theta$ of the Negative Exponential distribution;

iv) $(\Pi_{j=1}^n X_j, \Pi_{j=1}^n (1 - X_j))'$ is a sufficient statistic for $(\theta_1, \theta_2)' = (\alpha, \beta)'$ if the X's are distributed as Beta. In particular, $\Pi_{j=1}^n X_j$ or $-\sum_{j=1}^n \log X_j$ is a sufficient statistic for $\alpha = \theta$ if β is known, and $\Pi_{j=1}^n (1 - X_j)$ is a sufficient statistic for $\beta = \theta$ if α is known.

11.1.3 (Truncated Poisson r.v.'s) Let X_1, X_2 be i.i.d. r.v.'s with p.d.f. $f(\cdot; \theta)$ given by:

$$f(0; \theta) = e^{-\theta}, \quad f(1; \theta) = \theta e^{-\theta}, \quad f(2; \theta) = 1 - e^{-\theta} - \theta e^{-\theta},$$
$$f(x; \theta) = 0, \quad x \neq 0, 1, 2,$$

where $\theta > 0$. Then show that $X_1 + X_2$ is *not* a sufficient statistic for θ.

11.1.4 Let $X_1, \ldots, X_n$ be i.i.d. r.v.'s with the Double Exponential p.d.f. $f(\cdot; \theta)$ given in Exercise 3.3.13(iii) of Chapter 3. Then show that $\sum_{j=1}^{n}|X_j|$ is a sufficient statistic for θ.

11.1.5 If $\mathbf{X}_j = (X_{1j}, X_{2j})', j = 1, \ldots, n$, is a random sample of size n from the Bivariate Normal distribution with parameter $\boldsymbol{\theta}$ as described in Example 4, then, by using Theorem 1, show that:

$$\left(\bar{X}_1, \bar{X}_2, \sum_{j=1}^{n} X_{1j}^2, \sum_{j=1}^{n} X_{2j}^2, \sum_{j=1}^{n} X_{1j} X_{2j}\right)'$$

is a sufficient statistic for $\boldsymbol{\theta}$.

11.1.6 If $X_1, \ldots, X_n$ is a random sample of size n from $U(-\theta, \theta)$, $\theta \in (0, \infty)$, show that $(X_{(1)}, X_{(n)})'$ is a sufficient statistic for θ. Furthermore, show that this statistic is not minimal by establishing that $T = \max(|X_1|, \ldots, |X_n|)$ is also a sufficient statistic for θ.

11.1.7 If $X_1, \ldots, X_n$ is a random sample of size n from $N(\theta, \theta^2)$, $\theta \in \mathbb{R}$, show that

$$\left(\sum_{j=1}^{n} X_j, \sum_{j=1}^{n} X_j^2\right)' \quad \text{or} \quad \left(\bar{X}, \sum_{j=1}^{n} X_j^2\right)'$$

is a sufficient statistic for θ.

11.1.8 If $X_1, \ldots, X_n$ is a random sample of size n with p.d.f.

$$f(x; \theta) = e^{-(x-\theta)} I_{(\theta, \infty)}(x), \quad \theta \in \mathbb{R},$$

show that $X_{(1)}$ is a sufficient statistic for θ.

11.1.9 Let $X_1, \ldots, X_n$ be a random sample of size n from the Bernoulli distribution, and set T_1 for the number of X's which are equal to 0 and T_2 for the number of X's which are equal to 1. Then show that $\mathbf{T} = (T_1, T_2)'$ is a sufficient statistic for θ.

11.1.10 If $X_1, \ldots, X_n$ are i.i.d. r.v.'s with p.d.f. $f(\cdot; \theta)$ given below, find a sufficient statistic for θ.

i) $f(x; \theta) = \theta x^{\theta-1} I_{(0, 1)}(x), \quad \theta \in (0, \infty)$;

ii) $f(x; \theta) = \frac{2}{\theta^2}(\theta - x) I_{(0, \theta)}(x), \quad \theta \in (0, \infty)$;

iii) $f(x; \theta) = \frac{1}{6\theta^4} x^3 e^{-x/\theta} I_{(0, \infty)}(x), \quad \theta \in (0, \infty)$;

iv) $f(x; \theta) = \left(\frac{\theta}{c}\right)\left(\frac{c}{x}\right)^{\theta+1} I_{(c, \infty)}(x), \quad \theta \in (0, \infty)$.

11.2 Completeness

In this section, we introduce the (technical) concept of completeness which we also illustrate by a number of examples. Its usefulness will become apparent in the subsequent sections. To this end, let $\mathbf{X}$ be a k-dimensional random vector with p.d.f. $f(\cdot; \boldsymbol{\theta})$, $\boldsymbol{\theta} \in \Omega \subseteq \mathcal{R}^r$, and let $g: \mathbb{R}^k \to \mathbb{R}$ be a (measurable) function, so that $g(\mathbf{X})$ is an r.v. We assume that $E_\theta g(X)$ exists for all $\theta \in \Omega$ and set $\mathcal{F} = \{f(.; \theta); \theta \in \Omega\}$.

DEFINITION 2 With the above notation, we say that the family $\mathcal{F}$ (or the random vector $\mathbf{X}$) is *complete* if for every g as above, $E_\theta g(\mathbf{X}) = 0$ for all $\boldsymbol{\theta} \in \Omega$ implies that $g(\mathbf{x}) = 0$ except possibly on a set N of $\mathbf{x}$'s such that $P_\theta(\mathbf{X} \in N) = 0$ for all $\boldsymbol{\theta} \in \Omega$.

The examples which follow illustrate the concept of completeness. Meanwhile let us recall that if $\sum_{j=0}^n c_{n-j} x^{n-j} = 0$ for more than n values of x, then $c_j = 0$, $j = 0, \ldots, n$. Also, if $\sum_{n=0}^\infty c_n x^n = 0$ for all values of x in an interval for which the series converges, then $c_n = 0$, $n = 0, 1, \ldots$.

EXAMPLE 9 Let

$$\mathcal{F} = \left\{ f(\cdot;\ \theta);\ f(x;\ \theta) = \binom{n}{x} \theta^x (1-\theta)^{n-x} I_A(x),\ \theta \in (0,\ 1) \right\},$$

where $A = \{0, 1, \ldots, n\}$. Then $\mathcal{F}$ is complete. In fact,

$$E_\theta g(X) = \sum_{x=0}^n g(x) \binom{n}{x} \theta^x (1-\theta)^{n-x} = (1-\theta)^n \sum_{x=0}^n g(x) \binom{n}{x} \rho^x,$$

where $\rho = \theta/(1 - \theta)$. Thus $E_\theta g(X) = 0$ for all $\theta \in (0, 1)$ is equivalent to

$$\sum_{x=0}^n g(x) \binom{n}{x} \rho^x = 0$$

for every $\rho \in (0, \infty)$, hence for more than n values of ρ, and therefore

$$g(x)\binom{n}{x} = 0,\ x = 0, 1, \ldots, n$$

which is equivalent to $g(x) = 0$, $x = 0, 1, \ldots, n$.

EXAMPLE 10 Let

$$\mathcal{F} = \left\{ f(\cdot;\ \theta);\ f(x;\ \theta) = e^{-\theta} \frac{\theta^x}{x!} I_A(x),\ \theta \in (0,\ \infty) \right\},$$

where $A = \{0, 1, \ldots\}$. Then $\mathcal{F}$ is complete. In fact,

$$E_\theta g(X) = \sum_{x=0}^\infty g(x) e^{-\theta} \frac{\theta^x}{x!} = e^{-\theta} \sum_{x=0}^\infty \frac{g(x)}{x!} \theta^x = 0$$

for $\theta \in (0, \infty)$ implies $g(x)/x! = 0$ for $x = 0, 1, \ldots$ and this is equivalent to $g(x) = 0$ for $x = 0, 1, \ldots$.

EXAMPLE 11 Let

$$\mathcal{F} = \left\{ f(\cdot; \theta); f(x; \theta) = \frac{1}{\theta - \alpha} I_{[\alpha, \theta]}(x), \theta \in (\alpha, \infty) \right\}.$$

Then $\mathcal{F}$ is complete. In fact,

$$E_\theta g(X) = \frac{1}{\theta - \alpha} \int_\alpha^\theta g(x) dx.$$

Thus, if $E_\theta g(X) = 0$ for all $\theta \in (\alpha, \infty)$, then $\int_\alpha^\theta g(x) dx = 0$ for all $\theta > \alpha$ which intuitively implies (and that can be rigorously justified) that $g(x) = 0$ except possibly on a set N of x's such that $P_\theta(X \in N) = 0$ for all $\theta \in \Omega$, where X is an r.v. with p.d.f. $f(\cdot; \theta)$. The same is seen to be true if $f(\cdot; \theta)$ is $U(\theta, \beta)$.

EXAMPLE 12 Let $X_1, \ldots, X_n$ be i.i.d. r.v.'s from $N(\mu, \sigma^2)$. If σ is known and $\mu = \theta$, it can be shown that

$$\mathcal{F} = \left\{ f(\cdot; \theta); f(x; \theta) = \frac{1}{\sqrt{2\pi}\sigma} \exp\left[-\frac{(x-\theta)^2}{2\sigma^2}\right], \theta \in \mathbb{R} \right\}$$

is complete. If μ is known and $\sigma^2 = \theta$, then

$$\mathcal{F} = \left\{ f(\cdot; \theta); f(x; \theta) = \frac{1}{\sqrt{2\pi\theta}} \exp\left[-\frac{(x-\mu)^2}{2\theta}\right], \theta \in (0, \infty) \right\}$$

is *not* complete. In fact, let $g(x) = x - \mu$. Then $E_\theta g(X) = E_\theta(X - \mu) = 0$ for all $\theta \in (0, \infty)$, while $g(x) = 0$ only for $x = \mu$. Finally, if both μ and σ^2 are unknown, it can be shown that $(\bar{X}, S^2)'$ is complete.

In the following, we establish two theorems which are useful in certain situations.

THEOREM 2 Let $X_1, \ldots, X_n$ be i.i.d. r.v.'s with p.d.f. $f(\cdot; \boldsymbol{\theta})$, $\boldsymbol{\theta} \in \Omega \subseteq \mathbb{R}^r$ and let $\mathbf{T} = (T_1, \ldots, T_m)'$ be a sufficient statistic for $\boldsymbol{\theta}$, where $T_j = T_j(X_1, \ldots, X_n)$, $j = 1, \ldots, m$. Let $g(\cdot; \boldsymbol{\theta})$ be the p.d.f. of $\mathbf{T}$ and assume that the set S of positivity of $g(\cdot; \boldsymbol{\theta})$ is the same for all $\boldsymbol{\theta} \in \Omega$. Let $\mathbf{V} = (V_1, \ldots, V_k)'$, $V_j = V_j(X_1, \ldots, X_n)$, $j = 1, \ldots, k$, be any other statistic which is assumed to be (stochastically) independent of $\mathbf{T}$. Then the distribution of $\mathbf{V}$ does not depend on $\boldsymbol{\theta}$.

PROOF We have that for $\mathbf{t} \in S$, $g(\mathbf{t}; \boldsymbol{\theta}) > 0$ for all $\boldsymbol{\theta} \in \Omega$ and so $f(\mathbf{v}|\mathbf{t})$ is well defined and is also independent of $\boldsymbol{\theta}$, by sufficiency. Then

$$f_{\mathbf{V}, \mathbf{T}}(\mathbf{v}, \mathbf{t}; \boldsymbol{\theta}) = f(\mathbf{v}|\mathbf{t}) g(\mathbf{t}; \boldsymbol{\theta})$$

for all $\mathbf{v}$ and $\mathbf{t} \in S$, while by independence

$$f_{\mathbf{V}, \mathbf{T}}(\mathbf{v}, \mathbf{t}; \boldsymbol{\theta}) = f_{\mathbf{V}}(\mathbf{v}; \boldsymbol{\theta}) g(\mathbf{t}; \boldsymbol{\theta})$$

for all $\mathbf{v}$ and $\mathbf{t}$. Therefore

$$f_{\mathbf{V}}(\mathbf{v};\boldsymbol{\theta})g(\mathbf{t};\boldsymbol{\theta}) = f(\mathbf{v}|\mathbf{t})g(\mathbf{t};\boldsymbol{\theta})$$

for all $\mathbf{v}$ and $\mathbf{t} \in S$. Hence $f_{\mathbf{V}}(\mathbf{v};\boldsymbol{\theta}) = f(\mathbf{v}/\mathbf{t})$ for all $\mathbf{v}$ and $\mathbf{t} \in S$; that is, $f_{\mathbf{V}}(\mathbf{v};\boldsymbol{\theta}) = f_{\mathbf{V}}(\mathbf{v})$ is independent of $\boldsymbol{\theta}$. ▲

REMARK 4 The theorem need not be true if S depends on θ.

Under certain regularity conditions, the converse of Theorem 2 is true and also more interesting. It relates sufficiency, completeness, and stochastic independence.

THEOREM 3 (Basu) Let $X_1, \ldots, X_n$ be i.i.d. r.v.'s with p.d.f. $f(\cdot;\boldsymbol{\theta})$, $\boldsymbol{\theta} \in \Omega \subseteq \mathbb{R}^r$ and let $\mathbf{T} = (T_1, \ldots, T_m)'$ be a sufficient statistic of $\boldsymbol{\theta}$, where $T_j = T_j(X_1, \ldots, X_n)$, $j = 1, \ldots, m$. Let $g(\cdot;\boldsymbol{\theta})$ be the p.d.f. of $\mathbf{T}$ and assume that $C = \{g(\cdot;\boldsymbol{\theta}); \boldsymbol{\theta} \in \Omega\}$ is complete. Let $\mathbf{V} = (V_1, \ldots, V_k)'$, $V_j = V_j(X_1, \ldots, X_n)$, $j = 1, \ldots, k$ be any other statistic. Then, if the distribution of $\mathbf{V}$ does not depend on $\boldsymbol{\theta}$, it follows that $\mathbf{V}$ and $\mathbf{T}$ are independent.

PROOF It suffices to show that for every $\mathbf{t} \in \mathbb{R}^m$ for which $f(\mathbf{v}|\mathbf{t})$ is defined, one has $f_{\mathbf{V}}(\mathbf{v}) = f(\mathbf{v}|\mathbf{t})$, $\mathbf{v} \in \mathbb{R}^k$. To this end, for an arbitrary but fixed $\mathbf{v}$, consider the statistic $\phi(\mathbf{T}; \mathbf{v}) = f_{\mathbf{V}}(\mathbf{v}) - f(\mathbf{v}|\mathbf{T})$ which is defined for all $\mathbf{t}$'s except perhaps for a set N of $\mathbf{t}$'s such that $P_{\boldsymbol{\theta}}(\mathbf{T} \in N) = 0$ for all $\boldsymbol{\theta} \in \Omega$. Then we have for the continuous case (the discrete case is treated similarly)

$$E_{\boldsymbol{\theta}}\phi(\mathbf{T}; \mathbf{v}) = E_{\boldsymbol{\theta}}\left[f_{\mathbf{V}}(\mathbf{v}) - f(\mathbf{v}|\mathbf{T})\right] = f_{\mathbf{V}}(\mathbf{v}) - E_{\boldsymbol{\theta}}f(\mathbf{v}|\mathbf{T})$$

$$= f_{\mathbf{V}}(\mathbf{v}) - \int_{-\infty}^{\infty}\cdots\int_{-\infty}^{\infty} f(\mathbf{v}|t_1, \ldots, t_m)g(t_1, \ldots, t_m; \boldsymbol{\theta})dt_1 \cdots dt_m$$

$$= f_{\mathbf{V}}(\mathbf{v}) - \int_{-\infty}^{\infty}\cdots\int_{-\infty}^{\infty} f(\mathbf{v}, t_1, \ldots, t_m; \boldsymbol{\theta})dt_1 \cdots dt_m$$

$$= f_{\mathbf{V}}(\mathbf{v}) - f_{\mathbf{V}}(\mathbf{v}) = 0;$$

that is, $E_{\boldsymbol{\theta}}\phi(\mathbf{T}; \mathbf{v}) = 0$ for all $\boldsymbol{\theta} \in \Omega$ and hence $\phi(\mathbf{t}; \mathbf{v}) = 0$ for all $\mathbf{t} \in N^c$ by completeness (N is independent of $\mathbf{v}$ by the definition of completeness). So $f_{\mathbf{V}}(\mathbf{v}) = f(\mathbf{v}/\mathbf{t})$, $\mathbf{t} \in N^c$, as was to be seen. ▲

Exercises

11.2.1 If $\mathcal{F}$ is the family of all Negative Binomial p.d.f.'s, then show that $\mathcal{F}$ is complete.

11.2.2 If $\mathcal{F}$ is the family of all $U(-\theta, \theta)$ p.d.f.'s, $\theta \in (0, \infty)$, then show that $\mathcal{F}$ is *not* complete.

11.2.3 (Basu) Consider an urn containing 10 identical balls numbered $\theta + 1$, $\theta + 2, \ldots, \theta + 10$, where $\theta \in \Omega = \{0, 10, 20, \ldots\}$. Two balls are drawn one by one with replacement, and let X_j be the number on the jth ball, $j = 1, 2$. Use this example to show that Theorem 2 need not be true if the set S in that theorem does depend on θ.

11.3 Unbiasedness—Uniqueness

In this section, we shall restrict ourselves to the case that the parameter is real-valued. We shall then introduce the concept of unbiasedness and we shall establish the existence and uniqueness of uniformly minimum variance unbiased statistics.

DEFINITION 3 Let $X_1, \ldots, X_n$ be i.i.d. r.v.'s with p.d.f. $f(\cdot; \theta)$, $\theta \in \Omega \subseteq \mathbb{R}$ and let $U = U(X_1, \ldots, X_n)$ be a statistic. Then we say that U is an *unbiased statistic* for θ if $E_\theta U = \theta$ for every $\theta \in \Omega$, where by $E_\theta U$ we mean that the expectation of U is calculated by using the p.d.f. $f(\cdot; \theta)$.

We can now formulate the following important theorem.

THEOREM 4 (Rao–Blackwell) Let $X_1, \ldots, X_n$ be i.i.d. r.v.'s with p.d.f. $f(\cdot; \theta)$, $\theta \in \Omega \subseteq \mathbb{R}$, and let $\mathbf{T} = (T_1, \ldots, T_m)'$, $T_j = T_j(X_1, \ldots, X_n)$, $j = 1, \ldots, m$, be a sufficient statistic for θ. Let $U = U(X_1, \ldots, X_n)$ be an unbiased statistic for θ which is not a function of $\mathbf{T}$ alone (with probability 1). Set $\phi(\mathbf{t}) = E_\theta(U|\mathbf{T} = \mathbf{t})$. Then we have that:

i) The r.v. $\phi(\mathbf{T})$ is a function of the sufficient statistic $\mathbf{T}$ alone.
ii) $\phi(\mathbf{T})$ is an unbiased statistic for θ.
iii) $\sigma_\theta^2[\phi(\mathbf{T})] < \sigma_\theta^2(U)$, $\theta \in \Omega$, provided $E_\theta U^2 < \infty$.

PROOF

i) That $\phi(\mathbf{T})$ is a function of the sufficient statistic $\mathbf{T}$ alone and does not depend on θ is a consequence of the sufficiency of $\mathbf{T}$.
ii) That $\phi(\mathbf{T})$ is unbiased for θ, that is, $E_\theta\phi(\mathbf{T}) = \theta$ for every $\theta \in \Omega$, follows from (CE1), Chapter 5, page 123.
iii) This follows from (CV), Chapter 5, page 123. ▲

The interpretation of the theorem is the following: If for some reason one is interested in finding a statistic with the smallest possible variance within the class of unbiased statistics of θ, then one may restrict oneself to the subclass of the unbiased statistics which depend on $\mathbf{T}$ alone (with probability 1). This is so because, if an unbiased statistic U is not already a function of $\mathbf{T}$ alone (with probability 1), then it becomes so by conditioning it with respect to $\mathbf{T}$. The variance of the resulting statistic will be smaller than the variance of the statistic we started out with by (iii) of the theorem. It is further clear that the variance does not decrease any further by conditioning again with respect to $\mathbf{T}$, since the resulting statistic will be the same (with probability 1) by (CE2′), Chapter 5, page 123. The process of forming the conditional expectation of an unbiased statistic of θ, given $\mathbf{T}$, is known as *Rao–Blackwellization*.

The concept of completeness in conjunction with the Rao–Blackwell theorem will now be used in the following theorem.

THEOREM 5 (Uniqueness theorem: Lehmann–Scheffé) Let $X_1, \ldots, X_n$ be i.i.d. r.v.'s with p.d.f. $f(\cdot; \theta)$, $\theta \in \Omega \subseteq \mathbb{R}$, and let $\mathcal{F} = \{f(\cdot; \theta); \theta \in \Omega\}$. Let $\mathbf{T} = (T_1, \ldots, T_m)'$, $T_j = T_j(X_1, \ldots, X_n)$, $j = 1, \ldots, m$, be a sufficient statistic for θ and let $g(\cdot; \theta)$ be its p.d.f. Set $C = \{g(\cdot; \theta); \theta \in \Omega\}$ and assume that C is complete. Let $U = U(\mathbf{T})$ be an unbiased statistic for θ and suppose that $E_\theta U^2 < \infty$ for all $\theta \in \Omega$. Then U is the *unique* unbiased statistic for θ with the smallest variance in the class of all unbiased statistics for θ in the sense that, if $V = V(\mathbf{T})$ is another unbiased statistic for θ, then $U(\mathbf{t}) = V(\mathbf{t})$ (except perhaps on a set N of $\mathbf{t}$'s such that $P_\theta(\mathbf{T} \in N) = 0$ for all $\theta \in \Omega$).

PROOF By the Rao–Blackwell theorem, it suffices to restrict ourselves in the class of unbiased statistics of θ which are functions of $\mathbf{T}$ alone. By the unbiasedness of U and V, we have then $E_\theta U(\mathbf{T}) = E_\theta V(\mathbf{T}) = \theta$, $\theta \in \Omega$; equivalently,

$$E_\theta[U(\mathbf{T}) - V(\mathbf{T})] = 0, \quad \theta \in \Omega, \quad \text{or} \quad E_\theta \phi(\mathbf{T}) = 0, \quad \theta \in \Omega,$$

where $\phi(\mathbf{T}) = U(\mathbf{T}) - V(\mathbf{T})$. Then by completeness of C, we have $\phi(\mathbf{t}) = 0$ for all $\mathbf{t} \in \mathcal{R}^m$ except possibly on a set N of $\mathbf{t}$'s such that $P_\theta(\mathbf{T} \in N) = 0$ for all $\theta \in \Omega$. ▲

DEFINITION 4 An unbiased statistic for θ which is of minimum variance in the class of all unbiased statistics of θ is called a *uniformly minimum variance* (UMV) unbiased statistic of θ (the term "uniformly" referring to the fact that the variance is minimum for all $\theta \in \Omega$).

Some illustrative examples follow.

EXAMPLE 13 Let $X_1, \ldots, X_n$ be i.i.d. r.v.'s from B for r = 2 or by Example 5, $(1, \theta)$, $\theta \in (0, 1)$. Then $T = \sum_{j=1}^n X_j$ is a sufficient statistic for θ, by Example 6, and also complete, by Example 9. Now $\overline{X} = (1/n)T$ is an unbiased statistic for θ and hence, by Theorem 5, UMV unbiased for θ.

EXAMPLE 14 Let $X_1, \ldots, X_n$ be i.i.d. r.v.'s from $N(\mu, \sigma^2)$. Then if σ is known and $\mu = \theta$, we have that $T = \sum_{j=1}^n X_j$ is a sufficient statistic for θ, by Example 8. It is also complete, by Example 12. Then, by Theorem 5, $\overline{X} = (1/n)T$ is UMV unbiased for θ, since it is unbiased for θ. Let μ be known and without loss of generality set $\mu = 0$ and $\sigma^2 = \theta$. Then $T = \sum_{j=1}^n X_j^2$ is a sufficient statistic for θ, by Example 8. Since T is also complete (by Theorem 8 below) and $S^2 = (1/n)T$ is unbiased for θ, it follows, by Theorem 5, that it is UMV unbiased for θ.

Here is another example which serves as an application to both Rao–Blackwell and Lehmann–Scheffé theorems.

EXAMPLE 15 Let X_1, X_2, X_3 be i.i.d. r.v.'s from the Negative Exponential p.d.f. with parameter λ. Setting $\theta = 1/\lambda$, the p.d.f. of the X's becomes $f(x; \theta) = 1/\theta e^{-x/\theta}$, $x > 0$. We have then that $E_\theta(X_j) = \theta$ and $\sigma_\theta^2(X_j) = \theta^2$, $j = 1, 2, 3$. Thus X_1, for example, is an unbiased statistic for θ with variance θ^2. It is further easily seen (by Theorem

8 below) that $T = X_1 + X_2 + X_3$ is a sufficient statistic for θ and it can be shown that it is also complete. Since X_1 is *not* a function of T, one then knows that X_1 is not the UMV unbiased statistic for θ. To actually find the UMV unbiased statistic for θ, it suffices to Rao–Blackwellize X_1. To this end, it is clear that, by symmetry, one has $E_\theta(X_1|T) = E_\theta(X_2|T) = E_\theta(X_3|T)$. Since also their sum is equal to $E_\theta(T|T) = T$, one has that their common value is $T/3$. Thus $E_\theta(X_1|T) = T/3$ which is what we were after. (One, of course, arrives at the same result by using transformations.) Just for the sake of verifying the Rao–Blackwell theorem, one sees that

$$E_\theta\left(\frac{T}{3}\right) = \theta \quad \text{and} \quad \sigma_\theta^2\left(\frac{T}{3}\right) = \frac{\theta^2}{3}(<\theta^2), \quad \theta \in (0, \infty).$$

Exercises

11.3.1 If $X_1, \ldots, X_n$ is a random sample of size n from $P(\theta)$, then show that $\overline{X}$ is the unique UMV unbiased statistic for θ.

11.3.2 Refer to Example 15 and, by utilizing the appropriate transformation, show that $\overline{X}$ is the unique UMV unbiased statistic for θ.

11.4 The Exponential Family of p.d.f.'s: One-Dimensional Parameter Case

A large class of p.d.f.'s depending on a real-valued parameter θ is of the following form:

$$f(x; \theta) = C(\theta)e^{Q(\theta)T(x)}h(x), \quad x \in \mathbb{R}, \quad \theta \in \Omega(\subseteq \mathbb{R}), \tag{1}$$

where $C(\theta) > 0$, $\theta \in \Omega$ and also $h(x) > 0$ for $x \in S$, the set of positivity of $f(x; \theta)$, which is independent of θ. It follows that

$$C^{-1}(\theta) = \sum_{x \in S} e^{Q(\theta)T(x)}h(x)$$

for the discrete case, and

$$C^{-1}(\theta) = \int_S e^{Q(\theta)T(x)}h(x)dx$$

for the continuous case. If $X_1, \ldots, X_n$ are i.i.d. r.v.'s with p.d.f. $f(\cdot; \theta)$ as above, then the joint p.d.f. of the X's is given by

$$f(x_1, \ldots, x_n; \theta) = C^n(\theta)\exp\left[Q(\theta)\sum_{j=1}^n T(x_j)\right]h(x_1) \cdots h(x_n),$$

$$x_j \in \mathbb{R}, j = 1, \ldots, n, \theta \in \Omega. \tag{2}$$

Some illustrative examples follow.

EXAMPLE 16 Let

$$f(x;\theta) = \binom{n}{x}\theta^x(1-\theta)^{n-x}I_A(x),$$

where $A = \{0, 1, \ldots, n\}$. This p.d.f. can also be written as follows,

$$f(x;\theta) = (1-\theta)^n \exp\left[\left(\log\frac{\theta}{1-\theta}\right)x\right]\binom{n}{x}I_A(x), \quad \theta \in (0, 1),$$

and hence is of the exponential form with

$$C(\theta) = (1-\theta)^n, \quad Q(\theta) = \log\frac{\theta}{1-\theta}, \quad T(x) = x, \quad h(x) = \binom{n}{x}I_A(x).$$

EXAMPLE 17 Let now the p.d.f. be $N(\mu, \sigma^2)$. Then if σ is known and $\mu = \theta$, we have

$$f(x;\theta) = \frac{1}{\sqrt{2\pi}\sigma}\exp\left(-\frac{\theta^2}{2\sigma^2}\right)\exp\left(\frac{\theta}{\sigma^2}x\right)\exp\left(-\frac{1}{2\sigma^2}x^2\right), \quad \theta \in \mathbb{R},$$

and hence is of the exponential form with

$$C(\theta) = \frac{1}{\sqrt{2\pi}\sigma}\exp\left(-\frac{\theta^2}{2\sigma^2}\right), \quad Q(\theta) = \frac{\theta}{\sigma^2},$$

$$T(x) = x, \quad h(x) = \exp\left(-\frac{1}{2\sigma^2}x^2\right).$$

If now μ is known and $\sigma^2 = \theta$, then we have

$$f(x;\theta) = \frac{1}{\sqrt{2\pi\theta}}\exp\left(-\frac{1}{2\theta}(x-\mu)^2\right), \quad \theta \in (0, \infty),$$

and hence it is again of the exponential form with

$$C(\theta) = \frac{1}{\sqrt{2\pi\theta}}, \quad Q(\theta) = -\frac{1}{2\theta}, \quad T(x) = (x-\mu)^2 \quad \text{and} \quad h(x) = 1.$$

If the parameter space Ω of a one-parameter exponential family of p.d.f.'s contains a non-degenerate interval, it can be shown that the family is complete. More precisely, the following result can be proved.

THEOREM 6 Let X be an r.v. with p.d.f. $f(\cdot;\theta)$, $\theta \in \Omega \subseteq \mathbb{R}$ given by (1) and set $C = \{g(\cdot;\theta); \theta \in \Omega\}$, where $g(\cdot;\theta)$ is the p.d.f. of $T(X)$. Then C is complete, provided Ω contains a non-degenerate interval.

Then the completeness of the families established in Examples 9 and 10 and the completeness of the families asserted in the first part of Example 12 and the last part of Example 14 follow from the above theorem.

In connection with families of p.d.f.'s of the one-parameter exponential form, the following theorem holds true.

THEOREM 7 Let $X_1, \ldots, X_n$ be i.i.d. r.v.'s with p.d.f. of the one-parameter exponential form. Then

i) $T^* = \sum_{j=1}^n T(X_j)$ is a sufficient statistic for θ.

ii) The p.d.f. of T^* is of the form

$$g(t;\, \theta) = C^n(\theta) e^{Q(\theta) t} h^*(t),$$

where the set of positivity of $h^*(t)$ is independent of θ.

PROOF

i) This is immediate from (2) and Theorem 1.

ii) First, suppose that the X's are discrete, and then so is T^*. Then we have $g(t;\, \theta) = P_\theta(T^* = t) = \sum f(x_1, \ldots, x_n;\, \theta)$, where the summation extends over all $(x_1, \ldots, x_n)'$ for which $\sum_{j=1}^n T(x_j) = t$. Thus

$$g(t;\, \theta) = \sum C^n(\theta) \exp\left[Q(\theta) \sum_{j=1}^n T(x_j)\right] \prod_{j=1}^n h(x_j)$$

$$= C^n(\theta) e^{Q(\theta) t} \sum \left[\prod_{j=1}^n h(x_j)\right] = C^n(\theta) e^{Q(\theta) t} h^*(t),$$

where

$$h^*(t) = \sum \left[\prod_{j=1}^n h(x_j)\right].$$

Next, let the X's be of the continuous type. Then the proof is carried out under certain regularity conditions to be spelled out. We set $Y_1 = \sum_{j=1}^n T(X_j)$ and let $Y_j = X_j, j = 2, \ldots, n$. Then consider the transformation

$$\begin{cases} y_1 = \sum_{j=1}^n T(x_j) \\ y_j = x_j,\, j = 2, \ldots, n; \end{cases} \text{hence} \begin{cases} T(x_1) = y_1 - \sum_{j=2}^n T(y_j) \\ x_j = y_j,\, j = 2, \ldots, n, \end{cases}$$

and thus

$$\begin{cases} x_1 = T^{-1}\left[y_1 - \sum_{j=2}^n T(y_j)\right] \\ x_j = y_j,\, j = 2, \cdots, n, \end{cases}$$

where we *assume* that $y = T(x)$ is one-to-one and hence the inverse T^{-1} exists. Next,

11.4 The Exponential Family of p.d.f.'s: One-Dimensional Parameter Case

$$\frac{\partial x_1}{\partial y_1} = \frac{1}{T'[T^{-1}(z)]}, \quad \text{where} \quad z = y_1 - \sum_{j=2}^{n} T(y_j),$$

provided we *assume* that the derivative T' of T exists and $T'[T^{-1}(z)] \neq 0$. Since for $j = 2, \ldots, n$, we have

$$\frac{\partial x_1}{\partial y_j} = \frac{1}{T'[T^{-1}(z)]} \frac{\partial z}{\partial y_j} = -\frac{T'(y_j)}{T'[T^{-1}(z)]}, \quad \text{and} \quad \frac{\partial x_j}{\partial y_j} = 1$$

for $j = 2, \ldots, n$ and $\partial x_j / \partial y_i = 0$ for $1 < i, j$, $i \neq j$, we have that

$$J = \frac{1}{T'[T^{-1}(z)]} = \frac{1}{T'\{T^{-1}[y_1 - T(y_2) - \cdots - T(y_n)]\}}.$$

Therefore, the joint p.d.f. of $Y_1, \ldots, Y_n$ is given by

$$g(y_1, \ldots, y_n; \theta) = C^n(\theta) \exp\{Q(\theta)[y_1 - T(y_2) - \cdots - T(y_n)$$
$$+ T(y_2) + \cdots + T(y_n)]\}$$
$$\times h\{T^{-1}[y_1 - T(y_2) - \cdots - T(y_n)]\} \prod_{j=2}^{n} h(y_j) |J|$$
$$= C^n(\theta) e^{Q(\theta)y_1} h\{T^{-1}[y_1 - T(y_2) - \cdots - T(y_n)]\}$$
$$\times \prod_{j=2}^{n} h(y_j) |J|.$$

So if we integrate with respect to $y_2, \ldots y_n$, set

$$h^*(y_1) = \int_{-\infty}^{\infty} \cdots \int_{-\infty}^{\infty} h\{T^{-1}[y_1 - T(y_2) - \cdots - T(y_n)]\}$$
$$\times \prod_{j=2}^{n} h(y_j) |J| dy_2 \cdots dy_n,$$

and replace y_1, by t, we arrive at the desired result. ▲

REMARK 5 The above proof goes through if $y = T(x)$ is one-to-one on each set of a finite partition of $\mathbb{R}$.

We next set $C = \{g(\cdot; \theta \in \Omega\}$, where $g(\cdot; \theta)$ is the p.d.f. of the sufficient statistic T^*. Then the following result concerning the completeness of C follows from Theorem 6.

THEOREM 8 The family $C = \{g(\cdot; \theta \in \Omega\}$ is complete, provided Ω contains a non-degenerate interval.

Now as a consequence of Theorems 2, 3, 7 and 8, we obtain the following result.

THEOREM 9 Let the r.v. $X_1, \ldots, X_n$ be i.i.d. from a p.d.f. of the one-parameter exponential form and let T^* be defined by (i) in Theorem 7. Then, if V is any other statistic, it follows that **V** and T^* are independent if and only if the distribution of **V** does not depend on θ.

PROOF In the first place, T^* is sufficient for θ, by Theorem 7(i), and the set of positivity of its p.d.f. is independent of θ, by Theorem 7(ii). Thus the assumptions of Theorem 2 are satisfied and therefore, if **V** is any statistic which is independent of T^*, it follows that the distribution of **V** is independent of θ. For the converse, we have that the family C of the p.d.f.'s of T^* is complete, by Theorem 8. Thus, if the distribution of a statistic **V** does not depend on θ, it follows, by Theorem 3, that **V** and T^* are independent. The proof is completed. ▲

APPLICATION Let $X_1, \ldots, X_n$ be i.i.d. r.v.'s from $N(\mu, \sigma^2)$. Then

$$\bar{X} = \frac{1}{n}\sum_{j=1}^{n} X_j \quad \text{and} \quad S^2 = \frac{1}{n}\sum_{j=1}^{n}(X_j - \bar{X})^2$$

are independent.

PROOF We treat μ as the unknown parameter θ and let σ^2 be arbitrary (>0) but fixed. Then the p.d.f. of the X's is of the one-parameter exponential form and $T = \bar{X}$ is both sufficient for θ and complete. Let

$$V = V(X_1, \ldots, X_n) = \sum_{j=1}^{n}(X_j - \bar{X})^2.$$

Then V and T will be independent, by Theorem 9, if and only if the distribution of V does not depend on θ. Now X_j being $N(\theta, \sigma^2)$ implies that $Y_j = X_j - \theta$ is $N(0, \sigma^2)$. Since $\bar{Y} = \bar{X} - \theta$, we have

$$\sum_{j=1}^{n}(X_j - \bar{X})^2 = \sum_{j=1}^{n}(Y_j - \bar{Y})^2.$$

But the distribution of $\sum_{j=1}^{n}(Y_j - \bar{Y})^2$ does not depend on θ, because $P[\sum_{j=1}^{n}(Y_j - \bar{Y})^2 \in B]$ is equal to the integral of the joint p.d.f. of the Y's over B and this p.d.f. does not depend on θ. ▲

Exercises

11.4.1 In each one of the following cases, show that the distribution of the r.v. X is of the one-parameter exponential form and identify the various quantities appearing in a one-parameter exponential family.

  i) X is distributed as Poisson;
  ii) X is distributed as Negative Binomial;
  iii) X is distributed as Gamma with β known;

iii′) X is distributed as Gamma with α known;

iv) X is distributed as Beta with β known;

iv′) X is distributed as Beta with α known.

11.4.2 Let $X_1, \ldots, X_n$ be i.i.d. r.v.'s with p.d.f. $f(\cdot; \theta)$ given by

$$f(x; \theta) = \frac{\gamma}{\theta} x^{\gamma-1} \exp\left(-\frac{x^\gamma}{\theta}\right) I_{(0, \infty)}(x), \quad \theta > 0, \gamma > 0 \text{ (known)}.$$

i) Show that $f(\cdot; \theta)$ is indeed a p.d.f.;

ii) Show that $\sum_{j=1}^{n} X_j^\gamma$ is a sufficient statistic for θ;

iii) Is $f(\cdot; \theta)$ a member of a one-parameter exponential family of p.d.f.'s?

11.4.3 Use Theorems 6 and 7 to discuss:

i) The completeness established or asserted in Examples 9, 10, 12 (for $\mu = \theta$ and σ known), 15;

ii) Completeness in the Beta and Gamma distributions when one of the parameters is unknown and the other is known.

11.5 Some Multiparameter Generalizations

Let $X_1, \ldots, X_k$ be i.i.d. r.v.'s and set $\mathbf{X} = (X_1, \ldots, X_k)'$. We say that the joint p.d.f. of the X's, or that the p.d.f. of $\mathbf{X}$, belongs to the *r-parameter exponential family* if it is of the following form:

$$f(\mathbf{x}; \boldsymbol{\theta}) = C(\boldsymbol{\theta}) \exp\left[\sum_{j=1}^{r} Q_j(\boldsymbol{\theta}) T_j(\mathbf{x})\right] h(\mathbf{x}),$$

where $\mathbf{x} = (x_1, \ldots, x_k)'$, $x_j \in \mathbb{R}$, $j = 1, \ldots, k$, $k \geq 1$, $\boldsymbol{\theta} = (\theta_1, \ldots, \theta_r)' \in \Omega \subseteq \mathcal{R}^r$, $C(\boldsymbol{\theta}) > 0$, $\boldsymbol{\theta} \in \Omega$ and $h(\mathbf{x}) > 0$ for $\mathbf{x} \in S$, the set of positivity of $f(\cdot; \boldsymbol{\theta})$, which is independent of $\boldsymbol{\theta}$.

The following are examples of multiparameter exponential families.

EXAMPLE 18 Let $\mathbf{X} = (X_1, \ldots, X_r)'$ have the multinomial p.d.f. Then

$$f(x_1, \ldots, x_r; \theta_1, \ldots, \theta_{r-1}) = (1 - \theta_1 - \cdots - \theta_{r-1})^n$$

$$\times \exp\left(\sum_{j=1}^{r} x_j \log \frac{\theta_j}{1 - \theta_1 - \cdots - \theta_{r-1}}\right) \times \frac{n!}{x_1! \cdots x_r!} I_A(x_1, \ldots, x_r),$$

where $A = \{(x_1, \cdots, x_r)' \in \mathbb{R}^r; x_j \geq 0, j = 1, \ldots, r \text{ and } \sum_{j=1}^{r} x_j = n\}$. Thus this p.d.f. is of exponential form with

$$C(\boldsymbol{\theta}) = (1 - \theta_1 - \cdots - \theta_{r-1})^n,$$

$$Q_j(\boldsymbol{\theta}) = \log\frac{\theta_j}{1 - \theta_1 - \cdots - \theta_{r-1}}, \quad T_j(x_1, \ldots, x_r) = x_j, \, j = 1, \ldots, r-1,$$

and

$$h(x_1, \ldots, x_r) = \frac{n!}{x_1! \cdots x_r!} I_A(x_1, \ldots, x_r).$$

EXAMPLE 19 Let X be $N(\theta_1, \theta_2)$. Then,

$$f(x; \theta_1, \theta_2) = \frac{1}{\sqrt{2\pi\theta_2}} \exp\left(-\frac{\theta_1^2}{2\theta_2}\right) \exp\left(\frac{\theta_1}{\theta_2} x - \frac{1}{2\theta_2} x^2\right),$$

and hence this p.d.f. is of exponential form with

$$C(\boldsymbol{\theta}) = \frac{1}{\sqrt{2\pi\theta_2}} \exp\left(-\frac{\theta_1^2}{2\theta_2}\right), \quad Q_1(\boldsymbol{\theta}) = \frac{\theta_1}{\theta_2}, \quad Q_2 = \frac{1}{2\theta_2}, \quad T_1(x) = x,$$

$$T_2(x) = -x^2 \quad \text{and} \quad h(x) = 1.$$

For multiparameter exponential families, appropriate versions of Theorems 6, 7 and 8 are also true. This point will not be pursued here, however.

Finally, if $X_1, \ldots, X_n$ are i.i.d. r.v.'s with p.d.f. $f(\cdot; \boldsymbol{\theta})$, $\boldsymbol{\theta} = (\theta_1, \ldots, \theta_r)' \in \Omega \subseteq \mathbb{R}^r$, not necessarily of an exponential form, the r-dimensional statistic $\mathbf{U} = (U_1, \ldots, U_r)'$, $U_j = U_j(X_1, \ldots, X_n)$, $j = 1, \ldots, r$, is said to be *unbiased* if $E_{\boldsymbol{\theta}} U_j = \theta_j$, $j = 1, \ldots, r$ for all $\boldsymbol{\theta} \in \Omega$. Again, multiparameter versions of Theorems 4–9 may be formulated but this matter will not be dealt with here.

Exercises

11.5.1 In each one of the following cases, show that the distribution of the r.v. X and the random vector $\mathbf{X}$ is of the multiparameter exponential form and identify the various quantities appearing in a multiparameter exponential family.

i) X is distributed as Gamma;
ii) X is distributed as Beta;
iii) $\mathbf{X} = (X_1, X_2)'$ is distributed as Bivariate Normal with parameters as described in Example 4.

11.5.2 If the r.v. X is distributed as $U(\alpha, \beta)$, show that the p.d.f. of X is *not* of an exponential form regardless of whether one or both of α, β are unknown.

11.5.3 Use the not explicitly stated multiparameter versions of Theorems 6 and 7 to discuss:

i) The completeness asserted in Example 15 when both parameters are unknown;

ii) Completeness in the Beta and Gamma distributions when both parameters are unknown.

11.5.4 (A bio-assay problem) Suppose that the probability of death $p(x)$ is related to the dose x of a certain drug in the following manner

$$p(x) = \frac{1}{1 + e^{-(\alpha + \beta x)}},$$

where $\alpha > 0$, $\beta \in \mathbb{R}$ are unknown parameters. In an experiment, k different doses of the drug are considered, each dose is applied to a number of animals and the number of deaths among them is recorded. The resulting data can be presented in a table as follows.

Dose	x_1	x_2	...	x_k
Number of animals used (n)	n_1	n_2	...	n_k
Number of deaths (Y)	Y_1	Y_2	...	Y_k

$x_1, x_2, \ldots, x_k$ and $n_1, n_2, \ldots, n_k$ are known constants, $Y_1, Y_2, \ldots, Y_k$ are independent r.v.'s; Y_j is distributed as $B(n_j, p(x_j))$. Then show that:

i) The joint distribution of $Y_1, Y_2, \ldots, Y_k$ constitutes an exponential family;

ii) The statistic

$$\left(\sum_{j=1}^{k} Y_j, \sum_{j=1}^{k} x_j Y_j \right)'$$

is sufficient for $\boldsymbol{\theta} = (\alpha, \beta)'$.

(**REMARK** In connection with the probability $p(x)$ given above, see also Exercise 4.1.8 in Chapter 4.)

Chapter 12

Point Estimation

12.1 Introduction

Let X be an r.v. with p.d.f. $f(\cdot; \boldsymbol{\theta})$, where $\boldsymbol{\theta} \in \Omega \subseteq \mathbb{R}^r$. If $\boldsymbol{\theta}$ is known, we can calculate, in principle, all probabilities we might be interested in. In practice, however, $\boldsymbol{\theta}$ is generally unknown. Then the problem of estimating $\boldsymbol{\theta}$ arises; or more generally, we might be interested in estimating some function of $\boldsymbol{\theta}$, $g(\boldsymbol{\theta})$, say, where g is (measurable and) usually a real-valued function. We now proceed to define what we mean by an estimator and an estimate of $g(\boldsymbol{\theta})$. Let $X_1, \ldots, X_n$ be i.i.d. r.v.'s with p.d.f. $f(\cdot; \boldsymbol{\theta})$. Then

DEFINITION 1 Any statistic $U = U(X_1, \ldots, X_n)$ which is used for estimating the unknown quantity $g(\boldsymbol{\theta})$ is called an *estimator* of $g(\boldsymbol{\theta})$. The value $U(x_1, \ldots, x_n)$ of U for the observed values of the X's is called an *estimate* of $g(\boldsymbol{\theta})$.

For simplicity and by slightly abusing the notation, the terms estimator and estimate are often used interchangeably.

Exercise

12.1.1 Let $X_1, \ldots, X_n$ be i.i.d. r.v.'s having the Cauchy distribution with $\sigma = 1$ and μ unknown. Suppose you were to estimate μ; which one of the estimators $X_1, \bar{X}$ would you choose? Justify your answer.

(Hint: Use the distributions of X_1 and $\bar{X}$ as a criterion of selection.)

12.2 Criteria for Selecting an Estimator: Unbiasedness, Minimum Variance

From Definition 1, it is obvious that in order to obtain a meaningful estimator of $g(\theta)$, one would have to choose that estimator from a specified class of estimators having some optimal properties. Thus the question arises as to how a class of estimators is to be selected. In this chapter, we will devote ourselves to discussing those criteria which are often used in selecting a class of estimators.

DEFINITION 2 Let g be as above and suppose that it is real-valued. Then the estimator $U = U(X_1, \ldots, X_n)$ is called an *unbiased* estimator of $g(\theta)$ if $E_\theta U(X_1, \ldots, X_n) = g(\theta)$ for all $\theta \in \Omega$.

DEFINITION 3 Let g be as above and suppose it is real-valued. $g(\theta)$ is said to be *estimable* if it has an unbiased estimator.

According to Definition 2, one could restrict oneself to the class of unbiased estimators. The interest in the members of this class stems from the interpretation of the expectation as an average value. Thus if $U = U(X_1, \ldots, X_n)$ is an unbiased estimator of $g(\theta)$, then, no matter what $\theta \in \Omega$ is, the average value (expectation under θ) of U is equal to $g(\theta)$.

Although the criterion of unbiasedness does specify a class of estimators with a certain property, this class is, as a rule, too large. This suggests that a second desirable criterion (that of variance) would have to be superimposed on that of unbiasedness. According to this criterion, among two estimators of $g(\theta)$ which are both unbiased, one would choose the one with smaller variance. (See Fig. 12.1.) The reason for doing so rests on the interpretation of variance as a measure of concentration about the mean. Thus, if $U = U(X_1, \ldots, X_n)$ is an unbiased estimator of $g(\theta)$, then by Tchebichev's inequality,

$$P_\theta\left[|U - g(\theta)| \leq \varepsilon\right] \geq 1 - \frac{\sigma_\theta^2 U}{\varepsilon^2}.$$

Therefore the smaller $\sigma_\theta^2 U$ is, the larger the lower bound of the probability of concentration of U about $g(\theta)$ becomes. A similar interpretation can be given by means of the CLT when applicable.

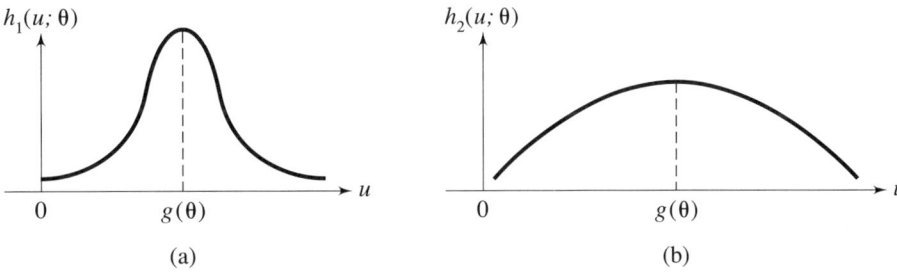

Figure 12.1 (a) p.d.f. of U_1 (for a fixed θ). (b) p.d.f. of U_2 (for a fixed θ).

Following this line of reasoning, one would restrict oneself first to the class of all unbiased estimators of $g(\theta)$ and next to the subclass of unbiased estimators which have finite variance under all $\theta \in \Omega$. Then, within this restricted class, one would search for an estimator with the smallest variance. Formalizing this, we have the following definition.

DEFINITION 4 Let g be estimable. An estimator $U = U(X_1, \ldots, X_n)$ is said to be a *uniformly minimum variance unbiased* (UMVU) estimator of $g(\theta)$ if it is unbiased and has the smallest variance within the class of all unbiased estimators of $g(\theta)$ under all $\theta \in \Omega$. That is, if $U_1 = U_1(X_1, \ldots, X_n)$ is any other unbiased estimator of $g(\theta)$, then $\sigma_\theta^2 U_1 \geq \sigma_\theta^2 U$ for all $\theta \in \Omega$.

In many cases of interest a UMVU estimator does exist. Once one decides to restrict oneself to the class of all unbiased estimators with finite variance, the problem arises as to how one would go about searching for a UMVU estimator (if such an estimator exists). There are two approaches which may be used. The first is appropriate when complete sufficient statistics are available and provides us with a UMVU estimator. Using the second approach, one would first determine a lower bound for the variances of all estimators in the class under consideration, and then would try to determine an estimator whose variance is equal to this lower bound. In the second method just described, the Cramér–Rao inequality, to be established below, is instrumental.

The second approach is appropriate when a complete sufficient statistic is not readily available. (Regarding sufficiency see, however, the corollary to Theorem 2.) It is more effective, in that it does provide a lower bound for the variances of all unbiased estimators regardless of the existence or not of a complete sufficient statistic.

Lest we give the impression that UMVU estimators are all-important, we refer the reader to Exercises 12.3.11 and 12.3.12, where the UMVU estimators involved behave in a rather ridiculous fashion.

Exercises

12.2.1 Let X be an r.v. distributed as $B(n, \theta)$. Show that there is no unbiased estimator of $g(\theta) = 1/\theta$ based on X.

In discussing Exercises 12.2.2–12.2.4 below, refer to Example 3 in Chapter 10 and Example 7 in Chapter 11.

12.2.2 Let $X_1, \ldots, X_n$ be independent r.v.'s distributed as $U(0, \theta)$, $\theta \in \Omega = (0, \infty)$. Find unbiased estimators of the mean and variance of the X's depending only on a sufficient statistic for θ.

12.2.3 Let $X_1, \ldots, X_n$ be i.i.d. r.v.'s from $U(\theta_1, \theta_2)$, $\theta_1 < \theta_2$ and find unbiased estimators for the mean $(\theta_1 + \theta_2)/2$ and the range $\theta_2 - \theta_1$ depending only on a sufficient statistic for $(\theta_1, \theta_2)'$.

12.2.4 Let $X_1, \ldots, X_n$ be i.i.d. r.v.'s from the $U(\theta, 2\theta)$, $\theta \in \Omega = (0, \infty)$ distribution and set

$$U_1 = \frac{n+1}{2n+1} X_{(n)} \quad \text{and} \quad U_2 = \frac{n+1}{5n+4}\left[2X_{(n)} + X_{(1)}\right].$$

Then show that both U_1 and U_2 are unbiased estimators of θ and that U_2 is uniformly better than U_1 (in the sense of variance).

12.2.5 Let $X_1, \ldots, X_n$ be i.i.d. r.v.'s from the Double Exponential distribution $f(x; \theta) = \frac{1}{2} e^{-|x-\theta|}$, $\theta \in \Omega = \mathbb{R}$. Then show that $(X_{(1)} + X_{(n)})/2$ is an unbiased estimator of θ.

12.2.6 Let $X_1, \ldots, X_m$ and $Y_1, \ldots, Y_n$ be two independent random samples with the same mean θ and known variances σ_1^2 and σ_2^2, respectively. Then show that for every $c \in [0, 1]$, $U = c\bar{X} + (1-c)\bar{Y}$ is an unbiased estimator of θ. Also find the value of c for which the variance of U is minimum.

12.2.7 Let $X_1, \ldots, X_n$ be i.i.d. r.v.'s with mean μ and variance σ^2, both unknown. Then show that $\bar{X}$ is the minimum variance unbiased linear estimator of μ.

12.3 The Case of Availability of Complete Sufficient Statistics

The first approach described above will now be looked into in some detail. To this end, let $\mathbf{T} = (T_1, \ldots, T_m)'$, $T_j = T_j(X_1, \ldots, X_n)$, $j = 1, \ldots, m$, be a statistic which is sufficient for $\boldsymbol{\theta}$ and let $U = U(X_1, \ldots, X_n)$ be an unbiased estimator of $g(\boldsymbol{\theta})$, where g is assumed to be real-valued. Set $\phi(T) = E_{\boldsymbol{\theta}}(U|\mathbf{T})$. Then by the Rao–Blackwell theorem (Theorem 4, Chapter 11) (or more precisely, an obvious modification of it), $\phi(\mathbf{T})$ is also an unbiased estimator of $g(\boldsymbol{\theta})$ and furthermore $\sigma_{\boldsymbol{\theta}}^2(\phi) \leq \sigma_{\boldsymbol{\theta}}^2 U$ for all $\boldsymbol{\theta} \in \Omega$ with equality holding only if U is a function of $\mathbf{T}$ (with $P_{\boldsymbol{\theta}}$-probability 1). Thus in the presence of a sufficient statistic, the Rao–Blackwell theorem tells us that, in searching for a UMVU estimator of $g(\boldsymbol{\theta})$, it suffices to restrict ourselves to the class of those unbiased estimators which depend on $\mathbf{T}$ alone. Next, assume that $\mathbf{T}$ is also complete. Then, by the Lehmann–Scheffé theorem (Theorem 5, Chapter 11) (or rather, an obvious modification of it), the unbiased estimator $\phi(\mathbf{T})$ is the one with uniformly minimum variance in the class of all unbiased estimators. Notice that the method just described not only secures the existence of a UMVU estimator, provided an unbiased estimator with finite variance exists, but also produces it. Namely, one starts out with any unbiased estimator of $g(\boldsymbol{\theta})$ with finite variance, U say, assuming that such an estimator exists. Then Rao–Blackwellize it and obtain $\phi(\mathbf{T})$. This is the required estimator. It is essentially unique in the sense that any other UMVU estimators will differ from $\phi(\mathbf{T})$ only on a set of $P_{\boldsymbol{\theta}}$-probability zero for all $\boldsymbol{\theta} \in \Omega$. Thus we have the following result.

12 Point Estimation

THEOREM 1 Let g be as in Definition 2 and assume that there exists an unbiased estimator $U = U(X_1, \ldots, X_n)$ of $g(\boldsymbol{\theta})$ with finite variance. Furthermore, let $\mathbf{T} = (T_1, \ldots, T_m)'$, $T_j = T_j(X_1, \ldots, X_n)$, $j = 1, \ldots, m$ be a sufficient statistic for $\boldsymbol{\theta}$ and suppose that it is also complete. Set $\phi(\mathbf{T}) = E_{\boldsymbol{\theta}}(U|\mathbf{T})$. Then $\phi(\mathbf{T})$ is a UMVU estimator of $g(\boldsymbol{\theta})$ and is essentially unique.

This theorem will be illustrated by a number of concrete examples.

EXAMPLE 1 Let $X_1, \ldots, X_n$ be i.i.d. r.v.'s from $B(1, p)$ and suppose we wish to find a UMVU estimator of the variance of the X's.

The variance of the X's is equal to pq. Therefore, if we set $p = \theta$, $\theta \in \Omega = (0, 1)$ and $g(\theta) = \theta(1 - \theta)$, the problem is that of finding a UMVU estimator for $g(\theta)$. We know that, if

$$U = \frac{1}{n-1} \sum_{j=1}^{n} (X_j - \overline{X})^2,$$

then $E_\theta U = g(\theta)$. Thus U is an unbiased estimator of $g(\theta)$. Furthermore,

$$\sum_{j=1}^{n}(X_j - \overline{X})^2 = \sum_{j=1}^{n} X_j^2 - n\overline{X}^2 = \sum_{j=1}^{n} X_j - n\left(\frac{1}{n}\sum_{j=1}^{n} X_j\right)^2$$

because X_j takes on the values 0 and 1 only and hence $X_j^2 = X_j$. By setting $T = \sum_{j=1}^{n} X_j$, we have then

$$\sum_{j=1}^{n}(X_j - \overline{X})^2 = T - \frac{T^2}{n}, \quad \text{so that} \quad U = \frac{1}{n-1}\left(T - \frac{T^2}{n}\right).$$

But T is a complete, sufficient statistic for θ by Examples 6 and 9 in Chapter 11. Therefore U is a UMVU estimator of the variance of the X's according to Theorem 1.

EXAMPLE 2 Let X be an r.v. distributed as $B(n, \theta)$ and set

$$g(\theta) = P_\theta(X \leq 2) = \sum_{x=0}^{2} \binom{n}{x} \theta^x (1-\theta)^{n-x}$$

$$= (1-\theta)^n + n\theta(1-\theta)^{n-1} + \binom{n}{2}\theta^2(1-\theta)^{n-2}.$$

On the basis of r independent r.v.'s $X_1, \ldots, X_r$ distributed as X, we would like to find a UMVU estimator of $g(\theta)$, if it exists. For example, θ may represent the probability of an item being defective, when chosen at random from a lot of such items. Then $g(\theta)$ represents the probability of accepting the entire lot, if the rule for rejection is this: Choose at random n (≥ 2) items from the lot and then accept the entire lot if the number of observed defective items is ≤ 2. The problem is that of finding a UMVU estimator of $g(\theta)$, if it exists, if the experiment just described is repeated independently r times.

Now the r.v.'s X_j, $j = 1, \ldots, r$ are independent $B(n, \theta)$, so that $T = \sum_{j=1}^{r} X_j$ is $B(nr, \theta)$. T is a complete, sufficient statistic for θ. Set

12.3 The Case of Availability of Complete Sufficient Statistics

$$U = \begin{cases} 1 & \text{if } X_1 \leq 2 \\ 0 & \text{if } X_1 > 2. \end{cases}$$

Then $E_\theta U = g(\theta)$ but it is not a function of T. Then one obtains the required estimator by Rao–Blackwellization of U.

To this end, we have

$$E_\theta(U|T=t) = P_\theta(U=1|T=t)$$

$$= P_\theta(X_1 \leq 2|T=t) = \frac{P_\theta(X_1 \leq 2, X_1 + \cdots + X_r = t)}{P_\theta(T=t)}$$

$$= \frac{1}{P_\theta(T=t)} \Big[P_\theta(X_1 = 0, X_2 + \cdots + X_r = t)$$

$$+ P_\theta(X_1 = 1, X_2 + \cdots + X_r = t-1)$$

$$+ P_\theta(X_1 = 2, X_2 + \cdots + X_r = t-2) \Big]$$

$$= \frac{1}{P_\theta(T=t)} \Big[P_\theta(X_1 = 0) P_\theta(X_2 + \cdots + X_r = t)$$

$$+ P_\theta(X_1 = 1) P_\theta(X_2 + \cdots + X_r = t-1)$$

$$+ P_\theta(X_1 = 2) P_\theta(X_2 + \cdots + X_r = t-2) \Big]$$

$$= \left[\binom{nr}{t} \theta^t (1-\theta)^{nr-t} \right]^{-1} \Bigg[(1-\theta)^n \binom{n(r-1)}{t} \theta^t (1-\theta)^{n(r-1)-t}$$

$$+ n\theta(1-\theta)^{n-1} \binom{n(r-1)}{t-1} \theta^{t-1} (1-\theta)^{n(r-1)-t+1}$$

$$+ \binom{n}{2} \theta^2 (1-\theta)^{n-2} \binom{n(r-1)}{t-2} \theta^{t-2} (1-\theta)^{n(r-1)-t+2} \Bigg]$$

$$= \left[\binom{nr}{t} \theta^t (1-\theta)^{nr-t} \right]^{-1} \theta^t (1-\theta)^{nr-t} \Bigg[\binom{n(r-1)}{t}$$

$$+ n\binom{n(r-1)}{t-1} + \binom{n}{2}\binom{n(r-1)}{t-2} \Bigg].$$

Therefore

$$\phi(T) = \binom{nr}{T}^{-1} \left[\binom{n(r-1)}{T} + n\binom{n(r-1)}{T-1} + \binom{n}{2}\binom{n(r-1)}{T-2} \right]$$

is a UMVU estimator of $g(\theta)$ by Theorem 1.

EXAMPLE 3 Consider certain events which occur according to the distribution $P(\lambda)$. Then the probability that no event occurs is equal to $e^{-\lambda}$. Let now $X_1, \ldots, X_n$ $(n \geq 2)$ be i.i.d. r.v.'s from $P(\lambda)$. Then the problem is that of finding a UMVU estimator of $e^{-\lambda}$.

Set

$$T = \sum_{j=1}^{n} X_j, \quad \lambda = \theta, \quad g(\theta) = e^{-\theta}$$

and define U by

$$U = \begin{cases} 1 & \text{if } X_1 = 0 \\ 0 & \text{if } X_1 \geq 1. \end{cases}$$

Then

$$E_\theta U = P_\theta(U = 1) = P_\theta(X_1 = 0) = g(\theta);$$

that is, U is an unbiased estimator of $g(\theta)$. However, it does not depend on T which is a complete, sufficient statistic for θ, according to Exercise 11.1.2(i) and Example 10 in Chapter 11. It remains then for us to Rao–Blackwellize U. For this purpose we use the fact that the conditional distribution of X_1, given $T = t$, is $B(t, 1/n)$. (See Exercise 12.3.1.) Then

$$E_\theta(U|T = t) = P_\theta(X_1 = 0|T = t) = \left(1 - \frac{1}{n}\right)^t,$$

so that

$$\phi(T) = \left(1 - \frac{1}{n}\right)^T$$

is a UMVU estimator of $e^{-\lambda}$.

EXAMPLE 4 Let $X_1, \ldots, X_n$ be i.i.d. r.v.'s from $N(\mu, \sigma^2)$ with σ^2 unknown and μ known. We are interested in finding a UMVU estimator of σ.

Set $\sigma^2 = \theta$ and let $g(\theta) = \sqrt{\theta}$. By Corollary 5, Chapter 7, we have that $1/\theta \sum_{j=1}^{n}(X_j - \mu)^2$ is χ_n^2. So, if we set

$$S^2 = \frac{1}{n}\sum_{j=1}^{n}(X_j - \mu)^2,$$

then nS^2/θ is χ_n^2, so that $\sqrt{n}S/\sqrt{\theta}$ is distributed as χ_n. Then the expectation $E_\theta(\sqrt{n}S/\sqrt{\theta})$ can be calculated and is independent of θ; call it c'_n (see Exercise 12.3.2). That is,

$$E_\theta\left(\frac{\sqrt{n}S}{\sqrt{\theta}}\right) = c'_n, \quad \text{so that} \quad E_\theta\left(\frac{\sqrt{n}S}{c'_n}\right) = \sqrt{\theta},$$

Setting finally $c_n = c'_n/\sqrt{n}$, we obtain

12.3 The Case of Availability of Complete Sufficient Statistics

$$E_\theta\left(\frac{S}{c_n}\right) = \sqrt{\theta};$$

that is, S/c_n is an unbiased estimator of $g(\theta)$. Since this estimator depends on the complete, sufficient statistic (see Example 8 and Exercise 11.5.3(ii), Chapter 11) S^2 alone, it follows that S/c_n is a UMVU estimator of σ.

EXAMPLE 5 Let again $X_1, \ldots, X_n$ be i.i.d. r.v.'s from $N(\mu, \sigma^2)$ with both μ and σ^2 unknown. We are interested in finding UMVU estimators for each one of μ and σ^2.

Here $\boldsymbol{\theta} = (\mu, \sigma^2)'$ and let $g_1(\boldsymbol{\theta}) = \mu$, $g_2(\boldsymbol{\theta}) = \sigma^2$. By setting

$$S^2 = \frac{1}{n}\sum_{j=1}^{n}(X_j - \overline{X})^2,$$

we have that $(\overline{X}, S^2)'$ is a sufficient statistic for $\boldsymbol{\theta}$. (See Example 8, Chapter 11.) Furthermore, it is complete. (See Example 12, Chapter 11.) Let $U_1 = \overline{X}$ and $U_2 = nS^2/(n-1)$. Clearly, $E_\boldsymbol{\theta} U_1 = \mu$. By Remark 5 in Chapter 7,

$$E_\boldsymbol{\theta}\left(\frac{nS^2}{\sigma^2}\right) = n-1.$$

Therefore

$$E_\boldsymbol{\theta}\left(\frac{nS^2}{n-1}\right) = \sigma^2.$$

So U_1 and U_2 are unbiased estimators of μ and σ^2, respectively. Since they depend only on the complete, sufficient statistic $(\overline{X}, S^2)'$, it follows that they are UMVU estimators.

EXAMPLE 6 Let $X_1, \ldots, X_n$ be i.i.d. r.v.'s from $N(\mu, \sigma^2)$ with both μ and σ^2 unknown, and set ζ_p for the upper pth quantile of the distribution ($0 < p < 1$). The problem is that of finding a UMVU estimator of ζ_p.

Set $\boldsymbol{\theta} = (\mu, \sigma^2)'$. From the definition of ζ_p, one has $P_\boldsymbol{\theta}(X_1 \geq \zeta_p) = p$. But

$$P_\boldsymbol{\theta}(X_1 \geq \zeta_p) = P_\boldsymbol{\theta}\left(\frac{X_1 - \mu}{\sigma} \geq \frac{\zeta_p - \mu}{\sigma}\right) = 1 - \Phi\left(\frac{\zeta_p - \mu}{\sigma}\right),$$

so that

$$\Phi\left(\frac{\zeta_p - \mu}{\sigma}\right) = 1 - p.$$

Hence

$$\frac{\zeta_p - \mu}{\sigma} = \Phi^{-1}(1-p) \quad \text{and} \quad \zeta_p = \mu + \sigma\Phi^{-1}(1-p).$$

Of course, since p is given, $\Phi^{-1}(1-p)$ is a uniquely determined number. Then by setting $g(\boldsymbol{\theta}) = \mu + \sigma\Phi^{-1}(1-p)$, our problem is that of finding a UMVU estimator of $g(\boldsymbol{\theta})$. Let

$$U = \overline{X} + \frac{S}{c_n}\Phi^{-1}(1-p),$$

where c_n is defined in Example 4. Then by the fact that $E_\theta \overline{X} = \mu$ and $E_\theta(S/c_n) = \sigma$ (see Example 4), we have that $E_\theta U = g(\theta)$. Since U depends only on the complete, sufficient statistic $(\overline{X}, S^2)'$, it follows that U is a UMVU estimator of ξ_p.

Exercises

12.3.1 Let $X_1, \ldots, X_n$ be i.i.d. r.v.'s from $P(\lambda)$ and set $T = \sum_{j=1}^{n} X_j$. Then show that the conditional p.d.f. of X_1, given $T = t$, is that of $B(t, 1/n)$. Furthermore, observe that the same is true if X_1 is replaced by any one of the remaining X's.

12.3.2 Refer to Example 4 and evaluate the quantity c'_n mentioned there.

12.3.3 If $X_1, \ldots, X_n$ are i.i.d. r.v.'s from $B(1, \theta)$, $\theta \in \Omega = (0, 1)$, by using Theorem 1, show that $\overline{X}$ is the UMVU estimator of θ.

12.3.4 If $X_1, \ldots, X_n$ are i.i.d. r.v.'s from $P(\theta)$, $\theta \in \Omega = (0, \infty)$, use Theorem 1 in order to determine the UMVU estimator of θ.

12.3.5 Let $X_1, \ldots, X_n$ be i.i.d. r.v.'s from the Negative Exponential distribution with parameter $\theta \in \Omega = (0, \infty)$. Use Theorem 1 in order to determine the UMVU estimator of θ.

12.3.6 Let X be an r.v. having the Negative Binomial distribution with parameter $\theta \in \Omega = (0, 1)$. Find the UMVU estimator of $g(\theta) = 1/\theta$ and determine its variance.

12.3.7 Let $X_1, \ldots, X_n$ be independent r.v.'s distributed as $N(\theta, 1)$. Show that $\overline{X}^2 - (1/n)$ is the UMVU estimator of $g(\theta) = \theta^2$.

12.3.8 Let $X_1, \ldots, X_n$ be independent r.v.'s distributed as $N(\mu, \sigma^2)$, where both μ and σ^2 are unknown. Find the UMVU estimator of μ/σ.

12.3.9 Let $(X_j, Y_j)'$, $j = 1, \ldots, n$ be independent random vectors having the Bivariate Normal distribution with parameter $\theta = (\mu_1, \mu_2, \sigma_1, \sigma_2, \rho)'$. Find the UMVU estimators of the following quantities: $\rho\sigma_1\sigma_2$, $\mu_1\mu_2$, $\rho\sigma_2/\sigma_1$.

12.3.10 Let X be an r.v. denoting the life span of a piece of equipment. Then the *reliability* of the equipment at time x, $R(x)$, is defined as the probability that $X > x$. If X has the Negative Exponential distribution with parameter $\theta \in \Omega = (0, \infty)$, find the UMVU estimator of the reliability $R(x; \theta)$ on the basis of n observations on X.

12.3.11 Let X be an r.v. having the Geometric distribution; that is,

$$f(x; \theta) = \theta(1-\theta)^x, \quad x = 0, 1, \ldots, \quad \theta \in \Omega = (0, 1),$$

and let $U(X)$ be defined as follows: $U(X) = 1$ if $X = 0$ and $U(X) = 0$ if $X \neq 0$. By using Theorem 1, show that $U(X)$ is a UMVU estimator of θ and conclude that it is an unreasonable one.

12.3.12 Let X be an r.v. denoting the number of telephone calls which arrive at a given telephone exchange, and suppose that X is distributed as $P(\theta)$, where $\theta \in \Omega = (0, \infty)$ is the number of calls arriving at the telephone exchange under consideration within a 15 minute period. Then the number of calls which arrive at the given telephone exchange within 30 minutes is an r.v. Y distributed as $P(2\theta)$, as can be shown. Thus $P_\theta(Y = 0) = e^{-2\theta} = g(\theta)$. Define $U(X)$ by $U(X) = (-1)^X$. Then show that $U(X)$ is the UMVU estimator of $g(\theta)$ and conclude that it is an entirely unreasonable estimator. (Hint: Use Theorem 1.)

12.3.13 Use Example 11, Chapter 11, in order to show that the unbiased estimator constructed in Exercise 12.2.2 is actually UMVU.

12.3.14 Use Exercise 11.1.4, Chapter 11, in order to conclude that the unbiased estimator constructed in Exercise 12.2.5 is not UMVU.

12.4 The Case Where Complete Sufficient Statistics Are Not Available or May Not Exist: Cramér–Rao Inequality

When complete, sufficient statistics are available, the problem of finding a UMVU estimator is settled as in Section 3. When such statistics do not exist, or it is not easy to identify them, one may use the approach described here in searching for a UMVU estimator. According to this method, we first establish a lower bound for the variances of all unbiased estimators and then we attempt to identify an unbiased estimator with variance equal to the lower bound found. If that is possible, the problem is solved again. At any rate, we do have a lower bound of the variances of a class of estimators, which may be useful for comparison purposes.

The following regularity conditions will be employed in proving the main result in this section. We assume that $\Omega \subseteq \mathbb{R}$ and that g is real-valued and differentiable for all $\theta \in \Omega$.

12.4.1 Regularity Conditions

Let X be an r.v. with p.d.f. $f(\cdot; \theta)$, $\theta \in \Omega \subseteq \mathbb{R}$. Then it is assumed that

  i) $f(x; \theta)$ is positive on a set S independent of $\theta \in \Omega$.
  ii) Ω is an open interval in $\mathbb{R}$ (finite or not).
  iii) $(\partial/\partial\theta) f(x; \theta)$ exists for all $\theta \in \Omega$ and all $x \in S$ except possibly on a set $N \subset S$ which is independent of θ and such that $P_\theta(X \in N) = 0$ for all $\theta \in \Omega$.
  iv) $$\int_S \cdots \int_S f(x_1; \theta) \cdots f(x_n; \theta) dx_1 \cdots dx_n$$

or
$$\sum_S \cdots \sum_S f(x_1; \theta) \cdots f(x_n; \theta)$$
may be differentiated under the integral or summation sign, respectively.

v) $E_\theta[(\partial/\partial\theta)\log f(X; \theta)]^2$, to be denoted by $I(\theta)$, is >0 for all $\theta \in \Omega$.

vi) $\int_S \cdots \int_S U(x_1, \ldots, x_n) f(x_1; \theta) \cdots f(x_n; \theta) dx_1 \cdots dx_n$

or
$$\sum_S \cdots \sum_S U(x_1, \ldots, x_n) f(x_1; \theta) \cdots f(x_n; \theta)$$
may be differentiated under the integral or summation sign, respectively, where $U(X_1, \ldots, X_n)$ is any unbiased estimator of $g(\theta)$. Then we have the following theorem.

THEOREM 2 (Cramér–Rao inequality.) Let $X_1, \ldots, X_n$ be i.i.d. r.v.'s with p.d.f. $f(\cdot; \theta)$ and assume that the regularity conditions (i)–(vi) are fulfilled. Then for any unbiased estimator $U = U(X_1, \ldots, X_n)$ of $g(\theta)$, one has

$$\sigma_\theta^2 U \geq \frac{[g'(\theta)]^2}{nI(\theta)}, \quad \theta \in \Omega, \quad \text{where} \quad g'(\theta) = \frac{dg(\theta)}{d\theta}.$$

PROOF If $\sigma_\theta^2 U = \infty$ or $I(\theta) = \infty$ for some $\theta \in \Omega$, the inequality is trivially true for those θ's. Hence we need only consider the case where $\sigma_\theta^2 U < \infty$ and $I(\theta) < \infty$ for all $\theta \in \Omega$. Also it suffices to discuss the continuous case only, since the discrete case is treated entirely similarly with integrals replaced by summation signs.

We have
$$E_\theta U(X_1, \ldots, X_n)$$
$$= \int_S \cdots \int_S U(x_1, \ldots, x_n) f(x_1; \theta) \cdots f(x_n; \theta) dx_1 \cdots dx_n = g(\theta). \quad (1)$$
Now restricting ourselves to S, we have

$$\frac{\partial}{\partial \theta}[f(x_1; \theta) \cdots f(x_n; \theta)]$$
$$= \left[\frac{\partial}{\partial \theta} f(x_1; \theta)\right] \prod_{j \neq 1} f(x_j; \theta) + \left[\frac{\partial}{\partial \theta} f(x_2; \theta)\right]$$
$$\times \prod_{j \neq 2} f(x_j; \theta) + \cdots + \left[\frac{\partial}{\partial \theta} f(x_n; \theta)\right] \prod_{j \neq n} f(x_j; \theta)$$
$$= \sum_{j=1}^n \left[\frac{\partial}{\partial \theta} f(x_j; \theta) \prod_{i \neq j} f(x_i; \theta)\right]$$
$$= \sum_{j=1}^n \left[\frac{1}{f(x_j; \theta)} \frac{\partial}{\partial \theta} f(x_j; \theta)\right] \prod_{i=1}^n f(x_i; \theta)$$
$$= \left[\sum_{j=1}^n \frac{\partial}{\partial \theta} \log f(x_j; \theta)\right] \prod_{i=1}^n f(x_i; \theta). \quad (2)$$

12.4 The Case Where Complete Sufficient Statistics Are Not Available or May Not Exist

Differentiating with respect to θ both sides of (1) on account of (vi) and utilizing (2), we obtain

$$g'(\theta) = \int_S \cdots \int_S U(x_1, \ldots, x_n) \left[\sum_{j=1}^n \frac{\partial}{\partial \theta} \log f(x_j; \theta) \right] \prod_{i=1}^n f(x_i; \theta) dx_1 \cdots dx_n$$

$$= E_\theta \left\{ U(X_1, \ldots, X_n) \left[\sum_{j=1}^n \frac{\partial}{\partial \theta} \log f(X_j; \theta) \right] \right\} = E_\theta(UV_\theta), \qquad (3)$$

where we set

$$V_\theta = V_\theta(X_1, \ldots, X_n) = \sum_{j=1}^n \frac{\partial}{\partial \theta} \log f(X_j; \theta).$$

Next,

$$\int_S \cdots \int_S f(x_1; \theta) \cdots f(x_n; \theta) dx_1 \cdots dx_n = 1.$$

Therefore differentiating both sides with respect to θ by virtue of (iv), and employing (2),

$$0 = \int_S \cdots \int_S \left[\sum_{j=1}^n \frac{\partial}{\partial \theta} \log f(x_j; \theta) \right] \prod_{i=1}^n f(x_i; \theta) dx_1 \cdots dx_n = E_\theta V_\theta. \qquad (4)$$

From (3) and (4), it follows that

$$Cov_\theta(U, V_\theta) = E_\theta(UV_\theta) - (E_\theta U)(E_\theta V_\theta) = E_\theta(UV_\theta) = g'(\theta). \qquad (5)$$

From (4) and the definition of V_θ, it further follows that

$$0 = E_\theta V_\theta = E_\theta \left[\sum_{j=1}^n \frac{\partial}{\partial \theta} \log f(X_j; \theta) \right] = \sum_{j=1}^n E_\theta \left[\frac{\partial}{\partial \theta} \log f(X_j; \theta) \right]$$

$$= n E_\theta \left[\frac{\partial}{\partial \theta} \log f(X_1; \theta) \right],$$

so that

$$E_\theta \left[\frac{\partial}{\partial \theta} \log f(X_1; \theta) \right] = 0.$$

Therefore

$$\sigma_\theta^2 V_\theta = \sigma_\theta^2 \left[\sum_{j=1}^n \frac{\partial}{\partial \theta} \log f(X_j; \theta) \right] = \sum_{j=1}^n \sigma_\theta^2 \left[\frac{\partial}{\partial \theta} \log f(X_j; \theta) \right]$$

$$= n \sigma_\theta^2 \left[\frac{\partial}{\partial \theta} \log f(X_1; \theta) \right]$$

$$= n E_\theta \left[\frac{\partial}{\partial \theta} \log f(X_1; \theta) \right]^2 = n E_\theta \left[\frac{\partial}{\partial \theta} \log f(X; \theta) \right]^2. \qquad (6)$$

But
$$\rho_\theta(U, V_\theta) = \frac{Cov(U, V_\theta)}{(\sigma_\theta U)(\sigma_\theta V_\theta)}$$

and $\rho_\theta^2(U, V_\theta) \leq 1$, which is equivalent to

$$C\hat{o}v_\theta(U, V_\theta) \leq (\sigma_\theta^2 U)(\sigma_\theta^2 V_\theta). \qquad (7)$$

Taking now into consideration (5) and (6), relation (7) becomes

$$[g'(\theta)]^2 \leq (\sigma_\theta^2 U) n E_\theta \left[\frac{\partial}{\partial \theta} \log f(X; \theta)\right]^2,$$

or by means of (v),

$$\sigma_\theta^2 U \geq \frac{[g'(\theta)]^2}{n E_\theta\left[(\partial/\partial \theta)\log f(X; \theta)\right]^2} = \frac{[g'(\theta)]^2}{nI(\theta)}. \qquad (8)$$

The proof of the theorem is completed. ▲

DEFINITION 5 The expression $E_\theta[(\partial/\partial\theta)\log f(X; \theta)]^2$, denoted by $I(\theta)$, is called *Fisher's information* (about θ) *number*; $nE_\theta[(\partial/\partial\theta)\log f(X; \theta)]^2$ is the information (about θ) contained in the sample $X_1, \ldots, X_n$.

(For an alternative way of calculating $I(\theta)$, see Exercises 12.4.6 and 12.4.7.)

Returning to the proof of Theorem 2, we have that equality holds in (8) if and only if $C\hat{o}v_\theta(U, V_\theta) = (\sigma_\theta^2 U)(\sigma_\theta^2 V_\theta)$ because of (7). By Schwarz inequality (Theorem 2, Chapter 5), this is equivalent to

$$V_\theta = E_\theta V_\theta + k(\theta)(U - E_\theta U) \text{ with } P_\theta\text{-probability } 1, \qquad (9)$$

where

$$k(\theta) = \pm \frac{\sigma_\theta V_\theta}{\sigma_\theta U}.$$

Furthermore, because of (i), the exceptional set for which (9) does not hold is independent of θ and has P_θ-probability 0 for all $\theta \in \Omega$. Taking into consideration (4), the fact that $E_\theta U = g(\theta)$ and the definition of V_θ, equation (9) becomes as follows:

$$\frac{\partial}{\partial \theta} \log \prod_{j=1}^n f(X_j; \theta) = k(\theta) U(X_1, \ldots, X_n) - g(\theta) k(\theta) \qquad (10)$$

outside a set N in $\mathbb{R}^n$ such that $P_\theta[(X_1, \ldots, X_n) \in N] = 0$ for all $\theta \in \Omega$. Integrating (10) (with respect to θ) and assuming that the indefinite integrals $\int k(\theta) d\theta$ and $\int g(\theta) k(\theta) d\theta$ exist, we obtain

$$\log \prod_{j=1}^n f(X_j; \theta) = U(X_1, \ldots, X_n) \int k(\theta) d\theta - \int g(\theta) k(\theta) d\theta + \tilde{h}(X_1, \ldots, X_n),$$

where $\tilde{h}(X_1, \ldots, X_n)$ is the "constant" of the integration, or

$$\log \prod_{j=1}^{n} f(x_j; \theta) = U(x_1, \ldots, x_n) \int k(\theta) d\theta - \int g(\theta) k(\theta) d\theta + \tilde{h}(x_1, \ldots, x_n). \quad (11)$$

Exponentiating both sides of (11), we obtain

$$\prod_{j=1}^{n} f(x_j; \theta) = C(\theta) \exp\left[Q(\theta) U(x_1, \ldots, x_n)\right] h(x_1, \ldots, x_n), \quad (12)$$

where

$$C(\theta) = \exp\left[-\int g(\theta) k(\theta) d\theta\right], \quad Q(\theta) = \int k(\theta) d\theta$$

and

$$h(x_1, \ldots, x_n) = \exp\left[\tilde{h}(x_1, \ldots, x_n)\right].$$

Thus, if equality occurs in the Cramér–Rao inequality for some unbiased estimator, then the joint p.d.f. of the X's is of the one-parameter exponential form, provided certain conditions are met. More precisely, we have the following result.

COROLLARY If in Theorem 2 equality occurs for some unbiased estimator $U = U(X_1, \ldots, X_n)$ of $g(\theta)$ and if the indefinite integrals $\int k(\theta) d\theta, \int g(\theta) k(\theta) d\theta$ exist, where

$$k(\theta) = \pm \frac{\sigma_\theta V_\theta}{\sigma_\theta U},$$

then

$$\prod_{j=1}^{n} f(x_j; \theta) = C(\theta) \exp\left[Q(\theta) U(x_1, \ldots, x_n)\right] h(x_1, \ldots, x_n)$$

outside a set N in $\mathbb{R}^n$ such that $P_\theta[(X_1, \ldots, X_n) \in N] = 0$ for all $\theta \in \Omega$; here $C(\theta) = \exp[-\int g(\theta) k(\theta) d\theta]$ and $Q(\theta) = \int k(\theta) d\theta$. That is, the joint p.d.f. of the X's is of the one-parameter exponential family (and hence U is sufficient for θ).

REMARK 1 Theorem 2 has a certain generalization for the multiparameter case, but this will not be discussed here.

In connection with the Cramér–Rao bound, we also have the following important result.

THEOREM 3 Let $X_1, \ldots, X_n$ be i.i.d. r.v.'s with p.d.f. $f(\cdot; \theta)$ and let g be an estimable real-valued function of θ. For an unbiased estimator $U = U(X_1, \cdots, X_n)$ of $g(\theta)$, we assume that regularity conditions (i)–(vi) are satisfied. Then $\sigma_\theta^2 U$ is equal to the Cramér–Rao bound if and only if there exists a real-valued function of θ, $d(\theta)$, such that $U = g(\theta) + d(\theta) V_\theta$ except perhaps on a set of P_θ-probability zero for all $\theta \in \Omega$.

PROOF Under the regularity conditions (i)–(vi), we have that

$$\sigma_\theta^2 U \geq \frac{[g'(\theta)]^2}{nI(\theta)}, \quad \text{or} \quad \sigma_\theta^2 U \geq \frac{[g'(\theta)]^2}{\sigma_\theta^2 V_\theta},$$

since $nI(\theta) = \sigma_\theta^2 V_\theta$ by (6). Then $\sigma_\theta^2 U$ is equal to the Cramér–Rao bound if and only if

$$[g'(\theta)]^2 = (\sigma_\theta^2 U)(\sigma_\theta^2 V_\theta).$$

But

$$[g'(\theta)]^2 = C\hat{o}v_\theta(U, V_\theta) \quad \text{by} \quad (5).$$

Thus $\sigma_\theta^2 U$ is equal to the Cramér–Rao bound if and only if $C_\theta^2(U, V_\theta) = (\sigma_\theta^2 U) \times (\sigma_\theta^2 V_\theta)$, or equivalently, if and only if $U = a(\theta) + d(\theta)V_\theta$ with P_θ-probability 1 for some functions of θ, $a(\theta)$ and $d(\theta)$. Furthermore, because of (i), the exceptional set for which this relationship does not hold is independent of θ and has P_θ-probability 0 for all $\theta \in \Omega$. Taking expectations and utilizing the unbiasedness of U and relation (4), we get that $U = g(\theta) + d(\theta)V_\theta$ except perhaps on a set of P_θ-probability 0 for all $\theta \in \Omega$. The proof of the theorem is completed. ▲

The following three examples serve to illustrate Theorem 2. The checking of the regularity conditions is left as an exercise.

EXAMPLE 7 Let $X_1, \ldots, X_n$ be i.i.d. r.v.'s from $B(1, p)$, $p \in (0, 1)$. By setting $p = \theta$, we have

$$f(x; \theta) = \theta^x(1-\theta)^{1-x}, \quad x = 0, 1$$

so that

$$\log f(x; \theta) = x \log \theta + (1-x)\log(1-\theta).$$

Then

$$\frac{\partial}{\partial \theta} \log f(x; \theta) = \frac{x}{\theta} - \frac{1-x}{1-\theta}$$

and

$$\left[\frac{\partial}{\partial \theta} \log f(x; \theta)\right]^2 = \frac{1}{\theta^2}x^2 + \frac{1}{(1-\theta)^2}(1-x)^2 - \frac{2}{\theta(1-\theta)}x(1-x).$$

Since

$$E_\theta X^2 = \theta, \quad E_\theta(1-X)^2 = 1-\theta \quad \text{and} \quad E_\theta[X(1-X)] = 0$$

(see Chapter 5), we have

$$E\left[\frac{\partial}{\partial \theta} \log f(X; \theta)\right]^2 = \frac{1}{\theta(1-\theta)},$$

so that the Cramér–Rao bound is equal to $\theta(1-\theta)/n$.

12.4 The Case Where Complete Sufficient Statistics Are Not Available or May Not Exist

Now $\bar{X}$ is an unbiased extimator of θ and its variance is $\sigma_\theta^2(\bar{X}) = \theta(1-\theta)/n$, that is, equal to the Cramér–Rao bound. Therefore $\bar{X}$ is a UMVU estimator of θ.

EXAMPLE 8 Let $X_1, \ldots, X_n$ be i.i.d. r.v.'s from $P(\lambda)$, $\lambda > 0$. Again by setting $\lambda = \theta$, we have

$$f(x;\theta) = e^{-\theta}\frac{\theta^x}{x!}, \quad x = 0,1,\ldots \quad \text{so that} \quad \log f(x;\theta) = -\theta + x\log\theta - \log x!.$$

Then

$$\frac{\partial}{\partial\theta}\log f(x;\theta) = -1 + \frac{x}{\theta}$$

and

$$\left[\frac{\partial}{\partial\theta}\log f(x;\theta)\right]^2 = 1 + \frac{1}{\theta^2}x^2 - \frac{2}{\theta}x.$$

Since $E_\theta X = \theta$ and $E_\theta X^2 = \theta(1+\theta)$ (see Chapter 5), we obtain

$$E_\theta\left[\frac{\partial}{\partial\theta}\log f(X;\theta)\right]^2 = \frac{1}{\theta},$$

so that the Cramér–Rao bound is equal to θ/n. Since again $\bar{X}$ is an unbiased estimator of θ with variance θ/n, we have that $\bar{X}$ is a UMVU estimator of θ.

EXAMPLE 9 Let $X_1, \ldots, X_n$ be i.i.d. r.v.'s from $N(\mu, \sigma^2)$. Assume first that σ^2 is known and set $\mu = \theta$. Then

$$f(x;\theta) = \frac{1}{\sqrt{2\pi}\sigma}\exp\left[-\frac{(x-\theta)^2}{2\sigma^2}\right], \quad x \in \mathbb{R}$$

and hence

$$\log f(x;\theta) = \log\left(\frac{1}{\sqrt{2\pi}\sigma}\right) - \frac{(x-\theta)^2}{2\sigma^2}.$$

Next,

$$\frac{\partial}{\partial\theta}\log f(x;\theta) = \frac{1}{\sigma}\frac{x-\theta}{\sigma},$$

so that

$$\left[\frac{\partial}{\partial\theta}\log f(x;\theta)\right]^2 = \frac{1}{\sigma^2}\left(\frac{x-\theta}{\sigma}\right)^2.$$

Then

$$E_\theta\left[\frac{\partial}{\partial\theta}\log f(X;\theta)\right]^2 = \frac{1}{\sigma^2},$$

since $(X - \theta)/\sigma$ is $N(0, 1)$ and hence

$$E_\theta = \left(\frac{X - \theta}{\sigma}\right)^2 = 1. \quad \text{(See Chapter 5.)}$$

Thus the Cramér–Rao bound is σ^2/n. Once again, $\bar{X}$ is an unbiased estimate of θ and its variance is equal to σ^2/n, that is, the Cramér–Rao bound. Therefore, $\bar{X}$ is a UMVU estimator. This was also shown in Example 5.

Suppose now that μ is known and set $\sigma^2 = \theta$. Then

$$f(x; \theta) = \frac{1}{\sqrt{2\pi\theta}} \exp\left[-\frac{(x-\mu)^2}{2\theta}\right],$$

so that

$$\log f(x; \theta) = -\frac{1}{2}\log(2\pi) - \frac{1}{2}\log\theta - \frac{(x-\mu)^2}{2\theta}$$

and

$$\frac{\partial}{\partial \theta}\log f(x; \theta) = -\frac{1}{2\theta} + \frac{(x-\mu)^2}{2\theta^2}.$$

Then

$$\left[\frac{\partial}{\partial\theta}\log f(x; \theta)\right]^2 = \frac{1}{4\theta^2} - \frac{1}{2\theta^2}\left(\frac{x-\mu}{\sqrt{\theta}}\right)^2 + \frac{1}{4\theta^2}\left(\frac{x-\mu}{\sqrt{\theta}}\right)^4$$

and since $(X - \mu)/\sqrt{\theta}$ is $N(0, 1)$, we obtain

$$E_\theta\left(\frac{X-\mu}{\sqrt{\theta}}\right)^2 = 1, \quad E_\theta\left(\frac{X-\mu}{\sqrt{\theta}}\right)^4 = 3. \quad \text{(See Chapter 5.)}$$

Therefore

$$E_\theta\left[\frac{\partial}{\partial\theta}\log f(X; \theta)\right]^2 = \frac{1}{2\theta^2}$$

and the Cramér–Rao bound is $2\theta^2/n$. Next,

$$\sum_{j=1}^{n}\left(\frac{X_j - \mu}{\sqrt{\theta}}\right)^2 \text{ is } \chi_n^2$$

(see first corollary to Theorem 5, Chapter 7), so that

$$E_\theta\left[\sum_{j=1}^{n}\left(\frac{X_j-\mu}{\sqrt{\theta}}\right)^2\right] = n \quad \text{and} \quad \sigma_\theta^2\left[\sum_{j=1}^{n}\left(\frac{X_j-\mu}{\sqrt{\theta}}\right)^2\right] = 2n$$

(see Remark 5 in Chapter 7). Therefore $(1/n)\sum_{j=1}^{n}(X_j - \mu)^2$ is an unbiased estimator of θ and its variance is $2\theta^2/n$, equal to the Cramér–Rao bound. Thus $(1/n)\sum_{j=1}^{n}(X_j - \mu)^2$ is a UMVU estimator of θ.

Finally, we assume that both μ and σ^2 are unknown and set $\mu = \theta_1$, $\sigma^2 = \theta_2$. Suppose that we are interested in finding a UMVU estimator of θ_2. By using the generalization we spoke of in Remark 1, it can be seen that the Cramér–Rao bound is again equal to $2\theta_2^2/n$. As a matter of fact, we arrive at the same conclusion by treating θ_1 as a constant and θ_2 as the (unknown) parameter θ and calculating the Cramér–Rao bound, provided by Theorem 2. Now it has been seen in Example 5 that

$$\frac{1}{n-1}\sum_{j=1}^{n}\left(X_j - \overline{X}\right)^2$$

is a UMVU estimator of θ_2. Since

$$\sum_{j=1}^{n}\left(\frac{X_j - \overline{X}}{\sqrt{\theta_2}}\right)^2 \text{ is } \chi^2_{n-1}$$

(see second corollary to Theorem 5, Chapter 7), it follows that

$$\sigma^2_\theta\left[\frac{1}{n-1}\sum_{j=1}^{n}\left(X_j - \overline{X}\right)^2\right] = \frac{2\theta_2^2}{n-1} > \frac{2\theta_2^2}{n},$$

the Cramér–Rao bound.

This then is an example of a case where a UMVU estimator does exist but its variance is larger than the Cramér–Rao bound.

A UMVU estimator of $g(\theta)$ is also called an *efficient* estimator of $g(\theta)$ (in the sense of variance). Thus if U is a UMVU estimator of $g(\boldsymbol{\theta})$ and U^* is any other unbiased estimator of $g(\boldsymbol{\theta})$, then the quantity $\sigma^2_\theta U/(\sigma^2_\theta U^*)$ may serve as a measure of expressing the efficiency of U^* relative to that of U. It is known as *relative efficiency* (r.eff.) of U^* and, clearly, takes values in $(0, 1]$.

REMARK 2 Corollary D in Chapter 6 indicates the sort of conditions which would guarantee the fulfillment of the regularity conditions (iv) and (vi).

Exercises

12.4.1 Let $X_1, \ldots, X_n$ be i.i.d. r.v.'s from the Gamma distribution with α known and $\beta = \theta \in \Omega$ $(0, \infty)$ unknown. Then show that the UMVU estimator of θ is

$$U(X_1, \ldots, X_n) = \frac{1}{n\alpha}\sum_{j=1}^{n}X_j$$

and its variance attains the Cramér–Rao bound.

12.4.2 Refer to Exercise 12.3.5 and investigate whether the Cramér–Rao bound is attained.

12.4.3 Refer to Exercise 12.3.6 and investigate whether the Cramér–Rao bound is attained.

12.4.4 Refer to Exercise 12.3.7 and show that the Cramér–Rao bound is not attained for the UMVU estimator of $g(\theta) = \theta^2$.

12.4.5 Refer to Exercise 12.3.11 and investigate whether the Cramér–Rao bound is attained.

12.4.6 Assume conditions (i) and (ii) listed just before Theorem 2, and also suppose that the $\frac{\partial^2}{\partial \theta^2} f(x; \theta)$ exists for all $\theta \in \Omega$ and all $x \in S$ except, perhaps, on a set $N \subset S$ with $P_\theta(X \in N) = 0$ for all $\theta \in \Omega$. Furthermore, suppose that, respectively,

$$\int_S \frac{\partial^2}{\partial \theta^2} f(x; \theta) dx = 0 \quad \text{or} \quad \sum_S \frac{\partial^2}{\partial \theta^2} f(x; \theta) = 0.$$

Then show that $I(\theta) = -E_\theta \left[\frac{\partial^2}{\partial \theta^2} \log f(X; \theta) \right]$.

12.4.7 In Exercises 12.4.1–12.4.4, recalculate $I(\theta)$ and the Cramér–Rao bound by utilizing Exercise 12.4.5 where appropriate.

12.4.8 Let $X_1, \ldots, X_n$ be i.i.d. r.v.'s with p.d.f. $f(\cdot; \theta)$, $\theta \in \Omega \subseteq \mathbb{R}$. For an estimator $V = V(X_1, \ldots, X_n)$ of θ for which $E_\theta V$ is finite, write $E_\theta V = \theta + b(\theta)$. Then $b(\theta)$ is called the *bias* of V. Show that, under the regularity conditions (i)–(vi) preceding Theorem 2—where (vi) is assumed to hold true for all estimators for which the integral (sum) is finite—one has

$$\sigma_\theta^2 V \geq \frac{[1 + b'(\theta)]^2}{nE_\theta[(\partial/\partial \theta) \log f(X; \theta)]^2}, \quad \theta \in \Omega.$$

Here X is an r.v. with p.d.f. $f(\cdot; \theta)$ and $b'(\theta) = db(\theta)/d\theta$. (This inequality is established along the same lines as those used in proving Theorem 2.)

12.5 Criteria for Selecting an Estimator: The Maximum Likelihood Principle

So far we have concerned ourselves with the problem of finding an estimator on the basis of the criteria of unbiasedness and minimum variance. Another principle which is very often used is that of the *maximum likelihood*.

Let $X_1, \ldots, X_n$ be i.i.d. r.v.'s with p.d.f. $f(\cdot; \boldsymbol{\theta})$, $\boldsymbol{\theta} \in \Omega \subseteq \mathbb{R}^r$ and consider the

12.5 Criteria for Selecting an Estimator: The Maximum Likelihood Principle

joint p.d.f. of the X's $f(x_1; \boldsymbol{\theta}) \cdots f(x_n; \boldsymbol{\theta})$. Treating the **x**'s as if they were constants and looking at this joint p.d.f. as a function of $\boldsymbol{\theta}$, we denote it by $L(\boldsymbol{\theta}|x_1, \ldots, x_n)$ and call it the *likelihood function*.

DEFINITION 6 The estimate $\hat{\boldsymbol{\theta}} = \hat{\boldsymbol{\theta}}(x_1, \ldots, x_n)$ is called a *maximum likelihood estimate* (MLE) of $\boldsymbol{\theta}$ if

$$L(\hat{\boldsymbol{\theta}}|x_1, \ldots, x_n) = \max\left[L(\boldsymbol{\theta}|x_1, \ldots, x_n); \boldsymbol{\theta} \in \Omega\right];$$

$\hat{\boldsymbol{\theta}}(X_1, \ldots, X_n)$ is called an *ML estimator* (MLE for short) of $\boldsymbol{\theta}$.

REMARK 3 Since the function $y = \log x$, $x > 0$ is strictly increasing, in order to maximize (with respect to $\boldsymbol{\theta}$) $L(\boldsymbol{\theta}|x_1, \ldots, x_n)$ in the case that $\Omega \in \mathbb{R}$, it suffices to maximize $\log L(\boldsymbol{\theta}|x_1, \ldots, x_n)$. This is much more convenient to work with, as will become apparent from examples to be discussed below.

In order to give an intuitive interpretation of a MLE, suppose first that the X's are discrete. Then

$$L(\boldsymbol{\theta}|x_1, \ldots, x_n) = P_{\boldsymbol{\theta}}(X_1 = x_1, \ldots, X_n = x_n);$$

that is, $L(\boldsymbol{\theta}|x_1, \ldots, x_n)$ is the probability of observing the **x**'s which were acutally observed. Then it is intuitively clear that one should select as an estimate of $\boldsymbol{\theta}$ that $\boldsymbol{\theta}$ which maximizes the probability of observing the **x**'s which were actually observed, if such a $\boldsymbol{\theta}$ exists. A similar interpretation holds true for the case that the **X**'s are continuous by replacing $L(\boldsymbol{\theta}|x_1, \ldots, x_n)$ with the probability element $L(\boldsymbol{\theta}|x_1, \ldots, x_n)dx_1 \cdots dx_n$ which represents the probability (under $P_{\boldsymbol{\theta}}$) that X_j lies between x_j and $x_j + dx_j$, $j = 1, \ldots, n$.

In many important cases there is a unique MLE, which we then call *the* MLE and which is often obtained by differentiation.

Although the principle of maximum likelihood does not seem to be justifiable by a purely mathematical reasoning, it does provide a method for producing estimates in many cases of practical importance. In addition, an MLE is often shown to have several desirable properties. We will elaborate on this point later.

The method of maximum likelihood estimation will now be applied to a number of concrete examples.

EXAMPLE 10 Let $X_1, \ldots, X_n$ be i.i.d. r.v.'s from $P(\theta)$. Then

$$L(\theta|x_1, \ldots, x_n) = e^{-n\theta} \frac{1}{\prod_{j=1}^{n} x_j!} \theta^{\sum_{j=1}^{n} x_j}$$

and hence

$$\log L(\theta|x_1, \ldots, x_n) = -\log\left(\prod_{j=1}^{n} x_j!\right) - n\theta + \left(\sum_{j=1}^{n} x_j\right)\log\theta.$$

Therefore the likelihood equation

$$\frac{\partial}{\partial \theta} \log L(\theta | x_1, \ldots, x_n) = 0 \quad \text{becomes} \quad -n + n\bar{x}\frac{1}{\theta} = 0$$

which gives $\tilde{\theta} = \bar{x}$. Next,

$$\frac{\partial^2}{\partial \theta^2} L(\theta | x_1, \ldots, x_n) = -n\bar{x}\frac{1}{\theta^2} < 0 \quad \text{for all} \quad \theta > 0$$

and hence, in particular, for $\theta = \tilde{\theta}$. Thus $\theta = \bar{x}$ is the MLE of θ.

EXAMPLE 11 Let $X_1, \ldots, X_r$ be multinomially distributed r.v.'s with parameter $\theta = (p_1, \cdots, p_r)' \in \Omega$, where Ω is the $(r-1)$-dimensional hyperplane in $\mathbb{R}^r$ defined by

$$\Omega = \left\{ \theta = (p_1, \ldots, p_r)' \in \mathbb{R}^r;\ p_j > 0, j = 1, \ldots, r \ \text{ and } \ \sum_{j=1}^{r} p_j = 1 \right\}.$$

Then

$$L(\theta | x_1, \ldots, x_r) = \frac{n!}{\prod_{j=1}^{r} x_j!} p_1^{x_1} \cdots p_r^{x_r}$$

$$= \frac{n!}{\prod_{j=1}^{r} x_j!} p_1^{x_1} \cdots p_{r-1}^{x_{r-1}} (1 - p_1 - \cdots - p_{r-1})^{x_r},$$

where $n = \sum_{j=1}^{r} x_j$. Then

$$\log L(\theta | x_1, \ldots, x_r) = \log \frac{n!}{\prod_{j=1}^{r} x_j!} + x_1 \log p_1 + \cdots$$

$$+ x_{r-1} \log p_{r-1} + x_r \log(1 - p_1 - \cdots - p_{r-1}).$$

Differentiating with respect to p_j, $j = 1, \ldots, r-1$ and equating the resulting expressions to zero, we get

$$x_j \frac{1}{p_j} - x_r \frac{1}{p_r} = 0, \quad j = 1, \ldots, r-1.$$

This is equivalent to

$$\frac{x_j}{p_j} = \frac{x_r}{p_r}, \quad j = 1, \ldots, r-1;$$

that is,

$$\frac{x_1}{p_1} = \cdots = \frac{x_{r-1}}{p_{r-1}} = \frac{x_r}{p_r},$$

and this common value is equal to

$$\frac{x_1 + \cdots + x_{r-1} + x_r}{p_1 + \cdots + p_{r-1} + p_r} = \frac{n}{1}.$$

12.5 Criteria for Selecting an Estimator: The Maximum Likelihood Principle

Hence $x_j/p_j = n$ and $p_j = x_j/n$, $j = 1, \ldots, r$. It can be seen that these values of the p's actually maximize the likelihood function, and therefore $\hat{p}_j = x_j/n$, $j = 1, \ldots, r$ are the MLE's of the p's. (See Exercise 12.5.4.)

EXAMPLE 12 Let $X_1, \ldots, X_n$ be i.i.d. r.v.'s from $N(\mu, \sigma^2)$ with parameter $\boldsymbol{\theta} = (\mu, \sigma^2)'$. Then

$$L(\boldsymbol{\theta}|x_1, \ldots, x_n) = \left(\frac{1}{\sqrt{2\pi\sigma^2}}\right)^n \exp\left[-\frac{1}{2\sigma^2}\sum_{j=1}^{n}(x_j - \mu)^2\right],$$

so that

$$\log L(\boldsymbol{\theta}|x_1, \ldots, x_n) = -n\log\sqrt{2\pi} - n\log\sqrt{\sigma^2} - \frac{1}{2\sigma^2}\sum_{j=1}^{n}(x_j - \mu)^2.$$

Differentiating with respect to μ and σ^2 and equating the resulting expressions to zero, we obtain

$$\frac{\partial}{\partial \mu}\log L(\boldsymbol{\theta}|x_1, \ldots, x_n) = \frac{n}{\sigma^2}(\bar{x} - \mu) = 0$$

$$\frac{\partial}{\partial \sigma^2}\log L(\boldsymbol{\theta}|x_1, \ldots, x_n) = -\frac{n}{2\sigma^2} + \frac{1}{2\sigma^4}\sum_{j=1}^{n}(x_j - \mu)^2 = 0.$$

Then

$$\tilde{\mu} = \bar{x} \quad \text{and} \quad \tilde{\sigma}^2 = \frac{1}{n}\sum_{j=1}^{n}(x_j - \bar{x})^2$$

are the roots of these equations. It is further shown that $\tilde{\mu}$ and $\tilde{\sigma}^2$ actually maximize the likelihood function (see Exercise 12.5.5) and therefore

$$\hat{\mu} = \bar{x} \quad \text{and} \quad \hat{\sigma}^2 = \frac{1}{n}\sum_{j=1}^{n}(x_j - \bar{x})^2$$

are the MLE's of μ and σ^2, respectively.

Now, if we assume that σ^2 is known and set $\mu = \theta$, then we have again that $\tilde{\mu} = \bar{x}$ is the root of the equation

$$\frac{\partial}{\partial \theta}\log L(\theta|x_1, \ldots, x_n) = 0.$$

In this case it is readily seen that

$$\frac{\partial^2}{\partial \theta^2}\log L(\theta|x_1, \ldots, x_n) = -\frac{n}{\sigma^2} < 0$$

and hence $\hat{\mu} = \bar{x}$ is the MLE of μ.

On the other hand, if μ is known and we set $\sigma^2 = \theta$, then the root of

$$\frac{\partial}{\partial \theta}\log L(\theta|x_1, \ldots, x_n) = 0$$

is equal to

$$\frac{1}{n}\sum_{j=1}^{n}(x_j - \mu)^2.$$

Next,

$$\frac{\partial^2}{\partial \theta^2} \log L(\theta | x_1, \ldots, x_n) = \frac{1}{\sigma^4}\left[\frac{n}{2} - \frac{1}{\sigma^2}\sum_{j=1}^{n}(x_j - \mu)^2\right]$$

which, for σ^2 equal to

$$\frac{1}{n}\sum_{j=1}^{n}(x_j - \mu)^2,$$

becomes

$$\frac{1}{\sigma^4}\left(\frac{n}{2} - n\right) = -\frac{n}{2\sigma^4} < 0.$$

So

$$\hat{\sigma}^2 = \frac{1}{n}\sum_{j=1}^{n}(x_j - \mu)^2$$

is the MLE of σ^2 in this case.

EXAMPLE 13 Let $X_1, \ldots, X_n$ be i.i.d. r.v.'s from $U(\alpha, \beta)$. Here $\boldsymbol{\theta} = (\alpha, \beta)' \in \Omega$ which is the part of the plane above the main diagonal.
Then

$$L(\boldsymbol{\theta} | x_1, \ldots, x_n) = \frac{1}{(\beta - \alpha)^n} I_{[\alpha, \infty)}(x_{(1)}) I_{(-\infty, \beta]}(x_{(n)}).$$

Here the likelihood function is not differentiable with respect to α and β, but it is, clearly, maximized when $\beta - \alpha$ is minimum, subject to the conditions that $\alpha \leq x_{(1)}$ and $\beta \geq x_{(n)}$. This happens when $\hat{\alpha} = x_{(1)}$ and $\hat{\beta} = x_{(n)}$. Thus $\hat{\alpha} = x_{(1)}$ and $\hat{\beta} = x_{(n)}$ are the MLE's of α and β, respectively.

In particular, if $\alpha = \theta - c$, $\beta = \theta + c$, where c is a given positive constant, then

$$L(\boldsymbol{\theta} | x_1, \ldots, x_n) = \frac{1}{(2c)^n} I_{[\theta - c, \infty)}(x_{(1)}) I_{(-\infty, \theta + c]}(x_{(n)}).$$

The likelihood function is maximized, and its maximum is $1/(2c)^n$, for any θ such that $\theta - c \leq x_{(1)}$ and $\theta + c \geq x_{(n)}$; equivalently, $\theta \leq x_{(1)} + c$ and $\theta \geq x_{(n)} - c$. This shows that any statistic that lies between $X_{(1)} + c$ and $X_{(n)} - c$ is an MLE of θ. For example, $\frac{1}{2}[X_{(1)} + X_{(n)}]$ is such a statistic and hence an MLE of θ.

If β is known and $\alpha = \theta$, or if α is known and $\beta = \theta$, then, clearly, $x_{(1)}$ and $x_{(n)}$ are the MLE's of α and β, respectively.

REMARK 4

i) The MLE may be a UMVU estimator. This, for instance, happens in Example 10, for $\hat{\mu}$ in Example 12, and also for $\hat{\sigma}^2$ in the same example when μ is known.

ii) The MLE need not be UMVU. This happens, e.g., in Example 12 for $\hat{\sigma}^2$ when μ is unknown.

iii) The MLE is not always obtainable by differentiation. This is the case in Example 13.

iv) There may be more than one MLE. This case occurs in Example 13 when $\alpha = \theta - c$, $\beta = \theta + c$, $c > 0$

In the following, we present two of the general properties that an MLE enjoys.

THEOREM 4 Let $X_1, \ldots, X_n$ be i.i.d. r.v.'s with p.d.f. $f(\cdot; \boldsymbol{\theta})$, $\boldsymbol{\theta} \in \Omega \subseteq \mathbb{R}^r$, and let $\mathbf{T} = (T_1, \ldots, T_r)'$, $T_j = T_j(X_1, \ldots, X_n)$, $j = 1, \ldots, r$ be a sufficient statistic for $\boldsymbol{\theta} = (\theta_1, \ldots, \theta_r)'$. Then, if $\hat{\boldsymbol{\theta}} = (\hat{\theta}_1, \ldots, \hat{\theta}_r)'$ is the unique MLE $\boldsymbol{\theta}$, it follows that $\hat{\boldsymbol{\theta}}$ is a function of $\mathbf{T}$.

PROOF Since $\mathbf{T}$ is sufficient, Theorem 1 in Chapter 11 implies the following factorization:

$$f(x_1; \boldsymbol{\theta}) \cdots f(x_n; \boldsymbol{\theta}) = g[\mathbf{T}(x_1, \ldots, x_n); \boldsymbol{\theta}] h(x_1, \ldots, x_n),$$

where h is independent of $\boldsymbol{\theta}$.

Therefore

$$\max[f(x_1; \boldsymbol{\theta}) \cdots f(x_n; \boldsymbol{\theta}); \boldsymbol{\theta} \in \Omega]$$
$$= h(x_1, \ldots, x_n) \max\{g[\mathbf{T}(x_1, \ldots, x_n); \boldsymbol{\theta}]; \boldsymbol{\theta} \in \Omega\}.$$

Thus, if a unique MLE exists, it will have to be a function of $\mathbf{T}$, as it follows from the right-hand side of the equation above. ▲

REMARK 5 Notice that the conclusion of the theorem holds true in all Examples 10–13. See also Exercise 12.3.10.

Another optimal property of an MLE is *invariance*, as is proved in the following theorem.

THEOREM 5 Let $X_1, \ldots, X_n$ be i.i.d. r.v.'s with p.d.f. $f(x; \boldsymbol{\theta})$, $\boldsymbol{\theta} \in \Omega \subseteq \mathbb{R}^r$, and let ϕ be defined on Ω onto $\Omega^* \subseteq \mathbb{R}^m$ and let it be one-to-one. Suppose $\hat{\boldsymbol{\theta}}$ is an MLE of $\boldsymbol{\theta}$. Then $\phi(\hat{\boldsymbol{\theta}})$ is an MLE of $\phi(\boldsymbol{\theta})$. That is, an MLE is invariant under one-to-one transformations.

PROOF Set $\boldsymbol{\theta}^* = \phi(\boldsymbol{\theta})$, so that $\boldsymbol{\theta} = \phi^{-1}(\boldsymbol{\theta}^*)$. Then

$$L(\boldsymbol{\theta}|x_1, \ldots, x_n) = L[\phi^{-1}(\boldsymbol{\theta}^*)|x_1, \ldots, x_n],$$

call it $L^*(\boldsymbol{\theta}^*|x_1, \ldots, x_n)$. It follows that

$$\max[L(\boldsymbol{\theta}|x_1, \ldots, x_n); \boldsymbol{\theta} \in \Omega] = \max[L^*(\boldsymbol{\theta}^*|x_1, \ldots, x_n); \boldsymbol{\theta}^* \in \Omega^*].$$

By assuming the existence of an MLE, we have that the maximum at the left-hand side above is attained at an MLE $\hat{\boldsymbol{\theta}}$. Then, clearly, the right-hand side attains its maximum at $\hat{\boldsymbol{\theta}}^*$, where $\hat{\boldsymbol{\theta}}^* = \phi(\hat{\boldsymbol{\theta}})$. Thus $\phi(\hat{\boldsymbol{\theta}})$ is an MLE of $\phi(\boldsymbol{\theta})$. ▲

For instance, since

$$\frac{1}{n}\sum_{j=1}^{n}(x_j - \bar{x})^2$$

is the MLE of σ^2 in the normal case (see Example 12), it follows that

$$\sqrt{\frac{1}{n}\sum_{j=1}^{n}(x_j - \bar{x})^2}$$

is the MLE of σ.

Exercises

12.5.1 If $X_1, \ldots, X_n$ are i.i.d. r.v.'s from $B(m, \theta)$, $\theta \in \Omega = (0, \infty)$, show that $\bar{X}/m$ is the MLE of θ.

12.5.2 If $X_1, \ldots, X_n$ are i.i.d. r.v.'s from the Negative Binomial distribution with parameter $\theta \in \Omega = (0, 1]$, show that $r/(r + \bar{X})$ is the MLE of θ.

12.5.3 If $X_1, \ldots, X_n$ are i.i.d. r.v.'s from the Negative Exponential distribution with parameter $\theta \in \Omega = (0, \infty]$, show that $1/\bar{X}$ is the MLE of θ.

12.5.4 Refer to Example 11 and show that the quantities $\hat{p}_j = x_j/n$, $j = 1, \ldots, r$ indeed maximize the likelihood function.

12.5.5 Refer to Example 12 and consider the case that both μ and σ^2 are unknown. Then show that the quantities $\tilde{\mu} = \bar{x}$ and

$$\tilde{\sigma}^2 = \frac{1}{n}\sum_{j=1}^{n}(x_j - \bar{x})^2$$

indeed maximize the likelihood function.

12.5.6 Suppose that certain particles are emitted by a radioactive source (whose strength remains the same over a long period of time) according to a Poisson distribution with parameter θ during a unit of time. The source in question is observed for n time units, and let X be the r.v. denoting the number of times that no particles were emitted. Find the MLE of θ in terms of X.

12.5.7 Let $X_1, \ldots, X_n$ be i.i.d. r.v.'s with p.d.f. $f(\cdot; \theta_1, \theta_2)$ given by

$$f(x; \theta_1, \theta_2) = \frac{1}{\theta_2}\exp\left(-\frac{x - \theta_1}{\theta_2}\right), \quad x > \theta_1, \quad \boldsymbol{\theta} = (\theta_1, \theta_2)' \in \Omega = \mathbb{R} \times (0, \infty).$$

Find the MLE's of θ_1, θ_2.

12.5.8 Refer to Exercise 11.4.2, Chapter 11, and find the MLE of θ.

12.5.9 Refer to Exercise 12.3.10 and find the MLE of the reliability $\mathbb{R}(x; \theta)$.

12.5.10 Let $X_1, \ldots, X_n$ be i.i.d. r.v.'s from the $U(\theta - \frac{1}{2}, \theta + \frac{1}{2})$, $\theta \in \Omega \subseteq \mathbb{R}$ distribution, and let

$$\hat{\theta} = \hat{\theta}(X_1, \ldots, X_n) = \left(X_{(n)} - \frac{1}{2}\right) + \left(\cos^2 X_1\right)\left(X_{(1)} - X_{(n)} + 1\right).$$

Then show that $\hat{\theta}$ is an MLE of θ but it is not a function only of the sufficient statistic $(X_{(1)}, X_{(n)})'$. (Thus Theorem 4 need not be correct if there exists more than one MLE of the parameters involved. For this, see also the paper *Maximum Likelihood and Sufficient Statistics* by D. S. Moore in the *American Mathematical Monthly*, Vol. 78, No. 1, January 1971, pp. 42–45.)

12.6 Criteria for Selecting an Estimator: The Decision-Theoretic Approach

We will first develop the general theory underlying the decision-theoretic method of estimation and then we will illustrate the theory by means of concrete examples. In this section, we will restrict ourselves to a real-valued parameter. So let $X_1, \ldots, X_n$ be i.i.d. r.v.'s with p.d.f. $f(\cdot; \theta)$, $\theta \in \Omega \subseteq \mathbb{R}$. Our problem is that of estimating θ.

DEFINITION 7 A *decision function (or rule)* δ is a (measurable) function defined on $\mathbb{R}^n$ into $\mathbb{R}$. The value $\delta(x_1, \ldots, x_n)$ of δ at $(x_1, \ldots, x_n)'$ is called a *decision*.

DEFINITION 8 For estimating θ on the basis of $X_1, \ldots, X_n$ and by using the decision function δ, a *loss function* is a nonnegative function in the arguments θ and $\delta(x_1, \ldots, x_n)$ which expresses the (financial) loss incurred when θ is estimated by $\delta(x_1, \ldots, x_n)$.

The loss functions which are usually used are of the following form:

$$L[\theta; \delta(x_1, \ldots, x_n)] = |\theta - \delta(x_1, \ldots, x_n)|,$$

or more generally,

$$L[\theta; \delta(x_1, \ldots, x_n)] = v(\theta)|\theta - \delta(x_1, \ldots, x_n)|^k, \quad k > 0;$$

or $L[\cdot; \delta(x_1, \ldots, x_n)]$ is taken to be a convex function of θ. The most convenient form of a loss function is the *squared loss function*; that is,

$$L[\theta; \delta(x_1, \ldots, x_n)] = [\theta - \delta(x_1, \ldots, x_n)]^2.$$

DEFINITION 9 The *risk function* corresponding to the loss function $L(\cdot; \cdot)$ is denoted by $R(\cdot; \cdot)$ and is defined by

$$R(\theta; \delta) = E_\theta L[\theta; \delta(X_1, \ldots, X_n)]$$

$$= \begin{cases} \int_{-\infty}^{\infty} \cdots \int_{-\infty}^{\infty} L[\theta; \delta(x_1, \ldots, x_n)] f(x_1; \theta) \cdots f(x_n; \theta) dx_1 \cdots dx_n \\ \sum_{x_1} \cdots \sum_{x_n} L[\theta; \delta(x_1, \ldots, x_n)] f(x_1; \theta) \cdots f(x_n; \theta). \end{cases}$$

That is, the risk corresponding to a given decision function is simply the average loss incurred if that decision function is used.

Two decision functions δ and δ^* such that

$$R(\theta; \delta) = E_\theta L[\theta; \delta(X_1, \ldots, X_n)] = E_\theta L[\theta; \delta^*(X_1, \ldots, X_n)] = R(\theta; \delta^*)$$

for all $\theta \in \Omega$ are said to be *equivalent*.

In the present context of (point) estimation, the decision $\delta = \delta(x_1, \ldots, x_n)$ will be called *an estimate of* θ, and its goodness will be judged on the basis of its risk $R(\cdot; \delta)$. It is, of course, assumed that a certain loss function is chosen and then kept fixed throughout. To start with, we first rule out those estimates which are not admissible (inadmissible), where

DEFINITION 10 The estimator δ of θ is said to be *admissible* if there is no other estimator δ^* of θ such that $R(\theta; \delta^*) \leq R(\theta; \delta)$ for all $\theta \in \Omega$ with strict inequality for at least one θ.

Since for any two equivalent estimators δ and δ^* we have $R(\theta; \delta) = R(\theta; \delta^*)$ for all $\theta \in \Omega$, it suffices to restrict ourselves to an essentially complete class of estimators, where

DEFINITION 11 A class $\mathcal{D}$ of estimators of θ is said to be *essentially complete* if for any estimator δ^* of θ not in $\mathcal{D}$ one can find an estimator δ in $\mathcal{D}$ such that $R(\theta; \delta^*) = R(\theta; \delta)$ for all $\theta \in \Omega$.

Thus, searching for an estimator with some optimal properties, we confine our attention to an essentially complete class of admissible estimators. Once this has been done the question arises as to which member of this class is to be chosen as an estimator of θ. An apparently obvious answer to this question would be to choose an estimator δ such that $R(\theta; \delta) \leq R(\theta; \delta^*)$ for any other estimator δ^* within the class and for all $\theta \in \Omega$. Unfortunately, such estimators do not exist except in trivial cases. However, if we restrict ourselves only to the class of unbiased estimators with finite variance and take the loss function to be the squared loss function (see paragraph following Definition 8), then, clearly, $R(\theta; \delta)$ becomes simply the variance of $\delta(X_1, \ldots, X_n)$. The criterion proposed above for selecting δ then coincides with that of finding a UMVU estimator. This problem has already been discussed in Section 3 and Section 4. Actually, some authors discuss UMVU estimators as a special case within the decision-theoretic approach as just mentioned. However, we believe that the approach adopted here is more pedagogic and easier for the reader to follow.

Setting aside the fruitless search for an estimator which would uniformly (in θ) minimize the risk within the entire class of admissible estimators, there are two principles on which our search may be based. The first is to look for an estimator which minimizes the worst which could happen to us, that is, to minimize the maximum (over θ) risk. Such an estimator, if it exists, is called a *minimax* (from *mini*mizing the *max*imum) *estimator*. However, in this case, while we may still confine ourselves to the essentially complete class of estimators, we may not rule out inadmissible estimators, for it might so happen that

12.6 Criteria for Selecting an Estimator: The Decision-Theoretic Approach

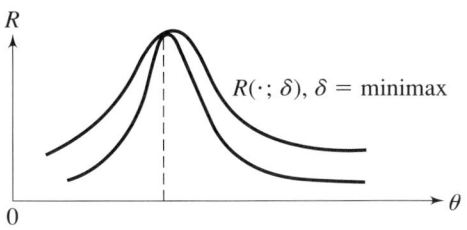

Figure 12.2

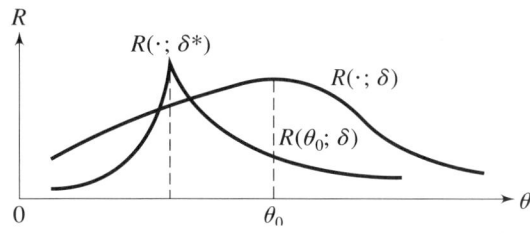

Figure 12.3

a minimax estimator is inadmissible. (See Fig. 12.2.) Instead, we restrict our attention to the class $\mathcal{D}_1$ of all estimators for which $R(\theta; \delta)$ is finite for all $\theta \in \Omega$. Then we have the following definition:

DEFINITION 12 Within the class $\mathcal{D}_1$, the estimator δ is said to be *minimax* if for any other estimator δ^*, one has

$$\sup\left[R(\theta; \delta);\ \theta \in \Omega\right] \leq \sup\left[R(\theta; \delta^*);\ \theta \in \Omega\right].$$

Figure 12.2 illustrates the fact that a minimax estimator may be inadmissible.

Now one may very well object to the minimax principle on the grounds that it gives too much weight to the maximum risk and entirely neglects its other values. For example, in Fig. 12.3, whereas the minimax estimate δ is slightly better at its maximum $R(\theta_0; \delta)$, it is much worse than δ^* at almost all other points.

Legitimate objections to minimax principles like the one just cited prompted the advancement of the concept of a *Bayes estimate*. To see what this is, some further notation is required. Recall that $\Omega \subseteq \mathbb{R}$, and suppose now that θ is an r.v. itself with p.d.f. λ, to be called a *prior* p.d.f. Then set

$$R(\delta) = E_\lambda R(\theta; \delta) = \begin{cases} \int_\Omega R(\theta; \delta)\lambda(\theta)d\theta \\ \sum_\Omega R(\theta; \delta)\lambda(\theta). \end{cases}$$

Assuming that the quantity just defined is finite, it is clear that $R(\delta)$ is simply the *average* (with respect to λ) *risk* over the entire parameter space Ω when the estimator δ is employed. Then it makes sense to choose that δ for which $R(\delta) \leq R(\delta^*)$ for any other δ^*. Such a δ is called a Bayes estimator of θ, provided it exists. Let $\mathcal{D}_2$ be the class of all estimators for which $R(\delta)$ is finite for a given prior p.d.f. λ on Ω. Then

DEFINITION 13 Within the class $\mathcal{D}_2$, the estimator δ is said to be a *Bayes estimator* (in the decision-theoretic sense and with respect to the prior p.d.f. λ on Ω) if $R(\delta) \leq R(\delta^*)$ for any other estimator δ^*.

It should be pointed out at the outset that the Bayes approach to estimation poses several issues that we have to reckon with. First, the assumption of θ being an r.v. might be entirely unreasonable. For example, θ may denote the (unknown but fixed) distance between Chicago and New York City, which is to be determined by repeated measurements. This difficulty may be circumvented by pretending that this assumption is only a mathematical device, by means of which we expect to construct estimates with some tangible and mathematically optimal properties. This granted, there still is a problem in choosing the prior λ on Ω. Of course, in principle, there are infinitely many such choices. However, in concrete cases, choices do suggest themselves. In addition, when choosing λ we have the flexibility to weigh the parameters the way we feel appropriate, and also incorporate in it any prior knowledge we might have in connection with the true value of the parameter. For instance, prior experience might suggest that it is more likely that the true parameter lies in a given subset of Ω rather than in its complement. Then, in choosing λ, it is sensible to assign more weight in the subset under question than to its complement. Thus we have the possibility of incorporating prior information about θ or expressing our prior opinion about θ. Another decisive factor in choosing λ is that of mathematical convenience; we are forced to select λ so that the resulting formulas can be handled.

We should like to mention once and for all that the results in the following two sections are derived by employing squared loss functions. It should be emphasized, however, that the same results may be discussed by using other loss functions.

12.7 Finding Bayes Estimators

Let $X_1, \ldots, X_n$ be i.i.d. r.v.'s with p.d.f. $f(\cdot; \theta)$, $\theta \in \Omega \subseteq \mathbb{R}$, and consider the squared loss function. That is, for an estimate

$$\delta = \delta(x_1, \ldots, x_n), \quad L(\theta; \delta) = L[\theta; \delta(x_1, \ldots, x_n)] = [\theta - \delta(x_1, \ldots, x_n)]^2.$$

Let θ be an r.v. with prior p.d.f. λ. Then we are interested in determining δ so that it will be a Bayes estimate (of θ in the decision-theoretic sense). We consider the continuous case, since the discrete case is handled similarly with the integrals replaced by summation signs. We have

$$R(\theta; \delta) = E_\theta [\theta - \delta(X_1, \ldots, X_n)]^2$$

$$= \int_{-\infty}^{\infty} \cdots \int_{-\infty}^{\infty} [\theta - \delta(x_1, \ldots, x_n)]^2 f(x_1; \theta) \cdots f(x_n; \theta) dx_1 \cdots dx_n.$$

Therefore

$$R(\delta) = \int_\Omega R(\theta; \delta) \lambda(\theta) d\theta$$

$$= \int_\Omega \left\{ \int_{-\infty}^{\infty} \cdots \int_{-\infty}^{\infty} [\theta - \delta(x_1, \ldots, x_n)]^2 \right.$$
$$\left. \times f(x_1; \theta) \cdots f(x_n; \theta) dx_1 \cdots dx_n \right\} \lambda(\theta) d\theta$$

$$= \int_{-\infty}^{\infty} \cdots \int_{-\infty}^{\infty} \left\{ \int_{\Omega} [\theta - \delta(x_1, \ldots, x_n)]^2 \right.$$
$$\left. \times \lambda(\theta) f(x_1; \theta) \cdots f(x_n; \theta) d\theta \right\} dx_1 \cdots dx_n. \tag{13}$$

(As can be shown, the interchange of the order of integration is valid here because the integrand is nonnegative. The theorem used is known as the Fubini theorem.)

From (13), it follows that if δ is chosen so that

$$\int_{\Omega} [\theta - \delta(x_1, \ldots, x_n)]^2 \lambda(\theta) f(x_1; \theta) \cdots f(x_n; \theta) d\theta$$

is minimized for each $(x_1, \ldots, x_n)'$, then $R(\delta)$ is also minimized. But

$$\int_{\Omega} [\theta - \delta(x_1, \ldots, x_n)]^2 \lambda(\theta) f(x_1; \theta) \cdots f(x_n; \theta) d\theta$$
$$= \delta^2(x_1, \ldots, x_n) \int_{\Omega} f(x_1; \theta) \cdots f(x_n; \theta) \lambda(\theta) d\theta - 2\delta(x_1, \ldots, x_n)$$
$$\times \int_{\Omega} \theta f(x_1; \theta) \cdots f(x_n; \theta) \lambda(\theta) d\theta + \int_{\Omega} \theta^2 f(x_1; \theta) \cdots f(x_n; \theta) \lambda(\theta) d\theta, \tag{14}$$

and the right-hand side of (14) is of the form

$$g(t) = at^2 - 2bt + c \quad (a > 0)$$

which is minimized for $t = b/a$. (In fact, $g'(t) = 2at - 2b = 0$ implies $t = b/a$ and $g''(t) = 2a > 0$.)

Thus the required estimate is given by

$$\delta(x_1, \ldots, x_n) = \frac{\int_{\Omega} \theta f(x_1; \theta) \cdots f(x_n; \theta) \lambda(\theta) d\theta}{\int_{\Omega} f(x_1; \theta) \cdots f(x_n; \theta) \lambda(\theta) d\theta}.$$

Formalizing this result, we have the following theorem:

THEOREM 6 A Bayes estimate $\delta(x_1, \ldots, x_n)$ (of θ) corresponding to a prior p.d.f. λ on Ω for which

$$\left| \int_{\Omega} \theta f(x_1; \theta) \cdots f(x_n; \theta) \lambda(\theta) d\theta \right| < \infty,$$
$$0 < \int_{\Omega} f(x_1; \theta) \cdots f(x_n; \theta) \lambda(\theta) d\theta < \infty,$$

and

$$\int_{\Omega} \theta^2 f(x_1; \theta) \cdots f(x_n; \theta) \lambda(\theta) d\theta < \infty,$$

for each $(x_1, \ldots, x_n)'$, is given by

$$\delta(x_1, \ldots, x_n) = \frac{\int_{\Omega} \theta f(x_1; \theta) \cdots f(x_n; \theta) \lambda(\theta) d\theta}{\int_{\Omega} f(x_1; \theta) \cdots f(x_n; \theta) \lambda(\theta) d\theta}, \tag{15}$$

provided λ is of the continuous type. Integrals in (15) are to be replaced by summation signs if λ is of the discrete type.

Now, if the observed value of X_j is x_j, $j = 1, \ldots, n$, we determine the conditional p.d.f. of θ, given $X_1 = x_1, \ldots, X_n = x_n$. This is called the *posterior* p.d.f. of θ and represents our revised opinion about θ after new evidence (the observed X's) has come in. Setting $\mathbf{x} = (x_1, \ldots, x_n)'$ and denoting by $h(\cdot|\mathbf{x})$ the posterior p.d.f. of θ, we have then

$$h(\theta|\mathbf{x}) = \frac{f(\theta, \mathbf{x})}{h(\mathbf{x})} = \frac{f(\mathbf{x}; \theta)\lambda(\theta)}{h(\mathbf{x})} = \frac{f(x_1; \theta) \cdots f(x_n; \theta)\lambda(\theta)}{h(\mathbf{x})}, \quad (16)$$

where

$$h(\mathbf{x}) = \int_\Omega f(\mathbf{x}; \theta)\lambda(\theta)d\theta = \int_\Omega f(x_1; \theta) \cdots f(x_n; \theta)\lambda(\theta)d\theta$$

for the case that λ is of the continuous type. By means of (15) and (16), it follows then that the Bayes estimate of θ (in the decision-theoretic sense) $\delta(x_1, \ldots, x_n)$ is the expectation of θ with respect to its posterior p.d.f., that is,

$$\delta(x_1, \ldots, x_n) = \int_\Omega \theta h(\theta|\mathbf{x})d\theta.$$

Another Bayesian estimate of θ could be provided by the median of $h(\cdot|\mathbf{x})$, or the mode of $h(\cdot|\mathbf{x})$, if it exists.

REMARK 6 At this point, let us make the following observation regarding the maximum likelihood and the Bayesian approach to estimation problems. As will be seen, this observation establishes a link between maximum likelihood and Bayes estimates and provides insight into each other. To this end, let $h(\cdot|\mathbf{x})$ be the posterior p.d.f. of θ given by (16) and corresponding to the prior p.d.f. λ. Since $f(\mathbf{x}; \theta) = L(\theta|\mathbf{x})$, $h(\cdot|\mathbf{x})$ may be written as follows:

$$h(\theta|\mathbf{x}) = \frac{L(\theta|\mathbf{x})\lambda(\theta)}{h(\mathbf{x})}. \quad (17)$$

Now let us suppose that Ω is bounded and let λ be constant on Ω, $\lambda(\theta) = c$, say, $\theta \in \Omega$. Then it follows from (17) that the MLE of θ, if it exists, is simply that value of $\boldsymbol{\theta}$ which maximizes $h(\cdot|\mathbf{x})$. Thus when no prior knowledge about θ is available (which is expressed by taking $\lambda(\theta) = c$, $\theta \in \Omega$), the likelihood function is maximized if and only if the posterior p.d.f. is.

Some examples follow.

EXAMPLE 14 Let $X_1, \ldots, X_n$ be i.i.d. r.v.'s from $B(1, \theta)$, $\theta \in \Omega = (0, 1)$. We choose λ to be the Beta density with parameters α and β; that is,

$$\lambda(\theta) = \begin{cases} \dfrac{\Gamma(\alpha + \beta)}{\Gamma(\alpha)\Gamma(\beta)} \theta^{\alpha-1}(1-\theta)^{\beta-1}, & \text{if } \theta \in (0, 1) \\ 0, & \text{otherwise.} \end{cases}$$

Now, from the definition of the p.d.f. of a Beta distribution with parameters α and β, we have

12.7 Finding Bayes Estimators

$$\int_0^1 x^{\alpha-1}(1-x)^{\beta-1}dx = \frac{\Gamma(\alpha)\Gamma(\beta)}{\Gamma(\alpha+\beta)}, \tag{18}$$

and, of course $\Gamma(\gamma) = (\gamma-1)\Gamma(\gamma-1)$. Then, for simplicity, writing $\sum_j^n x_j$ rather than $\sum_{j=1}^n x_j$ when this last expression appears as an exponent, we have

$$\begin{aligned}I_1 &= \int_\Omega f(x_1;\theta)\cdots f(x_n;\theta)\lambda(\theta)d\theta \\ &= \frac{\Gamma(\alpha+\beta)}{\Gamma(\alpha)\Gamma(\beta)}\int_0^1 \theta^{\Sigma_j x_j}(1-\theta)^{n-\Sigma_j x_j}\theta^{\alpha-1}(1-\theta)^{\beta-1}d\theta \\ &= \frac{\Gamma(\alpha+\beta)}{\Gamma(\alpha)\Gamma(\beta)}\int_0^1 \theta^{(\alpha+\Sigma_j x_j)-1}(1-\theta)^{(\beta+n-\Sigma_j x_j)-1}d\theta,\end{aligned}$$

which by means of (18) becomes as follows:

$$I_1 = \frac{\Gamma(\alpha+\beta)}{\Gamma(\alpha)\Gamma(\beta)}\cdot\frac{\Gamma\left(\alpha+\sum_{j=1}^n x_j\right)\Gamma\left(\beta+n-\sum_{j=1}^n x_j\right)}{\Gamma(\alpha+\beta+n)}. \tag{19}$$

Next,

$$\begin{aligned}I_1 &= \int_\Omega \theta f(x_1;\theta)\cdots f(x_n;\theta)\lambda(\theta)d\theta \\ &= \frac{\Gamma(\alpha+\beta)}{\Gamma(\alpha)\Gamma(\beta)}\int_0^1 \theta\theta^{\Sigma_j x_j}(1-\theta)^{n-\Sigma_j x_j}\theta^{\alpha-1}(1-\theta)^{\beta-1}d\theta \\ &= \frac{\Gamma(\alpha+\beta)}{\Gamma(\alpha)\Gamma(\beta)}\int_0^1 \theta^{(\alpha+\Sigma_j x_j+1)-1}(1-\theta)^{(\beta+n-\Sigma_j x_j)-1}d\theta.\end{aligned}$$

Once more relation (18) gives

$$I_2 = \frac{\Gamma(\alpha+\beta)}{\Gamma(\alpha)\Gamma(\beta)}\cdot\frac{\Gamma\left(\alpha+\sum_{j=1}^n x_j+1\right)\Gamma\left(\beta+n-\sum_{j=1}^n x_j\right)}{\Gamma(\alpha+\beta+n+1)}. \tag{20}$$

Relations (19) and (20) imply, by virtue of (15),

$$\delta(x_1,\ldots,x_n) = \frac{\Gamma(\alpha+\beta+n)\Gamma\left(\alpha+\sum_{j=1}^n x_j+1\right)}{\Gamma(\alpha+\beta+n+1)\Gamma\left(\alpha+\sum_{j=1}^n x_j\right)} = \frac{\alpha+\sum_{j=1}^n x_j}{\alpha+\beta+n};$$

that is,

$$\delta(x_1,\ldots,x_n) = \frac{\sum_{j=1}^n x_j + \alpha}{n+\alpha+\beta}. \tag{21}$$

REMARK 7 We know (see Remark 4 in Chapter 3) that if $\alpha = \beta = 1$, then the Beta distribution becomes $U(0, 1)$. In this case the corresponding Bayes estimate is

$$\delta(x_1, \ldots, x_n) = \frac{\sum_{j=1}^{n} x_j + 1}{n+2},$$

as follows from (21).

EXAMPLE 15 Let $X_1, \ldots, X_n$ be i.i.d. r.v.'s from $N(\theta, 1)$. Take λ to be $N(\mu, 1)$, where μ is known. Then

$$I_1 = \int_\Omega f(x_1; \theta) \cdots f(x_n; \theta) \lambda(\theta) d\theta$$

$$= \frac{1}{(\sqrt{2\pi})^{n+1}} \int_{-\infty}^{\infty} \exp\left[-\frac{1}{2} \sum_{j=1}^{n} (x_j - \theta)^2\right] \exp\left[-\frac{(\theta-\mu)^2}{2}\right] d\theta$$

$$= \frac{1}{(\sqrt{2\pi})^{n+1}} \exp\left[-\frac{1}{2}\left(\sum_{j=1}^{n} x_j^2 + \mu^2\right)\right]$$

$$\times \int_{-\infty}^{\infty} \exp\left\{-\frac{1}{2}\left[(n+1)\theta^2 - 2(n\bar{x} + \mu)\theta\right]\right\} d\theta.$$

But

$$(n+1)\theta^2 - 2(n\bar{x} + \mu)\theta = (n+1)\left(\theta^2 - 2\frac{n\bar{x} + \mu}{n+1}\theta\right)$$

$$= (n+1)\left[\theta^2 - 2\frac{n\bar{x} + \mu}{n+1}\theta + \left(\frac{n\bar{x} + \mu}{n+1}\right)^2 - \left(\frac{n\bar{x} + \mu}{n+1}\right)^2\right]$$

$$= (n+1)\left[\left(\theta - \frac{n\bar{x} + \mu}{n+1}\right)^2 - \left(\frac{n\bar{x} + \mu}{n+1}\right)^2\right].$$

Therefore

$$I_1 = \frac{1}{\sqrt{n+1}} \frac{1}{(\sqrt{2\pi})^n} \exp\left\{-\frac{1}{2}\left[\sum_{j=1}^{n} x_j^2 + \mu^2 - \frac{(n\bar{x} + \mu)^2}{n+1}\right]\right\}$$

$$\times \frac{1}{\sqrt{2\pi}(1/\sqrt{n+1})} \int_{-\infty}^{\infty} \exp\left[-\frac{1}{2(1/\sqrt{n+1})^2}\left(\theta - \frac{n\bar{x} + \mu}{n+1}\right)^2\right] d\theta$$

$$= \frac{1}{\sqrt{n+1}} \frac{1}{(\sqrt{2\pi})^n} \exp\left\{-\frac{1}{2}\left[\sum_{j=1}^{n} x_j^2 + \mu^2 - \frac{(n\bar{x} + \mu)^2}{n+1}\right]\right\}. \tag{22}$$

Next,

$$I_2 = \int_\Omega \theta f(x_1; \theta) \cdots f(x_n; \theta)\lambda(\theta)d\theta$$

$$= \frac{1}{\left(\sqrt{2\pi}\right)^{n+1}} \int_{-\infty}^{\infty} \theta \exp\left[-\frac{1}{2}\sum_{j=1}^{n}(x_j - \theta)^2\right]\exp\left[-\frac{(\theta - \mu)^2}{2}\right]d\theta$$

$$= \frac{1}{\sqrt{n+1}} \frac{1}{\left(\sqrt{2\pi}\right)^n} \exp\left\{-\frac{1}{2}\left[\sum_{j=1}^{n} x_j^2 + \mu^2 - \frac{(n\bar{x} + \mu)^2}{n+1}\right]\right\}$$

$$\times \frac{1}{\sqrt{2\pi}\left(1/\sqrt{n+1}\right)} \int_{-\infty}^{\infty} \theta \exp\left[-\frac{1}{2\left(1/\sqrt{n+1}\right)^2}\left(\theta - \frac{n\bar{x}+\mu}{n+1}\right)^2\right]d\theta$$

$$= \frac{1}{\sqrt{n+1}} \frac{1}{\left(\sqrt{2\pi}\right)^n} \exp\left\{-\frac{1}{2}\left[\sum_{j=1}^{n} x_j^2 + \mu^2 - \frac{(n\bar{x}+\mu)^2}{n+1}\right]\right\}\frac{n\bar{x}+\mu}{n+1}. \quad (23)$$

By means of (22) and (23), one has, on account of (15),

$$\delta(x_1, \ldots, x_n) = \frac{n\bar{x} + \mu}{n+1}. \quad (24)$$

Exercises

12.7.1 Refer to Example 14 and:

i) Determine the posterior p.d.f. $h(\theta|\mathbf{x})$;

ii) Construct a $100(1 - \alpha)\%$ Bayes confidence interval for θ; that is, determine a set $\{\theta \in (0, 1); h(\theta|\mathbf{x}) \geq c(\mathbf{x})\}$, where $c(\mathbf{x})$ is determined by the requirement that the P_λ-probability of this set is equal to $1 - \alpha$;

iii) Derive the Bayes estimate in (21) as the mean of the posterior p.d.f. $h(\theta|\mathbf{x})$.

(Hint: For simplicity, assign equal probabilities to the two tails.)

12.7.2 Refer to Example 15 and:

i) Determine the posterior p.d.f. $h(\theta|\mathbf{x})$;

ii) Construct the equal-tail $100(1 - \alpha)\%$ Bayes confidence interval for θ;

iii) Derive the Bayes estimate in (24) as the mean of the posterior p.d.f. $h(\theta|\mathbf{x})$.

12.7.3 Let X be an r.v. distributed as $P(\theta)$, and let the prior p.d.f. λ of θ be Negative Exponential with parameter τ. Then, on the basis of X:

i) Determine the posterior p.d.f. $h(\theta|\mathbf{x})$;

ii) Construct the equal-tail $100(1 - \alpha)\%$ Bayes confidence interval for θ;

iii) Derive the Bayes estimates $\delta(x)$ for the loss functions $L(\theta; \delta) = [\theta - \delta(x)]^2$ as well as $L(\theta; \delta) = [\theta - \delta(x)]^2/\theta$;

iv) Do parts (i)–(iii) for any sample size n.

12.7.4 Let X be an r.v. having the Beta p.d.f. with parameters $\alpha = \theta$ and $\beta = 1$, and let the prior p.d.f. λ of θ be the Negative Exponential with parameter τ. Then, on the basis of X:

i) Determine the posterior p.d.f. $h(\theta|\mathbf{x})$;

ii) Construct the equal-tail $100(1 - \alpha)\%$ Bayes confidence interval for θ;

iii) Derive the Bayes estimates $\delta(x)$ for the loss functions $L(\theta; \delta) = [\theta - \delta(x)]^2$ as well as $L(\theta; \delta) = [\theta - \delta(x)]^2/\theta$;

iv) Do parts (i)–(iii) for any sample size n;

v) Do parts (i)–(iv) for any sample size n when λ is Gamma with parameters k (positive integer) and β.

$\Big($Hint: If Y is distributed as Gamma with parameters k and β, then it is easily seen that $\frac{2Y}{\beta} \sim \chi^2_{2k}.\Big)$

12.8 Finding Minimax Estimators

Although there is no general method for deriving minimax estimates, this can be achieved in many instances by means of the Bayes method described in the previous section.

Let $X_1, \ldots, X_n$ be i.i.d. r.v.'s with p.d.f. $f(\cdot; \theta)$, $\theta \in \Omega$ ($\subseteq \mathbb{R}$) and let λ be a prior p.d.f. on Ω. Then the posterior p.d.f. of θ, given $\mathbf{X} = (X_1, \ldots, X_n)' = (x_1, \ldots, x_n)' = \mathbf{x}$, $h(\cdot|\mathbf{x})$, is given by (16), and as has been already observed, the Bayes estimate of θ (in the decision-theoretic sense) is given by

$$\delta(x_1, \ldots, x_n) = \int_\Omega \theta h(\theta|\mathbf{x}) d\theta,$$

provided λ is of the continuous type. Then we have the following result.

THEOREM 7 Suppose there is a prior p.d.f. λ on Ω such that for the Bayes estimate δ defined by (15) the risk $R(\theta; \delta)$ is independent of θ. Then δ is minimax.

PROOF By the fact that δ is the Bayes estimate corresponding to the prior λ, one has

$$\int_\Omega R(\theta;\delta)\lambda(\theta)d\theta \leq \int_\Omega R(\theta;\delta^*)\lambda(\theta)d\theta$$

for any estimate δ^*. But $R(\theta;\delta) = c$ by assumption. Hence

$$\sup[R(\theta;\delta);\ \theta\in\Omega] = c \leq \int_\Omega R(\theta;\delta^*)\lambda(\theta)d\theta \leq \sup[R(\theta;\delta^*);\ \theta\in\Omega]$$

for any estimate δ^*. Therefore δ is minimax. The case that λ is of the discrete type is treated similarly. ▲

The theorem just proved is illustrated by the following example.

EXAMPLE 16 Let $X_1, \ldots, X_n$ and λ be as in Example 14. Then the corresponding Bayes estimate δ is given by (21). Now by setting $X = \sum_{j=1}^n X_j$ and taking into consideration that $E_\theta X = n\theta$ and $E_\theta X^2 = n\theta(1 - \theta + n\theta)$, we obtain

$$R(\theta;\delta) = E_\theta\left(\theta - \frac{X+\alpha}{n+\alpha+\beta}\right)^2$$

$$= \frac{1}{(n+\alpha+\beta)^2}\left\{\left[(\alpha+\beta)^2 - n\right]\theta^2 - (2\alpha^2 + 2\alpha\beta - n)\theta + \alpha^2\right\}.$$

By taking $\alpha = \beta = \frac{1}{2}\sqrt{n}$ and denoting by δ^* the resulting estimate, we have

$$(\alpha+\beta)^2 - n = 0, \quad 2\alpha^2 + 2\alpha\beta - n = 0,$$

so that

$$R(\theta;\delta^*) = \frac{\alpha^2}{(n+\alpha+\beta)^2} = \frac{n}{4(n+\sqrt{n})^2} = \frac{1}{4(1+\sqrt{n})^2}.$$

Since $R(\theta;\delta^*)$ is independent of θ, Theorem 6 implies that

$$\delta^*(x_1,\ldots,x_n) = \frac{\sum_{j=1}^n x_j + \frac{1}{2}\sqrt{n}}{n+\sqrt{n}} = \frac{2\sqrt{n}\bar{x}+1}{2(1+\sqrt{n})}$$

is minimax.

EXAMPLE 17 Let $X_1, \ldots, X_n$ be i.i.d. r.v.'s from $N(\mu, \sigma^2)$, where σ^2 is known and $\mu = \theta$.

It was shown (see Example 9) that the estimator $\bar{X}$ of θ was UMVU. It can be shown that it is also minimax and admissible. The proof of these latter two facts, however, will not be presented here.

Now a UMVU estimator has uniformly (in θ) smallest risk when its competitors lie in the class of unbiased estimators with finite variance. However, outside this class there might be estimators which are better than a UMVU estimator. In other words, a UMVU estimator need not be admissible. Here is an example.

EXAMPLE 18 Let $X_1, \ldots, X_n$ be i.i.d. r.v.'s from $N(0, \sigma^2)$. Set $\sigma^2 = \theta$. Then the UMVU estimator of θ is given by

$$U = \frac{1}{n} \sum_{j=1}^{n} X_j^2.$$

(See Example 9.) Its variance (risk) was seen to be equal to $2\theta^2/n$; that is, $R(\theta; U) = 2\theta^2/n$. Consider the estimator $\delta = \alpha U$. Then its risk is

$$R(\theta; \delta) = E_\theta (\alpha U - \theta)^2 = E_\theta [\alpha(U - \theta) + (\alpha - 1)\theta]^2 = \frac{\theta^2}{n}\left[(n+2)\alpha^2 - 2n\alpha + n\right].$$

The value $\alpha = n/(n + 2)$ minimizes this risk and the minimum risk is equal to $2\theta^2/(n + 2) < 2\theta^2/n$ for all θ. Thus U is *not* admissible.

Exercise

12.8.1 Let $X_1, \ldots, X_n$ be independent r.v.'s from the $P(\theta)$ distribution, and consider the loss function $L(\theta; \delta) = [\theta - \delta(\mathbf{x})]^2/\theta$. Then for the estimate $\delta(\mathbf{x}) = \bar{x}$, calculate the risk $R(\theta; \delta) = E_\theta[\theta - \delta(\mathbf{X})]^2$, and conclude that $\delta(\mathbf{x})$ is minimax.

12.9 Other Methods of Estimation

Minimum chi-square method. This method of estimation is applicable in situations which can be described by a Multinomial distribution. Namely, consider n independent repetitions of an experiment whose possible outcomes are the k pairwise disjoint events $A_j, j = 1, \ldots, k$. Let X_j be the number of trials which result in A_j and let p_j be the probability that any one of the trials results in A_j. The probabilities p_j may be functions of r parameters; that is,

$$p_j = p_j(\boldsymbol{\theta}), \quad \boldsymbol{\theta} = (\theta_1, \ldots, \theta_r)', \quad j = 1, \ldots, k.$$

Then the present method of estimating $\boldsymbol{\theta}$ consists in minimizing some measure of discrepancy between the observed X's and the expected values of them. One such measure is the following:

$$\chi^2 = \sum_{j=1}^{k} \frac{[X_j - np_j(\boldsymbol{\theta})]^2}{np_j(\boldsymbol{\theta})}.$$

Often the p's are differentiable with respect to the θ's, and then the minimization can be achieved, in principle, by differentiation. However, the actual solution of the resulting system of r equations is often tedious. The solution may be easier by minimizing the following modified χ^2 expression:

$$\chi^2_{\text{mod}} = \sum_{j=1}^{k} \frac{\left[X_j - np_j(\boldsymbol{\theta})\right]^2}{X_j},$$

provided, of course, all $X_j > 0$, $j = 1, \ldots, k$.

Under suitable regularity conditions, the resulting estimators can be shown to have some asymptotic optimal properties. (See Section 12.10.)

The method of moments. Let $X_1, \ldots, X_n$ be i.i.d. r.v.'s with p.d.f. $f(\cdot; \boldsymbol{\theta})$ and for a positive integer r, assume that $EX^r = m_r$ is finite. The problem is that of estimating m_r. According to the present method, m_r will be estimated by the corresponding *sample moment*

$$\frac{1}{n}\sum_{j=1}^{n} X_j^r,$$

The resulting moment estimates are always unbiased and, under suitable regularity conditions, they enjoy some asymptotic optimal properties as well.

On the other hand the theoretical moments are also functions of $\boldsymbol{\theta} = (\theta_1, \ldots, \theta_r)'$. Then we consider the following system

$$\frac{1}{n}\sum_{j=1}^{n} X_j^k = m_k(\theta_1, \ldots, \theta_r), \quad k = 1, \ldots, r,$$

the solution of which (if possible) will provide estimators for θ_j, $j = 1, \ldots, r$.

EXAMPLE 19 Let $X_1, \ldots, X_n$ be i.i.d. r.v.'s from $N(\mu, \sigma^2)$, where both μ and σ^2 are unknown. By the method of moments, we have

$$\begin{cases} \overline{X} = \mu \\ \frac{1}{n}\sum_{j=1}^{n} X_j^2 = \sigma^2 + \mu^2, \end{cases} \quad \text{hence} \quad \hat{\mu} = \overline{X}, \quad \hat{\sigma}^2 = \frac{1}{n}\sum_{j=1}^{n}(X_j - \overline{X})^2.$$

EXAMPLE 20 Let $X_1, \ldots, X_n$ be i.i.d. r.v.'s from $U(\alpha, \beta)$, where both α and β are unknown. Since

$$EX_1 = \frac{\alpha + \beta}{2} \quad \text{and} \quad \sigma^2(X_1) = \frac{(\alpha - \beta)^2}{12}$$

(see Chapter 5), we have

$$\begin{cases} \overline{X} = \frac{\alpha + \beta}{2} \\ \frac{1}{n}\sum_{j=1}^{n} X_j^2 = \frac{(\alpha - \beta)^2}{12} + \frac{(\alpha + \beta)^2}{4}, \end{cases} \quad \text{or} \quad \begin{cases} \beta + \alpha = 2\overline{X} \\ \beta - \alpha = S\sqrt{12}, \end{cases}$$

where

$$S = \sqrt{\frac{1}{n}\sum_{j=1}^{n}(X_j - \overline{X})^2}.$$

Hence $\hat{\alpha} = \overline{X} - S\sqrt{3}$, $\hat{\beta} = \overline{X} + S\sqrt{3}$.

REMARK 8 In Example 20, we see that the moment estimators $\hat{\alpha}, \hat{\beta}$ of α, β, respectively, are not functions of the sufficient statistic $(X_{(1)}, X_{(n)})'$ of $(\alpha, \beta)'$. This is a drawback of the method of moment estimation. Another obvious disadvantage of this method is that it fails when no moments exist (as in the case of the Cauchy distribution), or when not enough moments exist.

Least square method. This method is applicable when the underlying distribution is of a certain special form and it will be discussed in detail in Chapter 16.

Exercises

12.9.1 Let $X_1, \ldots, X_n$ be independent r.v.'s distributed as $U(\theta - a, \theta + b)$, where $a, b > 0$ are known and $\theta \in \Omega = \mathbb{R}$. Find the moment estimator of θ and calculate its variance.

12.9.2 If $X_1, \ldots, X_n$ are independent r.v.'s distributed as $U(-\theta, \theta)$, $\theta \in \Omega = (0, \infty)$, does the method of moments provide an estimator for θ?

12.9.3 If $X_1, \ldots, X_n$ are i.i.d. r.v.'s from the Gamma distribution with parameters α and β, show that $\hat{\alpha} = \overline{X}^2/S^2$ and $\hat{\beta} = S^2/\overline{X}$ are the moment estimators of α and β, respectively, where

$$S^2 = \frac{1}{n}\sum_{j=1}^{n}(X_j - \overline{X})^2.$$

12.9.4 Let X_1, X_2 be independent r.v.'s with p.d.f. $f(\cdot; \theta)$ given by

$$f(x; \theta) = \frac{2}{\theta^2}(\theta - x)I_{(0,\theta)}(x), \quad \theta \in \Omega = (0, \infty).$$

Find the moment estimator of θ.

12.9.5 Let $X_1, \ldots, X_n$ be i.i.d. r.v.'s from the Beta distribution with parameters α, β and find the moment estimators of α and β.

12.9.6 Refer to Exercise 12.5.7 and find the moment estimators of θ_1 and θ_2.

12.10 Asymptotically Optimal Properties of Estimators

So far we have occupied ourselves with the problem of constructing an estimator on the basis of a sample of fixed size n, and having one or more of the following properties: Unbiasedness, (uniformly) minimum variance, minimax, minimum average risk (Bayes), the (intuitively optimal) property associated with an MLE. If however, the sample size n may increase indefinitely, then some additional, asymptotic properties can be associated with an estimator. To this effect, we have the following definitions.

Let $X_1, \ldots, X_n$ be i.i.d. r.v.'s with p.d.f. $f(\cdot; \theta)$, $\theta \in \Omega \subseteq \mathbb{R}$.

DEFINITION 14 The sequence of estimators of θ, $\{V_n\} = \{V(X_1, \ldots, X_n)\}$, is said to be *consistent in probability* (or *weakly consistent*) if $V_n \xrightarrow[]{P_\theta} \theta$ as $n \to \infty$, for all $\theta \in \Omega$. It is said to be *a.s. consistent* (or *strongly consistent*) if $V_n \xrightarrow[P_\theta]{a.s.} \theta$ as $n \to \infty$, for all $\theta \in \Omega$. (See Chapter 8.)

From now on, the term "consistent" will be used in the sense of "weakly consistent."

The following theorem provides a criterion for a sequence of estimates to be consistent.

THEOREM 8 If, as $n \to \infty$, $E_\theta V_n \to \theta$ and $\sigma_\theta^2 V_n \to 0$, then $V_n \xrightarrow[]{P_\theta} \theta$.

PROOF For the proof of the theorem the reader is referred to Remark 5, Chapter 8. ▲

DEFINITION 15 The sequence of estimators of θ, $\{V_n\} = \{V(X_1, \ldots, X_n)\}$, properly normalized, is said to be *asymptotically normal* $N(0, \sigma^2(\theta))$, if, as $n \to \infty$, $\sqrt{n}(V_n - \theta) \xrightarrow[(P_\theta)]{d} X$ for all $\theta \in \Omega$, where X is distributed (under P_θ) as $N(0, \sigma^2(\theta))$. (See Chapter 8.)

This is often expressed (loosely) by writing $V_n \approx N(\theta, \sigma^2(\theta)/n)$. If

$$\sqrt{n}(V_n - \theta) \xrightarrow[(P_\theta)]{d} N(0, \sigma^2(\theta)), \text{ as } n \to \infty,$$

it follows that $V_n \xrightarrow[n \to \infty]{P_\theta} \theta$ (see Exercise 12.10.1).

DEFINITION 16 The sequence of estimators of θ, $\{V_n\} = \{V(X_1, \ldots, X_n)\}$, is said to be *best asymptotically normal* (BAN) if:

i) It is asymptotically normal and

ii) The variance $\sigma^2(\theta)$ of its limiting normal distribution is smallest for all $\theta \in \Omega$ in the class of all sequences of estimators which satisfy (i).

A BAN sequence of estimators is also called *asymptotically efficient* (with respect to the variance). The *relative asymptotic efficiency* of any other sequence of estimators which satisfies (i) only is expressed by the quotient of the smallest variance mentioned in (ii) to the variance of the asymptotic normal distribution of the sequence of estimators under consideration.

In connection with the concepts introduced above, we have the following result.

12 Point Estimation

THEOREM 9 Let $X_1, \ldots, X_n$ be i.i.d. r.v.'s with p.d.f. $f(\cdot; \theta)$, $\theta \in \Omega \subseteq \mathbb{R}$. Then, if certain suitable regularity conditions are satisfied, the likelihood equation

$$\frac{\partial}{\partial \theta} \log L(\theta | X_1, \ldots, X_n) = 0$$

has a root $\theta_n^* = \theta^*(X_1, \ldots, X_n)$, for each n, such that the sequence $\{\theta_n^*\}$ of estimators is BAN and the variance of its limiting normal distribution is equal to the inverse of Fisher's information number

$$I(\theta) = E_\theta \left[\frac{\partial}{\partial \theta} \log f(X; \theta) \right]^2,$$

where X is an r.v. distributed as the X's above.

In smooth cases, θ_n^* will be an MLE or *the* MLE. Examples have been constructed, however, for which $\{\theta_n^*\}$ does not satisfy (ii) of Definition 16 for some exceptional θ's. Appropriate regularity conditions ensure that these exceptional θ's are only "a few" (in the sense of their set having Lebesgue measure zero). The fact that there can be exceptional θ's, along with other considerations, has prompted the introduction of other criteria of asymptotic efficiency. However, this topic will not be touched upon here. Also, the proof of Theorem 9 is beyond the scope of this book, and therefore it will be omitted.

EXAMPLE 21

i) Let $X_1, \ldots, X_n$ be i.i.d. r.v.'s from $B(1, \theta)$. Then, by Exercise 12.5.1, the MLE of θ is $\bar{X}$, which we denote by $\bar{X}_n$ here. The weak and strong consistency of $\bar{X}_n$ follows by the WLLN and SLLN, respectively (see Chapter 8). That $\sqrt{n}(\bar{X}_n - \theta)$ is asymptotically normal $N(0, I^{-1}(\theta))$, where $I(\theta) = 1/[\theta(1-\theta)]$ (see Example 7), follows from the fact that $\sqrt{n}(\bar{X}_n - \theta) / \sqrt{\theta(1-\theta)}$ is asymptotically $N(0, 1)$ by the CLT (see Chapter 8).

ii) If $X_1, \ldots, X_n$ are i.i.d. r.v.'s from $P(\theta)$, then the MLE $\bar{X} = \bar{X}_n$ of θ (see Example 10) is both (strongly) consistent and asymptotically normal by the same reasoning as above, with the variance of limiting normal distribution being equal to $I^{-1}(\theta) = \theta$ (see Example 8).

iii) The same is true of the MLE $\bar{X} = \bar{X}_n$ of μ and $(1/n)\sum_{j=1}^n (X_j - \mu)^2$ of σ^2 if $X_1, \ldots, X_n$ are i.i.d. r.v.'s from $N(\mu, \sigma^2)$ with one parameter known and the other unknown (see Example 12). The variance of the (normal) distribution of $\sqrt{n}(\bar{X}_n - \mu)$ is $I^{-1}(\mu) = \sigma^2$, and the variance of the limiting normal distribution of

$$\sqrt{n}\left[\frac{1}{n}\sum_{j=1}^n (X_j - \mu)^2 - \sigma^2 \right] \text{ is } I^{-1}(\sigma^2) = 2\sigma^4 \text{ (see Example 9)}.$$

It can further be shown that in all cases (i)–(iii) just considered the regularity conditions not explicitly mentioned in Theorem 9 are satisfied, and therefore the above sequences of estimators are actually BAN.

Exercise

12.10.1 Let $X_1, \ldots, X_n$ be i.i.d. r.v.'s with p.d.f. $f(\cdot; \theta)$; $\theta \in \Omega \subseteq \mathbb{R}$ and let $\{V_n\} = \{V_n(X_1, \ldots, X_n)\}$ be a sequence of estimators of θ such that $\sqrt{n}(V_n - \theta) \xrightarrow[(P_\theta)]{d} Y$ as $n \to \infty$, where Y is an r.v. distributed as $N(0, \sigma^2(\theta))$. Then show that $V_n \xrightarrow[n \to \infty]{P_\theta} \theta$. (That is, asymptotic normality of $\{V_n\}$ implies its consistency in probability.)

12.11 Closing Remarks

The following definition serves the purpose of asymptotically comparing two estimators.

DEFINITION 17 Let $X_1, \ldots, X_n$ be i.i.d. r.v.'s with p.d.f. $f(\cdot; \theta)$, $\theta \in \Omega \subseteq \mathbb{R}$ and let

$$\{U_n\} = \{U_n(X_1, \ldots, X_n)\} \quad \text{and} \quad \{V_n\} = \{V_n(X_1, \ldots, X_n)\}$$

be two sequences of estimators of θ. Then we say that $\{U_n\}$ and $\{V_n\}$ are *asymptotically equivalent* if for every $\theta \in \Omega$,

$$\sqrt{n}(U_n - V_n) \xrightarrow[n \to \infty]{P_\theta} 0.$$

For an example, suppose that the X's are from $B(1, \theta)$. It has been shown (see Exercise 12.3.3) that the UMVU estimator of θ is $U_n = \bar{X}_n \, (= \bar{X})$ and this coincides with the MLE of θ (Exercise 12.5.1). However, the Bayes estimator of θ, corresponding to a Beta p.d.f. λ, is given by

$$V_n = \frac{\sum_{j=1}^{n} X_j + \alpha}{n + \alpha + \beta}, \tag{25}$$

and the minimax estimator is

$$W_n = \frac{\sum_{j=1}^{n} X_j + \sqrt{n}/2}{n + \sqrt{n}}. \tag{26}$$

That is, four different methods of estimation of the same parameter θ provided three different estimators. This is not surprising, since the criteria of optimality employed in the four approaches were different. Next, by the CLT, $\sqrt{n}(U_n - \theta) \xrightarrow[(P_\theta)]{d} Z$, as $n \to \infty$, where Z is an r.v. distributed as $N(0, \theta(1 - \theta))$, and it can also be shown (see Exercise 11.1), that $\sqrt{n}(V_n - \theta) \xrightarrow[(P_\theta)]{d} Z$, as $n \to \infty$, for any arbitrary but fixed (that is, not functions of n) values of α and β. It can also be shown (see Exercise 12.11.2) that $\sqrt{n}(U_n - V_n) \xrightarrow[n \to \infty]{P_\theta} 0$. Thus $\{U_n\}$ and $\{V_n\}$ are asymptotically equivalent according to Definition 17. As for W_n, it can be established (see Exercise 12.11.3) that $\sqrt{n}(W_n - \theta) \xrightarrow[(P_\theta)]{d} W$, as $n \to \infty$, where W is an r.v. distributed as $N(\frac{1}{2} - \theta, \theta(1 - \theta))$.

Thus $\{U_n\}$ and $\{W_n\}$ or $\{V_n\}$ and $\{W_n\}$ are not even comparable on the basis of Definition 17.

Finally, regarding the question as to which estimator is to be selected in a given case, the answer would be that this would depend on which kind of optimality is judged to be most appropriate for the case in question.

Although the preceding comments were made in reference to the Binomial case, they are of a general nature, and were used for the sake of definiteness only.

Exercises

12.11.1 In reference to Example 14, the estimator V_n given by (25) is the Bayes estimator of θ, corresponding to a prior Beta p.d.f. Then show that $\sqrt{n}(V_n - \theta) \xrightarrow[(P_\theta)]{d} Z$ as $n \to \infty$, where Z is an r.v. distributed as $N(0, \theta(1 - \theta))$.

12.11.2 In reference to Example 14, $U_n = \bar{X}_n$ is the UMVU (and also the ML) estimator of θ, whereas the estimator V_n is given by (25). Then show that $\sqrt{n}(U_n - V_n) \xrightarrow[n \to \infty]{P_\theta} 0$.

12.11.3 In reference to Example 14, W_n, given by (26), is the minimax estimator of θ. Then show that $\sqrt{n}(W_n - \theta) \xrightarrow[(P_\theta)]{d} W$ as $n \to \infty$, where W is an r.v. distributed as $(N_{\frac{1}{2}} - \theta, \theta(1 - \theta).)$

Chapter 13

Testing Hypotheses

Throughout this chapter, $X_1, \ldots, X_n$ will be i.i.d. r.v.'s defined on a probability space (S, class of events, $\mathbf{P}_\theta$), $\boldsymbol{\theta} \in \boldsymbol{\Omega} \subseteq \mathbb{R}^r$ and having p.d.f. $f(\cdot; \boldsymbol{\theta})$.

13.1 General Concepts of the Neyman–Pearson Testing Hypotheses Theory

In this section, we introduce the basic concepts of testing hypotheses theory.

DEFINITION 1 A statement regarding the parameter $\boldsymbol{\theta}$, such as $\boldsymbol{\theta} \in \boldsymbol{\omega} \subset \boldsymbol{\Omega}$, is called a (statistical) *hypothesis* (about $\boldsymbol{\theta}$) and is usually denoted by H (or H_0). The statement that $\boldsymbol{\theta} \in \boldsymbol{\omega}^c$ (the complement of $\boldsymbol{\omega}$ with respect to $\boldsymbol{\Omega}$) is also a (statistical) hypothesis about $\boldsymbol{\theta}$, which is called the *alternative* to H (or H_0) and is usually denoted by A. Thus

$$H(H_0): \boldsymbol{\theta} \in \boldsymbol{\omega}$$
$$A: \boldsymbol{\theta} \in \boldsymbol{\omega}^c.$$

Often hypotheses come up in the form of a claim that a new product, a new technique, etc., is more efficient than existing ones. In this context, H (or H_0) is a statement which nullifies this claim and is called a *null hypothesis*.

If $\boldsymbol{\omega}$ contains only one point, that is, $\boldsymbol{\omega} = \{\boldsymbol{\theta}_0\}$, then H is called a *simple* hypothesis, otherwise it is called a *composite* hypothesis. Similarly for alternatives.

Once a hypothesis H is formulated, the problem is that of *testing* H on the basis of the observed values of the X's.

DEFINITION 2 A *randomized* (statistical) *test* (or *test function*) for testing H against the alternative A is a (measurable) function ϕ defined on $\mathbb{R}^n$, taking values in $[0, 1]$ and having the following interpretation: If $(x_1, \ldots, x_n)'$ is the observed value of $(X_1, \ldots, X_n)'$ and $\phi(x_1, \ldots, x_n) = y$, then a coin, whose probability of falling heads is y, is tossed and H is rejected or accepted when heads or tails appear, respectively. In the particular case where y can be either 0 or 1 for all $(x_1, \ldots, x_n)'$, then the test ϕ is called a *nonrandomized* test.

Thus a nonrandomized test has the following form:

$$\phi(x_1, \ldots, x_n) = \begin{cases} 1 & \text{if } (x_1, \ldots, x_n)' \in B \\ 0 & \text{if } (x_1, \ldots, x_n)' \in B^c. \end{cases}$$

In this case, the (Borel) set B in $\mathbb{R}^n$ is called the *rejection* or *critical region* and B^c is called the *acceptance region*.

In testing a hypothesis H, one may commit either one of the following two kinds of errors: to reject H when actually H is true, that is, the (unknown) parameter θ does lie in the subset ω specified by H; or to accept H when H is actually false.

DEFINITION 3 Let $\beta(\theta) = P_\theta$ (rejecting H), so that $1 - \beta(\theta) = P_\theta$ (accepting H), $\theta \in \Omega$. Then $\beta(\theta)$ with $\theta \in \omega$ is the probability of rejecting H, calculated under the assumption that H is true. Thus for $\theta \in \omega$, $\beta(\theta)$ is the probability of an error, namely, the probability of *type-I error*. $1 - \beta(\theta)$ with $\theta \in \omega^c$ is the probability of accepting H, calculated under the assumption that H is false. Thus for $\theta \in \omega^c$, $1 - \beta(\theta)$ represents the probability of an error, namely, the probability of *type-II error*. The function β restricted to ω^c is called the *power function* of the test and $\beta(\theta)$ is called the *power of the test at* $\theta \in \omega^c$. The sup $[\beta(\theta); \theta \in \omega]$ is denoted by α and is called the *level of significance* or *size* of the test.

Clearly, α is the smallest upper bound of the type-I error probabilities. It is also plain that one would desire to make α as small as possible (preferably 0) and at the same time to make the power as large as possible (preferably 1). Of course, maximizing the power is equivalent to minimizing the type-II error probability. Unfortunately, with a fixed sample size, this cannot be done, in general. What the classical theory of testing hypotheses does is to fix the size α at a desirable level (which is usually taken to be 0.005, 0.01, 0.05, 0.10) and then derive tests which maximize the power. This will be done explicitly in this chapter for a number of interesting cases. The reason for this course of action is that the roles played by H and A are not at all symmetric. From the consideration of potential losses due to wrong decisions (which may or may not be quantifiable in monetary terms), the decision maker is somewhat conservative for holding the null hypothesis as true unless there is overwhelming evidence from the data that it is false. He/she believes that the consequence of wrongly rejecting the null hypothesis is much more severe to him/her than that of wrongly accepting it. For example, suppose a pharmaceutical company is considering the marketing of a newly developed drug for treatment of a disease for which the best available drug in the market has a cure rate of 60%. On the basis of limited experimentation, the research division claims that the new drug is more effective. If, in fact, it fails to be more effective or if it has harmful side effects, the loss sustained by the company due to an immediate obsolescence of the product, decline of the company's image, etc., will be quite severe. On the other hand, failure to market a truly better drug is an opportunity loss, but that may not be considered to be as serious as the other loss. If a decision is to be made on the basis of a number of clinical trials, the null hypothesis H should be that the cure rate of the new drug is *no more than* 60% and A should be that this cure rate *exceeds* 60%.

We notice that for a nonrandomized test with critical region B, we have

$$\beta(\theta) = P_\theta\left[(X_1, \ldots, X_n)' \in B\right] = 1 \cdot P_\theta\left[(X_1, \ldots, X_n)' \in B\right]$$
$$+ 0 \cdot P_\theta\left[(X_1, \ldots, X_n)' \in B^c\right] = E_\theta \phi(X_1, \ldots, X_n),$$

and the same can be shown to be true for randomized tests (by an appropriate application of property (CE1) in Section 3 of Chapter 5). Thus

$$\beta_\phi(\theta) = \beta(\theta) = E_\theta \phi(X_1, \ldots, X_n), \qquad \theta \in \Omega. \tag{1}$$

DEFINITION 4 A level-α test which maximizes the power among all tests of level α is said to be *uniformly most powerful* (UMP). Thus ϕ is a UMP, level-α test if (i) sup $[\beta_\phi(\theta); \theta \in \omega] = \alpha$ and (ii) $\beta_\phi(\theta) \geq \beta_{\phi^*}(\theta)$, $\theta \in \omega^c$ for any other test ϕ^* which satisfies (i).

If ω^c consists of a single point only, a UMP test is simply called *most powerful* (MP). In many important cases a UMP test does exist.

Exercise

13.1.1 In the following examples indicate which statements constitute a simple and which a composite hypothesis:

i) X is an r.v. whose p.d.f. f is given by $f(x) = 2e^{-2x}I_{(0,\infty)}(x)$;

ii) When tossing a coin, let X be the r.v. taking the value 1 if head appears and 0 if tail appears. Then the statement is: The coin is biased;

iii) X is an r.v. whose expectation is equal to 5.

13.2 Testing a Simple Hypothesis Against a Simple Alternative

In the present case, we take Ω to consist of two points only, which can be labeled as θ_0 and θ_1; that is, $\Omega = \{\theta_0, \theta_1\}$. In actuality, Ω may consist of more than two points but we focus attention only on two of its points. Let f_{θ_0} and f_{θ_1} be two given p.d.f.'s. We set $f_0 = f(\cdot; \theta_0)$, $f_1 = f(\cdot; \theta_1)$ and let $X_1, \ldots, X_n$ be i.i.d. r.v.'s with p.d.f., $f(\cdot; \theta)$, $\theta \in \Omega$. The problem is that of testing the hypothesis $H: \theta \in \omega = \{\theta_0\}$ against the alternative $A: \theta \in \omega^c = \{\theta_1\}$ at level α. In other words, we want to test the hypothesis that the underlying p.d.f. of the X's is f_0 against the alternative that it is f_1. In such a formulation, the p.d.f.'s f_0 and f_1 need not even be members of a parametric family of p.d.f.'s; they may be any p.d.f.'s which are of interest to us.

In connection with this testing problem, we are going to prove the following result.

THEOREM 1 (Neyman–Pearson Fundamental Lemma) Let $X_1, \ldots, X_n$ be i.i.d. r.v.'s with p.d.f. $f(\cdot; \theta)$, $\theta \in \Omega = \{\theta_0, \theta_1\}$. We are interested in testing the hypothesis H: $\theta = \theta_0$ against the alternative $A: \theta = \theta_1$ at level α $(0 < \alpha < 1)$. Let ϕ be the test defined as follows:

$$\phi(x_1, \ldots, x_n) = \begin{cases} 1, & \text{if } f(x_1; \theta_1) \cdots f(x_n; \theta_1) > Cf(x_1; \theta_0) \cdots f(x_n; \theta_0) \\ \gamma, & \text{if } f(x_1; \theta_1) \cdots f(x_n; \theta_1) = Cf(x_1; \theta_0) \cdots f(x_n; \theta_0) \\ 0, & \text{otherwise,} \end{cases} \quad (2)$$

where the constants γ $(0 \leq \gamma \leq 1)$ and C (>0) are determined so that

$$E_{\theta_0} \phi(X_1, \ldots, X_n) = \alpha. \quad (3)$$

Then, for testing H against A at level α, the test defined by (2) and (3) is MP within the class of all tests whose level is $\leq \alpha$.

The proof is presented for the case that the X's are of the continuous type, since the discrete case is dealt with similarly by replacing integrals by summation signs.

PROOF For convenient writing, we set

$$\mathbf{z} = (x_1, \ldots, x_n)', \quad d\mathbf{z} = dx_1 \cdots dx_n, \quad \mathbf{Z} = (X_1, \ldots, X_n)'$$

and $f(\mathbf{z}; \theta)$, $f(\mathbf{Z}; \theta)$ for $f(x_1; \theta) \cdots f(x_n; \theta)$, $f(X_1; \theta) \cdots f(X_n; \theta)$, respectively. Next, let T be the set of points $\mathbf{z}$ in $\mathbb{R}^n$ such that $f_0(\mathbf{z}) > 0$ and let $D^c = \mathbf{Z}^{-1}(T^c)$. Then

$$P_{\theta_0}(D^c) = P_{\theta_0}(\mathbf{Z} \in T^c) = \int_{T^c} f_0(\mathbf{z}) d\mathbf{z} = 0,$$

and therefore in calculating P_{θ_0}-probabilities we may redefine and modify r.v.'s on the set D^c. Thus we have, in particular,

$$\begin{aligned} E_{\theta_0} \phi(\mathbf{Z}) &= P_{\theta_0}[f_1(\mathbf{Z}) > Cf_0(\mathbf{Z})] + \gamma P_{\theta_0}[f_1(\mathbf{Z}) = Cf_0(\mathbf{Z})] \\ &= P_{\theta_0}\{[f_1(\mathbf{Z}) > Cf_0(\mathbf{Z})] \cap D\} + \gamma P_{\theta_0}\{[f_1(\mathbf{Z}) = Cf_0(\mathbf{Z})] \cap D\} \\ &= P_{\theta_0}\left\{\left[\frac{f_1(\mathbf{Z})}{f_0(\mathbf{Z})} > C\right] \cap D\right\} + \gamma P_{\theta_0}\left\{\left[\frac{f_1(\mathbf{Z})}{f_0(\mathbf{Z})} = C\right] \cap D\right\} \\ &= P_{\theta_0}[(Y > C) \cap D] + \gamma P_{\theta_0}[(Y = C) \cap D] \\ &= P_{\theta_0}(Y > C) + \gamma P_{\theta_0}(Y = C), \end{aligned} \quad (4)$$

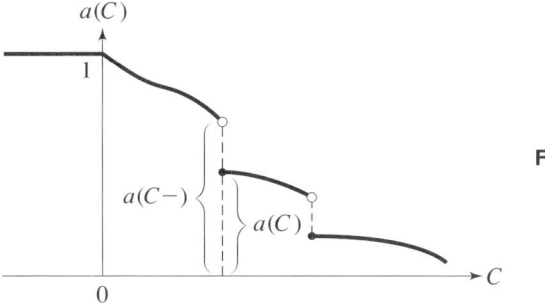

Figure 13.1

where $Y = f_1(\mathbf{Z})/f_0(\mathbf{Z})$ on D and let Y be arbitrary (but measurable) on D^c. Now let $a(C) = P_{\theta_0}(Y > C)$, so that $G(C) = 1 - a(C) = P_{\theta_0}(Y \leq C)$ is the d.f. of the r.v. Y. Since G is a d.f., we have $G(-\infty) = 0$, $G(\infty) = 1$, G is nondecreasing and continuous from the right. These properties of G imply that the function a is such that $a(-\infty) = 1$, $a(\infty) = 0$, a is nonincreasing and continuous from the right. Futhermore,

$$P_{\theta_0}(Y = C) = G(C) - G(C-) = [1 - a(C)] - [1 - a(C-)] = a(C-) - a(C),$$

and $a(C) = 1$ for $C < 0$, since $P_{\theta_0}(Y \geq 0) = 1$

Figure 13.1 represents the graph of a typical function a. Now for any α ($0 < \alpha < 1$) there exists C_0 (≥ 0) such that $a(C_0) \leq \alpha \leq a(C_0 -)$. (See Fig. 13.1.) At this point, there are two cases to consider. First, $a(C_0) = a(C_0 -)$; that is, C_0 is a continuity point of the function a. Then, $\alpha = a(C_0)$ and if in (2) C is replaced by C_0 and $\gamma = 0$, the resulting test is of level α. In fact, in this case (4) becomes

$$E_{\theta_0} \phi(\mathbf{Z}) = P_{\theta_0}(Y > C_0) = a(C_0) = \alpha,$$

as was to be seen.

Next, we assume that C_0 is a discontinuity point of a. In this case, take again $C = C_0$ in (2) and also set

$$\gamma = \frac{\alpha - a(C_0)}{a(C_0 -) - a(C_0)}$$

(so that $0 \leq \gamma \leq 1$). Again we assert that the resulting test is of level α. In the present case, (4) becomes as follows:

$$E_{\theta_0} \phi(\mathbf{Z}) = P_{\theta_0}(Y > C_0) + \gamma P_{\theta_0}(Y = C_0)$$
$$= a(C_0) + \frac{\alpha - a(C_0)}{a(C_0 -) - a(C_0)}[a(C_0 -) - a(C_0)] = \alpha.$$

Summarizing what we have done so far, we have that with $C = C_0$, as defined above, and

$$\gamma = \frac{\alpha - a(C_0)}{a(C_0-) - a(C_0)}$$

(which it is to be interpreted as 0 whenever is of the form 0/0), the test defined by (2) is of level α. That is, (3) is satisfied.

Now it remains for us to show that the test so defined is MP, as described in the theorem. To see this, let $\phi*$ be any test of level $\leq \alpha$ and set

$$B^+ = \{\mathbf{z} \in \mathbb{R}^n;\ \phi(\mathbf{z}) - \phi*(\mathbf{z}) > 0\} = (\phi - \phi* > 0),$$
$$B^- = \{\mathbf{z} \in \mathbb{R}^n;\ \phi(\mathbf{z}) - \phi*(\mathbf{z}) < 0\} = (\phi - \phi* < 0).$$

Then $B^+ \cap B^- = \emptyset$ and, clearly,

$$\left.\begin{array}{l} B^+ = (\phi > \phi*) \subseteq (\phi = 1) \cup (\phi = \gamma) = (f_1 \geq Cf_0) \\ B^- = (\phi < \phi*) \subseteq (\phi = 0) \cup (\phi = \gamma) = (f_1 \leq Cf_0). \end{array}\right\} \quad (5)$$

Therefore

$$\int_{\mathbb{R}^n} [\phi(\mathbf{z}) - \phi*(\mathbf{z})][f_1(\mathbf{z}) - Cf_0(\mathbf{z})]d\mathbf{z}$$
$$= \int_{B^+} [\phi(\mathbf{z}) - \phi*(\mathbf{z})][f_1(\mathbf{z}) - Cf_0(\mathbf{z})]d\mathbf{z}$$
$$+ \int_{B^-} [\phi(\mathbf{z}) - \phi*(\mathbf{z})][f_1(\mathbf{z}) - Cf_0(\mathbf{z})]d\mathbf{z}$$

and this is ≥ 0 on account of (5). That is,

$$\int_{\mathbb{R}^n} [\phi(\mathbf{z}) - \phi*(\mathbf{z})][f_1(\mathbf{z}) - Cf_0(\mathbf{z})]d\mathbf{z} \geq 0,$$

which is equivalent to

$$\int_{\mathbb{R}^n} [\phi(\mathbf{z}) - \phi*(\mathbf{z})]f_1(\mathbf{z})d\mathbf{z} \geq C \int_{\mathbb{R}^n} [\phi(\mathbf{z}) - \phi*(\mathbf{z})]f_0(\mathbf{z})d\mathbf{z}. \quad (6)$$

But

$$\int_{\mathbb{R}^n} [\phi(\mathbf{z}) - \phi*(\mathbf{z})]f_0(\mathbf{z})d\mathbf{z} = \int_{\mathbb{R}^n} \phi(\mathbf{z})f_0(\mathbf{z})d\mathbf{z} - \int_{\mathbb{R}^n} \phi*(\mathbf{z})f_0(\mathbf{z})d\mathbf{z}$$
$$= E_{\theta_0}\phi(\mathbf{Z}) - E_{\theta_0}\phi*(\mathbf{Z}) = \alpha - E_{\theta_0}\phi*(\mathbf{Z}) \geq 0, \quad (7)$$

and similarly,

$$\int_{\mathbb{R}^n} [\phi(\mathbf{z}) - \phi*(\mathbf{z})]f_1(\mathbf{z})d\mathbf{z} = E_{\theta_1}\phi(\mathbf{Z}) - E_{\theta_1}\phi*(\mathbf{Z}) = \beta_\phi(\theta_1) - \beta_{\phi*}(\theta_1). \quad (8)$$

Relations (6), (7) and (8) yield $\beta_\phi(\theta_1) - \beta_{\phi*}(\theta_1) \geq 0$, or $\beta_\phi(\theta_1) \geq \beta_{\phi*}(\theta_1)$. This completes the proof of the theorem. ▲

The theorem also guarantees that the power $\beta_\phi(\theta_1)$ is at least α. That is,

COROLLARY Let ϕ be defined by (2) and (3). Then $\beta_\phi(\theta_1) \geq \alpha$.

PROOF The test $\phi^*(z) = \alpha$ is of level α, and since ϕ is most powerful, we have $\beta_\phi(\theta_1) \geq \beta_{\phi^*}(\theta_1) = \alpha$. ▲

REMARK 1

i) The determination of C and γ is essentially unique. In fact, if $C = C_0$ is a discontinuity point of a, then both C and γ are uniquely defined the way it was done in the proof of the theorem. Next, if the (straight) line through the point $(0, \alpha)$ and parallel to the C-axis has only one point in common with the graph of a, then $\gamma = 0$ and C is the unique point for which $a(C) = \alpha$. Finally, if the above (straight) line coincides with part of the graph of a corresponding to an interval $(b_1, b_2]$, say, then $\gamma = 0$ again and any C in $(b_1, b_2]$ can be chosen without affecting the level of the test. This is so because

$$P_{\theta_0}\left[Y \in (b_1, b_2]\right] \leq G(b_2) - G(b_1)$$
$$= [1 - a(b_2)] - [1 - a(b_1)] = a(b_2) - a(b_1) = 0.$$

ii) The theorem shows that there is always a test of the structure (2) and (3) which is MP. The converse is also true, namely, if ϕ is an MP level α test, then ϕ necessarily has the form (2) unless there is a test of size $< \alpha$ with power 1.

This point will not be pursued further here.

The examples to be discussed below will illustrate how the theorem is actually used in concrete cases. In the examples to follow, $\Omega = \{\theta_0, \theta_1\}$ and the problem will be that of testing a simple hypothesis against a simple alternative at level of significance α. It will then prove convenient to set

$$R(z; \theta_0, \theta_1) = \frac{f(x_1; \theta_1) \cdots f(x_n; \theta_1)}{f(x_1; \theta_0) \cdots f(x_n; \theta_0)}$$

whenever the denominator is greater than 0. Also it is often more convenient to work with $\log R(z; \theta_0; \theta_1)$ rather than $R(z; \theta_0, \theta_1)$ itself, provided, of course, $R(z; \theta_0, \theta_1) > 0$.

EXAMPLE 1 Let $X_1, \ldots, X_n$ be i.i.d. r.v.'s from $B(1, \theta)$ and suppose $\theta_0 < \theta_1$. Then

$$\log R(z; \theta_0, \theta_1) = x \log \frac{\theta_1}{\theta_0} + (n - x) \log \frac{1 - \theta_1}{1 - \theta_0},$$

where $x = \sum_{j=1}^n x_j$ and therefore, by the fact that $\theta_0 < \theta_1$, $R(z; \theta_0, \theta_1) > C$ is equivalent to

$$x > C_0, \quad \text{where} \quad C_0 = \left(\log C - n \log \frac{1 - \theta_1}{1 - \theta_0}\right) \bigg/ \log \frac{\theta_1(1 - \theta_0)}{\theta_0(1 - \theta_1)}.$$

Thus the MP test is given by

$$\phi(\mathbf{z}) = \begin{cases} 1, & \text{if } \sum_{j=1}^{n} x_j > C_0 \\ \gamma, & \text{if } \sum_{j=1}^{n} x_j = C_0 \\ 0, & \text{otherwise,} \end{cases} \qquad (9)$$

where C_0 and γ are determined by

$$E_{\theta_0}\phi(\mathbf{Z}) = P_{\theta_0}(X > C_0) + \gamma P_{\theta_0}(X = C_0) = \alpha, \qquad (10)$$

and $X = \sum_{j=1}^{n} X_j$ is $B(n, \theta_i)$, $i = 0, 1$. If $\theta_0 > \theta_1$, the inequality signs in (9) and (10) are reversed.

For the sake of definiteness, let us take $\theta_0 = 0.50$, $\theta_1 = 0.75$, $\alpha = 0.05$ and $n = 25$. Then

$$0.05 = P_{0.5}(X > C_0) + \gamma P_{0.5}(X = C_0) = 1 - P_{0.5}(X \leq C_0) + \gamma P_{0.5}(X = C_0)$$

is equivalent to

$$P_{0.5}(X \leq C_0) - \gamma P_{0.5}(X = C_0) = 0.95.$$

For $C_0 = 17$, we have, by means of the Binomial tables, $P_{0.5}(X \leq 17) = 0.9784$ and $P_{0.5}(X = 17) = 0.0323$. Thus γ is defined by $0.9784 - 0.0323\gamma = 0.95$, whence $\gamma = 0.8792$. Therefore the MP test in this case is given by (9) with $C_0 = 17$ and $\gamma = 0.8792$. The power of the test is $P_{0.75}(X > 17) + 0.8792\, P_{0.75}(X = 17) = 0.8356$.

EXAMPLE 2 Let $X_1, \ldots, X_n$ be i.i.d. r.v.'s from $P(\theta)$ and suppose $\theta_0 < \theta_1$. Then

$$\log R(\mathbf{z}; \theta_0, \theta_1) = x \log \frac{\theta_1}{\theta_0} - n(\theta_1 - \theta_0),$$

where

$$x = \sum_{j=1}^{n} x_j$$

and hence, by using the assumption that $\theta_0 < \theta_1$, one has that $R(\mathbf{z}; \theta_0, \theta_1) > C$ is equivalent to $x > C_0$, where

$$C_0 = \frac{\log\left[Ce^{n(\theta_1 - \theta_0)}\right]}{\log(\theta_1/\theta_0)}.$$

Thus the MP test is defined by

$$\phi(\mathbf{z}) = \begin{cases} 1, & \text{if } \sum_{j=1}^{n} x_j > C_0 \\ \gamma, & \text{if } \sum_{j=1}^{n} x_j = C_0 \\ 0, & \text{otherwise,} \end{cases} \qquad (11)$$

where C_0 and γ are determined by

$$E_{\theta_0}\phi(\mathbf{Z}) = P_{\theta_0}(X > C_0) + \gamma P_{\theta_0}(X = C_0) = \alpha, \quad (12)$$

and $X = \sum_{j=1}^{n} X_j$ is $P(n\theta_i)$, $i = 0, 1$. If $\theta_0 > \theta_1$, the inequality signs in (11) and (12) are reversed.

As an application, let us take $\theta_0 = 0.3$, $\theta_1 = 0.4$, $\alpha = 0.05$ and $n = 20$. Then (12) becomes

$$P_{0.3}(X \leq C_0) - \gamma P_{0.3}(X = C_0) = 0.95.$$

By means of the Poisson tables, one has that for $C_0 = 10$, $P_{0.3}(X \leq 10) = 0.9574$ and $P_{0.3}(X = 10) = 0.0413$. Therefore γ is defined by $0.9574 - 0.0413\gamma = 0.95$, whence $\gamma = 0.1791$.

Thus the test is given by (11) with $C_0 = 10$ and $\gamma = 0.1791$. The power of the test is

$$P_{0.4}(X > 10) + 0.1791\, P_{0.4}(X = 10) = 0.2013.$$

EXAMPLE 3 Let $X_1, \ldots, X_n$ be i.i.d. r.v.'s from $N(\theta, 1)$ and suppose $\theta_0 < \theta_1$. Then

$$\log R(\mathbf{z}; \theta_0, \theta_1) = \frac{1}{2}\sum_{j=1}^{n}\left[(x_j - \theta_0)^2 - (x_j - \theta_1)^2\right]$$

and therefore $R(\mathbf{z}; \theta_0, \theta_1) > C$ is equivalent to $\bar{x} > C_0$, where

$$C_0 = \frac{1}{n}\left[\frac{\log C}{\theta_1 - \theta_0} + \frac{n(\theta_0 + \theta_1)}{2}\right]$$

by using the fact that $\theta_0 < \theta_1$.

Thus the MP test is given by

$$\phi(\mathbf{z}) = \begin{cases} 1, & \text{if } \bar{x} > C_0 \\ 0, & \text{otherwise,} \end{cases} \quad (13)$$

where C_0 is determined by

$$E_{\theta_0}\phi(\mathbf{Z}) = P_{\theta_0}(\bar{X} > C_0) = \alpha, \quad (14)$$

and $\bar{X}$ is $N(\theta_i, 1/n)$, $i = 0, 1$. If $\theta_0 > \theta_1$, the inequality signs in (13) and (14) are reversed.

Let, for example, $\theta_0 = -1$, $\theta_1 = 1$, $\alpha = 0.001$ and $n = 9$. Then (14) gives

$$P_{-1}(\bar{X} > C_0) = P_{-1}[3(\bar{X} + 1) > 3(C_0 + 1)] = P[N(0, 1) > 3(C_0 + 1)] = 0.001,$$

whence $C_0 = 0.03$. Therefore the MP test in this case is given by (13) with $C_0 = 0.03$. The power of the test is

$$P_1(\bar{X} > 0.03) = P_1[3(\bar{X} - 1) > -2.91] = P[N(0, 1) > -2.91] = 0.9982.$$

EXAMPLE 4 Let $X_1, \ldots, X_n$ be i.i.d. r.v.'s from $N(0, \theta)$ and suppose $\theta_0 < \theta_1$. Here

$$\log R(\mathbf{z}; \theta_0, \theta_1) = \frac{\theta_1 - \theta_0}{2\theta_0\theta_1} x + \frac{1}{2}\log\frac{\theta_0}{\theta_1},$$

where $x = \sum_{j=1}^{n} x_j^2$, so that, by means of $\theta_0 < \theta_1$, one has that $R(\mathbf{z}; \theta_0, \theta_1) > C$ is equivalent to $x > C_0$, where

$$C_0 = \frac{2\theta_0\theta_1}{\theta_1 - \theta_0}\log\left(C\sqrt{\frac{\theta_1}{\theta_0}}\right).$$

Thus the MP test in the present case is given by

$$\phi(\mathbf{z}) = \begin{cases} 1, & \text{if } \sum_{j=1}^{n} x_j^2 > C_0 \\ 0, & \text{otherwise,} \end{cases} \qquad (15)$$

where C_0 is determined by

$$E_{\theta_0}\phi(\mathbf{P}) = E_{\theta_0}\left(\sum_{j=1}^{n} X_j^2 > C_0\right) = \alpha, \qquad (16)$$

and $\frac{X}{\theta_i}$ is distributed as χ_n^2, $i = 0, 1$, where $X = \sum_{j=1}^{n} X_j^2$. If $\theta_0 > \theta_1$, the inequality signs in (15) and (16) are reversed. For an example, let $\theta_0 = 4$, $\theta_1 = 16$, $\alpha = 0.01$ and $n = 20$. Then (16) becomes

$$P_4(X > C_0) = P_4\left(\frac{X}{4} > \frac{C_0}{4}\right) = P\left(\chi_{20}^2 > \frac{C_0}{4}\right) = 0.01,$$

whence $C_0 = 150.264$. Thus the test is given by (15) with $C_0 = 150.264$. The power of the test is

$$P_{16}(X > 150.264) = P_{16}\left(\frac{X}{16} > \frac{150.264}{16}\right) = P\left(\chi_{20}^2 > 9.3915\right) = 0.977.$$

Exercises

13.2.1 If $X_1, \ldots, X_{16}$ are independent r.v.'s, construct the MP test of the hypothesis H that the common distribution of the X's is $N(0, 9)$ against the alternative A that it is $N(1, 9)$ at level of significance $\alpha = 0.05$. Also find the power of the test.

13.2.2 Let $X_1, \ldots, X_n$ be independent r.v.'s distributed as $N(\mu, \sigma^2)$, where μ is unknown and σ is known. Show that the sample size n can be determined so that when testing the hypothesis $H: \mu = 0$ against the alternative $A: \mu = 1$, one has predetermined values for α and β. What is the numerical value of n if $\alpha = 0.05$, $\beta = 0.9$ and $\sigma = 1$?

13.2.3 Let $X_1, \ldots, X_n$ be independent r.v.'s distributed as $N(\mu, \sigma^2)$, where μ is unknown and σ is known. For testing the hypothesis $H: \mu = \mu_1$ against the alternative $A: \mu = \mu_2$, show that α can get arbitrarily small and β arbitrarily large for sufficiently large n.

13.2.4 Let $X_1, \ldots, X_{100}$ be independent r.v.'s distributed as $N(\mu, \sigma^2)$. If $\bar{x} = 3.2$, construct the MP test of the hypothesis $H: \mu = 3$, $\sigma^2 = 4$ against the alternative $A: \mu = 3.5$, $\sigma^2 = 4$ at level of significance $\alpha = 0.01$.

13.2.5 Let $X_1, \ldots, X_{30}$ be independent r.v.'s distributed as Gamma with $\alpha = 10$ and β unknown. Construct the MP test of the hypothesis $H: \beta = 2$ against the alternative $A: \beta = 3$ at level of significance 0.05.

13.2.6 Let X be an r.v. whose p.d.f. is either the $U(0, 1)$ p.d.f. denoted by f_0, or the Triangular p.d.f. over the $[0, 1]$ interval, denoted by f_1 (that is, $f_1(x) = 4x$ for $0 \leq x < \frac{1}{2}$, $f_1(x) = 4 - 4x$ for $\frac{1}{2} \leq x \leq 1$ and 0 otherwise). On the basis of one observation on X, construct the MP test of the hypothesis $H: f = f_0$ against the alternative $A: f = f_1$ at level of significance $\alpha = 0.05$.

13.2.7 Let X be an r.v. with p.d.f. f which can be either f_0 or else f_1, where f_0 is $P(1)$ and f_1 is the Geometric p.d.f. with $p = \frac{1}{2}$. For testing the hypothesis $H: f = f_0$ against the alternative $A: f = f_1$:

i) Show that the rejection region is defined by: $\{x \geq 0 \text{ integer}; 1.36 \times \frac{x!}{2^x} \geq C\}$ for some positive number C;

ii) Determine the level of the test α when $C = 3$.

(Hint: Observe that the function $x!/2^x$ is nondecreasing for x integer ≥ 1.)

13.3 UMP Tests for Testing Certain Composite Hypotheses

In the previous section an MP test was constructed for the problem of testing a simple hypothesis against a simple alternative. However, in most problems of practical interest, at least one of the hypotheses H or A is composite. In cases like this it so happens that for certain families of distributions and certain H and A, UMP tests do exist. This will be shown in the present section.

Let $X_1, \ldots, X_n$ be i.i.d. r.v.'s with p.d.f. $f(\cdot; \theta)$, $\theta \in \Omega \subseteq \mathbb{R}$. It will prove convenient to set

$$g(\mathbf{z}; \theta) = f(x_1; \theta) \cdots f(x_1; \theta), \quad \mathbf{z} = (x_1, \ldots, x_n)'. \tag{17}$$

Also $\mathbf{Z} = (X_1, \ldots, X_n)'$.

In the following, we give the definition of a family of p.d.f.'s having the monotone likelihood ratio property. This definition is somewhat more restrictive than the one found in more advanced textbooks but it is sufficient for our purposes.

DEFINITION 5 The family $\{g(\cdot; \theta); \theta \in \Omega\}$ is said to have the *monotone likelihood ratio* (MLR) property in V if the set of $\mathbf{z}$'s for which $g(\mathbf{z}; \theta) > 0$ is independent of θ and there exists a (measurable) function V defined in $\mathbb{R}^n$ into $\mathbb{R}$ such that whenever $\theta, \theta' \in \Omega$ with $\theta < \theta'$ then: (i) $g(\cdot; \theta)$ and $g(\cdot; \theta')$ are distinct and (ii) $g(\mathbf{z}; \theta')/g(\mathbf{z}; \theta)$ is a monotone function of $V(\mathbf{z})$.

Note that the likelihood ratio (LR) in (ii) is well defined except perhaps on a set N of $\mathbf{z}$'s such that $P_\theta(\mathbf{Z} \in N) = 0$ for all $\theta \in \Omega$. In what follows, we will always work outside such a set.

An important family of p.d.f.'s having the MLR property is a one-parameter exponential family.

PROPOSITION 1 Consider the exponential family

$$f(x; \theta) = C(\theta) e^{Q(\theta)T(x)} h(x),$$

where $C(\theta) > 0$ for all $\theta \in \Omega \subseteq \mathbb{R}$ and the set of positivity of h is independent of θ. Suppose that Q is increasing. Then the family $\{g(\cdot; \theta); \theta \in \Omega\}$ has the MLR property in V, where $V(\mathbf{z}) = \sum_{j=1}^{n} T(x_j)$ and $g(\cdot; \theta)$ is given by (17). If Q is decreasing, the family has the MLR property in $V' = -V$.

PROOF We have

$$g(\mathbf{z}; \theta) = C_0(\theta) e^{Q(\theta)V(\mathbf{z})} h^*(\mathbf{z}),$$

where $C_0(\theta) = C^n(\theta)$, $V(\mathbf{z}) = \sum_{j=1}^{n} T(x_j)$ and $h^*(\mathbf{z}) = h(x_1) \cdots h(x_n)$. Therefore on the set of $\mathbf{z}$'s for which $h^*(\mathbf{z}) > 0$ (which set has P_θ-probability 1 for all θ), one has

$$\frac{g(\mathbf{z}; \theta')}{g(\mathbf{z}; \theta)} = \frac{C_0(\theta') e^{Q(\theta')V(\mathbf{z})}}{C_0(\theta) e^{Q(\theta)V(\mathbf{z})}} = \frac{C_0(\theta')}{C_0(\theta)} e^{[Q(\theta') - Q(\theta)]V(\mathbf{z})}.$$

Now for $\theta < \theta'$, the assumption that Q is increasing implies that $g(\mathbf{z}; \theta')/g(\mathbf{z}; \theta)$ is an increasing function of $V(\mathbf{z})$. This completes the proof of the first assertion. The proof of the second assertion follows from the fact that

$$[Q(\theta') - Q(\theta)]V(\mathbf{z}) = [Q(\theta) - Q(\theta')]V'(\mathbf{z}). \blacktriangle$$

From examples and exercises in Chapter 11, it follows that all of the following families have the MLR property: Binomial, Poisson, Negative Binomial, $N(\theta, \sigma^2)$ with σ^2 known and $N(\mu, \theta)$ with μ known, Gamma with $\alpha = \theta$ and β known, or $\beta = \theta$ and α known. Below we present an example of a family which has the MLR property, but it is not of a one-parameter exponential type.

EXAMPLE 5 Consider the Logistic p.d.f. (see also Exercise 4.1.8(i), Chapter 4) with parameter θ; that is,

$$f(x; \theta) = \frac{e^{-x-\theta}}{\left(1 + e^{-x-\theta}\right)^2}, \quad x \in \mathbb{R}, \; \theta \in \Omega = \mathbb{R}. \tag{18}$$

13.3 UMP Tests for Testing Certain Composite Hypotheses

Then

$$\frac{f(x;\,\theta')}{f(x;\,\theta)} = e^{\theta-\theta'}\left(\frac{1+e^{-x-\theta}}{1+e^{-x-\theta'}}\right)^2 \quad \text{and} \quad \frac{f(x;\,\theta')}{f(x;\,\theta)} < \frac{f(x';\,\theta')}{f(x';\,\theta)}$$

if and only if

$$e^{\theta-\theta'}\left(\frac{1+e^{-x-\theta}}{1+e^{-x-\theta'}}\right)^2 < e^{\theta-\theta'}\left(\frac{1+e^{-x'-\theta}}{1+e^{-x'-\theta'}}\right)^2.$$

However, this is equivalent to $e^{-x}(e^{-\theta} - e^{-\theta'}) < e^{-x'}(e^{-\theta} - e^{-\theta'})$. Therefore if $\theta < \theta'$, the last inequality is equivalent to $e^{-x} < e^{-x'}$ or $-x < -x'$. This shows that the family $\{f(\cdot;\,\theta);\,\theta \in \mathbb{R}\}$ has the MLR property in $-x$.

For families of p.d.f.'s having the MLR property, we have the following important theorem.

THEOREM 2 Let $X_1, \ldots, X_n$ be i.i.d. r.v.'s with p.d.f. $f(x;\,\theta)$, $\theta \in \Omega \subseteq \mathbb{R}$ and let the family $\{g(\cdot;\,\theta);\,\theta \in \Omega\}$ have the MLR property in V, where $g(\cdot;\,\theta)$ is defined in (17). Let $\theta_0 \in \Omega$ and set $\omega = \{\theta \in \Omega;\,\theta \leq \theta_0\}$. Then for testing the (composite) hypothesis $H: \theta \in \omega$ against the (composite) alternative $A: \theta \in \omega^c$ at level of significance α, there exists a test ϕ which is UMP within the class of all tests of level $\leq \alpha$. In the case that the LR is increasing in $V(\mathbf{z})$, the test is given by

$$\phi(\mathbf{z}) = \begin{cases} 1, & \text{if } V(\mathbf{z}) > C \\ \gamma, & \text{if } V(\mathbf{z}) = C \\ 0, & \text{otherwise}, \end{cases} \quad (19)$$

where C and γ are determined by

$$E_{\theta_0}\phi(\mathbf{Z}) = P_{\theta_0}[V(\mathbf{Z}) > C] + \gamma P_{\theta_0}[V(\mathbf{Z}) = C] = \alpha. \quad (19')$$

If the LR is decreasing in $V(\mathbf{z})$, the test is taken from (19) and (19') with the inequality signs reversed.

The proof of the theorem is a consequence of the following two lemmas.

LEMMA 1 Under the assumptions made in Theorem 2, the test ϕ defined by (19) and (19') is MP (at level α) for testing the (simple) hypothesis $H_0: \theta = \theta_0$ against the (composite) alternative $A: \theta \in \omega^c$ among all tests of level $\leq \alpha$.

PROOF Let θ' be an arbitrary but fixed point in ω^c and consider the problem of testing the above hypothesis H_0 against the (simple) alternative $A': \theta = \theta'$ at level α. Then, by Theorem 1, the MP test ϕ' is given by

$$\phi'(\mathbf{z}) = \begin{cases} 1, & \text{if } g(\mathbf{z};\,\theta') > C'g(\mathbf{z};\,\theta_0) \\ \gamma, & \text{if } g(\mathbf{z};\,\theta') = C'g(\mathbf{z};\,\theta_0) \\ 0, & \text{otherwise}, \end{cases}$$

where C' and γ' are defined by

$$E_{\theta_0}\phi'(\mathbf{Z}) = \alpha.$$

Let $g(\mathbf{z}; \theta')/g(\mathbf{z}; \theta_0) = \psi[V(\mathbf{z})]$. Then in the case under consideration ψ is defined on $\mathbb{R}$ into itself and is increasing. Therefore

$$\left.\begin{array}{ll} \psi[V(\mathbf{z})] > C' & \text{if and only if} \quad V(\mathbf{z}) > \psi^{-1}(C') = C_0 \\ \psi[V(\mathbf{z})] = C' & \text{if and only if} \quad V(\mathbf{z}) = C_0. \end{array}\right\} \quad (20)$$

In addition,

$$E_{\theta_0}\phi'(\mathbf{Z}) = P_{\theta_0}\{\psi[V(\mathbf{Z})] > C'\} + \gamma P_{\theta_0}\{\psi[V(\mathbf{Z})] = C'\}$$
$$= P_{\theta_0}[V(\mathbf{Z}) > C_0] + \gamma P_{\theta_0}[V(\mathbf{Z}) = C_0].$$

Therefore the test ϕ' defined above becomes as follows:

$$\phi'(\mathbf{z}) = \begin{cases} 1, & \text{if } V(\mathbf{z}) > C_0 \\ \gamma, & \text{if } V(\mathbf{z}) = C_0 \\ 0, & \text{otherwise,} \end{cases} \quad (21)$$

and

$$E_{\theta_0}\phi'(\mathbf{Z}) = P_{\theta_0}[V(\mathbf{Z}) > C_0] + \gamma' P_{\theta_0}[V(\mathbf{Z}) = C_0] = \alpha, \quad (21')$$

so that $C_0 = C$ and $\gamma' = \gamma$ by means of (19) and (19').

It follows from (21) and (21') that the test ϕ' is *independent of* $\theta' \in \omega^c$. In other words, we have that $C = C_0$ and $\gamma = \gamma'$ and the test given by (19) and (19') is UMP for testing $H_0: \theta = \theta_0$ against $A: \theta \in \omega^c$ (at level α). ▲

LEMMA 2 Under the assumptions made in Theorem 2, and for the test function ϕ defined by (19) and (19'), we have $E_\theta\phi(\mathbf{Z}) \leq \alpha$ for all $\theta' \in \omega$.

PROOF Let θ' be an arbitrary but fixed point in ω and consider the problem of testing the (simple) hypothesis $H': \theta = \theta'$ against the (simple) alternative $A_0(= H_0): \theta = \theta_0$ at level $\alpha(\theta') = E_{\theta'}\phi(\mathbf{Z})$. Once again, by Theorem 1, the MP test ϕ' is given by

$$\phi'(\mathbf{z}) = \begin{cases} 1, & \text{if } g(\mathbf{z}; \theta_0) > C'g(\mathbf{z}; \theta') \\ \gamma, & \text{if } g(\mathbf{z}; \theta_0) = C'g(\mathbf{z}; \theta') \\ 0, & \text{otherwise,} \end{cases}$$

where C' and γ are determined by

$$E_{\theta'}\phi'(\mathbf{Z}) = P_{\theta'}\{\psi[V(\mathbf{Z})] > C'\} + \gamma P_{\theta'}\{\psi[V(\mathbf{Z})] = C'\} = \alpha(\theta').$$

On account of (20), the test ϕ' above also becomes as follows:

$$\phi'(\mathbf{z}) = \begin{cases} 1, & \text{if } V(\mathbf{z}) > C_0' \\ \gamma, & \text{if } V(\mathbf{z}) = C_0' \\ 0, & \text{otherwise,} \end{cases} \quad (22)$$

$$E_{\theta'}\phi'(\mathbf{Z}) = P_{\theta'}\left[V(\mathbf{Z}) > C_0'\right] + \gamma P_{\theta'}\left[V(\mathbf{Z}) = C_0'\right] = \alpha(\theta'), \quad (22')$$

where $C_0' = \psi^{-1}(C')$.

Replacing θ_0 by θ' in the expression on the left-hand side of (19') and comparing the resulting expression with (22'), one has that $C_0' = C$ and $\gamma' = \gamma$. Therefore the tests ϕ' and ϕ are identical. Furthermore, by the corollary to Theorem 1, one has that $\alpha(\theta') \leq \alpha$, since α is the power of the test ϕ'. ▲

PROOF OF THEOREM 2 Define the classes of test C and C_0 as follows:

$$C = \left\{\text{all level } \alpha \text{ tests for testing } H : \theta \leq \theta_0\right\},$$
$$C_0 = \left\{\text{all level } \alpha \text{ tests for testing } H_0 : \theta = \theta_0\right\}.$$

Then, clearly, $C \subseteq C_0$. Next, the test ϕ, defined by (19) and (19'), belongs in C by Lemma 2, and is MP among all tests in C_0, by Lemma 1. Hence it is MP among tests in C. The desired result follows. ▲

REMARK 2 For the symmetric case where $\omega = \{\theta \in \Omega; \theta \geq \theta_0\}$, under the assumptions of Theorem 2, a UMP test also exists for testing $H : \theta \in \omega$ against $A : \theta \in \omega^c$. The test is given by (19) and (19') if the LR is decreasing in $V(z)$ and by those relationships with the inequality signs reversed if the LR is increasing in $V(\mathbf{z})$. The relevant proof is entirely analogous to that of Theorem 2.

COROLLARY Let $X_1, \ldots, X_n$ be i.i.d. r.v.'s with p.d.f. $f(\cdot; \theta)$ given by

$$f(x; \theta) = C(\theta)e^{Q(\theta)T(x)}h(x),$$

where Q is strictly monotone. Then for testing $H : \theta \in \omega = \{\theta \in \Omega; \theta \leq \theta_0\}$ against $A : \theta \in \omega^c$ at level of significance α, there is a test ϕ which is UMP within the class of all tests of level $\leq \alpha$. This test is given by (19) and (19') if Q is increasing and by (19) and (19') with reversed inequality signs if Q is decreasing.

Also for testing $H : \theta \in \omega = \{\theta \in \Omega; \theta \leq \theta_0\}$ against $A : \theta \in \omega^c$ at level α, there is a test ϕ which is UMP within the class of all tests of level $\leq \alpha$. This test is given by (19) and (19') if Q is decreasing and by those relationships with reversed inequality signs if Q is increasing.

In all tests, $V(\mathbf{z}) = \sum_{j=1}^{n} T(x_j)$.

PROOF It is immediate on account of Proposition 1 and Remark 2. ▲

It can further be shown that the function $\beta(\theta) = E_\theta \phi(\mathbf{Z})$, $\theta \in \Omega$, for the problem discussed in Theorem 2 and also the symmetric situation mentioned

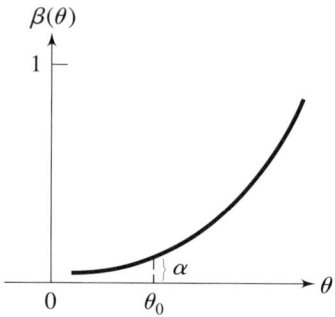

Figure 13.2 $H: \theta \leq \theta_0$, $A: \theta > \theta_0$ **Figure 13.3** $H: \theta \geq \theta_0$, $A: \theta < \theta_0$

in Remark 2, is increasing for those θ's for which it is less than 1 (see Figs. 13.2 and 13.3, respectively).

Another problem of practical importance is that of testing

$$H: \theta \in \omega = \{\theta \in \Omega;\ \theta \leq \theta_1 \ \text{or}\ \theta \geq \theta_2\}$$

against $A: \theta \in \omega^c$, where $\theta_1, \theta_2 \in \Omega$ and $\theta_1 < \theta_2$. For instance, θ may represent a dose of a certain medicine and θ_1, θ_2 are the limits within which θ is allowed to vary. If $\theta \leq \theta_1$ the dose is rendered harmless but also useless, whereas if $\theta \geq \theta_2$ the dose becomes harmful. One may then hypothesize that the dose in question is either useless or harmful and go about testing the hypothesis.

If the underlying distribution of the relevant measurements is assumed to be of a certain exponential form, then a UMP test for the testing problem above does exist. This result is stated as a theorem below but its proof is not given, since this would rather exceed the scope of this book.

THEOREM 3 Let $X_1, \ldots, X_n$ be i.i.d. r.v.'s with p.d.f. $f(\cdot;\ \theta)$, given by

$$f(x;\ \theta) = C(\theta) e^{Q(\theta)T(x)} h(x), \quad (23)$$

where Q is assumed to be strictly monotone and $\theta \in \Omega = \mathbb{R}$.

Set $\omega = \{\theta \in \Omega;\ \theta \leq \theta_1 \ \text{or}\ \theta \geq \theta_2\}$, where $\theta_1, \theta_2 \in \Omega$ and $\theta_1 < \theta_2$. Then for testing the (composite) hypothesis $H: \theta \in \omega$ against the (composite) alternative $A: \theta \in \omega^c$ at level of significance α, there exists a UMP test ϕ. In the case that Q is increasing, ϕ is given by

$$\phi(\mathbf{z}) = \begin{cases} 1, & \text{if}\ C_1 < V(\mathbf{z}) < C_2 \\ \gamma_i, & \text{if}\ V(\mathbf{z}) = C_i\ (i = 1, 2)(C_1 < C_2) \\ 0, & \text{otherwise}, \end{cases} \quad (24)$$

where C_1, C_2 and γ_1, γ_2 are determined by

$$E_{\theta_i} \phi(\mathbf{Z}) = P_{\theta_i}\left[C_1 < V(\mathbf{Z}) < C_2\right] + \gamma_1 P_{\theta_i}\left[V(\mathbf{Z}) = C_1\right]$$
$$+ \gamma_2 P_{\theta_i}\left[V(\mathbf{Z}) = C_2\right] = \alpha, \quad i = 1, 2, \ \text{and}\ V(\mathbf{z}) = \sum_{j=1}^n T(x_j). \quad (25)$$

13.3 UMP Tests for Testing Certain Composite Hypotheses

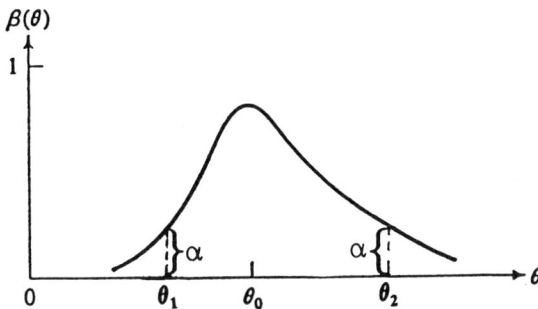

Figure 13.4 $H: \theta \leq \theta_1$ or $\theta \geq \theta_2$, $A: \theta_1 < \theta < \theta_2$.

If Q is decreasing, the test is given again by (24) and (25) with $C_1 < V(\mathbf{z}) < C_2$ replaced by $V(\mathbf{z}) < C_1$ or $V(\mathbf{z}) > C_2$.

It can also be shown that (in nondegenerate cases) the function $\beta(\theta) = E_\theta \phi(\mathbf{Z})$, $\theta \in \Omega$ for the problem discussed in Theorem 3, increases for $\theta \leq \theta_0$ and decreases for $\theta \geq \theta_0$ for some $\theta_1 < \theta_0 < \theta_2$ (see also Fig. 13.4).

Theorems 2 and 3 are illustrated by a number of examples below. In order to avoid trivial repetitions, we mention once and for all that the hypotheses to be tested are $H: \theta \in \omega = \{\theta \in \Omega; \theta \leq \theta_0\}$ against $A: \theta \in \omega^c$ and $H': \theta \in \omega = \{\theta \in \Omega; \theta \leq \theta_1 \text{ or } \theta \geq \theta_2\}$ against $A': \theta \in \omega^c$; $\theta_0, \theta_1, \theta_2 \in \Omega$ and $\theta_1 < \theta_2$. The level of significance is $\alpha (0 < \alpha < 1)$.

EXAMPLE 6 Let $X_1, \ldots, X_n$ be i.i.d. r.v.'s from $B(1, \theta)$, $\theta \in \Omega = (0, 1)$. Here

$$V(\mathbf{z}) = \sum_{j=1}^{n} x_j \quad \text{and} \quad Q(\theta) = \log \frac{\theta}{1-\theta}$$

is increasing since $\theta/(1 - \theta)$ is so. Therefore, on account of the corollary to Theorem 2, the UMP test for testing H is given by

$$\phi(\mathbf{z}) = \begin{cases} 1, & \text{if } \sum_{j=1}^{n} x_j > C \\ \gamma, & \text{if } \sum_{j=1}^{n} x_j = C \\ 0, & \text{otherwise,} \end{cases} \quad (26)$$

where C and γ are determined by

$$E_{\theta_0} \phi(\mathbf{Z}) = P_{\theta_0}(X > C) + \gamma P_{\theta_0}(X = C) = \alpha, \quad (27)$$

and

$$X = \sum_{j=1}^{n} X_j \quad \text{is} \quad B(n, \theta).$$

For a numerical application, let $\theta_0 = 0.5$, $\alpha = 0.01$ and $n = 25$. Then one has

$$P_{0.5}(X > C) + \gamma P_{0.5}(X = C) = 0.01.$$

The Binomial tables provided the values $C = 18$ and $\gamma = \frac{27}{143}$. The power of the test at $\theta = 0.75$ is

$$\beta_\phi(0.75) = P_{0.75}(X > 18) + \frac{27}{143} P_{0.75}(X = 18) = 0.5923.$$

By virtue of Theorem 3, for testing H' the UMP test is given by

$$\phi(\mathbf{z}) = \begin{cases} 1 & \text{if } C_1 < \sum_{j=1}^{n} x_j < C_2 \\ \gamma_i & \text{if } \sum_{j=1}^{n} x_j = C_i \quad (i = 1, 2) \\ 0, & \text{otherwise,} \end{cases}$$

with C_1, C_2 and γ_1, γ_2 defined by

$$E_{\theta_i}\phi(\mathbf{Z}) = P_{\theta_i}(C_1 < X < C_2) + \gamma_1 P_{\theta_i}(X = C_1) + \gamma_2 P_{\theta_i}(X = C_2) = \alpha, \quad i = 1, 2.$$

Again for a numerical application, take $\theta_1 = 0.25$, $\theta_2 = 0.75$, $\alpha = 0.05$ and $n = 25$. One has then

$$P_{0.25}(C_1 < X < C_2) + \gamma_1 P_{0.25}(X = C_1) + \gamma_2 P_{0.25}(X = C_2) = 0.05$$
$$P_{0.75}(C_1 < X < C_2) + \gamma_1 P_{0.75}(X = C_1) + \gamma_2 P_{0.75}(X = C_2) = 0.05.$$

For $C_1 = 10$ and $C_2 = 15$, one has after some simplifications

$$416\gamma_1 + 2\gamma_2 = 205$$
$$2\gamma_1 + 416\gamma_2 = 205,$$

from which we obtain

$$\gamma_1 = \gamma_2 = \frac{205}{418} \approx 0.4904.$$

The power of the test at $\theta = 0.5$ is

$$\beta_\phi(0.5) = P_{0.5}(10 < X < 15) + \frac{205}{418}\left[P_{0.5}(X = 10) + P_{0.5}(X = 15)\right] = 0.6711.$$

EXAMPLE 7 Let $X_1, \ldots, X_n$ be i.i.d. r.v.'s from $P(\theta)$, $\theta \in \Omega = (0, \infty)$. Here $V(\mathbf{z}) = \sum_{j=1}^{n} x_j$ and $Q(\theta) = \log \theta$ is increasing. Therefore the UMP test for testing H is again given by (26) and (27), where now X is $P(n\theta)$.

For a numerical example, take $\theta_0 = 0.5$, $\alpha = 0.05$ and $n = 10$. Then, by means of the Poisson tables, we find $C = 9$ and

$$\gamma = \frac{182}{363} \approx 0.5014.$$

The power of the test at $\theta = 1$ is $\beta_\phi(1) = 0.6048$.

EXAMPLE 8 Let $X_1, \ldots, X_n$ be i.i.d. r.v.'s from $N(\theta, \sigma^2)$ with σ^2 known. Here

$$V(\mathbf{z}) = \sum_{j=1}^{n} x_j \quad \text{and} \quad Q(\theta) = \frac{1}{\sigma^2}\theta$$

is increasing. Therefore for testing H the UMP test is given by (dividing by n)

$$\phi(\mathbf{z}) = \begin{cases} 1, & \text{if } \bar{x} > C \\ 0, & \text{otherwise,} \end{cases}$$

where C is determined by

$$E_{\theta_0}\phi(\mathbf{z}) = P_{\theta_0}(\bar{X} > C) = \alpha,$$

and $\bar{X}$ is $N(\theta, \sigma^2/n)$. (See also Figs. 13.5 and 13.6.)

The power of the test, as is easily seen, is given by

$$\beta_\phi(\theta) = 1 - \Phi\left[\frac{\sqrt{n}(C - \theta)}{\sigma}\right].$$

For instance, for $\sigma = 2$ and $\theta_0 = 20$, $\alpha = 0.05$ and $n = 25$, one has $C = 20.66$. For $\theta = 21$, the power of the test is

$$\beta_\phi(21) = 0.8023.$$

On the other hand, for testing H' the UMP test is given by

$$\phi(\mathbf{z}) = \begin{cases} 1, & \text{if } C_1 < \bar{x} < C_2 \\ 0, & \text{otherwise,} \end{cases}$$

where C_1, C_2 are determined by

$$E_{\theta_i}\phi(\mathbf{z}) = P_{\theta_i}(C_1 < \bar{X} < C_2) = \alpha, \quad i = 1, 2.$$

(See also Fig. 13.7.)

The power of the test is given by

$$\beta_\phi(\theta) = \Phi\left[\frac{\sqrt{n}(C_2 - \theta)}{\sigma}\right] - \Phi\left[\frac{\sqrt{n}(C_1 - \theta)}{\sigma}\right].$$

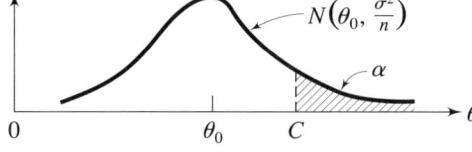

Figure 13.5 $H: \theta \leq \theta_0$, $A: \theta > \theta_0$.

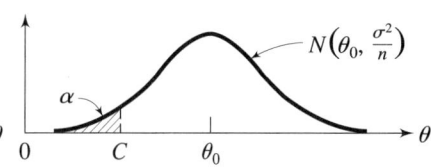

Figure 13.6 $H: \theta \geq \theta_0$, $A: \theta < \theta_0$.

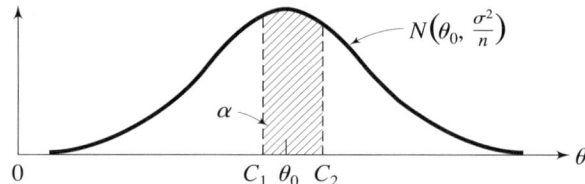

Figure 13.7 $H': \theta \leq \theta_1$ or $\theta \geq \theta_2$, $A': \theta_1 < \theta < \theta_2$.

For instance, for $\sigma = 2$ and $\theta_1 = -1$, $\theta_2 = 1$, $\alpha = 0.05$ and $n = 25$, one has $C_1 = -0.344$, $C_2 = 0.344$, and for $\theta = 0$, the power of the test is $\beta_\phi(0) = 0.610$.

EXAMPLE 9 Let $X_1, \ldots, X_n$ be i.i.d. r.v.'s from $N(\mu, \theta)$ with μ known. Then $V(\mathbf{z}) = \sum_{j=1}^n (x_j - \mu)^2$ and $Q(\theta) = -1/(2\theta)$ is increasing. Therefore for testing H, the UMP test is given by

$$\phi(\mathbf{z}) = \begin{cases} 1, & \text{if } \sum_{j=1}^n (x_j - \mu)^2 > C \\ 0, & \text{otherwise,} \end{cases}$$

where C is determined by

$$E_{\theta_0} \phi(\mathbf{Z}) = P_{\theta_0}\left[\sum_{j=1}^n (X_j - \mu)^2 > C\right] = \alpha.$$

The power of the test, as is easily seen, is given by

$$\beta_\phi(\theta) = 1 - P(\chi_n^2 < C/\theta) \quad \text{(independent of } \mu\text{!)}.$$

(See also Figs. 13.8 and 13.9; χ_n^2 stands for an r.v. distribution as χ_n^2.)

For a numerical example, take $\theta_0 = 4$, $\alpha = 0.05$ and $n = 25$. Then one has $C = 150.608$, and for $\theta = 12$, the power of the test is $\beta_\phi(12) = 0.980$.

On the other hand, for testing H' the UMP test is given by

$$\phi(\mathbf{z}) = \begin{cases} 1, & \text{if } C_1 < \sum_{j=1}^n (x_j - \mu)^2 < C_2 \\ 0, & \text{otherwise,} \end{cases}$$

where C_1, C_2 are determined by

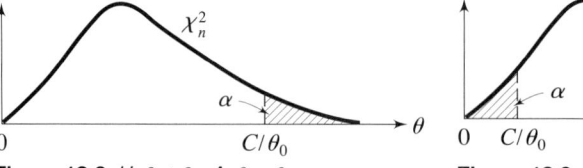

Figure 13.8 $H: \theta \leq \theta_0$, $A: \theta > \theta_0$.

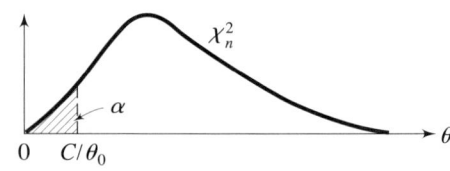

Figure 13.9 $H: \theta \geq \theta_0$, $A: \theta < \theta_0$.

$$E_{\theta_i}\phi(\mathbf{Z}) = P_{\theta_i}\left[C_1 < \sum_{j=1}^{n}(X_j - \mu)^2 < C_2\right] = \alpha, \quad i = 1, 2.$$

The power of the test, as is easily seen, is given by

$$\beta_\phi(\theta) = P\left(\chi_n^2 < \frac{C_2}{\theta}\right) - P\left(\chi_n^2 < \frac{C_1}{\theta}\right) \quad \text{(independent of } \mu\text{!)}.$$

For instance, for $\theta_1 = 1$, $\theta_2 = 3$, $\alpha = 0.01$ and $n = 25$, we have

$$P(\chi_{25}^2 < C_2) - P(\chi_{25}^2 < C_1) = 0.01, \quad P\left(\chi_{25}^2 < \frac{C_2}{3}\right) - P\left(\chi_{25}^2 < \frac{C_1}{3}\right) = 0.01$$

and C_1, C_2 are determined from the Chi-square tables (by trial and error).

Exercises

13.3.1 Let $X_1, \ldots, X_n$ be i.i.d. r.v.'s with p.d.f. f given below. In each case, show that the joint p.d.f. of the X's has the MLR property in $V = V(x_1, \ldots, x_n)$ and identity V.

i) $f(x; \theta) = \dfrac{\theta^\alpha}{\Gamma(\alpha)} x^{\alpha-1} e^{-\theta x} I_{(0,\infty)}(x)$, $\theta \in \Omega = (0, \infty)$, $\alpha = \text{known}$ (>0);

ii) $f(x; \theta) = \theta^r \dbinom{r+x-1}{x}(1-\theta)^x I_A(x)$, $A = \{0, 1, \ldots\}$, $\theta \in \Omega = (0, 1)$.

13.3.2 Refer to Example 8 and show that, for testing the hypotheses H and H' mentioned there, the power of the respective tests is given by

$$\beta_\phi(\theta) = 1 - \Phi\left[\frac{\sqrt{n}(C - \theta)}{\sigma}\right]$$

and

$$\beta_\phi(\theta) = \Phi\left[\frac{\sqrt{n}(C_2 - \theta)}{\sigma}\right] - \Phi\left[\frac{\sqrt{n}(C_1 - \theta)}{\sigma}\right]$$

as asserted.

13.3.3 The length of life X of a 50-watt light bulb of a certain brand may be assumed to be a normally distributed r.v. with unknown mean μ and known s.d. $\sigma = 150$ hours. Let $X_1, \ldots, X_{25}$ be independent r.v.'s distributed as X and suppose that $\bar{x} = 1{,}730$ hours. Test the hypothesis $H: \mu = 1{,}800$ against the alternative $A: \mu < 1{,}800$ at level of significance $\alpha = 0.01$.

13.3.4 The rainfall X at a certain station during a year may be assumed to be a normally distributed r.v. with s.d. $\sigma = 3$ inches and unknown mean μ. For the past 10 years, the record provides the following rainfalls: $x_1 = 30.5$, $x_2 = 34.1$, $x_3 = 27.9$, $x_4 = 29.4$, $x_5 = 35.0$, $x_6 = 26.9$, $x_7 = 30.2$, $x_8 = 28.3$, $x_9 = 31.7$, $x_{10} = 25.8$. Test the hypothesis $H: \mu = 30$ against the alternative $A: \mu < 30$ at level of significance $\alpha = 0.05$.

13.3.5 Refer to Example 9 and show that, for testing the hypotheses H and H' mentioned there, the power of the respective tests is given by

$$\beta_\phi(\theta) = 1 - P\left(\chi_n^2 < \frac{C}{\theta}\right) \quad \text{and} \quad \beta_\phi(\theta) = P\left(\chi_n^2 < \frac{C_2}{\theta}\right) - P\left(\chi_n^2 < \frac{C_1}{\theta}\right)$$

as asserted.

13.3.6 Let $X_1, \ldots, X_{25}$ be independent r.v.'s distributed as $N(0, \sigma^2)$. Test the hypothesis $H: \sigma \leq 2$ against the alternative $A: \sigma > 2$ at level of significance $\alpha = 0.05$. What does the relevant test become for $\sum_{j=1}^{25} x_j^2 = 120$, where x_j is the observed value of X_j, $j = 1, \ldots, 25$.

13.3.7 In a certain university 400 students were chosen at random and it was found that 95 of them were women. On the basis of this, test the hypothesis H that the proportion of women is 25% against the alternative A that is less than 25% at level of significance $\alpha = 0.05$. Use the CLT in order to determine the cut-off point.

13.3.8 Let $X_1, \ldots, X_n$ be independent r.v.'s distributed as $B(1, p)$. For testing the hypothesis $H: p \leq \frac{1}{2}$ against the alternative $A: p > \frac{1}{2}$, suppose that $\alpha = 0.05$ and $\beta(\frac{7}{8}) = 0.95$. Use the CLT in order to determine the required sample size n.

13.3.9 Let X be an r.v. distributed as $B(n, \theta)$, $\theta \in \Omega = (0, 1)$.

i) Derive the UMP test for testing the hypothesis $H: \theta \leq \theta_0$ against the alternative $A: \theta > \theta_0$ at level of significance α.

ii) What does the test in (i) become for $n = 10$, $\theta_0 = 0.25$ and $\alpha = 0.05$?

iii) Compute the power at $\theta_1 = 0.375, 0.500, 0.625, 0.750, 0.875$.

Now let $\theta_0 = 0.125$ and $\alpha = 0.1$ and suppose that we are interested in securing power at least 0.9 against the alternative $\theta_1 = 0.25$.

iv) Determine the minimum sample size n required by using the Binomial tables (if possible) and also by using the CLT.

13.3.10 The number X of fatal traffic accidents in a certain city during a year may be assumed to be an r.v. distributed as $P(\lambda)$. For the latest year $x = 4$, whereas for the past several years the average was 10. Test whether it has been an improvement, at level of significance $\alpha = 0.01$. First, write out the expression for the exact determination of the cut-off point, and secondly, use the CLT for its numerical determination.

13.3.11 Let X be the number of times that an electric light switch can be turned on and off until failure occurs. Then X may be considered to be an r.v. distributed as Negative Binomial with $r = 1$ and unknown p. Let $X_1, \ldots, X_{15}$ be independent r.v.'s distributed as X and suppose that $\bar{x} = 15{,}150$. Test the hypothesis $H: p = 10^{-4}$ against the alternative $A: p > 10^{-4}$ at level of significance $\alpha = 0.05$.

13.3.12 Let $X_1, \ldots, X_n$ be independent r.v.'s with p.d.f. f given by

$$f(x;\ \theta) = \frac{1}{\theta} e^{-x/\theta} I_{(0,\infty)}(x), \qquad \theta \in \Omega = (0,\ \infty).$$

i) Derive the UMP test for testing the hypothesis $H: \theta \geq \theta_0$ against the alternative $A: \theta < \theta_0$ at level of significance α;

ii) Determine the minimum sample size n required to obtain power at least 0.95 against the alternative $\theta_1 = 500$ when $\theta_0 = 1{,}000$ and $\alpha = 0.05$.

13.4 UMPU Tests for Testing Certain Composite Hypotheses

In Section 13.3, it was stated that under the assumptions of Theorem 3, for testing $H: \theta \in \omega = \{\theta \in \Omega;\ \theta \leq \theta_1 \text{ or } \theta \geq \theta_2\}$ against $A: \theta \in \omega^c$, a UMP test exists. It is then somewhat surprising that, if the roles of H and A are interchanged, a UMP test does not exist any longer. Also under the assumptions of Theorem 2, for testing $H_0: \theta = \theta_0$ against $A'': \theta \neq \theta_0$ a UMP does not exist. This is so because the test given by (19) and (19′) is UMP for $\theta > \theta_0$ but is worse than the trivial test $\phi(\mathbf{z}) = \alpha$ for $\theta < \theta_0$. Thus there is no unique test which is UMP for all $\theta \neq \theta_0$.

The above observations suggest that in order to find a test with some optimal property, one would have to restrict oneself to a smaller class of tests. This leads us to introducing the concept of an unbiased test.

DEFINITION 6 Let $X_1, \ldots, X_n$ be i.i.d. r.v.'s with p.d.f. $f(\cdot;\ \boldsymbol{\theta})$, $\boldsymbol{\theta} \in \boldsymbol{\Omega}$ and let $\boldsymbol{\omega} \subset \boldsymbol{\Omega} \subseteq \mathbb{R}^r$. Then for testing the hypothesis $H: \boldsymbol{\theta} \in \boldsymbol{\omega}$ against the alternative $A: \boldsymbol{\theta} \in \boldsymbol{\omega}^c$ at level of significance α, a test ϕ based on $X_1, \ldots, X_n$ is said to be *unbiased* if $E_{\boldsymbol{\theta}}\phi(X_1, \ldots, X_n) \leq \alpha$ for all $\boldsymbol{\theta} \in \boldsymbol{\omega}$ and $E_{\boldsymbol{\theta}}\phi(X_1, \ldots, X_n) \geq \alpha$ for all $\boldsymbol{\theta} \in \boldsymbol{\omega}^c$.

That is, the defining property of an unbiased test is that the type-I error probability is *at most* α and the power of the test is *at least* α.

DEFINITION 7 In the notation of Definition 6, a test is said to be *uniformly most powerful unbiased* (UMPU) if it is UMP within the class of all unbiased tests.

REMARK 3 A UMP test is always UMPU. In fact, in the first place it is unbiased because it is at least as powerful as the test which is identically equal to α. Next, it is UMPU because it is UMP within a class including the class of unbiased tests.

For certain important classes of distributions and certain hypotheses, UMPU tests do exist. The following theorem covers cases of this sort, but it will be presented without a proof.

THEOREM 4 Let $X_1, \ldots, X_n$ be i.i.d. r.v.'s with p.d.f. $f(\cdot; \theta)$ given by

$$f(x; \theta) = C(\theta) e^{\theta T(x)} h(x), \quad \theta \in \Omega \subseteq \mathbb{R}. \tag{28}$$

Let $\omega = \{\theta \in \Omega; \theta_1 \leq \theta \leq \theta_2\}$ and $\omega_0 = \{\theta_0\}$, where $\theta_0, \theta_1, \theta_2 \in \Omega$ and $\theta_1 < \theta_2$.

Then for testing the hypothesis $H: \theta \in \omega$ against $A: \theta \in \omega^c$ and the hypothesis $H_0: \theta \in \omega_0$ against $A_0: \theta \in \omega_0^c$ at level of significance α, there exist UMPU tests which are given by

$$\phi(\mathbf{z}) = \begin{cases} 1, & \text{if } V(\mathbf{z}) < C_1 \text{ or } V(\mathbf{z}) > C_2 \\ \gamma_i, & \text{if } V(\mathbf{z}) = C_i, \; (i = 1, 2) \; (C_1 < C_2) \\ 0, & \text{otherwise,} \end{cases}$$

where the constants $C_i, \gamma_i, i = 1, 2$ are given by

$$E_{\theta_i} \phi(\mathbf{Z}) = \alpha, \quad i = 1, 2 \quad \text{for} \quad H,$$

and

$$E_{\theta_0} \phi(\mathbf{Z}) = \alpha, \; E_{\theta_0}\left[V(\mathbf{Z})\phi(\mathbf{Z})\right] = \alpha E_{\theta_0} V(\mathbf{Z}) \quad \text{for} \quad H_0.$$

(Recall that $\mathbf{z} = (x_1, \ldots, x_n)'$, $\mathbf{Z} = (X_1, \ldots, X_n)'$ and $V(\mathbf{z}) = \sum_{j=1}^{n} T(x_j)$.)

Furthermore, it can be shown that the function $\beta_\phi(\theta) = E_\theta \phi(\mathbf{Z})$, $\theta \in \Omega$ (except for degenerate cases) is decreasing for $\theta \leq \theta_0$ and increasing for $\theta \geq \theta_0$ for some $\theta_1 < \theta_0 < \theta_2$ (see also Fig. 13.10).

REMARK 4 We would expect that cases like Binomial, Poisson and Normal would fall under Theorem 4, while they seemingly do not. However, a simple reparametrization of the families brings them under the form (28). In fact, by Examples and Exercises of Chapter 11 it can be seen that all these families are of the exponential form

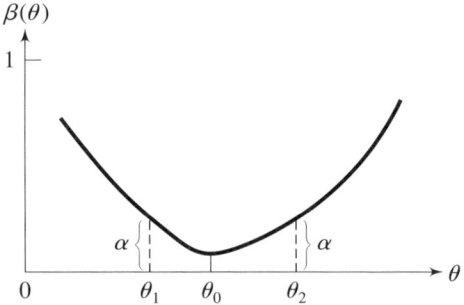

Figure 13.10 $H: \theta_1 \leq \theta \leq \theta_2$, $A: \theta < \theta_1$ or $\theta > \theta_2$.

13.4 UMPU Tests for Testing Certain Composite Hypotheses

$$f(x;\,\theta) = C(\theta)e^{Q(\theta)T(x)}h(x).$$

i) *For the Binomial case*, $Q(\theta) = \log[\theta/(1-\theta)]$. Then by setting $\log[\theta/(1-\theta)] = \tau$, the family is brought under the form (28). From this transformation, we get $\theta = e^{\tau}/(1 + e^{\tau})$ and the hypotheses $\theta_1 \le \theta \le \theta_2$, $\theta = \theta_0$ become equivalently, $\tau_1 \le \tau \le \tau_2$, $\tau = \tau_0$, where

$$\tau_i = \log\frac{\theta_i}{1-\theta_i}, \quad i = 0, 1, 2.$$

ii) *For the Poisson case*, $Q(\theta) = \log\theta$ and the transformation $\log\theta = \tau$ brings the family under the form (28). The transformation implies $\theta = e^{\tau}$ and the hypotheses $\theta_1 \le \theta \le \theta_2$, $\theta = \theta_0$ become, equivalently, $\tau_1 \le \tau \le \tau_2$, $\tau = \tau_0$ with $\tau_i = \log\theta_i$, $i = 0, 1, 2$.

iii) *For the Normal case with σ known and $\mu = \theta$*, $Q(\theta) = (1/\sigma^2)\theta$ and the factor $1/\sigma^2$ may be absorbed into $T(x)$.

iv) *For the Normal case with μ known and $\sigma^2 = \theta$*, $Q(\theta) = -1/(2\theta)$ and the transformation $-1/(2\theta) = \tau$ brings the family under the form (28) again. Since $\theta = -1/(2\tau)$, the hypotheses $\theta_1 \le \theta \le \theta_2$ and $\theta = \theta_0$ become, equivalently, $\tau_1 \le \tau \le \tau_2$ and $\tau = \tau_0$, where $\tau_i = -1/(2\theta_i)$, $i = 0, 1, 2$.

As an application to Theorem 4 and for later reference, we consider the following example. The level of significance will be α.

EXAMPLE 10 Suppose $X_1, \ldots, X_n$ are i.i.d. r.v.'s from $N(\mu, \sigma^2)$. Let σ be known and set $\mu = \theta$. Suppose that we are interested in testing the hypothesis $H: \theta = \theta_0$ against the alternative $A: \theta \ne \theta_0$. In the present case,

$$T(x) = \frac{1}{\sigma^2}x,$$

so that

$$V(\mathbf{z}) = \sum_{j=1}^{n} T(x_j) = \frac{1}{\sigma^2}\sum_{j=1}^{n} x_j = \frac{n}{\sigma^2}\bar{x}.$$

Therefore, by Theorem 4, the UMPU test is as follows:

$$\phi(\mathbf{z}) = \begin{cases} 1, & \text{if } \frac{n}{\sigma^2}\bar{x} < C_1 \text{ or } \frac{n}{\sigma^2}\bar{x} > C_2 \\ 0, & \text{otherwise,} \end{cases}$$

where C_1, C_2 are determined by

$$E_{\theta_0}\phi(\mathbf{Z}) = \alpha, \quad E_{\theta_0}\left[V(\mathbf{Z})\phi(\mathbf{Z})\right] = \alpha E_{\theta_0}V(\mathbf{Z}).$$

Now ϕ can be expressed equivalently as follows:

$$\phi(\mathbf{z}) = \begin{cases} 1, & \text{if } \dfrac{\sqrt{n}(\bar{x}-\theta_0)}{\sigma} < C_1' \text{ or } \dfrac{\sqrt{n}(\bar{x}-\theta_0)}{\sigma} > C_2' \\ 0, & \text{otherwise,} \end{cases}$$

where

$$C_1' = \frac{\sigma C_1}{\sqrt{n}} - \frac{\sqrt{n}\theta_0}{\sigma}, \quad C_2' = \frac{\sigma C_2}{\sqrt{n}} - \frac{\sqrt{n}\theta_0}{\sigma}.$$

On the other hand, under H, $\sqrt{n}(\bar{X}-\theta_0)/\sigma$ is $N(0, 1)$. Therefore, because of symmetry $C_1' = -C_2' = -C$, say ($C > 0$). Also

$$\frac{\sqrt{n}(\bar{x}-\theta_0)}{\sigma} < -C \quad \text{or} \quad \frac{\sqrt{n}(\bar{x}-\theta_0)}{\sigma} > C$$

is equivalent to

$$\left[\frac{\sqrt{n}(\bar{x}-\theta_0)}{\sigma}\right]^2 > C$$

and, of course, $[\sqrt{n}(\bar{X}-\theta_0)/\sigma]^2$ is χ_1^2, under H. By summarizing then, we have

$$\phi(\mathbf{z}) = \begin{cases} 1, & \text{if } \left[\dfrac{\sqrt{n}(\bar{x}-\theta_0)}{\sigma}\right]^2 > C \\ 0, & \text{otherwise,} \end{cases}$$

where C is determined by

$$P(\chi_1^2 > C) = \alpha.$$

In many situations of practical importance, the underlying p.d.f. involves a real-valued parameter θ in which we are exclusively interested, and in addition some other real-valued parameters $\vartheta_1, \ldots, \vartheta_k$ in which we have no interest. These latter parameters are known as *nuisance parameters*. More explicitly, the p.d.f. is of the following exponential form:

$$f(x; \theta, \vartheta_1, \ldots, \vartheta_k) = C(\theta, \vartheta_1, \ldots, \vartheta_k) \exp[\theta T(x) \\ + \vartheta_1 T_1(x) + \cdots + \vartheta_k T_k(x_k)] h(x), \tag{29}$$

where $\theta \in \Omega \subseteq \mathbb{R}$, $\vartheta_1, \ldots, \vartheta_k$ are real-valued and $h(x) > 0$ on a set independent of all parameters involved.

Let $\theta_0, \theta_1, \theta_2 \in \Omega$ with $\theta_1 < \theta_2$. Then the (composite) hypotheses of interest are the following ones, where $\vartheta_1, \ldots, \vartheta_k$ are left unspecified.

$$\begin{cases} H_1: \theta \in \omega = \{\theta \in \Omega;\ \theta \leq \theta_0\} \\ H_1': \theta \in \omega = \{\theta \in \Omega;\ \theta \geq \theta_0\} \\ H_2: \theta \in \omega = \{\theta \in \Omega;\ \theta \leq \theta_1\ \text{or}\ \theta \geq \theta_2\} \\ H_3: \theta \in \omega = \{\theta \in \Omega;\ \theta_1 \leq \theta \leq \theta_2\} \\ H_4: \theta \in \omega = \{\theta_0\} \end{cases} A_i(A_i'): \theta \in \omega^c,\ i = 1, \ldots, 4.$$

(30)

We may now formulate the following theorem, whose proof is omitted.

THEOREM 5 Let $X_1, \ldots, X_n$ be i.i.d. r.v.'s with p.d.f. given by (29). Then, under some additional regularity conditions, there exist UMPU tests with level of significance α for testing any one of the hypotheses $H_i(H_i')$ against the alternatives $A_i(A_i')$, $i = 1, \ldots, 4$, respectively.

Because of the special role that normal populations play in practice, the following two sections are devoted to presenting simple tests for the hypotheses specified in (30). Some of the tests will be arrived at again on the basis of the principle of likelihood ratio to be discussed in Section 7. However, the optimal character of the tests will not become apparent by that process.

Exercises

13.4.1 A coin, with probability p of falling heads, is tossed independently 100 times and 60 heads are observed. Use the UMPU test for testing the hypothesis $H: p = \frac{1}{2}$ against the alternative $A: p \neq \frac{1}{2}$ at level of significance $\alpha = 0.1$.

13.4.2 Let X_1, X_2, X_3 be independent r.v.'s distributed as $B(1, p)$. Derive the UMPU test for testing $H: p = 0.25$ against $A: p \neq 0.25$ at level of significance α. Determine the test for $\alpha = 0.05$.

13.5 Testing the Parameters of a Normal Distribution

In the present section, $X_1, \ldots, X_n$ are assumed to be i.i.d. r.v.'s from $N(\mu, \sigma^2)$, where both μ and σ^2 are unknown. One of the parameters at a time will be the parameter of interest, the other serving as a nuisance parameter. Under appropriate reparametrization, as indicated in Remark 5, the family is brought under the form (29). Also the remaining (unspecified) regularity conditions in Theorem 5 can be shown to be satisfied here, and therefore the conclusion of the theorem holds.

All tests to be presented below are UMPU, except for the first one which is UMP. This is a consequence of Theorem 5 (except again for the UMP test).

Whenever convenient, we will also use the notation $\mathbf{z}$ and $\mathbf{Z}$ instead of $(x_1, \ldots, x_n)'$ and $(X_1, \ldots, X_n)'$, respectively. Finally, all tests will be of level α.

13.5.1 Tests about the Variance

PROPOSITION 2 For testing $H_1: \sigma \leq \sigma_0$ against $A_1: \sigma > \sigma_0$, the test given by

$$\phi(\mathbf{z}) = \begin{cases} 1, & \text{if } \sum_{j=1}^{n}(x_j - \bar{x})^2 > C \\ 0, & \text{otherwise,} \end{cases} \quad (31)$$

where C is determined by

$$P\left(\chi_{n-1}^2 > C/\sigma_0^2\right) = \alpha, \quad (32)$$

is UMP. The test given by (31) and (32) with reversed inequalities is UMPU for testing $H_1': \sigma \geq \sigma_0$ against $A_1': \sigma < \sigma_0$.

The power of the tests is easily determined by the fact that $(1/\sigma^2) \sum_{j=1}^{n} (X_j - \bar{X})^2$ is χ_{n-1}^2 when σ obtains (that is, σ is the true s.d.). For example, for $n = 25$, $\sigma_0 = 3$ and $\alpha = 0.05$, we have for H_1, $C/9 = 36.415$, so that $C = 327.735$. The power of the test at $\sigma = 5$ is equal to $P(\chi_{24}^2 > 13.1094) = 0.962$.

For H_1', $C/9 = 13.848$, so that $C = 124.632$, and the power at $\sigma = 2$ is $P\chi_{24}^2$ $(< 31.158) = 0.8384$.

PROPOSITION 3 For testing $H_2: \sigma \leq \sigma_1$ or $\sigma \geq \sigma_2$, against $A_2: \sigma_1 < \sigma < \sigma_2$, the test given by

$$\phi(\mathbf{z}) = \begin{cases} 1, & \text{if } C_1 < \sum_{j=1}^{n}(x_j - \bar{x})^2 < C_2 \\ 0, & \text{otherwise,} \end{cases} \quad (33)$$

where C_1, C_2 are determined by

$$P\left(C_1/\sigma_i^2 < \chi_{n-1}^2 < C_2/\sigma_i^2\right) = \alpha, \quad i = 1, 2, \quad (34)$$

is UMPU. The test given by (33) and (34), where the inequalities $C_1 < \sum_{j=1}^{n} (x_j - \bar{x})^2 < C_2$ are replaced by

$$\sum_{j=1}^{n}(x_j - \bar{x})^2 < C_1 \quad \text{or} \quad \sum_{j=1}^{n}(x_j - \bar{x})^2 > C_2,$$

and similarly for (34), is UMPU for testing $H_3: \sigma_1 \leq \sigma \leq \sigma_2$ against $A_3: \sigma < \sigma_1$ or $\sigma > \sigma_2$. Again, the power of the tests is determined by the fact that $(1/\sigma^2) \sum_{j=1}^{n} (X_j - \bar{X})^2$ is χ_{n-1}^2 when σ obtains.

For example, for H_2 and for $n = 25$, $\sigma_1 = 2$, $\sigma_2 = 3$ and $\alpha = 0.05$, C_1, C_2 are determined by

$$\begin{cases} P\left(\chi_{24}^2 > \dfrac{C_1}{4}\right) - P\left(\chi_{24}^2 > \dfrac{C_2}{4}\right) = 0.05 \\ P\left(\chi_{24}^2 > \dfrac{C_1}{9}\right) - P\left(\chi_{24}^2 > \dfrac{C_2}{9}\right) = 0.05 \end{cases}$$

from the Chi-square tables (by trial and error).

PROPOSITION 4 For testing $H_4: \sigma = \sigma_0$ against $A_4: \sigma \neq \sigma_0$, the test given by

$$\phi(\mathbf{z}) = \begin{cases} 1, & \text{if } \sum_{j=1}^{n}(x_j - \bar{x})^2 < C_1 \text{ or } \sum_{j=1}^{n}(x_j - \bar{x})^2 > C_2 \\ 0, & \text{otherwise,} \end{cases}$$

where C_1, C_2 are determined by

$$\int_{c_1/\sigma_0^2}^{c_2/\sigma_0^2} g(t)\,dt = \frac{1}{n-1}\int_{c_1/\sigma_0^2}^{c_2/\sigma_0^2} tg(t)\,dt = 1 - \alpha,$$

g being the p.d.f. of a χ_{n-1}^2 distribution, is UMPU.

The power of the test is determined as in the previous cases.

REMARK 5 The popular *equal tail* test is not UMPU; it is a close approximation to the UMPU test when n is large.

13.5.2 Tests about the Mean

In connection with the problem of testing the mean, UMPU tests exist in a simple form and are explicitly given for the following three cases: $\mu \leq \mu_0, \mu \geq \mu_0$ and $\mu = \mu_0$.

To facilitate the writing, we set

$$t(\mathbf{z}) = \frac{\sqrt{n}(\bar{x} - \mu_0)}{\sqrt{\dfrac{1}{n-1}\sum_{j=1}^{n}(x_j - \bar{x})^2}}. \tag{35}$$

PROPOSITION 5 For testing $H_1: \mu \leq \mu_0$ against $A_1: \mu > \mu_0$, the test given by

$$\phi(\mathbf{z}) = \begin{cases} 1, & \text{if } t(\mathbf{z}) > C \\ 0, & \text{otherwise,} \end{cases} \tag{36}$$

where C is determined by

$$P(t_{n-1} > C) = \alpha, \tag{37}$$

is UMPU. The test given by (36) and (37) with reversed inequalities is UMPU for testing $H_1': \mu \geq \mu_0$ against $A_1': \mu < \sigma_0$; $t(\mathbf{z})$ is given by (35). (See also Figs. 13.11 and 13.12; t_{n-1} stands for an r.v. distributed as t_{n-1}.)

For $n = 25$ and $\alpha = 0.05$, we have $P(t_{24} > C) = 0.05$; hence $C = 1.7109$ for H_1, and $C = -1.7109$ for H_1'.

PROPOSITION 6 For testing $H_4: \mu = \mu_0$ against $A_4: \mu \neq \mu_0$, the test given by

$$\phi(\mathbf{z}) = \begin{cases} 1, & \text{if } t(\mathbf{z}) < -C \text{ or } t(\mathbf{z}) > C \quad (C > 0) \\ 0, & \text{otherwise,} \end{cases}$$

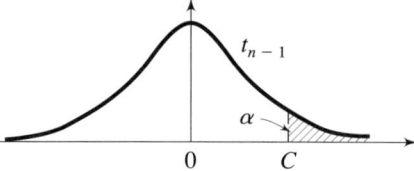

Figure 13.11 $H_1: \mu \leq \mu_0$, $A_1: \mu > \mu_0$.   **Figure 13.12** $H'_1: \mu \geq \mu_0$, $A'_1: \mu < \mu_0$.

where C is determined by

$$P(t_{n-1} > C) = \alpha/2,$$

is UMPU; $t(\mathbf{z})$ is given by (35). (See also Fig. 13.13.)

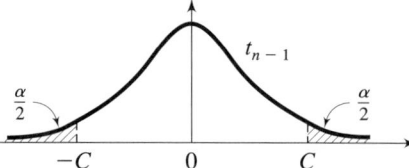

Figure 13.13 $H_4: \mu = \mu_0$, $A_4: \mu \neq \mu_0$.

For example, for $n = 25$ and $\alpha = 0.05$, we have $C = 2.0639$.

In both these last two propositions, the determination of the power involves what is known as *non-central t-distribution*, which is defined in Appendix II.

Exercises

13.5.1 The diameters of bolts produced by a certain machine are r.v.'s distributed as $N(\mu, \sigma^2)$. In order for the bolts to be usable for the intended purpose, the s.d. σ must not exceed 0.04 inch. A sample of size 16 is taken and is found that $s = 0.05$ inch. Formulate the appropriate testing hypothesis problem and carry out the test if $\alpha = 0.05$.

13.5.2 Let $X_1, \ldots, X_n$ be i.i.d. r.v.'s from $N(\mu, \sigma^2)$, where both μ and σ are unknown.

i) Derive the UMPU test for testing the hypothesis $H: \sigma = \sigma_0$ against the alternative $A: \sigma \neq \sigma_0$ at level of significance α;

ii) Carry out the test if $n = 25$, $\sigma_0 = 2$, $\sum_{j=1}^{25} (x_j - \bar{x})^2 = 24.8$, and $\alpha = 0.05$.

13.5.3 Discuss the testing hypothesis problem in Exercise 13.3.4 if both μ and σ are unknown.

13.5.4 A manufacturer claims that packages of certain goods contain 18 ounces. In order to check his claim, 100 packages are chosen at random from a large lot and it is found that

$$\sum_{j=1}^{100} x_j = 1{,}752 \quad \text{and} \quad \sum_{j=1}^{100} x_j^2 = 31{,}157.$$

Make the appropriate assumptions and test the hypothesis H that the manufacturer's claim is correct against the appropriate alternative A at level of significance $\alpha = 0.01$.

13.5.5 The diameters of certain cylindrical items produced by a machine are r.v.'s distributed as $N(\mu, 0.01)$. A sample of size 16 is taken and is found that $\bar{x} = 2.48$ inches. If the desired value for μ is 2.5 inches, formulate the appropriate testing hypothesis problem and carry out the test if $\alpha = 0.05$.

13.6 Comparing the Parameters of Two Normal Distributions

Let $X_1, \ldots, X_m$ be i.i.d. r.v.'s from $N(\mu_1, \sigma_1^2)$ and let $Y_1, \ldots, Y_n$ be i.i.d. r.v.'s from $N(\mu_2, \sigma_2^2)$. It is assumed that the two random samples are independent and that all four parameters involved are unknown. Set $\mu = \mu_1 - \mu_2$ and $\tau = \sigma_2^2/\sigma_1^2$. The problem to be discussed in this section is that of testing certain hypotheses about μ and τ. Each time either μ or τ will be the parameter of interest, the remaining parameters serving as nuisance parameters.

Writing down the joint p.d.f. of the X's and Y's and reparametrizing the family along the lines suggested in Remark 4 reveals that this joint p.d.f. has the form (29), in either one of the parameters μ or τ. Furthermore, it can be shown that the additional (but unspecified) regularity conditions of Theorem 5 are satisfied and therefore there exist UMPU tests for the hypotheses specified in (30). For some of these hypotheses, the tests have a simple form to be explicitly mentioned below. For convenient writing, we shall employ the notation

$$\mathbf{Z} = (X_1, \ldots, X_m)', \quad \mathbf{W} = (Y_1, \ldots, Y_n)'$$

for the X's and Y's, respectively, and

$$\mathbf{z} = (x_1, \ldots, x_m)', \quad \mathbf{w} = (y_1, \ldots, y_n)'$$

for their observed values.

13.6.1 Comparing the Variances of Two Normal Densities

PROPOSITION 7 For testing $H_1: \tau \leq \tau_0$ against $A_1: \tau > \tau_0$, the test given by

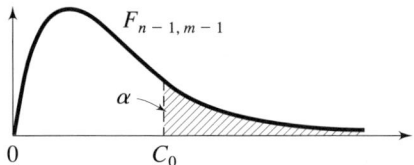

Figure 13.14 $H_1: \tau \leq \tau_0$, $A_1: \tau > \tau_0$.

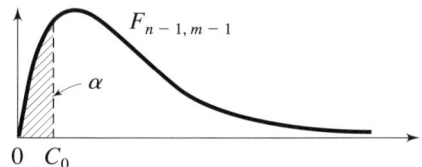

Figure 13.15 $H'_1: \tau \geq \tau_0$, $A'_1: \tau < \tau_0$.

$$\phi(\mathbf{z}, \mathbf{w}) = \begin{cases} 1, & \text{if } \dfrac{\sum_{j=1}^{n}(y_j - \bar{y})^2}{\sum_{i=1}^{m}(x_i - \bar{x})^2} > C \\ 0, & \text{otherwise,} \end{cases} \quad (38)$$

where C is determined by

$$P(F_{n-1,m-1} > C_0) = \alpha, \quad C_0 = \frac{(m-1)C}{(n-1)\tau_0}, \quad (39)$$

is UMPU. The test given by (38) and (39) with reversed inequalities is UMPU for testing $H': \tau \geq \tau_0$ against $A'_1: \tau < \tau_0$. (See also Figs. 13.14 and 13.15; $F_{n-1,m-1}$ stands for an r.v. distributed as $F_{n-1,m-1}$.)

The power of the test is easily determined by the fact that

$$\frac{\dfrac{1}{\sigma_2^2}\sum_{j=1}^{n}(Y_j - \bar{Y})^2 / n - 1}{\dfrac{1}{\sigma_1^2}\sum_{i=1}^{m}(X_j - \bar{X})^2 / m - 1} = \frac{1}{\tau} \frac{m-1}{n-1} \frac{\sum_{j=1}^{n}(Y_j - \bar{Y})^2}{\sum_{i=1}^{m}(X_i - \bar{X})^2}$$

is $F_{n-1,m-1}$ distributed when τ obtains. Thus the power of the test depends only on τ. For $m = 25$, $n = 21$, $\tau_0 = 2$ and $\alpha = 0.05$, one has $P(F_{20,24} > 5C/12) = 0.05$, hence $5C/12 = 2.0267$ and $C = 4.8640$ for H_1; for H'_1,

$$P\left(F_{20,24} < \frac{5C}{12}\right) = P\left(F_{24,20} > \frac{12}{5C}\right) = 0.05$$

implies $12/5C = 2.0825$ and hence $C = 1.1525$.

Now set

$$V(\mathbf{z}, \mathbf{w}) = \frac{\dfrac{1}{\tau_0}\sum_{j=1}^{n}(y_j - \bar{y})^2}{\sum_{i=1}^{m}(x_i - \bar{x})^2 + \dfrac{1}{\tau_0}\sum_{j=1}^{n}(y_j - \bar{y})^2}. \quad (40)$$

Then we have the following result.

PROPOSITION 8 For testing $H_4: \tau = \tau_0$ against $A_4: \tau \neq \tau_0$, the test given by

$$\phi(\mathbf{z}, \mathbf{w}) = \begin{cases} 1, & \text{if } V(\mathbf{z}, \mathbf{w}) < C_1 \text{ or } V(\mathbf{z}, \mathbf{w}) > C_2 \\ 0, & \text{otherwise,} \end{cases}$$

where C_1, C_2 are determined by

$$P\left[C_1 < B_{\frac{1}{2}(n-1), \frac{1}{2}(m-1)} < C_2\right] = P\left[C_1 < B_{\frac{1}{2}(n+1), \frac{1}{2}(m-1)} < C_2\right] = 1 - \alpha,$$

is UMPU; $V(\mathbf{z}, \mathbf{w})$ is defined by (40). (B_{r_1, r_2} stands for an r.v. distributed as Beta with r_1, r_2 degrees of freedom.) For the actual determination of C_1, C_2, we use the incomplete Beta tables. (See, for example, *New Tables of the Incomplete Gamma Function Ratio and of Percentage Points of the Chi-Square and Beta Distributions* by H. Leon Harter, Aerospace Research Laboratories, Office of Aerospace Research; also, *Tables of the Incomplete Beta-Function* by Karl Pearson, Cambridge University Press.)

13.6.2 Comparing the Means of Two Normal Densities

In the present context, it will be convenient to set

$$t(\mathbf{z}, \mathbf{w}) = \frac{\bar{y} - \bar{x}}{\sqrt{\sum_{i=1}^{m}(x_i - \bar{x})^2 + \sum_{j=1}^{n}(y_j - \bar{y})^2}}. \tag{41}$$

We shall also assume that $\sigma_1^2 = \sigma_2^2 = \sigma^2$ (unspecified).

PROPOSITION 9 For testing $H_1: \mu \leq 0$ against $A_1: \mu > 0$, the test given by

$$\phi(\mathbf{z}, \mathbf{w}) = \begin{cases} 1, & \text{if } t(\mathbf{z}, \mathbf{w}) > C \\ 0, & \text{otherwise,} \end{cases} \tag{42}$$

where C is determined by

$$P(t_{m+n-2} > C_0) = \alpha, \quad C_0 = C\sqrt{\frac{m+n-2}{(1/m)+(1/n)}}, \tag{43}$$

is UMPU. The test given by (42) and (43) with reversed inequalities is UMPU for testing $H_1': \mu \geq 0$ against $A_1': \mu < 0$; $t(\mathbf{z}, \mathbf{w})$ is given by (41). The determination of the power of the test involves a non-central t-distribution, as was also the case in Propositions 5 and 6.

For example, for $m = 15$, $n = 10$ and $\alpha = 0.05$, one has for H_1: $P(t_{23} > C\sqrt{23 \times 6}) = 0.05$; hence $C\sqrt{23 \times 6} = 1.7139$ and $C = 0.1459$. For H_1', $C = -0.1459$.

PROPOSITION 10 For testing $H_4: \mu = 0$ against $A_4: \mu \neq 0$, the test given by

$$\phi(\mathbf{z}, \mathbf{w}) = \begin{cases} 1, & \text{if } t(\mathbf{z}, \mathbf{w}) < -C \text{ or } t(\mathbf{z}, \mathbf{w}) > C \\ 0, & \text{otherwise,} \end{cases}$$

where C is determined by

$$P(t_{m+n-2} > C_0) = \alpha/2,$$

C_0 as above, is UMPU.

Again with $m = 15$, $n = 10$ and $\alpha = 0.05$, one has $P(t_{23} > C\sqrt{23 \times 6}) = 0.025$ and hence $C\sqrt{23 \times 6} = 2.0687$ and $C = 0.1762$.

Once again the determination of the power of the test involves the non-central t-distribution.

REMARK 6 In Propositions 9 and 10, if the variances are not equal, the tests presented above are not UMPU. The problem of comparing the means of two normal densities when the variances are unequal is known as the *Behrens–Fisher* problem. For this case, various tests have been proposed but we will not discuss them here.

Exercises

13.6.1 Let X_i, $i = 1, \ldots, 9$ and Y_j, $j = 1, \ldots, 10$ be independent random samples from the distributions $N(\mu_1, \sigma_1^2)$ and $N(\mu_2, \sigma_2^2)$, respectively. Suppose that the observed values of the sample s.d.'s are $s_X = 2$, $s_Y = 3$. At level of significance $\alpha = 0.05$, test the hypothesis: $H: \sigma_1 = \sigma_2$ against the alternative $A: \sigma_1 \neq \sigma_2$ and find the power of the test at $\sigma_1 = 2$, $\sigma_2 = 3$.

13.6.2 Let X_j, $j = 1, \ldots, 4$ and Y_j, $j = 1, \ldots, 4$ be two independent random samples from the distributions $N(\mu_1, \sigma_1^2)$ and $N(\mu_2, \sigma_2^2)$, respectively. Suppose that the observed values of the X's and Y's are as follows:

$x_1 = 10.1$, $x_2 = 8.4$, $x_3 = 14.3$, $x_4 = 11.7$,
$y_1 = 9.0$, $y_2 = 8.2$, $y_3 = 12.1$, $y_4 = 10.3$.

Test the hypothesis $H: \sigma_1 = \sigma_2$ against the alternative $A: \sigma_1 \neq \sigma_2$ at level of significance $\alpha = 0.05$.

13.6.3 Five resistance measurements are taken on two test pieces and the observed values (in ohms) are as follows:

$x_1 = 0.118$, $x_2 = 0.125$, $x_3 = 0.121$, $x_4 = 0.117$, $x_5 = 0.120$
$y_1 = 0.114$, $y_2 = 0.115$, $y_3 = 0.119$, $y_4 = 0.120$, $y_5 = 0.110$.

Make the appropriate assumptions and test the hypothesis $H: \sigma_1 = \sigma_2$ against the alternative $A: \sigma_1 \neq \sigma_2$ at level of significance $\alpha = 0.05$.

13.6.4 Refer to Exercise 13.6.2 and suppose it is known that $\sigma_1 = 4$ and $\sigma_2 = 3$. Test the hypothesis H that the two means do not differ by more than 1 at level of significance $\alpha = 0.05$.

13.6.5 The breaking powers of certain steel bars produced by processes A and B are r.v.'s distributed as normal with possibly different means but the same variance. A random sample of size 25 is taken from bars produced by each one of the processes, and it is found that $\bar{x} = 60$, $s_X = 6$, $\bar{y} = 65$, $s_Y = 7$. Test whether there is a difference between the two processes at the level of significance $\alpha = 0.05$.

13.6.6 Refer to Exercise 13.6.3, make the appropriate assumptions, and test the hypothesis $H: \mu_1 = \mu_2$ against the alternative $A: \mu_1 \neq \mu_2$ at level of significance $\alpha = 0.05$.

13.6.7 Let X_i, $i = 1, \ldots, n$ and Y_i, $i = 1, \ldots, n$ be independent random samples from the distributions $N(\mu_1, \sigma_1^2)$ and $N(\mu_2, \sigma_2^2)$, respectively, and suppose that all four parameters are unknown. By setting $Z_i = X_i - Y_i$, we have that the *paired* r.v.'s Z_i, $i = 1, \ldots, n$, are independent and distributed as $N(\mu, \sigma^2)$ with $\mu = \mu_1 - \mu_2$ and $\sigma^2 = \sigma_1^2 + \sigma_2^2$. Then one may use Propositions 5 and 6 to test hypotheses about μ.

Test the hypotheses $H_1: \mu \leq 0$ against $A_1: \mu > 0$ and $H_2: \mu = 0$ against $A_2: \mu \neq 0$ at level of significance $\alpha = 0.05$ for the data given in (i) Exercise 13.6.2; (ii) Exercise 13.6.3.

13.7 Likelihood Ratio Tests

Let $X_1, \ldots, X_n$ be i.i.d. r.v.'s with p.d.f. $f(\cdot; \boldsymbol{\theta})$, $\boldsymbol{\theta} \in \boldsymbol{\Omega} \subseteq \mathbb{R}^r$ and let $\boldsymbol{\omega} \subset \boldsymbol{\Omega}$. Set $L(\boldsymbol{\omega}) = f(x_1; \boldsymbol{\theta}) \cdots f(x_n; \boldsymbol{\theta})$ whenever $\boldsymbol{\theta} \in \boldsymbol{\omega}$, and $L(\boldsymbol{\omega}^c) = f(x_1; \boldsymbol{\theta}) \cdots f(x_n; \boldsymbol{\theta})$ when $\boldsymbol{\theta}$ is varying over $\boldsymbol{\omega}^c$. Now, when both $\boldsymbol{\omega}$ and $\boldsymbol{\omega}^c$ consist of a single point, then $L(\boldsymbol{\omega})$ and $L(\boldsymbol{\omega}^c)$ are completely determined and for testing $H: \boldsymbol{\theta} \in \boldsymbol{\omega}$ against $A: \boldsymbol{\theta} \in \boldsymbol{\omega}^c$, the MP test rejects when the *likelihood ratio* (LR) $L(\boldsymbol{\omega}^c)/L(\boldsymbol{\omega})$ is too large (greater than or equal to a constant C determined by the size of the test.) However, if $\boldsymbol{\omega}$ and $\boldsymbol{\omega}^c$ contain more than one point each, then neither $L(\boldsymbol{\omega})$ nor $L(\boldsymbol{\omega}^c)$ is determined by H and A and the above method of testing does not apply. The problem can be reduced to it though by the following device. $L(\boldsymbol{\omega})$ is to be replaced by $L(\hat{\boldsymbol{\omega}}) = \max[L(\boldsymbol{\theta}); \boldsymbol{\theta} \in \boldsymbol{\omega}]$ and $L(\boldsymbol{\omega}^c)$ is to be replaced by $L(\hat{\boldsymbol{\omega}}^c) = \max[L(\boldsymbol{\theta}); \boldsymbol{\theta} \in \boldsymbol{\omega}^c]$. Then for setting up a test, one would compare the quantities $L(\hat{\boldsymbol{\omega}})$ and $L(\hat{\boldsymbol{\omega}}^c)$. In practice, however, the statistic $L(\hat{\boldsymbol{\omega}}^c)/L(\hat{\boldsymbol{\Omega}})$ is used rather than $L(\hat{\boldsymbol{\omega}}^c)/L(\hat{\boldsymbol{\omega}})$, where, of course, $L(\hat{\boldsymbol{\Omega}}) = \max[L(\boldsymbol{\theta}); \boldsymbol{\theta} \in \boldsymbol{\Omega}]$. (When we talk about a statistic, it will be understood that the observed values have been replaced by the corresponding r.v.'s although the same notation will be employed.) In terms of this statistic, one rejects H if $L(\hat{\boldsymbol{\omega}})/L(\hat{\boldsymbol{\Omega}})$ is too small, that is, $\leq C$, where C is specified by the desired size of the test. For obvious reasons, the test is called a *likelihood ratio* (LR) *test*. Of course, the test specified by the Neyman–Pearson fundamental lemma is also a likelihood ratio test.

Now the likelihood ratio test which rejects H whenever $L(\hat{\omega})/L(\hat{\Omega})$ is too small has an intuitive interpretation, as follows: The quantity $L(\hat{\omega})$ and the probability element $L(\hat{\omega})dx_1 \cdots dx_n$ for the discrete and continuous case, respectively, is the maximum probability of observing $x_1, \ldots, x_n$ if θ lies in ω. Similarly, $L(\hat{\Omega})$ and $L(\hat{\Omega})dx_1 \cdots dx_n$ represent the maximum probability for the discrete and continuous case, respectively, of observing $x_1, \ldots, x_n$ without restrictions on θ. Thus, if $\theta \in \omega$, as specified by H, the quantities $L(\hat{\omega})$ and $L(\hat{\Omega})$ will tend to be close together (by an assumed continuity (in θ) of the likelihood function $L(\theta | x_1, \ldots, x_n)$), and therefore λ will be close to 1. Should λ be too far away from 1, the data would tend to discredit H, and therefore H is to be rejected.

The notation $\lambda = L(\hat{\omega})/L(\hat{\Omega})$ has been in wide use. (Notice that $0 < \lambda \leq 1$.) Also the statistic $-2\log\lambda$ rather than λ itself is employed, the reason being that, under certain regularity conditions, the asymptotic distribution of $-2\log\lambda$, under H, is known. Then in terms of this statistic, one rejects H whenever $-2\log\lambda > C$, where C is determined by the desired level of the test. Of course, this test is equivalent to the LR test. In carrying out the likelihood ratio test in actuality, one is apt to encounter two sorts of difficulties. First is the problem of determining the cut-off point C and second is the problem of actually determining $L(\hat{\omega})$ and $L(\hat{\Omega})$. The first difficulty is removed at the asymptotic level, in the sense that we may use as an approximation (for sufficiently large n) the limiting distribution of $-2\log\lambda$ for specifying C. The problem of finding $L(\hat{\Omega})$ is essentially that of finding the MLE of θ. Calculating $L(\hat{\omega})$ is a much harder problem. In many cases, however, H is simple and then no problem exists.

In spite of the apparent difficulties that a likelihood ratio test may present, it does provide a unified method for producing tests. Also in addition to its intuitive interpretation, in many cases of practical interest and for a fixed sample size, the likelihood ratio test happens to coincide with or to be close to other tests which are known to have some well defined optimal properties such as being UMP or being UMPU. Furthermore, under suitable regularity conditions, it enjoys some asymptotic optimal properties as well.

In the following, a theorem referring to the asymptotic distribution of $-2\log\lambda$ is stated (but not proved) and then a number of illustrative examples are discussed.

THEOREM 6 Let $X_1, \ldots, X_n$ be i.i.d. r.v.'s with p.d.f. $f(\cdot; \theta)$, $\theta \in \Omega$, where Ω is an r-dimensional subset of $\mathbb{R}^r$ and let ω be an m-dimensional subset of Ω. Suppose also that the set of positivity of the p.d.f. does not depend on θ. Then under some additional regularity conditions, the asymptotic distribution of $-2\log\lambda$ is χ^2_{r-m}, provided $\theta \in \omega$; that is, as $n \to \infty$,

$$P_\theta(-2\log\lambda \leq x) \to G(x), \quad x \geq 0 \quad \text{for all} \quad \theta \in \omega,$$

where G is the d.f. of a χ^2_{r-m} distribution.

Since in using the LR test, or some other test equivalent to it, the alternative A specifies that $\theta \in \omega^c$, this will not have to be mentioned explicitly in the sequel. Also the level of significance will always be α.

EXAMPLE 11 (Testing the mean of a normal distribution) Let $X_1, \ldots, X_n$ be i.i.d. r.v.'s from $N(\mu, \sigma^2)$, and consider the following testing hypotheses problems.

i) Let σ be known and suppose we are interested in testing the hypothesis $H: \mu \in \omega = \{\mu_0\}$.
Since the MLE of μ is $\hat{\mu}_\Omega = \bar{x}$ (see Example 12, Chapter 12), we have

$$L(\hat{\Omega}) = \frac{1}{(\sqrt{2\pi}\sigma)^n} \exp\left[-\frac{1}{2\sigma^2}\sum_{j=1}^{n}(x_j - \bar{x})^2\right]$$

and

$$L(\hat{\omega}) = \frac{1}{(\sqrt{2\pi}\sigma)^n} \exp\left[-\frac{1}{2\sigma^2}\sum_{j=1}^{n}(x_j - \mu_0)^2\right].$$

In this example, it is much easier to determine the distribution of $-2\log\lambda$ rather than that of λ. In fact,

$$-2\log\lambda = \frac{n}{\sigma^2}(\bar{x} - \mu_0)^2$$

and the LR test is equivalent to

$$\phi(\mathbf{z}) = \begin{cases} 1, & \text{if } \left[\frac{\sqrt{n}(\bar{x} - \mu_0)}{\sigma}\right]^2 > C \\ 0, & \text{otherwise,} \end{cases}$$

where C is determined by

$$P(\chi^2_1 > C) = \alpha.$$

(Recall that $\mathbf{z} = (x_1, \ldots, x_n)'$.)

Notice that this is consistent with Theorem 6. It should also be pointed out that this test is the same as the test found in Example 10 and therefore the present test is also UMPU.

ii) Consider the same problem as in (i) but suppose now that σ is also unknown. We are still interested in testing the hypothesis $H: \mu = \mu_0$ which now is composite, since σ is unspecified.

Now the MLE's of σ^2, under $\Omega = \{\theta = (\mu, \sigma)'; \mu \in \mathbb{R}, \sigma > 0\}$ and $\omega = \{\theta = (\mu, \sigma)'; \mu = \mu_0, \sigma > 0\}$ are, respectively,

$$\hat{\sigma}_\Omega^2 = \frac{1}{n}\sum_{j=1}^n (x_j - \bar{x})^2 \quad \text{and} \quad \hat{\sigma}_\omega^2 = \frac{1}{n}\sum_{j=1}^n (x_j - \mu_0)^2$$

(see Example 12, Chapter 12). Therefore

$$L(\hat{\Omega}) = \frac{1}{\left(\sqrt{2\pi}\hat{\sigma}_\Omega\right)^n} \exp\left[-\frac{1}{2\hat{\sigma}_\Omega^2}\sum_{j=1}^n (x_j - \bar{x})^2\right] = \frac{1}{\left(\sqrt{2\pi}\hat{\sigma}_\Omega\right)^n} e^{-n/2}$$

and

$$L(\hat{\omega}) = \frac{1}{\left(\sqrt{2\pi}\hat{\sigma}_\omega\right)^n} \exp\left[-\frac{1}{2\hat{\sigma}_\omega^2}\sum_{j=1}^n (x_j - \mu_0)^2\right] = \frac{1}{\left(\sqrt{2\pi}\hat{\sigma}_\omega\right)^n} e^{-n/2}.$$

Then

$$\lambda = \left(\frac{\hat{\sigma}_\Omega}{\hat{\sigma}_\omega}\right)^n \quad \text{or} \quad \lambda^{2/n} = \frac{\sum_{j=1}^n (x_j - \bar{x})^2}{\sum_{j=1}^n (x_j - \mu_0)^2}.$$

But

$$\sum_{j=1}^n (x_j - \mu_0)^2 = \sum_{j=1}^n (x_j - \bar{x})^2 + n(\bar{x} - \mu_0)^2$$

and therefore

$$\lambda^{2/n} = \left[1 + \frac{1}{n-1}\frac{n(\bar{x} - \mu_0)^2}{\frac{1}{n-1}\sum_{j=1}^n (x_j - \bar{x})^2}\right]^{-1} = \left(1 + \frac{t^2}{n-1}\right)^{-1},$$

where

$$t = t(\mathbf{z}) = \frac{\sqrt{n}(\bar{x} - \mu_0)}{\sqrt{\frac{1}{n-1}\sum_{j=1}^n (x_j - \bar{x})^2}}.$$

Then $\lambda < \lambda_0$ is equivalent to $t^2 > C$ for a certain constant C. That is, the LR test is equivalent to the test

$$\phi(\mathbf{z}) = \begin{cases} 1, & \text{if } t < -C \text{ or } t > C \\ 0, & \text{otherwise,} \end{cases}$$

where C is determined by
$$P(t_{n-1} > C) = \alpha/2.$$

Notice that, by Proposition 6, the test just derived is UMPU.

EXAMPLE 12 (Comparing the parameters of two normal distributions) Let $X_1, \ldots, X_m$ be i.i.d. r.v.'s from $N(\mu_1, \sigma_1^2)$ and $Y_1, \ldots, Y_n$ be i.i.d. r.v.'s from $N(\mu_2, \sigma_2^2)$. Suppose that the X's and Y's are independent and consider the following testing hypotheses problems. In the present case, the joint p.d.f. of the X's and Y's is given by

$$\frac{1}{\left(\sqrt{2\pi}\right)^{m+n}} \frac{1}{\sigma_1^m \sigma_2^n} \exp\left[-\frac{1}{2\sigma_1^2}\sum_{i=1}^{m}(x_i - \mu_1)^2 - \frac{1}{2\sigma_2^2}\sum_{j=1}^{m}(y_j - \mu_2)^2\right].$$

i) Assume that $\sigma_1 = \sigma_2 = \sigma$ unknown and we are interested in testing the hypothesis $H: \mu_1 = \mu_2 \, (= \mu \text{ unspecified})$. Under $\Omega = \{\theta = (\mu_1, \mu_2, \sigma)'; \mu_1, \mu_2 \in \mathbb{R}, \sigma > 0\}$, the MLE's of the parameters involved are given by

$$\hat{\mu}_{1,\Omega} = \bar{x}, \quad \hat{\mu}_{2,\Omega} = \bar{y}, \quad \hat{\sigma}_\Omega^2 = \frac{1}{m+n}\left[\sum_{i=1}^{m}(x_i - \bar{x})^2 + \sum_{j=1}^{n}(y_j - \bar{y})^2\right],$$

as is easily seen. Therefore

$$L(\hat{\Omega}) = \frac{1}{\left(\sqrt{2\pi}\hat{\sigma}_\Omega\right)^{m+n}} e^{-(m+n)/2}.$$

Under $\omega = \{\theta = (\mu_1, \mu_2, \sigma)'; \mu_1 = \mu_2 \in \mathbb{R}, \sigma > 0\}$, we have

$$\hat{\mu}_\omega = \frac{1}{m+n}\left(\sum_{i=1}^{m} x_i + \sum_{j=1}^{n} y_j\right) = \frac{m\bar{x} + n\bar{y}}{m+n},$$

and by setting $v_k = x_k$, $k = 1, \ldots, m$ and $v_{m+k} = y_k$, $k = 1, \ldots, n$, one has

$$\bar{v} = \frac{1}{m+n}\sum_{k=1}^{m+n} v_k = \frac{1}{m+n}\left(\sum_{i=1}^{m} x_i + \sum_{j=1}^{n} y_j\right) = \hat{\mu}_\omega$$

and

$$\hat{\sigma}_\omega^2 = \frac{1}{m+n}\sum_{k=1}^{m+n}(v_k - \bar{v})^2 = \frac{1}{m+n}\left[\sum_{i=1}^{m}(x_i - \hat{\mu}_\omega)^2 + \sum_{j=1}^{n}(y_j - \hat{\mu}_\omega)^2\right].$$

Therefore

$$L(\hat{\omega}) = \frac{1}{\left(\sqrt{2\pi}\hat{\sigma}_\omega\right)^{m+n}} e^{-(m+n)/2}.$$

It follows that

13 Testing Hypotheses

$$\lambda = \left(\frac{\hat{\sigma}_\Omega}{\hat{\sigma}_\omega}\right)^{m+n} \quad \text{and} \quad \lambda^{2/(m+n)} = \frac{\hat{\sigma}_\Omega^2}{\hat{\sigma}_\omega^2}.$$

Next

$$\sum_{i=1}^m (x_i - \hat{\mu}_\omega)^2 = \sum_{i=1}^m \left[(x_i - \bar{x}) + (\bar{x} - \hat{\mu}_\omega)\right]^2 = \sum_{i=1}^m (x_i - \bar{x})^2 + m(\bar{x} - \hat{\mu}_\omega)^2$$

$$= \sum_{i=1}^m (x_i - \bar{x})^2 + \frac{mn^2}{(m+n)^2}(\bar{x} - \bar{y})^2,$$

and in a similar manner

$$\sum_{j=1}^n (y_j - \hat{\mu}_\omega)^2 = \sum_{j=1}^n (y_j - \bar{y})^2 + \frac{m^2 n}{(m+n)^2}(\bar{x} - \bar{y})^2,$$

so that

$$(m+n)\hat{\sigma}_\omega^2 = (m+n)\hat{\sigma}_\Omega^2 + \frac{mn}{m+n}(\bar{x}-\bar{y})^2 = \sum_{i=1}^m (x_i - \bar{x})^2 + \sum_{j=1}^n (y_j - \bar{y})^2$$

$$+ \frac{mn}{m+n}(\bar{x}-\bar{y})^2.$$

It follows then that

$$\lambda^{2/(m+n)} = \left(1 + \frac{t^2}{m+n-2}\right)^{-1},$$

where

$$t = \sqrt{\frac{mn}{m+n}}(\bar{x}-\bar{y}) \bigg/ \sqrt{\frac{1}{m+n-2}\left[\sum_{i=1}^m (x_i - \bar{x})^2 + \sum_{j=1}^n (y_j - \bar{y})^2\right]}.$$

Therefore the LR test which rejects H whenever $\lambda < \lambda_0$ is equivalent to the following test:

$$\phi(\mathbf{z}, \mathbf{w}) = \begin{cases} 1, & \text{if } t < -C \text{ or } t > C \quad (C > 0) \\ 0, & \text{otherwise,} \end{cases}$$

where C is determined by

$$P(t_{m+n-2} > C) = \alpha/2,$$

and $\mathbf{z} = (x_1, \ldots, x_m)'$, $\mathbf{w} = (y_1, \ldots, y_n)'$, because, under H, t is distributed as t_{m+n-2}. We notice that the test ϕ above is the same as the UMPU test found in Proposition 10.

ii) Now we are interested in testing the hypothesis $H: \sigma_1 = \sigma_2$ ($= \sigma$ unspecified). Under $\Omega = \{\theta = (\mu_1, \mu_2, \sigma_1, \sigma_2)'; \mu_1, \mu_2 \in \mathbb{R}, \sigma_1, \sigma_2 > 0\}$, we have

$$\hat{\mu}_{1,\Omega} = \bar{x}, \quad \hat{\mu}_{2,\Omega} = \bar{y}, \quad \hat{\sigma}^2_{1,\Omega} = \frac{1}{m} \sum_{i=1}^{m} (x_i - \bar{x})^2$$

and

$$\hat{\sigma}^2_{2,\Omega} = \frac{1}{n} \sum_{j=1}^{n} (y_j - \bar{y})^2,$$

while under $\omega = \{\theta = (\mu_1, \mu_2, \sigma_1, \sigma_2)'; \mu_1, \mu_2 \in \mathbb{R}, \sigma_1 = \sigma_2 > 0\}$,

$$\hat{\mu}_{1,\omega} = \hat{\mu}_{1,\Omega}, \quad \hat{\mu}_{2,\omega} = \hat{\mu}_{2,\Omega}$$

and

$$\hat{\sigma}^2_{\omega} = \frac{1}{m+n} \left[\sum_{i=1}^{m} (x_i - \bar{x})^2 + \sum_{j=1}^{n} (y_j - \bar{y})^2 \right].$$

Therefore

$$L(\hat{\Omega}) = \frac{1}{\left(\sqrt{2\pi}\right)^{m+n}} \cdot \frac{1}{\left(\hat{\sigma}^2_{1,\Omega}\right)^{m/2} \left(\hat{\sigma}^2_{2,\Omega}\right)^{n/2}} e^{-(m+n)/2}$$

and

$$L(\hat{\omega}) = \frac{1}{\left(\sqrt{2\pi}\right)^{m+n}} \cdot \frac{1}{\left(\hat{\sigma}^2_{\omega}\right)^{-(m+n)/2}} e^{-(m+n)/2},$$

so that

$$\lambda = \frac{\left(\hat{\sigma}^2_{1,\Omega}\right)^{m/2} \left(\hat{\sigma}^2_{2,\Omega}\right)^{n/2}}{\left(\hat{\sigma}^2_{\omega}\right)^{(m+n)/2}}$$

$$= \frac{(m+n)^{(m+n)/2} \left[\sum_{i=1}^{m}(x_i-\bar{x})^2 / \sum_{j=1}^{n}(y_j-\bar{y})^2\right]^{m/2}}{m^{m/2} n^{n/2} \left\{\left[\sum_{i=1}^{m}(x_i-\bar{x})^2 + \sum_{j=1}^{n}(y_j-\bar{y})^2\right] / \sum_{j=1}^{n}(y_j-\bar{y})^2\right\}^{(m+n)/2}}$$

$$= \frac{(m+n)^{(m+n)/2}}{m^{m/2} n^{n/2}} \cdot \frac{\left[\frac{m-1}{n-1} \cdot \frac{\sum_{i=1}^{m}(x_i-\bar{x})^2 / m-1}{\sum_{j=1}^{n}(y_j-\bar{y})^2 / n-1}\right]^{m/2}}{\left[1 + \frac{m-1}{n-1} \cdot \frac{\sum_{i=1}^{m}(x_i-\bar{x})^2 / m-1}{\sum_{j=1}^{n}(y_j-\bar{y})^2 / n-1}\right]^{(m+n)/2}}$$

$$= \frac{(m+n)^{(n+m)/2}}{m^{m/2}n^{n/2}} \cdot \frac{\left(\frac{m-1}{n-1}f\right)^{m/2}}{\left(1+\frac{m-1}{n-1}f\right)^{(m+n)/2}},$$

where $f = [\sum_{i=1}^{m}(x_i - \bar{x})^2/m - 1]/[\sum_{j=1}^{n}(y_j - \bar{y})^2/n - 1]$.

Therefore the LR test, which rejects H whenever $\lambda < \lambda_0$, is equivalent to the test based on f and rejecting H if

$$\frac{\left(\frac{m-1}{n-1}f\right)^{m/2}}{\left(1+\frac{m-1}{n-1}f\right)^{(m+n)/2}} < C \quad \text{for a certain constant } C.$$

Setting $g(f)$ for the left hand side of this last inequality, we have that $g(0) = 0$ and $g(f) \to 0$ as $f \to \infty$. Furthermore, it can be seen (see Exercise 13.7.4) that $g(f)$ has a maximum at the point

$$f_{\max} = \frac{m(n-1)}{n(m-1)};$$

it is increasing between 0 and $f_{\max}$ and decreasing in $(f_{\max}, \infty)$. Therefore

$$g(f) < C \quad \text{if and only if} \quad f < C_1 \quad \text{or} \quad f > C_2$$

for certain specified constants C_1 and C_2.

Now, if in the expression of f the x's and y's are replaced by X's and Y's, respectively, and denote by F the resulting statistic, it follows that, under H, F is distributed as $F_{m-1,n-1}$. Therefore the constants C_1 and C_2 are uniquely determined by the following requirements:

$$P(F_{m-1,n-1} < C_1 \quad \text{or} \quad F_{m-1,n-1} > C_2) = \alpha \quad \text{and} \quad g(C_1) = g(C_2).$$

However, in practice the C_1 and C_2 are determined so as to assign probability $\alpha/2$ to each one of the two tails of the $F_{m-1,n-1}$ distribution; that is, such that

$$P(F_{m-1,n-1} < C_1) = P(F_{m-1,n-1} > C_2) = \alpha/2.$$

(See also Fig. 13.16.)

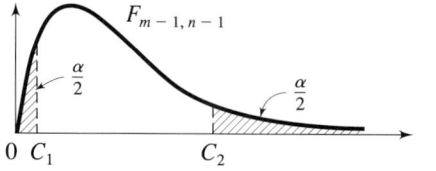

Figure 13.16

Exercises

13.7.1 Refer to Exercise 13.4.2 and use the LR test to test the hypothesis $H: p = 0.25$ against the alternative $A: p \neq 0.25$. Specifically, set $\lambda(t)$ for the likelihood function, where $t = x_1 + x_2 + x_3$, and:

i) Calculate the values $\lambda(t)$ for $t = 0, 1, 2, 3$ as well as the corresponding probabilities under H;

ii) Set up the LR test, in terms of both $\lambda(t)$ and t;

iii) Specify the (randomized) test of level of significance $\alpha = 0.05$;

iv) Compare the test in part (iii) with the UMPU test constructed in Exercise 13.4.2.

13.7.2 A coin, with probability p of falling heads, is tossed 100 times and 60 heads are observed. At level of significance $\alpha = 0.1$:

i) Test the hypothesis $H: p = \frac{1}{2}$ against the alternative $A: p \neq \frac{1}{2}$ by using the LR test and employ the appropriate approximation to determine the cut-off point;

ii) Compare the cut-off point in part (i) with that found in Exercise 13.4.1.

13.7.3 If $X_1, \ldots, X_n$ are i.i.d. r.v.'s from $N(\mu, \sigma^2)$, derive the LR test and the test based on $-2 \log \lambda$ for testing the hypothesis $H: \sigma = \sigma_0$ first in the case that μ is known and secondly in the case that μ is unknown. In the first case, compare the test based on $-2 \log \lambda$ with that derived in Example 11.

13.7.4 Consider the function

$$g(t) = \frac{\left(\frac{m-1}{n-1} t\right)^{m/2}}{\left(1 + \frac{m-1}{n-1} t\right)^{(m+n)/2}},$$

$t \geq 0$, $m, n \geq 2$, integers, and show that $g(t) \to 0$ as $t \to \infty$,

$$\max\left[g(t); t \geq 0\right] = \frac{m^{m/2}}{\left[1 + (m/n)\right]^{(m+n)/2}}$$

and that g is increasing in

$$\left[0, \frac{m(n-1)}{n(m-1)}\right] \quad \text{and decreasing in} \quad \left[\frac{m(n-1)}{n(m-1)}, \infty\right).$$

13.8 Applications of LR Tests: Contingency Tables, Goodness-of-Fit Tests

Now we turn to a slightly different testing hypotheses problem, where the LR is also appropriate. We consider an r. experiment which may result in k possibly different outcomes denoted by O_j, $j = 1, \ldots, k$. In n independent repetitions of the experiment, let p_j be the (constant) probability that each one of the trials will result in the outcome O_j and denote by X_j the number of trials which result in O_j, $j = 1, \ldots, k$. Then the joint distribution of the X's is the Multinomial distribution, that is,

$$P(X_1 = x_1, \ldots, X_k = x_k) = \frac{n!}{x_1! \cdots x_k!} p_1^{x_1} \cdots p_k^{x_k},$$

where $x_j \geq 0$, $j = 1, \ldots, k$, $\sum_{j=1}^{k} x_j = n$ and

$$\Omega = \left\{ \theta = (p_1, \ldots, p_k)'; p_j > 0, j = 1, \ldots, k, \sum_{j=1}^{k} p_j = 1 \right\}.$$

We may suspect that the p's have certain specified values; for example, in the case of a die, the die may be balanced. We then formulate this as a hypothesis and proceed to test it on the basis of the data. More generally, we may want to test the hypothesis that θ lies in a subset ω of Ω.

Consider the case that $H: \theta \in \omega = \{\theta_0\} = \{(p_{10}, \ldots, p_{k0})'\}$. Then, under ω,

$$L(\hat{\omega}) = \frac{n!}{x_1! \cdots x_k!} p_{10}^{x_1} \cdots p_{k0}^{x_k},$$

while, under Ω,

$$L(\hat{\Omega}) = \frac{n!}{x_1! \cdots x_k!} \hat{p}_1^{x_1} \cdots \hat{p}_k^{x_k},$$

where $\hat{p}_j = x_j/n$ are the MLE's of p_j, $j = 1, \ldots, k$ (see Example 11, Chapter 12). Therefore

$$\lambda = n^n \prod_{j=1}^{k} \left(\frac{p_{j0}}{x_j} \right)^{x_j}$$

and H is rejected if $-2 \log \lambda > C$. The constant C is determined by the fact that $-2 \log \lambda$ is asymptotically χ_{k-1}^2 distributed under H, as it can be shown on the basis of Theorem 6, and the desired level of significance α.

Now consider r events A_i, $i = 1, \ldots, r$ which form a partition of the sample space S and let $\{B_j, j = 1, \ldots, s\}$ be another partition of S. Let $p_{ij} = P(A_i \cap B_j)$ and let

$$p_{i \cdot} = \sum_{j=1}^{s} p_{ij}, \quad p_{\cdot j} = \sum_{i=1}^{r} p_{ij}.$$

13.8 Applications of LR Tests: Contingency Tables, Goodness-of-Fit Tests

Then, clearly, $p_{i.} = P(A_i)$, $p_{.j} = P(B_j)$ and

$$\sum_{i=1}^{r} p_{i.} = \sum_{j=1}^{s} p_{.j} = \sum_{i=1}^{r} \sum_{j=1}^{s} p_{ij} = 1.$$

Furthermore, the events $\{A_1, \ldots, A_r\}$ and $\{B_1, \ldots, B_s\}$ are independent if and only if $p_{ij} = p_{i.}p_{.j}$, $i = 1, \ldots, r$, $j = 1, \ldots, s$.

A situation where this set-up is appropriate is the following: Certain experimental units are classified according to two characteristics denoted by A and B and let $A_1, \ldots, A_r$ be the r levels of A and $B_1, \ldots, B_r$ be the J levels of B. For instance, A may stand for gender and A_1, A_2 for male and female, and B may denote educational status comprising the levels B_1 (elementary school graduate), B_2 (high school graduate), B_3 (college graduate), B_4 (beyond).

We may think of the rs events $A_i \cap B_j$ being arranged in an $r \times s$ rectangular array which is known as a *contingency table*; the event $A_i \cap B_j$ is called the (i, j)th cell.

Again consider n experimental units classified according to the characteristics A and B and let X_{ij} be the number of those falling into the (i, j)th cell. We set

$$X_{i.} = \sum_{j=1}^{s} X_{ij} \quad \text{and} \quad X_{.j} = \sum_{i=1}^{r} X_{ij}.$$

It is then clear that

$$\sum_{i=1}^{r} X_{i.} = \sum_{j=1}^{s} X_{.j} = n.$$

Let $\boldsymbol{\theta} = (p_{ij}, i = 1, \ldots, r, j = 1, \ldots, s)'$. Then the set $\boldsymbol{\Omega}$ of all possible values of $\boldsymbol{\theta}$ is an $(rs - 1)$-dimensional hyperplane in $\mathbb{R}^{rs}$. Namely, $\boldsymbol{\Omega} = \{\boldsymbol{\theta} = (p_{ij}, i = 1, \ldots, r, j = 1, \ldots, s)' \in \mathbb{R}^{rs}; p_{ij} > 0, i = 1, \ldots, r, j = 1, \ldots, s, \sum_{i=1}^{r} \sum_{j=1}^{s} p_{ij} = 1\}$.

Under the above set-up, the problem of interest is that of testing whether the characteristics A and B are independent. That is, we want to test the existence of probabilities p_i, q_j, $i = 1, \ldots, r$, $j = 1, \ldots, s$ such that $H: p_{ij} = p_i q_j$, $i = 1, \ldots, r$, $j = 1, \ldots, s$. Since for $i = 1, \ldots, r - 1$ and $j = 1, \ldots, s - 1$ we have the $r + s - 2$ independent linear relationships

$$\sum_{j=1}^{s} p_{ij} = p_i, \quad \sum_{i=1}^{r} p_{ij} = q_j,$$

it follows that the set $\boldsymbol{\omega}$, specified by H, is an $(r + s - 2)$-dimensional subset of $\boldsymbol{\Omega}$.

Next, if x_{ij} is the observed value of X_{ij} and if we set

$$x_{i.} = \sum_{j=1}^{s} x_{ij}, \quad x_{.j} = \sum_{i=1}^{r} x_{ij},$$

the likelihood function takes the following forms under Ω and ω, respectively. Writing $\Pi_{i,j}$ instead of $\Pi_{i=1}^r \Pi_{j=1}^s$, we have

$$L(\Omega) = \frac{n!}{\prod_{i,j} x_{ij}!} \prod_{i,j} p_{ij}^{x_{ij}},$$

$$L(\omega) = \frac{n!}{\prod_{i,j} x_{ij}!} \prod_{i,j} (p_i q_j)^{x_{ij}} = \frac{n!}{\prod_{i,j} x_{ij}!} \prod_{i,j} p_i^{x_{ij}} q_j^{x_{ij}} = \frac{n!}{\prod_{i,j} x_{ij}!} \left(\prod_i p_i^{x_{i\cdot}} \right) \left(\prod_j q_j^{x_{\cdot j}} \right)$$

since

$$\prod_{i,j} p_i^{x_{ij}} q_j^{x_{ij}} = \prod_i \prod_j p_i^{x_{ij}} q_j^{x_{ij}} = \prod_i p_i^{x_{i\cdot}} q_1^{x_{i1}} \cdots q_s^{x_{is}}$$

$$= \left(\prod_i p_i^{x_{i\cdot}} \right) \left(\prod_i q_1^{x_{i1}} \cdots q_s^{x_{is}} \right) = \left(\prod_i p_i^{x_{i\cdot}} \right) \left(\prod_j q_j^{x_{\cdot j}} \right).$$

Now the MLE's of p_{ij}, p_i and q_i are, under Ω and ω, respectively,

$$\hat{p}_{ij,\Omega} = \frac{x_{ij}}{n}, \quad \hat{p}_{i,\omega} = \frac{x_{i\cdot}}{n}, \quad \hat{q}_{j,\omega} = \frac{x_{\cdot j}}{n},$$

as is easily seen (see also Exercise 13.8.1). Therefore

$$L(\hat{\Omega}) = \frac{n!}{\prod_{i,j} x_{ij}!} \prod_{i,j} \left(\frac{x_{ij}}{n} \right)^{x_{ij}}, \quad L(\hat{\omega}) = \frac{n!}{\prod_{i,j} x_{ij}!} \left[\prod_i \left(\frac{x_{i\cdot}}{n} \right)^{x_{i\cdot}} \right] \left[\prod_j \left(\frac{x_{\cdot j}}{n} \right)^{x_{\cdot j}} \right]$$

and hence

$$\lambda = \frac{\left[\prod_i \left(\frac{x_{i\cdot}}{n} \right)^{x_{i\cdot}} \right] \left[\prod_j \left(\frac{x_{\cdot j}}{n} \right)^{x_{\cdot j}} \right]}{\prod_{i,j} \left(\frac{x_{ij}}{n} \right)^{x_{ij}}} = \frac{\left(\prod_i x_{i\cdot}^{x_{i\cdot}} \right) \left(\prod_j x_{\cdot j}^{x_{\cdot j}} \right)}{n^n \prod_{i,j} x_{ij}^{x_{ij}}}.$$

It can be shown that the (unspecified) assumptions of Theorem 6 are fulfilled in the present case and therefore $-2 \log \lambda$ is asymptotically χ_f^2 under ω, where $f = (rs - 1) - (r + s - 2) = (r - 1)(s - 1)$ according to Theorem 6. Hence the test for H can be carried out explicitly.

Now in a multinomial situation, as described at the beginning of this section and in connection with the estimation problem, it was seen (see Section 12.9, Chapter 12) that certain chi-square statistics were appropriate, in a sense. Recall that

$$\chi^2 = \sum_{j=1}^k \frac{(X_j - np_j)^2}{np_j}.$$

This χ^2 r.v. can be used for testing the hypothesis

$$H: \boldsymbol{\theta} \in \boldsymbol{\omega} = \{\boldsymbol{\theta}_0\} = \left\{ \left(p_{10}, \ldots, p_{k0}\right)' \right\},$$

where $\boldsymbol{\theta} = (p_1, \ldots, p_k)'$. That is, we consider

$$\chi_\omega^2 = \sum_{j=1}^k \frac{(x_j - np_{j0})^2}{np_{j0}}$$

and reject H if χ_ω^2 is too large, in the sense of being greater than a certain constant C which is specified by the desired level of the test. It can further be shown that, under $\boldsymbol{\omega}$, χ_ω^2 is asymptotically distributed as χ_{k-1}^2. In fact, the present test is asymptotically equivalent to the test based on $-2 \log \lambda$.

For the case of contingency tables and the problem of testing independence there, we have

$$\chi_\omega^2 = \sum_{i,j} \frac{(x_{ij} - np_i q_j)^2}{np_i q_j},$$

where $\boldsymbol{\omega}$ is as in the previous case in connection with the contingency tables. However, χ_ω^2 is not a statistic since it involves the parameters p_i, q_j. By replacing them by their MLE's, we obtain the statistic

$$\chi_{\hat{\omega}}^2 = \sum_{i,j} \frac{(x_{ij} - n\hat{p}_{i,\omega}\hat{p}_{j,\omega})^2}{n\hat{p}_{i,\omega}\hat{q}_{j,\omega}}.$$

By means of $\chi_{\hat{\omega}}^2$, one can test H by rejecting it whenever $\chi_{\hat{\omega}}^2 > C$. The constant C is to be determined by the significance level and the fact that the asymptotic distribution of $\chi_{\hat{\omega}}^2$, under $\boldsymbol{\omega}$, is χ_f^2 with $f = (r-1)(s-1)$, as can be shown. Once more this test is asymptotically equivalent to the corresponding test based on $-2 \log \lambda$.

Tests based on chi-square statistics are known as *chi-square tests* or *goodness-of-fit tests* for obvious reasons.

Exercises

13.8.1 Show that $\hat{p}_{ij,\Omega} = \frac{x_{ij}}{n}$, $\hat{p}_{i,\omega} = \frac{x_{i\cdot}}{n}$, $\hat{q}_{j,\omega} = \frac{x_{\cdot j}}{n}$ as claimed in the discussion in this section.

In Exercises 13.8.2–13.8.9 below, the test to be used will be the appropriate χ^2 test.

13.8.2 Refer to Exercise 13.7.2 and test the hypothesis formulated there at the specified level of significance. Also, compare the cut-off point with that found in Exercise 13.7.2(i).

13.8.3 A die is cast 600 times and the numbers 1 through 6 appear with the frequencies recorded below.

1	2	3	4	5	6
100	94	103	89	110	104

At the level of significance $\alpha = 0.1$, test the fairness of the die.

13.8.4 In a certain genetic experiment, two different varieties of a certain species are crossed and a specific characteristic of the offspring can only occur at three levels A, B and C, say. According to a proposed model, the probabilities for A, B and C are $\frac{1}{12}$, $\frac{3}{12}$ and $\frac{8}{12}$, respectively. Out of 60 offsprings, 6, 18, and 36 fall into levels A, B and C, respectively. Test the validity of the proposed model at the level of significance $\alpha = 0.05$.

13.8.5 Course work grades are often assumed to be normally distributed. In a certain class, suppose that letter grades are given in the following manner: A for grades in [90, 100], B for grades in [75, 89], C for grades in [60, 74], D for grades in [50, 59] and F for grades in [0, 49]. Use the data given below to check the assumption that the data is coming from an $N(75, 9^2)$ distribution. For this purpose, employ the appropriate χ^2 test and take $\alpha = 0.05$.

A	B	C	D	F
3	12	10	4	1

13.8.6 It is often assumed that I.Q. scores of human beings are normally distributed. Test this claim for the data given below by choosing appropriately the Normal distribution and taking $\alpha = 0.05$.

$x \leq 90$	$90 < x \leq 100$	$100 < x \leq 110$	$110 < x \leq 120$	$120 < x \leq 130$	$x > 130$
10	18	23	22	18	9

(Hint: Estimate μ and σ^2 from the grouped data; take the midpoints for the finite intervals and the points 65 and 160 for the leftmost and rightmost intervals, respectively.)

13.8.7 Consider a group of 100 people living and working under very similar conditions. Half of them are given a preventive shot against a certain disease and the other half serve as control. Of those who received the treatment, 40 did not contract the disease whereas the remaining 10 did so. Of those not treated, 30 did contract the disease and the remaining 20 did not. Test effectiveness of the vaccine at the level of significance $\alpha = 0.05$.

13.8.8 On the basis of the following scores, appropriately taken, test whether there are gender-associated differences in mathematical ability (as is often claimed!). Take $\alpha = 0.05$.

Boys: 80 96 98 87 75 83 70 92 97 82
Girls: 82 90 84 70 80 97 76 90 88 86

(Hint: Group the grades into the following six intervals: [70, 75), [75, 80), [80, 85), [85, 90), [90, 100).)

13.8.9 From each of four political wards of a city with approximately the same number of voters, 100 voters were chosen at random and their opinions were asked regarding a certain legislative proposal. On the basis of the data given below, test whether the fractions of voters favoring the legislative proposal under consideration differ in the four wards. Take $\alpha = 0.05$.

	WARD				Totals
	1	2	3	4	
Favor Proposal	37	29	32	21	119
Do not favor proposal	63	71	68	79	281
Totals	100	100	100	100	400

13.8.10 Let $X_1, \ldots, X_n$ be independent r.v.'s with p.d.f. $f(\cdot; \boldsymbol{\theta})$, $\boldsymbol{\theta} \in \Omega \subseteq \mathbb{R}^r$. For testing a hypothesis H against an alternative A at level of significance α, a test ϕ is said to be *consistent* if its power β_ϕ, evaluated at any fixed $\boldsymbol{\theta} \in \Omega$, converges to 1 as $n \to \infty$. Refer to the previous exercises and find at least one test which enjoys the property of consistency. Specifically, check whether the consistency property is satisfied with regards to Exercises 13.2.3 and 13.3.2.

13.9 Decision-Theoretic Viewpoint of Testing Hypotheses

For the definition of a decision, loss and risk function, the reader is referred to Section 6, Chapter 12.

Let $X_1, \ldots, X_n$ be i.i.d. r.v.'s with p.d.f. $f(\cdot; \boldsymbol{\theta})$, $\boldsymbol{\theta} \in \Omega \subseteq \mathbb{R}^r$, and let ω be a (measurable) subset of Ω. Then the hypothesis to be tested is $H : \boldsymbol{\theta} \in \omega$ against the alternative $A : \boldsymbol{\theta} \in \omega^c$. Let B be a critical region. Then by setting $\mathbf{z} = (x_1, \ldots, x_n)'$, in the present context a non-randomized decision function $\delta = \delta(\mathbf{z})$ is defined as follows:

$$\delta(\mathbf{z}) = \begin{cases} 1, & \text{if } \mathbf{z} \in B \\ 0, & \text{otherwise.} \end{cases}$$

We shall confine ourselves to non-randomized decision functions only. Also an appropriate loss function, corresponding to δ, is of the following form:

$$L(\theta; \delta) = \begin{cases} 0, & \text{if } \theta \in \omega \text{ and } \delta = 0, \text{ or } \theta \in \omega^c \text{ and } \delta = 1. \\ L_1, & \text{if } \theta \in \omega \text{ and } \delta = 1 \\ L_2, & \text{if } \theta \in \omega^c \text{ and } \delta = 0, \end{cases}$$

where $L_1, L_2 > 0$.

Clearly, a decision function in the present framework is simply a test function. The notation ϕ instead of δ could be used if one wished.

By setting $\mathbf{Z} = (X_1, \ldots, X_n)'$, the corresponding risk function is

$$R(\theta; \delta) = L(\theta; 1)P_\theta(\mathbf{Z} \in B) + L(\theta; 0)P_\theta(\mathbf{Z} \in B^c),$$

or

$$R(\theta; \delta) = \begin{cases} L_1 P_\theta(\mathbf{Z} \in B), & \text{if } \theta \in \omega \\ L_2 P_\theta(\mathbf{Z} \in B^c), & \text{if } \theta \in \omega^c. \end{cases} \quad (44)$$

In particular, if $\omega = \{\theta_0\}$, $\omega^c = \{\theta_1\}$ and $P_{\theta_0}(\mathbf{Z} \in B) = \alpha$, $P_{\theta_1}(\mathbf{Z} \in B) = \beta$, we have

$$R(\theta; \delta) = \begin{cases} L_1 \alpha, & \text{if } \theta = \theta_0 \\ L_2(1 - \beta), & \text{if } \theta = \theta_1. \end{cases} \quad (45)$$

As in the point estimation case, we would like to determine a decision function δ for which the corresponding risk would be uniformly (in θ) smaller than the risk corresponding to any other decision function δ^*. Since this is not feasible, except for trivial cases, we are led to *minimax* decision and *Bayes* decision functions corresponding to a given prior p.d.f. on Ω. Thus in the case that $\omega = \{\theta_0\}$ and $\omega^c = \{\theta_1\}$, δ is minimax if

$$\max[R(\theta_0; \delta), R(\theta_1; \delta)] \leq \max[R(\theta_0; \delta^*), R(\theta_1; \delta^*)]$$

for any other decision function δ^*.

Regarding the existence of minimax decision function, we have the result below. The r.v.'s $X_1, \ldots, X_n$ is a sample whose p.d.f. is either $f(\cdot; \theta_0)$ or else $f(\cdot; \theta_1)$. By setting $f_0 = f(\cdot; \theta_0)$ and $f_1 = f(\cdot; \theta_1)$, we have

THEOREM 7 Let $X_1, \ldots, X_n$ be i.i.d. r.v.'s with p.d.f. $f(\cdot; \theta)$, $\theta \in \Omega = \{\theta_0, \theta_1\}$. We are interested in testing the hypothesis $H: \theta = \theta_0$ against the alternative $A: \theta = \theta_1$ at level α. Define the subset B of $\mathbb{R}^n$ as follows: $B = \{\mathbf{z} \in \mathbb{R}^n; f(\mathbf{z}; \theta_1) > Cf(\mathbf{z}; \theta_0)\}$ and assume that there is a determination of the constant C such that

$$L_1 P_{\theta_0}(\mathbf{Z} \in B) = L_2 P_{\theta_1}(\mathbf{Z} \in B^c) \quad (\text{equivalently, } R(\theta_0; \delta) = R(\theta_1; \delta)). \quad (46)$$

Then the decision function δ defined by

$$\delta(\mathbf{z}) = \begin{cases} 1, & \text{if } \mathbf{z} \in B \\ 0, & \text{otherwise,} \end{cases} \quad (47)$$

is minimax.

PROOF For simplicity, set P_0 and P_1 for P_{θ_0} and P_{θ_1}, respectively, and similarly R(0; δ), R(1; δ) for R(θ_0; δ) and R(θ_1; δ). Also set $P_0(\mathbf{Z} \in B) = \alpha$ and $P_1(\mathbf{Z} \in B^c) = 1 - \beta$. The relation (45) implies that

$$R(0; \delta) = L_1\alpha \quad \text{and} \quad R(1; \delta) = L_2(1-\beta).$$

Let A be any other (measurable) subset of $\mathbb{R}^n$ and let δ^* be the corresponding decision function. Then

$$R(0; \delta^*) = L_1 P_0(\mathbf{Z} \in A) \quad \text{and} \quad R(1; \delta^*) = L_2 P_1(\mathbf{Z} \in A^c).$$

Consider R(0; δ) and R(0; δ^*) and suppose that R(0; δ^*) $\leq$ R(0; δ). This is equivalent to $L_1 P_0(\mathbf{Z} \in A) \leq L_1 P_0(\mathbf{Z} \in B)$, or

$$P_0(\mathbf{Z} \in A) \leq \alpha.$$

Then Theorem 1 implies that $P_1(\mathbf{Z} \in A) \leq P_1(\mathbf{Z} \in B)$ because the test defined by (47) is MP in the class of all tests of level $\leq \alpha$. Hence

$$P_1(\mathbf{Z} \in A^c) \geq P_1(\mathbf{Z} \in B^c), \quad \text{or} \quad L_2 P_1(\mathbf{Z} \in A^c) \geq L_2 P_1(\mathbf{Z} \in B^c),$$

or equivalently, R(1; δ^*) $\geq$ R(1; δ). That is, if

$$R(0; \delta^*) \leq R(0; \delta), \quad \text{then} \quad R(1; \delta) \leq R(1; \delta^*). \quad (48)$$

Since by assumption $R(0; \delta) = R(1; \delta)$, we have

$$\max[R(0; \delta^*), R(1; \delta^*)] = R(1; \delta^*) \geq R(1; \delta) = \max[R(0; \delta), R(1; \delta)], \quad (49)$$

whereas if R(0; δ) < R(0; δ^*), then

$$\max[R(0; \delta^*), R(1; \delta^*)] \geq R(0; \delta^*) > R(0; \delta) = \max[R(0; \delta), R(1; \delta)]. \quad (50)$$

Relations (49) and (50) show that δ is minimax, as was to be seen. ▲

REMARK 7 It follows that the minimax decision function defined by (46) is an LR test and, in fact, is the MP test of level $P_0(\mathbf{Z} \in B)$ constructed in Theorem 1.

We close this section with a consideration of the Bayesian approach. In connection with this it is shown that, corresponding to a given p.d.f. on $\Omega = \{\theta_0, \theta_1\}$, there is always a Bayes decision function which is actually an LR test. More precisely, we have

THEOREM 8 Let $X_1, \ldots, X_n$ be i.i.d. r.v.'s with p.d.f. $f(\cdot; \boldsymbol{\theta})$, $\boldsymbol{\theta} \in \Omega = \{\boldsymbol{\theta}_0, \boldsymbol{\theta}_1\}$ and let $\lambda_0 = \{p_0, p_1\}$ $(0 < p_0 < 1)$ be a probability distribution on Ω. Then for testing the hypothesis $H: \boldsymbol{\theta} = \boldsymbol{\theta}_0$ against the alternative $A: \boldsymbol{\theta} = \boldsymbol{\theta}_1$, there exists a *Bayes decision function* δ_{λ_0} corresponding to $\lambda_0 = \{p_0, p_1\}$, that is, a decision rule which minimizes the average risk $R(\boldsymbol{\theta}_0; \delta)p_0 + R(\boldsymbol{\theta}_1; \delta)p_1$, and is given by

$$\delta_{\lambda_0}(\mathbf{z}) = \begin{cases} 1, & \text{if } \mathbf{z} \in B \\ 0, & \text{otherwise,} \end{cases}$$

where $B = \{\mathbf{z} \in \mathbb{R}^n; f(\mathbf{z}; \boldsymbol{\theta}_1) > Cf(\mathbf{z}; \boldsymbol{\theta}_0)\}$ and $C = p_0 L_1 / p_1 L_2$.

PROOF Let $R_{\lambda_0}(\delta)$ be the average risk corresponding to λ_0. Then by virtue of (44), and by employing the simplified notation used in the proof of Theorem 7, we have

$$\begin{aligned} R_{\lambda_0}(\delta) &= L_1 P_0(\mathbf{Z} \in B) p_0 + L_2 P_1(\mathbf{Z} \in B^c) p_1 \\ &= p_0 L_1 P_0(\mathbf{Z} \in B) + p_1 L_2 \left[1 - P_1(\mathbf{Z} \in B)\right] \\ &= p_1 L_2 + \left[p_0 L_1 P(\mathbf{Z} \in B) - p_1 L_2 P(\mathbf{Z} \in B)\right] \end{aligned} \quad (51)$$

and this is equal to

$$p_1 L_2 + \int_B \left[p_0 L_1 f(\mathbf{z}; \boldsymbol{\theta}_0) - p_1 L_2 f(\mathbf{z}; \boldsymbol{\theta}_1)\right] d\mathbf{z}$$

for the continuous case and equal to

$$p_1 L_2 + \sum_{\mathbf{z} \in B} \left[p_0 L_1 f(\mathbf{z}; \boldsymbol{\theta}_0) - p_1 L_2 f(\mathbf{z}; \boldsymbol{\theta}_1)\right]$$

for the discrete case. In either case, it follows that the δ which minimizes $R_{\lambda_0}(\delta)$ is given by

$$\delta_{\lambda_0}(\mathbf{z}) = \begin{cases} 1, & \text{if } p_0 L_1 f(\mathbf{z}; \boldsymbol{\theta}_0) - p_1 L_2 f(\mathbf{z}; \boldsymbol{\theta}_1) < 0 \\ 0, & \text{otherwise;} \end{cases}$$

equivalently,

$$\delta_{\lambda_0}(\mathbf{z}) = \begin{cases} 1, & \text{if } \mathbf{z} \in B \\ 0, & \text{otherwise,} \end{cases}$$

where

$$B = \left\{\mathbf{z} \in \mathbb{R}^n; f(\mathbf{z}; \boldsymbol{\theta}_1) > \frac{p_0 L_1}{p_1 L_2} f(\mathbf{z}; \boldsymbol{\theta}_0)\right\},$$

as was to be seen. ▲

REMARK 8 It follows that the Bayesian decision function is an LR test and is, in fact, the MP test for testing H against A at the level $P_0(\mathbf{Z} \in B)$, as follows by Theorem 1.

13.9 Decision-Theoretic Viewpoint of Testing Hypotheses

The following examples are meant as illustrations of Theorems 7 and 8.

EXAMPLE 13 Let $X_1, \ldots, X_n$ be i.i.d. r.v.'s from $N(\theta, 1)$. We are interested in determining the minimax decision function δ for testing the hypothesis $H: \theta = \theta_0$ against the alternative $A: \theta = \theta_1$. We have

$$\frac{f(\mathbf{z}; \theta_1)}{f(\mathbf{z}; \theta_0)} = \frac{\exp\left[n(\theta_1 - \theta_0)\bar{x}\right]}{\exp\left[\frac{1}{2}n(\theta_1^2 - \theta_0^2)\right]},$$

so that $f(\mathbf{z}; \theta_1) > Cf(\mathbf{z}; \theta_0)$ is equivalent to

$$\exp\left[n(\theta_1 - \theta_0)\bar{x}\right] > C \exp\left[\frac{1}{2}n(\theta_1^2 - \theta_0^2)\right] \quad \text{or} \quad \bar{x} > C_0,$$

where

$$C_0 = \frac{1}{2}(\theta_1 + \theta_0) + \frac{\log C}{n(\theta_1 - \theta_0)}.$$

Then condition (46) becomes

$$L_1 P_{\theta_0}(\bar{X} > C_0) = L_2 P_{\theta_1}(\bar{X} < C_0).$$

As a numerical example, take $\theta_0 = 0$, $\theta_1 = 1$, $n = 25$ and $L_1 = 5$, $L_2 = 2.5$. Then

$$L_1 P_{\theta_0}(\bar{X} > C_0) = L_2 P_{\theta_1}(\bar{X} < C_0)$$

becomes

$$P_{\theta_1}(\bar{X} < C_0) = 2 P_{\theta_0}(\bar{X} > C_0),$$

or

$$P_{\theta_1}\left[\sqrt{n}(\bar{X} - \theta_1) < 5(C_0 - 1)\right] = 2 P_{\theta_0}\left[\sqrt{n}(\bar{X} - \theta_0) > 5C_0\right],$$

or

$$\Phi(5C_0 - 5) = 2\left[1 - \Phi(5C_0)\right], \quad \text{or} \quad 2\Phi(5C_0) - \Phi(5 - 5C_0) = 1$$

Hence $C_0 = 0.53$, as is found by the Normal tables.

Therefore the minimax decision function is given by

$$\delta(\mathbf{z}) = \begin{cases} 1, & \text{if } \bar{x} > 0.53 \\ 0, & \text{otherwise.} \end{cases}$$

The type-I error probability of this test is

$$P_{\theta_0}(\bar{X} > 0.53) = P[N(0, 1) > 0.53 \times 5] = 1 - \Phi(2.65) = 1 - 0.996 = 0.004$$

and the power of the test is

$$P_{\theta_1}(\overline{X} > 0.53) = P[N(0, 1) > 5(0.53 - 1)] = \Phi(2.35) = 0.9906.$$

Therefore relation (44) gives

$$R(\theta_0; \delta) = 5 \times 0.004 = 0.02 \quad \text{and} \quad R(\theta_1; \delta) = 2.5 \times 0.009 = 0.0235.$$

Thus

$$\max[R(\theta_0; \delta), R(\theta_1; \delta)] = 0.0235,$$

corresponding to the minimax δ given above.

EXAMPLE 14 Refer to Example 13 and determine the Bayes decision function corresponding to $\lambda_0 = \{p_0, p_1\}$.

From the discussion in the previous example it follows that the Bayes decision function is given by

$$\delta_{\lambda_0}(\mathbf{z}) = \begin{cases} 1, & \text{if } \overline{x} > C_0 \\ 0, & \text{otherwise,} \end{cases}$$

where

$$C_0 = \frac{1}{2}(\theta_1 + \theta_0) + \frac{\log C}{n(\theta_1 - \theta_0)} \quad \text{and} \quad C \frac{p_0 L_1}{p_1 L_2}.$$

Suppose $p_0 = \frac{2}{3}$, $p_1 = \frac{1}{3}$. Then $C = 4$ and $C_0 = 0.555451$ (≈ 0.55). Therefore the Bayes decision function corresponding to $\lambda_0' = \{\frac{2}{3}, \frac{1}{3}\}$ is given by

$$\delta_{\lambda_0}'(\mathbf{z}) = \begin{cases} 1, & \text{if } \overline{x} > 0.55 \\ 0, & \text{otherwise.} \end{cases}$$

The type-I error probability of this test is $P_{\theta_0}(\overline{X} > 0.55) = P[N(0, 1) > 2.75] = 1 - \Phi(2.75) = 0.003$ and the power of the test is $P_{\theta_1}(\overline{X} > 0.55) = P[N(1, 1) > -2.25] = \Phi(2.25) = 0.9878$. Therefore relation (51) gives that the Bayes risk corresponding to $\{\frac{2}{3}, \frac{1}{3}\}$ is equal to 0.0202.

EXAMPLE 15 Let $X_1, \ldots, X_n$ be i.i.d. r.v.'s from $B(1, \theta)$. We are interested in determining the minimax decision function δ for testing $H: \theta = \theta_0$ against $A: \theta = \theta_1$.

We have

$$\frac{f(\mathbf{z}; \theta_1)}{f(\mathbf{z}; \theta_0)} = \left(\frac{\theta_1}{\theta_0}\right)^x \left(\frac{1-\theta_1}{1-\theta_0}\right)^{n-x}, \quad \text{where } x = \sum_{j=1}^{n} x_j,$$

so that $f(\mathbf{z}; \theta_1) > Cf(\mathbf{z}; \theta_0)$ is equivalent to

$$x \log \frac{(1-\theta_0)\theta_1}{\theta_0(1-\theta_1)} > C_0',$$

where

$$C_0' = \log C - n\log\frac{1-\theta_1}{1-\theta_0}.$$

Let now $\theta_0 = 0.5$, $\theta_1 = 0.75$, $n = 20$ and $L_1 = 1071/577 \approx 1.856$, $L_2 = 0.5$. Then

$$\frac{(1-\theta_0)\theta_1}{\theta_0(1-\theta_1)} = 3 \quad (>1)$$

and therefore $f(\mathbf{z}; \theta_1) > Cf(\mathbf{z}; \theta_0)$ is equivalent to $x > C_0$, where

$$C_0 = \left(\log C - n\log\frac{1-\theta_1}{1-\theta_0}\right)\Big/ \log\frac{\theta_1(1-\theta_0)}{\theta_0(1-\theta_1)}.$$

Next, $X = \Sigma_{j=1}^n X_j$ is $B(n, \theta)$ and for $C_0 = 13$, we have $P_{0.5}(X > 13) = 0.0577$ and $P_{0.75}(X > 13) = 0.7858$, so that $P_{0.75}(X \leq 13) = 0.2142$. With the chosen values of L_1 and L_2, it follows then that relation (46) is satisfied. Therefore the minimax decision function is determined by

$$\delta(\mathbf{z}) = \begin{cases} 1, & \text{if } x > 13 \\ 0, & \text{otherwise.} \end{cases}$$

Furthermore, the minimax risk is equal to $0.5 \times 0.2142 = 0.1071$.

Chapter 14

Sequential Procedures

14.1 Some Basic Theorems of Sequential Sampling

In all of the discussions so far, the random sample $Z_1, \ldots, Z_n$, say, that we have dealt with was assumed to be of fixed size n. Thus, for example, in the point estimation and testing hypotheses problems the sample size n was fixed beforehand, then the relevant random experiment was supposed to have been independently repeated n times and finally, on the basis of the outcomes, a point estimate or a test was constructed with certain optimal properties.

Now, whereas in some situations the random experiment under consideration cannot be repeated at will, in many other cases this is, indeed, the case. In the latter case, as a rule, it is advantageous not to fix the sample size in advance, but to keep sampling and terminate the experiment according to a (random) stopping time.

DEFINITION 1 Let $\{Z_n\}$ be a sequence of r.v.'s. A *stopping time* (defined on this sequence) is a positive integer-valued r.v. N such that, for each n, the event $(N = n)$ depends on the r.v.'s $Z_1, \ldots, Z_n$ alone.

REMARK 1 In certain circumstances, a stopping time N is also allowed to take the value ∞ but with probability equal to zero. In such a case and when forming EN, the term $\infty \cdot 0$ appears, but that is interpreted as 0 and no problem arises.

Next, suppose we observe the r.v.'s $Z_1, Z_2, \ldots$ one after another, a single one at a time (sequentially), and we stop observing them after a specified event occurs. In connection with such a sampling scheme, we have the following definition.

DEFINITION 2 A sampling procedure which terminates according to a stopping time is called a *sequential procedure*.

14.1 Some Basic Theorems of Sequential Sampling

Thus a sequential procedure terminates with the r.v. Z_N, where Z_N is defined as follows:

$$\text{the value of } Z_N \text{ at } s \in S \text{ is equal to } Z_{N(s)}(s). \tag{1}$$

Quite often the partial sums $S_N = Z_1 + \cdots + Z_N$ defined by

$$S_N(s) = Z_1(s) + \cdots + Z_{N(s)}(s), \quad s \in S \tag{2}$$

are of interest and one of the problems associated with them is that of finding the expectation ES_N of the r.v. S_N. Under suitable regularity conditions, this expectation is provided by a formula due to Wald.

THEOREM 1 (Wald's lemma for sequential analysis) For $j \geq 1$, let Z_j be independent r.v.'s (not necessarily identically distributed) with identical first moments such that $E|Z_j| = M < \infty$, so that $EZ_j = \mu$ is also finite. Let N be a stopping time, defined on the sequence $\{Z_j\}, j \geq 1$, and assume that EN is finite. Then $E|S_N| < \infty$ and $ES_N = \mu EN$, where S_N is defined by (2) and Z_N is defined by (1).

The proof of the theorem is simplified by first formulating and proving a lemma. For this purpose, set $Y_j = Z_j - \mu$, $j \geq 1$. Then the r.v.'s $Y_1, Y_2, \ldots$ are independent, $EY_j = 0$ and have (common) finite absolute moment of first order to be denoted by m; that is, $E|Y_j| = m < \infty$. Also set $T_N = Y_1 + \cdots + Y_N$, where Y_N and T_N are defined in a way similar to the way Z_N and S_N are defined by (1) and (2), respectively. Then we will show that

$$E|T_N| < \infty \quad \text{and} \quad ET_N = 0. \tag{3}$$

In all that follows, it is assumed that all conditional expectations, given $N = n$, are finite for all n for which $P(N = n) > 0$. We set $E(Y_j|N = n) = 0$ (accordingly, $E(|Y_j||N = n) = 0$ for those n's for which $P(N = n) = 0$).

LEMMA 1 In the notation introduced above:

i) $\sum_{j=1}^{\infty}\sum_{n=j}^{\infty} E(|Y_j||N=n)P(N=n) = m\, EN(<\infty)$;

ii) $\sum_{n=1}^{\infty}\sum_{j=1}^{n} E(|Y_j||N=n)P(N=n) = \sum_{j=1}^{\infty}\sum_{n=j}^{\infty} E(|Y_j||N=n)P(N=n)$.

PROOF

i) For $j \geq 2$,

$$(\infty >)m = E|Y_j| = E\big[E(|Y_j|\|N)\big] = \sum_{n=1}^{\infty} E(|Y_j|\|N=n)P(N=n)$$

$$= \sum_{n=1}^{j-1} E(|Y_j|\|N=n)P(N=n) + \sum_{n=j}^{\infty} E(|Y_j|\|N=n)P(N=n). \tag{4}$$

The event $(N = n)$ depends only on $Y_1, \ldots, Y_n$ and hence, for $j > n$, $E(|Y_j||N = n) = E|Y_j| = m$. Therefore (4) becomes

$$m = m\sum_{n=1}^{j-1} P(N=n) + \sum_{n=j}^{\infty} E(|Y_j|\|N=n)P(N=n)$$

or

$$mP(N \geq j) = \sum_{n=j}^{\infty} E(|Y_j||N = n)P(N = n). \tag{5}$$

Equality (5) is also true for $j = 1$, as

$$mP(N \geq 1) = m = E|Y_1| = \sum_{n=1}^{\infty} E(|Y_1||N = n)P(N = n).$$

Therefore

$$\sum_{n=j}^{\infty} E(|Y_j||N = n)P(N = n) = mP(N \geq j), \quad j \geq 1,$$

and hence

$$\sum_{j=1}^{\infty} \sum_{n=j}^{\infty} E(|Y_j||N = n)P(N = n) = m \sum_{j=1}^{\infty} P(N \geq j) = m \sum_{j=1}^{\infty} jP(N = j) = mEN, \tag{6}$$

where the equality $\sum_{j=1}^{\infty} P(N \geq j) = \sum_{j=1}^{\infty} jP(N = j)$ is shown in Exercise 14.1.1. Relation (6) establishes part (i).

ii) By setting $p_{jn} = E(|Y_j||N = n)P(N = n)$, this part asserts that

$$\sum_{n=1}^{\infty} \sum_{j=1}^{n} p_{jn} = p_{11} + (p_{12} + p_{22}) + \cdots + (p_{1n} + p_{2n} + \cdots + p_{nn}) + \cdots,$$

and

$$\sum_{j=1}^{\infty} \sum_{n=j}^{\infty} p_{jn} = (p_{11} + p_{12} + \cdots) + (p_{22} + p_{23} + \cdots) + \cdots + (p_{nn} + p_{n,n+1} + \cdots) + \cdots$$

are equal. That this is, indeed, the case follows from part (i) and calculus results (see, for example, T.M. Apostol, Theorem 12–42, page 373, in *Mathematical Analysis*, Addison-Wesley, 1957). ▲

PROOF OF THEOREM 1 Since $T_N = S_N - \mu N$, it suffices to show (3). To this end, we have

$$E|T_N| = E[E(|T_N||N)] = \sum_{n=1}^{\infty} E(|T_N||N = n)P(N = n)(\ )$$

$$= \sum_{n=1}^{\infty} E\left(\left|\sum_{j=1}^{n} Y_j\right| \middle| N = n\right) P(N = n)$$

$$\leq \sum_{n=1}^{\infty} E\left(\sum_{j=1}^{n} |Y_j| \middle| N = n\right) P(N = n) = \sum_{n=1}^{\infty} \sum_{j=1}^{n} E(|Y_j||N = n)P(N = n)$$

$$= \sum_{j=1}^{\infty} \sum_{n=j}^{\infty} E(|Y_j||N = n)P(N = n) \quad \text{(by Lemma 1(ii))}$$

$$= mEN \ (< \infty) \quad \text{(by Lemma 1(i))}.$$

ii) Here

$$ET_N = E\left[E(T_N|N)\right] = \sum_{n=1}^{\infty} E(T_N|N=n)P(N=n)$$

$$= \sum_{n=1}^{\infty} E\left(\sum_{j=1}^{n} Y_j \bigg| N=n\right) P(N=n) = \sum_{n=1}^{\infty}\sum_{j=1}^{n} E(Y_j|N=n) P(N=n)$$

$$= \sum_{j=1}^{\infty}\sum_{n=j}^{\infty} E(Y_j|N=n) P(N=n). \tag{7}$$

This last equality holds by Lemma 1(ii), since

$$\sum_{j=1}^{\infty}\sum_{n=j}^{\infty} \left|E(Y_j|N=n)P(N=n)\right| \leq \sum_{j=1}^{\infty}\sum_{n=j}^{\infty} E(|Y_j||N=n)P(N=n) < \infty$$

by Lemma 1(i). Next, for $j \geq 1$,

$$0 = EY_j = E\left[E(Y_j|N)\right] = \sum_{n=1}^{\infty} E(Y_j|N=n)P(N=n), \tag{8}$$

whereas, for $j \geq 2$, relation (8) becomes as follows:

$$0 = \sum_{n=1}^{j-1} E(Y_j|N=n)P(N=n) + \sum_{n=j}^{\infty} E(Y_j|N=n)P(N=n)$$

$$= \sum_{n=j}^{\infty} E(Y_j|N=n)P(N=n). \tag{9}$$

This is so because the event $(N=n)$ depends only on $Y_1, \ldots, Y_n$, so that, for $j > n$, $E(Y_j|N=n) = EY_j = 0$. Therefore (9) yields

$$\sum_{n=j}^{\infty} E(Y_j|N=n)P(N=n) = 0, \quad j \geq 2. \tag{10}$$

By (8), this is also true for $j = 1$. Therefore

$$\sum_{n=j}^{\infty} E(Y_j|N=n)P(N=n) = 0, \quad j \geq 1. \tag{11}$$

Summing up over $j \geq 1$ in (11), we have then

$$\sum_{j=1}^{\infty}\sum_{n=j}^{\infty} E(Y_j|N=n)P(N=n) = 0. \tag{12}$$

Relations (7) and (12) complete the proof of the theorem. ▲

Now consider any r.v.'s $Z_1, Z_2, \ldots$ and let C_1, C_2 be two constants such that $C_1 < C_2$. Set $S_n = Z_1 + \cdots + Z_n$ and define the random quantity N as follows: N is the smallest value of n for which $S_n \leq C_1$ or $S_n \geq C_2$. If $C_1 < S_n < C_2$ for all n, then set $N = \infty$. In other words, for each $s \in S$, the value of N at s, $N(s)$, is assigned as follows: Look at $S_n(s)$ for $n \geq 1$, and find the first n, $N = N(s)$, say, for which $S_N(s) \leq C_1$ or $S_N(s) \geq C_2$. If $C_1 < S_n(s) < C_2$ for all n, then set $N(s) = \infty$. Then we have the following result.

THEOREM 2 Let $Z_1, Z_2, \ldots$ be i.i.d. r.v.'s such that $P(Z_j = 0) \neq 1$. Set $S_n = Z_1 + \cdots + Z_n$ and for two constants C_1, C_2 with $C_1 < C_2$, define the r. quantity N as the smallest n for which $S_n \leq C_1$ or $S_n \geq C_2$; set $N = \infty$ if $C_1 < S_n < C_2$ for all n. Then there exist $c > 0$ and $0 < r < 1$ such that

$$P(N \geq n) \leq cr^n \quad \text{for all} \quad n. \tag{13}$$

PROOF The assumption $P(Z_j = 0) \neq 1$ implies that $P(Z_j > 0) > 0$, or $P(Z_j < 0) > 0$. Let us suppose first that $P(Z_j > 0) > 0$. Then there exists $\varepsilon > 0$ such that $P(Z_j > \varepsilon) = \delta > 0$. In fact, if $P(Z_j > \varepsilon) = 0$ for every $\varepsilon > 0$, then, in particular, $P(Z_j > 1/n) = 0$ for all n. But $(Z_j > 1/n) \uparrow (Z_j > 0)$ and hence $0 = P(Z_j > 1/n) \to P(Z_j > 0) > 0$, a contradiction.

Thus for the case that $P(Z_j > 0) > 0$, we have that

$$\text{There exists } \varepsilon > 0 \quad \text{such that} \quad P(Z_j > \varepsilon) = \delta > 0. \tag{14}$$

With C_1, C_2 as in the theorem and ε as in (14), there exists a positive integer m such that

$$m\varepsilon > C_2 - C_1. \tag{15}$$

For such an m, we shall show that

$$P\left(\sum_{j=k+1}^{k+m} Z_j > C_2 - C_1\right) > \delta^m \quad \text{for} \quad k \geq 0. \tag{16}$$

We have

$$\bigcap_{j=k+1}^{k+m} (Z_j > \varepsilon) \subseteq \left(\sum_{j=k+1}^{k+m} Z_j > m\varepsilon\right) \subseteq \left(\sum_{j=k+1}^{k+m} Z_j > C_2 - C_1\right), \tag{17}$$

the first inclusion being obvious because there are m Z's, each one of which is greater than ε, and the second inclusion being true because of (15). Thus

$$P\left(\sum_{j=k+1}^{k+m} Z_j > C_2 - C_1\right) \geq P\left[\bigcap_{j=k+1}^{k+m} (Z_j > \varepsilon)\right] = \prod_{j=k+1}^{k+m} P(Z_j > \varepsilon) = \delta^m,$$

the inequality following from (17) and the equalities being true because of the independence of the Z's and (14). Clearly

$$S_{km} = \sum_{j=0}^{k-1} \left[Z_{jm+1} + \cdots + Z_{(j+1)m}\right].$$

Now we assert that

$$C_1 < S_i < C_2, \quad i = 1, \ldots, km$$

implies

$$Z_{jm+1} + \cdots + Z_{(j+1)m} \leq C_2 - C_1, \quad j = 0, 1, \ldots, k-1. \tag{18}$$

This is so because, if for some $j = 0, 1, \ldots, k-1$, we suppose that $Z_{jm+1} + \cdots + Z_{(j+1)m} > C_2 - C_1$, this inequality together with $S_{jm} > C_1$ would imply that $S_{(j+1)m} > C_2$, which is in contradiction to $C_1 < S_i < C_2$, $i = 1, \ldots, km$. Next,

$$(N \geq km+1) \subseteq (C_1 < S_j < C_2, j = 1, \ldots, km)$$
$$\subseteq \bigcap_{j=0}^{k-1} \left[Z_{jm+1} + \cdots + Z_{(j+1)m} \leq C_2 - C_1 \right],$$

the first inclusion being obvious from the definition of N and the second one following from (18). Therefore

$$P(N \geq km+1) \leq P\left\{ \bigcap_{j=0}^{k-1} \left[Z_{jm+1} + \cdots + Z_{(j+1)m} \leq C_2 - C_1 \right] \right\}$$
$$= \prod_{j=0}^{k-1} P\left[Z_{jm+1} + \cdots + Z_{(j+1)m} \leq C_2 - C_1 \right]$$
$$\leq \prod_{j=0}^{k-1} (1 - \delta^m) = (1 - \delta^m)^k,$$

the last inequality holding true because of (16) and the equality before it by the independence of the Z's. Thus

$$P(N \geq km+1) \leq (1 - \delta^m)^k. \tag{19}$$

Now set $c = 1/(1 - \delta^m)$, $r = (1 - \delta^m)^{1/m}$, and for a given n, choose k so that $km < n \leq (k+1)m$. We have then

$$P(N \geq n) \leq P(N \geq km+1) \leq (1 - \delta^m)^k$$
$$= \frac{1}{(1 - \delta^m)} (1 - \delta^m)^{k+1} = c \left[(1 - \delta^m)^{1/m} \right]^{(k+1)m}$$
$$= cr^{(k+1)m} \leq cr^n;$$

these inequalities and equalities are true because of the choice of k, relation (19) and the definition of c and r. Thus for the case that $P(Z_j > 0) > 0$, relation (13) is established. The case $P(Z_j < 0) > 0$ is treated entirely symmetrically, and also leads to (13). (See also Exercise 14.1.2.) The proof of the theorem is then completed. ▲

The theorem just proved has the following important corollary.

COROLLARY Under the assumptions of Theorem 2, we have (i) $P(N < \infty) = 1$ and (ii) $EN < \infty$.

PROOF

i) Set $A = (N = \infty)$ and $A_n = (N \geq n)$. Then, clearly, $A = \bigcap_{n=1}^{\infty} A_n$. Since also $A_1 \supseteq A_2 \supseteq \cdots$, we have $A = \lim_{n \to \infty} A_n$ and hence

$$P(A) = P\left(\lim_{n \to \infty} A_n \right) = \lim_{n \to \infty} P(A_n)$$

by Theorem 2 in Chapter 2. But $P(A_n) \leq cr^n$ by the theorem. Thus $\lim P(A_n) = 0$, so that $P(A) = 0$, as was to be shown.

ii) We have

$$EN = \sum_{n=1}^{\infty} nP(N = n) = \sum_{n=1}^{\infty} P(N \geq n) \leq \sum_{n=1}^{\infty} cr^n = c\sum_{n=1}^{\infty} r^n$$

$$= c\frac{r}{1-r} < \infty,$$

as was to be seen. ▲

REMARK 2 The r.v. N is positive integer-valued and it might also take on the value ∞ but with probability 0 by the first part of the corollary. On the other hand, from the definition of N it follows that for each n, the event $(N = n)$ depends only on the r.v.'s $Z_1, \ldots, Z_n$. Accordingly, N is a stopping time by Definition 1 and Remark 1.

Exercises

14.1.1 For a positive integer-valued r.v. N show that $EN = \sum_{n=1}^{\infty} P(N \geq n)$.

14.1.2 In Theorem 2, assume that $P(Z_j < 0) > 0$ and arrive at relation (13).

14.2 Sequential Probability Ratio Test

Although in the point estimation and testing hypotheses problems discussed in Chapter 12 and 13, respectively (as well as in the interval estimation problems to be dealt with in Chapter 15), sampling according to a stopping time is, in general, profitable, the mathematical machinery involved is well beyond the level of this book. We are going to consider only the problem of sequentially testing a simple hypothesis against a simple alternative as a way of illustrating the application of sequential procedures in a concrete problem.

To this end, let $X_1, X_2, \ldots$ be i.i.d. r.v.'s with p.d.f. either f_0 or else f_1, and suppose that we are interested in testing the (simple) hypothesis H: the true density is f_0 against the (simple) alternative A: the true density is f_1, at level of significance α $(0 < \alpha < 1)$ without fixing in advance the sample size n.

In order to simplify matters, we also assume that $\{x \in \mathbb{R}; f_0(x) > 0\} = \{x \in \mathbb{R}; f_1(x) > 0\}$.

Let a, b, be two numbers (to be determined later) such that $0 < a < b$, and for each n, consider the ratio

$$\lambda_n = \lambda_n(X_1, \ldots, X_n; 0, 1) = \frac{f_1(X_1) \cdots f_1(X_n)}{f_0(X_1) \cdots f_0(X_n)}.$$

We shall use the same notation λ_n for $\lambda_n(x_1, \ldots, x_n; 0, 1)$, where $x_1, \ldots, x_n$ are the observed values of $X_1, \ldots, X_n$.

For testing H against A, consider the following sequential procedure: As long as $a < \lambda_n < b$, take another observation, and as soon as $\lambda_n \leq a$, stop sampling and accept H and as soon as $\lambda_n \geq b$, stop sampling and reject H.

By letting N stand for the smallest n for which $\lambda_n \leq a$ or $\lambda_n \geq b$, we have that N takes on the values $1, 2, \ldots$ and possibly ∞, and, clearly, for each n, the event $(N = n)$ depends only on $X_1, \ldots, X_n$. Under suitable additional assumptions, we shall show that the value ∞ is taken on only with probability 0, so that N will be a stopping time.

Then the sequential procedure just described is called a *sequential probability ratio test* (SPRT) for obvious reasons.

In what follows, we restrict ourselves to the common set of positivity of f_0 and f_1, and for $j = 1, \ldots, n$, set

$$Z_j = Z_j(X_j; 0, 1) = \log \frac{f_1(X_j)}{f_0(X_j)}, \quad \text{so that} \quad \log \lambda_n = \sum_{j=1}^{n} Z_j.$$

Clearly, the Z_j's are i.i.d. since the X's are so, and if $S_n = \sum_{j=1}^{n} Z_j$, then N is redefined as the smallest n for which $S_n \leq \log a$ or $S_n \geq \log b$.

At this point, we also make the assumption that $P_i[f_0(X_1) \neq f_1(X_1)] > 0$ for $i = 0, 1$; equivalently, if C is the set over which f_0 and f_1 differ, then it is assumed that $\int_C f_0(x)dx > 0$ and $\int_C f_1(x)dx > 0$ for the continuous case. This assumption is equivalent to $P_i(Z_1 \neq 0) > 0$ under which the corollary to Theorem 2 applies.

Summarizing, we have the following result.

PROPOSITION 1 Let $X_1, X_2, \ldots$ be i.i.d. r.v.'s with p.d.f. either f_0 or else f_1, and suppose that

$$\{x \in \mathbb{R}; f_0(x) > 0\} = \{x \in \mathbb{R}; f_1(x) > 0\}$$

and that $P_i[f_0(X_1) \neq f_1(X_1)] > 0$, $i = 0, 1$. For each n, set

$$\lambda_n = \frac{f_1(X_1) \cdots f_1(X_n)}{f_0(X_1) \cdots f_0(X_n)}, \quad Z_j = \log \frac{f_1(X_j)}{f_0(X_j)}, \quad j = 1, \ldots, n$$

and

$$S_n = \sum_{j=1}^{n} Z_j = \log \lambda_n.$$

For two numbers a and b with $0 < a < b$, define the random quantity N as the smallest n for which $\lambda_n \leq a$ or $\lambda_n \geq b$; equivalently, the smallest n for which $S_n \leq \log a$ or $S_n \geq \log b$ for all n. Then

$$P_i(N < \infty) = 1 \quad \text{and} \quad E_i N < \infty, \quad i = 0, 1.$$

Thus, the proposition assures us that N is actually a stopping time with finite expectation, regardless of whether the true density is f_0 or f_1. The implication of $P_i(N < \infty) = 1$, $i = 0, 1$ is, of course, that the SPRT described above will terminate with probability one and acceptance or rejection of H, regardless of the true underlying density.

In the formulation of the proposition above, the determination of a and b was postponed until later. At this point, we shall see what is the exact determination of a and b, at least from theoretical point of view. However, the actual identification presents difficulties, as will be seen, and the use of approximate values is often necessary.

To start with, let α and $1 - \beta$ be prescribed first and second type of errors, respectively, in testing H against A, and let $\alpha < \beta < 1$. From their own definition, we have

$$\begin{aligned}
\alpha &= P(\text{rejecting } H \text{ when } H \text{ is true}) \\
&= P_0\big[(\lambda_1 \geq b) + (a < \lambda_1 < b, \lambda_2 \geq b) + \cdots \\
&\quad + (a < \lambda_1 < b, \ldots, a < \lambda_{n-1} < b, \lambda_n \geq b) + \cdots\big] \\
&= P_0(\lambda_1 \geq b) + P_0(a < \lambda_1 < b, \lambda_2 \geq b) + \cdots \\
&\quad + P_0(a < \lambda_1 < b, \ldots, a < \lambda_{n-1} < b, \lambda_n \geq b) + \cdots
\end{aligned} \quad (20)$$

and

$$\begin{aligned}
1 - \beta &= P(\text{accepting } H \text{ when } H \text{ is false}) \\
&= P_1\big[(\lambda_1 \leq a) + (a < \lambda_1 < b, \lambda_2 \leq a) + \cdots \\
&\quad + (a < \lambda_1 < b, \ldots, a < \lambda_{n-1} < b, \lambda_n \leq a) + \cdots\big] \\
&= P_1(\lambda_1 \leq a) + P_1(a < \lambda_1 < b, \lambda_2 \leq a) + \cdots \\
&\quad + P_1(a < \lambda_1 < b, \ldots, a < \lambda_{n-1} < b, \lambda_n \leq a) + \cdots.
\end{aligned} \quad (21)$$

Relations (20) and (21) allow us to determine theoretically the cut-off points a and b when α and β are given.

In order to find workable values of a and b, we proceed as follows. For each n, set

$$f_{in} = f(x_1, \ldots, x_n; i), \quad i = 0, 1$$

and in terms of them, define T'_n and T''_n as below; namely

$$T'_1 = \left\{ x_1 \in \mathbb{R};\ \frac{f_{11}}{f_{01}} \leq a \right\}, \quad T''_1 = \left\{ x_1 \in \mathbb{R};\ \frac{f_{11}(x_1)}{f_{01}(x_1)} \geq b \right\} \quad (22)$$

and for $n \geq 2$,

$$T'_n = \left\{ (x_1, \ldots, x_n)' \in \mathbb{R}^n;\ a < \frac{f_{1j}}{f_{0j}} < b, j = 1, \ldots, n-1 \text{ and } \frac{f_{1n}}{f_{0n}} \leq a \right\}, \quad (23)$$

$$T''_n = \left\{ (x_1, \ldots, x_n)' \in \mathbb{R}^n;\ a < \frac{f_{1j}}{f_{0j}} < b, j = 1, \ldots, n-1 \text{ and } \frac{f_{1n}}{f_{0n}} \geq b \right\}. \quad (24)$$

14.2 Sequential Probability Ratio Test

In other words, T'_n is the set of points in $\mathbb{R}^n$ for which the SPRT terminates with n observations and accepts H, while T''_n is the set of points in $\mathbb{R}^n$ for which the SPRT terminates with n observations and rejects H.

In the remainder of this section, the arguments will be carried out for the case that the X_j's are continuous, the discrete case being treated in the same way by replacing integrals by summation signs. Also, for simplicity, the differentials in the integrals will not be indicated.

From (20), (22) and (23), one has

$$\alpha = \sum_{n=1}^{\infty} \int_{T''_n} f_{0n}.$$

But on T''_n, $f_{1n}/f_{0n} \geq b$, so that $f_{0n} \leq (1/b)f_{1n}$. Therefore

$$\alpha = \sum_{n=1}^{\infty} \int_{T''_n} f_{0n} \leq \frac{1}{b}\sum_{n=1}^{\infty} \int_{T''_n} f_{1n}. \tag{25}$$

On the other hand, we clearly have

$$P_i(N = n) = \int_{T'_n} f_{in} + \int_{T''_n} f_{in}, \quad i = 0, 1,$$

and by Proposition 1,

$$1 = \sum_{n=1}^{\infty} P_i(N = n) = \sum_{n=1}^{\infty} \int_{T'_n} f_{in} + \sum_{n=1}^{\infty} \int_{T''_n} f_{in}, \quad i = 0, 1. \tag{26}$$

From (21), (22), (24) and (26) (with $i = 1$), we have

$$1 - \beta = \sum_{n=1}^{\infty} \int_{T'_n} f_{1n} = 1 - \sum_{n=1}^{\infty} \int_{T''_n} f_{1n}, \quad \text{so that} \quad \sum_{n=1}^{\infty} \int_{T''_n} f_{1n} = \beta.$$

Relation (25) becomes then

$$\alpha \leq \beta/b, \tag{27}$$

and in a very similar way (see also Exercise 14.2.1), we also obtain

$$1 - \alpha \geq (1 - \beta)/a. \tag{28}$$

From (27) and (28) it follows then that

$$a \geq \frac{1-\beta}{1-\alpha}, \quad b \leq \frac{\beta}{\alpha}. \tag{29}$$

Relation (29) provides us with a lower bound and an upper bound for the actual cut-off points a and b, respectively.

Now set

$$a' = \frac{1-\beta}{1-\alpha}, \quad \text{and} \quad b' = \frac{\beta}{\alpha}$$

$$\left(\text{so that } 0 < a' < b' \text{ by the assumption } \alpha < \beta < 1\right), \tag{30}$$

and suppose that the SPRT is carried out by employing the cut-off points a' and b' given by (30) rather than the original ones a and b. Furthermore, let α'

and $1 - \beta'$ be the two types of errors associated with a' and b'. Then replacing α, β, a and b by α', β', a' and b', respectively, in (29) and also taking into consideration (30), we obtain

$$\frac{1-\beta'}{1-\alpha'} \leq a' = \frac{1-\beta}{1-\alpha} \quad \text{and} \quad \frac{\beta}{\alpha} = b' \leq \frac{\beta'}{\alpha'}$$

and hence

$$1 - \beta' \leq \frac{1-\beta}{1-\alpha}(1-\alpha') \leq \frac{1-\beta}{1-\alpha} \quad \text{and} \quad \alpha' \leq \frac{\alpha}{\beta}\beta' \leq \frac{\alpha}{\beta}. \tag{31}$$

That is,

$$\alpha' \leq \frac{\alpha}{\beta} \quad \text{and} \quad 1 - \beta' \leq \frac{1-\beta}{1-\alpha}. \tag{32}$$

From (31) we also have

$$(1-\alpha)(1-\beta') \leq (1-\beta)(1-\alpha') \quad \text{and} \quad \alpha'\beta \leq \alpha\beta',$$

or

$$(1-\beta') - \alpha + \alpha\beta' \leq (1-\beta) - \alpha' + \alpha'\beta \quad \text{and} \quad -\alpha\beta' \leq -\alpha'\beta,$$

and by adding them up,

$$\alpha' + (1-\beta') \leq \alpha + (1-\beta). \tag{33}$$

Summarizing the main points of our derivations, we have the following result.

PROPOSITION 2 For testing H against A by means of the SPRT with prescribed error probabilities α and $1 - \beta$ such that $\alpha < \beta < 1$, the cut-off points a and b are determined by (20) and (21). Relation (30) provides approximate cut-off points a' and b' with corresponding error probabilities α' and $1 - \beta'$, say. Then relation (32) provides upper bounds for α' and $1 - \beta'$ and inequality (33) shows that their sum $\alpha' + (1 - \beta')$ is always bounded above by $\alpha + (1 - \beta)$.

REMARK 3 From (33) it follows that $\alpha' > \alpha$ and $1 - \beta' > 1 - \beta$ cannot happen simultaneously. Furthermore, the typical values of α and $1 - \beta$ are such as 0.01, 0.05 and 0.1, and then it follows from (32) that α' and $1 - \beta'$ lie close to α and $1 - \beta$, respectively. For example, for $\alpha = 0.01$ and $1 - \beta = 0.05$, we have $\alpha' < 0.0106$ and $1 - \beta' < 0.0506$. So there is no serious problem as far as α' and $1 - \beta'$ are concerned. The only problem which may arise is that, because a' and b' are used instead of a and b, the resulting α' and $1 - \beta'$ are too small compared to α and $1 - \beta$, respectively. As a consequence, we would be led to taking a much larger number of observations than would actually be needed to obtain α and β. It can be argued that this does not happen.

Exercise

14.2.1 Derive inequality (28) by using arguments similar to the ones employed in establishing relation (27).

14.3 Optimality of the SPRT-Expected Sample Size

An optimal property of the SPRT is stated in the following theorem, whose proof is omitted as being well beyond the scope of this book.

THEOREM 3 For testing H against A, the SPRT with error probabilities α and $1 - \beta$ minimizes the expected sample size under both H and A (that is, it minimizes $E_0 N$ and $E_1 N$) among all tests (sequential or not) with error probabilities bounded above by α and $1 - \beta$ and for which the expected sample size is finite under both H and A.

The remaining part of this section is devoted to calculating the expected sample size of the SPRT with given error probabilities, and also finding approximations to the expected sample size.

So consider the SPRT with error probabilities α and $1 - \beta$, and let N be the associated stopping time. Then we clearly have

$$E_i N = \sum_{n=1}^{\infty} n P_i(N = n) = 1 P_i(N = 1) + \sum_{n=2}^{\infty} n P_i(N = n)$$

$$= P_i(\lambda_1 \leq a \text{ or } \lambda_1 \geq b) + \sum_{n=2}^{\infty} n P_i(a < \lambda_j < b, j = 1, \ldots, n-1,$$

$$\lambda_n \leq a \text{ or } \lambda_n \geq b), \quad i = 0, 1. \quad (34)$$

Thus formula (34) provides the expected sample size of the SPRT under both H and A, but the actual calculations are tedious. This suggests that we should try to find an approximate value to $E_i N$, as follows. By setting $A = \log a$ and $B = \log b$, we have the relationships below:

$$\left(a < \lambda_j < b, j = 1, \ldots, n-1, \lambda_n \leq a \text{ or } \lambda \geq b\right)$$

$$= \left(A < \sum_{i=1}^{j} Z_i < B, j = 1, \ldots, n-1, \sum_{i=1}^{n} Z_i \leq A \text{ or } \sum_{i=1}^{n} Z_i \geq B\right), \quad n \geq 2 \quad (35)$$

and

$$(\lambda_1 \leq a \text{ or } \lambda_1 \geq b) = (Z_1 \leq A \text{ or } Z_1 \geq B). \quad (36)$$

From the right-hand side of (35), all partial sums $\sum_{i=1}^{j} Z_i$, $j = 1, \ldots, n - 1$ lie between A and B and it is only the $\sum_{i=1}^{n} Z_i$ which is either $\leq A$ or $\geq B$, and this is due to the nth observation Z_n. We would then expect that $\sum_{i=1}^{n} Z_i$ would not be too far away from either A or B. Accordingly, by letting $S_N = \sum_{i=1}^{N} Z_i$, we are led to assume as an approximation that S_N takes on the values A and B with respective probabilities

$$P_i(S_N \leq A) \text{ and } P_i(S_N \geq B), \quad i = 0, 1.$$

But

$$P_0(S_N \leq A) = 1 - \alpha, \quad P_0(S_N \geq B) = \alpha$$

and

$$P_1(S_N \leq A) = 1 - \beta, \quad P_1(S_N \geq B) = \beta.$$

Therefore we obtain

$$E_0 S_N \approx (1-\alpha)A + \alpha B \quad \text{and} \quad E_1 S_N \approx (1-\beta)A + \beta B. \tag{37}$$

On the other hand, by assuming that $E_i |Z_1| < \infty$, $i = 0, 1$, Theorem 1 gives $E_i S_N = (E_i N)(E_i Z_1)$. Hence, if also $E_i Z_1 \neq 0$, then $E_i N = (E_i S_N)/(E_i Z_1)$. By virtue of (37), this becomes

$$E_0 S_N \approx \frac{(1-\alpha)A + \alpha B}{E_0 Z_1}, \quad E_1 N \approx \frac{(1-\beta)A + \beta B}{E_1 Z_1}. \tag{38}$$

Thus we have the following result.

PROPOSITION 3 In the SPRT with error probabilities α and $1 - \beta$, the expected sample size $E_i N$, $i = 0, 1$ is given by (34). If furthermore $E_i|Z_1| < \infty$ and $E_i Z_1 \neq 0$, $i = 0, 1$, relation (38) provides approximations to $E_i N$, $i = 0, 1$.

REMARK 4 Actually, in order to be able to calculate the approximations given by (38), it is necessary to replace A and B by their approximate values taken from (30), that is,

$$A \approx \log a' = \log \frac{1-\beta}{1-\alpha} \quad \text{and} \quad B \approx \log b' = \frac{\beta}{\alpha}. \tag{39}$$

In utilizing (39), we also assume that $\alpha < \beta < 1$, since (30) was derived under this additional (but entirely reasonable) condition.

Exercises

14.3.1 Let $X_1, X_2, \ldots$ be independent r.v.'s distributed as $P(\theta)$, $\theta \in \Omega = (0, \infty)$. Use the SPRT for testing the hypothesis $H: \theta = 0.03$ against the alternative $A: \theta = 0.05$ with $\alpha = 0.1$, $1 - \beta = 0.05$. Find the expected sample sizes under both H and A and compare them with the fixed sample size of the MP test for testing H against A with the same α and $1 - \beta$ as above.

14.3.2 Discuss the same questions as in the previous exercise if the X_j's are independently distributed as Negative Exponential with parameter $\theta \in \Omega = (0, \infty)$.

14.4 Some Examples

This chapter is closed with two examples. In both, the r.v.'s $X_1, X_2, \ldots$ are i.i.d. with p.d.f. $f(\cdot; \theta)$, $\theta \in \Omega \subseteq \mathbb{R}$, and for $\theta_0, \theta_1 \in \Omega$ with $\theta_0 < \theta_1$, the problem is that of testing $H: \theta = \theta_0$ against $A: \theta = \theta_1$ by means of the SPRT with error probabilities α and $1 - \beta$. Thus in the present case $f_0 = f(\cdot; \theta_0)$ and $f_1 = f(\cdot; \theta_1)$.

14.4 Some Examples

What we explicitly do, is to set up the formal SPRT and for selected numerical values of α and $1 - \beta$, calculate a', b', upper bounds for α' and $1 - \beta'$, estimate $E_i N$, $i = 0, 1$, and finally compare the estimated $E_i N$, $i = 0, 1$ with the size of the fixed sample size test with the same error probabilities.

EXAMPLE 1 Let $X_1, X_2, \ldots$ be i.i.d. r.v.'s with p.d.f.

$$f(x; \theta) = \theta^x (1-\theta)^{1-x}, \quad x = 0, 1, \quad \theta \in \Omega = (0, 1).$$

Then the test statistic λ_n is given by

$$\lambda_n = \left(\frac{\theta_1}{\theta_2}\right)^{\sum_j X_j} \left(\frac{1-\theta_1}{1-\theta_0}\right)^{n-\sum_j X_j}$$

and we continue sampling as long as

$$\left(A - n\log\frac{1-\theta_1}{1-\theta_0}\right) \bigg/ \log\frac{\theta_1(1-\theta_0)}{\theta_0(1-\theta_1)}$$

$$< \sum_{j=1}^n X_j < \left(B - n\log\frac{1-\theta_1}{1-\theta_0}\right) \bigg/ \log\frac{\theta_1(1-\theta_0)}{\theta_0(1-\theta_1)}. \tag{40}$$

Next,

$$Z_1 = \log\frac{f_1(X_1)}{f_0(X_1)} = X_1 \log\frac{\theta_1(1-\theta_0)}{\theta_0(1-\theta_1)} + \log\frac{1-\theta_1}{1-\theta_0},$$

so that

$$E_i Z_1 = \theta_i \log\frac{\theta_1(1-\theta_0)}{\theta_0(1-\theta_1)} + \log\frac{1-\theta_1}{1-\theta_0}, \quad i = 0, 1. \tag{41}$$

For a numerical application, take $\alpha = 0.01$ and $1 - \beta = 0.05$. Then the cut-off points a and b are approximately equal to a' and b', respectively, where a' and b' are given by (30). In the present case,

$$a' = \frac{0.05}{1 - 0.01} = \frac{0.05}{0.99} \approx 0.0505 \quad \text{and} \quad b' = \frac{0.95}{0.01} = 95.$$

For the cut-off points a' and b', the corresponding error probabilities α' and $1 - \beta'$ are bounded as follows according to (32):

$$a' \leq \frac{0.01}{0.95} \approx 0.0105 \quad \text{and} \quad 1 - \beta' \leq \frac{0.05}{0.99} \approx 0.0505.$$

Next, relation (39) gives

$$A \approx \log\frac{5}{99} = -1.29667 \quad \text{and} \quad B \approx \log 95 = 1.97772. \tag{42}$$

At this point, let us suppose that $\theta_0 = \frac{3}{8}$ and $\theta_1 = \frac{4}{8}$. Then

$$\log\frac{\theta_1(1-\theta_0)}{\theta_0(1-\theta_1)}=\log\frac{5}{3}=0.22185 \quad \text{and} \quad \log\frac{1-\theta_1}{1-\theta_0}=\log\frac{4}{5}=-0.09691,$$

so that by means of (41), we have

$$E_0 Z_1 = -0.13716 \quad \text{and} \quad E_1 Z_1 = 0.014015. \tag{43}$$

Finally, by means of (42) and (43), relation (38) gives

$$E_0 N \approx 92.5 \quad \text{and} \quad E_1 N \approx 129.4$$

On the other hand, the MP test for testing H against A based on a fixed sample size n is given by (9) in Chapter 13. Using the normal approximation, we find that for the given $\alpha = 0.01$ and $\beta = 0.95$, n has to be equal to 244.05. Thus both $E_0 N$ and $E_1 N$ compare very favorably with it.

EXAMPLE 2 Let $X_1, X_2, \ldots$ be i.i.d. r.v.'s with p.d.f. that of $N(\theta, 1)$. Then

$$\lambda_n = \exp\left[(\theta_1 - \theta_0)\sum_{j=1}^{n} X_j - \frac{1}{2}n(\theta_1^2 - \theta_0^2)\right]$$

and we continue sampling as long as

$$\left[A + \frac{n}{2}(\theta_1^2 - \theta_0^2)\right]\bigg/(\theta_1 - \theta_0) < \sum_{j=1}^{n} X_j < \left[B + \frac{n}{2}(\theta_1^2 - \theta_0^2)\right]\bigg/(\theta_1 - \theta_0). \tag{44}$$

Next,

$$Z_1 = \log\frac{f_1(X_1)}{f_0(X_1)} = (\theta_1 - \theta_0)X_1 - \frac{1}{2}(\theta_1^2 - \theta_0^2),$$

so that

$$E_i Z_1 = \theta_i(\theta_1 - \theta_0) - \frac{1}{2}(\theta_1^2 - \theta_0^2), \quad i = 0, 1. \tag{45}$$

By using the same values of α and $1 - \beta$ as in the previous example, we have the same A and B as before. Taking $\theta_0 = 0$ and $\theta_1 = 1$, we have

$$E_0 Z_1 = -0.5 \quad \text{and} \quad E_1 Z_1 = 0.5.$$

Thus relation (38) gives

$$E_0 N \approx 2.53 \quad \text{and} \quad E_1 N \approx 3.63.$$

Now the fixed sample size MP test is given by (13) in Chapter 13. From this we find that $n \approx 15.84$. Again both $E_0 N$ and $E_1 N$ compare very favorably with the fixed value of n which provides the same protection.

Chapter 15

Confidence Regions—Tolerance Intervals

15.1 Confidence Intervals

Let $X_1, \ldots, X_n$ be i.i.d. r.v.'s with p.d.f. $f(\cdot; \boldsymbol{\theta})$ $\boldsymbol{\theta} \in \boldsymbol{\Omega} \subseteq \mathbb{R}^r$. In Chapter 12, we considered the problem of point estimation of a real-valued function of $\boldsymbol{\theta}$, $g(\boldsymbol{\theta})$. That is, we considered the problem of estimating $g(\boldsymbol{\theta})$ by a statistic (based on the X's) having certain optimality properties.

In the present chapter, we return to the estimation problem, but in a different context. First, we consider the case that θ is a real-valued parameter and proceed to define what is meant by a random interval and a confidence interval.

DEFINITION 1 A *random interval* is a finite or infinite interval, where at least one of the end points is an r.v.

DEFINITION 2 Let $L(X_1, \ldots, X_n)$ and $U(X_1, \ldots, X_n)$ be two statistics such that $L(X_1, \ldots, X_n) \leq U(X_1, \ldots, X_n)$. We say that the r. interval $[L(X_1, \ldots, X_n), U(X_1, \ldots, X_n)]$ is a *confidence interval* for θ with *confidence coefficient* $1 - \alpha$ $(0 < \alpha < 1)$ if

$$P_\theta\big[L(X_1, \ldots, X_n) \leq \theta \leq U(X_1, \ldots, X_n)\big] \geq 1 - \alpha \quad \text{for all} \quad \theta \in \Omega. \quad (1)$$

Also we say that $U(X_1, \ldots, X_n)$ and $L(X_1, \ldots, X_n)$ is an *upper* and a *lower confidence limit* for θ, respectively, with confidence coefficient $1 - \alpha$, if for all $\theta \in \Omega$,

$$P_\theta\big[-\infty < \theta \leq U(X_1, \ldots, X_n)\big] \geq 1 - \alpha$$

and

$$P_\theta\big[L(X_1, \ldots, X_n) \leq \theta < \infty\big] \geq 1 - \alpha. \quad (2)$$

Thus the r. interval $[L(X_1, \ldots, X_n), U(X_1, \ldots, X_n)]$ is a confidence interval for θ with confidence coefficient $1 - \alpha$, if the probability is at least $1 - \alpha$ that the

r. interval $[L(X_1, \ldots, X_n), U(X_1, \ldots, X_n)]$ covers the parameter θ no matter what $\theta \in \Omega$ is.

The interpretation of this statement is as follows: Suppose that the r. experiment under consideration is carried out independently n times, and if x_j is the observed value of X_j, $j = 1, \ldots, n$, construct the interval $[L(x_1, \ldots, x_n), U(x_1, \ldots, x_n)]$. Suppose now that this process is repeated independently N times, so that we obtain N intervals. Then, as N gets larger and larger, at least $(1 - \alpha)N$ of the N intervals will cover the true parameter θ.

A similar interpretation holds true for an upper and a lower confidence limit of θ.

REMARK 1 By relations (1) and (2) and the fact that

$$P_\theta[\theta \geq L(X_1, \ldots, X_n)] + P_\theta[\theta \leq U(X_1, \ldots, X_n)]$$
$$= P_\theta[L(X_1, \ldots, X_n) \leq \theta \leq U(X_1, \ldots, X_n)] + 1,$$

it follows that, if $L(X_1, \ldots, X_n)$ and $U(X_1, \ldots, X_n)$ is a lower and an upper confidence limit for θ, respectively, each with confidence coefficient $1 - \frac{1}{2}\alpha$, then $[L(X_1, \ldots, X_n), U(X_1, \ldots, X_n)]$ is a confidence interval for θ with confidence coefficient $1 - \alpha$. The *length* $l(X_1, \ldots, X_n)$ of this confidence interval is $l = l(X_1, \ldots, X_n) = U(X_1, \ldots, X_n) - L(X_1, \ldots, X_n)$ and the *expected length* is $E_\theta l$, if it exists.

Now it is quite possible that there exist more than one confidence interval for θ with the same confidence coefficient $1 - \alpha$. In such a case, it is obvious that we would be interested in finding the *shortest* confidence interval within a certain class of confidence intervals. This will be done explicitly in a number of interesting examples.

At this point, it should be pointed out that a general procedure for constructing a confidence interval is as follows: We start out with an r.v. $T_n(\theta) = T(X_1, \ldots, X_n; \theta)$ which depends on θ and on the X's only through a sufficient statistic of θ, and whose distribution, under P_θ, is completely determined. Then $L_n = L(X_1, \ldots, X_n)$ and $U_n = U(X_1, \ldots, X_n)$ are some rather simple functions of $T_n(\theta)$ which are chosen in an obvious manner.

The examples which follow illustrate the point.

Exercise

15.1.1 Establish the relation claimed in Remark 1 above.

15.2 Some Examples

We now proceed with the discussion of certain concrete cases. In all of the examples in the present section, the problem is that of constructing a confidence interval (and also the shortest confidence interval within a certain class) for θ with confidence coefficient $1 - \alpha$.

EXAMPLE 1 Let $X_1, \ldots, X_n$ be i.i.d. r.v.'s from $N(\mu, \sigma^2)$. First, suppose that σ is known, so that μ is the parameter, and consider the r.v. $T_n(\mu) = \sqrt{n}(\overline{X} - \mu)/\sigma$. Then $T_n(\mu)$ depends on the X's only through the sufficient statistic $\overline{X}$ of μ and its distribution is $N(0, 1)$ for all μ.

Next, determine two numbers a and b ($a < b$) such that

$$P[a \leq N(0, 1) \leq b] = 1 - \alpha. \tag{3}$$

From (3), we have

$$P_\mu\left[a \leq \frac{\sqrt{n}(\overline{X} - \mu)}{\sigma} \leq b\right] = 1 - \alpha$$

which is equivalent to

$$P_\mu\left(\overline{X} - b\frac{\sigma}{\sqrt{n}} \leq \mu \leq \overline{X} - a\frac{\sigma}{\sqrt{n}}\right) = 1 - \alpha.$$

Therefore

$$\left[\overline{X} - b\frac{\sigma}{\sqrt{n}}, \overline{X} - a\frac{\sigma}{\sqrt{n}}\right] \tag{4}$$

is a confidence interval for μ with confidence coefficient $1 - \alpha$. Its length is equal to $(b - a)\sigma/\sqrt{n}$. From this it follows that, among all confidence intervals with confidence coefficient $1 - \alpha$ which are of the form (4), the shortest one is that for which $b - a$ is smallest, where a and b satisfy (3). It can be seen (see also Exercise 15.2.1) that this happens if $b = c$ (> 0) and $a = -c$, where c is the upper $\alpha/2$ quantile of the $N(0, 1)$ distribution which we denote by $z_{\alpha/2}$. Therefore the shortest confidence interval for μ with confidence coefficient $1 - \alpha$ (and which is of the form (4)) is given by

$$\left[\overline{X} - z_{\alpha/2}\frac{\sigma}{\sqrt{n}}, \overline{X} + z_{\alpha/2}\frac{\sigma}{\sqrt{n}}\right]. \tag{5}$$

Next, assume that μ is known, so that σ^2 is the parameter, and consider the r.v.

$$\overline{T}_n(\sigma^2) = \frac{nS_n^2}{\sigma^2}, \quad \text{where} \quad S_n^2 = \frac{1}{n}\sum_{j=1}^{n}(X_j - \mu)^2.$$

Then $\overline{T}_n(\sigma^2)$ depends on the X's only through the sufficient statistic S_n^2 of σ^2 and its distribution is χ_n^2 for all σ^2.

Now determine two numbers a and b ($0 < a < b$) such that

$$P(a \leq \chi_n^2 \leq b) = 1 - \alpha. \tag{6}$$

From (6), we obtain

$$P_{\sigma^2}\left(a \leq \frac{nS_n^2}{\sigma^2} \leq b\right) = 1 - \alpha$$

which is equivalent to

$$P_{\sigma^2}\left(\frac{nS_n^2}{b} \leq \sigma^2 \leq \frac{nS_n^2}{a}\right) = 1 - \alpha.$$

Therefore

$$\left[\frac{nS_n^2}{b}, \frac{nS_n^2}{a}\right] \tag{7}$$

is a confidence interval for σ^2 with confidence coefficient $1 - \alpha$ and its length is equal to $(1/a - 1/b)nS_n^2$. The expected length is equal to $(1/a - 1/b)n\sigma^2$.

Now, although there are infinite pairs of numbers a and b satisfying (6), in practice they are often chosen by assigning mass $\alpha/2$ to each one of the tails of the χ_n^2 distribution. However, this is not the best choice because then the corresponding interval (7) is not the shortest one. For the determination of the shortest confidence interval, we work as follows. From (6), it is obvious that a and b are not independent of each other but the one is a function of the other. So let $b = b(a)$. Since the length of the confidence interval in (7) is $l = (1/a - 1/b)nS_n^2$, it clearly follows that that a for which l is shortest is given by $dl/da = 0$ which is equivalent to

$$\frac{db}{da} = \frac{b^2}{a^2}. \tag{8}$$

Now, letting G_n and g_n be the d.f. and the p.d.f. of the χ_n^2, relation (6) becomes $G_n(b) - G_n(a) = 1 - \alpha$. Differentiating it with respect to a, one obtains

$$g_n(b)\frac{db}{da} - g_n(a) = 0, \quad \text{or} \quad \frac{db}{da} = \frac{g_n(a)}{g_n(b)}.$$

Thus (8) becomes $a^2 g_n(a) = b^2 g_n(b)$. By means of this result and (6), it follows that a and b are determined by

$$a^2 g_n(a) = b^2 g_n(b) \quad \text{and} \quad \int_a^b g_n(t)\,dt = 1 - \alpha. \tag{9}$$

For the numerical solution of (9), tables are required. Such tables are available (see Table 678 in R. F. Tate and G. W. Klett, "Optimum confidence intervals for the variance of a normal distribution," *Journal of the American Statistical Association*, 1959, Vol. 54, pp. 674–682) for $n = 2(1)29$ and $1 - \alpha = 0.90, 0.95, 0.99, 0.995, 0.999)$.

To summarize then, the shortest (both in actual and expected length) confidence interval for σ^2 with confidence coefficient $1 - \alpha$ (and which is of the form (7)) is given by

$$\left[\frac{nS_n^2}{b}, \frac{nS_n^2}{a}\right],$$

where a and b are determined by (9).

As a numerical application, let $n = 25$, and $1 - \alpha = 0.95$. Then $z_{\alpha/2} = 1.96$, so that (5) gives $[\bar{X} - 0.392, \bar{X} + 0.392]$. Next, for the equal-tails confidence interval given by (7), we have $a = 13.120$ and $b = 40.646$, so that the equal-tails confidence interval itself is given by

$$\left[\frac{25 S_{25}^2}{40.646}, \frac{25 S_{25}^2}{13.120}\right].$$

On the other hand, the shortest confidence interval is equal to

$$\left[\frac{25 S_{25}^2}{45.7051}, \frac{25 S_{25}^2}{14.2636}\right]$$

and the ratio of their lengths is approximately 1.07.

EXAMPLE 2 Let $X_1, \ldots, X_n$ be i.i.d. r.v.'s from the Gamma distribution with parameter β and α a known positive integer, call it r. Then $\sum_{j=1}^{n} X_j$ is a sufficient statistic for β (see Exercise 11.1.2(iii), Chapter 11). Furthermore, for each $j = 1, \ldots, n$, the r.v. $2X_j/\beta$ is χ_{2r}^2, since

$$\phi_{2X_j/\beta}(t) = \phi_{X_j}\left(\frac{2t}{\beta}\right) = \frac{1}{(1-2it)^{2r/2}} \quad \text{(see Chapter 6)}.$$

Therefore

$$T_n(\beta) = \frac{2}{\beta} \sum_{j=1}^{n} X_j$$

is χ_{2rn}^2 for all $\beta > 0$. Now determine a and b ($0 < a < b$) such that

$$P(a \leq \chi_{2rn}^2 \leq b) = 1 - \alpha. \tag{10}$$

From (10), we obtain

$$P_\beta\left(a \leq \frac{2}{\beta} \sum_{j=1}^{n} X_j \leq b\right) = 1 - \alpha$$

which is equivalent to

$$P_\beta\left(2 \sum_{j=1}^{n} X_j / b \leq \beta \leq 2 \sum_{j=1}^{n} X_j / a\right) = 1 - \alpha.$$

Therefore a confidence interval with confidence coefficient $1 - \alpha$ is given by

15 Confidence Regions—Tolerance Intervals

$$\left[\frac{2\sum_{j=1}^{n} X_j}{b}, \frac{2\sum_{j=1}^{n} X_j}{a}\right]. \tag{11}$$

Its length and expected length are, respectively,

$$l = 2\left(\frac{1}{a} - \frac{1}{b}\right)\sum_{j=1}^{n} X_j, \quad E_\beta l = 2\beta rn\left(\frac{1}{a} - \frac{1}{b}\right).$$

As in the second part of Example 1, it follows that the equal-tails confidence interval, which is customarily employed, is not the shortest among those of the form (11).

In order to determine the shortest confidence interval, one has to minimize l subject to (10). But this is the same problem as the one we solved in the second part of Example 1. It follows then that the shortest (both in actual and expected length) confidence interval with confidence coefficient $1 - \alpha$ (which is of the form (11)) is given by (11) with a and b determined by

$$a^2 g_{2rn}(a) = b^2 g_{2rn}(b) \quad \text{and} \quad \int_a^b g_{2rn}(t)\,dt = 1 - \alpha.$$

For instance, for $n = 7$, $r = 2$ and $1 - \alpha = 0.95$, we have, by means of the tables cited in Example 1, $a = 16.5128$ and $b = 49.3675$. Thus the corresponding shortest confidence interval is then

$$\left[\frac{2\sum_{j=1}^{7} X_j}{49.3675}, \frac{2\sum_{j=1}^{7} X_j}{16.5128}\right].$$

The equal-tails confidence interval is

$$\left[\frac{2\sum_{j=1}^{7} X_j}{44.461}, \frac{2\sum_{j=1}^{7} X_j}{15.308}\right],$$

so that the ratio of their length is approximately equal to 1.075.

EXAMPLE 3 Let $X_1, \ldots, X_n$ be i.i.d. r.v.'s from the Beta distribution with $\beta = 1$ and $\alpha = \theta$ unknown.

Then $\prod_{j=1}^{n} X_j$, or $-\sum_{j=1}^{n} \log X_j$ is a sufficient statistic for θ. (See Exercise 11.1.2(iv) in Chapter 11.) Consider the r.v. $Y_j = -2\theta \log X_j$. It is easily seen that its p.d.f. is $\frac{1}{2}\exp(-y_j/2)$, $y_j > 0$, which is the p.d.f. of a χ_2^2. This shows that

$$T_n(\theta) = -2\theta \sum_{j=1}^{n} \log X_j = \sum_{j=1}^{n} Y_j$$

is distributed as χ_{2n}^2. Now determine a and b ($0 < a < b$) such that

$$P(a \leq \chi_{2n}^2 \leq b) = 1 - \alpha. \tag{12}$$

From (12), we obtain

$$P_\theta\left(a \le -2\theta \sum_{j=1}^{n} \log X_j \le b\right) = 1 - \alpha$$

which is equivalent to

$$P_\theta\left(a \bigg/ -2\sum_{j=1}^{n} \log X_j \le \theta \le b \bigg/ -2\sum_{j=1}^{n} \log X_j\right) = 1 - \alpha.$$

Therefore a confidence interval for θ with confidence coefficient $1 - \alpha$ is given by

$$\left[-\frac{a}{2\sum_{j=1}^{n} \log X_j}, -\frac{b}{2\sum_{j=1}^{n} \log X_j}\right]. \tag{13}$$

Its length is equal to

$$l = \frac{a-b}{2\sum_{j=1}^{n} \log X_j}.$$

Considering $dl/da = 0$ in conjunction with (12) in the same way as it was done in Example 2, we have that the shortest (both in actual and expected length) confidence interval (which is of the form (13)) is found by numerically solving the equations

$$g_{2n}(a) = g_{2n}(b) \quad \text{and} \quad \int_a^b g_{2n}(t)\,dt = 1 - \alpha.$$

However, no tables which would facilitate this solution are available.

For example, for $n = 25$ and $1 - \alpha = 0.95$, the equal-tails confidence interval for θ is given by (13) with $a = 32.357$ and $b = 71.420$.

EXAMPLE 4 Let $X_1, \ldots, X_n$ be i.i.d. r.v.'s from $U(0, \theta)$. Then $Y_n = X_{(n)}$ is a sufficient statistic for θ (see Example 7, Chapter 11) and its p.d.f. g_n is given by

$$g_n(y_n) = \frac{n}{\theta^n} y_n^{n-1}, \quad 0 \le y_n \le \theta \quad \text{(by Example 3, Chapter 10)}.$$

Consider the r.v. $T_n(\theta) = Y_n/\theta$. Its p.d.f. is easily seen to be given by

$$h_n(t) = n t^{n-1}, \quad 0 \le t \le 1.$$

Then define a and b with $0 \le a < b \le 1$ and such that

$$P_\theta\left[a \le T_n(\theta) \le b\right] = \int_a^b n t^{n-1}\,dt = b^n - a^n = 1 - \alpha. \tag{14}$$

From (14), we get $P_\theta(a \leq Y_n/\theta \leq b) = 1 - \alpha$ which is equivalent to $P_\theta[X_{(n)}/b \leq \theta \leq X_{(n)}/a] = 1 - \alpha$. Therefore a confidence interval for θ with confidence coefficient $1 - \alpha$ is given by

$$\left[\frac{X_{(n)}}{b}, \frac{X_{(n)}}{a}\right] \tag{15}$$

and its length is $l = (1/a - 1/b)X_{(n)}$. From this, we have

$$\frac{dl}{db} = X_{(n)}\left(-\frac{1}{a^2}\frac{da}{db} + \frac{1}{b^2}\right),$$

while by way of (14), $da/db = b^{n-1}/a^{n-1}$, so that

$$\frac{dl}{db} = X_{(n)}\frac{a^{n+1} - b^{n+1}}{b^2 a^{n+1}}.$$

Since this is less than 0 for all b, l is decreasing as a function of b and its minimum is obtained for $b = 1$, in which case $a = \alpha^{1/n}$, by means of (14). Therefore the shortest (both in actual and expected length) confidence interval with confidence coefficient $1 - \alpha$ (which is the form (15)) is given by

$$\left[X_{(n)}, \frac{X_{(n)}}{\alpha^{1/n}}\right].$$

For example, for $n = 32$ and $1 - \alpha = 0.95$, we have approximately $[X_{(32)}, 1.098 X_{(32)}]$.

Exercises 15.2.5–15.2.7 at the end of this section are treated along the same lines with the examples already discussed and provide additional interesting cases, where shortest confidence intervals exist. The inclusion of the discussions in relation to shortest confidence intervals in the previous examples, and the exercises just mentioned, has been motivated by a paper by W. C. Guenther on "Shortest confidence intervals" in *The American Statistican*, 1969, Vol. 23, Number 1.

Exercises

15.2.1 Let Φ be the d.f. of the $N(0, 1)$ distribution and let a and b with $a < b$ be such that $\Phi(b) - \Phi(a) = \gamma (0 < \gamma < 1)$. Show that $b - a$ is minimum if $b = c \, (> 0)$ and $a = -c$. (See also the discussion of the second part of Example 1.)

15.2.2 Let $X_1, \ldots, X_n$ be independent r.v.'s having the Negative Exponential distribution with parameter $\theta \in \Omega = (0, \infty)$, and set $U = \sum_{i=1}^n X_i$.

i) Show that the r.v. U is distributed as Gamma with parameters (n, θ) and that the r.v. $2U/\theta$ is distributed as χ^2_{2n};

ii) Use part (i) to construct a confidence interval for θ with confidence coefficient $1 - \alpha$.

15.2.3

i) If the r.v. X has the Negative Exponential distribution with parameter $\theta \in \Omega = (0, \infty)$, show that the reliability $R(x; \theta) = P_\theta(X > x)$ $(x > 0)$ is equal to $e^{-\theta x}$;

ii) If $X_1, \ldots, X_n$ is a random sample from the distribution in part (i) and $U = \sum_{i=1}^{n} X_i$, then (by Exercise 15.2.2(i)) $2U/\theta$ is distributed as χ^2_{2n}. Use this fact and part (i) of this exercise to construct a confidence interval for $R(x; \theta)$ with confidence coefficient $1 - \alpha$.

15.2.4 Refer to Example 4 and set $R = X_{(n)} - X_{(1)}$. Then:

i) Find the distribution of R;

(Hint: Take $c \in (0, 1)$ such that $P_\theta\left(c \leq \dfrac{R}{\theta} \leq 1\right) = 1 - \alpha$.)

ii) Show that a confidence interval for θ, based on R, with confidence coefficient $1 - \alpha$ is of the form $[R, R/c]$, where c is a root of the equation

$$c^{n-1}\left[n - (n-1)c\right] = \alpha$$

iii) Show that the expected length of the shortest confidence interval in Example 4 is shorter than that of the confidence interval in (ii) above.

15.2.5 Let $X_1, \ldots, X_n$ be i.i.d. r.v.'s with p.d.f. given by

$$f(x; \theta) = e^{-(x-\theta)} I_{(\theta, \infty)}(x), \qquad \theta \in \Omega = \mathbb{R}$$

and set $Y_1 = X_{(1)}$. Then show that:

i) The p.d.f. g of Y_1 is given by $g(y) = ne^{-n(y-\theta)} I_{(\theta, \infty)}(y)$

ii) The r.v. $T_n(\theta) = 2n(Y_1 - \theta)$ is distributed as χ^2_2;

iii) A confidence interval for θ, based on $T_n(\theta)$, with confidence coefficient $1 - \alpha$ is of the form $[Y_1 - (b/2n), Y_1 - (a/2n)]$;

iv) The shortest confidence interval of the form given in (iii) is provided by

$$\left[Y_1 - \dfrac{\chi^2_{2;\alpha}}{2n},\ Y_1\right],$$

where $\chi^2_2;\alpha$ is the upper αth quantile of the χ^2_2 distribution.

15.2.6 Let $X_1, \ldots, X_n$ be independent r.v.'s having the Weibull p.d.f. given in Exercise 11.4.2, Chapter 11. Then show that:

i) The r.v. $T_n(\theta) = 2Y/\theta$ is distributed as χ^2_{2n}, where $Y = \sum_{j=1}^{n} X_j^\gamma$;

ii) A confidence interval for θ, based on $T_n(\theta)$, with confidence coefficient $1 - \alpha$ is of the form $[2Y/b, 2Y/a]$;

iii) The shortest confidence interval of the form given in (ii) is taken for a and b satisfying the equations

$$\int_a^b g_{2n}(t)\,dt = 1-\alpha \quad \text{and} \quad a^2 g_{2n}(a) = b^2 g_{2n}(b),$$

where g_{2n} is the p.d.f. of the χ^2_{2n} distribution.

15.2.7 Let $X_1, \ldots, X_n$ be i.i.d. r.v.'s with p.d.f. given by

$$f(x;\theta) = \frac{1}{2\theta} e^{|x|/\theta}, \quad \theta \in \Omega = (0, \infty).$$

Then show that:

i) The r.v. $T_n(\theta) = \dfrac{2Y}{\theta}$ is distributed as χ^2_{2n}, where $Y = \sum_{j=1}^n |X_j|$;

ii) and **(iii)** as in Exercise 15.2.6.

15.2.8 Consider the independent random samples $X_1, \ldots, X_m$ from $N(\mu_1, \sigma_1^2)$ and $Y_1, \ldots, Y_n$ from $N(\mu_2, \sigma_2^2)$, where σ_1, σ_2 are known and μ_1, μ_2 are unknown, and let the r.v. $T_{m,n}(\mu_1 - \mu_2)$ be defined by

$$T_{m,n}(\mu_1 - \mu_2) = \frac{(\overline{X}_m - \overline{Y}_n) - (\mu_1 - \mu_2)}{\sqrt{(\sigma_1^2/m) + (\sigma_2^2/n)}}.$$

Then show that:

i) A confidence interval for $\mu_1 - \mu_2$, based on $T_{m,n}(\mu_1 - \mu_2)$, with confidence coefficient $1-\alpha$ is given by

$$\left[(\overline{X}_m - \overline{Y}_n) - b\sqrt{\frac{\sigma_1^2}{m} + \frac{\sigma_2^2}{n}},\ (\overline{X}_m - \overline{Y}_n) - a\sqrt{\frac{\sigma_1^2}{m} + \frac{\sigma_2^2}{n}} \right],$$

where a and b are such that $\Phi(b) - \Phi(a) = 1 - \alpha$;

ii) The shortest confidence interval of the aforementioned form is provided by the last expression above with $-a = b = z_{\frac{\alpha}{2}}$.

15.2.9 Refer to Exercise 15.2.8, but now suppose that μ_1, μ_2 are known and σ_1, σ_2 are unknown. Consider the r.v.

$$\overline{T}_{m,n}\left(\frac{\sigma_1}{\sigma_2}\right) = \frac{\sigma_1^2}{\sigma_2^2} \frac{S_n^2}{S_m^2}$$

and show that a confidence interval for σ_1^2/σ_2^2, based on $\overline{T}_{m,n}(\sigma_1/\sigma_2)$, with confidence coefficient $1-\alpha$ is given by

$$\left[a\frac{S_m^2}{S_n^2},\ b\frac{S_m^2}{S_n^2} \right],$$

where $0 < a < b$ are such that $P(a \leq F_{n,m} \leq b) = 1 - \alpha$. In particular, the equal-tails confidence interval is provided by the last expression above with $a = F'_{n,m;\alpha/2}$ and $b = F_{n,m;\alpha/2}$, where $F'_{n,m;\alpha/2}$ and $F_{n,m;\alpha/2}$ are the lower and the upper $\alpha/2$ quantiles, respectively, of $F_{n,m}$.

15.2.10 Let $X_1, \ldots, X_m$ and $Y_1, \ldots, Y_n$ be independent random samples from the Negative Exponential distributions with parameters θ_1 and θ_2, respectively, and set $U = \sum_{i=1}^{m} X_i$, $V = \sum_{i=1}^{n} Y_j$. Then (by Exercise 15.2.2(i)) the independent r.v.'s $2U/\theta_1$ and $2V/\theta_2$ are distributed as χ^2_{2m} and χ^2_{2n}, respectively, so that the r.v. $\frac{2V}{\theta_2} / \frac{2U}{\theta_1}$ is distributed as $F_{2n,2m}$. Use this result in order to construct a confidence interval for θ_1/θ_2 with confidence coefficient $1 - \alpha$.

15.3 Confidence Intervals in the Presence of Nuisance Parameters

So far we have been concerned with the problem of constructing a confidence interval for a real-valued parameter when no other parameters are present. However, in many interesting examples, in addition to the real-valued parameter of main interest, some other (*nuisance*) parameters do appear in the p.d.f. under consideration.

In such cases, we replace the nuisance parameters by appropriate estimators and then proceed as before.

The examples below illustrate the relevant procedure.

EXAMPLE 5 Refer to Example 1 and suppose that both μ and σ are unknown.

First, we suppose that we are interested in constructing a confidence interval for μ. For this purpose, consider the r.v. $T_n(\mu)$ of Example 1 and replace σ^2 by its usual estimator

$$S^2_{n-1} = \frac{1}{n-1} \sum_{j=1}^{n} (X_j - \overline{X}_n)^2.$$

Thus we obtain the new r.v. $T'_n(\mu) = \sqrt{n}(\overline{X}_n - \mu)/S_{n-1}$ which depends on the X's only through the sufficient statistic $(\overline{X}_n, S^2_{n-1})'$ of $(\mu, \sigma^2)'$. Basing the confidence interval in question on $T_n(\mu)$, which is t_{n-1} distributed, and working as in Example 1, we obtain a confidence interval of the form

$$\left[\overline{X}_n - b\frac{S_{n-1}}{\sqrt{n}}, \; \overline{X}_n - a\frac{S_{n-1}}{\sqrt{n}} \right]. \tag{16}$$

Furthermore, an argument similar to the one employed in Example 1 implies that the shortest (both in actual and expected length) confidence interval of the form (16) is given by

$$\left[\overline{X}_n - t_{n-1;\alpha/2}\frac{S_{n-1}}{\sqrt{n}}, \; \overline{X}_n + t_{n-1;\alpha/2}\frac{S_{n-1}}{\sqrt{n}} \right], \tag{17}$$

where $t_{n-1;\alpha/2}$ is the upper $\alpha/2$ quantile of the t_{n-1} distribution. For instance, for $n = 25$ and $1 - \alpha = 0.95$, the corresponding confidence interval for μ is taken from (17) with $t_{24;0.025} = 2.0639$. Thus we have approximately $[\bar{X}_n - 0.41278 S_{24}, \bar{X}_n + 0.41278 S_{24}]$.

Suppose now that we wish to construct a confidence interval for σ^2. To this end, modify the r.v. $\bar{T}_n(\sigma^2)$ of Example 1 as follows:

$$\bar{T}'_n(\sigma^2) = \frac{(n-1)S_{n-1}}{\sigma^2},$$

so that $\bar{T}'_n(\sigma^2)$ is χ^2_{n-1} distributed. Proceeding as in the corresponding case of Example 1, one has the following confidence interval for σ^2:

$$\left[\frac{(n-1)S^2_{n-1}}{b}, \frac{(n-1)S^2_{n-1}}{a} \right], \tag{18}$$

and the shortest confidence interval of this form is taken when a and b are numerical solutions of the equations

$$a^2 g_{n-1}(a) = b^2 g_{n-1}(b) \quad \text{and} \quad \int_a^b g_{n-1}(t)\,dt = 1 - \alpha.$$

Thus with n and $1 - \alpha$ as above, one has, by means of the tables cited in Example 1, $a = 13.5227$ and $b = 44.4802$, so that the corresponding interval approximately is equal to $[0.539 S^2_{24}, 1.775 S^2_{24}]$.

EXAMPLE 6 Consider the independent r. samples $X_1, \ldots, X_m$ from $N(\mu_1, \sigma_1^2)$ and $Y_1, \ldots, Y_n$ from $N(\mu_2, \sigma_2^2)$, where all μ_1, μ_2, σ_1 and σ_2 are unknown.

First, suppose that a confidence interval for $\mu_1 - \mu_2$ is desired. For this purpose, we have to assume that $\sigma_1 = \sigma_2 = \sigma$, say (unspecified).

Consider the r.v.

$$T_{m,n}(\mu_1 - \mu_2) = \frac{(\bar{X}_m - \bar{Y}_n) - (\mu_1 - \mu_2)}{\sqrt{\frac{(m-1)S^2_{m-1} + (n-1)S^2_{n-1}}{m+n-2} \left(\frac{1}{m} + \frac{1}{n} \right)}}.$$

Then $T_{m,n}(\mu_1 - \mu_2)$ is distributed as t_{m+n-2}. Thus, as in the first case of Example 1 (and also Example 5), the shortest (both in actual and expected length) confidence interval based on $T_{m,n}(\mu_1 - \mu_2)$ is given by

$$\left[(\bar{X}_m - \bar{Y}_n) - t_{m+n-2;\alpha/2} \sqrt{\frac{(m-1)S^2_{m-1} + (n-1)S^2_{n-1}}{m+n-2} \left(\frac{1}{m} + \frac{1}{n} \right)}, \right.$$

$$\left. (\bar{X}_m - \bar{Y}_n) + t_{m+n-2;\alpha/2} \sqrt{\frac{(m-1)S^2_{m-1} + (n-1)S^2_{n-1}}{m+n-2} \left(\frac{1}{m} + \frac{1}{n} \right)} \right].$$

15.3 Confidence Intervals in the Presence of Nuisance Parameters

For instance, for $m = 13$, $n = 14$ and $1 - \alpha = 0.95$, we have $t_{25;0.025} = 2.0595$, so that the corresponding interval approximately is equal to

$$\left[(\overline{X}_{13} - \overline{Y}_{14}) - 0.1586\sqrt{12S_{12}^2 + 13S_{13}^2}, \quad (\overline{X}_{13} - \overline{Y}_{14}) + 0.1586\sqrt{12S_{12}^2 + 13S_{13}^2} \right].$$

If our interest lies in constructing a confidence interval for σ_1^2/σ_2^2, we consider the r.v.

$$\overline{T}_{m,n}\left(\frac{\sigma_1}{\sigma_2}\right) = \frac{\sigma_1^2}{\sigma_2^2} \frac{S_{n-1}^2}{S_{m-1}^2}$$

which is distributed as $F_{n-1,m-1}$. Now determine two numbers a and b with $0 < a < b$ and such that

$$P(a \leq F_{n-1,m-1} \leq b) = 1 - \alpha.$$

Then

$$P_{\sigma_1/\sigma_2}\left(a \leq \frac{\sigma_1^2}{\sigma_2^2} \frac{S_{n-1}^2}{S_{m-1}^2} \leq b\right) = 1 - \alpha,$$

or

$$P_{\sigma_1/\sigma_2}\left(a \frac{S_{m-1}^2}{S_{n-1}^2} \leq \frac{\sigma_1^2}{\sigma_2^2} \leq b \frac{S_{m-1}^2}{S_{n-1}^2}\right) = 1 - \alpha.$$

Therefore a confidence interval for σ_1^2/σ_2^2 is given by

$$\left[a\frac{S_{m-1}^2}{S_{n-1}^2}, \quad b\frac{S_{m-1}^2}{S_{n-1}^2} \right].$$

In particular, the equal-tails confidence interval is provided by

$$\left[\frac{S_{m-1}^2}{S_{n-1}^2} F'_{n-1,m-1;\alpha/2}, \quad \frac{S_{m-1}^2}{S_{n-1}^2} F_{n-1,m-1;\alpha/2} \right],$$

where $F'_{n-1,m-1;\alpha/2}$ and $F_{n-1,m-1;\alpha/2}$ are the lower and the upper $\alpha/2$-quantiles of $F_{n-1,m-1}$. The point $F_{n-1,m-1;\alpha/2}$ is read off the F-tables and the point $F'_{n-1,m-1;\alpha/2}$ is given by

$$F'_{n-1,m-1;\alpha/2} = \frac{1}{F_{m-1,n-1;\alpha/2}}.$$

Thus, for the previous values of m, n and $1 - \alpha$, we have $F_{13,12;0.025} = 3.2388$ and $F_{12,13;0.025} = 3.1532$, so that the corresponding interval approximately is equal to

$$\left[0.3171 \frac{S_{12}^2}{S_{13}^2}, \quad 3.2388 \frac{S_{12}^2}{S_{13}^2} \right].$$

Exercise

15.3.1 Let $X_1, \ldots, X_n$ be independent r.v.'s distributed as $N(\mu, \sigma^2)$. Derive a confidence interval for σ with confidence coefficient $1 - \alpha$ when μ is unknown.

15.4 Confidence Regions—Approximate Confidence Intervals

The concept of a confidence interval can be generalized to that of a *confidence region* in the case that θ is a multi-dimensional parameter. This will be illustrated by means of the following example.

EXAMPLE 7 (Refer to Example 5.) Here the problem is that of constructing a confidence region in $\mathbb{R}^2$ for $(\mu, \sigma^2)'$. To this end, consider the r.v.'s

$$\frac{\sqrt{n}(\bar{X}_n - \mu)}{\sigma} \quad \text{and} \quad \frac{(n-1)S_{n-1}^2}{\sigma^2},$$

which are independently distributed as $N(0, 1)$ and χ_{n-1}^2, respectively. Next, define the constants $c\ (>0)$, a and b $(0 < a < b)$ by

$$P\left[-c \leq N(0, 1) \leq c\right] = \sqrt{1 - \alpha} \quad \text{and} \quad P\left(a \leq \chi_{n-1}^2 \leq b\right) = \sqrt{1 - \alpha}.$$

From these relationships, we obtain

$$P_{\mu,\sigma}\left[-c \leq \frac{\sqrt{n}(\bar{X}_n - \mu)}{\sigma} \leq c, \ a \leq \frac{(n-1)S_{n-1}^2}{\sigma^2} \leq b\right]$$

$$= P_{\mu,\sigma}\left[-c \leq \frac{\sqrt{n}(\bar{X}_n - \mu)}{\sigma} \leq c\right] \times P_{\mu,\sigma}\left[a \leq \frac{(n-1)S_{n-1}^2}{\sigma^2} \leq b\right] = 1 - \alpha.$$

Equivalently,

$$P_{\mu,\sigma}\left[(\mu - \bar{X}_n)^2 \leq \frac{c^2\sigma^2}{n}, \ \frac{(n-1)S_{n-1}^2}{b} \leq \sigma^2 \leq \frac{(n-1)S_{n-1}^2}{a}\right] = 1 - \alpha. \quad (19)$$

For the observed values of the X's, we have the confidence region for $(\mu, \sigma^2)'$ indicated in Fig. 15.1. The quantities a, b and c may be determined so that the resulting intervals are the shortest ones, both in actual and expected lengths.

Now suppose again that θ is real-valued. In all of the examples considered so far the r.v.'s employed for the construction of confidence intervals had an exact and known distribution. There are important examples, however, where this is not the case. That is, no suitable r.v. with known distribution is available which can be used for setting up confidence intervals. In cases like this, under

15.4 Confidence Regions—Approximate Confidence Intervals

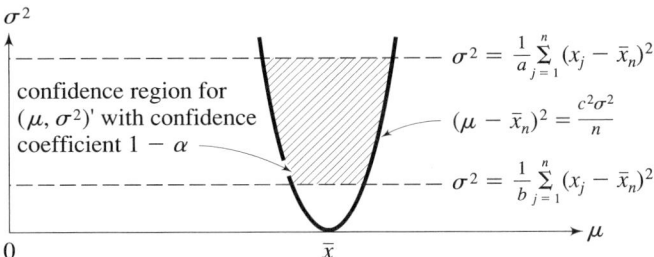

Figure 15.1

appropriate conditions, confidence intervals can be constructed by way of the CLT.

So let $X_1, \ldots, X_n$ be i.i.d. r.v.'s with finite mean and variance μ and σ^2, respectively. Then the CLT applies and gives that $\sqrt{n}(\overline{X}_n - \mu)/\sigma$ is approximately $N(0, 1)$ for large n. Thus, if we assume that σ is known, then a confidence interval for μ with approximate confidence coefficient $1 - \alpha$ is given by (5), provided n is sufficiently large. Suppose now that σ is also unknown. Then since

$$S_n^2 = \frac{1}{n}\sum_{j=1}^{n}\left(X_j - \overline{X}_n\right)^2 \xrightarrow[n\to\infty]{} \sigma^2$$

in probability, we have that $\sqrt{n}(\overline{X}_n - \mu)/S_n$ is again approximately $N(0, 1)$ for large n and therefore a confidence interval for μ with approximate confidence coefficient $1 - \alpha$ is given by (20) below, provided n is sufficiently large.

$$\left[\overline{X}_n - z_{\alpha/2}\frac{S_n}{\sqrt{n}},\ \overline{X}_n + z_{\alpha/2}\frac{S_n}{\sqrt{n}}\right]. \tag{20}$$

As an application, consider the Binomial and Poisson distributions.

EXAMPLE 8 Let $X_1, \ldots, X_n$ be i.i.d. r.v.'s from $B(1, p)$. The problem is that of constructing a confidence interval for p with approximate confidence coefficient $1 - \alpha$. Here $S_n^2 = \overline{X}_n(1 - \overline{X}_n)$, so that (20) becomes

$$\left[\overline{X}_n - z_{\alpha/2}\sqrt{\frac{\overline{X}_n(1-\overline{X}_n)}{n}},\ \overline{X}_n + z_{\alpha/2}\sqrt{\frac{\overline{X}_n(1-\overline{X}_n)}{n}}\right].$$

EXAMPLE 9 Let $X_1, \ldots, X_n$ be i.i.d. r.v.'s from $P(\lambda)$. Then a confidence interval for λ with approximate confidence coefficient $1 - \alpha$ is provided by

$$\overline{X}_n \pm z_{\alpha/2}\sqrt{\frac{\overline{X}_n}{n}}.$$

The two-sample problem also fits into this scheme, provided both means and variances (known or not) are finite.

We close this section with a result which shows that there is an intimate relationship between constructing confidence regions and testing hypotheses. Let $X_1, \ldots, X_n$ be i.i.d. r.v.'s with p.d.f. $f(x; \boldsymbol{\theta})$, $\boldsymbol{\theta} \in \Omega \subseteq \mathbb{R}^r$. For each $\boldsymbol{\theta}^* \in \Omega$ let us consider the problem of testing the hypothesis $H(\boldsymbol{\theta}^*)$: $\boldsymbol{\theta} = \boldsymbol{\theta}^*$ at level of significance α, and let $A(\boldsymbol{\theta}^*)$ stand for the acceptance region in $\mathbb{R}^n$. Set $\mathbf{Z} = (X_1, \ldots, X_n)'$, $\mathbf{z} = (x_1, \ldots, x_n)'$, and define the region $T(\mathbf{z})$ in Ω as follows:

$$T(\mathbf{z}) = \{\boldsymbol{\theta} \in \Omega : \mathbf{z} \in A(\boldsymbol{\theta})\}. \tag{21}$$

In other words, $T(\mathbf{z})$ is that subset of Ω with the following property: On the basis of $\mathbf{z}$, every $H(\boldsymbol{\theta})$ is accepted for $\boldsymbol{\theta} \in T(\mathbf{z})$. From (21), it is obvious that

$$\mathbf{z} \in A(\boldsymbol{\theta}) \quad \text{if and only if} \quad \boldsymbol{\theta} \in T(\mathbf{z}).$$

Therefore

$$P_{\boldsymbol{\theta}}[\boldsymbol{\theta} \in T(\mathbf{Z})] = P_{\boldsymbol{\theta}}[\mathbf{Z} \in A(\boldsymbol{\theta})] \geq 1 - \alpha,$$

so that $T(\mathbf{Z})$ is a confidence region for $\boldsymbol{\theta}$ with confidence coefficient $1 - \alpha$. Thus we have the following theorem.

THEOREM 1 Let $X_1, \ldots, X_n$ be i.i.d. r.v.'s with p.d.f. $f(x; \boldsymbol{\theta})$, $\boldsymbol{\theta} \in \Omega \subseteq \mathbb{R}^r$. For each $\boldsymbol{\theta}^* \in \Omega$, consider the problem of testing $H(\boldsymbol{\theta}^*): \boldsymbol{\theta} = \boldsymbol{\theta}^*$ at level α and let $A(\boldsymbol{\theta}^*)$ be the acceptance region. Set $\mathbf{Z} = (X_1, \ldots, X_n)'$, $\mathbf{z} = (x_1, \ldots, x_n)'$, and define $T(\mathbf{z})$ by (21). Then $T(\mathbf{Z})$ is a confidence region for $\boldsymbol{\theta}$ with confidence coefficient $1 - \alpha$.

Exercises

15.4.1 Let $X_1, \ldots, X_n$ be i.i.d. r.v.'s with (finite) unknown mean μ and (finite) known variance σ^2, and suppose that n is large.

i) Use the CLT to construct a confidence interval for μ with approximate confidence coefficient $1 - \alpha$;

ii) What does this interval become if $n = 100$, $\sigma = 1$ and $\alpha = 0.05$?

iii) Refer to part (i) and determine n so that the length of the confidence interval is 0.1, provided $\sigma = 1$ and $\alpha = 0.05$.

15.4.2 Refer to the previous problem and suppose that both μ and σ^2 are unknown. Then a confidence interval for μ with approximate confidence coefficient $1 - \alpha$ is given be relation (20).

i) What does this interval become for $n = 100$ and $\alpha = 0.05$?

ii) Show that the length of this confidence interval tends to 0 in probability (and also a.s.) as $n \to \infty$;

iii) Discuss part (i) for the case that the underlying distribution is $B(1, \theta)$, $\theta \in \Omega = (0, 1)$ or $P(\theta)$, $\theta \in \Omega = (0, \infty)$.

15.4.3 Let $X_1, \ldots, X_n$ be independent r.v.'s having the Negative Exponential distribution with parameter $\theta \in \Omega = (0, \infty)$ and suppose that n is large. Use the CLT to construct a confidence interval for θ with approximate confidence coefficient $1 - \alpha$. Compare this interval with that constructed in Exercise 15.2.2.

15.4.4 Construct confidence intervals as in Example 1 by utilizing Theorem 1.

15.4.5 Let $X_1, \ldots, X_n$ be i.i.d. r.v.'s with continuous d.f. F. Use Theorem 4 in Chapter 10 to construct a confidence interval for the pth quantile of F, where $p = 0.25, 0.50, 0.75$. Also identify the confidence coefficient $1 - \alpha$ if $n = 10$ for various values of the pair (i, j).

15.4.6 Refer to Example 14, Chapter 12, and show that the posterior p.d.f. of θ, given $x_1, \ldots, x_n$, is Beta with parameters $\alpha + \sum_{j=1}^{n} x_j$ and $\beta + n - \sum_{j=1}^{n} x_j$. Thus if x'_p and x_p are the lower and the upper pth quantiles, respectively, of the Beta p.d.f. mentioned above, it follows that $[x'_p, x_p]$ is a *prediction interval* for θ with confidence coefficient $1 - 2p$. (The term prediction interval rather than confidence interval is more appropriate here, since θ is considered to be an r.v. rather than a parameter. Thus the Bayes method of estimation considered in Section 7 of Chapter 12 also leads to the construction of prediction intervals for θ.)

15.4.7 Refer to Example 15, Chapter 12, and show that the posterior p.d.f. of θ, given $x_1, \ldots, x_n$, is $N((n\bar{x} + \mu)/(n + 1), 1/(n + 1))$. Then work as in Exercise 15.4.6 to find a prediction interval for θ with confidence coefficient $1 - p$. What does this interval become for $p = 0.05$, $n = 9$, $\mu = 1$ and $\bar{x} = 1.5$?

15.5 Tolerance Intervals

In the sections discussed so far, we assumed that $X_1, \ldots, X_n$ were an r. sample with a p.d.f. of known functional form and depending on a parameter θ. Then for the case that θ were real-valued, the problem was that of constructing a confidence interval for θ with a preassigned confidence coefficient. This problem was solved for certain cases.

Now we suppose that the p.d.f. f of the X's is not of known functional form; that is, we assume a *nonparametric* model. Then the concept of a confidence interval, as given in Definition 2, becomes meaningless in the present context. Instead, it is replaced by what is known as a tolerance interval. More precisely, we have the following definition.

DEFINITION 3 Let $X_1, \ldots, X_n$ be i.i.d. r.v.'s with a (nonparametric) d.f. F and let $T_1 = T_1(X_1, \ldots, X_n)$ and $T_2 = T_2(X_1, \ldots, X_n)$ be two statistics of the X's such that $T_1 \leq T_2$. For p and γ with $0 < p, \gamma < 1$, we say that the interval $(T_1, T_2]$ is a 100γ *percent tolerance interval* of $100p$ *percent* of F if $P[F(T_2) - F(T_1) \geq p] \geq \gamma$.

If we notice that for the observed values t_1 and t_2 of T_1 and T_2, respectively, $F(t_2) - F(t_1)$ is the portion of the distribution mass of F which lies in the interval $(t_1, t_2]$, the concept of a tolerance interval has an interpretation analogous to that of a confidence interval. Namely, suppose the r. experiment under consideration is carried out independently n times and let $(t_1, t_2]$ be the resulting interval for the observed values of the X's. Suppose now that this is repeated independently N times, so that we obtain N intervals $(t_1, t_2]$. Then as N gets larger and larger, at least 100γ of the N intervals will cover at least $100p$ percent of the distribution mass of F.

Now regarding the actual construction of tolerance intervals, we have the following result.

THEOREM 2 Let $X_1, \ldots, X_n$ be i.i.d. r.v.'s with p.d.f. f of the continuous type and let $Y_j = X_{(j)}$, $j = 1, \ldots, n$ be the order statistics. Then for any $p \in (0, 1)$ and $1 \leq i < j \leq n$, the r. interval $(Y_i, Y_j]$ is a 100γ percent tolerance interval of $100p$ percent of F, where γ is determined as follows:

$$\gamma = \int_p^1 g_{j-i}(v) dv,$$

g_{j-i} being the p.d.f. of a Beta distribution with parameters $\alpha = j - i$ and $\beta = n - j + i + 1$. (For selected values of p, α and β, $1 - \gamma$ is read off the Incomplete Beta tables.)

PROOF We wish to show that $P[F(Y_j) - F(Y_i) \geq p] = \gamma$. If we set $Z_k = F(Y_k)$, $k = 1, \ldots, n$, this becomes

$$P(Z_j - Z_i \geq p) = \gamma. \tag{22}$$

This suggests that we shall have to find the p.d.f. of $Z_j - Z_i$. Set

$$W_1 = Z_1 \quad \text{and} \quad W_k = Z_k - Z_{k-1}, \quad k = 2, \ldots, n.$$

Then the determinant of the transformation is easily seen to be 1 and therefore Theorem 3 in Chapter 10 gives

$$g(w_1, \ldots, w_n) = \begin{cases} n!, & 0 < w_k, \ k = 1, \ldots, n, \ w_1 + \cdots + w_n < 1 \\ 0, & \text{otherwise.} \end{cases}$$

From the transformation above, we also have

$$Z_j - Z_i = (W_1 + \cdots + W_j) - (W_1 + \cdots + W_i) = W_{i+1} + \cdots + W_j.$$

Thus it suffices to find the p.d.f. of $W_{i+1} + \cdots + W_j$. Actually, if we set $j - i = r$, then it is clear that the p.d.f. of the sum of any consecutive r W's is the same. Accordingly, it suffices to determine the p.d.f. of $W_1 + \cdots + W_r$. For this purpose, use the transformation $V_k = W_1 + \cdots W_k$, $k = 1, \ldots, n$. Then we see that formally we go back to the Z's and therefore

$$g(v_1, \ldots, v_k) = \begin{cases} n!, & 0 < v_1 < \cdots < v_k < 1 \\ 0, & \text{otherwise.} \end{cases}$$

It follows then from Theorem 2(i) in Chapter 10, that the marginal p.d.f. g_r is given by

$$g_r(v) = \begin{cases} \dfrac{n!}{(r-1)!(n-r)!} v^{r-1}(1-v)^{n-r}, & 0 < v < 1 \\ 0, & \text{otherwise.} \end{cases}$$

By taking into consideration that $\Gamma(m) = (m-1)!$, this can be rewritten as follows:

$$g_r(v) = \begin{cases} \dfrac{\Gamma(n+1)}{\Gamma(r)\Gamma(n-r+1)} v^{r-1}(1-v)^{n-r}, & 0 < v < 1 \\ 0, & \text{otherwise.} \end{cases}$$

But this is the p.d.f. of a Beta distribution with parameters $\alpha = r$ and $\beta = n - r + 1$. Since this is also the p.d.f. of $Z_j - Z_i$, it follows that (22) is true, provided γ is determined by

$$\int_p^1 g_r(v)\,dv = \gamma.$$

This completes the proof of the theorem. ▲

Let now f be positive in (a, b) with $-\infty \leq a < b \leq \infty$, so that F is strictly increasing. Then, if X is an r.v. with d.f. F, it follows that for any $p \in (0, 1)$, the r. interval $(-\infty, X]$ covers at most $100p$ percent of the distribution mass of F, and the r. interval (X, ∞) covers at least $100(1 - p)$ percent of the distribution mass of F, each with probability equal to p. In fact,

$$P[F(X) \leq p] = P[X \leq F^{-1}(p)] = F[F^{-1}(p)] = p,$$

so that $(-\infty, X]$ does cover at most p of the distribution mass of F with probability p, and

$$P[F(X) > 1 - p] = 1 - P[F(X) \leq 1 - p] = 1 - F[F^{-1}(1-p)] = 1 - (1-p) = p,$$

so that (X, ∞) does cover at least $1 - p$ of the distribution mass of F with probability p, as was to be seen.

Chapter 16

The General Linear Hypothesis

16.1 Introduction of the Model

In the present chapter, the reader is assumed to be familiar with the basics of linear algebra. However, for the sake of completeness, a brief exposition of the results employed herein is given in Appendix I.

For the introduction of the model, consider a certain chemical or physical experiment which is carried out at each one of the (without appreciable error) selected temperatures $x_j, j = 1, \ldots, n$ which need not be all distinct but they are not all identical either. Assume that a certain aspect of the experiment depends on the temperature and let y_j be some measurements taken at the temperatures $x_j, j = 1, \ldots, n$. Then one has the n pairs $(x_j, y_j), j = 1, \ldots, n$ which can be represented as points in the xy-plane. One question which naturally arises is how we draw a line in the xy-plane which would fit the data best in a certain sense; that is, which would pass through the pairs $(x_j, y_j), j = 1, \ldots, n$ as closely as possible. The reason that this line-fitting problem is important is twofold. First, it reveals the pattern according to which the y's depend on the x's, and secondly, it can be used for prediction purposes.

Quite often, as is seen by inspection, the pairs $(x_j, y_j), j = 1, \ldots, n$ are approximately linearly related; that is, they lie approximately on a straight line. In other cases, a polynomial of higher degree would seem to fit the data better, and still in others, the data is periodic and it is fit best by a trigonometric polynomial.

The underlying idea in all these cases is that, due to random errors in taking measurements, y_j is actually an observed value of a r.v. $Y_j, j = 1, \ldots, n$. If it were not for the r. errors, the pairs $(x_j, y_j), j = 1, \ldots, n$ would be (exactly) related as follows for the three cases considered above.

$$y_j = \beta_1 + \beta_2 x_j, \quad j = 1, \ldots, n \quad (n \geq 2), \tag{1}$$

for some values of the parameters β_1, β_2, or

16.1 Introduction of the Model

$$y_j = \beta_1 + \beta_2 x_j + \cdots + \beta_{k+1} x_j^k, \quad j=1,\ldots,n \quad (2 \le k \le n-1), \qquad (2)$$

for some values of the parameters $\beta_1, \ldots, \beta_{k+1}$, or, finally,

$$y_j = \beta_1 + \beta_2 \cos t_j + \beta_3 \sin t_j + \cdots + \beta_{2k} \cos(kt_j) + \beta_{2k+1} \sin(kt_j),$$
$$j = 1,\ldots,n \quad (n \ge 2k+1), \qquad (3)$$

for some values of the parameters $\beta_1, \ldots, \beta_{2k+1}$.

In the presence of r. errors e_j, $j=1,\ldots,n$, the y's appearing in (1)–(3) are observed values of the following r.v.'s, respectively:

$$Y_j = \beta_1 + \beta_2 x_j + e_j, \quad j=1,\ldots,n \quad (n \ge 2), \qquad (1')$$

$$Y_j = \beta_1 + \beta_2 x_j + \cdots + \beta_{k+1} x_j^k + e_j, \quad j=1,\ldots,n \quad (2 \le k \le n-1), \qquad (2')$$

and $\quad Y_j = \beta_1 + \beta_2 \cos t_j + \beta_3 \sin t_j + \cdots + \beta_{2k} \cos(kt_j) + \beta_{2k+1} \sin(kt_j), + e_j,$
$$j=1,\ldots,n \quad (n \ge 2k+1). \qquad (3')$$

At this point, one observes that the models appearing in relations (1')–(3') are special cases of the following general model:

$$Y_1 = x_{11}\beta_1 + x_{21}\beta_2 + \cdots + x_{p1}\beta_p + e_1$$
$$Y_2 = x_{12}\beta_1 + x_{22}\beta_2 + \cdots + x_{p2}\beta_p + e_2$$
$$\cdots\cdots\cdots\cdots\cdots\cdots\cdots\cdots\cdots$$
$$Y_n = x_{1n}\beta_1 + x_{2n}\beta_2 + \cdots + x_{pn}\beta_p + e_n,$$

or in a more compact form

$$Y_j = \sum_{i=1}^{p} x_{ij}\beta_i + e_j, \quad j=1,\ldots,n \quad \text{with} \quad p \le n \text{ and most often } p < n. \quad (4)$$

By setting

$$\mathbf{Y} = (Y_1,\ldots,Y_n)', \quad \boldsymbol{\beta} = (\beta_1,\ldots,\beta_p)', \quad \mathbf{e} = (e_1,\ldots,e_n)' \quad \text{and}$$

$$\mathbf{X} = \begin{pmatrix} x_{11} & x_{12} & \cdots & x_{1n} \\ x_{21} & x_{22} & \cdots & x_{2n} \\ \cdots & \cdots & \cdots & \cdots \\ x_{p1} & x_{p2} & \cdots & x_{pn} \end{pmatrix}, \quad \text{so that} \quad \mathbf{X}' = \begin{pmatrix} x_{11} & x_{21} & \cdots & x_{p1} \\ x_{12} & x_{22} & \cdots & x_{p2} \\ \cdots & \cdots & \cdots & \cdots \\ x_{1n} & x_{2n} & \cdots & x_{pn} \end{pmatrix}$$

relation (4) is written as follows in matrix notation:

$$\mathbf{Y} = \mathbf{X}'\boldsymbol{\beta} + \mathbf{e}. \qquad (5)$$

The model given by (5) is called the *general linear model* (linear because the parameters $\beta_1,\ldots,\beta_p$ enter the model in their first powers only). At this point, it should be noted that what one actually observes is the r. vector $\mathbf{Y}$, whereas the r. vector $\mathbf{e}$ is unobservable.

DEFINITION 1 Let $\mathbf{C} = (Z_{ij})$ be an $n \times k$ matrix whose elements Z_{ij} are r.v.'s. Then by assuming the EZ_{ij} are finite, the $E\mathbf{C}$ is defined as follows. $E\mathbf{C} = (EZ_{ij})$. In particular, for $\mathbf{Z} = (Z_1, \ldots, Z_n)'$, we have $E\mathbf{Z} = (EZ_1, \ldots, EZ_n)'$, and for $\mathbf{C} = (\mathbf{Z} - E\mathbf{Z})(\mathbf{Z} - E\mathbf{Z})'$, we have $E\mathbf{C} = E[(\mathbf{Z} - E\mathbf{Z})(\mathbf{Z} - E\mathbf{Z})']$. This last expression is denoted by $\mathbf{\Sigma}_z$ and is called the *variance–covariance* matrix of $\mathbf{Z}$, or just the *covariance matrix* of $\mathbf{Z}$. Clearly the (i, j)th element of the $n \times n$ matrix $\mathbf{\Sigma}_z$ is $\text{Cov}(Z_i, Z_j)$, the covariance of Z_i and Z_j, so that the diagonal elements are simply the variances of the Z's.

Since the r.v.'s e_j, $j = 1, \ldots, n$ are r. errors, it is reasonable to assume that $Ee_j = 0$ and that $\sigma^2(e_j) = \sigma^2$, $j = 1, \ldots, n$. Another assumption about the e's which is often made is that they are uncorrelated, that is, $\text{Cov}(e_i, e_j) = 0$, for $i \neq j$. These assumptions are summarized by writing $E(\mathbf{e}) = \mathbf{0}$ and $\mathbf{\Sigma}_e = \sigma^2 \mathbf{I}_n$, where $\mathbf{I}_n$ is the $n \times n$ unit matrix.

By then taking into consideration Definition 1 and the assumptions just made, our model in (5) becomes as follows:

$$\mathbf{Y} = \mathbf{X}'\boldsymbol{\beta} + \mathbf{e}, \quad E\mathbf{Y} = \mathbf{X}'\boldsymbol{\beta} = \boldsymbol{\eta}, \quad \mathbf{\Sigma}_Y = \sigma^2 \mathbf{I}_n, \tag{6}$$

where $\mathbf{e}$ is an $n \times 1$ r. vector, $\mathbf{X}'$ is an $n \times p$ $(p \leq n)$ matrix of known constants, and $\boldsymbol{\beta}$ is a $p \times 1$ vector of parameters, so that $\mathbf{Y}$ is an $n \times 1$ r. vector.

This is the model we are going to concern ourselves with from now on. It should also be mentioned in passing that the expectations η_j of the r.v.'s Y_j, $j = 1, \ldots, n$ are linearly related to the β's and are called *linear regression functions*. This motivates the title of the present chapter.

In the model represented by (6), there are $p + 1$ parameters $\beta_1, \ldots, \beta_p$, σ^2 and the problem is that of estimating these parameters and also testing certain hypotheses about the β's. This is done in the following sections.

16.2 Least Square Estimators—Normal Equations

According to the model assumed in (6), we would expect to have $\boldsymbol{\eta} = \mathbf{X}'\boldsymbol{\beta}$, whereas what we actually observe is $\mathbf{Y} = \mathbf{X}'\boldsymbol{\beta} + \mathbf{e} = \boldsymbol{\eta} + \mathbf{e}$ for some $\boldsymbol{\beta}$. Then the principle of *least squares* (LS) calls for determining $\boldsymbol{\beta}$, so that the difference between what we expect and what we actually observe is minimum. More precisely, $\boldsymbol{\beta}$ is to be determined so that the *sum of squares of errors*,

$$\|\mathbf{Y} - \boldsymbol{\eta}\|^2 = \|\mathbf{e}\|^2 = \sum_{j=1}^{n} e_j^2,$$

is minimum.

DEFINITION 2 Any value of $\boldsymbol{\beta}$ which minimizes the squared norm $\|\mathbf{Y} - \boldsymbol{\eta}\|^2$, where $\boldsymbol{\eta} = \mathbf{X}'\boldsymbol{\beta}$, is called a *least square estimator* (LSE) of $\boldsymbol{\beta}$ and is denoted by $\hat{\boldsymbol{\beta}}$.

The *norm* of an m-dimensional vector $\mathbf{v} = (v_1, \ldots, v_m)'$, denoted by $\|\mathbf{v}\|$, is the usual Euclidean norm, namely

16.2 Least Square Estimators—Normal Equations

$$\|\mathbf{v}\| = \left(\sum_{j=1}^{m} v_j^2\right)^{1/2}.$$

For the pictorial illustration of the principle of LS, let $p = 2$, $x_{1j} = 1$ and $x_{2j} = x_j$, $j = 1, \ldots, n$, so that $\eta_j = \beta_1 + \beta_2 x_j$, $j = 1, \ldots, n$. Thus $(x_j, \eta_j)'$, $j = 1, \ldots, n$ are n points on the straight line $\eta = \beta_1 + \beta_2 x$ and the LS principle specifies that β_1 and β_2 be chosen so that $\sum_{j=1}^{n}(Y_j - \eta_j)^2$ be minimum; Y_j is the (observable) r.v. corresponding to x_j, $j = 1, \ldots, n$. (See also Fig. 16.1.)

(The values of β_1 and β_2 are chosen in order to minimize the quantity $(Y_1 - \eta_1)^2 + \cdots + (Y_5 - \eta_5)^2$.)

From $(\eta_1, \ldots, \eta_n)' = \boldsymbol{\eta} = \mathbf{X}'\boldsymbol{\beta}$, we have that

$$\eta_j = \sum_{i=1}^{p} x_{ij}\beta_i, \quad j = 1, \ldots, n$$

and

$$\|\mathbf{Y} - \boldsymbol{\eta}\|^2 = \sum_{j=1}^{n}(Y_j - \eta_j)^2 = \sum_{j=1}^{n}\left(Y_j - \sum_{i=1}^{p} x_{ij}\beta_i\right)^2,$$

which we denote by $S(\mathbf{Y}, \boldsymbol{\beta})$. Then any LSE is a root of the equations

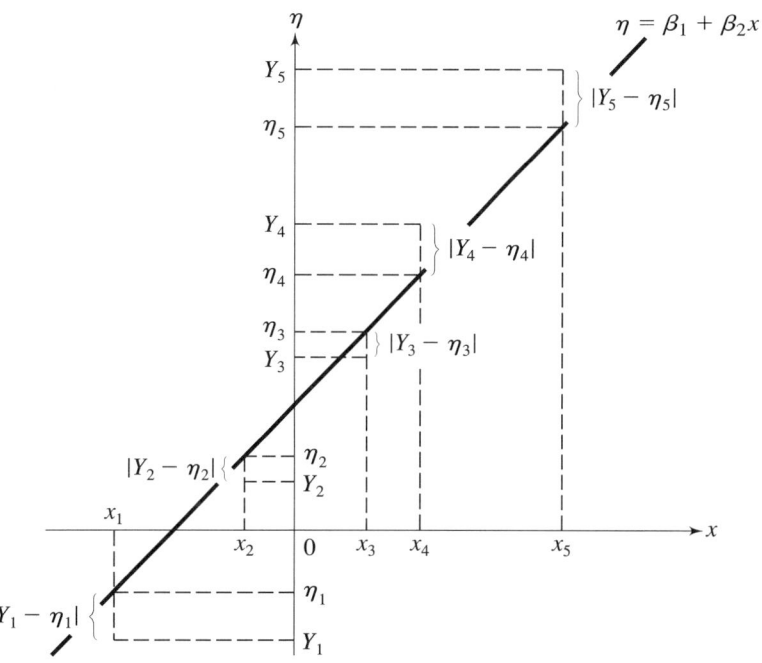

Figure 16.1

$$\frac{\partial}{\partial \beta_v} S(\mathbf{Y}, \boldsymbol{\beta}) = 0, \quad v = 1, \ldots, p$$

which are known as the *normal equations*.
Now

$$\frac{\partial}{\partial \beta_v} S(\mathbf{Y}, \boldsymbol{\beta}) = 2\sum_{j=1}^{n}\left(Y_j - \sum_{i=1}^{p} x_{ij}\beta_j\right)(1 - x_{vj}) = -2\sum_{j=1}^{n} x_{vj}Y_j + 2\sum_{j=1}^{n}\sum_{i=1}^{p} x_{vj}x_{ij}\beta_i,$$

so that the normal equations become

$$\sum_{j=1}^{n}\sum_{i=1}^{p} x_{vj} x_{ij} \beta_i = \sum_{j=1}^{n} x_{vj} Y_j, \quad v = 1, \cdots, p. \tag{7}$$

The equations in (7) are written as follows in matrix notation

$$\mathbf{X X}' \boldsymbol{\beta} = \mathbf{X Y}, \quad \text{or} \quad \mathbf{S} \boldsymbol{\beta} = \mathbf{X Y}, \quad \text{where} \quad \mathbf{S} = \mathbf{X X}'. \tag{7'}$$

Actually, the set of LSE's of $\boldsymbol{\beta}$ coincides with the set of solutions of the normal equations, as the following theorem shows. The normal equations provide a method for the actual calculation of LSE's.

THEOREM 1 Any LSE $\hat{\boldsymbol{\beta}}$ of $\boldsymbol{\beta}$ is a solution of the normal equations and any solution of the normal equations is an LSE.

PROOF We have

$$\boldsymbol{\eta} = \mathbf{X}'\boldsymbol{\beta} = \begin{pmatrix} x_{11} & x_{21} & \cdots & x_{p1} \\ x_{12} & x_{22} & \cdots & x_{p2} \\ \cdots & \cdots & \cdots & \cdots \\ x_{1n} & x_{2n} & \cdots & x_{pn} \end{pmatrix} \begin{pmatrix} \beta_1 \\ \beta_2 \\ \vdots \\ \beta_p \end{pmatrix}$$

$$= (x_{11}\beta_1 + x_{21}\beta_2 + \cdots + x_{p1}\beta_p, \, x_{12}\beta_1 + x_{22}\beta_2 + \cdots$$

$$+ x_{p2}\beta_p, \ldots, x_{1n}\beta_1 + x_{2n}\beta_2 + \cdots + x_{pn}\beta_p)'$$

$$= \beta_1(x_{11}, x_{12}, \ldots, x_{1n})' + \beta_2(x_{21}, x_{22}, \ldots, x_{2n})' + \cdots$$

$$+ \beta_p(x_{p1}, x_{p2}, \ldots, x_{pn})' = \beta_1 \boldsymbol{\xi}_1 + \beta_2 \boldsymbol{\xi}_2 + \cdots + \beta_p \boldsymbol{\xi}_p,$$

where $\boldsymbol{\xi}_j$ is the jth column of $\mathbf{X}'$, $j = 1, \ldots, n$. Thus

$$\boldsymbol{\eta} = \sum_{j=1}^{p} \beta_j \boldsymbol{\xi}_j \quad \text{with} \quad \boldsymbol{\xi}_j, \, j = 1, \ldots, p \quad \text{as above.} \tag{8}$$

Let V_n be the n-dimensional vector space $\mathbb{R}^n$ and let $r \,(\leq p)$ be the rank of $\mathbf{X}$ (= rank $\mathbf{X}'$). Then the vector space V_r generated by $\boldsymbol{\xi}_1, \ldots, \boldsymbol{\xi}_p$ is of dimension r

16.2 Least Square Estimators—Normal Equations

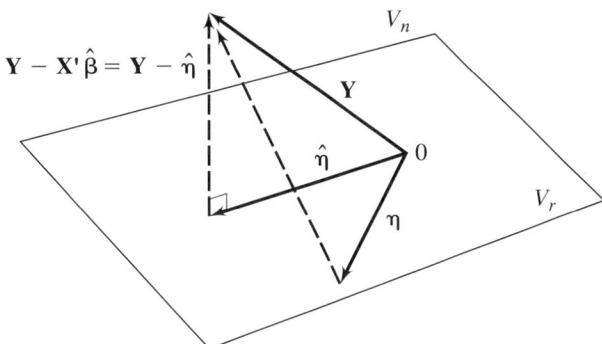

Figure 16.2 ($n = 3$, $r = 2$).

($\leq p$), and $V_r \subseteq V_n$. Of course, $\mathbf{Y} \in V_n$ and from (8), it follows that $\boldsymbol{\eta} \in V_r$. Let $\hat{\boldsymbol{\eta}}$ be the projection of $\mathbf{Y}$ into V_r. Then $\hat{\boldsymbol{\eta}} = \Sigma_{j=1}^{p} \hat{\beta}_j \boldsymbol{\xi}_j$, where $\hat{\beta}_j$, $j = 1, \ldots, p$ may not be uniquely determined ($\hat{\boldsymbol{\eta}}$ is, however) but may be chosen to be functions of $\mathbf{Y}$ since $\hat{\boldsymbol{\eta}}$ is a function of $\mathbf{Y}$. Now, as is well known, $\|\mathbf{Y} - \mathbf{X}'\boldsymbol{\beta}\|^2 = \|\mathbf{Y} - \boldsymbol{\eta}\|^2$ becomes minimum if $\boldsymbol{\eta} = \hat{\boldsymbol{\eta}}$. Thus $\hat{\boldsymbol{\beta}}$ is an LSE of $\boldsymbol{\beta}$ if and only if $\mathbf{X}'\hat{\boldsymbol{\beta}} = \hat{\boldsymbol{\eta}}$, and this is equivalent to saying that $\mathbf{Y} - \mathbf{X}'\hat{\boldsymbol{\beta}} \perp V_r$. Clearly, an equivalent condition to it is that $\mathbf{Y} - \mathbf{X}'\hat{\boldsymbol{\beta}} \perp \boldsymbol{\xi}_j$, $j = 1, \ldots, p$, or $\boldsymbol{\xi}_j'(\mathbf{Y} - \mathbf{X}'\hat{\boldsymbol{\beta}}) = 0$, $j = 1, \ldots, p$. From the definition of $\boldsymbol{\xi}_j$, $j = 1, \ldots, p$, this last condition is equivalent to $\mathbf{X}(\mathbf{Y} - \mathbf{X}'\hat{\boldsymbol{\beta}}) = 0$, or equivalently, $\mathbf{X}\mathbf{X}'\hat{\boldsymbol{\beta}} = \mathbf{X}\mathbf{Y}$, which is the matrix notation for the normal equations. This completes the proof of the theorem. (For a pictorial illustration of some of the arguments used in the proof of the theorem, see Fig. 16.2.) ▲

In the course of the proof of the last theorem, it was seen that there exists at least one LSE $\hat{\boldsymbol{\beta}}$ or $\boldsymbol{\beta}$ and by the theorem itself the totality of LSE's coincides with the set of the solutions of the normal equations. Now a special but important case is that where $\mathbf{X}$ is of full rank, that is, rank $\mathbf{X} = p$. Then $\mathbf{S} = \mathbf{X}\mathbf{X}'$ is a $p \times p$ symmetric matrix of rank p, so that $\mathbf{S}^{-1}$ exists. Therefore the normal equations in (7') provide a unique solution, namely $\hat{\boldsymbol{\beta}} = \mathbf{S}^{-1}\mathbf{X}\mathbf{Y}$. This is part of the following result.

THEOREM 2 If rank $\mathbf{X} = p$, then there exists a unique LSE $\hat{\boldsymbol{\beta}}$ of $\boldsymbol{\beta}$ given by the expression

$$\hat{\boldsymbol{\beta}} = \mathbf{S}^{-1}\mathbf{X}\mathbf{Y}, \quad \text{where} \quad \mathbf{S} = \mathbf{X}\mathbf{X}'. \tag{9}$$

Furthermore, this LSE is linear in Y, unbiased and has covariance matrix given by $\boldsymbol{\Sigma}_{\hat{\boldsymbol{\beta}}} = \sigma^2 \mathbf{S}^{-1}$.

PROOF The existence and uniqueness of the LSE $\hat{\boldsymbol{\beta}}$ and the fact that it is given by (9) have already been established. That it is linear in $\mathbf{Y}$ follows immediately from (9). Next, its unbiasedness is checked thus:

$$E\hat{\boldsymbol{\beta}} = E(\mathbf{S}^{-1}\mathbf{X}\mathbf{Y}) = \mathbf{S}^{-1}\mathbf{X}\, E\, \mathbf{Y} = \mathbf{S}^{-1}\mathbf{X}\mathbf{X}'\boldsymbol{\beta} = \mathbf{S}^{-1}\mathbf{S}\boldsymbol{\beta} = \mathbf{I}_p\boldsymbol{\beta} = \boldsymbol{\beta}.$$

Finally, for the calculation of the covariance of $\hat{\boldsymbol{\beta}}$, we need the following auxiliary result:

$$\boldsymbol{\Sigma}_{\mathbf{AV}} = \mathbf{A}\, \boldsymbol{\Sigma}_{\mathbf{V}} \mathbf{A}', \qquad (10)$$

where $\mathbf{V}$ is an $n \times 1$ r. vector with finite covariances and $\mathbf{A}$ is an $m \times n$ matrix of constants. Relation (10) is established as follows:

$$\boldsymbol{\Sigma}_{\mathbf{AV}} = E\left\{[\mathbf{AV} - E(\mathbf{AV})][\mathbf{AV} - E(\mathbf{AV})]'\right\} = E\left[\mathbf{A}(\mathbf{V} - E\mathbf{V})(\mathbf{V} - E\mathbf{V})'\mathbf{A}'\right]$$

$$= \mathbf{A}\, E\left[(\mathbf{V} - E\mathbf{V})(\mathbf{V} - E\mathbf{V})'\right]\mathbf{A}' = \mathbf{A}\, \boldsymbol{\Sigma}_{\mathbf{V}} \mathbf{A}'.$$

In the present case, $\mathbf{A} = \mathbf{S}^{-1}\mathbf{X}$ and $\mathbf{V} = \mathbf{Y}$ with $\boldsymbol{\Sigma}_{\mathbf{Y}} = \sigma^2 \mathbf{I}_n$, so that

$$\boldsymbol{\Sigma}_{\hat{\boldsymbol{\beta}}} = \sigma^2 \mathbf{S}^{-1}\mathbf{X}(\mathbf{S}^{-1}\mathbf{X})' = \sigma^2 \mathbf{S}^{-1}\mathbf{X}\mathbf{X}'(\mathbf{S}^{-1})' = \sigma^2 \mathbf{S}^{-1}\mathbf{S}(\mathbf{S}^{-1})'$$

$$= \sigma^2 \mathbf{I}_p \mathbf{S}^{-1} = \sigma^2 \mathbf{S}^{-1}$$

because $\mathbf{S}$ and hence $\mathbf{S}^{-1}$ is symmetric, so that $(\mathbf{S}^{-1})' = \mathbf{S}^{-1}$. This completes the proof of the theorem. ▲

The following definition will prove useful in the sequel.

DEFINITION 3 For a known $p \times 1$ vector $\mathbf{c}$, set $\psi = \mathbf{c}'\boldsymbol{\beta}$. Then ψ is called a *parametric function*. A parametric function ψ is called *estimable* if it has an unbiased, linear (in $\mathbf{Y}$) estimator; that is, if there exists a nonrandom $n \times 1$ vector $\mathbf{a}$ such that $E(\mathbf{a}'\mathbf{Y}) = \psi$ identically in $\boldsymbol{\beta}$.

In connection with estimable functions, the following result holds.

LEMMA 1 Let $\psi = \mathbf{c}'\boldsymbol{\beta}$ be an estimable function, so that there exists $\mathbf{a} \in V_n$ such that $E(\mathbf{a}'\mathbf{Y}) = \psi$ identically in $\boldsymbol{\beta}$. Furthermore, let $\mathbf{d}$ be the projection of $\mathbf{a}$ into V_r. Then:

 i) $E(\mathbf{d}'\mathbf{Y}) = \psi$.
 ii) $\sigma^2(\mathbf{a}'\mathbf{Y}) \geq \sigma^2(\mathbf{d}'\mathbf{Y})$.
 iii) $\mathbf{d}'\mathbf{Y} = \mathbf{c}'\hat{\boldsymbol{\beta}}$ for any LSE $\hat{\boldsymbol{\beta}}$ of $\boldsymbol{\beta}$.
 iv) If $\boldsymbol{\alpha}$ is any other nonrandom vector in V_r such that $E(\boldsymbol{\alpha}'\mathbf{Y}) = \psi$, then $\boldsymbol{\alpha} = \mathbf{d}$.

PROOF

 i) The vector $\mathbf{a}$ can be written uniquely as follows: $\mathbf{a} = \mathbf{d} + \mathbf{b}$, where $\mathbf{b} \perp V_r$. Hence

$$\mathbf{a}'\mathbf{Y} = (\mathbf{d} + \mathbf{b})'\mathbf{Y} = \mathbf{d}'\mathbf{Y} + \mathbf{b}'\mathbf{Y}$$

and therefore

$$\psi = E(\mathbf{a}'\mathbf{Y}) = E(\mathbf{d}'\mathbf{Y}) + \mathbf{b}'E\mathbf{Y} = E(\mathbf{d}'\mathbf{Y}) + \mathbf{b}'\mathbf{X}'\boldsymbol{\beta}.$$

But $\mathbf{b}'\mathbf{X}' = 0$ since $\mathbf{b} \perp V_r$ and thus $\mathbf{b} \perp \boldsymbol{\xi}_j$, $j = 1, \ldots, p$, the column vectors of $\overline{\mathbf{X}}'$. Hence $E(\mathbf{d}'\mathbf{Y}) = \psi$.

ii) From the decomposition $\mathbf{a} = \mathbf{d} + \mathbf{b}$ mentioned in (i), it follows that $\|\mathbf{a}\|^2 = \|\mathbf{d}\|^2 + \|\mathbf{b}\|^2$. Next, by means of (10),

$$\sigma^2(\mathbf{a}'\mathbf{Y}) = \mathbf{a}'\Sigma_Y \mathbf{a} = \sigma^2 \|\mathbf{a}\|^2 = \sigma^2 \|\mathbf{d}\|^2 + \sigma^2 \|\mathbf{b}\|^2.$$

Since also

$$\sigma^2 \|\mathbf{d}\|^2 = \sigma^2(\mathbf{d}'\mathbf{Y})$$

by (10) again, we have

$$\sigma^2(\mathbf{a}'\mathbf{Y}) = \sigma^2(\mathbf{d}'\mathbf{Y}) + \sigma^2 \|\mathbf{b}\|^2$$

from which we conclude that

$$\sigma^2(\mathbf{a}'\mathbf{Y}) \geq \sigma^2(\mathbf{d}'\mathbf{Y}).$$

iii) By (i), $E(\mathbf{d}'\mathbf{Y}) = \psi = \mathbf{c}'\boldsymbol{\beta}$ identically in β. But

$$E(\mathbf{d}'\mathbf{Y}) = \mathbf{d}'E\mathbf{Y} = \mathbf{d}'\mathbf{X}'\boldsymbol{\beta},$$

so that $\mathbf{d}'\mathbf{X}\boldsymbol{\beta} = \mathbf{c}'\boldsymbol{\beta}$ identically in β. Hence $\mathbf{d}'\mathbf{X}' = \mathbf{c}'$. Next, with $\hat{\boldsymbol{\eta}} = \mathbf{X}'\hat{\boldsymbol{\beta}}$, the projection of $\mathbf{Y}$ into V_r, one has $\mathbf{d}'(\mathbf{Y} - \hat{\boldsymbol{\eta}}) = 0$, since $\mathbf{d} \in V_r$. Therefore

$$\mathbf{d}'\mathbf{Y} = \mathbf{d}'\hat{\boldsymbol{\eta}} = \mathbf{d}'\mathbf{X}'\hat{\boldsymbol{\beta}} = \mathbf{c}'\hat{\boldsymbol{\beta}}.$$

iv) Finally, let $\boldsymbol{\alpha} \in V_r$ be such that $E(\boldsymbol{\alpha}'\mathbf{Y}) = \psi$. Then we have

$$0 = E(\boldsymbol{\alpha}'\mathbf{Y}) - E(\mathbf{d}'\mathbf{Y}) = E[(\boldsymbol{\alpha}' - \mathbf{d}')\mathbf{Y}] = (\boldsymbol{\alpha}' - \mathbf{d}')\mathbf{X}'\boldsymbol{\beta}.$$

That is, $(\boldsymbol{\alpha}' - \mathbf{d}')\mathbf{X}'\boldsymbol{\beta} = 0$ identically in β and hence $(\boldsymbol{\alpha}' - \mathbf{d}')\mathbf{X}' = \mathbf{0}$, or $(\boldsymbol{\alpha} - \mathbf{d})'\mathbf{X}' = \mathbf{0}$ which is equivalent to saying that $\boldsymbol{\alpha} - \mathbf{d} \perp V_r$. So, both $\boldsymbol{\alpha} - \mathbf{d} \in V_r$ and $\boldsymbol{\alpha} - \mathbf{d} \perp V_r$ and hence $\boldsymbol{\alpha} - \mathbf{d} = \mathbf{0}$. Thus $\boldsymbol{\alpha} = \mathbf{d}$, as was to be seen. ▲

Part (iii) of Lemma 1 justifies the following definition.

DEFINITION 4 Let $\psi = \mathbf{c}'\boldsymbol{\beta}$ be an estimable function. Thus there exists $\mathbf{a} \in V_n$ such that $E(\mathbf{a}'\mathbf{Y}) = \psi$ identically in β, and let $\mathbf{d}$ be the projection of $\mathbf{a}$ into V_r. Set $\hat{\psi} = \mathbf{c}'\hat{\boldsymbol{\beta}}(= \mathbf{d}'\mathbf{Y})$, where $\hat{\boldsymbol{\beta}}$ is any LSE of $\boldsymbol{\beta}$. Then the unbiased, linear (in $\mathbf{Y}$) estimator $\hat{\psi}$ of ψ is called the *LSE of ψ*.

We are now able to formulate and prove the following basic result.

THEOREM 3 (Gauss–Markov) Assume the model described in (6) and let ψ be an estimable function. Then its LSE $\hat{\psi}$ has the smallest variance in the class of all linear in $\mathbf{Y}$ and unbiased estimators of ψ.

PROOF Since ψ is estimable there exists $\mathbf{a} \in V_n$ such that $E(\mathbf{a}'\mathbf{Y}) = \psi$ identically in β, and let $\mathbf{d}$ be the projection of $\mathbf{a}$ into V_r. Then if $\mathbf{b}'\mathbf{Y}$ is any other linear in $\mathbf{Y}$ and unbiased estimator of ψ, it follows, by Lemma 1, that $\sigma^2(\mathbf{b}'\mathbf{Y}) \geq \sigma^2(\mathbf{d}'\mathbf{Y})$. Since $\mathbf{d}'\mathbf{Y} = \hat{\psi}$, the proof is complete. ▲

COROLLARY Suppose that rank $\mathbf{X} = p$. Then for any $\mathbf{c} \in V_p$, the function $\psi = \mathbf{c}'\boldsymbol{\beta}$ is estimable, and hence its LSE $\hat{\psi} = \mathbf{c}'\hat{\boldsymbol{\beta}}$ has the smallest variance in the class of all linear in $\mathbf{Y}$ and unbiased estimators of ψ. In particular, the same is true for each $\hat{\beta}_j$, $j = 1, \ldots, p$, where $\hat{\boldsymbol{\beta}} = (\hat{\beta}_1, \ldots, \hat{\beta}_p)'$.

PROOF The first part follows immediately by the fact that $\hat{\boldsymbol{\beta}} = \mathbf{S}^{-1}\mathbf{XY}$. The particular case follows from the first part by taking $\mathbf{c}$ to have all its components equal to zero except for the jth one which is equal to one, for $j = 1, \ldots, n$. ▲

16.3 Canonical Reduction of the Linear Model—Estimation of σ^2

Assuming the model described in (6), in the previous section we solved the problem of estimating $\boldsymbol{\beta}$ by means of the LS principle. In the present section, we obtain a certain reduction of the linear model under consideration, and as a by-product of it, we also obtain an estimator of the variance σ^2. For this, it will have to be assumed that $r < n$ as will become apparent in the sequel.

Recall that V_r is the r-dimensional vector space generated by the column vectors of $\mathbf{X}'$, where $r = $ rank $\mathbf{X}$, so that $V_r \subseteq V_n$. Let $\{\boldsymbol{\alpha}_1, \ldots, \boldsymbol{\alpha}_r\}$ be an orthonormal basis for V_r (that is, a basis for which $\boldsymbol{\alpha}_i'\boldsymbol{\alpha}_j = 0$, $i \neq j$ and $\|\boldsymbol{\alpha}_j\|^2 = 1$, $j = 1, \ldots, r$). Then this basis can be extended to an orthonormal basis $\{\boldsymbol{\alpha}_1, \ldots, \boldsymbol{\alpha}_r, \boldsymbol{\alpha}_{r+1}, \ldots, \boldsymbol{\alpha}_n\}$ for V_n. Now since $\mathbf{Y} \in V_n$, one has that $\mathbf{Y} = \sum_{j=1}^{n} Z_j \boldsymbol{\alpha}_j$ for certain r.v.'s Z_j, $j = 1, \ldots, n$ to be specified below. It follows that $\boldsymbol{\alpha}_i'\mathbf{Y} = \sum_{j=1}^{n} Z_j \boldsymbol{\alpha}_i'\boldsymbol{\alpha}_j$, so that $Z_i = \boldsymbol{\alpha}_i'\mathbf{Y}$, $i = 1, \ldots, n$. By letting $\mathbf{P}$ be the matrix with rows the vectors $\boldsymbol{\alpha}_i'$, $i = 1, \ldots, n$, the last n equations are summarized as follows: $\mathbf{Z} = \mathbf{PY}$, where $\mathbf{Z} = (Z_1, \ldots, Z_n)'$. From the definition of $\mathbf{P}$, it is clear that $\mathbf{PP}' = \mathbf{I}_n$, so that relation (10) gives

$$\Sigma_Z = \mathbf{P}\sigma^2 \mathbf{I}_n \mathbf{P}' = \sigma^2 \mathbf{I}_n.$$

Thus

$$\sigma^2(Z_j) = \sigma^2, \quad j = 1, \ldots, n. \tag{11}$$

Next, let $E\mathbf{Z} = \boldsymbol{\zeta} = (\zeta_1, \ldots, \zeta_n)'$. Then $\boldsymbol{\zeta} = E(\mathbf{PY}) = \mathbf{P}\boldsymbol{\eta}$, where $\boldsymbol{\eta} \in V_r$. It follows then that

$$\zeta_j = 0, \quad j = r+1, \ldots, n. \tag{12}$$

By recalling that $\hat{\boldsymbol{\eta}}$ is the projection of $\mathbf{Y}$ into V_r, we have

$$\mathbf{Y} = \sum_{j=1}^{n} Z_j \boldsymbol{\alpha}_j \quad \text{and} \quad \hat{\boldsymbol{\eta}} = \sum_{j=1}^{r} Z_j \boldsymbol{\alpha}_j,$$

so that

$$\|\mathbf{Y} - \hat{\boldsymbol{\eta}}\|^2 = \left\|\sum_{j=r+1}^{n} Z_j \boldsymbol{\alpha}_j\right\|^2 = \left(\sum_{j=r+1}^{n} Z_j \boldsymbol{\alpha}_j\right)' \left(\sum_{j=r+1}^{n} Z_j \boldsymbol{\alpha}_j\right) = \sum_{j=r+1}^{n} Z_j^2. \quad (13)$$

From (11) and (12), we get that $EZ_j^2 = \sigma^2$, $j = r + 1, \ldots, n$, so that (13) gives $E\|\mathbf{Y} - \hat{\boldsymbol{\eta}}\|^2 = (n - r)\sigma^2$. Hence $\|\mathbf{Y} - \hat{\boldsymbol{\eta}}\|^2/(n - r)$ is an unbiased estimator of σ^2. (Here is where we use the assumption that $r < n$ in order to ensure that $n - r > 0$.) Thus we have shown the following result.

THEOREM 4 In the model described in (6) with the assumption that $r < n$, an unbiased estimator for σ^2, $\tilde{\sigma}^2$, is provided by $\|\mathbf{Y} - \hat{\boldsymbol{\eta}}\|^2/(n - r)$, where $\hat{\boldsymbol{\eta}}$ is the projection of $\mathbf{Y}$ into V_r and r ($\leq p$) is the rank of $\mathbf{X}$. We may refer to $\tilde{\sigma}^2$ as the *LSE of* σ^2.

Now suppose that rank $\mathbf{X} = p$, so that $\hat{\boldsymbol{\beta}} = \mathbf{S}^{-1}\mathbf{X}\mathbf{Y}$ by Theorem 2. Next, the rows of $\mathbf{X}$ are $\boldsymbol{\xi}_j'$, where $\boldsymbol{\xi}_j$, $j = 1, \ldots, p$ are the column vectors of $\mathbf{X}'$, and $\boldsymbol{\xi}_j \in V_p$, $j = 1, \ldots, p$. Therefore $\boldsymbol{\xi}_j' \boldsymbol{\alpha}_i = 0$ for all $j = 1, \ldots, p$ and $i = p + 1, \ldots, n$. Since $\boldsymbol{\alpha}_j$, $j = 1, \ldots, n$ are the columns of $\mathbf{P}'$, it follows that the last $n - p$ elements in all p rows of the $p \times n$ matrix $\mathbf{XP}'$ are all equal to zero. Now from the transformation $\mathbf{Z} = \mathbf{PY}$ one has $\mathbf{Y} = \mathbf{P}^{-1}\mathbf{Z}$. But $\mathbf{P}^{-1} = \mathbf{P}'$ as follows from the fact that $\mathbf{PP}' = \mathbf{I}_n$. Thus $\mathbf{Y} = \mathbf{P}'\mathbf{Z}$ and therefore $\hat{\boldsymbol{\beta}} = \mathbf{S}^{-1}\mathbf{XP}'\mathbf{Z}$. Because of the special form of the matrix $\mathbf{XP}'$ mentioned above, it follows that $\hat{\boldsymbol{\beta}}$ is a function of Z_j, $j = 1, \ldots, p$ only (see also Exercise 16.3.1), whereas

$$\tilde{\sigma}^2 = \frac{1}{n-p} \sum_{j=p+1}^{n} Z_j^2 \quad \text{by (13)}.$$

By summarizing these results, we have then

COROLLARY Let rank $\mathbf{X} = p$ ($< n$). Then the LSE's $\hat{\boldsymbol{\beta}}$ and $\tilde{\sigma}^2$ are functions only of the transformed r.v.'s Z_j, $j = 1, \ldots, p$ and Z_j, $j = p + 1, \ldots, n$, respectively.

REMARK 1 From the last theorem above, it follows that in order for us to be able to actually calculate the LSE $\tilde{\sigma}^2$ of σ^2, we would have to rewrite $\|\mathbf{Y} - \hat{\boldsymbol{\eta}}\|^2$ in a form appropriate for calculation. To this end, we have

$$\|\mathbf{Y} - \hat{\boldsymbol{\eta}}\|^2 = \|\mathbf{Y} - \mathbf{X}'\hat{\boldsymbol{\beta}}\|^2 = (\mathbf{Y} - \mathbf{X}'\hat{\boldsymbol{\beta}})'(\mathbf{Y} - \mathbf{X}'\hat{\boldsymbol{\beta}})$$
$$= (\mathbf{Y}' - \hat{\boldsymbol{\beta}}'\mathbf{X})(\mathbf{Y} - \mathbf{X}'\hat{\boldsymbol{\beta}}) = \mathbf{Y}'\mathbf{Y} - \mathbf{Y}'\mathbf{X}'\hat{\boldsymbol{\beta}} - \hat{\boldsymbol{\beta}}'\mathbf{X}\mathbf{Y} + \hat{\boldsymbol{\beta}}'\mathbf{X}\mathbf{X}'\hat{\boldsymbol{\beta}}$$
$$= \mathbf{Y}'\mathbf{Y} - \mathbf{Y}'\mathbf{X}'\hat{\boldsymbol{\beta}} + \hat{\boldsymbol{\beta}}'(\mathbf{X}\mathbf{X}'\hat{\boldsymbol{\beta}} - \mathbf{X}\mathbf{Y}).$$

But $\mathbf{Y}'\mathbf{X}'\hat{\boldsymbol{\beta}}$ is $(1 \times n) \times (n \times p) \times (p \times 1) = 1 \times 1$, that is, a number. Hence $\mathbf{Y}'\mathbf{X}'\hat{\boldsymbol{\beta}} = (\mathbf{Y}'\mathbf{X}'\hat{\boldsymbol{\beta}})' = \hat{\boldsymbol{\beta}}'\mathbf{X}\mathbf{Y}$. On the other hand, $\mathbf{X}\mathbf{X}'\hat{\boldsymbol{\beta}} - \mathbf{X}\mathbf{Y} = 0$ since $\mathbf{X}\mathbf{X}'\hat{\boldsymbol{\beta}} = \mathbf{X}\mathbf{Y}$ by the normal equations (7'). Therefore

$$\|\mathbf{Y} - \hat{\boldsymbol{\eta}}\|^2 = \mathbf{Y}'\mathbf{Y} - \hat{\boldsymbol{\beta}}'\mathbf{X}\mathbf{Y} = \sum_{j=1}^{n} Y_j^2 - \hat{\boldsymbol{\beta}}'\mathbf{X}\mathbf{Y}. \quad (14)$$

16 The General Linear Hypothesis

Finally, denoting by r_v the vth element of the $p \times 1$ vector $\mathbf{XY}$, one has

$$r_v = \sum_{j=1}^{n} x_{vj} Y_j, \quad v = 1, \ldots, p \tag{15}$$

and therefore (14) becomes as follows:

$$\|\mathbf{Y} - \hat{\boldsymbol{\eta}}\|^2 = \sum_{j=1}^{n} Y_j^2 - \sum_{v=1}^{p} \hat{\beta}_v r_v. \tag{16}$$

As an application of some of the results obtained so far, consider the following example.

EXAMPLE 1 Let $Y_j = \beta_1 + \beta_2 x_j + e_j$, where $Ee_j = 0$ and $E(e_i e_j) = \sigma^2 \delta_{ij}$, $i, j = 1, \ldots, n$ ($\delta_{ij} = 1$, if $i = j$ and $\delta_{ij} = 0$ if $i \neq j$).

Clearly, this example fits the model described in (6) by taking

$$\mathbf{X}' = \begin{pmatrix} 1 & x_1 \\ 1 & x_2 \\ \cdot & \cdot \\ \cdot & \cdot \\ \cdot & \cdot \\ 1 & x_n \end{pmatrix} \quad \text{and} \quad \boldsymbol{\beta} = (\beta_1, \beta_2)'.$$

Next,

$$\mathbf{XX}' = \begin{pmatrix} 1 & 1 & \cdots & 1 \\ x_1 & x_2 & \cdots & x_n \end{pmatrix} \begin{pmatrix} 1 & x_1 \\ 1 & x_2 \\ \cdot & \cdot \\ \cdot & \cdot \\ \cdot & \cdot \\ 1 & x_n \end{pmatrix} = \begin{pmatrix} n & \sum_{j=1}^{n} x_j \\ \sum_{j=1}^{n} x_j & \sum_{j=1}^{n} x_j^2 \end{pmatrix} = \mathbf{S},$$

so that the normal equations are given by (7') with $\mathbf{S}$ as above and

$$\mathbf{XY} = \begin{pmatrix} 1 & 1 & \cdots & 1 \\ x_1 & x_2 & \cdots & x_n \end{pmatrix} (Y_1, Y_2, \ldots, Y_n)' = \left(\sum_{j=1}^{n} Y_j, \sum_{j=1}^{n} x_j Y_j \right)'. \tag{17}$$

Now

$$|\mathbf{S}| = n \sum_{j=1}^{n} x_j^2 - \left(\sum_{j=1}^{n} x_j \right)^2 = n \sum_{j=1}^{n} (x_j - \bar{x})^2,$$

so that

$$\sum_{j=1}^{n} (x_j - \bar{x})^2 \neq 0,$$

16.3 Canonical Reduction of the Linear Model—Estimation of σ^2

provided that not all x's are equal. Then $\mathbf{S}^{-1}$ exists and is given by

$$\mathbf{S}^{-1} = \frac{1}{n\sum_{j=1}^{n}(x_j-\bar{x})^2}\begin{pmatrix}\sum_{j=1}^{n}x_j^2 & -\sum_{j=1}^{n}x_j \\ -\sum_{j=1}^{n}x_j & n\end{pmatrix}. \tag{18}$$

It follows that

$$\hat{\boldsymbol{\beta}} = \begin{pmatrix}\hat{\beta}_1 \\ \hat{\beta}_2\end{pmatrix} = \mathbf{S}^{-1}\mathbf{X}\mathbf{Y} = \frac{1}{n\sum_{j=1}^{n}(x_j-\bar{x})^2}$$

$$\times \begin{pmatrix}\left(\sum_{j=1}^{n}x_j^2\right)\left(\sum_{j=1}^{n}Y_j\right) - \left(\sum_{j=1}^{n}x_j\right)\left(\sum_{j=1}^{n}x_jY_j\right) \\ \left(-\sum_{j=1}^{n}x_j\right)\left(\sum_{j=1}^{n}Y_j\right) + n\sum_{j=1}^{n}x_jY_j\end{pmatrix},$$

so that

$$\left.\begin{aligned}\hat{\beta}_1 &= \frac{\left(\sum_{j=1}^{n}x_j^2\right)\left(\sum_{j=1}^{n}Y_j\right) - \left(\sum_{j=1}^{n}x_j\right)\left(\sum_{j=1}^{n}x_jY_j\right)}{n\sum_{j=1}^{n}(x_j-\bar{x})^2}, \\ \hat{\beta}_2 &= \frac{n\sum_{j=1}^{n}x_jY_j - \left(\sum_{j=1}^{n}x_j\right)\left(\sum_{j=1}^{n}Y_j\right)}{n\sum_{j=1}^{n}(x_j-\bar{x})^2} \quad\text{and} \\ n\sum_{j=1}^{n}(x_j-\bar{x})^2 &= n\sum_{j=1}^{n}x_j^2 - \left(\sum_{j=1}^{n}x_j\right)^2.\end{aligned}\right\} \tag{19}$$

But

$$n\sum_{j=1}^{n}x_jY_j - \left(\sum_{j=1}^{n}x_j\right)\left(\sum_{j=1}^{n}Y_j\right) = n\sum_{j=1}^{n}(x_j-\bar{x})(Y_j-\bar{Y}),$$

as is easily seen, so that

$$\hat{\beta}_2 = \frac{\sum_{j=1}^{n}(x_j-\bar{x})(Y_j-\bar{Y})}{\sum_{j=1}^{n}(x_j-\bar{x})^2}. \tag{19'}$$

It is also verified (see also Exercise 16.3.2) that
$$\hat{\beta}_1 = \bar{Y} - \hat{\beta}_2 \bar{x}. \tag{19''}$$

The expressions of $\hat{\beta}_1$ and $\hat{\beta}_2$ given by (19'') and (19'), respectively, are more compact, but their expressions given by (19) are more appropriate for actual calculations.

In the present case, (15) gives in conjunction with (17)
$$r_1 = \sum_{j=1}^{n} Y_j \quad \text{and} \quad r_2 = \sum_{j=1}^{n} x_j Y_j,$$

so that (16) becomes
$$\|\mathbf{Y} - \hat{\boldsymbol{\eta}}\|^2 = \sum_{j=1}^{n} Y_j^2 - \hat{\beta}_1 \left(\sum_{j=1}^{n} Y_j \right) - \hat{\beta}_2 \left(\sum_{j=1}^{n} x_j Y_j \right). \tag{20}$$

Since also $r = p = 2$, the LSE of σ^2 is given by
$$\tilde{\sigma}^2 = \frac{\|\mathbf{Y} - \hat{\boldsymbol{\eta}}\|^2}{n-2}. \tag{21}$$

For a numerical example, take $n = 12$ and the x's and Y's as follows:

$x_1 = 30$	$x_7 = 70$	$Y_1 = 37$	$Y_7 = 20$
$x_2 = 30$	$x_8 = 70$	$Y_2 = 43$	$Y_8 = 26$
$x_3 = 30$	$x_9 = 70$	$Y_3 = 30$	$Y_9 = 22$
$x_4 = 50$	$x_{10} = 90$	$Y_4 = 32$	$Y_{10} = 15$
$x_5 = 50$	$x_{11} = 90$	$Y_5 = 27$	$Y_{11} = 19$
$x_6 = 50$	$x_{12} = 90$	$Y_6 = 34$	$Y_{12} = 20.$

Then relation (19) provides us with the estimates $\hat{\beta}_1 = 46.3833$ and $\hat{\beta}_2 = -0.3216$, and (20) and (21) give the estimate $\tilde{\sigma}^2 = 14.8939$.

Exercises

16.3.1 Referring to the proof of the corollary to Theorem 4, elaborate on the assertion that $\hat{\boldsymbol{\beta}}$ is a function of the r.v.'s Z_j, $j = 1, \ldots, r$ alone.

16.3.2 Verify relation (19'').

16.3.3 Show directly, by means of (19') and (19''), that
$$E\hat{\beta}_1 = \beta_1, \quad E\hat{\beta}_2 = \beta_2, \quad \sigma^2(\hat{\beta}_2) = \frac{\sigma^2}{\sum_{j=1}^{n}(x_j - \bar{x})^2}$$

and that $\hat{\beta}_2$ and $\hat{\beta}_1$ are normally distributed if the Y's are normally distributed.

16.3.4 i) Use relation (18) to show that

$$\sigma^2(\hat{\beta}_1) = \frac{\sigma^2 \sum_{j=1}^n x_j^2}{n \sum_{j=1}^n (x_j - \bar{x})^2}, \quad \sigma^2(\hat{\beta}_2) = \frac{\sigma^2}{\sum_{j=1}^n (x_j - \bar{x})^2}$$

and that, if $\bar{x} = 0$, then $\hat{\beta}_1$ and $\hat{\beta}_2$ are uncorrelated;
 Again refer to Example 1 and suppose that $\bar{x} = 0$. Then

ii) Conclude that $\hat{\beta}_1$ is normally distributed if the Y's are normally distributed;

iii) Show that

$$\sigma^2(\hat{\beta}_1) = \frac{\sigma^2}{n}, \quad \sigma^2(\hat{\beta}_2) = \frac{\sigma^2}{\sum_{j=1}^n x_j^2}$$

and, by assuming that n is even and $x_j \in [-x, x]$, $j = 1, \ldots, n$ for some $x > 0$, conclude that $\sigma^2(\hat{\beta}_2)$ becomes a minimum if half of the x's are chosen equal to x and the other half equal to $-x$ (if that is feasible). (It should be pointed out, however, that such a choice of the x's—when feasible—need not be "optimal." This is the case, for example, when there is doubt about the linearity of the model.)

16.3.5 Consider the linear model $Y_j = \beta_1 + \beta_2 x_j + \beta_3 x_j^2 + e_j$, $j = 1, \ldots, n$ under the usual assumptions and bring it under the form (6). Then for $n = 5$ and the data given in the table below find:

i) The LSE's of $\boldsymbol{\beta}$ and σ^2;

ii) The covariance of $\hat{\boldsymbol{\beta}}$ (the LSE of $\boldsymbol{\beta}$);

iii) An estimate of the covariance found in (ii).

x	1	2	3	4	5
y	1.0	1.5	1.3	2.5	1.7

16.4 Testing Hypotheses About $\eta = E(Y)$

The assumptions made in (6) were adequate for the derivation of the results obtained so far. Those assumptions did not specify any particular kind of distribution for the r. vector $\mathbf{e}$ and hence the r. vector $\mathbf{Y}$. However, in order to be able to carry out tests about $\boldsymbol{\eta}$, such an assumption will have to be made now. Since the r.v.'s e_j, $j = 1, \ldots, n$ represent errors in taking measurements, it is reasonable to assume that they are normally distributed. Denoting by (C) the set of conditions assumed so far, we have then

$$(C): \mathbf{Y} = \mathbf{X}'\boldsymbol{\beta} + \mathbf{e}, \quad \mathbf{e}: N(\mathbf{0}, \sigma^2 \mathbf{I}_n), \quad \text{rank } \mathbf{X} = r \, (\leq p < n). \tag{22}$$

The assumption that $\mathbf{e}: N(\mathbf{0}, \sigma^2 \mathbf{I}_n)$ simply states that the r.v.'s e_j, $j = 1, \ldots, n$ are uncorrelated normal (equivalently, independent normal) with mean zero and variance σ^2.

Now from (22), we have that $\boldsymbol{\eta} = E\mathbf{Y} = \mathbf{X}'\boldsymbol{\beta}$ and it was seen in (8) that $\boldsymbol{\eta} \in V_r$, the r-dimensional vector space generated by the column vectors of $\mathbf{X}'$. This means that the coordinates of $\boldsymbol{\eta}$ are not allowed to vary freely but they satisfy $n - r$ independent linear relationships. However, there might be some evidence that the components of $\boldsymbol{\eta}$ satisfy $0 < q \leq r$ additional independent linear relationships. This is expressed by saying that $\boldsymbol{\eta}$ actually belongs in V_{r-q}, that is, an $(r - q)$-dimensional vector subspace of V_r. Thus we hypothesize that

$$H: \boldsymbol{\eta} \in V_{r-q} \subset V_r \quad (q < r) \tag{23}$$

and denote by $c = C \cap H$, that is, the conditions that our model satisfies if in addition to (C), we also assume H.

For testing the hypothesis H, we are going to use the Likelihood Ratio (LR) test. For this purpose, denote by $f_\mathbf{Y}(\mathbf{y}; \boldsymbol{\beta}, \sigma^2)$, or $f_{Y_1,\ldots,Y_n}(y_1, \ldots, y_n; \boldsymbol{\beta}, \sigma^2)$ the joint p.d.f. of the Y's and let S_C and S_c stand for the minimum of

$$S(\mathbf{y}, \boldsymbol{\beta}) = \|\mathbf{y} - \mathbf{X}'\boldsymbol{\beta}\|^2 = \sum_{j=1}^{n}(y_j - EY_j)^2$$

under C and c, respectively. We have

$$f_\mathbf{Y}(\mathbf{y}: \boldsymbol{\beta}, \sigma^2) = \left(\frac{1}{\sqrt{2\pi\sigma^2}}\right)^n \exp\left[-\frac{1}{2\sigma^2} \sum_{j=1}^{n}(y_j - EY_j)^2\right]$$

$$= \left(\frac{1}{\sqrt{2\pi\sigma^2}}\right)^n \exp\left[-\frac{1}{2\sigma^2} S(\mathbf{y}, \boldsymbol{\beta})\right]. \tag{24}$$

From (24), it is obvious that for a fixed σ^2, the maximum of $f_\mathbf{Y}(\mathbf{Y}; \boldsymbol{\beta}, \sigma^2)$ with respect to $\boldsymbol{\beta}$, under C, is obtained when $S(\mathbf{y}, \boldsymbol{\beta})$ is replaced by S_C. Thus in order to maximize $f_\mathbf{Y}(\mathbf{y}; \boldsymbol{\beta}, \sigma^2)$ with respect to both $\boldsymbol{\beta}$ and σ^2, under C, it suffices to maximize with respect to σ^2 the quantity

$$\left(\frac{1}{\sqrt{2\pi\sigma^2}}\right)^n \exp\left(-\frac{1}{2\sigma^2} S_C\right),$$

or its logarithm

$$-\frac{n}{2}\log(2\pi) - \frac{n}{2}\log\sigma^2 - \frac{S_C}{2}\frac{1}{\sigma^2}.$$

Differentiating with respect to σ^2 this last expression and equating the derivative to zero, we obtain

$$-\frac{n}{2}\frac{1}{\sigma^2} + \frac{S_C}{2}\frac{1}{\sigma^4} = 0, \quad \text{so that} \quad \bar{\sigma}^2 = \frac{S_C}{n}.$$

The second derivative with respect to σ^2 is equal to $n/(2\sigma^4) - (S_C/\sigma^6)$ which for $\sigma^2 = \bar{\sigma}^2$ becomes $-n^3/(2S_C^2) < 0$. Therefore

$$\max_C f_Y(y; \beta, \sigma^2) = \left(\frac{n}{2\pi S_C}\right)^{n/2} \exp\left(-\frac{n}{2}\right). \tag{25}$$

In an entirely analogous way, one also has

$$\max_c f_Y(y; \beta, \sigma^2) = \left(\frac{n}{2\pi S_c}\right)^{n/2} \exp\left(-\frac{n}{2}\right), \tag{26}$$

so that the LR statistic λ is given by

$$\lambda = \left(\frac{S_c}{S_C}\right)^{-n/2} \tag{27}$$

where

$$S_C = \min_C S(Y, \beta) = \|Y - \hat{\eta}_C\|^2 = \|Y - X'\hat{\beta}_C\|^2, \tag{28}$$
$$\hat{\beta}_C = \text{LSE of } \beta \text{ under } C,$$

and

$$S_c = \min_c S(Y, \beta) = \|Y - \hat{\eta}_c\|^2 = \|Y - X'\hat{\beta}_c\|^2, \tag{29}$$
$$\hat{\beta}_c = \text{LSE of } \beta \text{ under } c.$$

The actual calculation of S_C and S_c is done by means of (16), where $\hat{\eta}$ is replaced by $\hat{\eta}_C$ and $\hat{\eta}_c$, respectively.

The LR test rejects H whenever $\lambda < \lambda_0$, where λ_0 is defined, so that the level of the test is α. Now the problem which arises is that of determining the distribution of λ (at least) under H. We will show in the following that the LR test is equivalent to a certain "F-test" based on a statistic whose distribution is the F-distribution, under H, with specified degrees of freedom. To this end, set

$$g(\lambda) = \lambda^{-2/n} - 1. \tag{30}$$

Then

$$\frac{dg(\lambda)}{d\lambda} = -\frac{2}{n}\lambda^{-(n+2)/n} < 0,$$

so that $g(\lambda)$ is decreasing. Thus $\lambda < \lambda_0$ if and only if $g(\lambda) > g(\lambda_0)$.

Taking into consideration relations (27) and (30), the last inequality becomes

$$\frac{n-r}{q}\frac{S_c - S_C}{S_C} > \mathcal{F}_0,$$

where $\mathcal{F}_0$ is determined by

$$P_H\left(\frac{n-r}{q}\frac{S_c - S_C}{S_C} > \mathcal{F}_0\right) = \alpha.$$

Therefore the LR test is equivalent to the test which rejects H whenever

$$\mathcal{F} > \mathcal{F}_0, \quad \text{where} \quad \mathcal{F} = \frac{n-r}{q}\frac{S_c - S_C}{S_C} \quad \text{and} \quad \mathcal{F}_0 = F_{q, n-r_{jx}}. \tag{31}$$

The statistics S_C and S_c are given by (28) and (29), respectively, and the distribution of $\mathcal{F}$, under H, is $F_{q,n-r}$, as is shown in the next section.

Now although the F-test in (31) is justified on the basis that it is equivalent to the LR test, its geometric interpretation illustrated by Fig. 16.3 below illuminates it even further. We have that $\hat{\eta}_C$ is the "best" estimator of η under C and $\hat{\eta}_c$ is the "best" estimator of η under c. Then the F-test rejects H whenever $\hat{\eta}_C$ and $\hat{\eta}_c$ differ by too much; equivalently, whenever $S_c - S_C$ is too large (when measured in terms of S_C).

Suppose now that rank $\mathbf{X} = p\ (< n)$, and let $\hat{\boldsymbol{\beta}}$ be the unique (and unbiased) LSE of $\boldsymbol{\beta}$. By the facet that $(\mathbf{Y} - \hat{\boldsymbol{\eta}})'(\hat{\boldsymbol{\eta}} - \boldsymbol{\eta}) = 0$ because $\mathbf{Y} - \hat{\boldsymbol{\eta}} \perp \hat{\boldsymbol{\eta}} - \boldsymbol{\eta}$, one has that the joint p.d.f. of the Y's is given by the following expression, where $\mathbf{y}$ has been replaced by $\mathbf{Y}$:

$$\left(\frac{1}{\sqrt{2\pi\sigma^2}}\right)^n \exp\left\{-\frac{1}{2\sigma^2}\left[(n-p)\tilde{\sigma}^2 + \left\|\mathbf{X}'\hat{\boldsymbol{\beta}} - \mathbf{X}'\boldsymbol{\beta}\right\|^2\right]\right\}.$$

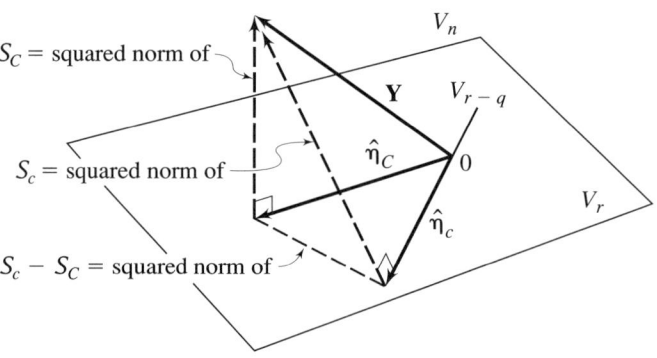

Figure 16.3 ($n = 3, r = 2, r - q = 1$).

This shows that $(\hat{\boldsymbol{\beta}}, \tilde{\sigma}^2)$ is a sufficient statistic for $(\boldsymbol{\beta}, \sigma^2)$. It can be shown that it is also complete (this follows from the multiparameter version of Theorem 6 in Chapter 11). By sufficiency and completeness, we have then the following result.

THEOREM 5 Under conditions (C) described in (22) and the assumption that rank $\mathbf{X} = p$ ($< n$), it follows that the LSE's $\hat{\beta}_j$ of β_j, $j = 1, \ldots, p$ have the smallest variance in the class of *all* unbiased estimators of β_j, $j = 1, \ldots, n$ (that is, regardless of whether they are linear in $\mathbf{Y}$ or not).

PROOF For $j = 1, \ldots, p$, $\hat{\beta}_j$ is an unbiased (and linear in $\mathbf{Y}$) estimate of β_j. It is also of minimum variance (by the Lehmann–Scheffé theorem) as it depends only on a sufficient and complete statistic. ▲

Exercises

16.4.1 Show that the MLE and the LSE of σ^2, $\hat{\sigma}^2$, and $\tilde{\sigma}^2$, respectively, are related as follows:

$$\hat{\sigma}^2 = \frac{n-r}{n} \tilde{\sigma}^2 \quad \text{and that} \quad \hat{\sigma}^2 = \bar{\sigma}^2$$

where $\bar{\sigma}^2$ is given in Section 16.4.

16.4.2 Let Y_j, $j = 1, \ldots, n$ be independent r.v.'s where Y_j is distributed as

$$N(\beta + \gamma(x_j - \bar{x}), \sigma^2); \quad x_j, \quad j = 1, \ldots, n$$

are known constants,

$$\bar{x} = \frac{1}{n} \sum_{j=1}^{n} x_j$$

and β, γ, σ^2 are parameters. Then:

i) Derive the LR test for testing the hypothesis $H: \gamma = \gamma_0$ against the alternative $A: \gamma \neq \gamma_0$ at level of significance α;

ii) Set up a confidence interval for γ with confidence coefficient $1 - \alpha$.

16.5 Derivation of the Distribution of the $\mathcal{F}$ Statistic

Consider the three vector spaces V_{r-q}, V_r and $V_n(r < n)$ which are related as follows: $V_{r-q} \subset V_r \subset V_n$. Following similar arguments to those in Section 16.3, let $\{\boldsymbol{\alpha}_{q+r}, \ldots, \boldsymbol{\alpha}_r\}$ be an orthonormal basis for V_{r-q} and extend it to an orthonormal basis $\{\boldsymbol{\alpha}_1, \ldots, \boldsymbol{\alpha}_q, \boldsymbol{\alpha}_{q+1}, \ldots, \boldsymbol{\alpha}_r\}$ for V_r and then to an orthonormal basis $\{\boldsymbol{\alpha}_1, \ldots, \boldsymbol{\alpha}_q, \boldsymbol{\alpha}_{q+1}, \ldots, \boldsymbol{\alpha}_r, \boldsymbol{\alpha}_{r+1}, \ldots, \boldsymbol{\alpha}_n\}$ for V_n. Let $\mathbf{P}$ be the $n \times n$ orthogonal matrix with rows the vectors $\boldsymbol{\alpha}'_j, j = 1, \ldots, n$, so that $\mathbf{PP}' = \mathbf{I}_n$. As in Section 16.3, consider the transformation $\mathbf{Z} = \mathbf{PY}$ and set $E\mathbf{Z} = \boldsymbol{\zeta} = (\zeta_1, \ldots, \zeta_n)'$. Then $\boldsymbol{\zeta} = \mathbf{P}\boldsymbol{\eta}$, where $\boldsymbol{\eta} \in V_r$. Thus $\zeta_j = 0, j = r+1, \ldots, n$. Now if H is true, we will further have that $\boldsymbol{\eta} \in V_{r-q}$, so that $\zeta_j = 0, j = 1, \ldots, q$, as follows from the transformation $\boldsymbol{\zeta} = \mathbf{P}\boldsymbol{\eta}$. The converse is also, clearly, true. Thus the hypothesis H is equivalent to the hypothesis

$$H': \zeta_j = 0, \quad j = 1, \ldots, q.$$

By (13) and (28),

$$S_C = \|\mathbf{Y} - \hat{\boldsymbol{\eta}}_C\|^2 = \sum_{j=r+1}^{n} Z_j^2.$$

On the other hand,

$$\mathbf{Y} = \sum_{j=1}^{n} \boldsymbol{\alpha}_j Z_j \quad \text{and} \quad \hat{\boldsymbol{\eta}}_c = \sum_{j=q+1}^{r} \boldsymbol{\alpha}_j Z_j,$$

since $\hat{\boldsymbol{\eta}}_c$ is the projection of $\hat{\boldsymbol{\eta}}$ into V_{r-q}. Therefore (29) gives

$$S_c = \|\mathbf{Y} - \hat{\boldsymbol{\eta}}_c\|^2 = \sum_{j=1}^{q} Z_j^2 + \sum_{j=r+1}^{n} Z_j^2.$$

Hence

$$S_c - S_C = \sum_{j=1}^{q} Z_j^2,$$

so that

$$\mathcal{F} = \frac{n-r}{q} \frac{\sum_{j=1}^{q} Z_j^2}{\sum_{j=r+1}^{n} Z_j^2} = \frac{\sum_{j=1}^{q} Z_j^2 / q}{\sum_{j=r+1}^{n} Z_j^2 / n-r}.$$

Now since the Y's are independent and the transformation $\mathbf{P}$ is orthogonal, it follows that Z's are also independent. (See Theorem 5, Chapter 9.) Since also $\sigma^2(Z_j) = \sigma^2, j = 1, \ldots, n$ by (11), it follows that, under H (or equivalently, H').

$$\sum_{j=1}^{q} Z_j^2 \text{ is } \sigma^2 \chi_q^2 \quad \text{and} \quad \sum_{j=r+1}^{n} Z_j^2 \text{ is } \sigma^2 \chi_{n-r}^2.$$

It follows that, under H (H'), the statistic $\mathcal{F}$ is distributed as $F_{q, n-r}$. The distribution of $\mathcal{F}$, under the alternatives, is *non-central* $F_{q, n-r}$ which is defined in terms of a χ_{n-r}^2 and a *non-central* χ_q^2 distribution. For these definitions, the reader is referred to Appendix II.

From the derivations in this section and previous results, one has the following theorem.

THEOREM 6 Assume the model described in (22) and let rank $\mathbf{X} = p\ (< n)$. Then the LSE's $\hat{\boldsymbol{\beta}}$ and $\tilde{\sigma}^2$ of $\boldsymbol{\beta}$ and σ^2, respectively, are independent.

PROOF It is an immediate consequence of the corollary to Theorem 4 and Theorem 5 in Chapter 9. ▲

Finally, we should like to emphasize that the transformation of the r.v.'s $Y_j, j = 1, \ldots, n$ to the r.v.'s $Z_j, j = 1, \ldots, n$ is only a technical device for deriving the distribution of $\mathcal{F}$ and also for proving unbiasedness of the LSE of σ^2. For actually carrying out the test and also for calculating the LSE of σ^2, the Y's rather than the Z's are used.

This section is closed with two examples.

EXAMPLE 2 Refer to Example 1 and suppose that the x's are not all equal, and that the Y's are normally distributed. It follows that rank $X = r = 2$, and the regression line is $y = \beta_1 + \beta_2 x$ in the xy-plane. Without loss of generality, we may assume that $x_1 \neq x_2$. Then $\eta_i = \beta_1 + \beta_2 x_i$, $i = 1, 2$ are linearly independent, and all $\eta_j, j = 3, \ldots, n$ are linear combinations of η_1 and η_2. Therefore $\eta \in V_2$.

Now, suppose we are interested in testing the following hypothesis about the slope of the regression line; namely $H_1: \beta_2 = \beta_{20}$, where β_{20} is a given number. Hypothesis H_1 is equivalent to the hypothesis $H_1': \eta_i = \beta_1 + \beta_{20} x_i$, $i = 1, 2$, from which it follows that, under H_1 (or H_1'), $\eta \in V_1$. Thus, $r - q = 1$, or $q = 1$. The LSE of β_1 and β_2 are $\hat{\beta}_{1C} = \bar{Y} - \hat{\beta}_{2C}\bar{x}$, $\hat{\beta}_{2C} = \sum_{j=1}^{n}(x_j - \bar{x})^2(Y_j - \bar{Y})/\sum_{j=1}^{n}(x_j - \bar{x})^2$, whereas under H_1 (or H_1'), the LSE become $\hat{\beta}_{1c} = \bar{Y} - \beta_{20}\bar{x}$, $\hat{\beta}_{2c} = \beta_{20}$. Then $\hat{\eta}_C = \mathbf{X}'\hat{\boldsymbol{\beta}}_C$ and $S_C = \|\mathbf{Y} - \hat{\boldsymbol{\eta}}_C\|^2 = \sum_{j=1}^{n}(Y_j - \bar{Y})^2 - \hat{\beta}_{2C}^2 \sum_{j=1}^{n}(x_j - \bar{x})^2$. Likewise, $\hat{\boldsymbol{\eta}}_c = \mathbf{X}'\hat{\boldsymbol{\beta}}_c$ and $S_c = \|\mathbf{Y} - \hat{\boldsymbol{\eta}}_c\|^2 = \sum_{j=1}^{n}(Y_j - \bar{Y})^2 + \beta_{20} \cdot (\beta_{20} - 2\hat{\beta}_{2C})\sum_{j=1}^{n}(x_j - \bar{x})^2$. It follows that $S_c - S_C = (\hat{\beta}_{2C} - \beta_{20})^2 \sum_{j=1}^{n}(x_j - \bar{x})^2$, and the test statistic is $\mathcal{F} = \frac{n-2}{1}\frac{S_c - S_C}{S_C} \sim F_{1, n-2}$, and the cut-off point is $\mathcal{F}_0 = F_{1, n-2; \alpha}$.

Next, suppose we are interested in testing the hypothesis $H_2: \beta_1 = \beta_{10}$, $\beta_2 = \beta_{20}$, where β_{10} and β_{20} are given numbers. Hypothesis H_2 is equivalent to the hypothesis $H_2': \eta_i = \beta_{10} + \beta_{20} x_i$, $i = 1, 2$, from which it follows that, under H_2 (or H_2'), $\boldsymbol{\eta} \in V_0$. Thus, $r - q = 0$, or $q = 2$. Clearly, $\hat{\beta}_{1c} = \beta_{10}$, $\hat{\beta}_{2c} = \beta_{20}$, so that $\hat{\boldsymbol{\eta}}_c = \mathbf{X}'\hat{\boldsymbol{\beta}}_c$ and $S_c = \sum_{j=1}^{n}(Y_j - \beta_{10} - \beta_{20} x_j)^2$. It follows that the test statistic here is $\mathcal{F} = \frac{n-2}{2}\frac{S_c - S_C}{S_C} \sim F_{2, n-2}$, where S_C was given above. The cut-off point is $\mathcal{F}_0 = F_{2, n-2; \alpha}$.

As mentioned earlier, the linear model adopted in this chapter can also be used for prediction purposes. The following example provides an instance of a prediction problem and its solution.

EXAMPLE 3 Refer to Example 2. Let $x_0 \neq x_j, j = 1, \ldots, n$ and suppose that the independent r.v.'s $Y_{0i} = \beta_1 + \beta_2 x_0 + e_i$, $i = 1, \ldots, m$ are to be observed at the point x_0. Let Y_0 be their sample mean. The problem is that of predicting Y_0 and also constructing a prediction interval for Y_0. Of course, it is also assumed that if the Y_{0i}'s were actually observed, they would be independent of the Y's.

The r.v. Y_0 is to be predicted by $\hat{Y}_0$, where $\hat{Y}_0 = \hat{\beta}_1 + \hat{\beta}_2 x_0$. Then it follows that $E\hat{Y}_0 = \beta_1 + \beta_2 x_0$. Thus, if we set $Z = Y_0 - \hat{Y}_0$, then $EZ = 0$. Furthermore, by means of Exercise 16.5.4(i) and (18), we find (see also Exercise 16.5.3) that

$$\sigma_Z^2 = \sigma^2 \left[\frac{1}{m} + \frac{1}{n} + \frac{(x_0 - \bar{x})^2}{\sum_{j=1}^{n}(x_j - \bar{x})^2} \right]. \qquad (32)$$

It follows by Theorem 6 that

$$\frac{Z/\sigma_Z}{\sqrt{\hat{\sigma}^2/\sigma^2}} = \frac{Y_0 - \hat{Y}_0}{\sqrt{\dfrac{\tilde{\sigma}^2}{n-2}\left[\dfrac{1}{m} + \dfrac{1}{n} + \dfrac{(x_0 - \bar{x})^2}{\sum_{j=1}^{n}(x_j - \bar{x})^2}\right]}}$$

is t_{n-2} distributed, so that a prediction interval for Y_0 with confidence coefficient $1 - \alpha$ is provided by

$$\left[\hat{Y}_0 - st_{n-2;\,\alpha/2},\ \hat{Y}_0 + st_{n-2;\,\alpha/2} \right],$$

where $\quad s = \sqrt{\dfrac{\tilde{\sigma}^2}{n-2}\left[\dfrac{1}{m} + \dfrac{1}{n} + \dfrac{(x_0 - \bar{x})^2}{\sum_{j=1}^{n}(x_j - \bar{x})^2}\right]}$

For a numerical example, refer to the data used in Example 1 and let $x_0 = 60$, $m = 1$ and $\alpha = 0.05$. Then $\hat{Y}_0 = 27.0873$, $s = 1.2703$ and $t_{10;0.025} = 2.2281$, so that the prediction interval for Y_0 is given by [24.2570, 29.9176].

Exercises

16.5.1 From the discussion in Section 16.5, it follows that the distribution of $[(n-r)\tilde{\sigma}^2]/\sigma^2$ is χ^2_{n-r}. Thus the statistic $\tilde{\sigma}^2$ can be used for testing hypotheses about σ^2 and also for constructing confidence intervals for σ^2.

i) Set up a confidence interval for σ^2 with confidence coefficient $1 - \alpha$;

ii) What is this confidence interval in the case of Example 2 when $n = 27$ and $\alpha = 0.05$?

16.5.2 Refer to Example 2 and:
i) Use Exercises 16.3.3 and 16.3.4 to show that

$$\frac{\sqrt{n}(\hat{\beta}_1 - \beta_1)}{\sigma} \quad \text{and} \quad \frac{\sqrt{\sum_{j=1}^{n} x_j^2}(\hat{\beta}_2 - \beta_2)}{\sigma}$$

are distributed as $N(0, 1)$, provided $\bar{x} = 0$;

ii) Use Theorem 6 and Exercise 16.5.1 to show that

$$\frac{\sqrt{n}(\hat{\beta}-\beta_1)}{\sqrt{\tilde{\sigma}^2}} \quad \text{and} \quad \frac{\sqrt{\sum_{j=1}^{n}(x_j-\bar{x})^2}(\hat{\beta}_2-\beta_2)}{\sqrt{\tilde{\sigma}^2}}$$

are distributed as t_{n-2}.

Thus the r.v.'s figuring in (ii) may be used for testing hypotheses about β_1 and β_2 and also for constructing confidence intervals for β_1 and β_2;

iii) Set up the test for testing the hypothesis $H: \beta_1 = 0$ (the regression line passes through the origin) against $A: \beta_1 \neq 0$ at level α and also construct a $1 - \alpha$ confidence interval for β_1;

iv) Set up the test for testing the hypothesis $H': \beta_2 = 0$ (the Y's are independent of the x's) against $A': \beta_2 \neq 0$ at level α and also construct a $1 - \alpha$ confidence interval for β_2;

v) What do the results in (iii) and (iv) become for $n = 27$ and $\alpha = 0.05$?

16.5.3 Verify relation (32).

16.5.4 Refer to Example 3 and suppose that the r.v.'s $Y_{0i} = \beta_1 + \beta_2 x_0 + e_i$, $i = 1, \ldots, m$ corresponding to an unknown point x_0 are observed. It is assumed that the r.v.'s Y_j, $j = 1, \ldots, n$ and Y_{0i}, $i = 1, \ldots, m$ are all independent.

i) Derive the MLE $\hat{x}_0$ of x_0;

ii) Set $V = \bar{Y}_0 - \hat{\beta}_1 - \hat{\beta}_2 x_0$, where $\bar{Y}_0$ is the sample mean of the Y_{0i}'s and show that the r.v.

$$\frac{V/\sigma_V}{\sqrt{(m+n)\hat{\sigma}^2/(m+n-3)\sigma^2}},$$

where

$$\sigma_V^2 = \sigma^2 \left[\frac{1}{m} + \frac{1}{n} + \frac{(x_0-\bar{x})^2}{\sum_{j=1}^{n}(x_j-\bar{x})^2}\right]$$

and

$$\hat{\sigma}^2 = \frac{1}{m+n}\left[\sum_{j=1}^{n}(Y_j-\hat{\beta}_1-\hat{\beta}_2 x_j)^2 + \sum_{i=1}^{m}(Y_{0i}-\bar{Y}_0)^2\right],$$

is distributed as t_{m+n-3}.

16.5.5 Refer to the model considered in Example 1 and suppose that the x's and the observed values of the Y's are given by the following table:

x	5	10	15	20	25	30
y	0.10	0.21	0.30	0.35	0.44	0.62

i) Find the LSE's of β_1, β_2 and σ^2 by utilizing the formulas (19'), (19'') and (21), respectively;

ii) Construct confidence intervals for β_1, β_2 and σ^2 with confidence coefficient $1 - \alpha = 0.95$ (see Exercises 16.5.1 and 16.5.2(ii));

iii) On the basis of the assumed model, predict Y_0 at $x_0 = 17$ and construct a prediction interval for Y_0 with confidence coefficient $1 - \alpha = 0.95$ (see Example 3).

16.5.6 The following table gives the reciprocal temperatures (x) and the corresponding observed solubilities of a certain chemical substance.

x	3.80	3.72	3.67	3.60	3.54
y	1.27 1.32 1.50	1.20 1.26	1.10 1.07	0.82 0.84 0.80	0.65 0.57 0.62

Assume the model considered in Example 1 and discuss questions (i) and (ii) of the previous exercise. Also discuss question (iii) of the same exercise for $x_0 = 3.77$.

16.5.7 Let Z_j, $j = 1, \ldots, n$ be independent r.v.'s, where Z_j is distributed as $N(\zeta_j, \sigma^2)$. Suppose that $\zeta_j = 0$ for $j = r + 1, \ldots, n$ whereas $\zeta_1, \ldots, \zeta_r$, σ^2 are parameters. Then derive the LR test for testing the hypothesis $H: \zeta_1 = \zeta_1'$ against the alternative $A: \zeta_1 \neq \zeta_1'$ at level of significance α.

16.5.8 Consider the r.v.'s of Exercise 16.4.2 and transform the Y's to Z's by means of an orthogonal transformation $\mathbf{P}$ whose first two rows are

$$\left(\frac{x_1 - \bar{x}}{s_x}, \ldots, \frac{x_n - \bar{x}}{s_x}\right), \quad \left(\frac{1}{\sqrt{n}}, \ldots, \frac{1}{\sqrt{n}}\right), \quad s_x^2 = \sum_{j=1}^n (x_j - \bar{x})^2.$$

Then:

i) Show that the Z's are as in Exercise 16.5.7 with $r = 2$, $\zeta_1 = \gamma s_x$, $\zeta_2 = \beta$;

ii) Set up the test mentioned in Exercise 16.5.7 and then transform the Z's back to Y's. Also compare the resulting test with the test mentioned in Exercise 16.4.2.

16.5.9 Let $Y_i = \beta_1 + \beta_2 x_i + e_i$, $i = 1, \ldots, m$, $Y_j^* = \beta_1^* + \beta_2^* x_j^* + e_j^*$, $j = 1, \ldots, n$, where the e's and e^*'s are independent r.v.'s distributed as $N(0, \sigma^2)$. Use Exercises 16.3.3, 16.3.4 and 16.5.1 to test the hypotheses $H_1: \beta_1 = \beta_1^*$, $H_2: \beta_2 = \beta_2^*$ against the corresponding alternatives $A_1: \beta_1 \neq \beta_1^*$, $A_2: \beta_2 \neq \beta_2^*$ at level of significance α.

Chapter 17

Analysis of Variance

The Analysis of Variance techniques discussed in this chapter can be used to study a great variety of problems of practical interest. Below we mention a few such problems.

Crop yields corresponding to different soil treatment.
Crop yields corresponding to different soils and fertilizers.
Comparison of a certain brand of gasoline with and without an additive by using it in several cars.
Comparison of different brands of gasoline by using them in several cars.
Comparison of the wearing of different materials.
Comparison of the effect of different types of oil on the wear of several piston rings, etc.
Comparison of the yields of a chemical substance by using different catalytic methods.
Comparison of the strengths of certain objects made of different batches of some material.
Identification of the melting point of a metal by using different thermometers.
Comparison of test scores from different schools and different teachers, etc.

Below, we discuss some statistical models which make the comparisons mentioned above possible.

17.1 One-way Layout (or One-way Classification) with the Same Number of Observations Per Cell

The models to be discussed in the present chapter are special cases of the general model which was studied in the previous chapter. In this section, we consider what is known as a one-way layout, or one-way classification, which we introduce by means of a couple of examples.

17.1 One-way Layout (or One-way Classification)

EXAMPLE 1 Consider I machines, each one of which is manufactured by I different companies but all intended for the same purpose. A purchaser who is interested in acquiring a number of these machines is then faced with the question as to which brand he should choose. Of course his decision is to be based on the productivity of each one of the I different machines. To this end, let a worker run each one of the I machines for J days each and always under the same conditions, and denote by Y_{ij} his output the jth day he is running the ith machine. Let μ_i be the average output of the worker when running the ith machine and let e_{ij} be his "error" (variation) the jth day when he is running the ith machine. Then it is reasonable to assume that the r.v.'s e_{ij} are normally distributed with mean 0 and variance σ^2. It is further assumed that they are independent. Therefore the Y_{ij}'s are r.v.'s themselves and one has the following model.

$$Y_{ij} = \mu_i + e_{ij} \quad \text{where } e_{ij} \text{ are independent } N(0,\sigma^2) \quad \text{for} \quad i=1,\ldots,I(\geq 2);$$
$$j=1,\ldots,J(\geq 2). \quad (1)$$

EXAMPLE 2 For an agricultural example, consider $I \cdot J$ identical plots arranged in an $I \times J$ orthogonal array. Suppose that the same agricultural commodity (some sort of a grain, tomatoes, etc.) is planted in all $I \cdot J$ plots and that the plants in the ith row are treated by the ith kind of I available fertilizers. All other conditions assumed to be the same, the problem is that of comparing the I different kinds of fertilizers with a view to using the most appropriate one on a large scale. Once again, we denote by μ_i the average yield of each one of the J plots in the ith row, and let e_{ij} stand for the variation of the yield from plot to plot in the ith row, $i = 1, \ldots, I$. Then it is again reasonable to assume that the r.v.'s e_{ij}, $i = 1, \ldots, I; j = 1, \ldots, J$ are independent $N(0, \sigma^2)$, so that the yield Y_{ij} of the jth plot treated by the ith kind of fertilizer is given by (1).

One may envision the I objects (machines, fertilizers, etc.) as being represented by the I spaces between $I+1$ horizontal (straight) lines and the J objects (days, plots, etc.) as being represented by the J spaces between $J + 1$ vertical (straight) lines. In such a case there are formed IJ rectangles in the resulting rectangular array which are also referred to as *cells* (see also Fig. 17.1). The same interpretation and terminology is used in similar situations throughout this chapter.

In connection with model (1), there are three basic problems we are interested in: Estimation of μ_i, $i = 1, \ldots, I$; testing the hypothesis: $H: \mu_1 = \cdots = \mu_I$ ($=\mu$, unspecified) (that is, there is no difference between the I machines, or the I kind of fertilizers) and estimation of σ^2. Set

$$\mathbf{Y} = (Y_{11},\ldots,Y_{1J};Y_{21},\ldots,Y_{2J};\ldots;Y_I,\ldots,Y_I)'$$
$$\mathbf{e} = (e_{11},\ldots,e_{1J};e_{21},\ldots,e_{2J};\ldots;e_I,\ldots,e_I)'$$
$$\boldsymbol{\beta} = (\mu_1,\ldots,\mu_I)'$$

	1	2	...	j	...	J−1	J
1							
2							
⋮							
i				(i, jth) cell			
⋮							
I−1							
I							

Figure 17.1

$$\mathbf{X}' = \begin{pmatrix} \overbrace{\begin{matrix} 1 & 0 & 0 & \cdot & \cdot & \cdot & 0 \\ \cdot & \cdot & \cdot & \cdot & \cdot & \cdot & \cdot \\ 1 & 0 & 0 & \cdot & \cdot & \cdot & 0 \end{matrix}}^{I} \left.\begin{matrix} \\ \\ \\ \end{matrix}\right\} J \\ \begin{matrix} 0 & 1 & 0 & \cdot & \cdot & \cdot & 0 \\ \cdot & \cdot & \cdot & \cdot & \cdot & \cdot & 0 \\ 0 & 1 & 0 & \cdot & \cdot & \cdot & 0 \end{matrix} \left.\begin{matrix} \\ \\ \\ \end{matrix}\right\} J \\ \begin{matrix} \cdot & \cdot & \cdot & \cdot & \cdot & \cdot & \cdot \\ \cdot & \cdot & \cdot & \cdot & \cdot & \cdot & \cdot \\ \cdot & \cdot & \cdot & \cdot & \cdot & \cdot & \cdot \end{matrix} \\ \begin{matrix} 0 & 0 & \cdot & \cdot & \cdot & 0 & 1 \\ \cdot & \cdot & \cdot & \cdot & \cdot & \cdot & \cdot \\ 0 & 0 & \cdot & \cdot & \cdot & 0 & 1 \end{matrix} \left.\begin{matrix} \\ \\ \\ \end{matrix}\right\} J \end{pmatrix}$$

Then it is clear that $\mathbf{Y} = \mathbf{X}'\boldsymbol{\beta} + \mathbf{e}$. Thus we have the model described in (6) of Chapter 16 with $n = IJ$ and $p = I$. Next, the I vectors $(1, 0, \cdots, 0)'$, $(0, 1, 0, \ldots, 0)', \ldots, (0, 0, \ldots, 0, 1)'$ are, clearly, independent and any other row vector in $\mathbf{X}'$ is a linear combination of them. Thus rank $\mathbf{X}' = I\ (= p)$, that is, $\mathbf{X}$

17.1 One-way Layout (or One-way Classification)

is of full rank. Then by Theorem 2, Chapter 16, $\mu_i = 1, \ldots, I$ have uniquely determined LSE's which have all the properties mentioned in Theorem 5 of the same chapter. In order to determine the explicit expression of them, we observe that

$$\mathbf{S} = \mathbf{XX}' = \begin{pmatrix} J & 0 & 0 & \cdots & 0 \\ 0 & J & 0 & \cdots & 0 \\ \cdot & \cdot & \cdot & \cdots & \cdot \\ 0 & 0 & \cdots & 0 & J \end{pmatrix} = J\mathbf{I}_p$$

and

$$\mathbf{XY} = \left(\sum_{j=1}^{J} Y_{1j}, \sum_{j=1}^{J} Y_{2j}, \ldots, \sum_{j=1}^{J} Y_{IJ} \right)',$$

so that, by (9), Chapter 16,

$$\hat{\boldsymbol{\beta}} = \mathbf{S}^{-1}\mathbf{XY} = \left(\frac{1}{J}\sum_{j=1}^{J} Y_{1j}, \frac{1}{J}\sum_{j=1}^{J} Y_{2j}, \ldots, \frac{1}{J}\sum_{j=1}^{J} Y_{IJ} \right)'.$$

Therefore the LSE's of the μ's are given by

$$\hat{\mu}_i = Y_{i.}, \quad \text{where} \quad Y_{i.} = \frac{1}{J}\sum_{j=1}^{J} Y_{ij}, \quad i = 1, \ldots, I. \tag{2}$$

Next, one has

$$\boldsymbol{\eta} = \mathbf{EY} = \left(\overbrace{\mu_1, \ldots, \mu_1}^{J}; \overbrace{\mu_2, \ldots, \mu_2}^{J}; \ldots; \overbrace{\mu_I, \ldots, \mu_I}^{J} \right)',$$

so that, under the hypothesis $H: \mu_1 = \cdots = \mu_I$ (= μ, unspecified), $\boldsymbol{\eta} \in V_1$. That is, $r - q = 1$ and hence $q = r - 1 = p - 1 = I - 1$. Therefore, according to (31) in Chapter 16, the $\mathcal{F}$ statistic for testing H is given by

$$\mathcal{F} = \frac{n-r}{q} \frac{S_c - S_C}{S_C} = \frac{I(J-1)}{I-1} \frac{S_c - S_C}{S_C}. \tag{3}$$

Now, under H, the model becomes $Y_{ij} = \mu + e_{ij}$ and the LSE of μ is obtained by differentiating with respect to μ the expression

$$\|\mathbf{Y} - \boldsymbol{\eta}_c\|^2 = \sum_{i=1}^{I}\sum_{j=1}^{J} (Y_{ij} - \mu)^2.$$

One has then the (unique) solution

$$\hat{\mu} = Y_{..}, \quad \text{where} \quad Y_{..} = \frac{1}{IJ}\sum_{i=1}^{I}\sum_{j=1}^{J} Y_{ij}. \tag{4}$$

Therefore relations (28) and (29) in Chapter 16 give

$$S_C = \|\mathbf{Y} - \hat{\boldsymbol{\eta}}_C\|^2 = \sum_{i=1}^{I}\sum_{j=1}^{J}(Y_{ij} - \hat{\eta}_{ij,C})^2 = \sum_{i=1}^{I}\sum_{j=1}^{J}(Y_{ij} - Y_{i.})^2$$

and

$$S_c = \|\mathbf{Y} - \hat{\boldsymbol{\eta}}_c\|^2 = \sum_{i=1}^{I}\sum_{j=1}^{J}(Y_{ij} - \hat{\eta}_{ij,c})^2 = \sum_{i=1}^{I}\sum_{j=1}^{J}(Y_{ij} - Y_{..})^2.$$

But for each fixed i,

$$\sum_{j=1}^{J}(Y_{ij} - Y_{i.})^2 = \sum_{j=1}^{J}Y_{ij}^2 - JY_{i.}^2,$$

so that

$$S_C = SS_e, \quad \text{where} \quad SS_e = \sum_{i=1}^{I}\sum_{j=1}^{J}(Y_{ij} - Y_{i.})^2 = \sum_{i=1}^{I}\sum_{j=1}^{J}Y_{ij}^2 - J\sum_{i=1}^{I}Y_{i.}^2. \quad (5)$$

Likewise,

$$S_c = SS_T, \quad \text{where} \quad SS_T = \sum_{i=1}^{I}\sum_{j=1}^{J}(Y_{ij} - Y_{..})^2 = \sum_{i=1}^{I}\sum_{j=1}^{J}Y_{ij}^2 - IJY_{..}^2, \quad (6)$$

so that, by means of (5) and (6), one has

$$S_c - S_C = J\sum_{i=1}^{I}Y_{i.}^2 - IJY_{..}^2 = J\left(\sum_{i=1}^{I}Y_{i.}^2 - IY_{..}^2\right) = J\sum_{i=1}^{I}(Y_{i.} - Y_{..})^2,$$

since

$$Y_{..} = \frac{1}{I}\sum_{i=1}^{I}\left(\frac{1}{J}\sum_{j=1}^{J}Y_{ij}\right) = \frac{1}{I}\sum_{i=1}^{I}Y_{i.}.$$

That is,

$$S_c - S_C = SS_H, \quad (7)$$

where

$$SS_H = J\sum_{i=1}^{I}(Y_{i.} - Y_{..})^2 = J\sum_{i=1}^{I}Y_{i.}^2 - IJY_{..}^2.$$

Therefore the $\mathcal{F}$ statistic given in (3) becomes as follows:

$$\mathcal{F} = \frac{I(J-1)}{I-1}\frac{SS_H}{SS_e} = \frac{MS_H}{MS_e}, \quad (8)$$

where

$$MS_H = \frac{SS_H}{I-1}, \quad MS_e = \frac{SS_e}{I(J-1)}$$

and SS_H and SS_e are given by (7) and (5), respectively. These expressions are also appropriate for actual calculations. Finally, according to Theorem 4 of Chapter 16, the LSE of σ^2 is given by

$$\tilde{\sigma}^2 = \frac{SS_e}{I(J-1)}. \quad (9)$$

Table 1 Analysis of Variance for One-Way Layout

source of variance	sums of squares	degrees of freedom	mean squares
between groups	$SS_H = J \sum_{i=1}^{I} (Y_{i.} - Y_{..})^2$	$I - 1$	$MS_H = \dfrac{SS_H}{I-1}$
within groups	$SS_e = \sum_{i=1}^{I} \sum_{j=1}^{J} (Y_{ij} - Y_{i.})^2$	$I(J - 1)$	$MS_e = \dfrac{SS_e}{I(J-1)}$
total	$SS_T = \sum_{i=1}^{I} \sum_{j=1}^{J} (Y_{ij} - Y_{..})^2$	$IJ - 1$	—

REMARK 1 From (5), (6) and (7) it follows that $SS_T = SS_H + SS_e$. Also from (6) it follows that SS_T stands for the sum of squares of the deviations of the Y_{ij}'s from the *grand (sample) mean* $Y_{..}$. Next, from (5) we have that, for each i, $\sum_{j=1}^{J}(Y_{ij} - Y_{i.})^2$ is the sum of squares of the deviations of Y_{ij}, $j = 1, \ldots, J$ within the ith group. For this reason, SS_e is called the *sum of squares within groups*. On the other hand, from (7) we have that SS_H represents the sum of squares of the deviations of the group means $Y_{i.}$ from the grand mean $Y_{..}$ (up to the factor J). For this reason, SS_H is called the *sum of squares between groups*. Finally, SS_T is called the *total sum of squares* for obvious reasons, and as mentioned above, it splits into SS_H and SS_e. Actually, the analysis of variance itself derives its name because of such a split of SS_T.

Now, as follows from the discussion in Section 5 of Chapter 16, the quantities SS_H and SS_e are independently distributed, under H, as $\sigma^2 \chi^2_{I-1}$ and $\sigma^2 \chi^2_{I(J-1)}$, respectively. Then SS_T is $\sigma^2 \chi^2_{IJ-1}$ distributed, under H. We may summarize all relevant information in a table (Table 1) which is known as an *Analysis of Variance Table*.

EXAMPLE 3 For a numerical example, take $I = 3$, $J = 5$ and let

$Y_{11} = 82 \quad Y_{21} = 61 \quad Y_{31} = 78$
$Y_{12} = 83 \quad Y_{22} = 62 \quad Y_{31} = 72$
$Y_{13} = 75 \quad Y_{23} = 67 \quad Y_{33} = 74$
$Y_{14} = 79 \quad Y_{24} = 65 \quad Y_{34} = 75$
$Y_{15} = 78 \quad Y_{25} = 64 \quad Y_{35} = 72$

We have then

$$\hat{\mu}_1 = 79.4, \quad \hat{\mu}_2 = 63.8, \quad \hat{\mu}_3 = 74.2$$

and $MS_H = 315.5392$, $MS_e = 7.4$, so that $\mathcal{F} = 42.6404$. Thus for $\alpha = 0.05$, $F_{2,12;0.05}$, $= 3.8853$ and the hypothesis $H : \mu_1 = \mu_2 = \mu_3$ is rejected. Of course, $\tilde{\sigma}^2 = MS_e = 7.4$.

Exercise

17.1.1 Apply the one-way layout analysis of variance to the data given in the table below.

A	B	C
10.0	9.1	9.2
11.5	10.3	8.4
11.7	9.4	9.4

17.2 Two-way Layout (Classification) with One Observation Per Cell

The model to be employed in this paragraph will be introduced by an appropriate modification of Examples 1 and 2.

EXAMPLE 4 Referring to Example 1, consider the I machines mentioned there and also J workers from a pool of available workers. Each one of the J workers is assigned to each one of the I machines which he runs for one day. Let μ_{ij} be the daily output of the jth worker when running the ith machine and let e_{ij} be his "error." His actual daily output is then an r.v. Y_{ij} such that $Y_{ij} = \mu_{ij} + e_{ij}$. At this point it is assumed that each μ_{ij} is equal to a certain quantity μ, the *grand mean*, plus a contribution α_i due to the ith row (ith machine), and called the *ith row effect*, plus a contribution β_j due to the jth worker, and called the *jth column effect*. It is further assumed that the I row effects and also the J column effects cancel out each other in the sense that

$$\sum_{i=1}^{I} \alpha_i = \sum_{j=1}^{J} \beta_j = 0.$$

Finally, it is assumed, as is usually the case, that the r. errors e_{ij}, $i = 1, \ldots, I$; $j = 1, \ldots, J$ are independent $N(0, \sigma^2)$. Thus the assumed model is then

$$Y_{ij} = \mu + \alpha_i + \beta_j + e_{ij}, \quad \text{where} \quad \sum_{i=1}^{I} \alpha_i = \sum_{j=1}^{J} \beta_j = 0 \qquad (10)$$

and e_{ij}, $i = 1, \ldots, I$ (≥ 2); $j = 1, \ldots, J$ (≥ 2) are independent $N(0, \sigma^2)$.

EXAMPLE 5 Consider the identical $I \cdot J$ plots described in Example 2, and suppose that J different varieties of a certain agricultural commodity are planted in each one of the I rows, one variety in each plot. Then all J plots in the ith row are treated by the ith of I different kinds of fertilizers. Then the yield of the jth variety of the commodity in question treated by the ith fertilizer is an r.v. Y_{ij} which is assumed again to have the structure described in (10). Here the ith row effect

17.2 Two-way Layout (Classification) with One Observation Per Cell

is the contribution of the ith fertilized and the jth column effect is the contribution of the jth variety of the commodity in question.

From the preceding two examples it follows that the outcome Y_{ij} is affected by two factors, machines and workers in Example 4 and fertilizers and varieties of agricultural commodity in Example 5. The I objects (machines or fertilizers) and the J objects (workers or varieties of an agricultural commodity) associated with these factors are also referred to as *levels* of the factors. The same interpretation and terminology is used in similar situations throughout this chapter.

In connection with model (10), there are the following three problems to be solved: Estimation of μ; α_i, $i = 1, \ldots, I$; β_j, $j = 1, \ldots, J$; testing the hypothesis $H_A: \alpha_1 = \cdots = \alpha_I = 0$ (that is, there is no row effect), $H_B: \beta_1 = \cdots = \beta_J = 0$ (that is, there is no column effect) and estimation of σ^2.

We first show that model (10) is a special case of the model described in (6) of Chapter 16. For this purpose, we set

$$\mathbf{Y} = (Y_{11}, \ldots, Y_{1J}; Y_{21}, \ldots, Y_{2J}; \ldots; Y_{I1}, \ldots, J_{IJ})'$$
$$\mathbf{e} = (e_{11}, \ldots, e_{1J}; e_{21}, \ldots, e_{2J}; \ldots; e_{I1}, \ldots, e_{IJ})'$$
$$\boldsymbol{\beta} = (\mu; \alpha_1, \ldots, \alpha_I; \beta_1, \ldots, \beta_J)'$$

and

$$\mathbf{X}' = \left(\begin{array}{cccccc|cccccc}
\multicolumn{6}{c}{\overbrace{\hspace{4em}}^{I}} & \multicolumn{6}{c}{\overbrace{\hspace{4em}}^{J}} \\
1 & 1 & 0 & 0 & \cdots & 0 & 1 & 0 & 0 & \cdots & & 0 \\
1 & 1 & 0 & 0 & \cdots & 0 & 0 & 1 & 0 & \cdots & & 0 \\
\cdot & \cdot & \cdot & \cdot & & \cdot & \cdot & \cdot & \cdot & & & \cdot \\
1 & 1 & 0 & 0 & \cdots & 0 & 0 & 0 & 0 & \cdots & 0 & 1 \\
1 & 0 & 1 & 0 & \cdots & 0 & 1 & 0 & 0 & \cdots & & 0 \\
1 & 0 & 1 & 0 & \cdots & 0 & 0 & 1 & 0 & \cdots & & 0 \\
\cdot & \cdot & \cdot & \cdot & & \cdot & \cdot & \cdot & \cdot & & & \cdot \\
1 & 0 & 1 & 0 & \cdots & 0 & 0 & 0 & 0 & \cdots & 0 & 1 \\
\cdot & \cdot & \cdot & \cdot & & \cdot & \cdot & \cdot & \cdot & & & \cdot \\
\cdot & \cdot & \cdot & \cdot & & \cdot & \cdot & \cdot & \cdot & & & \cdot \\
1 & 0 & 0 & 0 & \cdots & 0 \ 1 & 1 & 0 & 0 & \cdots & & 0 \\
1 & 0 & 0 & 0 & \cdots & 0 \ 1 & 0 & 1 & 0 & \cdots & & 0 \\
\cdot & \cdot & \cdot & \cdot & & \cdot & \cdot & \cdot & \cdot & & & \cdot \\
1 & 0 & 0 & 0 & \cdots & 0 \ 1 & 0 & 0 & 0 & \cdots & 0 & 1
\end{array}\right)\begin{array}{l} \\ \left.\vphantom{\begin{array}{c}1\\1\\\cdot\\1\end{array}}\right\}J \\ \\ \left.\vphantom{\begin{array}{c}1\\1\\\cdot\\1\end{array}}\right\}J \\ \\ \\ \\ \left.\vphantom{\begin{array}{c}1\\1\\\cdot\\1\end{array}}\right\}J \end{array}$$

and then we have

$$\mathbf{Y} = \mathbf{X}'\boldsymbol{\beta} + \mathbf{e} \quad \text{with} \quad n = IJ \quad \text{and} \quad p = I + J + 1.$$

It can be shown (see also Exercise 17.2.1) that $\mathbf{X}'$ is not of full rank but rank $\mathbf{X}' = r = I + J - 1$. However, because of the two independent restrictions

$$\sum_{i=1}^{I} \alpha_i = \sum_{j=1}^{J} \beta_j = 0,$$

imposed on the parameters, the normal equations still have a unique solution, as is found by differentiation.

In fact,

$$S(\mathbf{Y},\boldsymbol{\beta}) = \sum_{i=1}^{I}\sum_{j=1}^{J}\left(Y_{ij} - \mu - \alpha_i - \beta_j\right)^2 \quad \text{and} \quad \frac{\partial}{\partial \mu}S(\mathbf{Y},\boldsymbol{\beta}) = 0$$

implies $\hat{\mu} = Y_{..}$, where $Y_{..}$ is again given by (4);

$$\frac{\partial}{\partial \alpha_i}S(\mathbf{Y},\boldsymbol{\beta}) = 0$$

implies $\hat{\alpha}_i = Y_{i.} - Y_{..}$, where $Y_{i.}$ is given by (2) and $(\partial/\partial \beta_j)S(\mathbf{Y},\boldsymbol{\beta}) = 0$ implies $\hat{\beta}_j = Y_{.j} - Y_{..}$, where

$$Y_{.j} = \frac{1}{I}\sum_{i=1}^{I} Y_{ij}.$$

Summarizing these results, we have then that the LSE's of μ, α_i and β_j are, respectively,

$$\hat{\mu} = Y_{..}, \quad \hat{\alpha}_i = Y_{i.} - Y_{..}, \quad i = 1, \ldots, I, \quad \hat{\beta}_j = Y_{.j} - Y_{..}, \quad j = 1, \ldots, J. \quad (11)$$

where $Y_{i.}, i = 1, \ldots, I$ are given by (2), $Y_{..}$ is given by (4) and

$$Y_{.j} = \frac{1}{I}\sum_{i=1}^{I} Y_{ij}, j = 1, \ldots, J. \quad (12)$$

Now we turn to the testing hypotheses problems. We have

$$E\mathbf{Y} = \boldsymbol{\eta} = \mathbf{X}'\big(\mu : \alpha_1, \ldots, \alpha_I; \beta_1, \ldots, \beta_J\big)' \in V_r, \quad \text{where} \quad r = I + J - 1.$$

Consider the hypothesis

$$H_A: \alpha_1 = \cdots = \alpha_I = 0.$$

Then, under H_A, $\boldsymbol{\eta} \in V_{r-q}$, where $r - q_A = J$, so that $q_A = I - 1$.

Next, under H_A again, $S(\mathbf{Y}, \boldsymbol{\beta})$ becomes

$$\sum_{i=1}^{I}\sum_{j=1}^{J}\left(Y_{ij} - \mu - \beta_j\right)^2$$

from where by differentiation, we determine the LSE's of μ and β_j, to be denoted by $\hat{\mu}_A$ and $\hat{\beta}_{j,A}$, respectively. That is, one has

17.2 Two-way Layout (Classification) with One Observation Per Cell

$$\hat{\mu}_A = Y_{..} = \hat{\mu}, \quad \hat{\beta}_{j,A} = Y_{.j} - Y_{..} = \hat{\beta}_j, \quad j = 1, \ldots, J. \tag{13}$$

Therefore relations (28) and (29) in Chapter 16 give by means of (11) and (12)

$$S_C = \|\mathbf{Y} - \hat{\boldsymbol{\eta}}_C\|^2 = \sum_{i=1}^{I}\sum_{j=1}^{J}(Y_{ij} - \hat{\eta}_{ij,C})^2 = \sum_{i=1}^{I}\sum_{j=1}^{J}(Y_{ij} - Y_{i.} - Y_{.j} + Y_{..})^2$$

and

$$S_{C_A} = \|\mathbf{Y} - \hat{\boldsymbol{\eta}}_{C_A}\|^2 = \sum_{i=1}^{I}\sum_{j=1}^{J}(Y_{ij} - \hat{\eta}_{ij,C_A})^2 = \sum_{i=1}^{I}\sum_{j=1}^{J}(Y_{ij} - Y_{.j})^2.$$

Now S_C can be rewritten as follows:

$$S_C = SS_e = \sum_{i=1}^{I}\sum_{j=1}^{J}[(Y_{ij} - Y_{.j}) - (Y_{i.} - Y_{..})]^2$$

$$= \sum_{i=1}^{I}\sum_{j=1}^{J}(Y_{ij} - Y_{.j})^2 - J\sum_{i=1}^{I}(Y_{i.} - Y_{..})^2 \tag{14}$$

because

$$\sum_{i=1}^{I}\sum_{j=1}^{J}(Y_{ij} - Y_{.j})(Y_{i.} - Y_{..}) = \sum_{i=1}^{I}(Y_{i.} - Y_{..})\sum_{j=1}^{J}(Y_{ij} - Y_{.j})$$

$$= J\sum_{i=1}^{I}(Y_{i.} - Y_{..})^2.$$

Therefore

$$S_{C_A} - S_C = SS_A, \quad \text{where} \quad SS_A = J\sum_{i=1}^{I}\hat{\alpha}_i^2 = J\sum_{i=1}^{I}(Y_{i.} - Y_{..})^2$$

$$= J\sum_{i=1}^{I}Y_{i.}^2 - IJY_{..}^2. \tag{15}$$

It follows that for testing H_A, the $\mathcal{F}$ statistic, to be denoted here by $\mathcal{F}_A$, is given by

$$\mathcal{F}_A = \frac{(I-1)(J-1)}{I-1}\frac{SS_A}{SS_e} = \frac{MS_A}{MS_e}, \tag{16}$$

where

$$MS_A = \frac{SS_A}{I-1}, \quad MS_e = \frac{SS_e}{(I-1)(J-1)}$$

and SS_A, SS_e are given by (15) and (14), respectively. (However, for an expression of SS_e to be used in actual calculations, see (20) below.)

Next, for testing the hypothesis
$$H_B: \beta_1 = \cdots = \beta_J = 0,$$
we find in an entirely symmetric way that the $\mathcal{F}$ statistic, to be denoted here by $\mathcal{F}_B$, is given by
$$\mathcal{F}_B = \frac{(I-1)(J-1)}{J-1} \frac{SS_B}{SS_e} = \frac{MS_B}{MS_e}, \tag{17}$$
where $MS_B = SS_B/(J-1)$ and
$$SS_B = S_{c_B} - S_C = I \sum_{j=1}^{J} \hat{\beta}_j^2 = I \sum_{j=1}^{J} (Y_{.j} - Y_{..})^2 = I \sum_{j=1}^{J} Y_{.j}^2 - IJY_{..}^2. \tag{18}$$

The quantities SS_A and SS_B are known as sums of squares of *row effects* and *column effects*, respectively.

Finally, if we set
$$SS_T = \sum_{i=1}^{I} \sum_{j=1}^{J} (Y_{ij} - Y_{..})^2 = \sum_{i=1}^{I} \sum_{j=1}^{J} Y_{ij}^2 - IJY_{..}^2, \tag{19}$$
we show below that $SS_T = SS_e + SS_A + SS_B$ from where we get
$$SS_e = SS_T - SS_A - SS_B. \tag{20}$$

Relation (20) provides a way of calculating SS_e by way of (15), (18) and (19). Clearly,

$$SS_e = \sum_{i=1}^{I} \sum_{j=1}^{J} \left[(Y_{ij} - Y_{..}) - (Y_{i.} - Y_{..}) - (Y_{.j} - Y_{..}) \right]^2$$
$$= \sum_{i=1}^{I} \sum_{j=1}^{J} (Y_{ij} - Y_{..})^2 + J \sum_{i=1}^{I} (Y_{i.} - Y_{..})^2$$
$$+ I \sum_{j=1}^{J} (Y_{.j} - Y_{..})^2 - 2 \sum_{i=1}^{I} \sum_{j=1}^{J} (Y_{ij} - Y_{..})(Y_{i.} - Y_{..})$$
$$- 2 \sum_{i=1}^{I} \sum_{j=1}^{J} (Y_{ij} - Y_{..})(Y_{.j} - Y_{..})$$
$$+ 2 \sum_{i=1}^{I} \sum_{j=1}^{J} (Y_{i.} - Y_{..})(Y_{.j} - Y_{..}) = SS_{TT} - SS_A - SS_B$$

because

$$\sum_{i=1}^{I} \sum_{j=1}^{J} (Y_{ij} - Y_{..})(Y_{i.} - Y_{..}) = \sum_{i=1}^{I} (Y_{i.} - Y_{..}) \sum_{j=1}^{J} (Y_{ij} - Y_{..})$$
$$= J \sum_{i=1}^{I} (Y_{i.} - Y_{..})^2 = SS_A,$$

Table 2 Analysis of Variance for Two-way Layout with One Observation Per Cell

source of variance	sums of squares	degrees of freedom	mean squares
rows	$SS_A = J\sum_{i=1}^{I} \hat{\alpha}_i^2 = J\sum_{i=1}^{I}(Y_{i.} - Y_{..})^2$	$I-1$	$MS_A = \dfrac{SS_A}{I-1}$
columns	$SS_B = I\sum_{j=1}^{J} \hat{\beta}_j^2 = I\sum_{j=1}^{J}(Y_{.j} - Y_{..})^2$	$J-1$	$MS_B = \dfrac{SS_B}{J-1}$
residual	$SS_e = \sum_{i=1}^{I}\sum_{j=1}^{J}(Y_{ij} - Y_{i.} - Y_{.j} + Y_{..})^2$	$(I-1)\times(J-1)$	$MS_e = \dfrac{SS_e}{(I-1)(J-1)}$
total	$SS_T = \sum_{i=1}^{I}\sum_{j=1}^{J}(Y_{ij} - Y_{..})^2$	$IJ-1$	—

$$\sum_{i=1}^{I}\sum_{j=1}^{J}(Y_{ij} - Y_{..})(Y_{.j} - Y_{..}) = \sum_{j=1}^{J}(Y_{.j} - Y_{..})\sum_{i=1}^{I}(Y_{ij} - Y_{..})$$
$$= I\sum_{j=1}^{J}(Y_{.j} - Y_{..})^2 = SS_B$$

and

$$\sum_{i=1}^{I}\sum_{j=1}^{J}(Y_{i.} - Y_{..})(Y_{.j} - Y_{..}) = \sum_{i=1}^{I}(Y_{i.} - Y_{..})\sum_{j=1}^{J}(Y_{.j} - Y_{..}) = 0.$$

The pairs SS_e, SS_A and SS_e, SS_B are independent $\sigma^2\chi^2$ distributed r.v.'s with certain degrees of freedom, as a consequence of the discussion in Section 5 of Chapter 16. Finally, the LSE of σ^2 is given by

$$\tilde{\sigma}^2 = MS_e. \tag{21}$$

This section is closed by summarizing the basic results in Table 2 above.

Exercises

17.2.1 Show that rank $\mathbf{X}' = I + J - 1$, where $\mathbf{X}'$ is the matrix employed in Section 2.

17.2.2 Apply the two-way layout with one observation per cell analysis of variance to the data given in the following table (take $\alpha = 0.05$).

3	7	5	4
−1	2	0	2
1	2	4	0

17.3 Two-way Layout (Classification) with K (≥ 2) Observations Per Cell

In order to introduce the model of this section, consider Examples 4 and 5 and suppose that K (≥ 2) observations are taken in each one of the IJ cells. This amounts to saying that we observe the yields Y_{ijk}, $k = 1, \ldots, K$ of K identical plots with the (i, j)th plot, that is, the plot where the jth agricultural commodity was planted and it was treated by the ith fertilizer (in connection with Example 5); or we allow the jth worker to run the ith machine for K days instead of one day (Example 4). In the present case, the relevant model will have the form $Y_{ijk} = \mu_{ij} + e_{ijk}$. However, the means μ_{ij}, $i = 1, \ldots, I$; $j = 1, \ldots, J$ need not be additive any longer. In other words, except for the grand mean μ and the row and column effects α_i and β_j, respectively, which in the previous section added up to make μ_{ij}, we may now allow *interactions* γ_{ij} among the various factors involved, such as fertilizers and varieties of agricultural commodities, or workers and machines. It is not unreasonable to assume that, on the average, these interactions cancel out each other and we shall do so. Thus our present model is as follows:

$$Y_{ijk} = \mu + \alpha_i + \beta_j + \gamma_{ij} + e_{ijk}, \tag{22}$$

where

$$\sum_{i=1}^{I} \alpha_i = \sum_{j=1}^{J} \beta_j = \sum_{j=1}^{J} \gamma_{ij} = \sum_{i=1}^{I} \gamma_{ij} = 0$$

for all i and j and e_{ijk}, $i = 1, \ldots, I$ (≥ 2); $j = 1, \ldots, J$ (≥ 2); $k = 1, \ldots, K$ (≥ 2) are independent $N(0, \sigma^2)$.

Once again the problems of main interest are estimation of μ, α_i, β_j and γ_{ij}, $i = 1, \ldots, I$; $j = 1, \ldots, J$; testing the hypotheses: $H_A: \alpha_1 = \cdots = \alpha_I = 0$, $H_B: \beta_1 = \cdots = \beta_J = 0$ and $H_{AB}: \gamma_{ij} = 0$, $i = 1, \ldots, I$; $j = 1, \ldots, J$ (that is, there are no interactions present); and estimation of σ^2.

By setting

$$\mathbf{Y} = (Y_{111}, \ldots, Y_{11K}; \ldots; Y_{1J1}, \ldots, Y_{1JK}; \ldots; Y_{IJ1}, \ldots, Y_{IJK})'$$

$$\mathbf{e} = (e_{111}, \ldots, e_{11K}; \ldots; e_{1J1}, \ldots, e_{1JK}; \ldots; e_{IJ1}, \ldots, e_{IJK})'$$

$$\boldsymbol{\beta} = (\mu_{11}, \ldots, \mu_{1J}; \ldots; \mu_{I1}, \ldots, \mu_{IJ})'$$

and

$$\mathbf{X'} = \begin{pmatrix}
\overbrace{\begin{matrix} 1 & 0 & 0 & \cdots & & & \cdots & & 0 \\ \cdot & \cdot & \cdot & & & & & & \cdot \\ 1 & 0 & 0 & \cdots & & & \cdots & & 0 \end{matrix}}^{IJ} \Big\} K \\
\begin{matrix} 0 & 1 & 0 & \cdots & & & \cdots & & 0 \\ \cdot & \cdot & \cdot & & & & & & \cdot \\ 0 & 1 & 0 & \cdots & & & \cdots & & 0 \end{matrix} \Big\} K \\
\vdots \\
\overbrace{\begin{matrix} 0 & \cdots & \cdot & 0 & 1 & 0 & \cdots & \cdot & 0 \\ \cdot & & & \cdot & \cdot & \cdot & & & \cdot \\ 0 & \cdots & \cdot & 0 & 1 & 0 & \cdots & \cdot & 0 \end{matrix}}^{J} \Big\} K \\
\vdots \\
\overbrace{\begin{matrix} 0 & \cdots & & \cdot & 0 & 1 & 0 & \cdots & 0 \\ \cdot & & & & \cdot & \cdot & \cdot & & \cdot \\ 0 & \cdots & & \cdot & 0 & 1 & 0 & \cdots & 0 \end{matrix}}^{(I-1)J+1} \Big\} K \\
\vdots \\
\begin{matrix} 0 & \cdots & & & & & \cdot & 0 & 1 \\ \cdot & & & & & & & \cdot & \cdot \\ 0 & \cdots & & & & & \cdot & 0 & 1 \end{matrix} \Big\} K
\end{pmatrix},$$

it is readily seen that

$$\mathbf{Y} = \mathbf{X'}\boldsymbol{\beta} + \mathbf{e} \quad \text{with} \quad n = IJK \quad \text{and} \quad p = IJ, \tag{22'}$$

so that model (22′) is a special case of model (6) in Chapter 16. From the form of $\mathbf{X'}$ it is also clear that rank $\mathbf{X'} = r = p = IJ$; that is, $\mathbf{X'}$ is of full rank (see also Exercise 17.3.1). Therefore the unique LSE's of the parameters involved are obtained by differentiating with respect to μ_{ij} the expression

$$S(\mathbf{Y}, \boldsymbol{\beta}) = \sum_{i=1}^{I} \sum_{j=1}^{J} \sum_{k=1}^{K} (Y_{ijk} - \mu_{ij})^2.$$

We have then

$$\hat{\mu}_{ij} = Y_{ij.}, \quad i=1,\ldots,I; \quad j=1,\ldots,J. \tag{23}$$

Next, from the fact that $\mu_{ij} = \mu + \alpha_i + \beta_j + \gamma_{ij}$ and on the basis of the assumptions made in (22), we have

$$\mu = \mu_{..}, \quad \alpha_i = \mu_{i.} - \mu_{..}, \quad \beta_j = \mu_{.j} - \mu_{..}, \quad \gamma_{ij} = \mu_{ij} - \mu_{i.} - \mu_{.j} + \mu_{..}, \tag{24}$$

by employing the "dot" notation already used in the previous two sections. From (24) we have that μ, α_i, β_j and γ_{ij} are linear combinations of the parameters μ_{ij}. Therefore, by the corollary to Theorem 3 in Chapter 16, they are estimable, and their LSE's $\hat{\mu}$, $\hat{\alpha}_i$, $\hat{\beta}_j$, $\hat{\gamma}_{ij}$, are given by the above-mentioned linear combinations, upon replacing μ_{ij} by their LSE's. It is then readily seen that

$$\hat{\mu} = Y_{...}, \quad \hat{\alpha}_i = Y_{i..} - Y_{...}, \quad \hat{\beta}_j = Y_{.j.} - Y_{...},$$
$$\hat{\gamma}_{ij} = Y_{ij.} - Y_{i..} - Y_{.j.} + Y_{...}, \quad i=1,\ldots,I; \quad j=1,\ldots,J. \tag{25}$$

Now from (23) and (25) it follows that $\hat{\mu}_{ij} = \hat{\mu} + \hat{\alpha}_i + \hat{\beta}_j + \hat{\gamma}_{ij}$. Therefore

$$S_C = \sum_{i=1}^{I}\sum_{j=1}^{J}\sum_{k=1}^{K}\left(Y_{ijk} - \hat{\mu}_{ij}\right)^2 = \sum_{i=1}^{I}\sum_{j=1}^{J}\sum_{k=1}^{K}\left(Y_{ijk} - \hat{\mu} - \hat{\alpha}_i - \hat{\beta}_j - \hat{\gamma}_{ij}\right)^2.$$

Next,

$$Y_{ijk} - \mu_{ij} = \left(Y_{ijk} - \hat{\mu} - \hat{\alpha}_i - \hat{\beta}_j - \hat{\gamma}_{ij}\right) + \left(\hat{\mu} - \mu\right)$$
$$+ \left(\hat{\alpha}_i - \alpha_i\right) + \left(\hat{\beta}_j - \beta_j\right) + \left(\hat{\gamma}_{ij} + \gamma_{ij}\right)$$

and hence

$$S(\mathbf{Y},\boldsymbol{\beta}) = \sum_{i=1}^{I}\sum_{j=1}^{J}\sum_{k=1}^{K}\left(Y_{ijk} - \mu_{ij}\right)^2 = S_C + IJK\left(\hat{\mu}-\mu\right)^2$$
$$+ JK\sum_{i=1}^{I}\left(\hat{\alpha}_i - \alpha_i\right)^2 + IK\sum_{j=1}^{J}\left(\hat{\beta}_j - \beta_j\right)^2 + K\sum_{i=1}^{I}\sum_{j=1}^{J}\left(\hat{\gamma}_{ij} - \gamma_{ij}\right)^2, \tag{26}$$

because, as is easily seen, all other terms are equal to zero. (See also Exercise 17.3.2.)

From identity (26) it follows that, under the hypothesis

$$H_A: \alpha_1 = \cdots = \alpha_I = 0,$$

the LSE's of the remaining parameters remain the same as those given in (25). It follows then that

$$S_{C_A} = S_C + JK\sum_{i=1}^{I}\hat{\alpha}_i^2, \quad \text{so that} \quad S_{C_A} - S_C = JK\sum_{i=1}^{I}\hat{\alpha}_i^2.$$

Thus for testing the hypothesis H_A the sum of squares to be employed are

$$S_C = SS_e = \sum_{i=1}^{I}\sum_{j=1}^{J}\sum_{k=1}^{K}\left(Y_{ijk} - Y_{ij.}\right)^2$$
$$= \sum_{i=1}^{I}\sum_{j=1}^{J}\sum_{k=1}^{K}Y_{ijk}^2 - K\sum_{i=1}^{I}\sum_{j=1}^{J}Y_{ij.}^2, \tag{27}$$

and

17.3 Two-way Layout (Classification) with K (≥ 2) Observations Per Cell

$$S_{c_A} - S_C = SS_A = JK\sum_{i=1}^{I}\hat{\alpha}_i^2 = JK\sum_{i=1}^{I}\left(Y_{i..} - Y_{...}\right)^2$$

$$= JK\sum_{i=1}^{I}Y_{i..}^2 - IJKY_{...}^2. \tag{28}$$

For the purpose of determining the dimension $r - q_A$ of the vector space in which $\boldsymbol{\eta} = E\mathbf{Y}$ lies under H_A, we observe that $\mu_{i.} - \mu_{..} = \alpha_i$, so that, under H_A, $\mu_{i.} - \mu_{..} = 0$, $i = 1, \ldots, I$. For $i = 1, \ldots, I-1$, we get $I - 1$ independent linear relationships which the IJ components of $\boldsymbol{\eta}$ satisfy and hence $r - q_A = IJ - (I - 1)$. Thus $q_A = I - 1$ since $r = IJ$.

Therefore the $\mathcal{F}$ statistic in the present case is

$$\mathcal{F}_A = \frac{IJ(K-1)}{I-1}\frac{SS_A}{SS_e} = \frac{MS_A}{MS_e}, \tag{29}$$

where

$$MS_A = \frac{SS_A}{I-1}, \qquad MS_e = \frac{SS_e}{IJ(K-1)}$$

and SS_A, SS_e are given by (28) and (27), respectively.

For testing the hypothesis

$$H_B: \beta_1 = \cdots = \beta_J = 0,$$

we find in an entirely symmetric way that the $\mathcal{F}$ statistic to be employed is given by

$$\mathcal{F}_B = \frac{IJ(K-1)}{J-1}\frac{SS_B}{SS_e} = \frac{MS_B}{MS_e}, \tag{30}$$

where

$$MS_B = \frac{SS_B}{J-1} \quad \text{and} \quad SS_B = IK\sum_{j=1}^{J}\hat{\beta}_j^2 = IK\sum_{j=1}^{J}\left(Y_{.j.} - Y_{...}\right)^2$$

$$= IK\sum_{j=1}^{J}Y_{.j.}^2 - IJKY_{...}^2. \tag{31}$$

Also for testing the hypothesis

$$H_{AB}: \gamma_{ij} = 0, \qquad i = 1, \ldots, I; \quad j = 1, \ldots, J,$$

arguments similar to the ones used before yield the $\mathcal{F}$ statistic, which now is given by

$$\mathcal{F}_{AB} = \frac{IJ(K-1)}{(I-1)(J-1)}\frac{SS_{AB}}{SS_e} = \frac{MS_{AB}}{MS_e}, \tag{32}$$

where

$$MS_{AB} = \frac{SS_{AB}}{(I-1)(J-1)} \quad \text{and} \quad SS_{AB} = K\sum_{i=1}^{I}\sum_{j=1}^{J} \hat{\gamma}_{ij}^2$$

$$= K\sum_{i=1}^{I}\sum_{j=1}^{J}\left(Y_{ij.} - Y_{i..} - Y_{.j.} + Y_{...}\right)^2. \tag{33}$$

(However, for an expression of SS_{AB} suitable for calculations, see (35) below.)

Finally, by setting

$$SS_T = \sum_{i=1}^{I}\sum_{j=1}^{J}\sum_{k=1}^{K}\left(Y_{ijk} - Y_{...}\right)^2$$

$$= \sum_{i=1}^{I}\sum_{j=1}^{J}\sum_{k=1}^{K} Y_{ijk}^2 - IJKY_{...}^2, \tag{34}$$

we can show (see Exercise 17.3.3) that $SS_T = SS_e + SS_A + SS_B + SS_{AB}$, so that

$$SS_{AB} = SS_T - SS_e - SS_A - SS_B. \tag{35}$$

Relation (35) is suitable for calculating SS_{AB} in conjunction with (27), (28), (31) and (34).

Of course, the LSE of σ^2 is given by

$$\tilde{\sigma}^2 = MS_e. \tag{36}$$

Once again the main results of this section are summarized in a table, Table 3.

The number of degrees of freedom of SS_T is calculated by those of SS_A, SS_B, SS_{AB} and SS_e, which can be shown to be independently distributed as $\sigma^2\chi^2$ r.v.'s with certain degrees of freedom.

EXAMPLE 6 For a numerical application, consider two drugs ($I = 2$) administered in three dosages ($J = 3$) to three groups each of which consists of four ($K = 4$) subjects. Certain measurements are taken on the subjects and suppose they are as follows:

$$\begin{array}{lll}
X_{111} = 18 & X_{121} = 64 & X_{131} = 61 \\
X_{112} = 20 & X_{122} = 49 & X_{132} = 73 \\
X_{113} = 50 & X_{123} = 35 & X_{133} = 62 \\
X_{114} = 53 & X_{124} = 62 & X_{134} = 90 \\
X_{211} = 34 & X_{221} = 40 & X_{231} = 56 \\
X_{212} = 36 & X_{222} = 63 & X_{232} = 61 \\
X_{213} = 40 & X_{223} = 35 & X_{233} = 58 \\
X_{214} = 17 & X_{224} = 63 & X_{234} = 73
\end{array}$$

For this data we have

$\hat{\mu} = 50.5416; \quad \hat{\alpha}_1 = 2.5417, \quad \hat{\alpha}_2 = -2.5416; \quad \hat{\beta}_1 = -17.0416, \quad \hat{\beta}_2 = 0.8334,$
$\hat{\beta}_3 = 16.2084; \quad \hat{\gamma}_{11} = -0.7917, \quad \hat{\gamma}_{12} = -1.4167, \quad \hat{\gamma}_{13} = 2.2083, \quad \hat{\gamma}_{21} = 0.7916,$
$\hat{\gamma}_{22} = 1.4166, \quad \hat{\gamma}_{23} = -2.2084$

Table 3 Analysis of Variance for Two-way Layout with $K (\geq 2)$ Observations Per Cell

source of variance	sums of squares	degrees of freedom	mean squares
A main effects	$SS_A = JK\sum_{i=1}^{I}\hat{\alpha}_i^2 = JK\sum_{i=1}^{I}(Y_{i..} + Y_{...})^2$	$I - 1$	$MS_A = \dfrac{SS_A}{I-1}$
B main effects	$SS_B = IK\sum_{j=1}^{J}\hat{\beta}_j^2 = IK\sum_{j=1}^{J}(Y_{.j.} - Y_{...})^2$	$J - 1$	$MS_B = \dfrac{SS_B}{J-1}$
AB interactions	$SS_{AB} = K\sum_{i=1}^{I}\sum_{j=1}^{J}\hat{\gamma}_{ij}^2 = K\sum_{i=1}^{I}\sum_{j=1}^{J}(Y_{ij.} - Y_{i..} - Y_{.j.} + Y_{...})^2$	$(I-1)(J-1)$	$MS_{AB} = \dfrac{SS_{AB}}{(I-1)(J-1)}$
error	$SS_e = \sum_{i=1}^{I}\sum_{j=1}^{J}\sum_{k=1}^{K}(Y_{ijk} - Y_{ij.})^2$	$IJ(K-1)$	$MS_e = \dfrac{SS_e}{IJ(K-1)}$
total	$SS_T = \sum_{i=1}^{I}\sum_{j=1}^{J}\sum_{k=1}^{K}(Y_{ijk} - Y_{...})^2$	$IJK - 1$	—

and

$$\mathcal{F}_A = 0.8471, \quad \mathcal{F}_B = 12.1038, \quad \mathcal{F}_{AB} = 0.1641.$$

Thus for $\alpha = 0.05$, we have $F_{1,18;0.05} = 4.4139$ and $F_{2,18;0.05} = 3.5546$; we accept H_A, reject H_B and accept H_{AB}. Finally, we have $\hat{\sigma}^2 = 183.0230$.

The models analyzed in the previous three sections describe three experimental designs often used in practice. There are many others as well. Some of them are taken from the ones just described by allowing different numbers of observations per cell, by increasing the number of factors, by allowing the row effects, column effects and interactions to be r.v.'s themselves, by randomizing the levels of some of the factors, etc. However, even a brief study of these designs would be well beyond the scope of this book.

Exercises

17.3.1 Show that rank $\mathbf{X}' = IJ$, where $\mathbf{X}'$ is the matrix employed in Section 17.3.

17.3.2 Verify identity (26).

17.3.3 Show that $SS_T = SS_e + SS_A + SS_B + SS_{AB}$, where SS_e, SS_A, SS_B, SS_{AB} and SS_T are given by (27), (28), (31), (33) and (34), respectively.

17.3.4 Apply the two-way layout with two observations per cell analysis of variance to the data given in the table below (take $\alpha = 0.05$).

110	128	48	123	19
95	117	60	138	94
214	183	115	114	129
217	187	127	156	125
208	183	130	225	114
119	195	164	194	109

17.4 A Multicomparison Method

Consider again the one-way layout with $J (\geq 2)$ observations per cell described in Section 17.1 and suppose that in testing the hypothesis $H: \mu_1 = \cdots = \mu_I (= \mu,$ unspecified) we decided to reject it on the basis of the available data. In rejecting H, we simply conclude that the μ's are not all equal. No conclusions are reached as to which specific μ's may be unequal.

The multicomparison method described in this section sheds some light on this problem.

For the sake of simplicity, let us suppose that $I = 6$. After rejecting H, the natural quantities to look into are of the following sort:

$$\mu_i - \mu_j, i \neq j, \quad \text{or} \quad \frac{1}{3}(\mu_1 + \mu_2 + \mu_3) - \frac{1}{3}(\mu_4 + \mu_5 + \mu_6),$$

$$\text{or} \quad \frac{1}{3}(\mu_1 + \mu_3 + \mu_5) - \frac{1}{3}(\mu_2 + \mu_4 + \mu_6) \quad \text{etc.}$$

We observe that these quantities are all of the form

$$\sum_{i=1}^{6} c_i \mu_i \quad \text{with} \quad \sum_{i=1}^{6} c_i = 0.$$

This observation gives rise to the following definition.

DEFINITION 1 Any linear combination $\psi = \sum_{i=1}^{I} c_i \mu_i$ of the μ's, where c_i, $i = 1, \ldots, I$ are known constants such that $\sum_{i=1}^{I} c_i = 0$, is said to be a *contrast* among the parameters μ_i, $i = 1, \ldots, I$.

Let $\psi = \sum_{i=1}^{I} c_i \mu_i$ be a contrast among the μ's and let

$$\hat{\psi} = \sum_{i=1}^{I} c_i Y_{i\cdot}, \quad \hat{\sigma}^2(\hat{\psi}) = \frac{1}{J}\sum_{i=1}^{I} c_i^2 MS_e \quad \text{and} \quad S^2 = (I-1)F_{I-1,n-I;\alpha},$$

where $n = IJ$. We will show in the sequel that the interval $[\hat{\psi} - S\hat{\sigma}(\hat{\psi}), \hat{\psi} + S\hat{\sigma}(\hat{\psi})]$ is a confidence interval with confidence coefficient $1 - \alpha$ for all contrasts ψ. Next, consider the following definition.

DEFINITION 2 Let ψ and $\hat{\psi}$ be as above. We say that $\hat{\psi}$ is *significantly different from zero*, according to the S (for Scheffé) criterion, if the interval $[\hat{\psi} - S\hat{\sigma}(\hat{\psi}), \hat{\psi} + S\hat{\sigma}(\hat{\psi})]$ does not contain zero; equivalently, if $|\hat{\psi}| > S\hat{\sigma}(\hat{\psi})$.

Now it can be shown that the $\mathcal{F}$ test rejects the hypothesis H if and only if there is at least one contrast ψ such that $\hat{\psi}$ is significantly different from zero.

Thus following the rejection of H one would construct a confidence interval for each contrast ψ and then would proceed to find out which contrasts are responsible for the rejection of H starting with the simplest contrasts first.

The confidence intervals in question are provided by the following theorem.

THEOREM 1 Refer to the one-way layout described in Section 17.1 and let

$$\psi = \sum_{i=1}^{I} c_i \mu_i, \quad \sum_{i=1}^{I} c_i = 0,$$

so that

$$\hat{\sigma}^2(\hat{\psi}) = \frac{1}{J}\sum_{i=1}^{I} c_i^2 MS_e,$$

where MS_e is given in Table 1. Then the interval $[\hat{\psi} - S\hat{\sigma}(\hat{\psi}), \hat{\psi} + S\hat{\sigma}(\hat{\psi})]$ is a confidence interval simultaneously for all contrasts ψ with confidence coefficients $1 - \alpha$, where $S^2 = (I-1)F_{I-1,n-I;\alpha}$ and $n = IJ$.

PROOF Consider the problem of maximizing (minimizing) (with respect to c_i, $i = 1, \ldots, I$) the quantity

$$f(c_1, \ldots, c_I) = \frac{1}{\sqrt{\frac{1}{J}\sum_{i=1}^{I} c_i^2}} \sum_{i=1}^{I} c_i(Y_{i\cdot} - \mu_i)$$

subject to the contrast constraint

$$\sum_{i=1}^{I} c_i = 0.$$

Now, clearly, $f(c_1, \ldots, c_I) = f(\gamma c_1, \ldots, \gamma c_I)$ for any $\gamma > 0$. Therefore the maximum (minimum) of $f(c_1, \ldots, c_I)$, subject to the restraint

$$\sum_{i=1}^{I} c_i = 0,$$

is the same with the maximum (minimum) of $f(\gamma c_1, \ldots, \gamma c_I) = f(c_1', \ldots, c_I')$, $c_i' = \gamma c_i$, $i = 1, \ldots, I$ subject to the restraints

and
$$\sum_{i=1}^{I} c'_i = 0$$

$$\frac{1}{J}\sum_{i=1}^{I} c'^2_i = 1.$$

Hence the problem becomes that of maximizing (minimizing) the quantity

$$q(c_1,\ldots,c_I) = \sum_{i=1}^{I} c_i(Y_{i.} - \mu_i),$$

subject to the constraints

$$\sum_{i=1}^{I} c_i = 0 \quad \text{and} \quad \sum_{i=1}^{I} c_i^2 = J.$$

Thus the points which maximize (minimize) $q(c_1, \ldots, c_I)$ are to be found on the circumference of the circle which is the intersection of the sphere

$$\sum_{i=1}^{I} c_i^2 = J$$

and the plane

$$\sum_{i=1}^{I} c_i = 0$$

which passes through the origin. Because of this it is clear that $q(c_1, \ldots, c_I)$ has both a maximum and a minimum. The solution of the problem in question will be obtained by means of the Lagrange multipliers. To this end, one considers the expression

$$h = h(c_1, \ldots, c_I; \lambda_1, \lambda_2) = \sum_{i=1}^{I} c_i(Y_{i.} - \mu_i) + \lambda_1\left(\sum_{i=1}^{I} c_i\right) + \lambda_2\left(\sum_{i=1}^{I} c_i^2 - J\right)$$

and maximizes (minimizes) it with respect to c_i, $i = 1, \cdots, I$ and λ_1, λ_2. We have

$$\left.\begin{aligned}\frac{\partial h}{\partial c_k} &= Y_{k.} - \mu_k + \lambda_1 + 2\lambda_2 c_k = 0, \quad k = 1,\ldots,I \\ \frac{\partial h}{\partial \lambda_1} &= \sum_{i=1}^{I} c_i = 0 \\ \frac{\partial h}{\partial \lambda_2} &= \sum_{i=1}^{I} c_i^2 - J = 0.\end{aligned}\right\} \quad (37)$$

Solving for c_k in (37), we get

$$c_k = \frac{1}{2\lambda_2}(\mu_k - Y_{k.} - \lambda_1) \quad k = 1, \ldots, I. \quad (38)$$

Then the last two equations in (37) provide us with

17.4 A Multicomparison Method

$$\lambda_1 = \mu_{..} - Y_{...} \quad \text{and} \quad \lambda_2 = \pm \frac{1}{2\sqrt{J}} \sqrt{\sum_{i=1}^{I} (\mu_i - \mu_{..} - Y_{i.} + Y_{..})^2}.$$

Replacing these values in (38), we have

$$c_k = \frac{\pm J(\mu_k - \mu_{..} - Y_{k.} + Y_{..})}{\sqrt{J \sum_{i=1}^{I} (\mu_i - \mu_{..} - Y_{i.} + Y_{..})^2}}, \quad k = 1, \ldots, I.$$

Next,

$$\sum_{k=1}^{I} (Y_{k.} - \mu_k)(\mu_k - \mu_{..} - Y_{k.} + Y_{..}) = -\sum_{k=1}^{I} (\mu_k - Y_{k.})[(\mu_k - Y_{k.}) - (\mu_{..} - Y_{..})]$$

$$= -\sum_{k=1}^{I} (\mu_k - Y_{k.})^2 + (\mu_{..} - Y_{..}) \sum_{k=1}^{I} (\mu_k - Y_{k.})$$

$$= -\left[\sum_{k=1}^{I} (\mu_k - Y_{k.})^2 - I(\mu_{..} - Y_{..})^2 \right]$$

$$= -\sum_{k=1}^{I} [(\mu_k - Y_{k.}) - (\mu_{..} - Y_{..})]^2 \leq 0.$$

Therefore

$$-\sqrt{J \sum_{i=1}^{I} (\mu_i - \mu_{..} - Y_{i.} + Y_{..})^2} \leq \frac{\sum_{i=1}^{I} c_i (Y_{i.} - \mu_i)}{\sqrt{1/J \sum_{i=1}^{I} c_i^2}}$$

$$\leq \sqrt{J \sum_{i=1}^{I} (\mu_i - \mu_{..} - Y_{i.} + Y_{..})^2} \quad (39)$$

for *all* c_i, $i = 1, \ldots, I$ such that

$$\sum_{i=1}^{I} c_i = 0.$$

Now we observe that

$$J \sum_{i=1}^{I} (\mu_i - \mu_{..} - Y_{i.} + Y_{..})^2 = J \sum_{i=1}^{I} [(Y_{i.} - Y_{..}) - (\mu_i - \mu_{..})]^2$$

is $\sigma^2 \chi^2_{I-1}$ distributed (see also Exercise 17.4.1) and also independent of SS_e which is $\sigma^2 \chi^2_{n-I}$ distributed. (See Section 17.1.) Therefore

$$\frac{J \sum_{i=1}^{I} (\mu_i - \mu_{..} - Y_{i.} + Y_{..})^2 / (I-1)}{MS_e}$$

is $F_{I-1, n-I}$ distributed and thus

$$P\left[-\sqrt{(I-1)F_{I-1, n-I; \alpha} MS_e} \leq -\sqrt{J \sum_{i=1}^{I} (\mu_i - \mu_{..} - Y_{i.} + Y_{..})^2} \right]$$

$$= P\left[\sqrt{J \sum_{i=1}^{I} (\mu_i - \mu_{..} - Y_{i.} + Y_{..})^2} \leq \sqrt{(I-1)F_{I-1, n-I; \alpha} MS_e} \right] = 1 - \alpha. \quad (40)$$

From (40) and (39) it follows then that

$$P\left[-\sqrt{(I-1)F_{I-1,n-I;\alpha}\,1/J\sum_{i=1}^{I}c_i^2 MS_e} \leq \sum_{i=1}^{I}c_i(Y_{i.}-\mu_i)\right.$$
$$\left.\leq \sqrt{(I-1)F_{I-1,n-I;\alpha}\,1/J\sum_{i=1}^{I}c_i^2 MS_e}\right] = 1-\alpha,$$

for *all* c_i, $i = 1, \ldots, I$ such that $\sum_{i=1}^{I}c_i = 0$, or equivalently,

$$P\left[\hat{\psi} - S\hat{\sigma}(\hat{\psi}) \leq \psi < \hat{\psi} + S\hat{\sigma}(\hat{\psi})\right] = 1-\alpha,$$

for all contrasts ψ, as was to be seen. (This proof has been adapted from the paper "A simple proof of Scheffé's multiple comparison theorem for contrasts in the one-way layout" by Jerome Klotz in *The American Statistician*, 1969, Vol. 23, Number 5.) ▲

In closing, we would like to point out that a similar theorem to the one just proved can be shown for the two-way layout with ($K \geq 2$) observations per cell and as a consequence of it we can construct confidence intervals for all contrasts among the α's, or the β's, or the γ's.

Exercises

17.4.1 Show that the quantity $J\sum_{i=1}^{I}(\mu_i - \mu_. - Y_{i.} + Y_{..})^2$ mentioned in Section 17.4 is distributed as $\sigma^2\chi^2_{I-1}$, under the null hypothesis.

17.4.2 Refer to Exercise 17.1.1 and construct confidence intervals for all contrasts of the μ's (take $1 - \alpha = 0.95$).

Chapter 18

The Multivariate Normal Distribution

18.1 Introduction

In this chapter, we introduce the Multivariate Normal distribution and establish some of its fundamental properties. Also, certain estimation and independence testing problems closely connected with it are discussed.

Let $Y_j, j = 1, \ldots, m$ be i.i.d. r.v.'s with common distribution $N(0, 1)$. Then we know that for any constants $c_j, j = 1, \ldots, m$ and μ the r.v. $\sum_{j=1}^{m} c_j Y_j + \mu$ is distributed as $N(\mu, \sum_{j=1}^{m} c_j^2)$. Now instead of considering one (non-homogeneous) linear combination of the Y's, consider k such combinations; that is,

$$X_i = \sum_{j=1}^{m} c_{ij} Y_j + \mu_i, \qquad i = 1, \ldots, k, \tag{1}$$

or in matrix notation

$$\mathbf{X} = \mathbf{CY} + \boldsymbol{\mu}, \tag{2}$$

where

$$\mathbf{X} = (X_1, \ldots, X_k)', \qquad \mathbf{C} = (c_{ij})(k \times m),$$

$$\mathbf{Y} = (Y_1, \ldots, Y_m)', \quad \text{and} \quad \boldsymbol{\mu} = (\mu_1, \ldots, \mu_k)'.$$

Thus we can give the following definition.

DEFINITION 1 Let $Y_j, j = 1, \ldots, m$ be i.i.d. r.v.'s distributed as $N(0, 1)$ and let the r.v.'s $X_i, i = 1, \ldots, k$, or the r. vector $\mathbf{X}$, be defined by (1) or (2), respectively. Then the joint distribution of the r.v.'s $X_i, i = 1, \ldots, k$ or the distribution of the r. vector $\mathbf{X}$, is called *Multivariate* (or more specifically, *k-Variate*) *Normal*.

REMARK 1 From Definition 1, it follows that if $X_i, i = 1, \ldots, k$ are jointly normally distributed, then any subset of them also is a set of jointly normally distributed r.v.'s.

From (2) and relation (10), Chapter 16, it follows that $E\mathbf{X} = \boldsymbol{\mu}$ and $\boldsymbol{\Sigma}_x = \mathbf{C}\boldsymbol{\Sigma}_Y\mathbf{C}' = \mathbf{C}\mathbf{I}_m\mathbf{C}' = \mathbf{C}\mathbf{C}'$; that is,

$$E\mathbf{X} = \boldsymbol{\mu}, \qquad \boldsymbol{\Sigma}_x \text{ (or just } \boldsymbol{\Sigma}) = \mathbf{C}\mathbf{C}'. \tag{3}$$

We now proceed to finding the ch.f. $\phi_\mathbf{x}$ of the r. vector $\mathbf{X}$. For $\mathbf{t} = (t_1, \ldots, t_k)' \in \mathbb{R}^k$, we have

$$\phi_\mathbf{x}(\mathbf{t}) = E(\exp i\mathbf{t}'\mathbf{X}) = E[\exp i\mathbf{t}'(\mathbf{CY} + \boldsymbol{\mu})] = \exp i\mathbf{t}'\boldsymbol{\mu} E(\exp i\mathbf{t}'\mathbf{CY}). \tag{4}$$

But

$$\mathbf{t}'\mathbf{CY} = \left(\sum_{j=1}^k t_j c_{j1}, \ldots, \sum_{j=1}^k t_j c_{jm}\right)(Y_1, \ldots, Y_m)'$$

$$= \left(\sum_{j=1}^k t_j c_{j1}\right)Y_1 + \cdots + \left(\sum_{j=1}^k t_j c_{jm}\right)Y_m$$

and hence

$$E(\exp i\mathbf{t}'\mathbf{CY}) = \phi_{Y_1}\left(\sum_{j=1}^k t_j c_{j1}\right) + \cdots + \phi_{Y_m}\left(\sum_{j=1}^k t_j c_{jm}\right)$$

$$= \exp\left[-\frac{1}{2}\left(\sum_{j=1}^k t_j c_{j1}\right)^2 - \cdots - \frac{1}{2}\left(\sum_{j=1}^k t_j c_{jm}\right)^2\right]$$

$$= \exp\left(-\frac{1}{2}\mathbf{t}'\mathbf{CC}'\mathbf{t}\right) \tag{5}$$

because

$$\left(\sum_{j=1}^k t_j c_{j1}\right)^2 + \cdots + \left(\sum_{j=1}^k t_j c_{jm}\right)^2 = \mathbf{t}'\mathbf{CC}'\mathbf{t}.$$

Therefore by means of (3)–(5), we have the following result.

THEOREM 1 The ch.f. of the r. vector $\mathbf{X} = (X_1, \ldots, X_k)'$, which has the k-Variate Normal distribution with mean $\boldsymbol{\mu}$ and covariance matrix $\boldsymbol{\Sigma}$, is given by

$$\phi_\mathbf{x}(\mathbf{t}) = \exp\left(i\mathbf{t}'\boldsymbol{\mu} - \frac{1}{2}\mathbf{t}'\boldsymbol{\Sigma}\mathbf{t}\right). \tag{6}$$

From (6) it follows that $\phi_\mathbf{x}$, and therefore the distribution of $\mathbf{X}$, is completely determined by means of its mean $\boldsymbol{\mu}$ and covariance matrix $\boldsymbol{\Sigma}$, a fact analogous to that of a Univariate Normal distribution. This fact justifies the following notation:

$$\mathbf{X} \sim N(\boldsymbol{\mu}, \boldsymbol{\Sigma}).$$

where $\boldsymbol{\mu}$ and $\boldsymbol{\Sigma}$ are the *parameters* of the distribution.

Now we shall establish the following interesting result.

THEOREM 2 Let $Y_j, j = 1, \ldots, k$ be i.i.d. r.v.'s with distribution $N(0, 1)$ and set $\mathbf{X} = \mathbf{CY} + \boldsymbol{\mu}$, where $\mathbf{C}$ is a $k \times k$ non-singular matrix. Then the p.d.f. $f_\mathbf{x}$ of $\mathbf{X}$ exists and is given by

$$f_{\mathbf{X}}(\mathbf{x}) = (2\pi)^{-k/2} |\mathbf{\Sigma}|^{-1/2} \exp\left[-\frac{1}{2}(\mathbf{x}-\boldsymbol{\mu})' \mathbf{\Sigma}^{-1}(\mathbf{x}-\boldsymbol{\mu})\right], \quad \mathbf{x} \in \mathbb{R}^k, \tag{7}$$

where $\mathbf{\Sigma} = \mathbf{CC}'$ and $|\mathbf{\Sigma}|$ denotes the determinant of $\mathbf{\Sigma}$.

PROOF From $\mathbf{X} = \mathbf{CY} + \boldsymbol{\mu}$ we get $\mathbf{CY} = \mathbf{X} - \boldsymbol{\mu}$, which, since $\mathbf{C}$ is non-singular, gives

$$\mathbf{Y} = \mathbf{C}^{-1}(\mathbf{X}-\boldsymbol{\mu}).$$

Therefore

$$f_{\mathbf{X}}(\mathbf{x}) = f_{\mathbf{Y}}\left[\mathbf{C}^{-1}(\mathbf{x}-\boldsymbol{\mu})\right] \|\mathbf{C}^{-1}\| = (2\pi)^{-k/2} \exp\left(-\frac{1}{2}\sum_{j=1}^{k} y_j^2\right) \|\mathbf{C}^{-1}\|.$$

But

$$\sum_{j=1}^{k} y_j^2 = (\mathbf{x}-\boldsymbol{\mu})' (\mathbf{C}^{-1})' (\mathbf{C}^{-1})(\mathbf{x}-\boldsymbol{\mu}) \quad \text{(see also Exercise 18.1.2)},$$

$$|\mathbf{C}^{-1}| = |\mathbf{C}|^{-1} \quad \text{and} \quad (\mathbf{C}^{-1})'(\mathbf{C}^{-1}) = (\mathbf{C}')^{-1}\mathbf{C}^{-1} = (\mathbf{CC}')^{-1} = \mathbf{\Sigma}^{-1}. \tag{8}$$

Therefore

$$f_{\mathbf{X}}(\mathbf{x}) = (2\pi)^{-k/2} \|\mathbf{C}^{-1}\| \exp\left[-\frac{1}{2}(\mathbf{x}-\boldsymbol{\mu})' \mathbf{\Sigma}^{-1}(\mathbf{x}-\boldsymbol{\mu})\right].$$

Finally, from $\mathbf{\Sigma} = \mathbf{CC}'$, one has $|\mathbf{\Sigma}| = |\mathbf{C}| |\mathbf{C}'| = |\mathbf{C}|^2$, so that $\|\mathbf{C}\| = |\mathbf{\Sigma}|^{\frac{1}{2}}$. Thus

$$f_{\mathbf{X}}(\mathbf{x}) = (2\pi)^{-k/2} |\mathbf{\Sigma}|^{-1/2} \exp\left[-\frac{1}{2}(\mathbf{x}-\boldsymbol{\mu})' \mathbf{\Sigma}^{-1}(\mathbf{x}-\boldsymbol{\mu})\right],$$

as was to be seen. ▲

REMARK 2 A k-Variate Normal distribution with p.d.f. given by (7) is called a *non-singular k-Variate Normal*. The use of the term non-singular corresponds to the fact that $|\mathbf{\Sigma}| \neq 0$; that is, the fact that $\mathbf{\Sigma}$ is of full rank.

COROLLARY 1 In the theorem, let $k = 2$. Then $\mathbf{X} = (X_1, X_2)'$ and the joint p.d.f. of X_1, X_2 is the Bivariate Normal p.d.f.

PROOF By Remark 1, both X_1 and X_2 are normally distributed and let $X_1 \sim N(\mu_1, \sigma_1^2)$ and $X_2 \sim N(\mu_2, \sigma_2^2)$. Also let ρ be the correlation coefficient of X_1 and X_2. Then their covariance matrix $\mathbf{\Sigma}$ is given by

$$\mathbf{\Sigma} = \begin{pmatrix} \sigma_1^2 & \rho\sigma_1\sigma_2 \\ \rho\sigma_1\sigma_2 & \sigma_2^2 \end{pmatrix}$$

and hence $|\mathbf{\Sigma}| = \sigma_1^2 \sigma_2^2 (1 - \rho^2)$, so that

$$\mathbf{\Sigma}^{-1} = \frac{1}{\sigma_1^2 \sigma_2^2 (1-\rho^2)} \begin{pmatrix} \sigma_2^2 & -\rho\sigma_1\sigma_2 \\ -\rho\sigma_1\sigma_2 & \sigma_1^2 \end{pmatrix}.$$

Therefore

$$\sigma_1^2\sigma_2^2(1-\rho^2)(\mathbf{x}-\boldsymbol{\mu})'\boldsymbol{\Sigma}^{-1}(\mathbf{x}-\boldsymbol{\mu})$$

$$= (x_1-\mu_1, x_2-\mu_2)\begin{pmatrix} \sigma_2^2 & -\rho\sigma_1\sigma_2 \\ -\rho\sigma_1\sigma_2 & \sigma_1^2 \end{pmatrix}\begin{pmatrix} x_1-\mu_1 \\ x_2-\mu_2 \end{pmatrix}$$

$$= \left((x_1-\mu_1)\sigma_2^2 - (x_2-\mu_2)\rho\sigma_1\sigma_2,\ -(x_1-\mu_1)\rho\sigma_1\sigma_2 + (x_2-\mu_2)\sigma_1^2\right)\begin{pmatrix} x_1-\mu_1 \\ x_2-\mu_2 \end{pmatrix}$$

$$= (x_1-\mu_1)^2\sigma_2^2 - 2(x_1-\mu_1)(x_2-\mu_2)\rho\sigma_1\sigma_2 + (x_2-\mu_2)^2\sigma_1^2.$$

Hence

$$f_{X_1,X_2}(x_1,x_2) = \frac{1}{2\pi\sigma_1\sigma_2\sqrt{1-\rho^2}}\exp\left\{-\frac{1}{2(1-\rho^2)}\left[\left(\frac{x_1-\mu_1}{\sigma_1}\right)^2 - \frac{2\rho}{\sigma_1\sigma_2}(x_1-\mu_1)(x_2-\mu_2) + \left(\frac{x_2-\mu_2}{\sigma_2}\right)^2\right]\right\},$$

as was to be shown. ▲

COROLLARY 2   The (normal) r.v.'s X_i, $i = 1, \ldots, k$ are independent if and only if they are uncorrelated.

PROOF   The r.v.'s X_i, $i = 1, \ldots, k$ are uncorrelated if and only if $\boldsymbol{\Sigma}$ is a diagonal matrix and its diagonal elements are the variances of the X's. Then $|\boldsymbol{\Sigma}| = \sigma_1^2 \cdots \sigma_n^2$. On the other hand, $|\boldsymbol{\Sigma}|\boldsymbol{\Sigma}^{-1}$ is also a diagonal matrix with the jth diagonal element given by $\prod_{i\neq j}\sigma_i^2$, so that $\boldsymbol{\Sigma}^{-1}$ itself is a diagonal matrix with the jth diagonal element being given by $1/\sigma_j^2$. It follows that

$$f_{X_1,\ldots,X_k}(x_1,\ldots,x_k) = \prod_{i=1}^{k}\frac{1}{\sqrt{2\pi}\sigma_i}\exp\left[-\frac{1}{2\sigma_i^2}(x_i-\mu_i)^2\right]$$

and this establishes the independence of the X's. ▲

REMARK 3   The really important part of the corollary is that noncorrelation plus normality implies independence, since independence implies noncorrelation in any case. It is also to be noted that noncorrelation without normality need not imply independence, as it has been seen elsewhere.

Exercises

18.1.1   Use Definition 1 herein in order to conclude that the LSE $\hat{\boldsymbol{\beta}}$ of $\boldsymbol{\beta}$ in (9) of Chapter 16 has the n-Variate Normal distribution with mean $\boldsymbol{\beta}$ and covariance matrix $\sigma^2\mathbf{S}^{-1}$. In particular, $(\hat{\beta}_1, \hat{\beta}_2)'$, given by (19″) and (19′) of

Chapter 16, have the Bivariate Normal distribution with means and variances $E\hat{\beta}_1 = \beta_1$, $E\hat{\beta}_2 = \beta_2$ and

$$\sigma^2(\hat{\beta}_1) = \frac{\sigma^2 \sum_{j=1}^n x_j^2}{n \sum_{j=1}^n (x_j - \bar{x})^2}, \quad \sigma^2(\hat{\beta}_2) = \frac{\sigma^2}{\sum_{j=1}^n (x_j - \bar{x})^2}$$

and correlation coefficient equal to $-\frac{\sum_{j=1}^n x_j}{\sqrt{n \sum_{j=1}^n x_j^2}}$.

18.1.2 Verify relation (8).

18.1.3 Let the random vector $\mathbf{X} = (X_1, \ldots, X_k)'$ be distributed as $N(\boldsymbol{\mu}, \boldsymbol{\Sigma})$ and suppose that $\boldsymbol{\Sigma}$ is non-singular. Then show that the conditional joint distribution of $X_{i_1}, \ldots, X_{i_m}$, given $X_{j_1}, \ldots, X_{j_n}$ ($1 \leq m < k$, $m + n = k$, all $i_1, \ldots, i_m \neq$ from all $j_1, \ldots, j_n$), is Multivariate Normal and specify its parameters.

18.2 Some Properties of Multivariate Normal Distributions

In this section we establish some of the basic properties of a Multivariate Normal distribution.

THEOREM 3 Let $\mathbf{X} = (X_1, \ldots, X_k)'$ be $N(\boldsymbol{\mu}, \boldsymbol{\Sigma})$ (not necessarily non-singular). Then for any $m \times k$ constant matrix $\mathbf{A} = (\alpha_{ij})$, the r. vector $\mathbf{Y}$ defined by $\mathbf{Y} = \mathbf{AX}$ has the m-Variate Normal distribution with mean $\mathbf{A}\boldsymbol{\mu}$ and covariance matrix $\mathbf{A}\boldsymbol{\Sigma}\mathbf{A}'$. In particular, if $m = 1$, the r.v. Y is a linear combination of the X's, $Y = \boldsymbol{\alpha}'\mathbf{X}$, say, and Y has the Univariate Normal distribution with mean $\boldsymbol{\alpha}'\boldsymbol{\mu}$ and variance $\boldsymbol{\alpha}'\boldsymbol{\Sigma}\boldsymbol{\alpha}$.

PROOF For $\mathbf{t} \in \mathbb{R}^m$, we have

$$\phi_{\mathbf{Y}}(\mathbf{t}) = E[\exp(\mathbf{t}'\mathbf{Y})] = E(\exp \mathbf{t}'\mathbf{AX}) = E\left[\exp(\mathbf{A}'\mathbf{t})'\mathbf{X}\right] = \phi_{\mathbf{X}}(\mathbf{A}'\mathbf{t}),$$

so that by means of (6), we have

$$\phi_{\mathbf{Y}}(\mathbf{t}) = \exp\left[i(\mathbf{A}'\mathbf{t})'\boldsymbol{\mu} - \frac{1}{2}(\mathbf{A}'\mathbf{t})'\boldsymbol{\Sigma}(\mathbf{A}'\mathbf{t})\right] = \exp\left[i\mathbf{t}'(\mathbf{A}\boldsymbol{\mu}) - \frac{1}{2}\mathbf{t}'(\mathbf{A}\boldsymbol{\Sigma}\mathbf{A}')\mathbf{t}\right]$$

and this last expression is the ch.f. of the m-Variate Normal with mean $\mathbf{A}\boldsymbol{\mu}$ and covariance matrix $\mathbf{A}\boldsymbol{\Sigma}\mathbf{A}'$, as was to be seen. The particular case follows from the general one just established. ▲

THEOREM 4 For $j = 1, \ldots, n$, let $\mathbf{X}_j$ be independent $N(\boldsymbol{\mu}_j, \boldsymbol{\Sigma}_j)$ k-dimensional r. vectors and let c_j be constants. Then the r. vector

$$\mathbf{X} = \sum_{j=1}^n c_j \mathbf{X}_j \quad \text{is} \quad N\left(\sum_{j=1}^n c_j \boldsymbol{\mu}_j, \sum_{j=1}^n c_j^2 \boldsymbol{\Sigma}_j\right)$$

(a result parallel to a known one for r.v.'s).

PROOF For $\mathbf{t} \in \mathbb{R}^k$ and the independence of the $\mathbf{X}_j$'s, we have

$$\phi_{\mathbf{X}}(\mathbf{t}) = \prod_{j=1}^{n} \phi_{c_j \mathbf{X}_j}(\mathbf{t}) = \prod_{j=1}^{n} \phi_{\mathbf{X}_j}(c_j \mathbf{t}).$$

But

$$\phi_{\mathbf{X}_j}(c_j \mathbf{t}) = \exp\left[i(c_j \mathbf{t})' \boldsymbol{\mu}_j - \frac{1}{2}(c_j \mathbf{t})' \boldsymbol{\Sigma}_j (c_j \mathbf{t})\right]$$

$$= \exp\left[i\mathbf{t}'(c_j \boldsymbol{\mu}_j) - \frac{1}{2}\mathbf{t}'(c_j^2 \boldsymbol{\Sigma}_j)\mathbf{t}\right],$$

so that

$$\phi_{\mathbf{X}}(\mathbf{t}) = \exp\left[i\mathbf{t}'\left(\sum_{j=1}^{n} c_j \boldsymbol{\mu}_j\right) - \frac{1}{2}\mathbf{t}'\left(\sum_{j=1}^{n} c_j^2 \boldsymbol{\Sigma}_j\right)\mathbf{t}\right]. \blacktriangle$$

COROLLARY For $j = 1, \ldots, n$, let $\mathbf{X}_j$ be independent $N(\boldsymbol{\mu}, \boldsymbol{\Sigma})$ k-dimensional r. vectors and let

$$\overline{\mathbf{X}} = \frac{1}{n}\sum_{j=1}^{n} \mathbf{X}_j.$$

Then $\overline{\mathbf{X}}$ is $N(\boldsymbol{\mu}, (1/n)\boldsymbol{\Sigma})$.

PROOF In the theorem, taken $\boldsymbol{\mu}_j = \boldsymbol{\mu}$, $\boldsymbol{\Sigma}_j = \boldsymbol{\Sigma}$ and $c_j = 1/n$, $j = 1, \ldots, n$. $\blacktriangle$

THEOREM 5 Let $\mathbf{X} = (X_1, \ldots, X_k)'$ be non-singular $N(\boldsymbol{\mu}, \boldsymbol{\Sigma})$ and set $Q = (\mathbf{X} - \boldsymbol{\mu})' \boldsymbol{\Sigma}^{-1} (\mathbf{X} - \boldsymbol{\mu})$. Then Q is an r.v. distributed as χ_k^2.

PROOF For $t \in \mathbb{R}$, we have

$$\phi_Q(t) = E(\exp itQ) = \int_{\mathbb{R}^k} \exp\left[it(\mathbf{x} - \boldsymbol{\mu})' \boldsymbol{\Sigma}^{-1}(\mathbf{x} - \boldsymbol{\mu})\right](2\pi)^{-k/2}|\boldsymbol{\Sigma}|^{-1/2}$$

$$\times \exp\left[-\frac{1}{2}(\mathbf{x} - \boldsymbol{\mu})' \boldsymbol{\Sigma}^{-1}(\mathbf{x} - \boldsymbol{\mu})\right]d\mathbf{x}$$

$$= \int_{\mathbb{R}^k} (2\pi)^{-k/2}|\boldsymbol{\Sigma}|^{-1/2} \exp\left[-\frac{1}{2}(\mathbf{x} - \boldsymbol{\mu})' \boldsymbol{\Sigma}^{-1}(\mathbf{x} - \boldsymbol{\mu})(1 - 2it)\right]d\mathbf{x}$$

$$= (1 - 2it)^{-k/2} \int_{\mathbb{R}^k} (2\pi)^{-k/2}\left|\frac{\boldsymbol{\Sigma}}{1 - 2it}\right|^{-1/2}$$

$$\times \exp\left[-\frac{1}{2}(\mathbf{x} - \boldsymbol{\mu})'\left(\frac{\boldsymbol{\Sigma}}{1 - 2it}\right)^{-1}(\mathbf{x} - \boldsymbol{\mu})\right]d\mathbf{x},$$

since

$$\left|\frac{\boldsymbol{\Sigma}}{1 - 2it}\right| = (1 - 2it)^{-k}|\boldsymbol{\Sigma}|.$$

Now the integrand in the last integral above can be looked upon as the p.d.f. of a k-Variate Normal with mean $\boldsymbol{\mu}$ and covariance matrix $\boldsymbol{\Sigma}/(1 - 2it)$. Hence the integral is equal to one and we conclude that $\phi_Q(t) = (1-2it)^{-k/2}$ which is the ch.f. of χ_k^2. ▲

REMARK 4 Notice that Theorem 5 generalizes a known result for the one-dimensional case.

Exercise

18.2.1 Consider the k-dimensional random vectors $\mathbf{X}_n = (X_{1n}, \ldots, X_{kn})'$, $n = 1, 2, \ldots$ and $\mathbf{X} = (X_1, \ldots, X_k)'$ with d.f.'s F_n, F and ch.f.'s ϕ_n, ϕ, respectively. Then we say that $\{\mathbf{X}_n\}$ *converges in distribution* to $\mathbf{X}$ as $n \to \infty$, and we write $\mathbf{X}_n \xrightarrow[n\to\infty]{d} \mathbf{X}$, if $F_n(\mathbf{x}) \xrightarrow[n\to\infty]{} F(\mathbf{x})$ for all $\mathbf{x} \in \mathbb{R}^k$ for which F is continuous (see also Definition 1(iii) in Chapter 8). It can be shown that a multidimensional version of Theorem 2 in Chapter 8 holds true. Use this result (and also Theorem 3' in Chapter 6) in order to prove that $\mathbf{X}_n \xrightarrow[n\to\infty]{d} \mathbf{X}$, if and only if $\boldsymbol{\lambda}'\mathbf{X}_n \xrightarrow[n\to\infty]{d} \boldsymbol{\lambda}'\mathbf{X}$, for every $\boldsymbol{\lambda} = (\lambda, \ldots, \lambda_k)' \in \mathbb{R}^k$. In particular, $\mathbf{X}_n \xrightarrow[n\to\infty]{d} \mathbf{X}$, where $\mathbf{X}$ is distributed as $N(\boldsymbol{\mu}, \boldsymbol{\Sigma})$ if and only if $\{\boldsymbol{\lambda}'\mathbf{X}_n\}$ converges in distribution as $n \to \infty$, to an r.v. Y which is distributed as Normal with mean $\boldsymbol{\lambda}'\boldsymbol{\mu}$ and variance $\boldsymbol{\lambda}'\boldsymbol{\Sigma}\boldsymbol{\lambda}$ for every $\boldsymbol{\lambda} \in \mathbb{R}^k$.

18.3 Estimation of μ and Σ and a Test of Independence

First we formulate a theorem without proof, providing estimators for $\boldsymbol{\mu}$ and $\boldsymbol{\Sigma}$, and then we proceed with a certain testing hypothesis problem.

THEOREM 6 For $j = 1, \ldots, n$, let $\mathbf{X}_j = (X_{j1}, \ldots, X_{jk})'$ be independent, non-singular $N(\boldsymbol{\mu}, \boldsymbol{\Sigma})$ r. vectors and set

$$\overline{\mathbf{X}} = (\overline{X}_1, \ldots, \overline{X}_k)', \quad \text{where} \quad \overline{X}_i = \frac{1}{n}\sum_{j=1}^n X_{ji}, \quad i = 1, \ldots, k,$$

and

$$\mathbf{S} = (S_{ij}), \quad \text{where} \quad S_{ij} = \sum_{k=1}^n (X_{ki} - \overline{X}_i)(X_{kj} - \overline{X}_j), \quad i,j = 1, \ldots, k.$$

Then

i) $\overline{\mathbf{X}}$ and $\mathbf{S}$ are sufficient for $(\boldsymbol{\mu}, \boldsymbol{\Sigma})$;
ii) $\overline{\mathbf{X}}$ and $S/(n-1)$ are unbiased estimators of $\boldsymbol{\mu}$ and $\boldsymbol{\Sigma}$, respectively;
iii) $\overline{\mathbf{X}}$ and S/n are MLE's of $\boldsymbol{\mu}$ and $\boldsymbol{\Sigma}$, respectively.

Now suppose that the joint distribution of the r.v.'s X and Y is the Bivariate Normal distribution. That is,

$$f_{X,Y}(x,y) = \frac{1}{2\pi\sigma_1\sigma_2\sqrt{1-\rho^2}} e^{-q/2},$$

$$q = \frac{1}{1-\rho^2}\left[\left(\frac{x-\mu_1}{\sigma_1}\right)^2 - 2\rho\left(\frac{x-\mu_1}{\sigma_1}\right)\left(\frac{y-\mu_2}{\sigma_2}\right) + \left(\frac{y-\mu_2}{\sigma_2}\right)^2\right].$$

Then by Corollary 2 to Theorem 2, the r.v.'s X and Y are independent if and only if they are uncorrelated. Thus the problem of testing independence for X and Y becomes that of testing the hypothesis $H: \rho = 0$. For this purpose, consider an r. sample of size $n(X_j, Y_j), j = 1, \ldots, n$, from the Bivariate Normal under consideration. Then their joint p.d.f., f, is given by

$$\left(\frac{1}{2\pi\sigma_1\sigma_2\sqrt{1-\rho^2}}\right)^n e^{-Q/2},$$

where

$$Q = \sum_{j=1}^{n} q_j$$

and

$$q_j = \frac{1}{1-\rho^2}\left[\left(\frac{x_j-\mu_1}{\sigma_1}\right)^2 - 2\rho\left(\frac{x_j-\mu_1}{\sigma_1}\right)\left(\frac{y_j-\mu_2}{\sigma_2}\right) + \left(\frac{y_j-\mu_2}{\sigma_2}\right)^2\right],$$

$$j = 1, \ldots, n. \quad (9)$$

For testing H, we are going to employ the LR test. And although the MLE's of the parameters involved are readily given by Theorem 6, we choose to derive them directly. For this purpose, we set $g(\theta)$ for $\log f(\theta)$ considered as a function of the parameter $\theta \in \Omega$, where the parameter space Ω is given by

$$\Omega = \left\{\theta = (\mu_1, \mu_2, \sigma_1^2, \sigma_2^2, \rho)' \in \mathbb{R}^5; \mu_1, \mu_2 \in \mathbb{R}; \sigma_1^2, \sigma_2^2 > 0; -1 < \rho < 1\right\},$$

whereas under H, the parameter space ω becomes

$$\omega = \left\{\theta = (\mu_1, \mu_2, \sigma_1^2, \sigma_2^2, \rho)' \in \mathbb{R}^5; \mu_1, \mu_2 \in \mathbb{R}; \sigma_1^2, \sigma_2^2 > 0; \rho = 0\right\}.$$

We have

$$g = g(\theta) = g(\mu_1, \mu_2, \sigma_1^2, \sigma_2^2, \rho; x_1, \ldots, x_n, y_1, \ldots, y_n)$$

$$= n\log 2\pi - \frac{n}{2}\log\sigma_1^2 - \frac{n}{2}\log\sigma_2^2 - \frac{n}{2}\log(1-\rho^2) - \frac{1}{2}\sum_{j=1}^{n} q_j, \quad (10)$$

where q_j, $j = 1, \ldots, n$ are given by (9). Differentiating (10) with respect to μ_1 and μ_2 and equating the partial derivatives to zero, we get after some simplifications

$$\left.\begin{array}{l}\dfrac{\rho}{\sigma_2}\mu_2 - \dfrac{1}{\sigma_1}\mu_1 = \dfrac{\rho}{\sigma_2}\bar{y} - \dfrac{1}{\sigma_1}\bar{x} \\ \dfrac{\rho}{\sigma_1}\mu_1 - \dfrac{1}{\sigma_2}\mu_2 = \dfrac{\rho}{\sigma_1}\bar{x} - \dfrac{1}{\sigma_2}\bar{y}.\end{array}\right\} \quad \text{(See also Exercise 18.3.1.)} \quad (11)$$

Solving system (11) for μ_1 and μ_2, we get

$$\tilde{\mu}_1 = \bar{x}, \quad \tilde{\mu}_2 = \bar{y}. \quad (12)$$

Now let us set

$$S_x = \frac{1}{n}\sum_{j=1}^{n}(x_j - \bar{x})^2,$$
$$S_y = \frac{1}{n}\sum_{j=1}^{n}(y_j - \bar{y})^2 \quad \text{and} \quad S_{xy} = \frac{1}{n}\sum_{j=1}^{n}(x_j - \bar{x})(y_j - \bar{y}). \quad (13)$$

Then, differentiating g with respect to σ_1^2 and σ_2^2, equating the partial derivatives to zero and replacing μ_1 and μ_2 by $\tilde{\mu}_1$ and $\tilde{\mu}_2$, respectively, we obtain after some simplifications

$$\left.\begin{array}{l}\dfrac{1}{\sigma_1^2}S_x - \dfrac{\rho}{\sigma_1\sigma_2}S_{xy} = 1 - \rho^2 \\ \dfrac{1}{\sigma_2^2}S_y - \dfrac{\rho}{\sigma_1\sigma_2}S_{xy} = 1 - \rho^2.\end{array}\right\} \quad \text{(See also Exercise 18.3.2.)} \quad (14)$$

Next, differentiating g with respect to ρ and equating the partial derivative to zero, we obtain after some simplifications (see also Exercise 18.3.3)

$$\rho - \frac{\rho}{1-\rho^2}\left(\frac{1}{\sigma_1^2}S_x - \frac{2\rho}{\sigma_1\sigma_2}S_{xy} + \frac{1}{\sigma_2^2}S_y\right) + \frac{1}{\sigma_1\sigma_2}S_{xy} = 0. \quad (15)$$

In (14) and (15), solving for σ_1^2, σ_2^2 and ρ, we obtain (see also Exercise 18.3.4)

$$\tilde{\sigma}_1^2 = S_x, \quad \tilde{\sigma}_2^2 = S_y, \quad \tilde{\rho} = \frac{S_{xy}}{\sqrt{S_x S_y}}. \quad (16)$$

It can further be shown (see also Exercise 18.3.5) that the values of the parameters given by (12) and (16) actually maximize f (equivalently, g) and the maximum is given by

$$\max[f(\boldsymbol{\theta}); \boldsymbol{\theta} \in \Omega] = L(\hat{\Omega}) = \left(\frac{e^{-1}}{2\pi\sqrt{S_x S_y}\sqrt{1 - \dfrac{S_{xy}^2}{S_x S_y}}}\right)^n. \quad (17)$$

It follows that the MLE's of $\mu_1, \mu_2, \sigma_1^2, \sigma_2^2$ and ρ, under Ω, are given by (12) and (16), which we may now denote by $\hat{\mu}_{1,\Omega}, \hat{\mu}_{2,\Omega}, \hat{\sigma}_{1,\Omega}^2, \hat{\sigma}_{2,\Omega}^2$ and $\hat{\rho}_\Omega$. That is,

$$\hat{\mu}_{1,\Omega} = \bar{x}, \quad \hat{\mu}_{2,\Omega} = \bar{y}, \quad \hat{\sigma}_{1,\Omega}^2 = S_x, \quad \hat{\sigma}_{2,\Omega}^2 = S_y, \quad \hat{\rho}_\Omega = \frac{S_{xy}}{\sqrt{S_x S_y}}. \tag{18}$$

Under ω (that is, for $\rho = 0$), it is seen (see also Exercise 18.3.6) that the MLE's of the parameters involved are given by

$$\hat{\mu}_{1,\omega} = \bar{x}, \quad \hat{\mu}_{2,\omega} = \bar{y}, \quad \hat{\sigma}_{1,\omega}^2 = S_x, \quad \hat{\sigma}_{2,\omega}^2 = S_y \tag{19}$$

and

$$\max\left[f(\boldsymbol{\theta}); \boldsymbol{\theta} \in \omega\right] = L(\hat{\omega}) = \left(\frac{e^{-1}}{2\pi\sqrt{S_x S_y}}\right)^n. \tag{20}$$

Replacing the x's and y's by X's and Y's, respectively, in (17) and (20), we have that the LR statistic λ is given by

$$\lambda = \left(1 - \frac{S_{XY}^2}{S_X S_Y}\right)^{n/2} = \left(1 - \mathbf{R}^2\right)^{n/2}. \tag{21}$$

where R is the *sample correlation coefficient*, that is,

$$\mathbf{R} = \sum_{j=1}^{n}(X_j - \bar{X})(Y_j - \bar{Y}) \bigg/ \sqrt{\sum_{j=1}^{n}(X_j - \bar{X})^2 \sum_{j=1}^{n}(Y_j - \bar{Y})^2}. \tag{22}$$

From (22), it follows that $\mathbf{R}^2 \leq 1$. (See also Exercise 18.3.7.) Therefore by the fact that the LR test rejects H whenever $\lambda < \lambda_0$, where λ_0 is determined, so that $P_H(\lambda < \lambda_0) = \alpha$, we get by means of (21), that this test is equivalent to rejecting H whenever

$$\mathbf{R}^2 > c_0, \quad \text{equivalently,} \quad \mathbf{R} < -c_0 \text{ or } \mathbf{R} > c_0, \quad c_0 = 1 - \lambda_0^{2/n}. \tag{23}$$

In (23), in order to be able to determine the cut-off point c_0, we have to know the distribution of R under H. Now although the p.d.f. of the r.v. R can be derived, this p.d.f. is none of the usual ones. However, if we consider the function

$$W = W(\mathbf{R}) = \frac{\sqrt{n-2}\,\mathbf{R}}{\sqrt{1-\mathbf{R}^2}}, \tag{24}$$

it is easily seen, by differentiation, that W is an increasing function of R. Therefore, the test in (23) is equivalent to the following test

$$\text{Reject } H \quad \text{whenever} \quad W < -c \quad \text{or} \quad W > c, \tag{25}$$

where c is determined, so that $P_H(W < -c \text{ or } W > c) = \alpha$. It is shown in the sequel that the distribution of W under H is t_{n-2} and hence c is readily determined.

Suppose $X_j = x_j$, $j = 1, \ldots, n$ and that $\sum_{j=1}^{n}(x_j - \bar{x})^2 > 0$ and set

$$R_x = \sum_{j=1}^{n}(x_j - \bar{x})(Y_j - \bar{Y}) \bigg/ \sqrt{\sum_{j=1}^{n}(x_j - \bar{x})^2 \sum_{j=1}^{n}(Y_j - \bar{Y})^2}. \qquad (26)$$

Let also

$$v_j = (x_j - \bar{x}) \bigg/ \sqrt{\sum_{j=1}^{n}(x_j - \bar{x})^2},$$

so that

$$\sum_{j=1}^{n} v_j = 0 \quad \text{and} \quad \sum_{j=1}^{n} v_j^2 = 1. \qquad (27)$$

Let also $W_x = \sqrt{n-2}\, R_x \big/ \sqrt{n - R_x^2}$. It is readily seen that

$$R_x = R_v^* = \sum_{j=1}^{n} v_j Y_j \bigg/ \sqrt{\sum_{j=1}^{n} Y_j^2 - n\bar{Y}^2},$$

so that (see also Exercise 18.3.8)

$$W_x = W_v^* = \sum_{j=1}^{n} v_j Y_j \bigg/ \sqrt{\left[\sum_{j=1}^{n} Y_j^2 - n\bar{Y}^2 - \left(\sum_{j=1}^{n} v_j Y_j\right)^2\right] \bigg/ (n-2)}. \qquad (28)$$

We have that Y_j, $j = 1, \ldots, n$ are independent $N(\mu_2, \sigma_2^2)$. Now if we consider the $N(0, 1)$ r.v.'s $Y_j' = (Y_j - \mu_2)/\sigma_2$, $j = 1, \ldots, n$ and replace Y_j by Y_j' in (28), it is seen (see also Exercise 18.3.9) that $W_x = W_v^*$ remains unchanged. Therefore we may assume that the Y's are themselves independent $N(0, 1)$. Next consider the transformation

$$\begin{cases} Z_1 = \dfrac{1}{\sqrt{n}} Y_1 + \cdots + \dfrac{1}{\sqrt{n}} Y_n \\ Z_2 = v_1 Y_1 + \cdots + v_n Y_n. \end{cases}$$

Then because of $(1/\sqrt{n})^2 + \cdots + (1/\sqrt{n})^2 = n/n = 1$ and also because of (27), this transformation can be completed to an orthogonal transformation (see Theorem 8.I(i) in Appendix I) and let Z_j, $j = 3, \ldots, n$ be the remaining Z's. Then by Theorem 5, Chapter 9, it follows that the r.v.'s Z_j, $j = 1, \ldots, n$ are independent $N(0, 1)$. Also $\sum_{j=1}^{n} Y_j^2 = \sum_{j=1}^{n} Z_j^2$ by Theorem 4, Chapter 9. By means of the transformation in question, the statistic in (28) becomes $Z_2 \big/ \sqrt{\sum_{j=3}^{n} Z_j^2 / (n-2)}$. Therefore the distribution of W_x, equivalently the distribution of W, given $X_j = x_j$, $j = 1, \ldots, n$ is t_{n-2}. Since this distribution is independent of the x's, it follows that the unconditional distribution of W is t_{n-2}. Thus we have the following result.

THEOREM 7 For testing $H: \rho = 0$ against $A: \rho \neq 0$ at level of significance α, one rejects H whenever $W < -c$ or $W > c$, where W is given by (24), the sample correlation coefficient R is given by (22) and the cut-off point c is determined from $P(t_{n-2} > c) = \alpha/2$ by the fact that the distribution of W, under H, is t_{n-2}.

To this last theorem, one has the following corollary.

COROLLARY The p.d.f. of the correlation coefficient R is given by

$$f_R(r) = \frac{1}{\sqrt{\pi}} \frac{\Gamma\left[\frac{1}{2}(n-1)\right]}{\Gamma\left[\frac{1}{2}(n-2)\right]} (1-r^2)^{(n/2)-2}, \quad -1 < r < 1.$$

PROOF From $W = \sqrt{n-2}\, r/\sqrt{1-R^2}$, it follows that R and W have the same sign, that is, $RW \geq 0$. Solving for R, one has then $R = W/\sqrt{W^2 + n - 2}$. By setting $w = \sqrt{n-2}\, r/\sqrt{1-r^2}$, one has $dw/dr = \sqrt{n-2}(1-r^2)^{-3/2}$, whereas

$$f_W(w) = \frac{\Gamma\left[\frac{1}{2}(n-1)\right]}{\sqrt{\pi}\sqrt{n-2}\,\Gamma\left[\frac{1}{2}(n-2)\right]} \left(1 + \frac{w^2}{n-2}\right)^{-(n-1)/2}, \quad w \in \mathbb{R}.$$

Therefore

$$f_R(r) = f_W\left(\frac{\sqrt{n-2}\,r}{\sqrt{1-r^2}}\right) \frac{dw}{dr}$$

$$= \frac{\Gamma\left[\frac{1}{2}(n-1)\right]}{\sqrt{\pi}\sqrt{n-2}\,\Gamma\left[\frac{1}{2}(n-2)\right]} \left[1 + \frac{(n-2)r^2}{(n-2)(1-r^2)}\right]^{-(n-1)/2} \sqrt{n-2}(1-r^2)^{-3/2}$$

$$= \frac{\Gamma\left[\frac{1}{2}(n-1)\right]}{\sqrt{\pi}\,\Gamma\left[\frac{1}{2}(n-2)\right]} (1-r^2)^{(n/2)-2},$$

as was to be shown. ▲

The p.d.f. of R when $\rho \neq 0$ can also be obtained, but its expression is rather complicated and we choose not to go into it.

We close this chapter with the following comment. Let **X** be a k-dimensional random vector distributed as $N(\boldsymbol{\mu}, \boldsymbol{\Sigma})$. Then its ch.f. is given by (6). Furthermore, if $\boldsymbol{\Sigma}$ is non-singular, then the $N(\boldsymbol{\mu}, \boldsymbol{\Sigma})$ distribution has a p.d.f. which is given by (7). However, this is not the case if $\boldsymbol{\Sigma}$ is singular. In this latter case, the distribution is called *singular*, and it can be shown that it is concentrated in a hyperplane of dimensionality less than k.

Exercises

18.3.1 Verify relation (11).

18.3.2 Verify relation (14).

18.3.3 Verify relation (15).

18.3.4 Show that $\tilde{\sigma}_1^2$, $\tilde{\sigma}_2^2$, and $\tilde{\rho}$ given by (16) is indeed the solution of the system of the equations in (14) and (15).

18.3.5 Consider g given by (10) and set

$$d_{ij} = \frac{\partial^2}{\partial \theta_i \partial \theta_j} g(\boldsymbol{\theta}) \bigg|_{\boldsymbol{\theta} = \tilde{\boldsymbol{\theta}}},$$

where

$$\boldsymbol{\theta} = (\mu_1, \mu_2, \sigma_1^2, \sigma_2^2, \rho)' \quad \tilde{\boldsymbol{\theta}} = (\tilde{\mu}_1, \tilde{\mu}_2, \tilde{\sigma}_1^2, \tilde{\sigma}_2^2, \tilde{\rho})'$$

and $\tilde{\mu}_1$, $\tilde{\mu}_2$, $\tilde{\sigma}_1^2$, $\tilde{\sigma}_2^2$ and $\tilde{\rho}$ are given by (12) and (16). Let $\mathbf{D} = (d_{ij})$, $i, j = 1, \ldots, 5$ and denote by D_{5-k} the determinant obtained from $\mathbf{D}$ by deleting the last k rows and columns, $k = 1, \ldots, 5$, $D_0 = 1$. Then show that the six numbers D_0, $D_1, \ldots, D_5$ are alternately positive and negative. This result together with the fact that d_{ij}, $i, j = 1, \ldots, 5$ are continuous functions of $\boldsymbol{\theta}$ implies that the quantities given by (18) are, indeed, the MLE's of the parameters μ_1, μ_2, σ_1^2, σ_2^2 and ρ, under Ω. (See for example *Mathematical Analysis* by T. M. Apostol, Addison-Wesley, 1957, Theorem 7.9, pp. 151–152.)

18.3.6 Show that the MLE's of μ_1, μ_2, σ_1^2, and σ_2^2, under ω, are indeed given by (19).

18.3.7 Show that $\mathbb{R}^2 \leq 1$, where $\mathbb{R}$ is the sample correlation coefficient given by (22).

18.3.8 Verify relation (28).

18.3.9 Show that the statistic $W_x (= W_y^*)$ in (28) remains unchanged if the r.v.'s Y_j are replaced by the r.v.'s

$$Y_j' = \frac{Y_j - \mu_2}{\sigma^2}, \quad j = 1, \ldots, n.$$

18.3.10 Refer to the quantities $\overline{\mathbf{X}}$ and $\mathbf{S}$ defined in Theorem 6 and, by using Basu's theorem (Theorem 3, Chapter 11), show that they are independent.

Chapter 19

Quadratic Forms

19.1 Introduction

In this chapter, we introduce the concept of a quadratic form in the variables $x_j, j = 1, \ldots, n$ and then confine attention to quadratic forms in which the x_j's are replaced by independent normally distributed r.v.'s $X_j, j = 1, \ldots, n$. In this latter case, we formulate and prove a number of standard theorems referring to the distribution and/or independence of quadratic forms.

A quadratic form, Q, in the variables $x_j, j = 1, \ldots, n$ is a homogeneous quadratic (second degree) function of $x_j, j = 1, \ldots, n$. That is,

$$Q = \sum_{i=1}^{n} \sum_{j=1}^{n} c_{ij} x_i x_j,$$

where here and in the sequel the coefficients of the x's are always assumed to be real-valued constants. By setting $\mathbf{x} = (x_1, \ldots, x_n)'$ and $\mathbf{C} = (c_{ij})$, we can write $Q = \mathbf{x}'\mathbf{C}\mathbf{x}$. Now Q is a 1×1 matrix and hence $Q' = Q$, or $(\mathbf{x}'\mathbf{C}\mathbf{x})' = \mathbf{x}'\mathbf{C}'\mathbf{x} = \mathbf{x}'\mathbf{C}\mathbf{x}$. Therefore $Q = \frac{1}{2}(\mathbf{x}'\mathbf{C}'\mathbf{x} + \mathbf{x}'\mathbf{C}\mathbf{x}) = \mathbf{x}'\mathbf{A}\mathbf{x}$, where $\mathbf{A} = \frac{1}{2}(\mathbf{C} + \mathbf{C}')$; that is to say, if $\mathbf{A} = (a_{ij})$, then $a_{ij} = \frac{1}{2}(c_{ij} + c_{ji})$, so that $a_{ij} = a_{ji}$. Thus $\mathbf{A}$ is symmetric. We can then give the following definition.

DEFINITION 1 A (*real*) *quadratic form*, Q, in the variables $x_j, j = 1, \ldots, n$ is a homogeneous quadratic function of $x_j, j = 1, \ldots, n$,

$$Q = \sum_{i=1}^{n} \sum_{j=1}^{n} c_{ij} x_i x_j, \tag{1}$$

where $c_{ij} \in \mathbb{R}$ and $c_{ij} = c_{ji}, i, j = 1, \ldots, n$. In matrix notation, (1) becomes as follows:

$$Q = \mathbf{x}'\mathbf{C}\mathbf{x}, \tag{2}$$

where $\mathbf{x} = (x_1, \ldots, x_n)'$, $\mathbf{C} = (c_{ij})$ and $\mathbf{C}' = \mathbf{C}$ (which expresses the symmetry of $\mathbf{C}$).

DEFINITION 2 For an $n \times n$ matrix $\mathbf{C}$, the polynomial (in λ) $|\mathbf{C} - \lambda \mathbf{I}_n|$ is of degree n and is called the *characteristic polynomial* of $\mathbf{C}$. The n roots of the equation $|\mathbf{C} - \lambda \mathbf{I}_n| = 0$ are called *characteristic* or *latent roots* or *eigenvalues* of $\mathbf{C}$.

DEFINITION 3 The quadratic form $Q = \mathbf{x}'\mathbf{C}\mathbf{x}$ is called *positive definite* if $\mathbf{x}'\mathbf{C}\mathbf{x} > 0$ for every $\mathbf{x} \neq \mathbf{0}$; it is called *negative definite* if $\mathbf{x}'\mathbf{C}\mathbf{x} < 0$ for every $\mathbf{x} \neq \mathbf{0}$ and *positive semidefinite* if $\mathbf{x}'\mathbf{C}\mathbf{x} \geq 0$ for every $\mathbf{x}$. A symmetric $n \times n$ matrix C is called *positive definite*, *negative definite* or *positive semidefinite* if the quadratic form associated with it, $Q = \mathbf{x}'\mathbf{C}\mathbf{x}$, is positive definite, negative definite or positive semidefinite, respectively.

DEFINITION 4 If $Q = \mathbf{x}'\mathbf{C}\mathbf{x}$, then the rank of $\mathbf{C}$ is also called the *rank of Q*.

19.2 Some Theorems on Quadratic Forms

Throughout this section, it is assumed that the r.v.'s X_j, $j = 1, \ldots, n$ are independently distributed as $N(0, 1)$ and we set $\mathbf{X} = (X_1, \ldots, X_n)'$. We then replace $\mathbf{x}$ by $\mathbf{X}$ in (2) and obtain the following quadratic form in X_j, $j = 1, \ldots, n$, or X:

$$Q = \mathbf{X}'\mathbf{C}\mathbf{X}, \quad \text{where} \quad \mathbf{C}' = \mathbf{C}.$$

Some theorems related to such quadratic forms will now be established.

THEOREM 1 (Cochran) Let

$$\mathbf{X}'\mathbf{X} = \sum_{i=1}^{k} Q_i, \qquad (3)$$

where for $i = 1, \ldots, k$, Q_i are quadratic forms in $\mathbf{X}$ with rank $Q_i = r_i$. Then the r.v.'s Q_i are independent $\chi^2_{r_i}$ if and only if $\sum_{i=1}^{k} r_i = n$.

PROOF We have that $\mathbf{X}'\mathbf{X} = \sum_{j=1}^{n} X_j^2$ is χ^2_n. Therefore if for $i = 1, \ldots, k$, Q_i are independent $\chi^2_{r_i}$, then because of (3), $\sum_{i=1}^{k} r_i = n$.

Next, we suppose that $\sum_{i=1}^{k} r_i = n$ and show that for $i = 1, \ldots, k$, Q_i are independent $\chi^2_{r_i}$. To this end, one has that $Q_i = \mathbf{X}'\mathbf{C}_i\mathbf{X}$, where $\mathbf{C}_i$ is an $n \times n$ symmetric matrix with rank $\mathbf{C}_i = r_i$. Consider the matrix $\mathbf{C}_i$. By Theorem 11.I(ii) in Appendix I, there exist r_i linear forms in the $\mathbf{X}$'s such that

$$Q_i = \delta_1^{(i)} \left(b_{11}^{(i)} X_1 + \cdots + b_{1n}^{(i)} X_n \right)^2 + \cdots + \delta_{r_i}^{(i)} \left(b_{r_i 1}^{(i)} X_1 + \cdots + b_{r_i n}^{(i)} X_n \right)^2, \qquad (4)$$

where $\delta_1^{(i)}, \ldots, \delta_{r_i}^{(i)}$ are either 1 or -1. Now $\sum_{i=1}^{k} r_i = n$ and let $\mathbf{B}$ be the $n \times n$ matrix defined by

$$\mathbf{B} = \begin{Bmatrix} b_{11}^{(1)} & \cdots & b_{1n}^{(1)} \\ b_{r_11}^{(1)} & \cdots & b_{r_1n}^{(1)} \\ b_{11}^{(k)} & \cdots & b_{1n}^{(k)} \\ b_{r_k1}^{(k)} & \cdots & b_{r_kn}^{(k)} \end{Bmatrix}.$$

Then by (4) and the definition of **B**, it is clear that

$$\sum_{i=1}^{k} Q_i = (\mathbf{BX})'\mathbf{D}(\mathbf{BX}), \tag{5}$$

where **D** is an $n \times n$ diagonal matrix with diagonal elements equal to $\delta_1^{(i)}, \ldots, \delta_{r_i}^{(i)}$, $i = 1, \ldots, k$. On the other hand,

$$\sum_{i=1}^{k} Q_i = \mathbf{X}'\mathbf{X} \quad \text{and} \quad (\mathbf{BX})'\mathbf{D}(\mathbf{BX}) = \mathbf{X}'(\mathbf{B}'\mathbf{DB})\mathbf{X}.$$

Therefore (5) gives

$$\mathbf{X}'\mathbf{X} = \mathbf{X}'(\mathbf{B}'\mathbf{DB})\mathbf{X} \quad \text{identically in } \mathbf{X}.$$

Hence $\mathbf{B}'\mathbf{DB} = \mathbf{I}_n$. From the definition of **D**, it follows that $\|\mathbf{D}\| = 1$, so that rank $\mathbf{D} = n$. Let $r = $ rank **B**. Then, of course, $r \leq n$. Also $n = $ rank $\mathbf{I}_n = $ rank $(\mathbf{B}'\mathbf{DB}) \leq r$, so that $r = n$. It follows that B is nonsingular and therefore the relationship $\mathbf{B}'\mathbf{DB} = \mathbf{I}_n$ implies $\mathbf{D} = (\mathbf{B}')^{-1}\mathbf{B}^{-1} = (\mathbf{BB}')^{-1}$. On the other hand, for any nonsingular square matrix **M**, $\mathbf{MM}'$ is positive definite (by Theorem 10.I(ii) in Appendix I) and so is $(\mathbf{MM}')^{-1}$. Thus $(\mathbf{BB}')^{-1}$ is positive definite and hence so is **D**. From the form of **D**, it follows then that all diagonal elements of **D** are equal to 1, which implies that $\mathbf{D} = \mathbf{I}_n$ and hence $\mathbf{B}'\mathbf{B} = \mathbf{I}_n$; that is to say, **B** is orthogonal. Set $\mathbf{Y} = \mathbf{BX}$. By Theorem 5, Chapter 9, it follows that, if $\mathbf{Y} = (Y_1, \ldots, Y_n)'$, then the r.v.'s $Y_j, j = 1, \ldots, n$ are independent $N(0, 1)$. Also the fact that $\mathbf{D} = \mathbf{I}_n$ and the transformation $\mathbf{Y} = \mathbf{BX}$ imply, by means of (4), that Q_1 is equal to the sum of the squares of the first r_1, Y's, Q_2 is the sum of the squares of the next r_2 Y's, $\ldots$, Q_k is the sum of the squares of the last r_k Y's. It follows that, for $i = 1, \ldots, k$, Q_i are independent $\chi_{r_i}^2$. The proof is completed. ▲

APPLICATION 1 For $j = 1, \ldots, n$, let Z_j be independent r.v.'s distributed as $N(\mu, \sigma^2)$ and set $X_j = (Z_j - \mu)/\sigma$, so that the X's are i.i.d. distributed as $N(0, 1)$. It has been seen elsewhere that

$$\sum_{j=1}^{n}\left(\frac{Z_j - \mu}{\sigma}\right)^2 = \sum_{j=1}^{n}\left(\frac{Z_j - \bar{Z}}{\sigma}\right)^2 + \left[\frac{\sqrt{n}(\bar{Z} - \mu)}{\sigma}\right]^2;$$

equivalently,

$$\sum_{j=1}^{n} X_j^2 = \sum_{j=1}^{n}(X_j - \overline{X})^2 + \left(\sqrt{n^{\frac{1}{2}}}\overline{X}\right)^2.$$

Now

$$\left(\sqrt{n^{\frac{1}{2}}}\overline{X}\right)^2 = \frac{1}{n}\left(\sum_{j=1}^{n} X_j\right)^2 = \mathbf{X}'\mathbf{C}_2\mathbf{X},$$

where $\mathbf{C}_2$ has its elements identically equal to $1/n$, so that rank $\mathbf{C}_2 = 1$. Next it can be shown (see also Exercise 19.2.1) that

$$\sum_{j=1}^{n}(X_j - \overline{X})^2 = \mathbf{X}'\mathbf{C}_1\mathbf{X},$$

where $\mathbf{C}_1$ is given by

$$\mathbf{C}_1 = \begin{Bmatrix} (n-1)/n & -1/n & \cdots & -1/n \\ -1/n & (n-1)/n & \cdots & -1/n \\ \cdot & \cdot & \cdots & \cdot \\ -1/n & -1/n & \cdots & (n-1)/n \end{Bmatrix}$$

and that rank $\mathbf{C}_1 = n - 1$. Then Theorem 1 applies with $k = 2$ and gives that $\sum_{j=1}^{n}(X_j - \overline{X})^2$ and $(\sqrt{n}\overline{X})^2$ are independent distributed as χ^2_{n-1} and χ^2_1, respectively. Thus it follows that $(1/\sigma^2)\sum_{j=1}^{n}(Z_j - \overline{Z})^2$ is distributed as χ^2_{n-1} and is independent of $\overline{Z}$.

The following theorem refers to the distribution of a quadratic form in the independent $N(0, 1)$ r.v.'s X_j, $j = 1, \ldots, n$. Namely,

THEOREM 2 Consider the quadratic form $Q = \mathbf{X}'\mathbf{C}\mathbf{X}$. Then Q is distributed as χ^2_r if and only if $\mathbf{C}$ is idempotent (that is, $\mathbf{C}^2 = \mathbf{C}$) and rank $\mathbf{C} = r$.

PROOF Suppose that $\mathbf{C}$ is idempotent and that rank $\mathbf{C} = r$. Then by Theorem 12.I(iii) in Appendix I, we have

$$\text{rank } \mathbf{C} + \text{rank}(\mathbf{I}_n - \mathbf{C}) = n. \tag{6}$$

Also

$$\mathbf{X}'\mathbf{X} = \mathbf{X}'\mathbf{C}\mathbf{X} + \mathbf{X}'(\mathbf{I}_n - \mathbf{C})\mathbf{X}. \tag{7}$$

Then Theorem 1 applies with $k = 2$ and gives that $\mathbf{X}'\mathbf{C}\mathbf{X}$ is χ^2_r (and also $\mathbf{X}'(\mathbf{I}_n - \mathbf{C})\mathbf{X}$ is χ^2_{n-r}).

Assume now that $Q = \mathbf{X}'\mathbf{C}\mathbf{X}$ is χ^2_r. Then we first show that rank $\mathbf{C} = r$. By Theorem 11.I(iii) in Appendix I, there exists an orthogonal matrix $\mathbf{P}$ such that if $\mathbf{Y} = \mathbf{P}^{-1}\mathbf{X}$ (equivalently, $\mathbf{X} = \mathbf{P}\mathbf{Y}$), then

$$Q = \mathbf{X}'\mathbf{C}\mathbf{X} = (\mathbf{P}\mathbf{Y})'\mathbf{C}(\mathbf{P}\mathbf{Y}) = \mathbf{Y}'(\mathbf{P}'\mathbf{C}\mathbf{P})\mathbf{Y} = \sum_{j=1}^{m} \lambda_j Y_j^2, \tag{8}$$

where $(Y_1, \ldots, Y_n)' = \mathbf{Y}$ and λ_j, $j = 1, \ldots, m$ are the nonzero characteristic roots of $\mathbf{C}$.

By the orthogonality of $\mathbf{P}$, the Y's are independent $N(0, 1)$ (Theorem 5, Chapter 9), so that the Y^2's are independent χ_1^2. Therefore the ch.f. of $\sum_{j=1}^{m} \lambda_j Y_j^2$, evaluated at t, is given by

$$\left[(1 - 2i\lambda_1 t) \cdots (1 - 2i\lambda_m t)\right]^{-1/2}. \tag{9}$$

On the other hand, Q is χ_r^2 by assumption, so that its ch.f., evaluated at t, is given by

$$(1 - 2it)^{-r/2}. \tag{10}$$

From (8)–(10), one then has that (see also Exercise 19.2.2)

$$\lambda_1 = \cdots = \lambda_m = 1 \quad \text{and} \quad m = r. \tag{11}$$

It follows then that rank $\mathbf{C} = r$. We now show that $\mathbf{C}^2 = \mathbf{C}$. From (8), one has that $\mathbf{P'CP}$ is diagonal and, by (11), its diagonal elements are either 1 or 0. Hence $\mathbf{P'CP}$ is idempotent. Thus

$$\mathbf{P'CP} = (\mathbf{P'CP})^2 = (\mathbf{P'CP})(\mathbf{P'CP})$$
$$= \mathbf{P'C}(\mathbf{PP'})\mathbf{CP} = \mathbf{P'CI}_n\mathbf{CP} = \mathbf{P'C}^2\mathbf{P}.$$

That is,

$$\mathbf{P'CP} = \mathbf{P'C}^2\mathbf{P}. \tag{12}$$

Multiplying by $\mathbf{P'}^{-1}$ and $\mathbf{P}^{-1}$ on the left and right, respectively, both sides of (12), one concludes that $\mathbf{C} = \mathbf{C}^2$. This completes the proof of the theorem. ▲

APPLICATION 2 Refer to Application 1. It can be shown (see also Exercise 19.2.3) that $\mathbf{C}_1$ and $\mathbf{C}_2$ are idempotent. Then Theorem 2 implies that $\sum_{j=1}^{n}(X_j - \overline{X})^2$ and $(\sqrt{n}^{1/2}\overline{X})^2$, or equivalently

$$\frac{1}{\sigma^2}\sum_{j=1}^{n}(Z_j - \overline{Z})^2 \quad \text{and} \quad \left[\frac{\sqrt{n}(\overline{Z} - \mu)}{\sigma}\right]^2$$

are distributed as χ_{n-1}^2 and χ_1^2, respectively.

To this theorem there are the following two corollaries which will be employed in the sequel.

COROLLARY 1 If the quadratic form $Q = \mathbf{X'CX}$ is distributed as χ_r^2, then it is positive semidefinite.

PROOF From (8) and (10), one has that $Q = \mathbf{X'CX}$ is equal to $\sum_{j=1}^{r} Y_j^2$, so that $\mathbf{X'CX}$ is equal to $\sum_{j=1}^{r} Y_j^2$, where $\mathbf{X} = (X_1, \ldots, X_n)'$ and $(Y_1, \ldots, Y_n)' = \mathbf{Y} = \mathbf{P}^{-1}\mathbf{X}$. Thus $\mathbf{X'CX} \geq 0$ for every $\mathbf{X}$, as was to be seen. ▲

COROLLARY 2 Let $\mathbf{P}$ be an orthogonal matrix and consider the transformation $\mathbf{Y} = \mathbf{P}^{-1}\mathbf{X}$. Then if the quadratic form $Q = \mathbf{X}'\mathbf{CX}$ is χ_r^2, so is the quadratic form $Q^* = \mathbf{Y}'(\mathbf{P}'\mathbf{CP})\mathbf{Y}$.

PROOF By the theorem, it suffices to show that $\mathbf{P}'\mathbf{CP}$ is idempotent and that its rank is r. We have

$$(\mathbf{P}'\mathbf{CP})^2 = \mathbf{P}'\mathbf{C}(\mathbf{PP}')\mathbf{CP} = \mathbf{P}'\mathbf{CCP} = \mathbf{P}'\mathbf{CP}$$

since $\mathbf{C}^2 = \mathbf{C}$. That rank $\mathbf{P}'\mathbf{CP} = r$ follows from Theorem 9.I(iv) in Appendix I. Hence the result. ▲

THEOREM 3 Suppose that $\mathbf{X}'\mathbf{X} = Q_1 + Q_2$, where Q_1, Q_2 are quadratic forms in $\mathbf{X}$, and let Q_1 be $\chi_{r_1}^2$. Then Q_2 is $\chi_{n-r_1}^2$ and Q_1, Q_2 are independent.

PROOF Let $Q_1 = \mathbf{X}'\mathbf{C}_1\mathbf{X}$. Then the assumption that Q_1 is $\chi_{r_1}^2$ implies (by Theorem 2) that $\mathbf{C}_1$ is idempotent and rank $\mathbf{C}_1 = r_1$. Next

$$Q_2 = \mathbf{X}'\mathbf{X} - Q_1 = \mathbf{X}'\mathbf{X} - \mathbf{X}'\mathbf{C}_1\mathbf{X} = \mathbf{X}'(\mathbf{I}_n - \mathbf{C}_1)\mathbf{X}$$

and $(\mathbf{I}_n - \mathbf{C}_1)^2 = \mathbf{I}_n - 2\mathbf{C}_1 + \mathbf{C}_1^2 = \mathbf{I}_n - \mathbf{C}_1$, that is, $\mathbf{I}_n - \mathbf{C}_1$ is idempotent. Also rank $\mathbf{C}_1$ + rank $(\mathbf{I}_n - \mathbf{C}_1) = n$ by Theorem 12.I(iii) in Appendix I, so that rank $(\mathbf{I}_n - \mathbf{C}_1) = n - r_1$. We have then that rank Q_1 + rank $Q_2 = n$, and therefore Theorem 1 applies and gives the result. ▲

APPLICATION 3 Refer to Application 1. Since $\sqrt{n}\bar{X}$ is $N(0, 1)$, it follows that $(\sqrt{n}\bar{X})^2$ is χ_1^2. Then, by Theorem 3, $\sum_{j=1}^n (X_j - \bar{X})^2$ is distributed as χ_{n-1}^2 and is independent of $(\sqrt{n}\bar{X})^2$. Thus once again, $(1/\sigma^2)\sum_{j=1}^n (Z_j - \bar{Z})^2$ is distributed as χ_{n-1}^2 and is independent of $\bar{Z}$.

The following theorem is also of interest.

THEOREM 4 Suppose that $Q = Q_1 + Q_2$, where Q, Q_1 and Q_2 are quadratic forms in $\mathbf{X}$. Furthermore, let Q be χ_r^2, let Q_1 be $\chi_{r_1}^2$ and let Q_2 be positive semidefinite. Then Q_2 is $\chi_{r_2}^2$, where $r_2 = r - r_1$, and Q_1, Q_2 are independent.

PROOF Let $Q = \mathbf{X}'\mathbf{CX}$. Then, by Theorem 2, $\mathbf{C}$ is idempotent and rank $\mathbf{C} = r$. By Theorem 11.I(iv) in Appendix I, it follows that there exists an orthogonal matrix $\mathbf{P}$ such that if $\mathbf{Y} = \mathbf{P}^{-1}\mathbf{X}$ (equivalently, $\mathbf{X} = \mathbf{PY}$), then Q is transformed into $\mathbf{Y}'(\mathbf{P}'\mathbf{CP})\mathbf{Y} = \sum_{j=1}^r Y_j^2$. For $i = 1, 2$, let $Q_i = \mathbf{X}'\mathbf{C}_i\mathbf{X}$ and let Q_i^* be the quadratic form in $\mathbf{Y}$ into which Q_i is transformed under $\mathbf{P}$; that is,

$$Q_i^* = \mathbf{Y}'\mathbf{B}_i\mathbf{Y}, \quad \text{where} \quad \mathbf{B}_i = \mathbf{P}'\mathbf{C}_i\mathbf{P}, \quad i = 1, 2.$$

The equation $Q = Q_1 + Q_2$ implies

$$(Y_1, \ldots, Y_r)'(Y_1, \ldots, Y_r) = \sum_{j=1}^r Y_j^2 = Q_1^* + Q_2^*. \tag{13}$$

By Corollary 1 to Theorem 2, it follows that $\mathbf{C}_1$ is positive semidefinite and so is $\mathbf{C}_2$ by assumption. Therefore by Theorem 10.I in Appendix I, $\mathbf{B}_1$ and $\mathbf{B}_2$, or equivalently, Q_1^* and Q_2^* are positive semidefinite. From this result and (13), it follows that Q_1^* and Q_2^* are functions of Y_j, $j = 1, \ldots, r$ only. From the orthogonality of **P**, we have that the r.v.'s Y_j, $j = 1, \ldots, r$ are independent $N(0, 1)$. On the other hand, Q_1^* is $\chi_{r_1}^2$ by Corollary 2 to Theorem 2. These facts together with (13) imply that Theorem 3 applies (with $n = r$) and provides the desired result. ▲

This last theorem generalizes as follows.

THEOREM 5 Suppose that $Q = \sum_{i=1}^{k} Q_i$, where Q and Q_i, $i = 1, \ldots, k$ (≥ 2) are quadratic forms in **X**. Furthermore, let Q be χ_r^2, let Q_i be $\chi_{r_i}^2$, $i = 1, \ldots, k-1$ and let Q_k be positive semidefinite. Then Q_k is $\chi_{r_k}^2$, where

$$r_k = r - \sum_{i=1}^{k-1} r_i, \quad \text{and} \quad Q_i, i = 1, \ldots, k$$

are independent.

PROOF The proof is by induction. For $k = 2$ the conclusion is true by Theorem 4. Let the theorem hold for $k = m$ and show that it also holds for $m + 1$. We write

$$Q = \sum_{i=1}^{m-1} Q_i + Q_m^*, \quad \text{where} \quad Q_m^* = Q_m + Q_{m+1}.$$

By our assumptions and Corollary 1 to Theorem 2, it follows that Q_m^* is positive semidefinite. Hence Q_m^* is $\chi_{r_m^*}^2$,

$$r_m^* = r - \sum_{i=1}^{m-1} r_i, \quad \text{and} \quad Q_1, \ldots, Q_{m-1}, Q_m^*$$

are independent, by the induction hypothesis. Thus $Q_m^* = Q_m + Q_{m+1}$, where Q_m^* is $\chi_{r_m^*}^2$, Q_m is $\chi_{r_m}^2$ and Q_{m+1} is positive semidefinite. Once again Theorem 4 applies and gives that Q_{m+1} is $\chi_{r_{m+1}}^2$, where

$$r_{m+1} = r_m^* - r_m = r - \sum_{i=1}^{m} r_i,$$

and that Q_m and Q_{m+1} are independent. It follows that Q_i, $i = 1, \ldots, m+1$ are also independent and the proof is concluded. ▲

The theorem below gives a necessary and sufficient condition for independence of two quadratic forms. More precisely, we have the following result.

THEOREM 6 Consider the independent r.v.'s Y_j, $j = 1, \ldots, n$, where Y_j is distributed as $N(\mu_j, \sigma^2)$, and for $i = 1, 2$, let Q_i be quadratic forms in $\mathbf{Y} = (Y_1, \ldots, Y_n)'$; that is, $Q_i = \mathbf{Y}'\mathbf{C}_i\mathbf{Y}$. Then Q_1 and Q_2 are independent if and only if $\mathbf{C}_1\mathbf{C}_2 = \mathbf{0}$.

PROOF The proof is presented only for the special case that $Y_j = X_j \sim N(0, 1)$ and $Q_i \sim \chi_{r_i}^2$, $i = 1, 2$. To this end, suppose that $\mathbf{C}_1\mathbf{C}_2 = \mathbf{0}$. By the fact that

Q_i is distributed as $\chi^2_{r_i}$, it follows (by Theorem 2) that $\mathbf{C}_i$, $i = 1, 2$ are idempotent; that is, $\mathbf{C}_i^2 = \mathbf{C}_i$, $i = 1, 2$. Next by the symmetry of $\mathbf{C}_i$, one has $\mathbf{C}_2\mathbf{C}_1 = \mathbf{C}_2'\mathbf{C}_1' = (\mathbf{C}_1\mathbf{C}_2)' = \mathbf{0}' = \mathbf{0}$. Therefore

$$\mathbf{C}_1(\mathbf{I}_n - \mathbf{C}_1 - \mathbf{C}_2) = \mathbf{C}_2(\mathbf{I}_n - \mathbf{C}_1 - \mathbf{C}_2) = \mathbf{0}.$$

Then Theorem 12.I(iii), in Appendix I, implies that

$$\operatorname{rank} \mathbf{C}_1 + \operatorname{rank} \mathbf{C}_2 + \operatorname{rank}(\mathbf{I}_n - \mathbf{C}_1 - \mathbf{C}_2) = n. \tag{14}$$

On the other hand, clearly, we have

$$\mathbf{X}'\mathbf{X} = \mathbf{X}'\mathbf{C}_1\mathbf{X} + \mathbf{X}'\mathbf{C}_2\mathbf{X} + \mathbf{X}'(\mathbf{I}_n - \mathbf{C}_1 - \mathbf{C}_2)\mathbf{X}. \tag{15}$$

Then relations (14), (15) and Theorem 1 imply that $\mathbf{X}'\mathbf{C}_1\mathbf{X} = Q_1$, $\mathbf{X}'\mathbf{C}_2\mathbf{X} = Q_2$ (and $\mathbf{X}'(\mathbf{I}_n - \mathbf{C}_1 - \mathbf{C}_2)\mathbf{X}$) are independent.

Let now Q_1, Q_2 be independent. Since Q_1 is $\chi^2_{r_1}$ and Q_2 is $\chi^2_{r_2}$, it follows that $Q_1 + Q_2 = \mathbf{X}'(\mathbf{C}_1 + \mathbf{C}_2)\mathbf{X}$ is $\chi^2_{r_1+r_2}$. That is, $\mathbf{X}'(\mathbf{C}_1 + \mathbf{C}_2)\mathbf{X}$ is a quadratic form in $\mathbf{X}$ distributed as $\chi^2_{r_1+r_2}$. Thus $\mathbf{C}_1 + \mathbf{C}_2$ is idempotent by Theorem 2. So the matrices $\mathbf{C}_1$, $\mathbf{C}_2$ and $\mathbf{C}_1 \pm \mathbf{C}_2$ are all idempotent. Then Theorem 12.I(iv), in Appendix I, applies and gives that $\mathbf{C}_1\mathbf{C}_2 (= \mathbf{C}_2\mathbf{C}_1) = \mathbf{0}$. This concludes the proof of the theorem. ▲

REMARK 1 Consider the quadratic forms $\mathbf{X}'\mathbf{C}_1\mathbf{X}$ and $\mathbf{X}'\mathbf{C}_2\mathbf{X}$ figuring in Applications 1–3. Then, according to the conclusion reached in discussing those applications, $\mathbf{X}'\mathbf{C}_1\mathbf{X}$ and $\mathbf{X}'\mathbf{C}_2\mathbf{X}$ are distributed as χ^2_{n-1} and χ^2_1, respectively. This should imply that $\mathbf{C}_1\mathbf{C}_2 = \mathbf{0}$, by Theorem 6. This is, indeed, the case as is easily seen.

Exercises

19.2.1 Refer to Application 1 and show that

$$\sum_{j=1}^{n}(X_j - \overline{X})^2 = \mathbf{X}'\mathbf{C}_1\mathbf{X}$$

as asserted there.

19.2.2 Justify the equalities asserted in (11).

19.2.3 Refer to Application 2 and show that the matrices $\mathbf{C}_1$ and $\mathbf{C}_2$ are both idempotent.

19.2.4 Consider the usual linear model $\mathbf{Y} = \mathbf{X}'\boldsymbol{\beta} + \mathbf{e}$, where X is of full rank p, and let $\hat{\boldsymbol{\beta}} = \mathbf{S}^{-1}\mathbf{X}\mathbf{Y}$ be the LSE of $\boldsymbol{\beta}$. Write $\mathbf{Y}$ as follows: $\mathbf{Y} = \mathbf{X}'\hat{\boldsymbol{\beta}} + (\mathbf{Y} - \mathbf{X}'\hat{\boldsymbol{\beta}})$ and show that:

i) $\|\mathbf{Y}\|^2 = \mathbf{Y}'\mathbf{X}'\mathbf{S}^{-1}\mathbf{X}\mathbf{Y} + \|\mathbf{Y} - \mathbf{X}'\hat{\boldsymbol{\beta}}\|^2$;

ii) The r.v.'s $\mathbf{Y}'\mathbf{X}'\mathbf{S}^{-1}\mathbf{X}\mathbf{Y}$ and $\|\mathbf{Y} - \mathbf{X}'\hat{\boldsymbol{\beta}}\|^2$ are independent, the first being distributed as noncentral χ^2_p and the second as χ^2_{n-p}.

19.2.5 Let X_1, X_2, X_3 be independent r.v.'s distributed as $N(0, 1)$ and let the r.v. Q be defined by

$$Q = \frac{1}{6}\left(5X_1^2 + 2X_2^2 + 5X_3^2 + 4X_1X_2 - 2X_1X_3 + 4X_2X_3\right).$$

Then find the distribution of Q and show that Q is independent of the r.v. $\sum_{j=1}^{3} X_j^2 - Q$.

19.2.6 Refer to Example 1 in Chapter 16 and (by using Theorem 6 herein) show that the r.v.'s $\hat{\beta}_1, \tilde{\sigma}^2$, as well as the r.v.'s $\hat{\beta}_2, \tilde{\sigma}^2$, are independent, where $\tilde{\sigma}^2$ is the LSE of σ^2.

19.2.7 For $j = 1, \ldots, n$, let Y_j be independent r.v.'s, Y_j being distributed as $N(\mu_j, 1)$, and set $\mathbf{Y} = (Y_1, \ldots, Y_n)'$. Let $\mathbf{Y}'\mathbf{Y} = \sum_{i=1}^{k} Q_i$, where for $i = 1, \ldots, k$, Q_i are quadratic forms in $\mathbf{Y}$, $Q_i = \mathbf{Y}'\mathbf{C}_i\mathbf{Y}$, with rank $Q_i = r_i$. Then show that the r.v.'s Q_i are independent $\chi^2_{r_i;\delta_i}$ if and only if $\sum_{i=1}^{k} r_i = n$, where the noncentrality parameter $\delta_i = \boldsymbol{\mu}'\mathbf{C}_i\boldsymbol{\mu}$, $\boldsymbol{\mu} = (\mu_1, \ldots, \mu_n)'$, $i = 1, \ldots, k$. [Hint: The proof is presented along the same lines as that of Theorem 1.]

Chapter 20

Nonparametric Inference

In this chapter, we discuss briefly some instances of *nonparametric*, or more properly, *distribution-free* inference. That is, inferences which are made without any assumptions regarding the functional form of the underlying distributions. The first part of the chapter is devoted to nonparametric estimation and the remaining part of it to nonparametric testing of hypotheses.

20.1 Nonparametric Estimation

At the beginning, we should like to mention a few cases of nonparametric estimation which have already been discussed in previous chapters although the term "nonparametric" was not employed there. To this end, let X_j, $j = 1, \ldots, n$ be i.i.d. r.v.'s with certain distribution about which no functional form is stipulated. The only assumption made is that the X's have a finite (unknown) mean μ. Let $\overline{X}_n$ be the sample mean of the X's; that is,

$$\overline{X}_n = \frac{1}{n}\sum_{j=1}^{n} X_j.$$

Then it has been shown that $\overline{X}_n$, viewed as an estimator of μ, is *weakly consistent*, that is, consistent in the probability sense. Thus $\overline{X}_n \xrightarrow[n\to\infty]{P} \mu$. This is so by the WLLN's. It has also been mentioned that $\overline{X}_n$ is *strongly consistent*, namely, $\overline{X}_n \xrightarrow[n\to\infty]{a.s.} \mu$. This is justified on the basis of the SLLN's.

Let us suppose now that the X's also have finite (and positive) variance σ^2 which presently is assumed to be known. Then, according to the CLT,

$$\frac{\sqrt{n}(\overline{X}_n - \mu)}{\sigma} \xrightarrow[n\to\infty]{d} Z \sim N(0, 1).$$

Thus if $z_{\alpha/2}$ is the upper $\alpha/2$ quantile of the $N(0, 1)$ distribution, then

$$P\left[-z_{\alpha/2} \leq \frac{\sqrt{n}(\overline{X}_n - \mu)}{\sigma} \leq z_{\alpha/2}\right]$$

$$= P\left[\overline{X}_n - z_{\alpha/2}\frac{\sigma}{\sqrt{n}} \leq \mu \leq \overline{X}_n + z_{\alpha/2}\frac{\sigma}{\sqrt{n}}\right] \xrightarrow[n \to \infty]{} 1 - \alpha,$$

so that $[L_n, U_n]$ is a *confidence interval* for μ with *asymptotic confidence coefficient* $1 - \alpha$; here

$$L_n = L(X_1, \ldots, X_n) = \overline{X}_n - z_{\alpha/2}\frac{\sigma}{\sqrt{n}}$$

and

$$U_n = U(X_1, \ldots, X_n) = \overline{X}_n + z_{\alpha/2}\frac{\sigma}{\sqrt{n}}.$$

Next, suppose that σ^2 is unknown and set S_n^2 for the sample variance of the X's; namely,

$$S_n^2 = \frac{1}{n}\sum_{j=1}^{n}(X_j - \overline{X}_n)^2.$$

Then the WLLN's and the SLLN's, properly applied, ensured that S_n^2, viewed as an estimator of σ^2, was both a *weakly* and *strongly consistent* estimator of σ^2. Also by the corollary to Theorem 9 of Chapter 8, it follows that

$$\frac{\sqrt{n}(\overline{X}_n - \mu)}{S_n} \xrightarrow[n \to \infty]{d} Z \sim N(0, 1).$$

By setting

$$L_n^* = L^*(X_1, \ldots, X_n) = \overline{X}_n - z_{\alpha/2}\frac{S_n}{\sqrt{n}}$$

and

$$U_n^* = U^*(X_1, \ldots, X_n) = \overline{X}_n + z_{\alpha/2}\frac{S_n}{\sqrt{n}},$$

we have that $[L_n^*, U_n^*]$ is a *confidence interval* for μ with *asymptotic confidence coefficient* $1 - \alpha$.

Clearly, the examples mentioned so far are cases of nonparametric point and interval estimation. A further instance of point nonparametric estimation is provided by the following example. Let F be the (common and unknown) d.f. of the X_i's and set F_n for their sample or empirical d.f.; that is,

$$F_n(x; s) = \frac{1}{n}\big[\text{the number of } X_1(s), \ldots, X_n(s) \le x\big], \quad x \in \mathbb{R}, \quad s \in S. \quad (1)$$

We often omit the random element s and write $F_n(x)$ rather than $F_n(x; s)$. Then it was stated in Chapter 8 (see Theorem 6) that

$$F_n(x; \cdot) \xrightarrow[n\to\infty]{\text{a.s.}} F(x) \quad \text{uniformly in} \quad x \in \mathbb{R}. \quad (2)$$

Thus $F_n(x; \cdot)$ is a *strongly consistent* estimator of $F(x)$ and for almost all $s \in S$ and every $\varepsilon > 0$, we have

$$F_n(x; s) - \varepsilon \le F(x) \le F_n(x; s) + \varepsilon,$$

provided $n \ge n(\varepsilon, s)$ independent of $x \in \mathbb{R}$.

We close this section by observing that Section 5 of Chapter 15 is concerned with another nonparametric aspect, namely that of constructing tolerance intervals.

20.2 Nonparametric Estimation of a p.d.f.

At the end of the previous section, an unknown d.f. F was estimated by the sample d.f. F_n based on the i.i.d. r.v.'s $X_j, j = 1, \ldots, n$ whose (common) d.f. is assumed to be the unknown one F. In the present section, we shall consider the problem of estimation of an unknown p.d.f. To this end, let $X_j, j = 1, \ldots, n$ be i.i.d. r.v.'s with (the common) p.d.f. f which is assumed to be of the continuous type. A significant amount of work has been done regarding the estimation of $f(x)$, $x \in \mathbb{R}$, which, of course, is assumed to be unknown. In this section, we report some of these results without proofs. The relevant proofs can be found in the paper "On estimation of a probability density function and mode," by E. Parzen, which appeared in *The Annals of Mathematical Statistics*, Vol. 33 (1962), pp. 1065–1076.

First we shall try to give a motivation to the estimates to be employed in the sequel. To this end, recall that if F is the d.f. corresponding to the p.d.f. f, then

$$f(x) = \lim_{h \to 0} \frac{F(x+h) - F(x-h)}{2h}.$$

Thus, for $(0 <) h$ sufficiently small, the quantity $[F(x + h) - F(x - h)]/2h$ should be close to $f(x)$. This suggests estimating $f(x)$ by the (known) quantity

$$\hat{f}_n(x) = \frac{F_n(x+h) - F_n(x-h)}{2h}.$$

However,

$$\hat{f}_n(x) = \frac{F_n(x+h) - F_n(x-h)}{2h}$$

$$= \frac{1}{2h}\left(\frac{\text{the number of } X_1, \ldots, X_n \leq x+h}{n}\right.$$

$$\left. - \frac{\text{the number of } X_1, \ldots, X_n \leq x-h}{n}\right)$$

$$= \frac{1}{2h} \frac{\text{the number of } X_1, \ldots, X_n \text{ in } (x-h, x+h]}{n};$$

that is

$$\hat{f}_n(x) = \frac{1}{2h} \frac{\text{the number of } X_1, \ldots, X_n \text{ in } (x-h, x+h]}{n}$$

and it can be further easily seen (see Exercise 20.2.1) that

$$\hat{f}_n(x) = \frac{1}{nh} \sum_{j=1}^{n} K\left(\frac{x - X_j}{h}\right), \tag{3}$$

where K is the following p.d.f.:

$$K(x) = \begin{cases} \frac{1}{2}, & \text{if } x \in (-1, 1] \\ 0, & \text{otherwise.} \end{cases}$$

Thus the proposed estimator $\hat{f}_n(x)$ of $f(x)$ is expressed in terms of a known p.d.f. K by means of (3). This expression also suggests an entire class of estimators to be introduced below. For this purpose, let K be any p.d.f. defined on $\mathbb{R}$ into itself and satisfying the following properties:

$$\left.\begin{array}{l} \sup\{K(x); x \in \mathbb{R}\} < \infty \\ \lim |xK(x)| = 0 \quad \text{as} \quad |x| \to \infty \\ K(-x) = K(x), \quad x \in \mathbb{R}. \end{array}\right\} \tag{4}$$

Next, let $\{h_n\}$ be a sequence of positive constants such that

$$h_n \xrightarrow[n \to \infty]{} 0. \tag{5}$$

For each $x \in \mathbb{R}$ and by means of K and $\{h_n\}$, define the r.v. $\hat{f}_n(x; s)$, to be shortened to $\hat{f}_n(x)$, as follows:

$$\hat{f}_n(x) = \frac{1}{nh_n} \sum_{j=1}^{n} K\left(\frac{x - X_j}{h_n}\right). \tag{6}$$

Then we may formulate the following results.

THEOREM 1 Let $X_j, j = 1, \ldots, n$ be i.i.d. r.v.'s with (unknown) p.d.f. f and let K be a p.d.f. satisfying conditions (4). Also, let $\{h_n\}$ be a sequence of positive constants satisfying (5) and for each $x \in \mathbb{R}$ let $\hat{f}_n(x)$ be defined by (6). Then for any $x \in \mathbb{R}$ at which f is continuous, the r.v. $\hat{f}_n(x)$, viewed as an estimator of $f(x)$, is *asymptotically unbiased* in the sense that

$$E\hat{f}_n(x) \xrightarrow[n \to \infty]{} f(x).$$

Now let $\{h_n\}$ be as above and also satisfying the following requirement:

$$nh_n \xrightarrow[n \to \infty]{} \infty. \qquad (7)$$

Then the following results hold true.

THEOREM 2 Under the same assumptions as those in Theorem 1 and the additional condition (7), for each $x \in \mathbb{R}$ at which f is continuous, the estimator $\hat{f}_n(x)$ of $f(x)$ is *consistent in quadratic mean* in the sense that

$$E\left|\hat{f}_n(x) - f(x)\right|^2 \xrightarrow[n \to \infty]{} 0.$$

The estimator $\hat{f}_n(x)$, when properly normalized, is also asymptotically normal, as the following theorem states.

THEOREM 3 Under the same assumptions as those in Theorem 2, for each $x \in \mathbb{R}$ at which f is continuous,

$$\frac{\hat{f}_n(x) - E[\hat{f}_n(x)]}{\sigma[\hat{f}_n(x)]} \xrightarrow[n \to \infty]{d} Z \sim N(0, 1).$$

Finally, if it happens to be known that f belongs to a class of p.d.f.'s which are *uniformly continuous*, then by choosing the sequence $\{h_n\}$ of positive constants to tend to zero and also such that

$$nh_n^2 \xrightarrow[n \to \infty]{} \infty, \qquad (8)$$

we may show the following result.

THEOREM 4 Under the same assumptions as those in Theorem 1 and also condition (8),

$$\hat{f}_n(x) \xrightarrow[n \to \infty]{a.\,s.} f(x) \quad \text{uniformly in } x \in \mathbb{R},$$

provided f is uniformly continuous.

In closing this section, it should be pointed out that there are many p.d.f.'s of the type K satisfying conditions (4). For example, if K is taken to be the p.d.f. of the $N(0, 1)$, or the $U(-\frac{1}{2}, \frac{1}{2})$ p.d.f., these conditions are, clearly, satisfied. As for the sequence $\{h_n\}$, there is plenty of flexibility in choosing it. As an illustration, consider the following example.

EXAMPLE 1 Consider the i.i.d. r.v.'s X_j, $j = 1, \ldots, n$ with (unknown) p.d.f. f. Take

$$K(x) = \frac{1}{\sqrt{2\pi}} e^{-x^2/2}.$$

Then, clearly,

$$|K(x)| = \frac{1}{\sqrt{2\pi}} e^{-x^2/2} \leq \frac{1}{\sqrt{2\pi}}, \quad \text{so that} \quad \sup\{K(x); x \in \mathbb{R}\} < \infty.$$

Next, for $x > 1$, one has $e^x < e^{x^2}$, so that $e^{-x^2/2} < e^{-x/2}$ and hence

$$xe^{-x^2/2} < xe^{-x/2} = x/e^{x/2}.$$

Now consider the expansion $e^t = 1 + te^{\theta t}$ for some $0 < \theta < 1$, and replace t by $x/2$. We get then

$$\frac{x}{e^{x/2}} = \frac{x}{1 + (x/2)e^{\theta x/2}} = \frac{1}{(1/x) + \frac{1}{2}e^{\theta x/2}} \xrightarrow[x \to \infty]{} 0$$

and therefore $xe^{-x^2/2} \xrightarrow[x \to \infty]{} 0$. In a similar way $xe^{-x^2/2} \xrightarrow[x \to -\infty]{} 0$, so that $\lim |xK(x)| = 0$ as $|x| \to \infty$. Since also $K(-x) = K(x)$, condition (4) is satisfied. Let us now take $h_n = 1/n^{1/4}$. Then $0 < h_n \xrightarrow[n \to \infty]{} 0$, $nh_n^2 = n/n^{1/2} = n^{1/2} \xrightarrow[n \to \infty]{} \infty$ and $nh_n = n^{3/4} \xrightarrow[n \to \infty]{} \infty$. Thus the estimator given by (6) has all properties stated in Theorems 1–4. This estimator here becomes as follows:

$$\hat{f}_n(x) = \frac{1}{\sqrt{2\pi} n^{3/4}} \sum_{j=1}^n \exp\left[-\frac{(x - X_j)^2}{2n^{1/2}}\right].$$

Exercise

20.2.1 Let $X_j, j = 1, \ldots, n$ be i.i.d. r.v.'s and for some $h > 0$ and any $x \in \mathbb{R}$, define $\hat{f}_n(x)$ as follows:

$$\hat{f}_n(x) = \frac{1}{2h} \frac{\text{the number of } X_1, \ldots, X_n \text{ in } (x - h, x + h]}{n}.$$

Then show that

$$\hat{f}_n(x) = \frac{1}{nh} \sum_{j=1}^n K\left(\frac{x - X_j}{h}\right),$$

where $K(x)$ is $\frac{1}{2}$ if $x \in [-1, 1)$ and 0 otherwise.

20.3 Some Nonparametric Tests

Let $X_j, j = 1, \ldots, n$ be i.i.d. r.v.'s with unknown d.f. F. As was seen in Section 20.1, the sample d.f. F_n may be used for the purpose of estimating F. However, testing hypotheses problems about F also arise and are of practical importance. Thus we may be interested in testing the hypothesis $H: F = F_0$, a given d.f., against all possible alternatives. This hypothesis can be tested by utilizing the chi-square test for goodness of fit discussed in Chapter 13, Section 8. The chi-square test is the oldest nonparametric test regarding d.f.'s. Alternatively, the sample d.f. F_n may also be used for testing the same hypothesis as above. In order to be able to employ the test proposed below, we have to make the supplementary (but mild) assumption that F is *continuous*. Thus the hypothesis to be tested here is

$$H: F = F_0, \quad \text{a given continuous d.f.,}$$

against the alternative

$$A: F \neq F_0 \quad \left(\text{in the sense that } F(x) \neq F_0(x) \text{ for at least one } x \in \mathbb{R}\right).$$

Let α be the level of significance. Define the r.v. D_n as follows,

$$D_n = \sup\{|F_n(x) - F_0(x)|; \, x \in \mathbb{R}\}, \tag{9}$$

where F_n is the sample d.f. defined by (1). Then, under H, it follows from (2) that $D_n \xrightarrow[n \to \infty]{\text{a.s.}} 0$. Therefore we would reject H if $D_n > C$ and would accept it otherwise. The constant C is to be determined through the relationship

$$P(D_n > C | H) = \alpha. \tag{10}$$

In order for this determination to be possible, we would have to know the distribution of D_n, under H, or of some known multiple of it. It has been shown in the literature that

$$P(\sqrt{n} D_n \leq x | H) \xrightarrow[n \to \infty]{} \sum_{j=-\infty}^{\infty} (-1)^j e^{-2j^2 x^2}, \quad x \geq 0. \tag{11}$$

Thus for large n, the right-hand side of (11) may be used for the purpose of determining C by way of (10). For moderate values of n ($n \leq 100$) and selected α's ($\alpha = 0.10, 0.05, 0.025, 0.01, 0.005$), there are tables available which facilitate the calculation of C. (See, for example, *Handbook of Statistical Tables* by D. B. Owen, Addison-Wesley, 1962.) The test employed above is known as the *Kolmogorov one-sample test*.

The testing hypothesis problem just described is of limited practical importance. What arise naturally in practice are problems of the following type: Let $X_i, i = 1, \ldots, m$ be i.i.d. r.v.'s with *continuous* but unknown d.f. F and let $Y_j, j = 1, \ldots, n$ be i.i.d. r.v.'s with *continuous* but unknown d.f. G. The two random samples are assumed to be independent and the hypothesis of interest here is

$$H: F = G.$$

One possible alternative is the following:

$$A: F \neq G \tag{12}$$

(in the sense that $F(x) \neq G(x)$ for at least one $x \in \mathbb{R}$).

The hypothesis is to be tested at level α. Define the r.v. $D_{m,n}$ as follows:

$$D_{m,n} = \sup\{|F_m(x) - G_n(x)|; x \in \mathbb{R}\}, \tag{13}$$

where F_m, G_n are the sample d.f.'s of the X's and Y's, respectively. Under H, $F = G$, so that

$$|F_m(x) - G_n(x)| = |[F_m(x) - F(x)] - [G_n(x) - G(x)]|$$
$$\leq |F_m(x) - F(x)| + |G_n(x) - G(x)|.$$

Hence

$$D_{m,n} \leq \sup\{|F_m(x) - F(x)|; x \in \mathbb{R}\} + \sup\{|G_n(x) - G(x)|; x \in \mathbb{R}\},$$

whereas

$$\sup\{|F_m(x) - F(x)|; x \in \mathbb{R}\} \xrightarrow[m \to \infty]{a.s.} 0, \quad \sup\{|G_n(x) - G(x)|; x \in \mathbb{R}\} \xrightarrow[m \to \infty]{a.s.} 0.$$

In other words, we have that $D_{m,n} \xrightarrow{a.s.} 0$ as $m, n \to \infty$, and this suggests rejecting H if $D_{m,n} > C$ and accepting it otherwise. The constant C is determined by means of the relation

$$P(D_{m,n} > C|H) = \alpha. \tag{14}$$

Once again the actual determination of C requires the knowledge of the distribution of $D_{m,n}$ under H, or some known multiple of it. In connection with this it has been shown in the literature that

$$P(\sqrt{N} D_{m,n} \leq x|H) \to \sum_{j=-\infty}^{\infty} (-1)^j e^{-2j^2 x^2} \quad \text{as} \quad m, n \to \infty, \quad x \geq 0, \tag{15}$$

where $N = mn/(m + n)$.

Thus, for large m and n, the right-hand side of (15) may be used for the purpose of determining C by way of (14). For moderate values of m and n (such as $m = n \leq 40$), there are tables available which facilitate the calculation of C. (See reference cited above in connection with the one-sample Kolmogorov test.)

In addition to the alternative $A : F \neq G$ just considered, the following two alternatives are also of interest; namely,

$$A' : F > G, \tag{16}$$

in the sense that $F(x) \geq G(x)$ with strict inequality for at least one $x \in \mathbb{R}$, and

$$A'' : F < G, \tag{17}$$

in the sense that $F(x) \leq G(x)$ with strict inequality for at least one $x \in \mathbb{R}$.

For testing H against A', we employ the statistic $D_{m,n}^+$ defined by

$$D_{m,n}^+ = \sup\{F_m(x) - G_n(x); x \in \mathbb{R}\}$$

and reject H if $D_{m,n}^+ > C^+$. The cut-off point C^+ is determined through the relation

$$P\left(D_{m,n}^+ > C^+ \middle| H\right) = \alpha$$

by utilizing the fact that

$$P\left(\sqrt{N} D_{m,n}^+ \leq x\right) \to 1 - e^{-2x^2} \quad \text{as} \quad m, n \to \infty, \quad x \in \mathbb{R},$$

as can be shown. Here N is as before, that is, $N = mn/(m + n)$. Similarly, for testing H against A'', we employ the statistic $D_{m,n}^-$ defined by

$$D_{m,n}^- = \sup\{G_n(x) - F_m(x); \; x \in \mathbb{R}\}$$

and reject H if $D_{m,n}^- < C^-$. The cut-off point C^- is determined through the relation

$$P\left(D_{m,n}^- < C^- \middle| H\right) = \alpha$$

by utilizing the fact that

$$P\left(\sqrt{N} D_{m,n}^- \leq x\right) \to 1 - e^{-2x^2} \quad \text{as} \quad m, n \to \infty, \quad x \in \mathbb{R}.$$

For relevant tables, the reader is referred to the reference cited earlier in this section. The last three tests based on the statistics $D_{m,n}$, $D_{m,n}^+$ and $D_{m,n}^-$ are known as *Kolmogorov–Smirnov two-sample tests*.

20.4 More About Nonparametric Tests: Rank Tests

Consider again the two-sample problem discussed in the latter part of the previous section. Namely, let X_i, $i = 1, \ldots, m$ and Y_j, $j = 1, \ldots, n$ be two independent random samples with *continuous* d.f.'s F and G, respectively. The problem is that of testing the hypothesis $H: F = G$ against various alternatives at level of significance α.

Now it seems reasonable that in testing H on the basis of the X's and Y's, we should reach the same conclusion regarding the rejection or acceptance of H regardless of the scale used in measuring the X's and Y's. (That is, the conclusion should be the same if the X's and Y's are multiplied by the same positive constant. This is a special case of what is known as *invariance under monotone transformations*.) This is done by employing the ranks of the X's and Y's in the combined sample rather than their actual values. The *rank* of X_i in the combined sample of X's and Y's, to be denoted by $R(X_i)$, is that integer among the numbers $1, \ldots, N (= m + n)$ which corresponds to the position of X_i after the X's and Y's have been ordered according to their size. Of course, the *rank* $R(Y_j)$ of Y_j in the combined sample of the X's and Y's is defined in a similar fashion. By the assumption of continuity of F and G, it follows that in ordering the X's and Y's, we have strict inequalities with probability equal to one.

For testing the hypothesis H specified above, we are going to use either one of the *rank sum* statistics R_X, R_Y defined by

$$R_X = \sum_{i=1}^{m} R(X_i), \quad R_Y = \sum_{j=1}^{n} R(Y_j) \tag{18}$$

because $R_X + R_Y = N(N+1)/2$ (fixed), as is easily seen. (See Exercise 20.4.1.)

In the present case, and for reasons to become apparent soon, it is customary to take the level of significance α as follows:

$$\alpha = \frac{k}{\binom{N}{m}}, \quad 1 < k < \binom{N}{m}.$$

There are three alternatives of interest to consider, namely, A, A' and A'', as they are specified by (12), (16) and (17), respectively. As an example, let the r.v.'s X, Y be distributed as the X's and Y's, respectively, and let us consider alternative A'. Under A', $P(X \leq x) \geq P(Y \leq x)$, $x \in \mathbb{R}$, so that R_X would tend to take on small values; accordingly, we would reject H in favor of A' if

$$\mathbb{R}_X < C', \tag{19}$$

where C' is defined, so that

$$P(\mathbb{R}_X < C' | H) = \alpha. \tag{20}$$

Theoretically the determination of C' is a simple matter; under H, all $\binom{N}{m}$ values of $(R(X_1), \ldots, R(X_m))$ are equally likely each having probability $1/\binom{N}{m}$. The rejection region then is defined as follows: Consider all these $\binom{N}{m}$ values and for each one of them form the rank sum R_X. Then the rejection region consists of the k smallest values of these rank sums. For small values of m and n ($n \leq m \leq 10$), this procedure is facilitated by tables (see reference cited in previous section), whereas for large values of m and n it becomes unmanageable; for this latter case, the normal approximation to be discussed below may be employed. The remaining two alternatives are treated in a similar fashion.

Next, consider the function u defined as follows:

$$u(z) = \begin{cases} 1, & \text{if } z > 0 \\ 0, & \text{if } z < 0 \end{cases} \tag{21}$$

and set

$$U = \sum_{i=1}^{m} \sum_{j=1}^{n} u(X_i - Y_j). \tag{22}$$

Then U is, clearly, the number of times a Y precedes an X and it can be shown (see Exercise 20.4.2) that

$$U = mn + \frac{n(n+1)}{2} - R_Y = R_X - \frac{m(m+1)}{2}. \tag{23}$$

Therefore the test in (19) can be expressed equivalently in terms of the U statistic. This test is known as the *two-sample Wilcoxon–Mann–Whitney test*.

Now it can be shown (see Exercise 20.4.3) that under H,

$$EU = \frac{mn}{2}, \quad \sigma^2(U) = \frac{mn(m+n+1)}{12}.$$

Then the r.v. $(U - EU)/(\sigma(U))$ converges in distribution to an r.v. Z distributed as $N(0, 1)$ as $m, n \to \infty$ and therefore, for large m, n, the limiting distribution (along with the continuity correction for better precision) may be used for determining the cut-off point C' by means of (20).

A special interesting case, where the *rank sum* tests of the present section are appropriate, is that where the d.f. G of the Y's is assumed to be of the form

$$G(x) = F(x - \Delta), \quad x \in \mathbb{R} \quad \text{for some unknown} \quad \Delta \in \mathbb{R}.$$

As before, F is assumed to be unknown but continuous. In this case, we say that G is a *shift* of F (to the right if $\Delta > 0$ and to the left if $\Delta < 0$). Then the hypothesis $H: F = G$ is equivalent to testing $\Delta = 0$ and the alternatives $A: F \neq G$, $A': F > G$ and $A'': F < G$ are equivalent to $\Delta \neq 0$, $\Delta > 0$ and $\Delta < 0$, respectively.

In closing this section, we should like to mention that there is also the *one-sample Wilcoxon–Mann–Whitney test*, as well as other one-sample and two-sample rank tests available. However, their discussion here would be beyond the purposes of the present chapter.

As an illustration, consider the following numerical example.

EXAMPLE 2 Let $m = 5$, $n = 4$ and suppose that $X_1 = 78$, $X_2 = 65$, $X_3 = 74$, $X_4 = 45$, $X_5 = 82$; $Y_1 = 110$, $Y_2 = 71$, $Y_3 = 53$, $Y_4 = 50$. Combining these values and ordering them according to their size, we obtain

$$\begin{array}{ccccccccc} 45 & 50 & 53 & 65 & 71 & 74 & 78 & 82 & 110 \\ (X) & (Y) & (Y) & (X) & (Y) & (X) & (X) & (X) & (Y), \end{array}$$

where an X or Y below a number means that the number is coming from the X or Y sample, respectively. From this, we find that

$$R(X_1) = 7, \quad R(X_2) = 4, \quad R(X_3) = 6, \quad R(X_4) = 1, \quad R(X_5) = 8; \quad R(Y_1) = 9,$$
$$R(Y_2) = 5, \quad R(Y_3) = 3, \quad R(Y_4) = 2, \quad \text{so that} \quad R_X = 26, \quad R_Y = 19.$$

We also find that

$$U = 4 + 4 + 3 + 0 = 11.$$

(Incidentally, these results check with (23) and Exercise 20.4.1.) Now

$$N = 5 + 4 = 9 \quad \text{and} \quad \frac{1}{\binom{N}{m}} = \frac{1}{\binom{9}{5}} = \frac{1}{126},$$

and let us take

$$\alpha = \frac{5}{126}(\approx 0.04).$$

Then for testing H against A' (given by (16)), we would reject for small values of R_X, or equivalently (by means of (23)), for small values of U. For the given m, n, α and for the observed value of R_X (or U), H is accepted. (See tables on p. 341 of the reference cited in Section 20.3.)

Exercises

20.4.1 Consider the two independent random samples X_i, $i = 1, \ldots, m$ and Y_j, $j = 1, \ldots, n$ and let $R(X_i)$ and $\mathbb{R}(Y_j)$ be the ranks of X_i and Y_j, respectively, in the combined sample of the X's and Y's. Furthermore, let R_X and R_Y be defined by (18). Then show that

$$R_X + R_Y = \frac{N(N+1)}{2},$$

where $N = m + n$.

20.4.2 Let R_X and R_Y be as in the previous exercise and let U be defined by (22). Then establish (23).

20.4.3 Let X_i, $i = 1, \ldots, m$ and Y_j, $j = 1, \ldots, n$ be two independent random samples and let U be defined by (22). Then show that, under H,

$$EU = \frac{mn}{2}, \quad \sigma^2(U) = \frac{mn(m+n-1)}{12}.$$

20.5 Sign Test

In this section, we briefly mention another nonparametric test—the *two-sample Sign test*, which is easily applicable in many situations of practical importance. In order to avoid distribution related difficulties, we assume, as we have also done in previous sections, that the underlying distributions are continuous. More precisely, we suppose that X_j, $j = 1, \ldots, n$ are i.i.d. r.v.'s with *continuous* d.f. F and that Y_j, $j = 1, \ldots, n$ are also i.i.d. r.v.'s with *continuous* d.f. G. The two random samples are assumed to be independent and the hypothesis H to be tested is

$$H: F = G.$$

To this end, consider the n pairs (X_j, Y_j), $j = 1, \ldots, n$ and set

$$Z_j = \begin{cases} 1, & \text{if } X_j < Y_j \\ 0, & \text{if } X_j > Y_j. \end{cases}$$

Also set $Z = \sum_{j=1}^{n} Z_j$ and $p = P(X_j < Y_j)$. Then, clearly, Z is distributed as $B(n, p)$ and the hypothesis H above is equivalent to testing $p = \frac{1}{2}$. Depending on the type of the alternatives, one would use the two-sided or the appropriate one-sided test.

Some cases where the sign test just described is appropriate is when one is interested in comparing the effectiveness of two different drugs used for the treatment of the same disease, the efficiency of two manufacturing processes producing the same item, the response of n customers regarding their preferences towards a certain consumer item, etc.

Of course, there is also the *one-sample Sign test* available, but we will not discuss it here.

For the sake of an illustration, consider the following numerical example.

EXAMPLE 3 Let $n = 10$ and suppose that

$$X_1 = 73, \quad X_2 = 68, \quad X_3 = 64, \quad X_4 = 90, \quad X_5 = 83,$$
$$X_6 = 48, \quad X_7 = 100, \quad X_8 = 75, \quad X_9 = 90, \quad X_{10} = 85$$

and

$$Y_1 = 50, \quad Y_2 = 100, \quad Y_3 = 70, \quad Y_4 = 96, \quad Y_5 = 74,$$
$$Y_6 = 64, \quad Y_7 = 76, \quad Y_8 = 83, \quad Y_9 = 98, \quad Y_{10} = 40.$$

Then

$$Z_1 = 0, \quad Z_2 = 1, \quad Z_3 = 1, \quad Z_4 = 1, \quad Z_5 = 0, \quad Z_6 = 1, \quad Z_7 = 0,$$
$$Z_8 = 1, \quad Z_9 = 1, \quad Z_{10} = 0, \quad \text{so that} \quad \sum_{j=1}^{10} Z_j = 6.$$

Thus, if $\alpha = 0.1$ and if we are interested in testing $H : F = G$ against $A : F \neq G$ (equivalently, $p = \frac{1}{2}$ against $p \neq \frac{1}{2}$), we would accept H.

20.6 Relative Asymptotic Efficiency of Tests

Consider again the testing hypothesis problem discussed in the previous sections; namely, let X_i, $i = 1, \ldots, m$ and Y_j, $j = 1, \ldots, n$ be i.i.d. r.v.'s with continuous d.f.'s F and G, respectively. The hypothesis to be tested is $H : F = G$ and the alternative may be either $A : F \neq G$, or $A' : F > G$, or $A'' : F < G$. In employing either the Wilcoxon–Mann–Whitney test or the Sign test in the problem just described, we would like to have some measure on the basis of which we could judge the performance of the test in question at least in an asymptotic sense. This is obtained by introducing what is known as the Pitman asymptotic relative efficiency of tests. For the precise definition of this concept, suppose that the two sample sizes are the same and let n be the sample size needed in order to obtain a given power β, say, against a specified alternative when one of the above-mentioned tests is employed. The level of significance is α. Formally, we may also employ the t-test (see (36), Chapter 13) for testing the same hypothesis against the same specified alternative at the same level α. Let n^* be the (common) sample size required in order to achieve a power equal to β by employing the t-test. We further assume that the limit of n^*/n, as $n \to \infty$, exists and is independent of α and β. Denote this limit by e. Then this quantity e is the *Pitman asymptotic relative efficiency of the Wilcoxon–Mann–Whitney test* (or of the *Sign test*, depending on which one is used) *relative to the t-test*. Thus, if we use the Wilcoxon–Mann–Whitney test and if it so happens that $e = \frac{1}{3}$, then this means that the Wilcoxon–Mann–Whitney test requires approximately three times as many observations as the t-test in order to achieve the same power. However, if $e = 5$, then the Wilcoxon–Mann–Whitney test requires approximately only one-fifth as many observations as the t-test in order to achieve the same power.

It has been found in the literature that the asymptotic efficiency of the Wilcoxon–Mann–Whitney test relative to the t-test is $3/\pi \approx 0.95$ when the underlying distribution is Normal, 1 when the underlying distribution is Uniform and ∞ when the underlying distribution is Cauchy.

Appendix I

Topics from Vector and Matrix Algebra

I.1 Basic Definitions in Vector Spaces

For a positive integer n, $\mathbf{x}$ is said to be an *n-dimensional vector with real components* if it is an *n*-tuple of real numbers. All vectors will be column vectors but for typographical convenience, they will be written in the form of a row with a prime (′) to indicate *transpose*. Thus $\mathbf{x} = (x_1, \ldots, x_n)'$, $x_j \in \mathbb{R}, j = 1, \ldots, n$. Only vectors with real components will be considered. The set of all *n*-dimensional vectors is denoted by V_n. Thus $V_n = \mathbb{R}^n (= \mathbb{R} \times \ldots \times \mathbb{R}$, n factors) in our previous notation, and V_n is called *the (real) n-dimensional vector space*. The *zero vector*, to be denoted by $\mathbf{0}$, is the vector all of whose components are equal to 0. Two vectors $\mathbf{x} = (x_1, \ldots, x_n)'$, $\mathbf{y} = (y_1, \ldots, y_n)'$ are said to be *equal* if $x_j = y_j$, $j = 1, \ldots, n$. The sum $\mathbf{x} + \mathbf{y}$ of two vectors $\mathbf{x} = (x_1, \ldots, x_n)'$, $\mathbf{y} = (y_1, \ldots, y_n)'$ is the vector defined by $\mathbf{x} + \mathbf{y} = (x_1 + y_1, \ldots, x_n + y_n)'$. This definition is extended in an obvious manner to any finite number of vectors. For any three vectors $\mathbf{x}$, $\mathbf{y}$ and $\mathbf{z}$ in V_n, the following properties are immediate:

$$\mathbf{x}+\mathbf{y}=\mathbf{y}+\mathbf{x}, \quad (\mathbf{x}+\mathbf{y})+\mathbf{z}=\mathbf{x}+(\mathbf{y}+\mathbf{z})=\mathbf{x}+\mathbf{y}+\mathbf{z}.$$

The *product* $\alpha \mathbf{x}$ of the vector $\mathbf{x} = (x_1, \ldots, x_n)'$ by the real number α (*scalar*) is the vector defined by $\alpha \mathbf{x} = (\alpha x_1, \ldots, \alpha x_n)'$. For any two vectors $\mathbf{x}$, $\mathbf{y}$ in V_n and any two scalars α, β, the following properties are immediate:

$$\alpha(\mathbf{x}+\mathbf{y})=\alpha\mathbf{x}+\alpha\mathbf{y}, \quad (\alpha+\beta)\mathbf{x}=\alpha\mathbf{x}+\beta\mathbf{x}, \quad \alpha(\beta\mathbf{x})=\beta(\alpha\mathbf{x})=\alpha\beta\mathbf{x},$$

$$(\alpha\mathbf{x}+\beta\mathbf{y})' = \alpha\mathbf{x}'+\beta\mathbf{y}', \quad 1\mathbf{x}=\mathbf{x}.$$

The *inner* (or *scalar*) product $\mathbf{x}'\mathbf{y}$ of any two vectors $\mathbf{x} = (x_1, \ldots, x_n)'$, $\mathbf{y} = (y_1, \ldots, y_n)'$ is a scalar and is defined as follows:

$$\mathbf{x'y} = \sum_{j=1}^{n} x_j y_j.$$

For any three vectors $\mathbf{x}$, $\mathbf{y}$ and $\mathbf{z}$ in V_n and any scalars α, β, the following properties are immediate:

$$\mathbf{x'y} = \mathbf{y'x}, \quad \mathbf{x'}(\alpha\mathbf{y}) = (\alpha\mathbf{x})'\mathbf{y} = \alpha(\mathbf{x'y}), \quad \mathbf{x'}(\alpha\mathbf{y} + \beta\mathbf{z}) = \alpha\mathbf{x'y} + \beta\mathbf{x'z},$$
$$\mathbf{x'x} \geq 0 \quad \text{and} \quad \mathbf{x'x} = 0 \quad \text{if and only if} \quad \mathbf{x} = 0;$$

also if

$$\mathbf{x'y} = 0 \quad \text{for every} \quad \mathbf{y} \in V_n, \quad \text{then} \quad \mathbf{x} = 0.$$

The *norm* (or *length*) $\|\mathbf{x}\|$ of a vector $\mathbf{x}$ in V_n is a non-negative number and is defined by $\|\mathbf{x}\| = (\mathbf{x'x})^{1/2}$. For any vector in V_n and any scalar α, the following property is immediate:

$$\|\alpha\mathbf{x}\| = |\alpha|\|\mathbf{x}\|.$$

Two vectors $\mathbf{x}$, $\mathbf{y}$ in V_n are said to be *orthogonal* (or *perpendicular*), and we write $\mathbf{x} \perp \mathbf{y}$, if $\mathbf{x'y} = 0$. A vector x in V_n is said to be *orthogonal* (or *perpendicular*) to a subset $\mathcal{U}$ of V_n, and we write $\mathbf{x} \perp \mathcal{U}$, if $\mathbf{x'y} = 0$ for every $\mathbf{y} \in \mathcal{U}$.

A (nonempty) subset $\mathcal{V}$ of V_n is a vector space, which is a *subspace of* V_n, denoted by $\mathcal{V} \subseteq V_n$, if for any vectors $\mathbf{x}$, $\mathbf{y}$ in $\mathcal{V}$ and any scalars α and β, $\alpha\mathbf{x} + \beta\mathbf{y}$ is also in $\mathcal{V}$. Thus, for example, the straight line $\alpha_1 x_1 + \alpha_2 x_2 = 0$ in the plane, being the set $\{\mathbf{x} = (x_1, x_2)' \in V_2; \alpha_1 x_1 + \alpha_2 x_2 = 0\}$ is a subspace of V_2. It is shown easily that for any given set of vectors $\mathbf{x}_j$, $j = 1, \ldots, r$ in V_n, $\mathcal{V}$ defined by

$$\mathcal{V} = \left\{ \mathbf{y} \in V_n; \mathbf{y} = \sum_{j=1}^{r} \alpha_j \mathbf{x}_j, \quad \alpha_j \in \mathbb{R}, j = 1, \ldots, r \right\}$$

is a subspace of V_n.

The vectors $\mathbf{x}_j$, $j = 1, \ldots, r$ in V_n are said to *span* (or *generate*) the subspace $\mathcal{V} \subseteq V_n$ if every vector $\mathbf{y}$ in $\mathcal{V}$ may be written as follows: $\mathbf{y} = \sum_{j=1}^{r} \alpha_j \mathbf{x}_j$ for some scalars α_j, $j = 1, \ldots, r$.

For any positive integer $m < n$, the m-dimensional vector space V_m may be considered as a subspace of V_n by enlarging the m-tuples to n-tuples and identifying the appropriate components with zero in the resulting n-tuples. Thus, for examples, the x-axis in the plane may be identified with the set $\{\mathbf{x} = (x_1, x_2)' \in V_2; x_1 \in \mathbb{R}, x_2 = 0\}$ which is a subspace of V_2. Similarly the y-axis in the plane may be identified with the set $\{\mathbf{y} = (y_1, y_2)' \in V_2; y_1 = 0, y_2 \in \mathbb{R}\}$ which is a subspace of V_2; the xy-plane in the three-dimensional space may be identified with the set $\{\mathbf{z} = (x_1, x_2, x_3)' \in V_3; x_1, x_2 \in \mathbb{R}, x_3 = 0\}$ which is a subspace of V_3, etc.

From now on, we shall assume that the above-mentioned identification has been made and we shall write $V_m \subseteq V_n$ to indicate that V_m is a subspace of V_n.

The vectors $\mathbf{x}_j, j = 1, \ldots, k$ in the subspace $\mathcal{V} \subseteq V_n$ are said to be *linearly independent* if there are no scalars $\alpha_j, j = 1, \ldots, k$ which are not all zero for which $\Sigma_{j=1}^{k} \alpha_j x_j = 0$; otherwise they are said to be *linearly dependent*. A *basis* for the subspace $\mathcal{V} \subseteq V_n$ is any set of linearly independent vectors which span $\mathcal{V}$. The vectors $\{\mathbf{x}_j, j = 1, \ldots, k\}$ are said to form an *orthonormal basis in* $\mathcal{V}$ if they form a basis in $\mathcal{V}$ and also are pairwise orthogonal and of norm one; that is, $\mathbf{x}'_i \mathbf{x}_j = 0$ for $i \neq j$,

$$\mathbf{x}'_i \mathbf{x}_i = \|\mathbf{x}_i\|^2 = 1, \qquad i = 1, \ldots, k.$$

For example, by taking

$$\mathbf{e}_1 = (1, 0, \ldots, 0)', \quad \mathbf{e}_2 = (0, 1, 0, \ldots, 0)', \ldots, \quad \mathbf{e}_n = (0, \ldots, 0, 1)',$$

it is clear that the vectors $\{\mathbf{e}_j, j = 1, \ldots, n\}$ form an orthonormal basis in V_n. It can be shown that the number of vectors in any basis of $\mathcal{V}$ is the same and this is the largest number of linearly independent vectors in $\mathcal{V}$. This number is called the *dimension of* $\mathcal{V}$.

I.2 Some Theorems on Vector Spaces

In this section, we gather together for easy reference those results about vector spaces used in this book.

THEOREM 1.I For any positive integer n, consider any subspace $\mathcal{V} \subseteq V_n$. Then $\mathcal{V}$ has a basis and any two bases in $\mathcal{V}$ have the same number of vectors, say, m (the dimension of $\mathcal{V}$). In particular, the dimension of V_n is n and $m \leq n$.

THEOREM 2.I Let m, n be any positive integers with $m < n$ and let $\{\mathbf{x}_j, j = 1, \ldots, m\}$ be an orthonormal basis for V_m. Then this basis can be extended to an orthonormal basis $\{\mathbf{x}_j, j = 1, \ldots, n\}$ for V_n.

THEOREM 3.I Let n be any positive integer, let $\mathbf{x}$ be a vector in V_n and let $\mathcal{V}$ be a subspace of V_n. Then $\mathbf{x} \perp \mathcal{V}$ if and only if $\mathbf{x}$ is orthogonal to the vectors of a basis for $\mathcal{V}$, or to the vectors of any set of vectors in $\mathcal{V}$ spanning $\mathcal{V}$.

THEOREM 4.I Let m, n be any positive integers with $m < n$ and let $\mathcal{V}_m$ be a subspace of V_n of dimension m. Let $\mathcal{U}$ be the set of vectors in V_n each of which is orthogonal to $\mathcal{V}_m$. Then $\mathcal{U}$ is an r-dimensional subspace $\mathcal{U}_r$ of V_n with $r = n - m$ and is called the *orthocomplement* (or *orthogonal complement*) of $\mathcal{V}_m$ in V_n. Furthermore, any vector $\mathbf{x}$ in V_n may be written (decomposed) uniquely as follows: $\mathbf{x} = \mathbf{v} + \mathbf{u}$ with $\mathbf{v} \in \mathcal{V}_m$, $\mathbf{u} \in \mathcal{U}_r$. The vectors $\mathbf{v}, \mathbf{u}$ are called the *projections* of $\mathbf{x}$ into $\mathcal{V}_m$ and $\mathcal{U}_r$, respectively, and $\|\mathbf{x}\|^2 = \|\mathbf{v}\|^2 + \|\mathbf{u}\|^2$. Finally, as $\mathbf{z}$ varies in $\mathcal{V}_m$, $\|\mathbf{x} - \mathbf{z}\|$ has a minimum value obtained for $\mathbf{z} = \mathbf{v}$, and as $\mathbf{w}$ varies in $\mathcal{U}_r$, $\|\mathbf{x} - \mathbf{w}\|$ has a minimum value obtained for $\mathbf{w} = \mathbf{u}$.

I.3 Basic Definitions About Matrices

Let m, n be any positive integers. Then a (*real*) $m \times n$ *matrix* $\mathbf{A}^{m \times n}$ is a rectangular array of mn real numbers arranged in m rows and n columns; m and n are called the *dimensions* of the matrix. Thus

$$\mathbf{A}^{m \times n} = \begin{pmatrix} a_{11} & a_{12} & \cdots & a_{1n} \\ a_{21} & a_{22} & \cdots & a_{2n} \\ \cdots & \cdots & \cdots & \cdots \\ a_{m1} & a_{m2} & \cdots & a_{mn} \end{pmatrix}$$

The numbers a_{ij}, $i = 1, \ldots, m$; $j = 1, \ldots, n$ are called the *elements* of the matrix. For brevity, we shall write $\mathbf{A} = (a_{ij})$, $i = 1, \ldots, m$; $j = 1, \ldots, n$ for an $m \times n$ matrix, and only real matrices will be considered in the sequel. It follows that a vector in V_n is simply an $n \times 1$ matrix. A matrix is said to be a *square matrix* if $m = n$ and then n is called the *order* of the matrix. The elements a_{ii}, $i = 1, \ldots, n$ of a square matrix of order n are called the elements of the main diagonal of $\mathbf{A}$, or just the *diagonal elements* of $\mathbf{A}$. If $a_{ij} = 0$ for all $i \neq j$ (that is, if all of the elements off the main diagonal are 0), then $\mathbf{A}$ is called *diagonal*. A *zero* matrix is one in which all the elements are equal to zero. A zero matrix will be denoted by $\mathbf{0}$ regardless of its dimensions. A *unit* (or *identity*) matrix is a square matrix in which all diagonal elements are equal to 1 and all other elements are equal to 0. The proper notation for a unit matrix of order n is $\mathbf{I}_n$. However, we shall often write simply $\mathbf{I}$ and the order is to be understood from the context. Thus $\mathbf{I} = (\delta_{ij})$, where $\delta_{ij} = 1$ if $i = j$ and equals 0 if $i \neq j$. Two $m \times n$ matrices are said to be *equal* if they have identical elements. The *sum* $\mathbf{A} + \mathbf{B}$ of two $m \times n$ matrices $\mathbf{A} = (a_{ij})$, $\mathbf{B} = (b_{ij})$ is the $m = n$ matrix defined by $\mathbf{A} + \mathbf{B} = (a_{ij} + b_{ij})$. This definition is extended in an obvious manner to any finite number of $m \times n$ matrices. For any $m \times n$ matrices $\mathbf{A}$, $\mathbf{B}$ and $\mathbf{C}$, the following properties are immediate:

$$\mathbf{A} + \mathbf{B} = \mathbf{B} + \mathbf{A}, \quad (\mathbf{A} + \mathbf{B}) + \mathbf{C} = \mathbf{A} + (\mathbf{B} + \mathbf{C}) = \mathbf{A} + \mathbf{B} + \mathbf{C}.$$

The *product* $\alpha \mathbf{A}$ of the matrix $\mathbf{A} = (a_{ij})$ by the scalar α is the matrix defined by $\alpha \mathbf{A} = (\alpha a_{ij})$. The transpose $\mathbf{A}'$ of the $m \times n$ matrix $\mathbf{A} = (a_{ij})$ is the $n \times m$ matrix defined by $\mathbf{A}' = (a_{ji})$. Thus the rows and columns of $\mathbf{A}'$ are equal to the columns and rows of $\mathbf{A}$, respectively. If $\mathbf{A}$ is a square matrix and $\mathbf{A}' = \mathbf{A}$, then $\mathbf{A}$ is called *symmetric*. Clearly, for a symmetric matrix the elements symmetric with respect to the main diagonal of $\mathbf{A}$ are equal; that is, $a_{ij} = a_{ji}$ for all i and j. For any $m \times n$ matrices $\mathbf{A}$, $\mathbf{B}$ and any scalars α, β, the following properties are immediate:

$$(\mathbf{A}')' = \mathbf{A}, \quad (\alpha \mathbf{A})' = \alpha \mathbf{A}', \quad (\alpha \mathbf{A} + \beta \mathbf{B})' = \alpha \mathbf{A}' + \beta \mathbf{B}'.$$

The *product* $\mathbf{AB}$ of the $m \times n$ matrix $\mathbf{A} = (a_{ij})$ by the $n = r$ matrix $\mathbf{B} = (b_{ij})$ is the $m \times r$ matrix defined as follows: $\mathbf{AB} = (c_{ij})$, where $c_{ij} = \Sigma_{k=1}^{n} a_{ik} b_{kj}$. The product

BA is not defined unless $r = m$ and even then, it is not true, in general, that **AB** = **BA**. For example, take

$$\mathbf{A} = \begin{pmatrix} 0 & 0 \\ 0 & 1 \end{pmatrix}, \quad \mathbf{B} = \begin{pmatrix} 1 & 1 \\ 0 & 0 \end{pmatrix}.$$

Then

$$\mathbf{AB} = \begin{pmatrix} 0 & 0 \\ 0 & 0 \end{pmatrix}, \quad \mathbf{BA} = \begin{pmatrix} 0 & 1 \\ 0 & 0 \end{pmatrix},$$

so that **AB** ≠ **BA**. The products **AB**, **BA** are always defined for all square matrices of the same order.

Let A be an $m \times n$ matrix, let **B**, **C** be two $n \times r$ matrices and let **D** be an $r \times k$ matrix. Then for any scalars α, β and γ, the following properties are immediate:

$$\mathbf{IA} = \mathbf{AI} = \mathbf{A}, \mathbf{0A} = \mathbf{A0} = \mathbf{0}, \quad \mathbf{A}(\beta\mathbf{B} + \gamma\mathbf{C}) = \beta\mathbf{AB} + \gamma\mathbf{AC},$$
$$(\beta\mathbf{B} + \gamma\mathbf{C})\mathbf{D} = \beta\mathbf{BD} + \gamma\mathbf{CD}, \quad (\alpha\mathbf{A})\mathbf{B} = \mathbf{A}(\alpha\mathbf{B}) = \alpha(\mathbf{AB}) = \alpha\mathbf{AB},$$
$$(\mathbf{AB})' = \mathbf{B'A'}, \quad (\mathbf{AB})\mathbf{D} = \mathbf{A}(\mathbf{BD}).$$

By means of the last property, we may omit the parentheses and set **ABD** for (**AB**)**D** = **A**(**BD**).

Let **A** be an $m \times n$ matrix and let $r_i, i = 1, \ldots, m, c_j, j = 1, \ldots, n$ stand for the row and column vectors of A, respectively. Then it can be shown that the largest number of independent r-vectors is the same as the largest number of independent c-vectors and this common number is called *the rank of the matrix* **A**. Thus the rank of **A**, to be denoted by rank **A**, is the common dimension of the two vector spaces spanned by the r-vectors and the c-vectors. Always rank **A** $\leq \min(m, n)$ and if equality occurs, we say that **A** is *non-singular* or of *full rank*; otherwise **A** is called *singular*.

Let now |**A**| stand for the *determinant* of the *square matrix* **A**, defined only for square matrices, say $m \times m$, by the expression

$$|\mathbf{A}| = \sum \pm a_{1i} a_{2j} \cdots a_{mp},$$

where the a_{ij} are the elements of **A** and the summation extends over all permutations $(i, j, \ldots, p)$ of $(1, 2, \ldots, m)$. The plus sign is chosen if the permutation is even and the minus sign if it is odd. For further elaboration, see any of the references cited at the end of this appendix. It can be shown that **A** is nonsingular if and only is |**A**| ≠ 0. It can also be shown that if |**A**| ≠ 0, there exists a unique matrix, to be denoted by $\mathbf{A}^{-1}$, such that $\mathbf{AA}^{-1} = \mathbf{A}^{-1}\mathbf{A} = \mathbf{I}$. The matrix $\mathbf{A}^{-1}$ is called the *inverse* of **A**. Clearly, $(\mathbf{A}^{-1})^{-1} = \mathbf{A}$.

Let **A** be a square matrix of order n such that $\mathbf{A'A} = \mathbf{AA'} = \mathbf{I}$. Then **A** is said to be *orthogonal*. Let $\mathbf{r}_i$ and $\mathbf{c}_i, i = 1, \ldots, n$ stand for the row and column vectors of the matrix $\mathbf{A}$ of order n. Then the orthogonality of $\mathbf{A}$ is equivalent to the following properties:

$$\mathbf{r}_i' \mathbf{r}_i = \|\mathbf{r}_i\|^2 = \|\mathbf{c}_i\|^2 = \mathbf{c}_i' \mathbf{c}_i = 1 \quad \text{and} \quad \mathbf{r}_i' \mathbf{r}_j = \mathbf{c}_i' \mathbf{c}_j = 0 \quad \text{for} \quad i \neq j.$$

That is, $\{\mathbf{r}_j, j = 1, \ldots, n\}$ and $\{\mathbf{c}_j, j = 1, \ldots, n\}$ are orthonormal bases of V_n.

For a square matrix $\mathbf{A}$ of order n, consider the determinant $|\mathbf{A} - \lambda \mathbf{I}|$, where λ is a scalar. Then it is immediate that $|\mathbf{A} - \lambda \mathbf{I}|$ is a polynomial in λ of degree n and is called the *characteristic polynomial* of $\mathbf{A}$. The n roots of the equation $|\mathbf{A} - \lambda \mathbf{I}| = 0$ are called the *characteristic* (or *latent*) *roots*, or *eigenvalues* of $\mathbf{A}$. The matrix $\mathbf{A}$ is said to be *positive definite*, *negative definite*, or *positive semidefinite* if its characteristic roots λ_j, $j = 1, \ldots, n$ satisfy the following inequalities $\lambda_j > 0$, $\lambda_j < 0$, $\lambda_j \geq 0$, $j = 1, \ldots, n$, respectively.

REMARK 1.1 Although all matrices considered here are matrices with real elements, it should be noted that their characteristic roots will, in general, be complex numbers. However, they are always real for symmetric matrices.

Finally, a square matrix $\mathbf{A}$ is said to be *idempotent* if $\mathbf{A}^2 = \mathbf{A}$.

I.4 Some Theorems About Matrices and Quadratic Forms

Those theorems about matrices used in this book are gathered together here for easy reference.

THEOREM 5.I Let $\mathbf{A}, \mathbf{B}, \mathbf{C}$ be any $m \times n$, $n \times r$, $r \times s$ matrices, respectively. Then $(\mathbf{ABC})' = \mathbf{C}'\mathbf{B}'\mathbf{A}'$ and, in particular (by taking $\mathbf{C} = \mathbf{I}_r$), $(\mathbf{AB})' = \mathbf{B}'\mathbf{A}'$.

THEOREM 6.I
i) Let $\mathbf{A}, \mathbf{B}$ be any two matrices of the same order. Then $|\mathbf{AB}| = |\mathbf{BA}| = |\mathbf{A}||\mathbf{B}|$.

ii) For any diagonal matrix $\mathbf{A}$ of order n, $|\mathbf{A}| = \prod_{j=1}^{n} a_j$, where $a_j, j = 1, \ldots, n$ are the diagonal elements of $\mathbf{A}$.

iii) For any (square) matrix $\mathbf{A}$, $|\mathbf{A}| = |\mathbf{A}'|$.

iv) For any orthogonal matrix $\mathbf{A}$, $|\mathbf{A}|$ is either 1 or -1.

v) Let $\mathbf{A}, \mathbf{B}$ be matrices of the same order and suppose that $\mathbf{B}$ is orthogonal, Then $|\mathbf{B}'\mathbf{AB}| = |\mathbf{BAB}'| = |\mathbf{A}|$.

vi) For any matrix $\mathbf{A}$ for which $|\mathbf{A}| \neq 0$, $|\mathbf{A}^{-1}| = |\mathbf{A}|^{-1}$.

THEOREM 7.I
i) A square matrix $\mathbf{A}$ is non-singular if and only if $|\mathbf{A}| \neq 0$.

ii) Every orthogonal matrix is non-singular. (See (iv) of Theorem 6.I.)

iii) Let $\mathbf{A}$ be a non-singular square matrix. Then $\mathbf{A}'$, $\mathbf{A}^{-1}$ are also non-singular. (See (iii), (vi) of Theorem 6.I.)

iv) If $\mathbf{A}$ is symmetric non-singular, then so is $\mathbf{A}^{-1}$.

v) Let $\mathbf{A}, \mathbf{B}$ be non-singular $m \times m$ matrices.

Then the $m \times m$ matrix $\mathbf{AB}$ is non-singular and $(\mathbf{AB})^{-1} = \mathbf{B}^{-1}\mathbf{A}^{-1}$.

I.4 Some Theorems About Matrices and Quadratic Forms

THEOREM 8.1 i) Let $\mathbf{r}_1, \mathbf{r}_2$ be two vectors in V_n such that $\mathbf{r}_1'\mathbf{r}_2 = 0$ and $\|\mathbf{r}_1\| = \|\mathbf{r}_2\| = 1$. Then there exists an $n \times n$ orthogonal matrix, the first two rows of which are equal to $\mathbf{r}_1', \mathbf{r}_2'$.

(For a concrete example, see the application after Theorem 5 in Chapter 9.)

ii) Let $\mathbf{x}$ be a vector in V_n, let $\mathbf{A}$ be an $n \times n$ orthogonal matrix and set $\mathbf{y} = \mathbf{A}\mathbf{x}$. Then $\mathbf{x}'\mathbf{x} = \mathbf{y}'\mathbf{y}$, so that $\|\mathbf{x}\| = \|\mathbf{y}\|$.

iii) For every symmetric matrix $\mathbf{A}$ there is an orthogonal matrix $\mathbf{B}$ (of the same order as that of $\mathbf{A}$) such that the matrix $\mathbf{B}'\mathbf{A}\mathbf{B}$ is diagonal (and its diagonal elements are the characteristic roots of $\mathbf{A}$).

THEOREM 9.1 i) For any square matrix $\mathbf{A}$,

$$\text{rank}(\mathbf{A}\mathbf{A}') = \text{rank}(\mathbf{A}'\mathbf{A}) = \text{rank}\,\mathbf{A} = \text{rank}\,\mathbf{A}'.$$

ii) Let $\mathbf{A}, \mathbf{B}$ and $\mathbf{C}$ be $m \times n$, $n \times r$ and $r \times k$ matrices, respectively. Then

$$\text{rank}(\mathbf{A}\mathbf{B}) \leq \min(\text{rank}\,\mathbf{A}, \text{rank}\,\mathbf{B})$$

and

$$\text{rank}(\mathbf{A}\mathbf{B}\mathbf{C}) \leq \min(\text{rank}\,\mathbf{A}, \text{rank}\,\mathbf{B}, \text{rank}\,\mathbf{C}).$$

iii) Let $\mathbf{A}, \mathbf{B}$ and $\mathbf{C}$ be $m \times n$, $m \times m$ and $n \times n$ matrices, respectively, and suppose that $\mathbf{B}, \mathbf{C}$ are non-singular. Then

$$\text{rank}(\mathbf{B}\mathbf{A}) = \text{rank}(\mathbf{A}\mathbf{C}) = \text{rank}(\mathbf{B}\mathbf{A}\mathbf{C}) = \text{rank}\,\mathbf{A}.$$

iv) Let $\mathbf{A}, \mathbf{B}$ and $\mathbf{C}$ be $m \times n$, $m \times m$ and $n \times n$ matrices, respectively, and suppose that $\mathbf{B}, \mathbf{C}$ are non-singular. Then rank $(\mathbf{B}\mathbf{A}\mathbf{C})$ = rank $\mathbf{A}$. In particular, rank $(\mathbf{B}'\mathbf{A}\mathbf{B})$ = rank $(\mathbf{B}\mathbf{A}\mathbf{B}')$ = rank $\mathbf{A}$ if $m = n$ and $\mathbf{B}$ is orthogonal.

v) For any matrix $\mathbf{A}$, rank $\mathbf{A}$ = number of nonzero characteristic roots of $\mathbf{A}$.

THEOREM 10.1 i) If $\mathbf{A}$ is positive definite, $\mathbf{A}^{-1}$ exists and is also positive definite.

ii) For any nonsingular square matrix $\mathbf{A}$, $\mathbf{A}\mathbf{A}'$ is positive definite (and symmetric).

iii) Let $\mathbf{A} = (a_{ij})$, $i, j = 1, \ldots, n$ and define $\mathbf{A}_j$ by

$$\mathbf{A}_j = \begin{pmatrix} a_{11} & \cdots & a_{1j} \\ a_{j1} & \cdots & a_{jj} \end{pmatrix}, \quad j = 1, \ldots, n.$$

Then $\mathbf{A}$ is positive definite if and only if $|\mathbf{A}_j| > 0$, $j = 1, \ldots, n$. In particular, a diagonal matrix is positive definite if and only if its diagonal elements are all positive.

iv) A matrix $\mathbf{A}$ of order n is positive definite (semidefinite, negative definite, respectively,) if and only if $\mathbf{x}'\mathbf{A}\mathbf{x} > 0$ (≥ 0, <0, respectively) for every $\mathbf{x} \in V_n$ with $\mathbf{x} \neq \mathbf{0}$.

v) If **A** is a positive semidefinite matrix of order n and **B** is a non-singular matrix of order n, then **B'AB** is positive semidefinite.

vi) The characteristic roots of a positive definite (semidefinite) matrix are positive (nonnegative).

The following theorem refers to quadratic forms. For the definition of a quadratic form, the reader is referred to Definition 1, Chapter 19.

THEOREM 11.1 **i)** Let **A** be a symmetric matrix of order n. If $\mathbf{x'Ax} = \mathbf{x'x}$ identically in $\mathbf{x} \in V_n$, then $\mathbf{A} = \mathbf{I}$.

ii) Consider the quadratic form $Q = \mathbf{x'Ax}$, where **A** is of order n, and suppose that rank $\mathbf{A} = r$. Then there exist r linear forms in the x's

$$\sum_{j=1}^{n} b_{ij} x_j, \quad i = 1, \ldots, r$$

such that

$$Q = \sum_{i=1}^{r} \delta_i \left(\sum_{j=1}^{n} b_{ij} x_j \right)^2,$$

where δ_i is either 1 or -1, $i = 1, \ldots, r$.

iii) Let Q be as in (ii). There exists an orthogonal matrix **B** such that if

$$\mathbf{y} = \mathbf{B}^{-1}\mathbf{x}, \quad \text{then} \quad Q = \sum_{j=1}^{m} \lambda_j y_j^2,$$

where λ_j, $j = 1, \ldots, m$ are the nonzero characteristic roots of **A**.

iv) Let Q be as in (ii) and suppose that **A** is idempotent and rank $\mathbf{A} = r$. There exists an orthogonal matrix **B** such that if $\mathbf{y} = \mathbf{B}^{-1}\mathbf{x}$, then

$$Q = \sum_{j=1}^{r} y_j^2.$$

Finally, we formulate the following results referring to idempotent matrices:

THEOREM 12.1 **i)** The characteristic roots of an idempotent matrix are either 1 or 0.

ii) A diagonal matrix whose (diagonal) elements are either 1 or 0 is idempotent.

iii) If $\mathbf{A}_j$, $j = 1, \ldots, m$ are symmetric idempotent matrices of order n, such that $\mathbf{A}_i \mathbf{A}_j = 0$ for $1 \leq i < j \leq m$, then $\sum_{j=1}^{m} \mathbf{A}_j$ is idempotent and

$$\sum_{j=1}^{m} \text{rank } \mathbf{A}_j = \text{rank} \left(\sum_{j=1}^{m} \mathbf{A}_j \right).$$

In particular,

$$\text{rank } \mathbf{A}_1 + \text{rank} \left(\mathbf{I} - \mathbf{A}_1 \right) = n$$

and
$$\text{rank } \mathbf{A}_1 + \text{rank } \mathbf{A}_2 + \text{rank } (\mathbf{I} - \mathbf{A}_1 - \mathbf{A}_2) = n.$$

iv) If $\mathbf{A}_j$, $j = 1, \ldots, m$ are symmetric idempotent matrices of the same order and $\Sigma_{j=1}^{m} \mathbf{A}_j$ is also idempotent, the $\mathbf{A}_i \mathbf{A}_j = \mathbf{0}$ for $1 \leq i < j \leq m$.

The proof of the theorems formulated in this appendix may be found in most books of linear algebra. For example, see Birkhoff and MacLane, *A Survey of Modern Algebra*, 3d ed., MacMillan, 1965; S. Lang, *Linear Algebra*, Addison-Wesley, 1968; D. C. Murdoch, *Linear Algebra for Undergraduates*, Wiley, 1957; S. Perlis, *Theory of Matrices*, Addison-Wesley, 1952. For a brief exposition of most results from linear algebra employed in statistics, see also C. R. Rao, *Linear Statistical Inference and Its Applications*, Chapter 1, Wiley, 1965; H. Scheffé, *The Analysis of Variance*, Appendices I and II, Wiley, 1959; and F. A. Graybill, *An Introduction to Linear Statistical Models*, Vol. I, Chapter 1, McGraw-Hill, 1961.

Appendix II

Noncentral t, χ^2 and F Distributions

II.1 Noncentral t-Distribution

It was seen in Chapter 9, Application 2, that if the independent r.v.'s X and Y were distributed as $N(0, 1)$ and χ_r^2, respectively, then the distribution of the r.v. $T = X/\sqrt{(Y/r)}$ was the (Student's) t-distribution with r d.f. Now let X and Y be independent r.v.'s distributed as $N(\delta, 1)$ and χ_r^2, respectively, and set $T' = X/\sqrt{(Y/r)}$. The r.v. T' is said to have the *noncentral t-distribution with r d.f. and noncentrality parameter* δ. This distribution, as well as an r.v. having this distribution, if often denoted by $t'_{r;\delta}$. Using the definition of a $t'_{r;\delta}$ r.v., it can be found by well known methods that its p.d.f. is given by

$$f_{t'_{r;\delta}}(t;\delta) = \frac{1}{2^{(r+1)/2}\Gamma(r/2)\sqrt{\pi r}} \int_0^\infty x^{(r-1)/2}$$

$$\times \exp\left\{-\frac{1}{2}\left[x + \left(t\sqrt{\frac{x}{r}} - \delta\right)^2\right]\right\} dx, \quad t \in \mathbb{R}.$$

II.2 Noncentral χ^2-Distribution

It was seen in Chapter 7 (see corollary to Theorem 5) that if $X_1, \ldots, X_r$ were independent normally distributed r.v.'s with variance 1 and mean 0, then the r.v. $X = \sum_{j=1}^r X_j^2$ was distributed as χ_r^2. Let now the r.v.'s $X_1, \ldots, X_r$ be independent normally distributed with variance 1 but means $\mu_1, \ldots, \mu_r$, respectively. Then the distribution of the r.v. $X^* = \sum_{j=1}^r X_j^2$ is said to be the *noncentral* chi-square distribution with r d.f. and noncentrality parameter δ, where $\delta^2 = \Sigma_{j=1}^{r} \mu_j^2$. This distribution, and also an r.v. having this distribution, is often denoted by $\chi^2_{r;\delta}$. Using the definition of a $\chi^2_{r;\delta}$ r.v., one can find its p.d.f. but it does not have any simple closed form. It can be seen that this p.d.f. is a mixture of χ^2-distributions with Poisson weights. More precisely, one has

$$f_{\chi^2_{r;\delta}}(x;\;\delta) = \sum_{j=0}^{\infty} P_j(\delta) f_{r+2j}(x), \quad x \geq 0,$$

where

$$P_j(\delta) = e^{-\delta^2/2} \frac{(\delta^2/2)^j}{j!} \quad \text{and} \quad f_{r+2j} \quad \text{is the p.d.f. of} \quad \chi^2_{r+2j}, \quad j = 0, 1, \ldots.$$

II.3 Noncentral *F*-Distribution

In Chapter 9, Application 2, the *F*-distribution with r_1 and r_2 d.f. was defined as the distribution of the r.v.

$$F = \frac{X/r_1}{Y/r_2},$$

where X and Y were independent r.v.'s distributed as $\chi^2_{r_1}$ and $\chi^2_{r_2}$, respectively. Suppose now that the r.v.'s X and Y are independent and distributed as $\chi^2_{r_1;\delta}$ and $\chi^2_{r_2}$, respectively, and set

$$F' = \frac{X/r_1}{Y/r_2}.$$

Then the distribution of F' is said to be *the noncentral F-distribution with r_1 and r_2 d.f. and noncentrality parameter δ*. This distribution, and also an r.v. having this distribution, is often denoted by $F'_{r_1,r_2;\delta}$, and its p.d.f., which does not have any simple closed form, is given by the following expression:

$$f_{F'_{r_1,r_2;\delta}}(f;\;\delta) = e^{-\delta^2/2} \sum_{j=0}^{\infty} c_j \frac{(\delta^2/2)^j}{j!} \frac{f^{\frac{1}{2}r_1 - 1 + j}}{(1+f)^{\frac{1}{2}(r_1 + r_2) + j}}, \quad f \geq 0,$$

where

$$c_j = \frac{\Gamma\left[\frac{1}{2}(r_1 + r_2) + j\right]}{\Gamma\left(\frac{1}{2}r_1 + j\right)\Gamma\left(\frac{1}{2}r_2\right)}, \quad j = 0, 1, \ldots.$$

REMARKS

(i) By setting $\delta = 0$ in the noncentral t, χ^2 and F-distributions, we obtain the t, χ^2 and F-distributions, respectively. In view of this, the latter distributions may also be called central t, χ^2 and F-distributions.

(ii) Tables for the noncentral t, χ^2 and F-distributions are given in a reference cited elsewhere, namely, *Handbook of Statistical Tables* by D. B. Owen. Addison-Wesley, 1962.

Appendix III

Tables

Table 1 The Cumulative Binomial Distribution

The tabulated quantity is

$$\sum_{j=0}^{k}\binom{n}{j}p^{j}(1-p)^{n-j}.$$

					p				
n	k	1/16	2/16	3/16	4/16	5/16	6/16	7/16	8/16
2	0	0.8789	0.7656	0.6602	0.5625	0.4727	0.3906	0.3164	0.2500
	1	0.9961	0.9844	0.9648	0.9375	0.9023	0.8594	0.8086	0.7500
	2	1.0000	1.0000	1.0000	1.0000	1.0000	1.0000	1.0000	1.0000
3	0	0.8240	0.6699	0.5364	0.4219	0.3250	0.2441	0.1780	0.1250
	1	0.9888	0.9570	0.9077	0.8437	0.7681	0.6836	0.5933	0.5000
	2	0.9998	0.9980	0.9934	0.9844	0.9695	0.9473	0.9163	0.8750
	3	1.0000	1.0000	1.0000	1.0000	1.0000	1.0000	1.0000	1.0000
4	0	0.7725	0.5862	0.4358	0.3164	0.2234	0.1526	0.1001	0.0625
	1	0.9785	0.9211	0.8381	0.7383	0.6296	0.5188	0.4116	0.3125
	2	0.9991	0.9929	0.9773	0.9492	0.9065	0.8484	0.7749	0.6875
	3	1.0000	0.9998	0.9988	0.9961	0.9905	0.9802	0.9634	0.9375
	4	1.0000	1.0000	1.0000	1.0000	1.0000	1.0000	1.0000	1.0000
5	0	0.7242	0.5129	0.3541	0.2373	0.1536	0.0954	0.0563	0.0312
	1	0.9656	0.8793	0.7627	0.6328	0.5027	0.3815	0.2753	0.1875
	2	0.9978	0.9839	0.9512	0.8965	0.8200	0.7248	0.6160	0.5000
	3	0.9999	0.9989	0.9947	0.9844	0.9642	0.9308	0.8809	0.8125
	4	1.0000	1.0000	0.9998	0.9990	0.9970	0.9926	0.9840	0.9687
	5	1.0000	1.0000	1.0000	1.0000	1.0000	1.0000	1.0000	1.0000

Table 1 *(continued)*

n	k	1/16	2/16	3/16	p 4/16	5/16	6/16	7/16	8/16
6	0	0.6789	0.4488	0.2877	0.1780	0.1056	0.0596	0.0317	0.0156
	1	0.9505	0.8335	0.6861	0.5339	0.3936	0.2742	0.1795	0.1094
	2	0.9958	0.9709	0.9159	0.8306	0.7208	0.5960	0.4669	0.3437
	3	0.9998	0.9970	0.9866	0.9624	0.9192	0.8535	0.7650	0.6562
	4	1.0000	0.9998	0.9988	0.9954	0.9868	0.9694	0.9389	0.8906
	5	1.0000	1.0000	1.0000	0.9998	0.9991	0.9972	0.9930	0.9844
	6	1.0000	1.0000	1.0000	1.0000	1.0000	1.0000	1.0000	1.0000
7	0	0.6365	0.3927	0.2338	0.1335	0.0726	0.0373	0.0178	0.0078
	1	0.9335	0.7854	0.6114	0.4449	0.3036	0.1937	0.1148	0.0625
	2	0.9929	0.9537	0.8728	0.7564	0.6186	0.4753	0.3412	0.2266
	3	0.9995	0.9938	0.9733	0.9294	0.8572	0.7570	0.6346	0.5000
	4	1.0000	0.9995	0.9965	0.9871	0.9656	0.9260	0.8628	0.7734
	5	1.0000	1.0000	0.9997	0.9987	0.9952	0.9868	0.9693	0.9375
	6	1.0000	1.0000	1.0000	0.9999	0.9997	0.9990	0.9969	0.9922
	7	1.0000	1.0000	1.0000	1.0000	1.0000	1.0000	1.0000	1.0000
8	0	0.5967	0.3436	0.1899	0.1001	0.0499	0.0233	0.0100	0.0039
	1	0.9150	0.7363	0.5406	0.3671	0.2314	0.1350	0.0724	0.0352
	2	0.9892	0.9327	0.8238	0.6785	0.5201	0.3697	0.2422	0.1445
	3	0.9991	0.9888	0.9545	0.8862	0.7826	0.6514	0.5062	0.3633
	4	1.0000	0.9988	0.9922	0.9727	0.9318	0.8626	0.7630	0.6367
	5	1.0000	0.9999	0.9991	0.9958	0.9860	0.9640	0.9227	0.8555
	6	1.0000	1.0000	0.9999	0.9996	0.9983	0.9944	0.9849	0.9648
	7	1.0000	1.0000	1.0000	1.0000	0.9999	0.9996	0.9987	0.9961
	8	1.0000	1.0000	1.0000	1.0000	1.0000	1.0000	1.0000	1.0000
9	0	0.5594	0.3007	0.1543	0.0751	0.0343	0.0146	0.0056	0.0020
	1	0.8951	0.6872	0.4748	0.3003	0.1747	0.0931	0.0451	0.0195
	2	0.9846	0.9081	0.7707	0.6007	0.4299	0.2817	0.1679	0.0898
	3	0.9985	0.9817	0.9300	0.8343	0.7006	0.5458	0.3907	0.2539
	4	0.9999	0.9975	0.9851	0.9511	0.8851	0.7834	0.6506	0.5000
	5	1.0000	0.9998	0.9978	0.9900	0.9690	0.9260	0.8528	0.7461
	6	1.0000	1.0000	0.9998	0.9987	0.9945	0.9830	0.9577	0.9102
	7	1.0000	1.0000	1.0000	0.9999	0.9994	0.9977	0.9926	0.9805
	8	1.0000	1.0000	1.0000	1.0000	1.0000	0.9999	0.9994	0.9980
	9	1.0000	1.0000	1.0000	1.0000	1.0000	1.0000	1.0000	1.0000
10	0	0.5245	0.2631	0.1254	0.0563	0.0236	0.0091	0.0032	0.0010
	1	0.8741	0.6389	0.4147	0.2440	0.1308	0.0637	0.0278	0.0107
	2	0.9790	0.8805	0.7152	0.5256	0.3501	0.2110	0.1142	0.0547
	3	0.9976	0.9725	0.9001	0.7759	0.6160	0.4467	0.2932	0.1719
	4	0.9998	0.9955	0.9748	0.9219	0.8275	0.6943	0.5369	0.3770
	5	1.0000	0.9995	0.9955	0.9803	0.9428	0.8725	0.7644	0.6230
	6	1.0000	1.0000	0.9994	0.9965	0.9865	0.9616	0.9118	0.8281
	7	1.0000	1.0000	1.0000	0.9996	0.9979	0.9922	0.9773	0.9453
	8	1.0000	1.0000	1.0000	1.0000	0.9998	0.9990	0.9964	0.9893

Table 1 *(continued)*

n	k	1/16	2/16	3/16	p 4/16	5/16	6/16	7/16	8/16
10	9	1.0000	1.0000	1.0000	1.0000	1.0000	0.9999	0.9997	0.9990
	10	1.0000	1.0000	1.0000	1.0000	1.0000	1.0000	1.0000	1.0000
11	0	0.4917	0.2302	0.1019	0.0422	0.0162	0.0057	0.0018	0.0005
	1	0.8522	0.5919	0.3605	0.1971	0.0973	0.0432	0.0170	0.0059
	2	0.9724	0.8503	0.6589	0.4552	0.2816	0.1558	0.0764	0.0327
	3	0.9965	0.9610	0.8654	0.7133	0.5329	0.3583	0.2149	0.1133
	4	0.9997	0.9927	0.9608	0.8854	0.7614	0.6014	0.4303	0.2744
	5	1.0000	0.9990	0.9916	0.9657	0.9068	0.8057	0.6649	0.5000
	6	1.0000	0.9999	0.9987	0.9924	0.9729	0.9282	0.8473	0.7256
	7	1.0000	1.0000	0.9999	0.9988	0.9943	0.9807	0.9487	0.8867
	8	1.0000	1.0000	1.0000	0.9999	0.9992	0.9965	0.9881	0.9673
	9	1.0000	1.0000	1.0000	1.0000	0.9999	0.9996	0.9983	0.9941
	10	1.0000	1.0000	1.0000	1.0000	1.0000	1.0000	0.9999	0.9995
	11	1.0000	1.0000	1.0000	1.0000	1.0000	1.0000	1.0000	1.0000
12	0	0.4610	0.2014	0.0828	0.0317	0.0111	0.0036	0.0010	0.0002
	1	0.8297	0.5467	0.3120	0.1584	0.0720	0.0291	0.0104	0.0032
	2	0.9649	0.8180	0.6029	0.3907	0.2240	0.1135	0.0504	0.0193
	3	0.9950	0.9472	0.8267	0.6488	0.4544	0.2824	0.1543	0.0730
	4	0.9995	0.9887	0.9429	0.8424	0.6900	0.5103	0.3361	0.1938
	5	1.0000	0.9982	0.9858	0.9456	0.8613	0.7291	0.5622	0.3872
	6	1.0000	0.9998	0.9973	0.9857	0.9522	0.8822	0.7675	0.6128
	7	1.0000	1.0000	0.9996	0.9972	0.9876	0.9610	0.9043	0.8062
	8	1.0000	1.0000	1.0000	0.9996	0.9977	0.9905	0.9708	0.9270
	9	1.0000	1.0000	1.0000	1.0000	0.9997	0.9984	0.9938	0.9807
	10	1.0000	1.0000	1.0000	1.0000	1.0000	0.9998	0.9992	0.9968
	11	1.0000	1.0000	1.0000	1.0000	1.0000	1.0000	1.0000	0.9998
	12	1.0000	1.0000	1.0000	1.0000	1.0000	1.0000	1.0000	1.0000
13	0	0.4321	0.1762	0.0673	0.0238	0.0077	0.0022	0.0006	0.0001
	1	0.8067	0.5035	0.2690	0.1267	0.0530	0.0195	0.0063	0.0017
	2	0.9565	0.7841	0.5484	0.3326	0.1765	0.0819	0.0329	0.0112
	3	0.9931	0.9310	0.7847	0.5843	0.3824	0.2191	0.1089	0.0461
	4	0.9992	0.9835	0.9211	0.7940	0.6164	0.4248	0.2565	0.1334
	5	0.9999	0.9970	0.9778	0.9198	0.8078	0.6470	0.4633	0.2905
	6	1.0000	0.9996	0.9952	0.9757	0.9238	0.8248	0.6777	0.5000
	7	1.0000	1.0000	0.9992	0.9944	0.9765	0.9315	0.8445	0.7095
	8	1.0000	1.0000	0.9999	0.9990	0.9945	0.9795	0.9417	0.8666
	9	1.0000	1.0000	1.0000	0.9999	0.9991	0.9955	0.9838	0.9539
	10	1.0000	1.0000	1.0000	1.0000	0.9999	0.9993	0.9968	0.9888
	11	1.0000	1.0000	1.0000	1.0000	1.0000	0.9999	0.9996	0.9983
	12	1.0000	1.0000	1.0000	1.0000	1.0000	1.0000	1.0000	0.9999
	13	1.0000	1.0000	1.0000	1.0000	1.0000	1.0000	1.0000	1.0000
14	0	0.4051	0.1542	0.0546	0.0178	0.0053	0.0014	0.0003	0.0001
	1	0.7833	0.4626	0.2312	0.1010	0.0388	0.0130	0.0038	0.0009

Table 1 *(continued)*

n	k	1/16	2/16	3/16	p 4/16	5/16	6/16	7/16	8/16
14	2	0.9471	0.7490	0.4960	0.2811	0.1379	0.0585	0.0213	0.0065
	3	0.9908	0.9127	0.7404	0.5213	0.3181	0.1676	0.0756	0.0287
	4	0.9988	0.9970	0.8955	0.7415	0.5432	0.3477	0.1919	0.0898
	5	0.9999	0.9953	0.9671	0.8883	0.7480	0.5637	0.3728	0.2120
	6	1.0000	0.9993	0.9919	0.9167	0.8876	0.7581	0.5839	0.3953
	7	1.0000	0.9999	0.9985	0.9897	0.9601	0.8915	0.7715	0.6047
	8	1.0000	1.0000	0.9998	0.9978	0.9889	0.9615	0.8992	0.7880
	9	1.0000	1.0000	1.0000	0.9997	0.9976	0.9895	0.9654	0.9102
	10	1.0000	1.0000	1.0000	1.0000	0.9996	0.9979	0.9911	0.9713
	11	1.0000	1.0000	1.0000	1.0000	1.0000	0.9997	0.9984	0.9935
	12	1.0000	1.0000	1.0000	1.0000	1.0000	1.0000	0.9998	0.9991
	13	1.0000	1.0000	1.0000	1.0000	1.0000	1.0000	1.0000	0.9999
	14	1.0000	1.0000	1.0000	1.0000	1.0000	1.0000	1.0000	1.0000
15	0	0.3798	0.1349	0.0444	0.0134	0.0036	0.0009	0.0002	0.0000
	1	0.7596	0.4241	0.1981	0.0802	0.0283	0.0087	0.0023	0.0005
	2	0.9369	0.7132	0.4463	0.2361	0.1069	0.0415	0.0136	0.0037
	3	0.9881	0.8922	0.6946	0.4613	0.2618	0.1267	0.0518	0.0176
	4	0.9983	0.9689	0.8665	0.6865	0.4729	0.2801	0.1410	0.0592
	5	0.9998	0.9930	0.9537	0.8516	0.6840	0.4827	0.2937	0.1509
	6	1.0000	0.9988	0.9873	0.9434	0.8435	0.6852	0.4916	0.3036
	7	1.0000	0.9998	0.9972	0.9827	0.9374	0.8415	0.6894	0.5000
	8	1.0000	1.0000	0.9995	0.9958	0.9799	0.9352	0.8433	0.6964
	9	1.0000	1.0000	0.9999	0.9992	0.9949	0.9790	0.9364	0.8491
	10	1.0000	1.0000	1.0000	0.9999	0.9990	0.9947	0.9799	0.9408
	11	1.0000	1.0000	1.0000	1.0000	0.9999	0.9990	0.9952	0.9824
	12	1.0000	1.0000	1.0000	1.0000	1.0000	0.9999	0.9992	0.9963
	13	1.0000	1.0000	1.0000	1.0000	1.0000	1.0000	0.9999	0.9995
	14	1.0000	1.0000	1.0000	1.0000	1.0000	1.0000	1.0000	1.0000
	15	1.0000	1.0000	1.0000	1.0000	1.0000	1.0000	1.0000	1.0000
16	0	0.3561	0.1181	0.0361	0.0100	0.0025	0.0005	0.0001	0.0000
	1	0.7359	0.3879	0.1693	0.0635	0.0206	0.0057	0.0014	0.0003
	2	0.9258	0.6771	0.3998	0.1971	0.0824	0.0292	0.0086	0.0021
	3	0.9849	0.8698	0.6480	0.4050	0.2134	0.0947	0.0351	0.0106
	4	0.9977	0.9593	0.8342	0.6302	0.4069	0.2226	0.1020	0.0384
	5	0.9997	0.9900	0.9373	0.8103	0.6180	0.4067	0.2269	0.1051
	6	1.0000	0.9981	0.9810	0.9204	0.7940	0.6093	0.4050	0.2272
	7	1.0000	0.9997	0.9954	0.9729	0.9082	0.7829	0.6029	0.4018
	8	1.0000	1.0000	0.9991	0.9925	0.9666	0.9001	0.7760	0.5982
	9	1.0000	1.0000	0.9999	0.9984	0.9902	0.9626	0.8957	0.7728
	10	1.0000	1.0000	1.0000	0.9997	0.9977	0.9888	0.9609	0.8949
	11	1.0000	1.0000	1.0000	1.0000	0.9996	0.9974	0.9885	0.9616
	12	1.0000	1.0000	1.0000	1.0000	0.9999	0.9995	0.9975	0.9894
	13	1.0000	1.0000	1.0000	1.0000	1.0000	0.9999	0.9996	0.9979

Table 1 (continued)

n	k	1/16	2/16	3/16	p 4/16	5/16	6/16	7/16	8/16
16	14	1.0000	1.0000	1.0000	1.0000	1.0000	1.0000	1.0000	0.9997
	15	1.0000	1.0000	1.0000	1.0000	1.0000	1.0000	1.0000	1.0000
	16	1.0000	1.0000	1.0000	1.0000	1.0000	1.0000	1.0000	1.0000
17	0	0.3338	0.1033	0.0293	0.0075	0.0017	0.0003	0.0001	0.0000
	1	0.7121	0.3542	0.1443	0.0501	0.0149	0.0038	0.0008	0.0001
	2	0.9139	0.6409	0.3566	0.1637	0.0631	0.0204	0.0055	0.0012
	3	0.9812	0.8457	0.6015	0.3530	0.1724	0.0701	0.0235	0.0064
	4	0.9969	0.9482	0.7993	0.5739	0.3464	0.1747	0.0727	0.0245
	5	0.9996	0.9862	0.9180	0.7653	0.5520	0.3377	0.1723	0.0717
	6	1.0000	0.9971	0.9728	0.8929	0.7390	0.5333	0.3271	0.1662
	7	1.0000	0.9995	0.9927	0.9598	0.8725	0.7178	0.5163	0.3145
	8	1.0000	0.9999	0.9984	0.9876	0.9484	0.8561	0.7002	0.5000
	9	1.0000	1.0000	0.9997	0.9969	0.9828	0.9391	0.8433	0.6855
	10	1.0000	1.0000	1.0000	0.9994	0.9954	0.9790	0.9323	0.8338
	11	1.0000	1.0000	1.0000	0.9999	0.9990	0.9942	0.9764	0.9283
	12	1.0000	1.0000	1.0000	1.0000	0.9998	0.9987	0.9935	0.9755
	13	1.0000	1.0000	1.0000	1.0000	1.0000	0.9998	0.9987	0.9936
	14	1.0000	1.0000	1.0000	1.0000	1.0000	1.0000	0.9998	0.9988
	15	1.0000	1.0000	1.0000	1.0000	1.0000	1.0000	1.0000	0.9999
	16	1.0000	1.0000	1.0000	1.0000	1.0000	1.0000	1.0000	1.0000
18	0	0.3130	0.0904	0.0238	0.0056	0.0012	0.0002	0.0000	0.0000
	1	0.6885	0.3228	0.1227	0.0395	0.0108	0.0025	0.0005	0.0001
	2	0.9013	0.6051	0.3168	0.1353	0.0480	0.0142	0.0034	0.0007
	3	0.9770	0.8201	0.5556	0.3057	0.1383	0.0515	0.0156	0.0038
	4	0.9959	0.9354	0.7622	0.5187	0.2920	0.1355	0.0512	0.0154
	5	0.9994	0.9814	0.8958	0.7175	0.4878	0.2765	0.1287	0.0481
	6	0.9999	0.9957	0.9625	0.8610	0.6806	0.4600	0.2593	0.1189
	7	1.0000	0.9992	0.9889	0.9431	0.8308	0.6486	0.4335	0.2403
	8	1.0000	0.9999	0.9973	0.9807	0.9247	0.8042	0.6198	0.4073
	9	1.0000	1.0000	0.9995	0.9946	0.9721	0.9080	0.7807	0.5927
	10	1.0000	1.0000	0.9999	0.9988	0.9915	0.9640	0.8934	0.7597
	11	1.0000	1.0000	1.0000	0.9998	0.9979	0.9885	0.9571	0.8811
	12	1.0000	1.0000	1.0000	1.0000	0.9996	0.9970	0.9860	0.9519
	13	1.0000	1.0000	1.0000	1.0000	0.9999	0.9994	0.9964	0.9846
	14	1.0000	1.0000	1.0000	1.0000	1.0000	0.9999	0.9993	0.9962
	15	1.0000	1.0000	1.0000	1.0000	1.0000	1.0000	0.9999	0.9993
	16	1.0000	1.0000	1.0000	1.0000	1.0000	1.0000	1.0000	0.9999
	17	1.0000	1.0000	1.0000	1.0000	1.0000	1.0000	1.0000	1.0000
19	0	0.2934	0.0791	0.0193	0.0042	0.0008	0.0001	0.0000	0.0000
	1	0.6650	0.2938	0.1042	0.0310	0.0078	0.0016	0.0003	0.0000
	2	0.8880	0.5698	0.2804	0.1113	0.0364	0.0098	0.0021	0.0004
	3	0.9722	0.7933	0.5108	0.2631	0.1101	0.0375	0.0103	0.0022
	4	0.9947	0.9209	0.7235	0.4654	0.2440	0.1040	0.0356	0.0096

Table 1 *(continued)*

n	k	1/16	2/16	3/16	p 4/16	5/16	6/16	7/16	8/16
19	5	0.9992	0.9757	0.8707	0.6678	0.4266	0.2236	0.0948	0.0318
	6	0.9999	0.9939	0.9500	0.8251	0.6203	0.3912	0.2022	0.0835
	7	1.0000	0.9988	0.9840	0.9225	0.7838	0.5779	0.3573	0.1796
	8	1.0000	0.9998	0.9957	0.9713	0.8953	0.7459	0.5383	0.3238
	9	1.0000	1.0000	0.9991	0.9911	0.9573	0.8691	0.7103	0.5000
	10	1.0000	1.0000	0.9998	0.9977	0.9854	0.9430	0.8441	0.0672
	11	1.0000	1.0000	1.0000	0.9995	0.9959	0.9793	0.9292	0.8204
	12	1.0000	1.0000	1.0000	0.9999	0.9990	0.9938	0.9734	0.9165
	13	1.0000	1.0000	1.0000	1.0000	0.9998	0.9985	0.9919	0.9682
	14	1.0000	1.0000	1.0000	1.0000	1.0000	0.9997	0.9980	0.9904
	15	1.0000	1.0000	1.0000	1.0000	1.0000	1.0000	0.9996	0.9978
	16	1.0000	1.0000	1.0000	1.0000	1.0000	1.0000	1.0000	0.9996
	17	1.0000	1.0000	1.0000	1.0000	1.0000	1.0000	1.0000	1.0000
	18	1.0000	1.0000	1.0000	1.0000	1.0000	1.0000	1.0000	1.0000
20	0	0.2751	0.0692	0.0157	0.0032	0.0006	0.0001	0.0000	0.0000
	1	0.6148	0.2669	0.0883	0.0243	0.0056	0.0011	0.0002	0.0000
	2	0.8741	0.5353	0.2473	0.0913	0.0275	0.0067	0.0013	0.0002
	3	0.9670	0.7653	0.4676	0.2252	0.0870	0.0271	0.0067	0.0013
	4	0.9933	0.9050	0.6836	0.4148	0.2021	0.0790	0.0245	0.0059
	5	0.9989	0.9688	0.8431	0.6172	0.3695	0.1788	0.0689	0.0207
	6	0.9999	0.9916	0.9351	0.7858	0.5598	0.3284	0.1552	0.0577
	7	1.0000	0.9981	0.9776	0.8982	0.7327	0.5079	0.2894	0.1316
	8	1.0000	0.9997	0.9935	0.9591	0.8605	0.6829	0.4591	0.2517
	9	1.0000	0.9999	0.9984	0.9861	0.9379	0.8229	0.6350	0.4119
	10	1.0000	1.0000	0.9997	0.9961	0.9766	0.9153	0.7856	0.5881
	11	1.0000	1.0000	0.9999	0.9991	0.9926	0.9657	0.8920	0.7483
	12	1.0000	1.0000	1.0000	0.9998	0.9981	0.9884	0.9541	0.8684
	13	1.0000	1.0000	1.0000	1.0000	0.9996	0.9968	0.9838	0.9423
	14	1.0000	1.0000	1.0000	1.0000	0.9999	0.9993	0.9953	0.9793
	15	1.0000	1.0000	1.0000	1.0000	1.0000	0.9999	0.9989	0.9941
	16	1.0000	1.0000	1.0000	1.0000	1.0000	1.0000	0.9998	0.9987
	17	1.0000	1.0000	1.0000	1.0000	1.0000	1.0000	1.0000	0.9998
	18	1.0000	1.0000	1.0000	1.0000	1.0000	1.0000	1.0000	1.0000
	19	1.0000	1.0000	1.0000	1.0000	1.0000	1.0000	1.0000	1.0000
21	0	0.2579	0.0606	0.0128	0.0024	0.0004	0.0001	0.0000	0.0000
	1	0.6189	0.2422	0.0747	0.0190	0.0040	0.0007	0.0001	0.0000
	2	0.8596	0.5018	0.2175	0.0745	0.0206	0.0046	0.0008	0.0001
	3	0.9612	0.7366	0.4263	0.1917	0.0684	0.0195	0.0044	0.0007
	4	0.9917	0.8875	0.6431	0.3674	0.1662	0.0596	0.0167	0.0036
	5	0.9986	0.9609	0.8132	0.5666	0.3172	0.1414	0.0495	0.0133
	6	0.9998	0.9888	0.9179	0.7436	0.5003	0.2723	0.1175	0.0392
	7	1.0000	0.9973	0.9696	0.8701	0.6787	0.4405	0.2307	0.0946
	8	1.0000	0.9995	0.9906	0.9439	0.8206	0.6172	0.3849	0.1917

Table 1 (continued)

n	k	1/16	2/16	3/16	p 4/16	5/16	6/16	7/16	8/16
21	9	1.0000	0.9999	0.9975	0.9794	0.9137	0.7704	0.5581	0.3318
	10	1.0000	1.0000	0.9995	0.9936	0.9645	0.8806	0.7197	0.5000
	11	1.0000	1.0000	0.9999	0.9983	0.9876	0.9468	0.8454	0.6682
	12	1.0000	1.0000	1.0000	0.9996	0.9964	0.9799	0.9269	0.8083
	13	1.0000	1.0000	1.0000	0.9999	0.9991	0.9936	0.9708	0.9054
	14	1.0000	1.0000	1.0000	1.0000	0.9998	0.9983	0.9903	0.9605
	15	1.0000	1.0000	1.0000	1.0000	1.0000	0.9996	0.9974	0.9867
	16	1.0000	1.0000	1.0000	1.0000	1.0000	0.9999	0.9994	0.9964
	17	1.0000	1.0000	1.0000	1.0000	1.0000	1.0000	0.9999	0.9993
	18	1.0000	1.0000	1.0000	1.0000	1.0000	1.0000	1.0000	0.9999
	19	1.0000	1.0000	1.0000	1.0000	1.0000	1.0000	1.0000	1.0000
	20	1.0000	1.0000	1.0000	1.0000	1.0000	1.0000	1.0000	1.0000
22	0	0.2418	0.0530	0.0104	0.0018	0.0003	0.0000	0.0000	0.0000
	1	0.5963	0.2195	0.0631	0.0149	0.0029	0.0005	0.0001	0.0000
	2	0.8445	0.4693	0.1907	0.0606	0.0154	0.0031	0.0005	0.0001
	3	0.9548	0.7072	0.3871	0.1624	0.0535	0.0139	0.0028	0.0004
	4	0.9898	0.8687	0.6024	0.3235	0.1356	0.0445	0.0133	0.0022
	5	0.9981	0.9517	0.7813	0.5168	0.2700	0.1107	0.0352	0.0085
	6	0.9997	0.9853	0.8983	0.6994	0.4431	0.2232	0.0877	0.0267
	7	1.0000	0.9963	0.9599	0.8385	0.6230	0.3774	0.1812	0.0669
	8	1.0000	0.9992	0.9866	0.9254	0.7762	0.5510	0.3174	0.1431
	9	1.0000	0.9999	0.9962	0.9705	0.8846	0.7130	0.4823	0.2617
	10	1.0000	1.0000	0.9991	0.9900	0.9486	0.8393	0.6490	0.4159
	11	1.0000	1.0000	0.9998	0.9971	0.9804	0.9220	0.7904	0.5841
	12	1.0000	1.0000	1.0000	0.9993	0.9936	0.9675	0.8913	0.7383
	13	1.0000	1.0000	1.0000	0.9999	0.9982	0.9885	0.9516	0.8569
	14	1.0000	1.0000	1.0000	1.0000	0.9996	0.9966	0.9818	0.9331
	15	1.0000	1.0000	1.0000	1.0000	0.9999	0.9991	0.9943	0.9739
	16	1.0000	1.0000	1.0000	1.0000	1.0000	0.9998	0.9985	0.9915
	17	1.0000	1.0000	1.0000	1.0000	1.0000	1.0000	0.9997	0.9978
	18	1.0000	1.0000	1.0000	1.0000	1.0000	1.0000	1.0000	0.9995
	19	1.0000	1.0000	1.0000	1.0000	1.0000	1.0000	1.0000	0.9999
	20	1.0000	1.0000	1.0000	1.0000	1.0000	1.0000	1.0000	1.0000
23	0	0.2266	0.0464	0.0084	0.0013	0.0002	0.0000	0.0000	0.0000
	1	0.5742	0.1987	0.0532	0.0116	0.0021	0.0003	0.0000	0.0000
	2	0.8290	0.4381	0.1668	0.0492	0.0115	0.0021	0.0003	0.0000
	3	0.9479	0.6775	0.3503	0.1370	0.0416	0.0099	0.0018	0.0002
	4	0.9876	0.8485	0.5621	0.2832	0.1100	0.0330	0.0076	0.0013
	5	0.9976	0.9413	0.7478	0.4685	0.2280	0.0859	0.0247	0.0053
	6	0.9996	0.9811	0.8763	0.6537	0.3890	0.1810	0.0647	0.0173
	7	1.0000	0.9949	0.9484	0.8037	0.5668	0.3196	0.1403	0.0466
	8	1.0000	0.9988	0.9816	0.9037	0.7283	0.4859	0.2578	0.1050
	9	1.0000	0.9998	0.9944	0.9592	0.8507	0.6522	0.4102	0.2024

Table 1 *(continued)*

n	k	1/16	2/16	3/16	p 4/16	5/16	6/16	7/16	8/16
23	10	1.0000	1.0000	0.9986	0.9851	0.9286	0.7919	0.5761	0.3388
	11	1.0000	1.0000	0.9997	0.9954	0.9705	0.8910	0.7285	0.5000
	12	1.0000	1.0000	0.9999	0.9988	0.9895	0.9504	0.8471	0.6612
	13	1.0000	1.0000	1.0000	0.9997	0.9968	0.9806	0.9252	0.7976
	14	1.0000	1.0000	1.0000	0.9999	0.9992	0.9935	0.9686	0.8950
	15	1.0000	1.0000	1.0000	1.0000	0.9998	0.9982	0.9888	0.9534
	16	1.0000	1.0000	1.0000	1.0000	1.0000	0.9996	0.9967	0.9827
	17	1.0000	1.0000	1.0000	1.0000	1.0000	0.9999	0.9992	0.9947
	18	1.0000	1.0000	1.0000	1.0000	1.0000	1.0000	0.9998	0.9987
	19	1.0000	1.0000	1.0000	1.0000	1.0000	1.0000	1.0000	0.9998
	20	1.0000	1.0000	1.0000	1.0000	1.0000	1.0000	1.0000	1.0000
	21	1.0000	1.0000	1.0000	1.0000	1.0000	1.0000	1.0000	1.0000
24	0	0.2125	0.0406	0.0069	0.0010	0.0001	0.0000	0.0000	0.0000
	1	0.5524	0.1797	0.0448	0.0090	0.0015	0.0002	0.0000	0.0000
	2	0.8131	0.4082	0.1455	0.0398	0.0086	0.0014	0.0002	0.0000
	3	0.9405	0.6476	0.3159	0.1150	0.0322	0.0070	0.0011	0.0001
	4	0.9851	0.8271	0.5224	0.2466	0.0886	0.0243	0.0051	0.0008
	5	0.9970	0.9297	0.7130	0.4222	0.1911	0.0661	0.0172	0.0033
	6	0.9995	0.9761	0.8522	0.6074	0.3387	0.1453	0.0472	0.0113
	7	0.9999	0.9932	0.9349	0.7662	0.5112	0.2676	0.1072	0.0320
	8	1.0000	0.9983	0.9754	0.8787	0.6778	0.4235	0.2064	0.0758
	9	1.0000	0.9997	0.9920	0.9453	0.8125	0.5898	0.3435	0.1537
	10	1.0000	0.9999	0.9978	0.9787	0.9043	0.7395	0.5035	0.2706
	11	1.0000	1.0000	0.9995	0.9928	0.9574	0.8538	0.6618	0.4194
	12	1.0000	1.0000	0.9999	0.9979	0.9835	0.9281	0.7953	0.5806
	13	1.0000	1.0000	1.0000	0.9995	0.9945	0.9693	0.8911	0.7294
	14	1.0000	1.0000	1.0000	0.9999	0.9984	0.9887	0.9496	0.8463
	15	1.0000	1.0000	1.0000	1.0000	0.9996	0.9964	0.9799	0.9242
	16	1.0000	1.0000	1.0000	1.0000	0.9999	0.9990	0.9932	0.9680
	17	1.0000	1.0000	1.0000	1.0000	1.0000	0.9998	0.9981	0.9887
	18	1.0000	1.0000	1.0000	1.0000	1.0000	1.0000	0.9996	0.9967
	19	1.0000	1.0000	1.0000	1.0000	1.0000	1.0000	0.9999	0.9992
	20	1.0000	1.0000	1.0000	1.0000	1.0000	1.0000	1.0000	0.9999
	21	1.0000	1.0000	1.0000	1.0000	1.0000	1.0000	1.0000	1.0000
	22	1.0000	1.0000	1.0000	1.0000	1.0000	1.0000	1.0000	1.0000
25	0	0.1992	0.0355	0.0056	0.0008	0.0001	0.0000	0.0000	0.0000
	1	0.5132	0.1623	0.0377	0.0070	0.0011	0.0001	0.0000	0.0000
	2	0.7968	0.3796	0.1266	0.0321	0.0064	0.0010	0.0001	0.0000
	3	0.9325	0.6176	0.2840	0.0962	0.0248	0.0049	0.0007	0.0001
	4	0.9823	0.8047	0.4837	0.2137	0.0710	0.0178	0.0033	0.0005
	5	0.9962	0.9169	0.6772	0.3783	0.1591	0.0504	0.0119	0.0028
	6	0.9993	0.9703	0.8261	0.5611	0.2926	0.1156	0.0341	0.0073
	7	0.9999	0.9910	0.9194	0.7265	0.4573	0.2218	0.0810	0.0216

Table 1 *(continued)*

n	k	1/16	2/16	3/16	p 4/16	5/16	6/16	7/16	8/16
25	8	1.0000	0.9977	0.9678	0.8506	0.6258	0.3651	0.1630	0.0539
	9	1.0000	0.9995	0.9889	0.9287	0.7704	0.5275	0.2835	0.1148
	10	1.0000	0.9999	0.9967	0.9703	0.8756	0.6834	0.4335	0.2122
	11	1.0000	1.0000	0.9992	0.9893	0.9408	0.8110	0.5926	0.3450
	12	1.0000	1.0000	0.9998	0.9966	0.9754	0.9003	0.7369	0.5000
	13	1.0000	1.0000	1.0000	0.9991	0.9911	0.9538	0.8491	0.6550
	14	1.0000	1.0000	1.0000	0.9998	0.9972	0.9814	0.9240	0.7878
	15	1.0000	1.0000	1.0000	1.0000	0.9992	0.9935	0.9667	0.8852
	16	1.0000	1.0000	1.0000	1.0000	0.9998	0.9981	0.9874	0.9462
	17	1.0000	1.0000	1.0000	1.0000	1.0000	0.9995	0.9960	0.9784
	18	1.0000	1.0000	1.0000	1.0000	1.0000	0.9999	0.9989	0.9927
	19	1.0000	1.0000	1.0000	1.0000	1.0000	1.0000	0.9998	0.9980
	20	1.0000	1.0000	1.0000	1.0000	1.0000	1.0000	1.0000	0.9995
	21	1.0000	1.0000	1.0000	1.0000	1.0000	1.0000	1.0000	0.9999
	22	1.0000	1.0000	1.0000	1.0000	1.0000	1.0000	1.0000	1.0000

Table 2 The Cumulative Poisson Distribution

The tabulated quantity is

$$\sum_{j=0}^{k} e^{-\lambda} \frac{\lambda^j}{j!}.$$

			λ			
k	0.001	0.005	0.010	0.015	0.020	0.025
0	0.9990 0050	0.9950 1248	0.9900 4983	0.9851 1194	0.9801 9867	0.9753 099
1	0.9999 9950	0.9999 8754	0.9999 5033	0.9998 8862	0.9998 0264	0.9996 927
2	1.0000 0000	0.9999 9998	0.9999 9983	0.9999 9945	0.9999 9868	0.9999 974
3		1.0000 0000	1.0000 0000	1.0000 0000	0.9999 9999	1.0000 000
4					1.0000 0000	1.0000 000

			λ			
k	0.030	0.035	0.040	0.045	0.050	0.055
0	0.970 446	0.965 605	0.960 789	0.955 997	0.951 229	0.946 485
1	0.999 559	0.999 402	0.999 221	0.999 017	0.998 791	0.998 542
2	0.999 996	0.999 993	0.999 990	0.999 985	0.999 980	0.999 973
3	1.000 000	1.000 000	1.000 000	1.000 000	1.000 000	1.000 000

			λ			
k	0.060	0.065	0.070	0.075	0.080	0.085
0	0.941 765	0.937 067	0.932 394	0.927 743	0.923 116	0.918 512
1	0.998 270	0.997 977	0.997 661	0.997 324	0.996 966	0.996 586
2	0.999 966	0.999 956	0.999 946	0.999 934	0.999 920	0.999 904
3	0.999 999	0.999 999	0.999 999	0.999 999	0.999 998	0.999 998
4	1.000 000	1.000 000	1.000 000	1.000 000	1.000 000	1.000 000

			λ			
k	0.090	0.095	0.100	0.200	0.300	0.400
0	0.913 931	0.909 373	0.904 837	0.818 731	0.740 818	0.670 320
1	0.996 185	0.995 763	0.995 321	0.982 477	0.963 064	0.938 448
2	0.999 886	0.999 867	0.999 845	0.998 852	0.996 401	0.992 074
3	0.999 997	0.999 997	0.999 996	0.999 943	0.999 734	0.999 224
4	1.000 000	1.000 000	1.000 000	0.999 998	0.999 984	0.999 939
5				1.000 000	0.999 999	0.999 996
6					1.000 000	1.000 000

			λ			
k	0.500	0.600	0.700	0.800	0.900	1.000
0	0.606 531	0.548 812	0.496 585	0.449 329	0.406 329	0.367 879
1	0.909 796	0.878 099	0.844 195	0.808 792	0.772 482	0.735 759
2	0.985 612	0.976 885	0.965 858	0.952 577	0.937 143	0.919 699

Table 2 *(continued)*

k	λ					
	0.500	0.600	0.700	0.800	0.900	1.000
3	0.998 248	0.996 642	0.994 247	0.990 920	0.986 541	0.981 012
4	0.999 828	0.999 606	0.999 214	0.998 589	0.997 656	0.996 340
5	0.999 986	0.999 961	0.999 910	0.999 816	0.999 657	0.999 406
6	0.999 999	0.999 997	0.999 991	0.999 979	0.999 957	0.999 917
7	1.000 000	1.000 000	0.999 999	0.999 998	0.999 995	0.999 990
8			1.000 000	1.000 000	1.000 000	0.999 999
9						1.000 000

k	λ							
	1.20	1.40	1.60	1.80	2.00	2.50	3.00	3.50
0	0.3012	0.2466	0.2019	0.1653	0.1353	0.0821	0.0498	0.0302
1	0.6626	0.5918	0.5249	0.4628	0.4060	0.2873	0.1991	0.1359
2	0.8795	0.8335	0.7834	0.7306	0.6767	0.5438	0.4232	0.3208
3	0.9662	0.9463	0.9212	0.8913	0.8571	0.7576	0.6472	0.5366
4	0.9923	0.9857	0.9763	0.9636	0.9473	0.8912	0.8153	0.7254
5	0.9985	0.9968	0.9940	0.9896	0.9834	0.9580	0.9161	0.8576
6	0.9997	0.9994	0.9987	0.9974	0.9955	0.9858	0.9665	0.9347
7	1.0000	0.9999	0.9997	0.9994	0.9989	0.9958	0.9881	0.9733
8		1.0000	1.0000	0.9999	0.9998	0.9989	0.9962	0.9901
9				1.0000	1.0000	0.9997	0.9989	0.9967
10						0.9999	0.9997	0.9990
11						1.0000	0.9999	0.9997
12							1.0000	0.9999
13								1.0000

k	λ							
	4.00	4.50	5.00	6.00	7.00	8.00	9.00	10.00
0	0.0183	0.0111	0.0067	0.0025	0.0009	0.0003	0.0001	0.0000
1	0.0916	0.0611	0.0404	0.0174	0.0073	0.0030	0.0012	0.0005
2	0.2381	0.1736	0.1247	0.0620	0.0296	0.0138	0.0062	0.0028
3	0.4335	0.3423	0.2650	0.1512	0.0818	0.0424	0.0212	0.0103
4	0.6288	0.5321	0.4405	0.2851	0.1730	0.0996	0.0550	0.0293
5	0.7851	0.7029	0.6160	0.4457	0.3007	0.1912	0.1157	0.0671
6	0.8893	0.8311	0.7622	0.6063	0.4497	0.3134	0.2068	0.1301
7	0.9489	0.9134	0.8666	0.7440	0.5987	0.4530	0.3239	0.2202
8	0.9786	0.9597	0.9319	0.8472	0.7291	0.5925	0.4577	0.3328
9	0.9919	0.9829	0.9682	0.9161	0.8305	0.7166	0.5874	0.4579
10	0.9972	0.9933	0.9863	0.9574	0.9015	0.8159	0.7060	0.5830
11	0.9991	0.9976	0.9945	0.9799	0.9467	0.8881	0.8030	0.6968
12	0.9997	0.9992	0.9980	0.9912	0.9730	0.9362	0.8758	0.7916
13	0.9999	0.9997	0.9993	0.9964	0.9872	0.9658	0.9261	0.8645
14	1.0000	0.9999	0.9998	0.9986	0.9943	0.9827	0.9585	0.9165

Table 2 (continued)

k	4.00	4.50	5.00	λ 6.00	7.00	8.00	9.00	10.00
15		1.0000	0.9999	0.9995	0.9976	0.9918	0.9780	0.9513
16			1.0000	0.9998	0.9990	0.9963	0.9889	0.9730
17				0.9999	0.9996	0.9984	0.9947	0.9857
18				1.0000	0.9999	0.9993	0.9976	0.9928
19						0.9997	0.9989	0.9965
20					1.0000	0.9999	0.9996	0.9984
21						1.0000	0.9998	0.9993
22							0.9999	0.9997
23							1.0000	0.9999
24								1.0000

Table 3 The Normal Distribution

The tabulated quantity is

$$\Phi(x) = \frac{1}{\sqrt{2\pi}} \int_{-\infty}^{x} e^{-t^2/2} dt.$$

$$[\Phi(-x) = 1 - \Phi(x)].$$

x	$\Phi(x)$	x	$\Phi(x)$	x	$\Phi(x)$	x	$\Phi(x)$
0.00	0.500000	0.35	0.636831	0.70	0.758036	1.05	0.853141
0.01	0.503989	0.36	0.640576	0.71	0.761148	1.06	0.855428
0.02	0.507978	0.37	0.644309	0.72	0.764238	1.07	0.857690
0.03	0.511966	0.38	0.648027	0.73	0.767305	1.08	0.859929
0.04	0.515953	0.39	0.651732	0.74	0.770350	1.09	0.862143
0.05	0.519939	0.40	0.655422	0.75	0.773373	1.10	0.864334
0.06	0.523922	0.41	0.659097	0.76	0.776373	1.11	0.866500
0.07	0.527903	0.42	0.662757	0.77	0.779350	1.12	0.868643
0.08	0.531881	0.43	0.666402	0.78	0.782305	1.13	0.870762
0.09	0.535856	0.44	0.670031	0.79	0.785236	1.14	0.872857
0.10	0.539828	0.45	0.673645	0.80	0.788145	1.15	0.874928
0.11	0.543795	0.46	0.677242	0.81	0.791030	1.16	0.876976
0.12	0.547758	0.47	0.680822	0.82	0.793892	1.17	0.879000
0.13	0.551717	0.48	0.684386	0.83	0.796731	1.18	0.881000
0.14	0.555670	0.49	0.687933	0.84	0.799546	1.19	0.882977
0.15	0.559618	0.50	0.691462	0.85	0.802337	1.20	0.884930
0.16	0.563559	0.51	0.694974	0.86	0.805105	1.21	0.886861
0.17	0.567495	0.52	0.698468	0.87	0.807850	1.22	0.888768
0.18	0.571424	0.53	0.701944	0.88	0.810570	1.23	0.890651
0.19	0.575345	0.54	0.705401	0.89	0.813267	1.24	0.892512
0.20	0.579260	0.55	0.708840	0.90	0.815940	1.25	0.894350
0.21	0.583166	0.56	0.712260	0.91	0.818589	1.26	0.896165
0.22	0.587064	0.57	0.715661	0.92	0.821214	1.27	0.897958
0.23	0.590954	0.58	0.719043	0.93	0.823814	1.28	0.899727
0.24	0.594835	0.59	0.722405	0.94	0.826391	1.29	0.901475
0.25	0.598706	0.60	0.725747	0.95	0.828944	1.30	0.903200
0.26	0.602568	0.61	0.279069	0.96	0.831472	1.31	0.904902
0.27	0.606420	0.62	0.732371	0.97	0.833977	1.32	0.906582
0.28	0.610261	0.63	0.735653	0.98	0.836457	1.33	0.908241
0.29	0.614092	0.64	0.738914	0.99	0.838913	1.34	0.909877
0.30	0.617911	0.65	0.742154	1.00	0.841345	1.35	0.911492
0.31	0.621720	0.66	0.745373	1.01	0.843752	1.36	0.913085
0.32	0.625516	0.67	0.748571	1.02	0.846136	1.37	0.914657
0.33	0.629300	0.68	0.751748	1.03	0.848495	1.38	0.916207
0.34	0.633072	0.69	0.754903	1.04	0.850830	1.39	0.917736

Table 3 (continued)

x	$\Phi(x)$	x	$\Phi(x)$	x	$\Phi(x)$	x	$\Phi(x)$
1.40	0.919243	1.85	0.967843	2.30	0.989276	2.75	0.997020
1.41	0.920730	1.86	0.968557	2.31	0.989556	2.76	0.997110
1.42	0.922196	1.87	0.969258	2.32	0.989830	2.77	0.997197
1.43	0.923641	1.88	0.969946	2.33	0.990097	2.78	0.997282
1.44	0.925066	1.89	0.970621	2.34	0.990358	2.79	0.997365
1.45	0.926471	1.90	0.971283	2.35	0.990613	2.80	0.997445
1.46	0.927855	1.91	0.971933	2.36	0.990863	2.81	0.997523
1.47	0.929219	1.92	0.972571	2.37	0.991106	2.82	0.997599
1.48	0.930563	1.93	0.973197	2.38	0.991344	2.83	0.997673
1.49	0.931888	1.94	0.973810	2.39	0.991576	2.84	0.997744
1.50	0.933193	1.95	0.974412	2.40	0.991802	2.85	0.997814
1.51	0.934478	1.96	0.975002	2.41	0.992024	2.86	0.997882
1.52	0.935745	1.97	0.975581	2.42	0.992240	2.87	0.997948
1.53	0.936992	1.98	0.976148	2.43	0.992451	2.88	0.998012
1.54	0.938220	1.99	0.976705	2.44	0.992656	2.89	0.998074
1.55	0.939429	2.00	0.977250	2.45	0.992857	2.90	0.998134
1.56	0.940620	2.01	0.977784	2.46	0.993053	2.91	0.998193
1.57	0.941792	2.02	0.978308	2.47	0.993244	2.92	0.998250
1.58	0.942947	2.03	0.978822	2.48	0.993431	2.93	0.998305
1.59	0.944083	2.04	0.979325	2.49	0.993613	2.54	0.998359
1.60	0.945201	2.05	0.979818	2.50	0.993790	2.95	0.998411
1.61	0.946301	2.06	0.980301	2.51	0.993963	2.96	0.998462
1.62	0.947384	2.07	0.980774	2.52	0.994132	2.97	0.998511
1.63	0.948449	2.08	0.981237	2.53	0.994297	2.98	0.998559
1.64	0.949497	2.09	0.981691	2.54	0.994457	2.99	0.998605
1.65	0.950529	2.10	0.982136	2.55	0.994614	3.00	0.998650
1.66	0.951543	2.11	0.982571	2.56	0.994766	3.01	0.998694
1.67	0.952540	2.12	0.982997	2.57	0.994915	3.02	0.998736
1.68	0.953521	2.13	0.983414	2.58	0.995060	3.03	0.998777
1.69	0.954486	2.14	0.983823	2.59	0.995201	3.04	0.998817
1.70	0.955435	2.15	0.984222	2.60	0.995339	3.05	0.998856
1.71	0.956367	2.16	0.984614	2.61	0.995473	3.06	0.998893
1.72	0.957284	2.17	0.984997	2.62	0.995604	3.07	0.998930
1.73	0.958185	2.18	0.985371	2.63	0.995731	3.08	0.998965
1.74	0.959070	2.19	0.985738	2.64	0.995855	3.09	0.998999
1.75	0.959941	2.20	0.986097	2.65	0.995975	3.10	0.999032
1.76	0.960796	2.21	0.986447	2.66	0.996093	3.11	0.999065
1.77	0.961636	2.22	0.986791	2.67	0.996207	3.12	0.999096
1.78	0.962462	2.23	0.987126	2.68	0.996319	3.13	0.999126
1.79	0.963273	2.24	0.987455	2.69	0.996427	3.14	0.999155
1.80	0.964070	2.25	0.987776	2.70	0.996533	3.15	0.999184
1.81	0.964852	2.26	0.988089	2.71	0.996636	3.16	0.999211
1.82	0.965620	2.27	0.988396	2.72	0.996736	3.17	0.999238
1.83	0.966375	2.28	0.988696	2.73	0.996833	3.18	0.999264
1.84	0.967116	2.29	0.988989	2.74	0.996928	3.19	0.999289

Table 3 *(continued)*

x	$\Phi(x)$	x	$\Phi(x)$	x	$\Phi(x)$	x	$\Phi(x)$
3.20	0.999313	3.40	0.999663	3.60	0.999841	3.80	0.999928
3.21	0.999336	3.41	0.999675	3.61	0.999847	3.81	0.999931
3.22	0.999359	3.42	0.999687	3.62	0.999853	3.82	0.999933
3.23	0.999381	3.43	0.999698	3.63	0.999858	3.83	0.999936
3.24	0.999402	3.44	0.999709	3.64	0.999864	3.84	0.999938
3.25	0.999423	3.45	0.999720	3.65	0.999869	3.85	0.999941
3.26	0.999443	3.46	0.999730	3.66	0.999874	3.86	0.999943
3.27	0.999462	3.47	0.999740	3.67	0.999879	3.87	0.999946
3.28	0.999481	3.48	0.999749	3.68	0.999883	3.88	0.999948
3.29	0.999499	3.49	0.999758	3.69	0.999888	3.89	0.999950
3.30	0.999517	3.50	0.999767	3.70	0.999892	3.90	0.999952
3.31	0.999534	3.51	0.999776	3.71	0.999896	3.91	0.999954
3.32	0.999550	3.52	0.999784	3.72	0.999900	3.92	0.999956
3.33	0.999566	3.53	0.999792	3.73	0.999904	3.93	0.999958
3.34	0.999581	3.53	0.999800	3.74	0.999908	3.94	0.999959
3.35	0.999596	3.55	0.999807	3.75	0.999912	3.95	0.999961
3.36	0.999610	3.56	0.999815	3.76	0.999915	3.96	0.999963
3.37	0.999624	3.57	0.999822	3.77	0.999918	3.97	0.999964
3.38	0.999638	3.58	0.999828	3.78	0.999922	3.98	0.999966
3.39	0.999651	3.59	0.999835	3.79	0.999925	3.99	0.999967

Table 4 Critical Values for Student's *t*-Distribution

Let t_r be a random variable having the Student's *t*-distribution with r degrees of freedom. Then the tabulated quantities are the numbers x for which

$$P(t_r \leq x) = \gamma.$$

			γ			
r	0.75	0.90	0.95	0.975	0.99	0.995
1	1.0000	3.0777	6.3138	12.7062	31.8207	63.6574
2	0.8165	1.8856	2.9200	4.3027	6.9646	9.9248
3	0.7649	1.6377	2.3534	3.1824	4.5407	5.8409
4	0.7407	1.5332	2.1318	2.7764	3.7649	4.6041
5	0.7267	1.4759	2.0150	2.5706	3.3649	4.0322
6	0.7176	1.4398	1.9432	2.4469	3.1427	3.7074
7	0.7111	1.4149	1.8946	2.3646	2.9980	3.4995
8	0.7064	1.3968	1.8595	3.3060	2.8965	3.3554
9	0.7027	1.3830	1.8331	2.2622	2.8214	3.2498
10	0.6998	1.3722	1.8125	2.2281	2.7638	3.1693
11	0.6974	1.3634	1.7959	2.2010	2.7181	3.1058
12	0.6955	1.3562	1.7823	2.1788	2.6810	3.0545
13	0.6938	1.3502	1.7709	1.1604	2.6503	3.0123
14	0.6924	1.3450	1.7613	2.1448	2.6245	2.9768
15	0.6912	1.3406	1.7531	2.1315	2.6025	2.9467
16	0.6901	1.3368	1.7459	2.1199	2.5835	2.9208
17	0.6892	1.3334	1.7396	2.1098	2.5669	2.8982
18	0.6884	1.3304	1.7341	2.1009	2.5524	2.8784
19	0.6876	1.3277	1.7291	2.0930	2.5395	2.8609
20	0.6870	1.3253	1.7247	2.0860	2.5280	2.8453
21	0.6864	1.3232	1.7207	2.0796	2.5177	2.8314
22	0.6858	1.3212	1.7171	2.0739	2.5083	2.8188
23	0.6853	1.3195	1.7139	2.0687	2.4999	2.8073
24	0.6848	1.3178	1.7109	2.0639	2.4922	2.7969
25	0.6844	1.3163	1.7081	2.0595	2.4851	2.7874
26	0.6840	1.3150	1.7056	2.0555	2.4786	2.7787
27	0.6837	1.3137	1.7033	2.0518	2.4727	2.7707
28	0.6834	1.3125	1.7011	2.0484	2.4671	2.7633
29	0.6830	1.3114	1.6991	2.0452	2.4620	2.7564
30	0.6828	1.3104	1.6973	2.0423	2.4573	2.7500
31	0.6825	1.3095	1.6955	2.0395	2.4528	2.7440
32	0.6822	1.3086	1.6939	2.0369	2.4487	2.7385
33	0.6820	1.3077	1.6924	2.0345	2.4448	2.7333
34	0.6818	1.3070	1.6909	2.0322	2.4411	2.7284
35	0.6816	1.3062	1.6896	2.0301	2.4377	2.7238
36	0.6814	1.3055	1.6883	2.0281	2.4345	2.7195
37	0.6812	1.3049	1.6871	2.0262	2.4314	1.7154
38	0.6810	1.3042	1.6860	2.0244	2.4286	2.7116

Table 4 *(continued)*

r	0.75	0.90	γ 0.95	0.975	0.99	0.995
39	0.6808	1.3036	1.6849	2.0227	2.4258	2.7079
40	0.6807	1.3031	1.6839	2.0211	2.4233	2.7045
41	0.6805	1.3025	1.6829	2.0195	2.4208	2.7012
42	0.6804	1.3020	1.6820	2.0181	2.4185	2.6981
43	0.6802	1.3016	1.6811	2.0167	2.4163	2.6951
44	0.6801	1.3011	1.6802	2.0154	2.4141	2.6923
45	0.6800	1.3006	1.6794	2.0141	2.4121	2.6896
46	0.6799	1.3002	1.6787	2.0129	2.4102	2.6870
47	0.6797	1.2998	1.6779	2.0117	2.4083	2.6846
48	0.6796	1.2994	1.6772	2.0106	2.4066	2.6822
49	0.6795	1.2991	1.6766	2.0096	2.4069	2.6800
50	0.6794	1.2987	1.6759	2.0086	2.4033	2.6778
51	0.6793	1.2984	1.6753	2.0076	2.4017	2.6757
52	0.6792	1.2980	1.6747	2.0066	2.4002	2.6737
53	0.6791	1.2977	1.6741	2.0057	2.3988	2.6718
54	0.6791	1.2974	1.6736	2.0049	2.3974	2.6700
55	0.6790	1.2971	1.6730	2.0040	2.3961	2.6682
56	0.6789	1.2969	1.6725	2.0032	2.3948	2.6665
57	0.6788	1.2966	1.6720	2.0025	2.3936	2.6649
58	0.6787	1.2963	1.6716	2.0017	2.3924	2.6633
59	0.6787	1.2961	1.6711	2.0010	2.3912	2.6618
60	0.6786	1.2958	1.6706	2.0003	2.3901	2.6603
61	0.6785	1.2956	1.6702	1.9996	2.3890	2.6589
62	0.6785	1.2954	1.6698	1.9990	2.3880	2.6575
63	0.6784	1.2951	1.6694	1.9983	2.3870	2.6561
64	0.6783	1.2949	1.6690	1.9977	2.3860	2.6549
65	0.6783	1.2947	1.6686	1.9971	2.3851	2.6536
66	0.6782	1.2945	1.6683	1.9966	2.3842	2.6524
67	0.6782	1.2943	1.6679	1.9960	2.3833	2.6512
68	0.6781	1.2941	1.6676	1.9955	2.3824	2.6501
69	0.6781	1.2939	1.6672	1.9949	2.3816	2.6490
70	0.6780	1.2938	1.6669	1.9944	2.3808	2.6479
71	0.6780	1.2936	1.6666	1.9939	2.3800	2.6469
72	0.6779	1.2934	1.6663	1.9935	2.3793	2.6459
73	0.6779	1.2933	1.6660	1.9930	2.3785	2.6449
74	0.6778	1.2931	1.6657	1.9925	2.3778	2.6439
75	0.6778	1.2929	1.6654	1.9921	2.3771	2.6430
76	0.6777	1.2928	1.6652	1.9917	2.3764	2.6421
77	0.6777	1.2926	1.6649	1.9913	2.3758	2.6412
78	0.6776	1.2925	1.6646	1.9908	2.3751	2.6403
79	0.6776	1.2924	1.6644	1.9905	2.3745	2.6395
80	0.6776	1.2922	1.6641	1.9901	2.3739	2.6387
81	0.6775	1.2921	1.6639	1.9897	2.3733	2.6379

Table 4 (continued)

r	0.75	0.90	γ 0.95	0.975	0.99	0.995
82	0.6775	1.2920	1.6636	1.9893	2.3727	2.6371
83	0.6775	1.2918	1.6634	1.9890	2.3721	2.6364
84	0.6774	1.2917	1.6632	1.9886	2.3716	2.6356
85	0.6774	1.2916	1.6630	1.9883	2.3710	2.6349
86	0.6774	1.2915	1.6628	1.9879	2.3705	2.6342
87	0.6773	1.2914	1.6626	1.9876	2.3700	2.6335
88	0.6773	1.2912	1.6624	1.9873	2.3695	2.6329
89	0.6773	1.2911	1.6622	1.9870	2.3690	2.6322
90	0.6772	1.2910	1.6620	1.9867	2.3685	2.6316

Table 5 Critical Values for the Chi-Square Distribution

Let χ_r^2 be a random variable having the chi-square distribution with r degrees of freedom. Then the tabulated quantities are the numbers x for which

$$P(\chi_r^2 \leq x) = \gamma.$$

r	0.005	0.01	γ 0.025	0.05	0.10	0.25
1	—	—	0.001	0.004	0.016	0.102
2	0.010	0.020	0.051	0.103	0.211	0.575
3	0.072	0.115	0.216	0.352	0.584	1.213
4	0.207	0.297	0.484	0.711	1.064	1.923
5	0.412	0.554	0.831	1.145	1.610	2.675
6	0.676	0.872	1.237	1.635	2.204	3.455
7	0.989	1.239	1.690	2.167	2.833	4.255
8	1.344	1.646	2.180	2.733	3.490	5.071
9	1.735	2.088	2.700	2.325	4.168	5.899
10	2.156	2.558	3.247	3.940	4.865	6.737
11	2.603	3.053	3.816	4.575	5.578	7.584
12	3.074	3.571	4.404	5.226	6.304	9.438
13	3.565	4.107	5.009	5.892	7.042	9.299
14	4.075	4.660	5.629	6.571	7.790	10.165
15	4.601	5.229	6.262	7.261	8.547	11.037
16	5.142	5.812	6.908	7.962	9.312	11.912
17	5.697	6.408	7.564	8.672	10.085	12.792
18	6.265	7.015	8.231	8.390	10.865	13.675
19	6.844	7.633	8.907	10.117	11.651	14.562
20	7.434	8.260	9.591	10.851	12.443	15.452
21	8.034	8.897	10.283	11.591	13.240	16.344
22	8.643	9.542	10.982	12.338	14.042	17.240
23	9.260	10.196	11.689	13.091	14.848	18.137
24	9.886	10.856	12.401	13.848	15.659	19.037
25	10.520	11.524	13.120	14.611	16.473	19.939
26	11.160	12.198	13.844	13.379	17.292	20.843
27	11.808	12.879	14.573	16.151	18.114	21.749
28	12.461	13.565	15.308	16.928	18.939	22.657
29	13.121	14.257	16.047	17.708	19.768	23.567
30	13.787	14.954	16.791	18.493	20.599	24.478
31	14.458	15.655	17.539	19.281	21.434	25.390
32	15.134	16.362	18.291	20.072	22.271	26.304
33	15.815	17.074	19.047	20.867	23.110	27.219
34	16.501	17.789	19.806	21.664	23.952	28.136
35	17.192	18.509	20.569	22.465	24.797	29.054
36	17.887	19.233	21.336	23.269	25.643	29.973
37	18.586	19.960	22.106	24.075	26.492	30.893
38	19.289	20.691	22.878	24.884	27.343	31.815

Table 5 (continued)

r	γ 0.005	0.01	0.025	0.05	0.10	0.25
39	19.996	21.426	23.654	25.695	28.196	32.737
40	20.707	22.164	24.433	26.509	29.051	33.660
41	21.421	22.906	25.215	27.326	29.907	34.585
42	22.138	23.650	25.999	28.144	30.765	35.510
43	22.859	24.398	26.785	28.965	31.625	36.436
44	23.584	25.148	27.575	29.787	32.487	37.363
45	24.311	25.901	28.366	30.612	33.350	38.291

r	γ 0.75	0.90	0.95	0.975	0.99	0.995
1	1.323	2.706	3.841	5.024	6.635	7.879
2	2.773	4.605	5.991	7.378	9.210	10.597
3	4.108	6.251	7.815	9.348	11.345	12.838
4	5.385	7.779	9.488	11.143	13.277	14.860
5	6.626	9.236	11.071	12.833	15.086	16.750
6	7.841	10.645	12.592	14.449	16.812	18.548
7	9.037	12.017	14.067	16.013	18.475	20.278
8	10.219	13.362	15.507	17.535	20.090	21.955
9	11.389	14.684	16.919	19.023	21.666	23.589
10	12.549	15.987	18.307	20.483	23.209	25.188
11	13.701	17.275	19.675	21.920	24.725	26.757
12	14.845	18.549	21.026	23.337	26.217	28.299
13	15.984	19.812	23.362	24.736	27.688	29.819
14	17.117	21.064	23.685	26.119	29.141	31.319
15	18.245	22.307	24.996	27.488	30.578	32.801
16	19.369	23.542	26.296	28.845	32.000	34.267
17	20.489	24.769	27.587	30.191	33.409	35.718
18	21.605	25.989	28.869	31.526	34.805	37.156
19	22.718	27.204	30.144	32.852	36.191	38.582
20	23.828	28.412	31.410	34.170	37.566	39.997
21	24.935	29.615	32.671	35.479	38.932	41.401
22	26.039	30.813	33.924	36.781	40.289	42.796
23	27.141	32.007	35.172	38.076	41.638	44.181
24	28.241	33.196	36.415	39.364	42.980	45.559
25	29.339	34.382	37.652	40.646	44.314	46.928
26	30.435	35.563	38.885	41.923	45.642	48.290
27	31.528	36.741	40.113	43.194	46.963	49.645
28	32.620	37.916	41.337	44.641	48.278	50.993
29	33.711	39.087	42.557	45.722	49.588	52.336
30	34.800	40.256	43.773	46.979	50.892	53.672
31	35.887	41.422	44.985	48.232	51.191	55.003
32	36.973	42.585	46.194	49.480	53.486	56.328

Table 5 *(continued)*

| | | | γ | | | |
r	0.75	0.90	0.95	0.975	0.99	0.995
33	38.058	43.745	47.400	50.725	54.776	57.648
34	39.141	44.903	48.602	51.966	56.061	58.964
35	40.223	46.059	49.802	53.203	57.342	60.275
36	41.304	47.212	50.998	54.437	58.619	61.581
37	42.383	48.363	52.192	55.668	59.892	62.883
38	43.462	49.513	53.384	56.896	61.162	64.181
39	44.539	50.660	54.572	58.120	62.428	65.476
40	45.616	51.805	55.758	59.342	63.691	66.766
41	46.692	52.949	56.942	60.561	64.950	68.053
42	47.766	54.090	58.124	61.777	66.206	69.336
43	48.840	55.230	59.304	62.990	67.459	70.616
44	49.913	56.369	60.481	64.201	68.710	71.893
45	50.985	57.505	61.656	65.410	69.957	73.166

Table 6 Critical Values for the *F*-Distribution

Let F_{r_1,r_2} be a random variable having the *F*-distribution with r_1, r_2 degrees of freedom. Then the tabulated quantities are the numbers x for which

$$P(F_{r_1,r_2} \leq x) = \gamma.$$

		\multicolumn{6}{c}{r_1}	
	γ	1	2	3	4	5	6	γ
1	0.500	1.0000	1.5000	1.7092	1.8227	1.8937	1.9422	0.500
	0.750	5.8285	7.5000	8.1999	8.5810	8.8198	8.9833	0.750
	0.900	39.864	49.500	53.593	55.833	57.241	58.204	0.900
	0.950	161.45	199.50	215.71	224.58	230.16	233.99	0.950
	0.975	647.79	799.50	864.16	899.58	921.85	937.11	0.975
	0.990	4052.2	4999.5	5403.3	5624.6	5763.7	5859.0	0.990
	0.995	16211	20000	21615	22500	23056	23437	0.995
2	0.500	0.66667	1.0000	1.1349	1.2071	1.2519	1.2824	0.500
	0.750	2.5714	3.0000	3.1534	3.2320	3.2799	3.3121	0.750
	0.900	8.5623	9.0000	9.1618	9.2434	9.2926	9.3255	0.900
	0.950	18.513	19.000	19.164	19.247	19.296	19.330	0.950
	0.975	38.506	39.000	39.165	39.248	39.298	39.331	0.975
	0.990	98.503	99.000	99.166	99.249	99.299	99.332	0.990
	0.995	198.50	199.00	199.17	199.25	199.30	199.33	0.995
3	0.500	0.58506	0.88110	1.0000	1.0632	1.1024	1.1289	0.500
	0.750	2.0239	2.2798	2.3555	2.3901	2.4095	2.4218	0.750
	0.900	5.5383	5.4624	5.3908	5.3427	5.3092	5.2847	0.900
	0.950	10.128	9.5521	9.2766	9.1172	9.0135	8.9406	0.950
	0.975	17.443	16.044	15.439	15.101	14.885	14.735	0.975
	0.990	34.116	30.817	29.457	28.710	28.237	27.911	0.990
	0.995	55.552	49.799	47.467	46.195	45.392	44.838	0.995
4	0.500	0.54863	0.82843	0.94054	1.0000	1.0367	1.0617	0.500
	0.750	1.8074	2.0000	2.0467	2.0642	2.0723	2.0766	0.750
	0.900	4.5448	4.3246	4.1908	4.1073	4.0506	4.0098	0.900
	0.950	7.7086	6.9443	6.5914	6.3883	6.2560	6.1631	0.950
	0.975	12.218	10.649	9.9792	9.6045	9.3645	9.1973	0.975
	0.990	21.198	18.000	16.694	15.977	15.522	15.207	0.990
	0.995	31.333	26.284	24.259	23.155	22.456	21.975	0.995
5	0.500	0.52807	0.79877	0.90715	0.96456	1.0000	1.0240	0.500
	0.750	1.6925	1.8528	1.8843	1.8927	1.8947	1.8945	0.750
	0.900	4.0604	3.7797	3.6195	3.5202	3.4530	3.4045	0.900
	0.950	6.6079	5.7861	5.4095	5.1922	5.0503	4.9503	0.950
	0.975	10.007	8.4336	7.7636	7.3879	7.1464	6.9777	0.975
	0.990	16.258	13.274	12.060	11.392	10.967	10.672	0.990
	0.995	22.785	18.314	16.530	15.556	14.940	14.513	0.995
6	0.500	0.51489	0.77976	0.88578	0.94191	0.97654	1.0000	0.500
	0.750	1.6214	1.7622	1.7844	1.7872	1.7852	1.7821	0.750
	0.900	3.7760	3.4633	3.2888	3.1808	3.1075	3.0546	0.900
	0.950	5.9874	5.1433	4.7571	4.5337	4.3874	4.2839	0.950
	0.975	8.8131	7.2598	6.5988	6.2272	5.9876	5.8197	0.975
	0.990	13.745	10.925	9.7795	9.1483	8.7459	8.4661	0.990
	0.995	18.635	14.544	12.917	12.028	11.464	11.073	0.995

Table 6 (continued)

		γ	7	8	9	10	11	12	γ	
					r_1					
		0.500	1.9774	2.0041	2.0250	2.0419	2.0558	2.0674	0.500	
		0.750	9.1021	9.1922	9.2631	9.3202	9.3672	9.4064	0.750	
		0.900	58.906	59.439	59.858	60.195	60.473	60.705	0.900	
	1	0.950	236.77	238.88	240.54	241.88	242.99	243.91	0.950	1
		0.975	948.22	956.66	963.28	968.63	973.04	976.71	0.975	
		0.990	5928.3	5981.1	6022.5	6055.8	6083.3	6106.3	0.990	
		0.995	23715	23925	24091	24224	24334	24426	0.995	
		0.500	1.3045	1.3213	1.3344	1.3450	1.3537	1.3610	0.500	
		0.750	3.3352	3.3526	3.3661	3.3770	3.3859	3.3934	0.750	
		0.900	9.3491	9.3668	9.3805	9.3916	9.4006	9.4081	0.900	
	2	0.950	19.353	19.371	19.385	19.396	19.405	19.413	0.950	2
		0.975	39.355	39.373	39.387	39.398	39.407	39.415	0.975	
		0.990	99.356	99.374	99.388	99.399	99.408	99.416	0.990	
		0.995	199.36	199.37	199.39	199.40	199.41	199.42	0.995	
		0.500	1.1482	1.1627	1.1741	1.1833	1.1909	1.1972	0.500	
		0.750	2.4302	2.4364	2.4410	2.4447	2.4476	2.4500	0.750	
		0.900	5.2662	5.2517	5.2400	5.2304	5.2223	5.2156	0.900	
	3	0.950	8.8868	8.8452	8.8123	8.7855	8.7632	8.7446	0.950	3
		0.975	14.624	14.540	14.473	14.419	14.374	14.337	0.975	
		0.990	27.672	27.489	27.345	27.229	27.132	27.052	0.990	
r_2		0.995	44.434	44.126	43.882	43.686	43.523	43.387	0.995	r_2
		0.500	1.0797	1.0933	1.1040	1.1126	1.1196	1.1255	0.500	
		0.750	2.0790	2.0805	2.0814	2.0820	2.0823	2.0826	0.750	
		0.900	3.9790	3.9549	3.9357	3.9199	3.9066	3.8955	0.900	
	4	0.950	6.0942	6.0410	5.9988	5.9644	5.9357	5.9117	0.950	4
		0.975	9.0741	8.9796	8.9047	8.8439	8.7933	8.7512	0.975	
		0.990	14.976	14.799	14.659	14.546	14.452	14.374	0.990	
		0.995	21.622	21.352	21.139	20.967	20.824	20.705	0.995	
		0.500	1.0414	1.0545	1.0648	1.0730	1.0798	1.0855	0.500	
		0.750	1.8935	1.8923	1.8911	1.8899	1.8887	1.8877	0.750	
		0.900	3.3679	3.3393	3.3163	3.2974	3.2815	3.2682	0.900	
	5	0.950	4.8759	4.8183	4.7725	4.7351	4.7038	4.6777	0.950	5
		0.975	6.8531	6.7572	6.6810	6.6192	6.5676	6.5246	0.975	
		0.990	10.456	10.289	10.158	10.051	9.9623	9.8883	0.990	
		0.995	14.200	13.961	13.772	13.618	13.490	13.384	0.995	
		0.500	1.0169	1.0298	1.0398	1.0478	1.0545	1.0600	0.500	
		0.750	1.7789	1.7760	1.7733	1.7708	1.7686	1.7668	0.750	
		0.900	3.0145	2.9830	2.9577	2.9369	2.9193	2.9047	0.900	
	6	0.950	4.2066	4.1468	4.0990	4.0600	4.0272	3.9999	0.950	6
		0.975	5.6955	5.5996	5.5234	5.4613	5.4094	5.3662	0.975	
		0.990	8.2600	8.1016	7.9761	7.8741	7.7891	7.7183	0.990	
		0.995	10.786	10.566	10.391	10.250	10.132	10.034	0.995	

Table 6 *(continued)*

			r_1							
		γ	13	14	15	18	20	24	γ	
		0.500	2.0773	2.0858	2.0931	2.1104	2.1190	2.1321	0.500	
		0.750	9.4399	9.4685	9.4934	9.5520	9.5813	9.6255	0.750	
		0.900	60.903	61.073	61.220	61.567	61.740	62.002	0.900	
	1	0.950	244.69	245.37	245.95	247.32	248.01	249.05	0.950	1
		0.975	979.85	982.54	984.87	990.36	993.10	997.25	0.975	
		0.990	6125.9	6142.7	6157.3	6191.6	6208.7	6234.6	0.990	
		0.995	24504	24572	24630	24767	24836	24940	0.995	
		0.500	1.3672	1.3725	1.3771	1.3879	1.3933	1.4014	0.500	
		0.750	3.3997	3.4051	3.4098	3.4208	3.4263	3.4345	0.750	
		0.900	9.4145	9.4200	9.4247	9.4358	9.4413	9.4496	0.900	
	2	0.950	19.419	19.424	19.429	19.440	19.446	19.454	0.950	2
		0.975	39.421	39.426	39.431	39.442	39.448	39.456	0.975	
		0.990	99.422	99.427	99.432	99.443	99.449	99.458	0.990	
		0.995	199.42	199.43	199.43	199.44	199.45	199.46	0.995	
		0.500	1.2025	1.2071	1.2111	1.2205	1.2252	1.2322	0.500	
		0.750	2.4520	2.4537	2.4552	2.4585	2.4602	2.4626	0.750	
		0.900	5.2097	5.2047	5.2003	5.1898	5.1845	5.1764	0.900	
	3	0.950	8.7286	8.7148	8.7029	8.6744	8.6602	8.6385	0.940	3
		0.975	14.305	14.277	14.253	14.196	14.167	14.124	0.975	
		0.990	26.983	26.923	26.872	26.751	26.690	26.598	0.990	
r_2		0.995	43.271	43.171	43.085	42.880	42.778	42.622	0.955	r_2
		0.500	1.1305	1.1349	1.1386	1.1473	1.1517	1.1583	0.500	
		0.750	2.0827	2.0828	2.0829	2.0828	2.0828	2.0827	0.750	
		0.900	3.8853	3.8765	3.8689	3.8525	3.8443	3.8310	0.900	
	4	0.950	5.8910	5.8732	5.8578	5.8209	5.8025	5.7744	0.950	4
		0.975	8.7148	8.6836	8.6565	8.5921	8.5599	8.5109	0.975	
		0.990	14.306	14.248	14.198	14.079	14.020	13.929	0.990	
		0.995	20.602	20.514	20.438	20.257	20.167	20.030	0.995	
		0.500	1.0903	1.0944	1.0980	1.1064	1.1106	1.1170	0.500	
		0.750	1.8867	1.8858	1.8851	1.8830	1.8820	1.8802	0.750	
		0.900	3.2566	3.2466	3.2380	3.2171	3.2067	3.1905	0.900	
	5	0.950	4.6550	4.6356	4.6188	4.5783	4.5581	4.5272	0.950	5
		0.975	6.4873	6.4554	6.4277	6.3616	6.3285	6.2780	0.975	
		0.990	9.8244	9.7697	9.7222	9.6092	9.5527	9.4665	0.990	
		0.995	13.292	13.214	13.146	12.984	12.903	12.780	0.995	
		0.500	1.0647	1.0687	1.0722	1.0804	1.0845	1.0907	0.500	
		0.750	1.7650	1.7634	1.7621	1.7586	1.7569	1.7540	0.750	
		0.900	2.8918	2.8808	2.8712	2.8479	2.8363	2.8183	0.900	
	6	0.950	3.9761	3.9558	3.9381	3.8955	3.8742	3.8415	0.950	6
		0.975	5.3287	5.2966	5.2687	5.2018	5.1684	5.1172	0.975	
		0.990	7.6570	7.6045	7.5590	7.4502	7.3958	7.3127	0.990	
		0.995	9.9494	9.8769	9.8140	9.6639	9.5888	9.4741	0.995	

Table 6 (continued)

	γ	30	40	48	r_1 60	120	∞	γ	
	0.500	2.1452	2.1584	2.1650	2.1716	2.1848	2.1981	0.500	
	0.750	9.6698	9.7144	9.7368	9.7591	9.8041	9.8492	0.750	
	0.900	62.265	62.529	62.662	62.794	63.061	63.328	0.990	
1	0.950	250.09	251.14	251.67	252.20	253.25	254.32	0.950	1
	0.975	1001.4	1005.6	1007.7	1009.8	1014.0	1018.3	0.975	
	0.990	6260.7	6286.8	6299.9	6313.0	6339.4	6366.0	0.990	
	0.995	25044	25148	25201	25253	25359	25465	0.995	
	0.500	1.4096	1.4178	1.4220	1.4261	1.4344	1.4427	0.500	
	0.750	3.4428	3.4511	3.4553	3.4594	3.4677	3.4761	0.750	
	0.900	9.4579	9.4663	9.4705	9.4746	9.4829	9.4913	0.900	
2	0.950	19.462	19.471	19.475	19.479	19.487	19.496	0.950	2
	0.975	39.465	39.473	39.477	39.481	39.490	39.498	0.975	
	0.990	99.466	99.474	99.478	99.483	99.491	99.499	0.990	
	0.995	199.47	199.47	199.47	199.48	199.49	199.51	0.995	
	0.500	1.2393	1.2464	1.2500	1.2536	1.2608	1.2680	0.500	
	0.750	2.4650	2.4674	2.4686	2.4697	2.4720	2.4742	0.750	
	0.900	5.1681	5.1597	5.1555	5.1512	5.1425	5.1337	0.900	
3	0.950	8.6166	8.5944	8.5832	8.5720	8.5494	8.5265	0.950	3
	0.975	14.081	14.037	14.015	13.992	13.947	13.902	0.975	
	0.990	26.505	26.411	26.364	26.316	26.221	26.125	0.990	
r_2	0.995	42.466	42.308	42.229	42.149	41.989	41.829	0.995	r_2
	0.500	1.1649	1.1716	1.1749	1.1782	1.1849	1.1916	0.500	
	0.750	2.0825	2.0821	2.0819	2.0817	2.0812	2.0806	0.750	
	0.900	3.8174	3.8036	3.7966	3.7896	3.7753	3.7607	0.900	
4	0.950	5.7459	5.7170	5.7024	5.6878	5.6581	5.6281	0.950	4
	0.975	8.4613	8.4111	8.3858	8.3604	8.3092	8.2573	0.975	
	0.990	13.838	13.745	13.699	13.652	13.558	13.463	0.990	
	0.995	19.892	19.752	19.682	19.611	19.468	19.325	0.995	
	0.500	1.1234	1.1297	1.1329	1.1361	1.1426	1.1490	0.500	
	0.750	1.8784	1.8763	1.8753	1.8742	1.8719	1.8694	0.750	
	0.900	3.1741	3.1573	3.1488	1.1402	3.1228	3.1050	0.900	
5	0.950	4.4957	4.4638	4.4476	4.4314	4.3984	4.3650	0.950	5
	0.975	6.2269	6.1751	6.1488	6.1225	6.0693	6.0153	0.975	
	0.990	9.3793	9.2912	9.2466	9.2020	9.1118	0.0204	0.990	
	0.995	12.656	12.530	12.466	12.402	12.274	12.144	0.995	
	0.500	1.0969	1.1031	1.1062	1.1093	1.1156	1.1219	0.500	
	0.750	1.7510	1.7477	1.7460	1.7443	1.7407	1.7368	0.750	
	0.900	2.8000	2.7812	2.7716	2.7620	2.7423	2.7222	0.900	
6	0.950	3.8082	3.7743	3.7571	3.7398	3.7047	3.6688	0.950	6
	0.975	5.0652	5.0125	4.9857	4.9589	4.9045	4.9491	0.975	
	0.990	7.2285	7.1432	7.1000	7.0568	6.9690	6.8801	0.990	
	0.995	9.3583	9.2408	9.1814	9.1219	9.0015	8.8793	0.995	

Table 6 *(continued)*

	γ	1	2	3	4	5	6	γ	
					r_1				
	0.500	0.50572	0.76655	0.87095	0.92619	0.96026	0.98334	0.500	
	0.750	1.5732	1.7010	1.7169	1.7157	1.7111	1.7059	0.750	
	0.900	3.5894	3.2574	3.0741	2.9605	2.8833	2.8274	0.900	
7	0.950	5.5914	4.7374	4.3468	4.1203	3.9715	3.8660	0.950	7
	0.975	8.0727	6.5415	5.8898	5.5226	5.2852	5.1186	0.975	
	0.990	12.246	9.5466	8.4513	7.8467	7.4604	7.1914	0.990	
	0.995	16.236	12.404	10.882	10.050	9.5221	9.1554	0.995	
	0.500	0.49898	0.75683	0.86004	0.91464	0.94831	0.97111	0.500	
	0.750	1.5384	1.6569	1.6683	1.6642	1.6575	1.6508	0.750	
	0.900	3.4579	3.1131	2.9238	2.8064	2.7265	2.6683	0.900	
8	0.950	5.3177	4.4590	4.0662	3.8378	3.6875	3.5806	0.950	8
	0.975	7.5709	6.0595	5.4160	5.0526	4.8173	4.6517	0.975	
	0.990	11.259	8.6491	7.5910	7.0060	6.6318	6.3707	0.990	
	0.995	14.688	11.042	9.5965	8.8051	8.3018	7.9520	0.995	
	0.500	0.49382	0.74938	0.85168	0.90580	0.93916	0.96175	0.500	
	0.750	1.5121	1.6236	1.6315	1.6253	1.6170	1.6091	0.750	
	0.900	3.3603	3.0065	2.8129	2.6927	2.6106	2.5509	0.900	
9	0.950	5.1174	4.2565	3.8626	3.6331	3.4817	3.3738	0.950	9
	0.975	7.2093	5.7147	5.0781	4.7181	4.4844	4.3197	0.975	
	0.990	10.561	8.0215	6.9919	6.4221	6.0569	5.8018	0.990	
r_2	0.995	13.614	10.107	8.7171	7.9559	7.4711	7.1338	0.995	r_2
	0.500	0.48973	0.74349	0.84508	0.89882	0.93193	0.95436	0.500	
	0.750	1.4915	1.5975	1.6028	1.5949	1.5853	1.5765	0.750	
	0.900	3.2850	2.9245	2.7277	2.6053	2.5216	2.4606	0.900	
10	0.950	4.9646	4.1028	3.7083	3.4780	3.3258	3.2172	0.950	10
	0.975	6.9367	5.4564	4.8256	4.4683	4.2361	4.0721	0.975	
	0.990	10.044	7.5594	6.5523	5.9943	5.6363	5.3858	0.990	
	0.995	12.826	9.4270	8.0807	7.3428	6.8723	6.5446	0.995	
	0.500	0.48644	0.73872	0.83973	0.89316	0.92608	0.94837	0.500	
	0.750	1.4749	1.5767	1.5798	1.5704	1.5598	1.5502	0.750	
	0.900	3.2252	2.8595	2.6602	2.5362	2.4512	2.3891	0.900	
11	0.950	4.8443	3.9823	3.5874	3.3567	3.2039	3.0946	0.950	11
	0.975	6.7241	5.2559	4.6300	4.2751	4.0440	3.8807	0.975	
	0.990	9.6460	7.2057	6.2167	5.6683	5.3160	5.0692	0.990	
	0.995	12.226	8.9122	7.6004	6.8809	6.4217	6.1015	0.995	
	0.500	0.48369	0.73477	0.83530	0.88848	0.92124	0.94342	0.500	
	0.750	1.4613	1.5595	1.5609	1.5503	1.5389	1.5286	0.750	
	0.900	3.1765	2.8068	2.6055	2.4801	2.3940	2.3310	0.900	
12	0.950	4.7472	3.8853	3.4903	3.2592	3.1059	2.9961	0.950	12
	0.975	6.5538	5.0959	4.4742	4.1212	3.8911	3.7283	0.975	
	0.990	9.3302	6.9266	5.9526	5.4119	5.0643	4.8206	0.990	
	0.995	11.754	8.5096	7.2258	6.5211	6.0711	5.7570	0.995	

Table 6 *(continued)*

		γ	7	8	9	r_1 10	11	12	γ	
		0.500	1.0000	1.0216	1.0224	1.0304	1.0369	1.0423	0.500	
		0.750	1.7011	1.6969	1.6931	1.6898	1.6868	1.6843	0.750	
		0.900	2.7849	2.7516	2.7247	2.7025	2.6837	2.6681	0.900	
	7	0.950	3.7870	3.7257	3.6767	3.6365	3.6028	3.5747	0.950	7
		0.975	4.9949	4.8994	4.8232	4.7611	4.7091	4.6658	0.975	
		0.990	6.9928	6.8401	6.7188	6.6201	6.5377	6.4691	0.990	
		0.995	8.8854	8.6781	8.5138	8.3803	8.2691	8.1764	0.995	
		0.500	0.98757	1.0000	1.0097	1.0175	1.0239	1.0293	0.500	
		0.750	1.6448	1.6396	1.6350	1.6310	1.6274	1.6244	0.750	
		0.900	2.6241	2.5893	2.5612	2.5380	2.5184	2.5020	0.900	
	8	0.950	3.5005	3.4381	3.3881	3.3472	3.3127	3.2840	0.950	8
		0.975	4.5286	4.4332	4.3572	4.2951	4.2431	4.1997	0.975	
		0.990	6.1776	6.0289	5.9106	5.8143	5.7338	5.6668	0.990	
		0.995	7.6942	7.4960	7.3386	7.2107	7.1039	7.0149	0.995	
		0.500	0.97805	0.99037	1.0000	1.0077	1.0141	1.0194	0.500	
		0.750	1.6022	1.5961	1.5909	1.5863	1.5822	1.5788	0.750	
		0.900	2.5053	2.4694	2.4403	2.4163	2.3959	2.3789	0.900	
	9	0.950	3.2927	3.2296	3.1789	3.1373	3.1022	3.0729	0.950	9
		0.975	4.1971	4.1020	4.0260	3.9639	3.9117	3.8682	0.975	
		0.990	5.6129	5.4671	5.3511	5.2565	5.1774	5.1114	0.990	
r_2		0.995	6.8849	6.6933	6.5411	6.4171	6.3136	6.2274	0.995	r_2
		0.500	0.97054	0.98276	0.99232	1.0000	1.0063	1.0166	0.500	
		0.750	1.5688	1.5621	1.5563	1.5513	1.5468	1.5430	0.750	
		0.900	2.4140	2.3772	2.3473	2.3226	2.3016	2.2841	0.900	
	10	0.950	3.1355	3.0717	3.0204	2.9782	2.9426	2.9130	0.950	10
		0.975	3.9498	3.8549	3.7790	3.7168	3.6645	3.6209	0.975	
		0.990	5.2001	5.0567	4.9424	4.8492	4.7710	4.7059	0.990	
		0.995	6.3025	6.1159	5.9676	5.8467	5.7456	5.6613	0.995	
		0.500	0.96445	0.97661	0.98610	0.99373	0.99999	1.0052	0.500	
		0.750	1.5418	1.5346	1.5284	1.5230	1.5181	1.5140	0.750	
		0.900	2.3416	2.3040	2.2735	2.2482	2.2267	2.2087	0.900	
	11	0.950	3.0123	2.9480	2.8962	2.8536	2.8176	2.7876	0.950	11
		0.975	3.7586	3.6638	3.5879	3.5257	3.4733	3.4296	0.975	
		0.990	4.8861	4.7445	4.6315	4.5393	4.4619	4.3974	0.990	
		0.995	5.8648	5.6821	5.5368	5.4182	5.3190	5.2363	0.995	
		0.500	0.95943	0.97152	0.98097	0.98856	0.99480	1.0000	0.500	
		0.750	1.5197	1.5120	1.5054	1.4996	1.4945	1.4902	0.750	
		0.900	2.2828	2.2446	2.2135	2.1878	1.1658	1.1474	0.900	
	12	0.950	2.9134	2.8486	2.7964	2.7534	2.7170	2.6866	0.950	12
		0.975	3.6065	3.5118	3.4358	3.3736	3.3211	3.2773	0.975	
		0.990	4.6395	4.4994	4.3875	4.2961	4.2193	4.1553	0.990	
		0.995	5.5245	5.3451	5.2021	5.0855	4.9878	4.9063	0.995	

Table 6 *(continued)*

	γ	13	14	15	r_1 18	20	24	γ	
	0.500	1.0469	1.0509	1.0543	1.0624	1.0664	1.0724	0.500	
	0.750	1.6819	1.6799	1.6781	1.6735	1.6712	1.6675	0.750	
	0.900	2.6543	2.6425	2.6322	2.6072	2.5947	2.5753	0.900	
7	0.950	3.5501	3.5291	3.5108	3.4666	3.4445	3.4105	0.950	7
	0.975	4.6281	4.5958	4.5678	4.5004	4.4667	4.4150	0.975	
	0.990	6.4096	6.3585	6.3143	6.2084	6.1554	6.0743	0.990	
	0.995	8.0962	8.0274	7.9678	7.8253	7.7540	7.6450	0.995	
	0.500	1.0339	1.0378	1.0412	1.0491	1.0531	1.0591	0.500	
	0.750	1.6216	1.6191	1.6170	1.6115	1.6088	1.6043	0.750	
	0.900	2.4875	2.4750	2.4642	2.4378	2.4246	2.4041	0.900	
8	0.950	3.2588	3.2371	3.2184	3.1730	3.1503	3.1152	0.950	8
	0.975	4.1618	4.1293	4.1012	4.0334	3.9995	3.9472	0.975	
	0.990	5.6085	5.5584	5.5151	5.4111	5.3591	5.2793	0.990	
	0.995	6.9377	6.8716	6.8143	6.6769	6.6082	6.5029	0.995	
	0.500	1.0239	1.0278	1.0311	1.0390	1.0429	1.0489	0.500	
	0.750	1.5756	1.5729	1.5705	1.5642	1.5611	1.5560	0.750	
	0.900	2.3638	2.3508	2.3396	2.3121	2.9893	2.2768	0.900	
9	0.950	3.0472	3.0252	3.0061	2.9597	2.9365	2.9005	0.950	9
	0.975	3.8302	3.7976	3.7694	3.7011	3.6669	3.6142	0.975	
	0.990	5.0540	5.0048	4.9621	4.8594	4.8080	4.7290	0.990	
r_2	0.995	6.1524	6.0882	6.0325	5.8987	5.8318	5.7292	0.995	r_2
	0.500	1.0161	1.0199	1.0232	1.0310	1.0349	1.0408	0.500	
	0.750	1.5395	1.5364	1.5338	1.5269	1.5235	1.5179	0.750	
	0.900	2.2685	2.2551	2.2435	2.2150	2.2007	2.1784	0.900	
10	0.950	2.8868	2.8644	2.8450	2.7977	2.7740	2.7372	0.950	10
	0.975	3.5827	3.5500	3.5217	3.4530	3.4186	3.3654	0.975	
	0.990	4.6491	4.6004	4.5582	4.4563	4.4054	4.3269	0.990	
	0.995	5.5880	5.5252	5.4707	5.3396	5.2740	5.1732	0.995	
	0.500	1.0097	1.0135	1.0168	1.0245	1.0284	1.0343	0.500	
	0.750	1.5102	1.5069	1.5041	1.4967	1.4930	1.4869	0.750	
	0.900	2.1927	2.1790	2.1671	2.1377	2.1230	2.1000	0.900	
11	0.950	2.7611	2.7383	2.7186	2.6705	2.6464	2.6090	0.950	11
	0.975	3.3913	3.3584	3.3299	3.2607	3.2261	3.1725	0.975	
	0.990	4.3411	4.2928	4.2509	4.1496	4.0990	4.0209	0.990	
	0.995	5.1642	5.1024	5.0489	4.9198	4.8552	4.7557	0.995	
	0.500	1.0044	1.0082	1.0115	1.0192	1.0231	1.0289	0.500	
	0.750	1.4861	1.4826	1.4796	1.4717	1.4678	1.4613	0.750	
	0.900	2.1311	2.1170	1.1049	2.0748	2.0597	2.0360	0.900	
12	0.950	2.6598	2.6368	2.6169	2.5680	2.5436	2.5055	0.950	12
	0.975	3.2388	3.2058	3.1772	3.1076	3.0728	3.0187	0.975	
	0.990	4.0993	4.0512	4.0096	3.9088	3.8584	3.7805	0.990	
	0.995	4.8352	4.7742	4.7214	4.5937	4.5299	4.4315	0.995	

Table 6 *(continued)*

					r_1				
	γ	30	40	48	60	120	∞	γ	
	0.500	1.0785	1.0846	1.0877	1.0908	1.0969	1.1031	0.500	
	0.750	1.6635	1.6593	1.6571	1.6548	1.6502	1.6452	0.750	
	0.900	2.5555	2.5351	2.5427	2.5142	2.4928	2.4708	0.900	
7	0.950	3.3758	3.3404	3.3224	3.3043	3.2674	3.2298	0.950	7
	0.975	4.3624	4.3089	4.2817	4.2544	4.1989	4.1423	0.975	
	0.990	5.9921	5.9084	5.8660	5.8236	5.7372	5.6495	0.990	
	0.995	7.5345	7.4225	7.3657	7.3088	7.1933	7.0760	0.995	
	0.500	1.0651	1.0711	1.0741	1.0771	1.0832	1.0893	0.500	
	0.750	1.5996	1.5945	1.5919	1.5892	1.5836	1.5777	0.750	
	0.900	2.3830	2.3614	2.3503	2.3391	2.3162	2.2926	0.900	
8	0.950	3.0794	3.0428	3.0241	3.0053	2.9669	2.9276	0.950	8
	0.975	3.8940	3.8398	3.8121	3.7844	3.7279	3.6702	0.975	
	0.990	5.1981	5.1156	5.0736	5.0316	4.9460	4.8588	0.990	
	0.995	6.3961	6.2875	6.2324	6.1772	6.0649	5.9505	0.995	
	0.500	1.0548	1.0608	1.0638	1.0667	1.0727	1.0788	0.500	
	0.750	1.5506	1.5450	1.5420	1.5389	1.5325	1.5257	0.750	
	0.900	2.2547	2.2320	2.2203	2.2085	2.1843	2.1592	0.900	
9	0.950	2.8637	2.8259	2.8066	2.7872	2.7475	2.7067	0.950	9
	0.975	3.5604	3.5055	3.4774	3.4493	3.3918	3.3329	0.975	
	0.990	4.6486	4.5667	4.5249	4.4831	4.3978	4.3105	0.990	
r_2	0.995	5.6248	5.5186	5.4645	5.4104	5.3001	5.1875	0.995	r_2
	0.500	1.0467	1.0526	1.0556	1.0585	1.0645	1.0705	0.500	
	0.750	1.5119	1.5056	1.5023	1.4990	1.4919	1.4843	0.750	
	0.900	2.1554	1.1317	2.1195	2.1072	2.0818	2.0554	0.900	
10	0.950	2.6996	2.6609	2.6410	2.6211	2.5801	2.5379	0.950	10
	0.975	3.3110	3.2554	3.2269	3.1984	3.1399	3.0798	0.975	
	0.990	4.2469	4.1653	4.1236	4.0819	3.9965	3.9090	0.990	
	0.995	5.0705	4.9659	4.9126	4.8592	4.7501	4.6385	0.995	
	0.500	1.0401	1.0460	1.0490	1.0519	1.0578	1.0637	0.500	
	0.750	1.4805	1.4737	1.4701	1.4664	1.4587	1.4504	0.750	
	0.900	2.0762	2.0516	2.0389	2.0261	1.9997	1.9721	0.900	
11	0.950	2.5705	2.5309	2.5105	2.4901	2.4480	2.4045	0.950	11
	0.975	3.1176	3.0613	3.0324	3.0035	2.9441	2.8828	0.975	
	0.990	3.9411	3.8596	3.8179	3.7761	3.6904	3.6025	0.990	
	0.995	4.6543	4.5508	4.4979	4.4450	4.3367	4.2256	0.995	
	0.500	1.0347	1.0405	1.0435	1.0464	1.0523	1.0582	0.500	
	0.750	1.4544	1.4471	1.4432	1.4393	1.4310	1.4221	0.750	
	0.900	2.0115	1.9861	1.9729	1.9597	1.9323	1.9036	0.900	
12	0.950	2.4663	2.4259	2.4051	2.3842	2.3410	2.2962	0.950	12
	0.975	2.9633	2.9063	2.8771	2.8478	2.7874	2.7249	0.975	
	0.990	3.7008	3.6192	3.5774	3.5355	3.4494	3.3608	0.990	
	0.995	4.3309	4.2282	4.1756	4.1229	4.0149	3.9039	0.995	

Table 6 (continued)

					r_1				
	γ	1	2	3	4	5	6	γ	
	0.500	0.48141	0.73145	0.83159	0.88454	0.91718	0.93926	0.500	
	0.750	1.4500	1.5452	1.5451	1.5336	1.5214	1.5105	0.750	
	0.900	3.1362	2.7632	2.5603	2.4337	2.3467	2.2830	0.900	
13	0.950	4.6672	3.8056	3.4105	3.1791	3.0254	2.9153	0.950	13
	0.975	6.4143	4.9653	4.3472	3.9959	3.7667	3.6043	0.975	
	0.990	9.0738	6.7010	5.7394	5.2053	4.8616	4.6204	0.990	
	0.995	11.374	8.1865	6.9257	6.2335	5.7910	5.4819	0.995	
	0.500	0.47944	0.72862	0.82842	0.88119	0.91371	0.93573	0.500	
	0.750	1.4403	1.5331	1.5317	1.5194	1.5066	1.4952	0.750	
	0.900	3.1022	2.7265	2.5222	2.3947	2.3069	2.2426	0.900	
14	0.950	4.6001	3.7389	3.3439	3.1122	2.9582	2.8477	0.950	14
	0.975	6.2979	4.8567	4.2417	3.8919	3.6634	3.5014	0.975	
	0.990	8.8616	6.5149	5.5639	5.0354	4.6950	4.4558	0.990	
	0.995	11.060	7.9216	6.6803	5.9984	5.5623	5.2574	0.995	
	0.500	0.47775	0.72619	0.82569	0.87830	0.91073	0.93267	0.500	
	0.750	1.4321	1.5227	1.5202	1.5071	1.4938	1.4820	0.750	
	0.900	3.0732	2.6952	2.4898	2.3614	2.2730	2.2081	0.900	
15	0.950	4.5431	3.6823	3.2874	3.0556	2.9013	2.7905	0.950	15
	0.975	6.1995	4.7650	4.1528	3.8043	3.5764	3.4147	0.975	
	0.990	8.6831	6.3589	5.4170	4.8932	4.5556	4.3183	0.990	
r_2	0.995	10.798	7.7008	6.4760	5.8029	5.3721	5.0708	0.995	r_2
	0.500	0.47628	0.72406	0.82330	0.87578	0.90812	0.93001	0.500	
	0.750	1.4249	1.5137	1.5103	1.4965	1.4827	1.4705	0.750	
	0.900	3.0481	2.6682	2.4618	2.3327	2.2438	2.1783	0.900	
16	0.950	4.4940	3.6337	3.2389	3.0069	2.8524	2.7413	0.950	16
	0.975	6.1151	4.6867	4.0768	3.7294	3.5021	3.3406	0.975	
	0.990	8.5310	6.2262	5.2922	4.7726	4.4374	4.2016	0.990	
	0.995	10.575	7.5138	6.3034	5.6378	5.2117	4.9134	0.995	
	0.500	0.47499	0.72219	0.82121	0.87357	0.90584	0.92767	0.500	
	0.750	1.4186	1.5057	1.5015	1.4873	1.4730	1.4605	0.750	
	0.900	3.0262	2.6446	2.4374	2.3077	2.2183	2.1524	0.900	
17	0.950	4.4513	3.5915	3.1968	2.9647	2.8100	2.6987	0.950	17
	0.975	6.0420	4.6189	4.0112	3.6648	3.4379	3.2767	0.975	
	0.990	8.3997	6.1121	5.1850	4.6690	4.3359	4.1015	0.990	
	0.995	10.384	7.3536	6.1556	5.4967	5.0746	5.7789	0.995	
	0.500	0.47385	0.72053	0.81936	0.87161	0.90381	0.92560	0.500	
	0.750	1.4130	1.4988	1.4938	1.4790	1.4644	1.4516	0.750	
	0.900	3.0070	2.6239	2.4160	2.2858	2.1958	1.1296	0.900	
18	0.950	4.4139	3.5546	3.1599	2.9277	2.7729	2.6613	0.950	18
	0.975	5.9781	4.5597	3.9539	3.6083	3.3820	3.2209	0.975	
	0.990	8.2854	6.0129	5.0919	4.5790	4.2479	4.0146	0.990	
	0.995	10.218	7.2148	6.0277	5.3746	4.9560	4.6627	0.995	

Table 6 *(continued)*

					r_1				
	γ	7	8	9	10	11	12	γ	
	0.500	0.95520	0.96724	0.97665	0.98421	0.99042	0.99560	0.500	
	0.750	1.5011	1.4931	1.4861	1.4801	1.4746	1.4701	0.750	
	0.900	2.2341	2.1953	2.1638	1.1376	1.1152	2.0966	0.900	
13	0.950	2.8321	2.7669	2.7144	2.6710	2.6343	2.6037	0.950	13
	0.975	3.4827	3.3880	3.3120	3.2497	3.1971	3.1532	0.975	
	0.990	4.4410	4.3021	4.1911	4.1003	4.0239	3.9603	0.990	
	0.995	5.2529	5.0761	4.9351	4.8199	4.7234	4.6429	0.995	
	0.500	0.95161	0.96360	0.97298	0.98051	0.98670	0.99186	0.500	
	0.750	1.4854	1.4770	1.4697	1.4634	1.4577	1.4530	0.750	
	0.900	2.1931	2.1539	2.1220	2.0954	2.0727	2.0537	0.900	
14	0.950	2.7642	2.6987	2.6548	2.6021	2.5651	2.5342	0.950	14
	0.975	3.3799	2.2853	3.2093	3.1469	3.0941	3.0501	0.975	
	0.990	4.2779	4.1399	4.0297	3.9394	3.8634	3.8001	0.990	
	0.995	5.0313	4.8566	4.7173	4.6034	4.5078	4.4281	0.995	
	0.500	0.94850	0.96046	0.96981	0.97732	0.98349	0.98863	0.500	
	0.750	1.4718	1.4631	1.4556	1.4491	1.4432	1.4383	0.750	
	0.900	2.1582	2.1185	2.0862	2.0593	2.0363	2.0171	0.900	
15	0.950	2.7066	2.6408	2.5876	2.5437	2.5064	2.4753	0.950	15
	0.975	3.2934	3.1987	3.1227	3.0602	3.0073	2.9633	0.975	
	0.990	4.1415	4.0045	3.8948	3.8049	3.7292	3.6662	0.990	
r_2	0.995	4.8473	4.6743	4.5364	4.4236	4.3288	4.2498	0.995	r_2
	0.500	0.94580	0.95773	0.96705	0.97454	0.98069	0.98582	0.500	
	0.750	1.4601	1.4511	1.4433	1.4366	1.4305	1.4255	0.750	
	0.900	2.1280	2.0880	2.0553	2.0281	2.0048	1.9854	0.900	
16	0.950	2.6572	2.5911	2.5377	2.4935	2.4560	2.4247	0.950	16
	0.975	3.2194	3.1248	3.0488	2.9862	2.9332	2.8890	0.975	
	0.990	4.0259	3.8896	3.7804	3.6909	3.6155	3.5527	0.990	
	0.995	4.6920	4.5207	4.3838	4.2719	4.1778	4.0994	0.995	
	0.500	0.94342	0.95532	0.96462	0.97209	0.97823	0.98334	0.500	
	0.750	1.4497	1.4405	1.4325	1.4256	1.4194	1.4142	0.750	
	0.900	2.1017	2.0613	2.0284	2.0009	1.9773	1.9577	0.900	
17	0.950	2.6143	2.5480	2.4943	2.4499	2.4122	2.3807	0.950	17
	0.975	3.1556	3.0610	2.9849	2.9222	2.8691	2.8249	0.975	
	0.990	3.9267	3.7910	3.6822	3.5931	3.5179	3.4552	0.990	
	0.995	4.5594	4.3893	4.2535	4.1423	4.0488	3.9709	0.995	
	0.500	0.94132	0.95319	0.96247	0.96993	0.97606	0.98116	0.500	
	0.750	1.4406	1.4312	1.4320	1.4159	1.4095	1.4042	0.750	
	0.900	2.0785	2.0379	2.0047	1.9770	1.9532	1.9333	0.900	
18	0.950	2.5767	2.5102	2.4563	2.4117	2.3737	2.3421	0.950	18
	0.975	3.0999	3.0053	2.9291	2.8664	2.8132	2.7689	0.975	
	0.990	3.8406	3.7054	3.5971	3.5082	3.4331	3.3706	0.990	
	0.995	4.4448	4.2759	4.1410	4.0305	3.9374	3.8599	0.995	

These tables have been adapted from Donald B. Owen's *Handbook of Statistical Tables*, published by Addison-Wesley, by permission of the publishers.

Table 7 Table of Selected Discrete and Continuous Distributions and Some of their Characteristics

Distribution	Probability density function	Mean	Variance
Binomial, $B(n, p)$	$f(x) = \binom{n}{x} p^x q^{n-x}, x = 0, 1, \ldots, n;$ $0 < p < 1, q = 1 - p$	np	npq
(Bernoulli, $B(1, p)$	$f(x) = p^x q^{1-x}, x = 0, 1$	p	$pq)$
Poisson $P(\lambda)$	$f(x) = e^{-\lambda} \dfrac{\lambda^x}{x!}, x = 0, 1, \ldots;$ $\lambda > 0$	λ	λ
Hypergeometric	$f(x) = \dfrac{\binom{m}{x}\binom{n}{r-x}}{\binom{m+n}{r}}$, where $x = 0, 1, \ldots, \min(r, m)$	$\dfrac{mr}{m+n}$	$\dfrac{mnr(m+n-r)}{(m+n)^2(m+n-1)}$
Negative Binomial	$f(x) = p^r \binom{r+x-1}{x} q^x, x = 0, 1, \ldots;$ $0 < p < 1, q = 1 - p$	$\dfrac{rq}{p}$	$\dfrac{rq}{p^2}$
(Geometric	$f(x) = pq^x, x = 0, 1, \ldots$	$\dfrac{q}{p}$	$\dfrac{q}{p^2})$
Multinomial	$f(x_1, \ldots, x_k) = \dfrac{n!}{x_1! x_2! \cdots x_k!} \times$ $p_1^{x_1} p_2^{x_2} \cdots p_k^{x_k}, x_j \geq 0$ integers, $x_1 + \cdots + x_k = n; p_j > 0, j = 1, 2, \ldots, k, p_1 + p_2 + \cdots + p_k = 1$	vector of expectations: $(np_1, \ldots, np_k)'$	vector of variances: $(np_1 q_1, \ldots, np_k q_k)'$ $q_j = 1 - p_j, j = 1, \ldots, k$
Normal, $N(\mu, \sigma^2)$	$f(x) = \dfrac{1}{\sqrt{2\pi}\sigma} \exp\left[-\dfrac{(x-\mu)^2}{2\sigma^2}\right],$ $x \in \mathbb{R}; \mu \in \mathbb{R}, \sigma > 0$	μ	σ^2
(Standard Normal, $N(0, 1)$	$f(x) = \dfrac{1}{\sqrt{2\pi}} \exp\left(-\dfrac{x^2}{2}\right), x \in \mathbb{R}$	0	$1)$
Gamma	$f(x) = \dfrac{1}{\Gamma(\alpha)\beta^\alpha} x^{\alpha-1} \exp\left(-\dfrac{x}{\beta}\right), x > 0;$ $\alpha, \beta > 0$	$\alpha\beta$	$\alpha\beta^2$
Chi-square	$f(x) = \dfrac{1}{\Gamma\left(\frac{r}{2}\right) 2^{r/2}} x^{\frac{r}{2}-1} \exp\left(-\dfrac{x}{2}\right), x > 0;$ $r > 0$ integer	r	$2r$

Table 7 (continued)

Distribution	Probability density function	Mean	Variance
Negative Exponential	$f(x) = \lambda \exp(-\lambda x),\ x > 0;\ \lambda > 0$	$\dfrac{1}{\lambda}$	$\dfrac{1}{\lambda^2}$
Uniform, U(α, β)	$f(x) = \dfrac{1}{\beta - \alpha},\ \alpha \leq x \leq \beta;$ $-\infty < \alpha < \beta < \infty$	$\dfrac{\alpha + \beta}{2}$	$\dfrac{(\alpha - \beta)^2}{12}$
Beta	$f(x) = \dfrac{\Gamma(\alpha+\beta)}{\Gamma(\alpha)\Gamma(\beta)} x^{\alpha-1}(1-x)^{\beta-1},\ 0 < x < 1;$ $\alpha, \beta > 0$	$\dfrac{\alpha}{\alpha+\beta}$	$\dfrac{\alpha\beta}{(\alpha+\beta)^2(\alpha+\beta+1)}$
Cauchy	$f(x) = \dfrac{\sigma}{\pi} \cdot \dfrac{1}{\sigma^2 + (x-\mu)^2},\ x \in \mathbb{R};$ $\mu \in \mathbb{R},\ \sigma > 0$	Does not exist	Does not exist
Bivariate Normal	$f(x_1, x_2) = \dfrac{1}{2\pi \sigma_1 \sigma_2 \sqrt{1-\rho^2}} \exp\left(-\dfrac{q}{2}\right),$ $q = \dfrac{1}{1-\rho^2}\left[\left(\dfrac{x_1-\mu_1}{\sigma_1}\right)^2 - 2\rho\left(\dfrac{x_1-\mu_1}{\sigma_1}\right)\right.$ $\left.\times\left(\dfrac{x_2-\mu_2}{\sigma_2}\right) + \left(\dfrac{x_2-\mu_2}{\sigma_2}\right)^2\right],$ $x_1, x_2 \in \mathbb{R};$ $\mu_1, \mu_2 \in \mathbb{R},\ \sigma_1, \sigma_2 > 0,\ -1 \leq \rho \leq 1$	vector of expectations: $(\mu_1, \mu_2)'$	vector of variances: $(\sigma_1^2, \sigma_2^2)'$.
k-Variate Normal, $N(\boldsymbol{\mu}, \boldsymbol{\Sigma})$	$f(\mathbf{x}) = (2\pi)^{-k/2}|\boldsymbol{\Sigma}|^{-1/2} \times$ $\exp\left[-\dfrac{1}{2}(\mathbf{x}-\boldsymbol{\mu})'\boldsymbol{\Sigma}^{-1}(\mathbf{x}-\boldsymbol{\mu})\right],$ $\mathbf{x} \in \mathbb{R}^k;\ \boldsymbol{\mu} \in \mathbb{R},\ \boldsymbol{\Sigma}: k \times k$ non-singular symmetric matrix	mean vector: $\boldsymbol{\mu}$	covariance matrix: $\boldsymbol{\Sigma}$

Distribution	Characteristic function	Moment generating function
Binomial, B(n, p)	$\varphi(t) = (pe^{it} + q)^n,\ t \in \mathbb{R}$	$M(t) = (pe^t + q)^n,\ t \in \mathbb{R}$
(Bernoulli, B($1, p$)	$\varphi(t) = pe^{it} + q,\ t \in \mathbb{R}$	$M(t) = pe^t + q,\ t \in \mathbb{R}$)
Poisson, $P(\lambda)$	$\varphi(t) = \exp(\lambda e^{it} - \lambda),\ t \in \mathbb{R}$	$M(t) = \exp(\lambda e^t - \lambda),\ t \in \mathbb{R}$
Negative Binomial	$\varphi(t) = \dfrac{p^r}{(1-qe^{it})^r},\ t \in \mathbb{R}$	$M(t) = \dfrac{p^r}{(1-qe^t)^r},\ t < -\log q$
(Geometric	$\varphi(t) = \dfrac{p}{1-qe^{it}},\ t \in \mathbb{R}$	$M(t) = \dfrac{p}{1-qe^t},\ t < -\log q$)

Table 7 *(continued)*

Distribution	Characteristic function	Moment generating function
Multinomial	$\varphi(t_1, \ldots, t_k) = \left(p_1 e^{it_1} + \cdots + p_k e^{it_k}\right)^n$, $t_1, \ldots, t_k \in \mathbb{R}$	$M(t_1, \ldots, t_k) = \left(p_1 e^{t_1} + \cdots + p_k e^{t_k}\right)^n$, $t_1, \ldots, t_k \in \mathbb{R}$
Normal, $N(\mu, \sigma^2)$	$\varphi(t) = \exp\left(i\mu t - \frac{\sigma^2 t^2}{2}\right)$, $t \in \mathbb{R}$	$M(t) = \exp\left(\mu t + \frac{\sigma^2 t^2}{2}\right)$, $t \in \mathbb{R}$
(Standard Normal	$\varphi(t) = \exp\left(-\frac{t^2}{2}\right)$, $t \in \mathbb{R}$	$M(t) = \exp\left(\frac{t^2}{2}\right)$, $t \in \mathbb{R}$)
Gamma	$\varphi(t) = \frac{1}{(1 - i\beta t)^\alpha}$, $t \in \mathbb{R}$	$M(t) = \frac{1}{(1 - \beta t)^\alpha}$, $t < \frac{1}{\beta}$
Chi-square	$\varphi(t) = \frac{1}{(1 - 2it)^{r/2}}$, $t \in \mathbb{R}$	$M(t) = \frac{1}{(1 - 2t)^{r/2}}$, $t < \frac{1}{2}$
Negative Exponential	$\varphi(t) = \frac{\lambda}{\lambda - it}$, $t \in \mathbb{R}$	$M(t) = \frac{\lambda}{\lambda - t}$, $t < \lambda$
Uniform, $U(\alpha, \beta)$	$\varphi(t) = \frac{e^{it\beta} - e^{it\alpha}}{it(\beta - \alpha)}$, $t \in \mathbb{R}$	$M(t) = \frac{e^{t\beta} - e^{t\alpha}}{t(\beta - \alpha)}$, $t \in \mathbb{R}$
Cauchy ($\mu = 0, \sigma = 1$)	$\varphi(t) = \exp(-|t|)$, $t \in \mathbb{R}$	Does not exist (for $t \neq 0$)
Bivariate Normal	$\varphi(t_1, t_2) = \exp\left[i\mu_1 t_1 + i\mu_2 t_2 - \frac{1}{2}\left(\sigma_1^2 t_1^2 + 2\rho\sigma_1\sigma_2 t_1 t_2 + \sigma_2^2 t_2^2\right)\right]$, $t_1, t_2 \in \mathbb{R}$	$M(t_1, t_2) = \exp\left[\mu_1 t_1 + \mu_2 t_2 + \frac{1}{2}\left(\sigma_1^2 t_1^2 + 2\rho\sigma_1\sigma_2 t_1 t_2 + \sigma_2^2 t_2^2\right)\right]$, $t_1, t_2 \in \mathbb{R}$
k-Variate Normal	$\varphi(\mathbf{t}) = \exp\left(i\mathbf{t}'\boldsymbol{\mu} - \frac{1}{2}\mathbf{t}'\boldsymbol{\Sigma}\mathbf{t}\right)$, $\mathbf{t} \in \mathbb{R}^k$	$M(\mathbf{t}) = \exp\left(\mathbf{t}'\boldsymbol{\mu} + \frac{1}{2}\mathbf{t}'\boldsymbol{\Sigma}\mathbf{t}\right)$, $\mathbf{t} \in \mathbb{R}^k$

Some Notation and Abbreviations

$\mathcal{F}, \mathcal{A}$	(usually) a field and sigma-field, respectively
$\mathcal{F}(C), \sigma(C)$	field and σ-field, respectively, generated by the class C
$\mathcal{A}_A$	σ-field of members of $\mathcal{A}$ which are subsets of A
$(S, \mathcal{A})$	measurable space
$\mathbb{R}^k, \mathcal{B}^k, k \geq 1$	k-dimensional Euclidean space and Borel σ-field, respectively
$(\mathbb{R}^1, \mathcal{B}^1) = (\mathbb{R}, \mathcal{B})$	Borel real line
$\uparrow, \downarrow$	increasing (non-decreasing) and decreasing (non-increasing), respectively
$P, (S, \mathcal{A}, P)$	probability measure (function) and probability space, respectively
I_A	indicator of the set A
$X^{-1}(B)$	inverse image of the set B under X
$\mathcal{A}_X$ or $X^{-1}(\sigma\text{-field})$	σ-field induced by X
$(X \in B) = [X \in B]$ $= X^{-1}(B)$	the set of points for which X takes values in B
r.v., r. vector,	random variable, random vector
r. experiment,	random experiment
r. sample, r. interval,	random sample, random interval
r. error	random error
$X(S)$	range of X
$X_{(j)}$ or Y_j	jth order statistic
$B(n, p)$	Binomial distribution (or r.v.) with parameters n and p
$P(\lambda)$	Poisson distribution (or r.v.) with parameter λ
$N(\mu, \sigma^2)$	Normal distribution (or r.v.) with parameters μ and σ^2
Φ	distribution function of $N(0, 1^2)$

Some Notation and Abbreviations

χ_r^2	Chi-square distribution (or r.v.) with r degrees of freedom (d.f.)
$U(\alpha, \beta)$ or $R(\alpha, \beta)$	Uniform or Rectangular distribution (or r.v.) with parameters α and β
t_r	(Student's) t distribution (or r.v.) with r d.f.
F_{r_1, r_2}	F distribution (or r.v.) with r_1 and r_2 d.f.
$\chi'^2_{r:\delta}$	noncentral Chi-square distribution with r d.f. and noncentrality parameter δ
$t'_{r:\delta}$	noncentral t distribution with r d.f. and noncentrality parameter δ
$F'_{r_1, r_2 : \delta}$	noncentral F distribution with r_1, r_2 d.f. and noncentrality parameter δ
$E(X)$ or EX or $\mu(X)$ or μ_X or just μ	expectation (mean value, mean) of X
$\sigma^2(X)$ ($\sigma(X)$) or σ_X^2 (σ_X) or just $\sigma^2(\sigma)$	variance (standard deviation) of X
$\mathrm{Cov}(X, Y), \rho(X, Y)$	Covariance and correlation coefficient, respectively, of X and Y
φ_X or $\varphi_{X_1, \ldots, X_n}$, φ_X or just φ	characteristic function (cf. f.)
M_X or $M_{X_1, \ldots, X_n}$, M_X or just M	moment generating function (m.g.f.)
η_X	factorial moment generating function
$\xrightarrow{\text{a.s.}}$	almost sure (a.s.) convergence or convergence with probability one
$\xrightarrow{P}, \xrightarrow{d}, \xrightarrow{\text{q.m.}}$	convergence in probability, distribution, quadratic mean, respectively
UMV (UMVU)	uniformly minimum variance (unbiased)
ML (MLE)	maximum likelihood (estimator or estimate)
(UMP) MP (UMPU)	(uniformly) most powerful (unbiased)
(MLR) LR	(monotone) likelihood ratio
SPRT	sequential probability ratio test
LE (LSE)	least square (estimator or estimate)

Answers to Selected Exercises

Chapter 1

1.1.1. (i), (ii) incorrect; (iii), (iv) correct.

1.1.2. $A_1 = \{(0, 0), (1, 1), (2, 2), (3, 3), (4, 4), (5, 5)\}$
$A_2 = \{(-5, 5), (-4, 4), (-3, 3), (-2, 2), (-1, 1), (0, 0)\}$
$A_5 = \{(0, 0), (0, 1), (0, 2), (1, 0), (1, 1), (2, 0), (-1, 0), (-1, 1), (-2, 0)\}$.

1.1.9. $\underline{A} = \bigcup_{n=1}^{\infty} \bigcap_{j=n}^{\infty} A_j$; (ii) $\overline{A} = \bigcap_{n=1}^{\infty} \bigcup_{j=n}^{\infty} A_j$; (iii)–(v) follow from (i), (ii).

1.1.11. $A_n \uparrow A = (-5, 20)$, $B_n \downarrow B = (0, 7]$.

1.2.2. Let $A_1, A_2 \in \mathcal{F}$. Then $A_1^c, A_2^c \in \mathcal{F}$ by ($\mathcal{F}2$). Also $A_1^c \cap A_2^c \in \mathcal{F}$ by ($\mathcal{F}3'$). But $A_1^c \cap A_2^c = (A_1 \cup A_2)^c$. Thus $(A_1 \cup A_2)^c \in \mathcal{F}$ and hence $A_1 \cup A_2 \in \mathcal{F}$ by ($\mathcal{F}2$).

1.2.7. C is not a field because, for example, $\{3\} \in C$ but $\{3\}^c = \{1, 2, 4\} \notin C$.

Chapter 2

2.1.1. $P(A_1^c \cap A_2) = P(A_2^c \cap A_3) = 1/6$, $P(A_1^c \cap A_3) = 1/3$,
$P(A_1 \cap A_2^c \cap A_3^c) = 0$, $P(A_1^c \cap A_2^c \cap A_3^c) = 5/12$.

2.1.2. (i) 1/9; (ii) 1/3.

2.1.3. (i) 3/190; (ii) 4/190.

2.1.4. $P(A) = 0.14$, $P(B) = 0.315$, $P(C) = 0.095$.

2.2.7. $P(A_j|A) = j(5 - j)/20$, $j = 1, \ldots, 5$.

2.2.8. (i) 2/5; (ii) 5/7.

2.2.9. (i) 15/26; (ii) 13/24.

2.2.11. (i) 7/9; (ii) 1/6.

2.2.12. 19/218.

2.2.17. $\dfrac{1}{3} \times \sum_{j=1}^{6}[m_j n_j /(m_j + n_j)(m_j + n_j - 1)]$.

2.3.1. $0 = P(\emptyset) = P(A \cap B)$. Thus $P(A \cap B) = 0$ if and only if $P(A) = 0$ or $P(B) = 0$ or $P(A) = P(B) = 0$.

2.3.6. $0.54 \times \sum_{j=1}^{n-1} (0.1)^{j-1}(0.4)^{n-j-1}$.

2.4.1. 720.

2.4.2. 2^n.

2.4.3. 900.

2.4.4. (i) 10^7; (ii) 10^4.

2.4.6. 1/360.

2.4.7. (i) $1/(24!)$; (ii) $1/(13!) \times (9!)$.

2.4.8. $n(n-1)$.

2.4.12. 29/56.

2.4.13. $(2n)!$.

2.4.14. $(1/2)^{2n} \times \sum_{j=n+1}^{2n} \binom{2n}{j}$.

2.4.15. $\sum_{j=0}^{n}\left[\binom{n}{j} p^j (1-p)^{n-j}\right]^2$.

2.4.21. $\sum_{j=5}^{10} \binom{10}{j}\left(\dfrac{1}{5}\right)^j \left(\dfrac{4}{5}\right)^{10-j}$.

2.4.29. With regard to order: $P(A_1) = 0.125$; $P(A_2) = 0.25$; $P(A_3) = \dfrac{4 \times 48^2 \times 3}{52^3} \approx 0.19663$; $P(A_4) = 0.015625$; $P[A_5] = 0.09375$.

Without regard to order: $P(A_1) = \dfrac{7}{53} \approx 0.13207$; $P(A_2) = 0.50$; $P(A_3) = \dfrac{392}{2067} \approx 0.18965$; $P(A_5) = \dfrac{507}{5724} \approx 0.08857$.

2.4.31. (i) $2 \times \binom{26}{3} \times \binom{26}{2}$; (ii) $4 \times \binom{13}{1}^3 \times \binom{13}{2}$; (iii) $\sum_{j=2}^{4} \binom{4}{j}\binom{48}{5-j}$; (iv) 384; (v) $4 \times \binom{13}{5}$.

2.6.3. 244/495.

Chapter 3

3.2.1. (i) {0, 1, 2, 3, 4}; (ii) $P(X=x) = \binom{4}{x} / 2^4$, $x = 0, 1, \ldots, 4$.

3.2.2. (i) $1 - (0.875)^{25} = 0.9645$; (ii) 1; (iii) 0.1953.

3.2.5. $\lambda = 2.3026$ and $P(X > 5) = 0.032$.

3.2.6. e^{-4}.

3.2.10. $1 - \sum_{x=0}^{9}\left[\binom{400}{x}\binom{1200}{25-x} / \binom{1600}{25}\right]$.

3.2.14. $c = 1 - \alpha$.

3.2.15. (i) 2/3; (ii) $(1/3)^{10}$; (iii) 0.25; (iv) 3/13.

3.3.5. $a = 3.94$, $b = 18.3$.

3.3.7. (i) $e^{-\lambda j}(1 - e^{-\lambda})$, $j = 0, 1, \ldots$; (ii) $e^{-\lambda s}$; (iii) $e^{-\lambda t}$; (iv) $\lambda = -\log \alpha / s$.

3.3.8. (i) $e^{-1.3} \approx 0.27$; (ii) $e^{-0.6} \approx 0.55$; (iii) $50 \log 2 \approx 34.65$.

3.3.9. (i) $\alpha = 2$; (ii) $\alpha = 3$.

3.3.12. $2 \tan \sqrt[4]{c} / \pi$.

3.3.15. (i) 1.5; (ii) 3.

3.3.16. $\exp(-x^3)$.

3.3.21. (i) 27/400; (ii) 12/25; (iii) 0; (iv) 2/25.

3.4.2. (i) $1 - e^{-20}$; (ii), (iii) 0.5591.

3.4.3. 0.0713.

Chapter 4

4.1.10. (i) 0; (ii) 0.159; (iii) 0.

4.1.11. (i), (ii) 0; (iii) 0.988.

4.1.13. (i) 0.584.

4.1.14. $c = \mu + 1.15\sigma$.

4.2.1. (i) $f_j(x) = \binom{21}{x}\left(\frac{1}{6}\right)^x\left(\frac{5}{6}\right)^{21-x}$, $j = 1, \ldots, 6$; (ii) $1 - \sum_{x=0}^{4}\binom{21}{x}\left(\frac{1}{6}\right)^x\left(\frac{5}{6}\right)^{21-x}$.

4.2.3. (i) c = any real > 0; (ii) $f_X(x) = 2(c-x)/c^2$, $x \in [0, c]$, $f_Y(y) = 2y/c^2$, $y \in [0, c]$; (iii) $f(x|y) = 1/y$, $x \in [0, y]$, $y \in (0, c]$, $f(y|x) = 1/(c-x)$, $0 \le x < y < c$; (iv) $(2c-1)/c^2$.

4.2.4. (i) $1 - e^{-x}$, $x > 0$; (ii) $1 - e^{-y}$, $y > 0$; (iii) $1/2$; (iv) $1 - 4e^{-3}$.

Chapter 5

5.1.6. 0, c^2.

5.1.12. (i) 2, 4; (ii) 2.

5.2.1. $n(n-1)\cdots(n-k+1)p^k$.

5.2.2. (i) 0.5, 0.25; (ii) 0.75; (iii) 500; (iv) $0.05\sqrt{2\theta}$.

5.2.3. $45(O)$, $40(A)$, $10(B)$, $5(AB)$.

5.2.4. λ^k.

5.2.10. 0.075.

5.2.13. $\exp\left(\frac{2\alpha + \beta^2}{2}\right)$, $(\exp\beta^2 - 1)\exp(2\alpha + \beta^2)$.

5.2.15. (ii) $\gamma_1 = [np(p-1)(2p-1)]/\sigma^3$, $\sigma^2 = np(1-p)$, so that $\gamma_1 < 0$ if $p < 1/2$ and $\gamma_1 > 0$ if $p > 1/2$; (iii) $\lambda^{-1/2}$, 2.

5.2.16. (i) $\gamma_2 = -1.2$; (ii) 9.

5.3.2. $[n(n+1) - y(y-1)]/2(n-y+1)$, $y = 1, \ldots, n$, $(x+1)/2$, $x = 1, \ldots, n$, $1/2$.

5.3.3. $7/12$, $11/144$, $7/12$, $11/144$, $(3y+2)/(6y+3)$, $y \in (0, 1)$, $(6y^2 + 6y + 1)/2(6y+3)^2$, $y \in (0, 1)$.

5.3.4. $1/\lambda$, $1/\lambda^2$, $1/\lambda$, $1/\lambda^2$, $1/\lambda$, $1/\lambda^2$.

5.4.5. $P(X = \mu) = P\left[\bigcap_{n=1}^{\infty}\left(|X - \mu| < \frac{1}{n}\right)\right] = P\left[\lim_{n\to\infty}\left(|X-\mu| < \frac{1}{n}\right)\right]$
$= \lim_{n\to\infty} P\left(|X-\mu| < \frac{1}{n}\right) = 1.$

5.5.1. 0, 2.5, 2.5, 2.25, 0.

Chapter 6

6.2.10. $\phi(t) = 1/(1 - it)^2$, $EX^n = (n + 1)!$.

6.2.11. $(1 - \cos x)/\pi x^2$ for $x \neq 0$, anything (for example, $1/(2\pi)$) for $x = 0$.

6.5.1. $\left(\sum_{j=1}^{6} e^{jt}\right)\bigg/6$, $t \in \mathbb{R}$.

6.5.2. $e^t/(2 - e^t)$, $t < \log 2$, $e^{it}/(2 - e^{it})$, 2, 4, 2.

6.5.4. $\lambda e^{\alpha t}/(\lambda - t)$, $t < \lambda$, $\lambda e^{i\alpha t}/(\lambda - it)$, $\alpha + (1/\lambda)$, $1/\lambda^2$.

6.5.5. $\eta(t) = [pt + (1 - p)]^n$, $t \in \mathbb{R}$.

6.5.12. $\gamma(t) = e^{\lambda t}$, $t \in \mathbb{R}$.

6.5.15. $M(t) = 1/(1 - t)$, $t \in (-1, 1)$, $\phi(t) = 1/(1 - it)$, $f(x) = e^{-x}$, $x > 0$.

6.5.17. $EX_1 = 1$, $\sigma^2(X_1) = 0.5$, $Cov(X_1, X_2) = 1/3$.

6.5.24. $M(t_1, t_2) = \exp(\boldsymbol{\mu}'\mathbf{t} + \frac{1}{2}\mathbf{t}'\mathbf{Ct})$, $\boldsymbol{\mu} = (\mu_1, \mu_2)'$, $\mathbf{t} = (t_1, t_2)' \in \mathbb{R}^2$,

$\mathbf{C} = \begin{pmatrix} \sigma_1^2 & \rho\sigma_1\sigma_2 \\ \rho\sigma_1\sigma_2 & \sigma_2^2 \end{pmatrix}$, $\phi(t_1, t_2) = \exp(i\boldsymbol{\mu}'\mathbf{t} - \frac{1}{2}\mathbf{t}'\mathbf{Ct})$, $E(X_1, X_2) = \rho\sigma_1\sigma_2 + \mu_1\mu_2$.

Chapter 7

7.1.1. $F_{X_{(1)}}(x) = 1 - [1 - F(x)]^n$, $F_{X_{(n)}}(x) = [F(x)]^n$. Then for the continuous case and continuity points of f, we have $f_{X_{(1)}}(x) = nf(x)[1 - F(x)]^{n-1}$, $f_{X_{(n)}}(x) = nf(x)[F(x)]^{n-1}$.

7.1.2. (i) $f_{X_1}(x_1) = I_{(0,1)}(x_1)$, $f_{X_2}(x_2) = I_{(0,1)}(x_2)$; (ii) $1/18$, $\pi/4$, $(1 - \log 2)/2$.

7.1.6. (i) For $j \neq 1$, $f_{X_1,X_j}(0, 0) = f_{X_1,X_j}(0, 1) = f_{X_1,X_j}(1, 0) = f_{X_1,X_j}(1, 1) = 1/4$ and $f_{X_i}(0) = f_{X_i}(1) = 1/2$, $i = 1, 2, 3$; (ii) Follows from (i).

7.1.8. (i) $\sum_{j=k}^{n} \binom{n}{j} p^j (1-p)^{n-j}$, $p = P(X_1 \in B)$; (ii) $p = 1/e$;

(iii) $\sum_{j=5}^{10} \binom{10}{j} e^{-j} (1 - e^{-1})^{10-j} \approx 0.3057$ independently of λ.

7.2.2. $1/c^2$, $1/c^3$, $2/c^2$, $3/c^2$.

7.2.4. (i) n = integral part of $\sigma^2/(1 - \alpha)c^2$ plus 1; (ii) 4,000.

7.3.3. (i) 200; (ii) $f_{X+Y}(z) = \lambda^2 z e^{-\lambda z}$, $z > 0$; (iii) $3.5e^{-2.5}$.

Chapter 8

8.1.1. $p_n \to 0$ as $n \to \infty$.

8.1.3. $E(\overline{X}_n - \mu)^2 = \sigma^2(\overline{X}_n) = \dfrac{\sigma^2}{n} \to 0$ as $n \to \infty$.

8.1.4. $E(Y_n - X)^2 = E(X_n - Y_n)^2 + E(X_n - X)^2 - 2E[(X_n - Y_n)(X_n - X)]$ and $|E[(X_n - Y_n)(X_n - X)]| \le E|(X_n - Y_n)(X_n - X)| \le E^{1/2}(X_n - Y_n)^2 \times E^{1/2}(X_n - X)^2$.

8.2.2. $\phi_{X_n}(t) = \left(1 - \frac{\lambda_n}{n}\right)^n \left(1 + \frac{\lambda_n}{n - \lambda_n} e^{it}\right)^n \xrightarrow{n \to \infty} \exp(-\lambda + \lambda e^{it}) = \phi_X(t)$,

where $X \sim P(\lambda)$.

8.3.2. $P(180 \le X \le 200) \approx 0.88$.

8.3.3. $P(150 \le X \le 200) \approx 0.96155$.

8.3.5. $P(65 \le X \le 90) \approx 0.87686$.

8.3.7. 0.99985.

8.3.8. $4,146$.

8.3.11. $c = 0.329$.

8.3.15. $n = 123$.

8.3.16. 26.

8.4.3. $E(\overline{X}_n - \overline{\mu}_n)^2 = \frac{1}{n^2} E\left[\sum_{j=1}^{n}(X_j - \mu_j)\right]^2 = \frac{1}{n^2} \sum_{j=1}^{n} E(X_j - \mu_j)^2 = \frac{1}{n^2} \sum_{j=1}^{n} \sigma_j^2$

$$\le \frac{1}{n^2} nM = \frac{M}{n} \xrightarrow{n \to \infty} 0.$$

8.4.6. $\sigma^2(X_j) = \sigma^2\left(\frac{1}{\sqrt{j}} \chi_j^2\right) = \frac{1}{j}\sigma^2(\chi_j^2) = 2$ and then Exercise 8.4.3 applies.

8.4.7. $\sigma^2(X_j) = \lambda_j$ so that $E(\overline{X}_n - \overline{\mu}_n)^2 = \frac{1}{n^2}\sum_{j=1}^{n} \sigma_j^2 = \frac{1}{n^2}\sum_{j=1}^{n} \lambda_j$

$$= \frac{1}{n}\frac{1}{n}\sum_{j=1}^{n} \lambda_j \xrightarrow{n \to \infty} 0 \cdot c = 0.$$

Chapter 9

9.1.2. (ii) $P(Y = c_1) = 0.9596$, $P(Y = c_2) = 0.0393$, $P(Y = c_3) = 0.0011$; (iii) $0.9596 c_1 + 0.0393 c_2 + 0.0011 c_3$.

9.1.3. $N(\frac{9}{5}\mu + 32, \frac{81}{25}\sigma^2)$.

9.1.6. $f_Z(z) = 1/\pi\sqrt{1 - z^2}$, $z \in (-1, 1)$.

9.1.8. $f(y) = \left[\Gamma\left(\frac{r}{2}\right)2^{r/2}\right]^{-1} y^{(r/2)-1}(1 - y)^{-[(r/2)+1]} \exp[-y/2(1 - y)]$, $y \in (0, 1)$.

9.2.1. $P(X_1 + X_2 = j) = (j-1)/36, j = 2, \ldots, 7$ and $P(X_1 + X_2 = j) = (13-j)/36$, $j = 8, \ldots, 12$.

9.2.6. $f_X(x) = \frac{1}{2} I_{(0,1)}(x) + \frac{1}{2x^2} I_{(1,\infty)}$.

9.2.7. (ii) $\frac{r}{r-2}, r \neq 2$; (iii) $\left(1 + \frac{t^2}{r}\right)^{-(r+1)/2} \xrightarrow[r \to \infty]{} e^{-t^2/2}$.

By using the approximation $\Gamma(n) \approx (2\pi)^{1/2} n^{(2n-1)/2} e^{-n}$ (see Stirling's formula in W. Feller's book *An Introduction to Probability Theory*, Vol. I, 3rd ed., 1968, page 50), we obtain

$$\left[\Gamma\left(\frac{r+1}{2}\right) \Big/ r^{1/2} \Gamma\left(\frac{r}{2}\right)\right] \xrightarrow[r \to \infty]{} 1/\sqrt{2}. \text{ Hence the result.}$$

9.2.8.

(i) Use $EX^k = \left(\frac{r_2}{r_1}\right)^k \frac{\Gamma\left[\left(\frac{1}{2}r_1\right) + k\right] \Gamma\left[\left(\frac{1}{2}r_2\right) - k\right]}{\Gamma\left(\frac{1}{2}r_1\right) \Gamma\left(\frac{1}{2}r_2\right)}$;

(ii) The transformation $y = \frac{1}{1 + \frac{r_1}{r_2} x}$ gives

$$x = \frac{r_2}{r_1} \cdot \frac{1-y}{y}, 0 < y < 1, \text{ and } \left|\frac{dx}{dy}\right| = \frac{r_2}{r_1} \cdot \frac{1}{y^2}.$$

Then the p.d.f. of F_{r_1, r_2}, on page 236 yields:

$$f_Y(y) = f_X\left(\frac{r_2}{r_1} \cdot \frac{1-y}{y}\right) \cdot \frac{r_2}{r_1} \cdot \frac{1}{y^2} = \frac{\Gamma\left(\frac{r_1+r_2}{2}\right)}{\Gamma\left(\frac{r_1}{2}\right) \Gamma\left(\frac{r_2}{2}\right)} y^{\frac{r_2}{2}-1}(1-y)^{\frac{r_1}{2}-1}$$

after cancellations. This last expression is the p.d.f. of the Beta distribution with degrees of freedom $\frac{r_2}{2}$ and $\frac{r_1}{2}$.

(iii) By (ii) and for $r_1 = (= r)$, $1/(1 + X)$ is $B(r/2, r/2)$ which is symmetric about $1/2$.

Hence $P(X \leq 1) = P\left(\frac{1}{1+X} \geq \frac{1}{2}\right) = \frac{1}{2}$; (iv) Set $Y = r_1 X$. Then

$$f_Y(y) = \frac{\Gamma\left[\frac{1}{2}(r_1 + r_2)\right]}{\Gamma\left(\frac{1}{2}r_1\right) \Gamma\left(\frac{1}{2}r_2\right) r_2^{r_1/2}} y^{(r_1/2)-1}\left(1 + \frac{y}{r_2}\right)^{-r_2/2}\left(1 + \frac{y}{r_2}\right)^{-r_1/2}$$

$$\xrightarrow[r_2\to\infty]{} \frac{1}{\Gamma\left(\frac{1}{2}r_1\right) 2^{(r_1/2)-1}} y^{(r_1/2)-1} e^{-y/2}$$

since $\left[\Gamma\left(\frac{r_1+r_2}{2}\right) \bigg/ \Gamma\left(\frac{r_2}{2}\right) r_2^{r_1/2}\right] \xrightarrow[r_2\to\infty]{} 1 / \left[2^{(r_1/2)-1}\right]$

by the approximation employed in Exercise 2.7(iii).

9.3.5. $\begin{pmatrix} Y_1 \\ Y_2 \end{pmatrix} : N\left(\begin{pmatrix} 0 \\ 0 \end{pmatrix}, \begin{pmatrix} 2+2p & 0 \\ 0 & 2-2p \end{pmatrix}\right)$.

Chapter 10

10.1.1. $P(X_j > m, j = 1, \ldots, n) = \frac{1}{2^n}$, $P(Y_n \le m) = 1/2^n$.

10.1.2. $c = \theta - \log[1 - (0.9)^{1/3}]$.

10.1.4.

(i) $\alpha + (\beta - \alpha)j/(n+1)$, $(\beta - \alpha)^2 j(n-j+1)/(n+1)^2 (n+2)$;

(ii) $f_Y(y) = \dfrac{n(n-1)}{(\beta-\alpha)^n} y^{n-2}(\beta - \alpha - y)$, $y \in (0, \beta - \alpha)$,

$P(a < Y < b) = \dfrac{n(n-1)}{(\beta-\alpha)^n}\left((\beta-\alpha)\dfrac{b^{n-1}-a^{n-1}}{n-1} - \dfrac{b^n - a^n}{n}\right)$,

(iii) $\rho(Y_i, Y_j) = i(n-j+1)/[i(n-i+1)j(n-j+1)]^{1/2}$.

10.1.9. For the converse, $e^{-n\lambda t} = P(Y_1 > t) = P(X_j > t, j = 1, \ldots, n) = [P(X_1 > t)]^n$ so that $P(X_1 > t) = e^{-\lambda t}$. Thus the common distribution of the X's is the Negative Exponential distribution with parameter λ.

10.1.14. With $k = \dfrac{n+1}{2}$, $\dfrac{(2k-1)!}{[(k-1)!]^2} \dfrac{1}{(\beta-\alpha)^{2k-1}}(y-\alpha)^{k-1}(\beta-y)^{k+1}$, $y \in (\alpha, \beta)$,

and $\dfrac{(2k-1)!}{[(k-1)!]^2} \lambda(1-e^{-\lambda y})^{k-1}(e^{-\lambda y})^k$, $y > 0$.

10.1.15. For $n = 2k-1$, $f_{S_M}(y) = \dfrac{(2k-1)!}{[(k-1)!]^2}[F(y)]^{k-1}[1-F(y)]^{k-1} f(y)$, $y \in \mathbb{R}$. But $f(\mu - y) = f(\mu + y)$ and $F(\mu + y) = 1 - F(\mu - y)$. Hence the result.

10.1.17. $f_{S_M}(y) = 3! e^{-2(y-\theta)}[1 - e^{-(y-\theta)}]$, $y > 0$.

10.2.1. Set $Z = F(Y_1)$. Then $f_Z(z) = n(1-z)^{n-1}$, $z \in (0, 1)$ and $EZ = 1/(n+1)$.

Chapter 11

11.1.10. (i) $\prod_{j=1}^{n} X_j$; (ii) $(X_1, \ldots, X_n)$; (iii) $\sum_{j=1}^{n} X_j$; (iv) $\prod_{j=1}^{n} X_j$.

11.2.2. Take $g(x) = x$. Then $E_\theta g(X) = \dfrac{1}{2\theta}\int_{-\theta}^{\theta} x\, dx = 0$ for every $\theta \in \Omega = (0, \infty)$.

11.2.3. $f(x;\theta) = P_\theta(X_j = x) = 1/10$, $x = \theta + 1, \ldots, \theta + 10$, $j = 1, 2$ and let $T(X_1, X_2) = X_1$, $V(X_1, X_2) = X_2$. Then T is sufficient for θ and T, V are independent. But the distribution of V does depend on θ. This is so because the set of positivity of the p.d.f. of T depends on θ (see Theorem 2).

11.3.1. Set $T = \sum_{j=1}^{n} X_j$. Then T is $P(n\theta)$, sufficient for θ (Exercise 11.1.2(i)) and complete (Example 10). Finally, $E_\theta(T/n) = \theta$ for every $\theta \in \Omega = (0, \infty)$.

11.5.2. $f_X(x) = \dfrac{1}{\beta - x} I_{(\alpha, \beta)}(x)$ so that the set of positivity of f does depend on the parameter(s).

Chapter 12

12.2.2. $\dfrac{n+1}{2n} X_{(n)}$, $\dfrac{n+2}{12n} X_{(n)}^2$.

12.2.3. $[X_{(1)} + X_{(n)}]/2$, $\dfrac{n+1}{n-1}[X_{(n)} - X_{(1)}]$.

12.3.2. $c'_n = \sqrt{2}\,\Gamma\!\left[\dfrac{1}{2}(n+1)\right] \Big/ \Gamma\!\left(\dfrac{1}{2}n\right)$.

12.3.5. $\overline{X}$.

12.3.6. $(X + r)/r$, $\sigma_\theta^2[(X+r)/r] = (1-\theta)/r\theta^2$.

12.3.8. $\sqrt{2}\,\Gamma\!\left(\dfrac{n-1}{2}\right)\overline{X} \Big/ \Gamma\!\left(\dfrac{n-2}{2}\right)\left[\sum_{j=1}^{n}(X_j - \overline{X})^2\right]$.

12.3.9. $\sum_{j=1}^{n}(X_j - \overline{X})(Y_j - \overline{Y})/(n-1)$, $\overline{X}\overline{Y} - \dfrac{1}{n(n-1)}\sum_{j=1}^{n}(X_j - \overline{X})(Y_j - \overline{Y})$,

$2\Gamma\!\left(\dfrac{n-1}{2}\right)\sum_{j=1}^{n}(X_j - \overline{X})(Y_j - \overline{Y})/(n-1)\Gamma\!\left(\dfrac{n-3}{2}\right)\sum_{j=1}^{n}(X_j - \overline{X})^2$.

12.5.6. $-\log(X/n)$.

12.5.7. $X_{(1)}$, $\overline{X} - X_{(1)}$.

12.5.8. $\left(\sum_{j=1}^{n} X_j^r\right)\!\Big/ n$.

12.5.9. $\exp(-x/\overline{X})$.

12.9.1. $\overline{X} - \dfrac{b-a}{2}$, $\sigma_\theta^2\left(\overline{X} - \dfrac{b-a}{2}\right) = (a+b)^2/12n$.

12.9.4. $3(X_1 + X_2)/2$.

12.9.6. $\overline{X} - S$ and S, where $S^2 = \dfrac{1}{n}\sum_{j=1}^{n}(X_j - \overline{X})^2$.

12.11.2. $\sqrt{n}(U_n - V_n) = \left[(\alpha+\beta)\sum_{j=1}^{n} X_j - \alpha n\right] / \sqrt{n}(\alpha+\beta+n)$ and

$E_\theta[\sqrt{n}(U_n - V_n)] = [(\alpha+\beta)n\theta - \alpha n]/\sqrt{n}(\alpha+\beta+n) \xrightarrow[n\to\infty]{} 0$,

$\sigma_\theta^2[\sqrt{n}(U_n - V_n)] = n\theta(1-\theta)[(\alpha+\beta)/\sqrt{n}(\alpha+\beta+n)]^2 \xrightarrow[n\to\infty]{} 0$.

Chapter 13

13.2.1. Reject H if $\overline{x} > 1.2338$. Power $= 0.378$.

13.2.2. $n = 9$.

12.2.4. Cut-off point $= 3.466$, H is accepted.

13.3.1. $V = -\sum_{j=1}^{n} x_j$ in both cases.

13.3.4. Cut-off point $= 28.44$, H is accepted.

13.3.7. Cut-off point ≈ 86, H is accepted.

13.3.8. $n = 14$.

13.3.10. Cut-off point $= 10$, $H:\lambda = 20$ (there is no improvement) is accepted.

13.3.12. (i) Reject H if $\sum_{j=1}^{n} x_j < C$, $C: P_{\theta_0}\left(\sum_{j=1}^{n} X_j < C\right) = \alpha$; (ii) $n = 23$.

13.4.1. H is rejected.

13.5.1. $H: \sigma \leq 0.04$, $A: \sigma > 0.04$. H is accepted.

13.5.4. Assume normality and independence. H is accepted.

13.5.5. $H: \mu = 2.5$, $A: \mu \neq 2.5$. H is accepted.

13.6.2. Cut-off point $= 5.9$, H is accepted.

13.6.3. Assume normality and independence. Both hypotheses are accepted.

13.7.2. H is rejected in both cases.

13.8.3. Cut-off point $= 2.82$, H is accepted.

Chapter 14

13.8.4. H (hypothesizing the validity of the model) is accepted.

13.8.7. H (the vaccine is not effective) is rejected.

14.3.1. $E_0(N) = 77.3545$, $E_1(N) = 97.20$, n (fixed sample size) = 869.90 ≈ 870.

14.3.2. $E_0(N) = 2.32$, $E_1(N) = 4.863$, n (fixed sample size) = 32.18 ≈ 33.

Chapter 15

15.2.4. (i) $f_R(r) = n(n-1)r^{n-2}(\theta - r)/\theta^n$, $r \in (0, \theta)$; (iii) The expected length of the shortest confidence interval in Example 4 is $= n\theta(\alpha^{-1/n} - 1)/(n+1)$. The expected length of the confidence interval in (ii) is $= (n-1) \times \theta(1-c)/c(n+1)$ and the required inequality may be seen to be true.

15.4.1.

(i) $[\overline{X}_n - z_{\alpha/2}\sigma/\sqrt{n}, \overline{X}_n + z_{\alpha/2}\sigma/\sqrt{n}]$;

(ii) $[\overline{X}_{100} - 0.196, \overline{X}_{100} + 0.196]$;

(iii) $n = 1537$.

15.4.2.

(i) $[\overline{X}_{100} - 0.0196 S_{100}, \overline{X}_{100} + 0.0196 S_{100}]$, $S_{100}^2 = \sum_{j=1}^{100}(X_j - \overline{X}_{100})^2/100$;

(ii) $S_n \xrightarrow[n\to\infty]{} \sigma$ in probability (and also a.s.).

15.4.4. σ = known, μ = unknown: $[\overline{X}_n - z_{\alpha/2}\sigma/\sqrt{n}, \overline{X}_n + z_{\alpha/2}\sigma/\sqrt{n}]$; μ = known, σ = unknown: $[nS_n^2/C_2, nS_n^2/C_1]$, $C_1, C_2 : P(\chi_n^2 < C_1 \text{ or } \chi_n^2 > C_2) = \alpha$.

15.4.5. $P(Y_i < x_p < Y_j) = \sum_{k=i}^{j-1}\binom{10}{k}p^k(1-p)^{10-k} = 1-\alpha$. Let $p = 0.25$ and $(i, j) = (2, 9), (3, 4), (4, 7)$. Then $1 - \alpha = 0.756, 0.474, 0.2237$, respectively. For $p = 0.50$ and (i, j) as above, $1 - \alpha = 0.9786, 0.8906, 0.7734$, respectively.

15.4.7. $[x'_{p/2}, x_{p/2}]$, $[0.8302, 2.0698]$.

Chapter 16

16.3.5.

(i) $\hat{\boldsymbol{\beta}} = \begin{pmatrix} 0.280 \\ 0.572 \\ -0.268 \end{pmatrix}$, $\tilde{\sigma}^2 = 7.9536$;

(ii) $\sigma^2 \begin{pmatrix} 4.6 & -3.30 & 0.50 \\ -3.3 & 2.67 & -0.43 \\ 0.5 & -0.43 & 0.07 \end{pmatrix}$;

(iii) $\begin{pmatrix} 36.5865 & -26.2469 & 3.9768 \\ -26.2469 & 21.2361 & -3.4200 \\ 3.9768 & -3.4200 & 0.5567 \end{pmatrix}$

16.4.2.

(i) (i) Reject H if $\left|(\hat{\gamma}-\gamma_0)\sqrt{\sum_{j=1}^{n}(x_j-\bar{x})^2}\ \sqrt{n\hat{\sigma}^2/(n-1)}\right| > t_{n-2;\alpha/2}$,

where $\hat{\gamma} = \sum_{j=1}^{n}(x_j-\bar{x})Y_j \Big/ \sum_{j=1}^{n}(x_j-\bar{x})^2$, $\hat{\sigma}^2$

$= \sum_{j=1}^{n}[Y_j-\hat{\beta}-\hat{\gamma}(x_j-\bar{x})]^2 \Big/ n$, $\hat{\beta} = \bar{Y}$;

(ii) $\left[\hat{\gamma}-t_{n-2;\alpha/2}\sqrt{n\hat{\sigma}^2 \Big/ (n-2)\sum_{j=1}^{n}(x_j-\bar{x})^2}\right.$,

$\left.\hat{\gamma}+t_{n-2;\alpha/2}\sqrt{n\hat{\sigma}^2 \Big/ (n-2)\sum_{j=1}^{n}(x_j-\bar{x})^2}\right]$.

16.5.1.

(i) $[(n-r)\tilde{\sigma}^2/b, (n-r)\tilde{\sigma}^2/a]$, $a, b: P(a \leq \chi^2_{n-r} \leq b) = 1 - \alpha$;

(ii) $[25\tilde{\sigma}^2/40.6, 25\tilde{\sigma}^2/13.1]$.

16.5.2.

(iii) Reject H if $\left|\sqrt{n}\hat{\beta}_1/\sqrt{\tilde{\sigma}^2}\right| > t_{n-2;\alpha/2}$,

$\left[\hat{\beta}_1-b\sqrt{\tilde{\sigma}^2/n},\ \hat{\beta}_1-a\sqrt{\tilde{\sigma}^2/n}\right]$, $a, b: P(a \leq t_{n-2} \leq b) = 1-\alpha$;

(iv) Reject H' if $\left|\hat{\beta}_2\sqrt{\sum_{j=1}^{n}x_j^2}\Big/\sqrt{\tilde{\sigma}^2}\right| > t_{n-2;\alpha/2}$,

$\left[\hat{\beta}_2-b\sqrt{\tilde{\sigma}^2}\Big/\sqrt{\sum_{j=1}^{n}x_j^2},\ \hat{\beta}_2-a\sqrt{\tilde{\sigma}^2}\Big/\sqrt{\sum_{j=1}^{n}x_j^2}\right]$, a, b as in (iii).

16.5.5.

(i) $\hat{\beta}_1 = -0.0125$, $\hat{\beta}_2 = 0.019$, $\tilde{\sigma}^2 = 0.01$;

(ii) β_1: [−0.1235, 0.0985], β_2: [0.134, 0.0246], σ^2: [0.003, 0.008];
(iii) [−0.2891, 0.9101].

16.5.9. Reject H_1 if

$$\left|(\hat{\beta}_1 - \hat{\beta}_1^*)\middle/ \sqrt{\tilde{\sigma}^2\left\{\left[\sum_{i=1}^{m} x_i^2 \middle/ m\sum_{i=1}^{m}(x_i - \bar{x})^2\right] + \left[\sum_{j=1}^{n} x_j^{*2} \middle/ n\sum_{j=1}^{n}(x_j^* - \bar{x}^*)^2\right]\right\}}\right| > t_{m+n-4;\alpha/2},$$

and reject H_2 if

$$\left|(\hat{\beta}_2 - \hat{\beta}_2^*)\middle/ \sqrt{\tilde{\sigma}^2\left\{\left[1\middle/\sum_{i=1}^{m}(x_i - \bar{x})^2\right] + \left[1\middle/\sum_{j=1}^{n}(x_j^* - \bar{x}_j^*)^2\right]\right\}}\right| > t_{m+n-4;\alpha/2}.$$

Chapter 17

17.1.1. $SS_H = 0.9609$ (d.f. = 2), $MS_H = 0.48045$, $SS_e = 8.9044$ (d.f. = 6), $MS_e = 1.48407$, $SS_T = 9.8653$ (d.f. = 8).

17.2.2. $SS_A = 34.6652$ (d.f. = 2), $MS_A = 17.3326$, $SS_B = 12.2484$ (d.f. = 3), $MS_B = 4.0828$, $SS_e = 12.0016$ (d.f. = 6), $MS_e = 2.0003$, $SS_T = 58.9152$ (d.f. = 11).

17.4.1. $Y_{ij} \approx N(\mu_i, \sigma^2)$, $i = 1, \ldots, I$; $j = 1, \ldots, J$ independent implies

$y_{i\cdot} - \mu_i \sim N\left(0, \dfrac{\sigma^2}{J}\right)$, $i = 1, \ldots, I$ independent. Since

$\dfrac{1}{I}\sum_{i=1}^{I}(Y_{i\cdot} - \mu_i) = Y_{\cdot\cdot} - \mu_{\cdot}$, we have that

$\sum_{i=1}^{I}[(Y_{i\cdot} - \mu_i) - (Y_{\cdot\cdot} - \mu_{\cdot})]^2 \middle/ \sigma^2/J \sim \chi^2_{I-1}$ Hence that result.

Chapter 18

18.1.3. Let $\mathbf{X}^{(1)} = (X_{i_1}, \ldots, X_{i_m})'$, $\mathbf{X}^{(2)} = (X_{j_1}, \ldots, X_{j_n})'$ and partition $\boldsymbol{\mu}$ and $\boldsymbol{\Sigma}$ as follows:

$$\boldsymbol{\mu} = \begin{pmatrix} \boldsymbol{\mu}^{(1)} \\ \boldsymbol{\mu}^{(2)} \end{pmatrix}, \boldsymbol{\Sigma} = \begin{pmatrix} \boldsymbol{\Sigma}_{11} & \boldsymbol{\Sigma}_{12} \\ \boldsymbol{\Sigma}_{21} & \boldsymbol{\Sigma}_{22} \end{pmatrix}.$$

Then the conditional distribution of $\mathbf{X}^{(1)}$, given $\mathbf{X}^{(2)} = \mathbf{x}^{(2)}$, is the m-variate Normal with parameters:

$$\boldsymbol{\mu}^{(1)} + \boldsymbol{\Sigma}_{12} + \boldsymbol{\Sigma}_{22}^{-1}\left[\mathbf{x}^{(2)} - \boldsymbol{\mu}^{(2)}\right], \boldsymbol{\Sigma}_{11} - \boldsymbol{\Sigma}_{12}\boldsymbol{\Sigma}_{22}^{-1}\boldsymbol{\Sigma}_{21}.$$

18.3.7. In the inequality $\left(\sum_{j=1}^{n}\alpha_j\beta_j\right)^2 \leq \left(\sum_{i=1}^{n}\alpha_i^2\right)\left(\sum_{j=1}^{n}\beta_j^2\right)$, $\alpha_i, \beta_j \in \mathbb{R}, i, j = 1, \ldots, n$, set $\alpha_i = X_i - \overline{X}$, $\beta_j = Y_j - \overline{Y}$.

Chapter 19

19.2.5. $Q = \mathbf{X}'\mathbf{C}\mathbf{X}$, where $\mathbf{X} = (X_1, X_2, X_3)'$, $\mathbf{C} = \begin{pmatrix} 5/6 & 1/3 & -1/6 \\ 1/3 & 1/3 & 1/3 \\ -1/6 & 1/3 & 5/6 \end{pmatrix}$

and $\mathbf{C}$ is idempotent and of full rank. Thus Q is χ_3^2. Furthermore, $\mathbf{X}'\mathbf{X} - Q = \frac{1}{6}(X_1 - X_2 + X_3)^2 \geq 0$, so that $\mathbf{X}'\mathbf{X} - Q$ is positive definite. Then, by Theorem 4, $\mathbf{X}'\mathbf{X} - Q$ and Q are independent.

Chapter 20

20.4.1. $R_X + R_Y = \sum_{i=1}^{m} R(X_i) + \sum_{j=1}^{n} R(Y_j) = 1 + 2 + \cdots + N = N(N+1)/2$.

20.4.3. $Eu(X_i - Y_j) = Eu^2(X_i - Y_j) = P(X_i > Y_j) = 1/2$, so that $\sigma^2 u(X_i - Y_j) = 1/4$, and $\text{Cov}[u(X_i - Y_j), u(X_k - Y_l)] = 0$, $i \neq k, j \neq l$, $\text{Cov}[u(X_i - Y_j), u(X_k - Y_l)] = \frac{1}{3}$, $i = k, j \neq l$ or $i \neq k, j = l$. The result follows.

INDEX

A

Absolutely continuous, 54, 55
Acceptance region, 332
Analysis of variance, 440
    one-way layout (classification) in, 440
    tables for, 445, 451, 457
    two-way layout (classification) in, 446, 452
Asymptotic efficiency (efficient), 327
Asymptotic relative efficiency, 327, 497
    Pittman, 498
        of the sign test, 498
        of the t-test, 498
        of the Wilcoxon-Mann-Whitney test, 498

B

Basis, 501
    orthonormal, 501
Bayes, decision function, 380, 382
    estimator (estimate), 315–317, 322
    formula, 24
Behrens-Fisher problem, 364
Best linear predictor, 134
Beta distribution, 71
    expectation of, 120, 543
    graphs of, 72
    moments of, 120
    parameters of, 71
    p.d.f. of, 70, 542
    variance of, 543
Beta function, 71
Bias, 306
Bienaymé equality, 171
Binomial distribution, 56
    approximation by Poisson, 58, 79
    as an approximation to Hypergeometric, 81
    ch.f. of, 144, 146, 543
    expectation of, 114, 542
    factorial moments of, 119
    graphs of, 56, 57
    m.g.f. of, 154, 543
    parameters of, 56
    Point, 56
    p.d.f. of, 55
    tables of, 511–519
    variance of, 114, 542
Binomial experiment, 55, 59
Bio-assay, 287
Bivariate Normal distribution, 74, 96, 543
    ch.f. of, 544
    conditional moments of, 122
    conditional p.d.f.'s of, 122
    conditional variance of, 122
    graph of, 75
    independence in, 168
    marginal p.d.f.'s of, 75
    m.g.f. of, 158, 544
    parameters of, 74
    p.d.f. of, 74, 97, 134, 465–466, 543
    test of independence in, 470
    uncorrelation in, 168
    vector of expectations of, 543
    vector of variances of, 543
Borel, real line, 11
    σ-field, 12
        k-dimensional, 13

C

Cauchy distribution, 72, 543
    ch.f. of, 149, 544
    graph of, 73
    m.g.f. of, 156, 544
    parameters of, 72

Index

Cauchy principle value integral, 118
Central Limit Theorem (CLT), 66, 189
   applications of, 191–196
   continuity correction in, 192
Center of gravity, 108
Ch.f.'s, 138, 140
   application of, 166, 173–176
   basic properties of, 140
   joint, 150
   of an r.v., 140
   of Binomial distribution, 144, 146, 543
   of Bivariate Normal distribution, 544
   of Cauchy distribution, 149, 544
   of Chi-square distribution, 148, 544
   of Gamma distribution, 148, 544
   of Geometric distribution, 150, 544
   of k-Variate Normal distribution, 464, 544
   of Multinomial distribution, 152, 544
   of Negative Binomial distribution, 150, 543
   of Negative Exponential distribution, 148, 544
   of Normal distribution, 145, 147, 544
   of Poisson distribution, 146, 544
   of Uniform distribution, 150, 544
   properties of, 140–141, 150–151
Characteristic polynomial, 477, 504
   roots, 477, 504
Chi-square distribution, 69
   critical values of, 529–531
   degrees of freedom of, 69
   expectation of, 118, 542
   noncentral, 434, 508
   p.d.f. of, 69
   variance of, 118, 542
Chi-square statistic, 377
   asymptotic distribution of, 377
Chi-square test, 377
Column effect, 446, 450
Combinatorial results, 34
Completeness, 275, 281, 283, 285
   lack of, 276
   of Beta distribution, 285, 287
   of Binomial distribution, 275
   of Exponential distibution, 281, 283
   of Gamma distribution, 285, 287
   of Negative Binomial distribution, 287
   of Negative Exponential distribution, 287
   of Normal N (θ, σ^2) distribution, 276
   of Poisson distribution, 275
   of Uniform distribution, 276
Conditional expectation, 122
   basic properties of, 122
Confidence coefficient, 397
   approximate, 410
   asymptotic, 486
Confidence interval(s), 397, 486
   expected length of, 398
   in Beta distribution, 402–403, 409
   in Binomial distribution, 411
   in Gamma distribution, 401–402
   in Normal distribution, 398–401, 407–409, 410–411
   in Poisson distribution, 411
   in the presence of nuisance parameters, 407
   in Uniform distribution, 403
   length (shortest) of, 398–400, 402, 404, 408
Confidence limits (lower, upper), 397
Confidence region(s), 397, 410
   and testing hypotheses, 412
   in Normal distribution, 411
Contingency tables, 374–375
Continuity correction, 192–196
Contrast, 458
Convergence of r.v.'s, 180
   almost sure, 180, 183
   in distribution, 181, 183, 469
   in probability, 180, 183
   in quadratic mean, 182–183
   modes of, 180–181
   strong, 180
   weak, 181
   with probability one, 180
Convolution of d.f.'s, 224
Correlation coefficient, 129, 134
   interpretation of, 129–133
   p.d.f. of sample, 474
   sample, 472
Covariance, 129
   of a matrix of r.v.'s, 418
   of LSE in case of full rank, 421
Cramér-Rao bound, 297–298
   in Binomial distribution, 302
   in Normal distribution, 303–305
   in Poisson distribution, 303
Critical region, 332
Cumulant generating function, 162
Cumulative d.f., 85
   basic properties of, 85

D

Data, 263
Decision, 313
   rule, 313
Decision function(s), 313
   Bayes, 380, 382, 384
   equivalent, 314
   minimax, 380, 383–385
   nonrandomized, 308
Degrees of freedom
   of Chi-square distribution, 69
   of F-distribution, 233
   of noncentral Chi-square distribution, 508–509
   of noncentral F-distribution, 509
   of noncentral t-distribution, 508
   of t-distribution, 233
De Morgan's laws, 4, 7–8
Dependent

negatively quadrant, 163
positively quadrant, 163
Dimension of a space, 501
Distinct, balls, 38
  cells, 38
Distribution
  Bernoulli (or Point Binomial), 56, 542–543
  Beta, 70, 543
  Binomial, 55–56, 542–543
  Bivariate Normal, 74, 96, 465–466, 543–544
  Cauchy, 72, 543–544
  central Chi-square, 511
  central F-, 511
  central t-, 511
  Chi-square, 69, 542, 544
  continuous, 54
  discrete, 54
  discrete Uniform, 60
  Double Exponential, 78
  F-, 73, 223
  -free, 485
  function(s), 85
    basic properties of, 85
    conditional, 91
    cumulative, 85
    empirical (or sample), 200, 486
    graphs of, 87
    joint, 91, 93
    marginal, 91–93
    variation of, 92
  Gamma, 67, 542, 544
  Gauss, 65
  Geometric, 60, 542–543
  Hypergeometric, 59, 542
    Multiple, 63
  kurtosis of a, 121
  k-Variate Normal, 462, 543–544
  lattice, 141
  leptokurtic, 121
  Logistic, 90
  Lognormal, 73
  Maxwell, 78
  memoryless, 77
  mixed, 79
  Multinomial, 60–61, 95–96, 542, 544
  Multivariate Normal, 463
  Negative Binomial, 59, 542–543
  Negative Exponential, 69, 543–544
  noncentral Chi-square, 434, 508–509
  noncentral F-, 434, 509
  noncentral t-, 360, 508
  Normal, 65, 542, 544
  of an r.v., 53
  of the F statistic, 433
  of the sample variance in Normal, 480
  Pareto, 78
  Pascal, 60
  platykurtic, 121
  Point Binomial, 56
  Poisson, 57, 542–543

Raleigh, 78
Rectangular, 70
skewed, 121
skewness, 120
Standard Normal, 65
t-, 223
Triangular, 228
Uniform, 60, 70, 543–544
Weibull, 78, 285

E

Effects
  column, 446, 450
  main, 457
  row, 446, 450
Efficiency, relative asymptotic, 327. *See also*
  Pittman asymptotic relative efficiency
Eigenvalues, 477, 504
Error(s), sum of squares of, 418
  type-I, 332
  type-II, 332
Estimable, function, 422
Estimation
  by the method of moments, 324
  decision-theoretic, 313
  least square, 326
  maximum likelihood, 307
  minimum chi-square, 324
  nonparametric, 485
  point, 288
Estimator(s) (estimate(s))
  admissible, 314
  a.s. consistent, 327
  asymptotically efficient, 327
  asymptotically equivalent, 329
  asymptotically normal, 327, 489
  asymptotically optimal properties of, 326
  asymptotically unbiased, 489
  Bayes, 315–317, 322
  best asymptotically normal, 327
  consistent in probability, 327
  consistent in quadratic mean (q.m.), 489
  criteria for selecting, 289, 306, 313
  efficient, 305
  essentially complete class of, 314
  inadmissible, 315
  least square, 418
  maximum likelihood, 307
  minimax, 314, 322–323
  strongly consistent, 327, 485–487, 489
  unbiased, 289
  uniformly minimum variance unbiased
    (UMVU), 292, 297
  weakly consistent, 327, 485–486
Event(s)
  certain (or sure), 14
  composite, 14
  dependent, 33
  happens (or occurs), 15

Event(s) *continued*
  impossible, 14
  independent, 28
    completely, 28
    in the probability sure, 28
    mutually, 28
    pairwise, 28
    statistically, 28
    stochastically, 28
  null, 15
  simple, 14
Expectation, 107–108
  basic properties of, 110
  conditional, 123
  mathematical, 107–108
  of a matrix of r.v's, 418
  of Beta distribution, 120, 543
  of Binomial distribution, 114, 542
  of Chi-square distribution, 118, 542
  of Gamma distribution, 117, 542
  of Geometric distribution, 119, 542
  of Hypergeometric distribution, 119, 542
  of Lognormal distribution, 120
  of Negative Binomial distribution, 119, 542
  of Negative Exponential distribution, 118, 543
  of Normal distribution, 117, 542
  of Poisson distribution, 115, 542
  of Standard Normal distribution, 116
  of Uniform distribution, 113, 543
  of Weibull distribution, 114
  properties of conditional, 123
Experiment(s)
  binomial, 55, 59
  composite (or compound), 32
  dependent, 33
  deterministic, 14
  independent, 32, 46
  multinomial, 60
  random, 14
Exponential distribution(s), 280
  multiparameter, 285–286
  Negative Exponential, 69
  one-dimensional parameter, 280, 282
  $U(\alpha, \beta)$ is not an, 286

F

F-distribution, 73, 233
  critical values of, 532–541
  expectation of, 238
  graph of, 236
  noncentral, 434, 509
  the p.d.f. of, 234
  variance of, 258
F-test, 432
  geometric interpretation of, 432
Factorial moment, 110
  generating function of a, 153
  of Binomial distribution, 157

  of Negative Binomial distribution, 161
  of Poisson distribution, 115, 119
Failure rate, 91
Field(s), 8
  consequences of the definition of a, 8
  discrete, 9
  examples of, 9
  generated by a class of sets, 9
  trivial, 9
Finitely additive, 15
Fisher's information number, 300
Fourier transform, 140
Function
  Bayes, 380, 382
  Beta, 71
  continuous, 103
  cumulative distribution, 85
  decision, 313, 379
  empirical distribution, 200, 486
  estimable, 422
  Gamma, 67
  indicator, 84, 135
  loss, 313, 380
  measurable, 103
  minimax, 380
  non-randomized, 379
  parametric, 422
  projection, 104
  risk, 313, 380
  sample distribution, 200, 486
  squared loss, 313
Fundamental Principle of Counting, 35

G

Gamma distribution, 67, 542
  ch.f. of, 148, 544
  expectation of, 117, 542
  function, 67
  graph of, 68
  m.g.f. of, 155, 544
  parameters of, 67
  p.d.f. of, 67, 542
  variance of, 117, 542
General linear
  hypothesis, 416
  model, 417
Generating function, 121
  cumulant, 162
  factorial moment, 153
  moment, 138, 153, 158
Geometric distribution, 60
  ch.f. of, 543
  expectation of, 119, 542
  m.g.f. of, 543
  p.d.f. of, 60, 542
  variance of, 119, 542
Grand
  mean, 446
  sample mean, 445

H

Hazard rate, 91
Hypergeometric distribution, 59, 542
  approximation to, 81
  expectation of, 119, 542
  Multiple, 63
  p.d.f. of, 59, 542
  variance of, 119, 542
Hypothesis(-es)
  alternative, 331
  composite, 331, 353
  linear, 416
  null, 331
  simple, 331
  statistical, 331
  testing a simple, 333, 337
  testing a composite, 341

I

Independence, 27
  complete (or mutual), 28
  criteria of, 164
  in Bivariate Normal distribution, 168
  in the sense of probability, 28
  of classes, 177–178
  of events, 28
  of random experiments, 32, 46
  of r.v.'s, 164, 178
  of sample mean and sample variance in the Normal distribution, 244, 284, 479, 481
  of σ-fields, 46, 178
  pairwise, 28
  statistical, 28
  stochastic, 28
Independent
  Binomial r.v.'s, 173
  Chi-square r.v.'s, 175
  classes, 178
  completely, 28
  events, 28
  in the sense of probability, 28
  mutually, 28
  Normal r.v.'s, 174–175
  pairwise, 28
  Poisson r.v.'s, 173
  random experiments, 32, 46
  r.v.'s, 164, 178
  σ-fields, 46, 178
  statistically, 28
  stochastically, 28
Indicator, 84
  function, 135
Indistinguishable balls, 38
Inequality(-ies)
  Cauchy-Schwarz, 127
  Cramér-Rao, 297–298
  Markov, 126
  moment, 125
  probability, 125
  Tchebichev's, 126
Inference(s), 263
Interaction(s), 452, 457
Intersection, 2
Invariance, 311, 493
Inverse image, 82
Inversion formula, 141, 151

J

Jacobian, 226, 237, 240
Joint, ch.f., 150
  conditional distribution, 95, 467
  conditional p.d.f., 95
  d.f., 91, 94
  moment, 107
  m.g.f., 158
  probability, 25
  p.d.f., 95
  p.d.f. of order statistics, 249

K

Kolmogorov
  one sample test, 491
  -Smirnov, two-sample test, 493
Kurtosis of a distribution, 121
  of Double Exponential distribution, 121
  of Uniform distribution, 121

L

Laplace transform, 153
Latent roots, 481, 504
Laws of Large Numbers (LLNs), 198
  Strong (SLLNs), 198, 200
  Weak (WLLNs), 198–200, 210
Least squares, 418
  estimator, 418, 420
  estimator in the case of full rank, 421
Lebesgue measure, 186, 328
Leptokurtic distribution, 121
Level, of factor, 447
  of significance, 332
Likelihood, function, 307
  ratio, 365
Likelihood ratio test, 365, 430, 432
  applications of, 374
  in Normal distribution(s), 367–372
  interpretation of, 366
Likelihood statistic, 365
  asymptotic distribution of, 366
Limit
  inferior, 5
  of a monotone sequence of sets, 5
  superior, 5
  theorems, basic, 180
  theorems, further, 202

Linear
  dependence, 129
  hypothesis, 416
  interpolation, 61
  model, 424
    canonical reduction of, 424
  regression functions, 418
Lognormal distribution, 73
  expectation of, 120
  graphs of, 74
  parameters of, 73
  p.d.f. of, 73
  variance of, 120
Loss function, 313, 380
  squared, 313
Least Square Estimate (LSE), of μ's, 443
  of μ, α_i, β_j, 448–449
  of μ_{ij} etc., 454
  of σ^2, 425, 444, 451, 456

M

Matching problem, 48
Matrix (-ices), 502
  characteristic polynomial of a, 504
  characteristic (latent) roots of a, 504–505
  determinant of a square, 503
  diagonal, 502, 504–506
  diagonal element of a, 502
  dimensions of a, 502
  eigenvalues of a, 504
  elements of a, 502
  equal, 502
  idempotent, 504, 506–507
  indentity, 502
  inverse, 503
  negative definite (semidefinite), 504–506
  nonsingular, 503–506
  of full rank, 503
  order of a, 502, 505–507
  orthogonal, 503–506
  positive definite (semidefinite), 504–506
  product by a scalar, 502
  product of, 502
  rank of a, 503, 505–507
  singular, 503
  some theorems about, 504
  square, 502, 504–505
  sum of, 502
  symmetric, 502, 505–507
  transpose, 502
  unit, 502
  zero, 502
Maximum likelihood, 306
  and Bayes estimation, 318
  estimator (estimate) (MIE), 307
    in Binomial distribution, 309
    in Bivariate Normal distribution, 472, 475
    in Multinomial distribution, 308–309, 312
    in Negative Binomial distribution, 312
    in Negative Exponential distribution, 312
    in Normal distribution, 309–310, 312
    in Poisson distribution, 307
    in Uniform distribution, 310
    interpretation of, 307
    invariance of, 311
  function, 307
  principle of, 306
Mean, 107–108. *See also*, expectation
  grand, 446
  grand sample, 445
  interpretation of, 108
Mean value, 107–108
Measurable, function, 103
  space, 11
Measurable spaces
  product of, 45–46
Median, 99
  sample, 259
Mellin (or Mellin-Stieltjes)
    transform, 156
Method(s) of estimation, other, 324
  least square, 326, 418
  minimum chi-square, 324
  moments, 325
    in Beta distribution, 326
    in Gamma distribution, 326
    in Normal distribution, 325
    in Uniform distribution, 325
M.g.f., 138, 153, 158
  factorial, 153, 156–157
  joint, 158
  of Binomial distribution, 154, 543
  of Bivariate Normal distribution, 158, 544
  of Cauchy distribution, 156, 544
  of Chi-square distribution, 544
  of Gamma distribution, 155, 544
  of Geometric distribution, 543
  of Multinomial distribution, 158, 544
  of Negative Binomial distribution, 161, 543
  of Negative Exponential distribution, 156, 544
  of Normal distribution, 155, 544
  of Poisson distribution, 154, 543
  of Uniform distribution, 161, 544
Mode(s), 100
  of Binomial distribution, 101
  of Poisson distribution, 101
Moment(s)
  about the mean, 109
  absolute, 107
  central, 107, 109
  central joint, 109
  conditional, 122
  factorial, 110, 115
    of Binomial distribution, 119
    of Poisson distribution, 119
  generating function (m.g.f.), 153–156
  inequalities, 125
  joint, 107

n-th, 106–107
of an r.v., 106, 108
of Beta distribution, 120
of inertia, 107
of Standard Normal, 116
r-th absolute, 107
sample, 325
Monotone likelihood ratio (MLR) property, 342
of Exponential distribution, 342
of Logistic distribution, 342–343
testing under, 343
Multicomparison method, 458
Multinomial distribution, 60–61, 64, 95–96, 542
ch.f. of, 152, 544
distribution, 60–61, 64, 95–96, 542
experiment, 60
m.g.f. of, 158, 544
parameters of, 60–61
p.d.f. of, 60
vector of expectations of, 542
vector of variances, 542
Multivariate (or k-Variate) Normal distribution, 463–465, 543–544
ch.f. of, 544
estimation of μ and Σ, 469
m.g.f. of, 544
nonsingular, 465
parameters of, 464
p.d.f. of, 465, 543
singular, 474
some properties of, 467

N

Negative Binomial distribution, 59
ch.f. of, 150, 543
expectation of, 119, 542
factorial m.g.f., 161
m.g.f., 161, 543
p.d.f. of, 59, 542
variance of, 119, 542
Negative Exponential distribution, 69
ch.f. of, 544
expectation of, 118, 543
m.g.f. of, 156, 544
parameter of, 69
p.d.f. of, 69, 543
variance of, 118, 543
Neyman-Pearson, 331, 366
fundamental lemma, 334
Noncentrality parameter, 508–509
Noncomplete distributions, 276–277
Nonexponential distribution, 286
Nonnormal Bivariate distribution, 99
Nonparametric, estimation, 485
estimation of a p.d.f., 487
inference, 485
tests, 490
Normal distribution, 65, 542

approximation by, 194–196
ch.f. of, 145, 147, 544
expectation of, 116–117, 542
graph of, 66
independence of sample mean and sample variance in, 244, 284, 479, 481
m.g.f. of, 155, 544
moments of Standard, 116
parameters of, 65
p.d.f. of, 65, 542
Standard, 65
tabulation of Standard, 523–525
variance of, 116–117, 542
Normal equations, 418, 420
Notation and abbreviations, 545–546

O

Occupancy numbers, 39
Order statistic(s), 249
joint p.d.f. of, 249
p.d.f. of j-th, 252
Orthocomplement (or orthogonal complement), 501

P

Parameters, 263
noncentrality, 508–509
nuisance, 356, 407
space, 263
Partition (finite, denumerably infinite), 23
Permutation(s), 35
Pittman asymptotic relative efficiency, 498
of the Sign test, 498
of the Wilcoxon-Mann-Whitney, 498
telative to the t-test, 498
Platikurtic distribution, 121
Poisson distribution, 57
approximation to Binomial distribution, 58, 79
ch.f. of, 146, 543
expectation of, 115, 542
factorial moments of, 119
graph of, 58
m.g.f. of, 154, 543
parameter of, 57
p.d.f. of, 57, 542
tabulation of, 520–522
variance of, 115, 542
Poisson, truncated r.v.('s), 62, 273
Polya's lemma, 210
urn scheme, 65
Power function, 332
Prediction interval, 413
Predictor, best linear, 134
Probability(-ies)
classical definition of, 17
conditional, 21–22

Probability(-ies) *continued*
  consequences of the definition of, 15–16
  defining properties of, 15
  densities, 85
  density function, 54
  distribution, 53, 84
  distribution function, 53, 84
  function, 15
    uniform, 17
  inequalities, 125
  integral transform, 246
  joint, 25
  Kolmogorov definition of, 17
  marginal, 25
  measure(s), 15
    product of, 46
  of coverage of a population quantile, 260
  of matchings, 47
  of type-I error, 332
  of type-II error, 332
  relative frequency definition of, 17
  space, 15
  statistical definition of, 17
Probability density function(s) (p.d.f.('s)), 53–54
  conditional, 91, 94
  convolution of, 224
  F-, 233–234
  joint, 93
  joint conditional, 95
  marginal, 91, 93, 95
  of Bernoulli (or Point Binomial) distribution, 56, 542
  of Beta distribution, 70, 543
  of Binomial distribution, 55, 542
  of Bivariate Normal distribution, 74, 97, 134, 465–466, 543
  of Cauchy distribution, 72, 543
  of Chi-square distribution, 69, 542
  of continuous Uniform distribution, 70, 543
  of discrete Uniform distribution, 60
  of Double Exponential distribution, 78
  of Gamma distribution, 67, 542
  of Geometric distribution, 60, 542
  of Hypergeometric distribution, 59, 542
  of Lognormal distribution, 73
  of Maxwell distribution, 78
  of Multinomial distribution, 60, 542
  of Multivariate Normal distribution, 465, 543
  of Negative Binomial distribution, 59, 542
  of Negative Exponential distribution, 69, 543
  of Normal distribution, 65, 542
  of Pareto distribution, 78
  of Pascal distribution, 60
  of Poisson distribution, 57, 542
  of Raleigh distribution, 78
  of sample range, 255
  of Standard Normal distribution, 65
  of t-distribution, 233
  of Triangular distribution, 228
  posterior, 318
  prior, 315, 380
Product probability
  measures, 46
  spaces, 45–46

Q

Quadratic form(s), 476, 504, 506
  negative definite (semidefinite), 477
  positive definite (semidefinite), 477
  rank of a, 477
  some theorems on, 477
Quantile, p-th, 99, 261

R

Random variable(s) (r.v.('s)), 53
  absolute moments of a, 107
  absolutely continuous, 54
  as a measurable function, 82–83
  Bernoulli (or Point Binomial), 56
  Beta, 70
  Binomial, 55
  Cauchy, 72
  central moment of a, 107
  ch.f. of a, 138, 140
  Chi-square, 69
  completely correlated, 129
  continuous, 53–54
  criteria of independence of, 164–167
  cumulative distribution function of a, 85
  degenerate, 183
  dependent, 164
  discrete, 53–55
  discrete Uniform, 60
  distribution function of a, 85
  distribution of a, 53
  expectation of a, 107–108
  functions of, 107–108
  Gamma, 66, 117
  Geometric, 119, 542
  Hypergeometric, 59, 119
  independent, 178
  linearly related, 129
  Lognormal, 72
  mathematical expectation of a, 107–108
  mean (or mean value) of a, 107–108
  m.g.f. of a, 138, 153, 158
  moments of a, 107–108
  Negative Binomial, 59
  Negative Exponential, 69
  negatively correlated, 129
  Normal, 65
  normalization of a, 89
  pairwise uncorrelated, 171
  Poisson, 57
  Poisson, truncated, 62

Index

positively correlated, 129
p.d.f. of a, 54
probability distribution function of a, 53
Standard Normal, 65
transformation of a, 216
uncorrelated, 129
Uniform, 60, 70
variance of a, 107
Random vector(s), 55
  absolutely continuous, 55
  as a measurable function, 82–83
  Bivariate Normal, 74
  ch.f. of, 150
  continuous, 55
  discrete, 55
  distribution (or probability distribution function) of, 55
  distribution function of, 85
  k-dimensional, 55
  linear transformation of, 239
  Multinomial, 60
  Multiple Hypergeometric, 63
  Multivariate (or k-variate) Normal, 463
  orthogonal transformation of, 239–240
  p.d.f. of a, 55
  transformation of, 216, 223
Range
  sample, 255
  studentized, 256
Rao-Blackwell, 278
  Rao-Blackwellization, 278
Rectangular distribution, 70
Recursive formula(s) for, 64
  Binomial probabilites, 64
  Hypergeometric probabilities, 64
  Poisson probabilities, 64
Region(s)
  acceptance, 332
  contidence, 397, 410
  critical, 332
  rejection, 332
Regression function(s)
  linear, 418
Regularity conditions, 297
Reliability (function), 91, 296, 312, 405
Risk
  average, 315
  function, 313, 380
Row effect(s), 446, 450

S

Sample(s)
  correlation coefficient, 472
  mean, 445
  median, 259
  moment, 325
  ordered, 35
  point, 14
  random, 263

range, 255
size, 35
space, 14
unordered, 35
Sampling
  sequential, 382
  with replacement, 35–36, 49, 82
  without replacement, 35–36, 48–49, 59, 63, 82
Scalar, 499, 502
Scheffé criterion, 459
Sequential
  analysis, 383
  probability ratio test (SPRT), 388–389
    expected sample size, 393–394
    for Binomial distribution, 395
    for Normal distribution, 396
    optimality of, 393
    with prescribed error probabilities, 392
  procedures, 382
  sampling, 383
Set(s)
  basic, 1
  complement of, 1
  decreasing sequence of, 5
  difference of, 3
  disjoint, 3
  element of, 1
  empty, 3
  equal, 3
  increasing sequence of, 5
  inferior limit of a sequence of, 5
  intersection of, 2
  limit of a sequence of, 5
  measure of a, 102
  member of a, 1
  monotone sequence of, 5
  mutually disjoint, 3
  of convergence of a sequence of r.v.'s, 183
  open, 102–103
  operations, 1
  pairwise disjoint, 3
  properties of operations on, 4
  superior limit of a sequence of, 5
  symmetric difference of, 3
  union of, 2
  universal, 1
Shift, 499
σ-additive, 15
σ-field(s)
  Borel, 12
  consequences of the definition of a, 10
  dependent, 46
  discrete, 10
  examples of, 10
  generated by, 11
  independent, 46
  induced by, 178
  of events, 15
  k-dimensional Borel, 13

σ-field(s) *continued*
  trivial, 10
  two-dimensional Borel, 13
Sign test
  one-sample, 497
  two-sample, 496
Significantly different from zero, 459
Skewed, to the left (to the right), 121
Skewness, 120
  of Binomial distribution, 121
  of Negative Exponential distribution, 121
  of Poisson distribution, 121
Space(s)
  measurable, 11
  probability, 15
  product probability, 45–46
  sample, 14
Standard
  deviation, 107, 109
  Normal distribution, 65
Statistic(s), 263
  complete, 292
  m-dimensional, 263
  minimal sufficient, 272
  of uniformly minimum variance (UMV), 279
  sufficient, 266, 270, 292
  U-, 495
  unbiased, 278
Stopping time, 382
Strong Law of Large Numbers (SLLNs), 198, 200
Studentized range, 256
Subadditive, 16
Subset, 1
  proper, 1
Sub-σ-fields
  independent, 46
Sufficiency, 263–264
  definition of, 264
  some basic results on, 264
Sufficient statistic(s), 266, 270, 282, 292
  complete, 291
  for Bernoulli distribution, 274
  for Beta distribution, 273
  for Bivariate Normal distribution, 274
  for Cauchy distribution, 272
  for Double Exponential distribution, 274
  for Exponential distribution, 282
  for Gamma distribution, 273
  for Multinomial distribution, 270
  for Multivariate Normal distribution, 469
  for Negative Binomial distribution, 273
  for Negative Exponential distribution, 273
  for Normal distribution, 271
  for Poisson distribution, 273
  for Uniform distribution, 271
  for Weibull distribution, 285
  m-dimensional, 263
  minimal, 272

Sums of squares
  between groups, 445
  total, 440
  within groups, 445

T

Tables, 511
  of critical values for Chi-square distribution, 529–531
  of critical values for F-distribution, 532–541
  of critical values for Student's t-distribution, 526–528
  of the cumulative Binomial distribution, 511–519
  of the cumulative Normal distribution, 523–525
  of the cumulative Poisson distribution, 520–522
t (Student's)-distribution, 73, 233–234
  critical values of, 526–528
  expectation of, 239
  graph of, 234
  noncentral, 360, 364, 508
  noncentrality parameter, 508
  p.d.f. of, 234
  variance of, 239
Test(s), 311
  about mean(s) in Normal distribution(s), 339, 349, 355, 359–360, 363–364, 383–384
  about variance(s) in Normal distribution(s), 340, 350, 358–359, 361–363
  Bayes, in Normal distribution, 384
  chi-square, 377
  consistent, 379
  equal-tail, 359, 372
  F-, 432
  function, 331
  goodness-of-fit, 374
  Kolmogorov one-sample, 491
  Kolmogorov-Smirnov two-sample test, 493
  level-α, 333
  level of significance of a, 332
  likelihood ratio (LR), 365, 367–372
  minimax in Binomial distribution, 385
  minimax in Normal distribution, 384
  most powerful (MP), 333, 338–340
  nonparametric, 490–493
  nonrandomized, 332
  of independence in Bivariate Normal distribution, 470
  one-sample Wilcoxon-Mann-Whitney, 495
  power of a, 332
  randomized, 331
  rank, 493
  rank sum, 495
  relative asymptotic efficiency of a, 497
  sign, 496–497
  size of a, 332
  statistical, 331

two-sample Wilcoxon-Mann-Whitney, 495
unbiased, 353
uniformly most powerful (UMP), 333, 343, 345, 346–350, 353, 357–358
uniformly most powerful unbiased (UMPU), 353–354, 357–360, 362–364
Testing hypothesis(-es), 331
   about η, 434
   composite, 343, 345–346, 353
   decision-theoretic viewpoint of, 379
   general concepts of, 331
   in Binomial distribution, 337, 347, 355, 384–385
   in Exponential family, 346, 354
   in Normal distribution, 339–340, 349–350, 355, 358–364, 383–384
   in Poisson distribution, 338, 348, 355
   theory of, 331
   under monotone likelihood ratio (MLR) property, 343, 345–346
Testing, composite hypotheses, 343, 345–346
   simple hypotheses, 333–334
Theorem(s)
   additive probability, 17
   basic limit, 180
   Basu, 277
   Carathéodory extension, 46
   Central Limit (CLT), 189
   Cochran, 477
   De Moivre, 192
   factorization, 165–166
   Fisher-Neyman factorization, 267
   Fubini, 317
   further limit, 202
   Gauss-Markov, 423
   Glivenko-Cantelli, 201
   Lehmann-Scheffé, 279
   multiplicative probability, 22
   P. Lévy's continuity, 187
   Rao-Blackwell, 278
   Slutsky's, 207
   total probability, 23
   uniqueness, 145, 151
   Wald's, 383
Tolerance interval(s), 413–414
   interpretation of, 414
Tolerance limits, 397
Transformation(s)
   invariance under monotone, 493
   linear, 239–244
   of continuous r.v.'s, 218–224, 226–237
   of discrete r.v.'s, 216–218, 225
   orthogonal, 239–240
   the multivariate case of, 223–237
   the univariate case of, 216–222
Transitive relationship, 6
Type-I error, 332
   probability of, 332
Type-II error, 332
   probability of, 332

U

U statistic, 495
Unbiasedness, 278, 286, 289
Unbiased statistic, 278
   asymptotically, 489
   unique, 279
Uniform distribution, 60, 70
   ch.f. of, 150, 544
   continuous, 70
   discrete, 60
   expectation of, 113, 543
   graphs of, 60, 70
   m.g.f. of, 161, 544
   parameters of, 70
   p.d.f. of, 60, 70, 544
   variance of, 113, 543
Uniformly minimum variance unbiasedness, 279, 290
Uniformly minimum variance unbiased (UMVU) estimator(s) (estimate(s)), 290–291
   in Binomial distribution, 292–294, 296
   in Bivariate Normal distribution, 296
   in Gamma distribution, 305
   in Geometric distribution, 297
   in Negative Binomial distribution, 296
   in Negative Exponential distribution, 296
   in Normal distribution, 294–295, 303–305
   in Poisson distribution, 294, 303
Uniformly minimum variance unbiased statistic(s), 279
   in Binomial distribution, 279
   in Negative Exponential distribution, 279–280
   in Normal distribution, 279
   in Poisson distribution, 280
Uniformly most powerful (UMP) test(s), 333, 343, 345
   in Binomial distribution, 347
   in distributions with the MLR property, 343
   in Exponential distributions, 345–346
   in Normal distribution, 349–350, 358
   in Poisson distribution, 348
Uniformly most powerful unbiased (UMPU) test(s), 353, 357
   in Exponential distributions, 354
   in Normal distributions, 355, 357–360
   in the presence of nuisance parameters, 357–364
Uniform probability measure, 60
Union of sets, 2

V

Variance, 107, 109
   basic properties of, 111
   covariance of a matrix of r.v.'s, 418
   interpretation of, 109

Variance *continued*
  minimum, 289
  of Beta distribution, 120, 543
  of Binomial distribution, 114, 542
  of Chi-square distribution, 118, 542
  of Gamma distribution, 117, 542
  of Geometric distribution, 119, 542
  of Hypergeometric distribution, 119, 542
  of Lognormal distribution, 120
  of Negative Binomial distribution, 119, 542
  of Negative Exponential distribution, 118, 543
  of Normal distribution, 117, 543
  of Poisson distribution, 115, 542
  of Standard Normal distribution, 116, 542
  of Uniform distribution, 113, 543
  of Weibull distribution, 114
Vector(s), 499
  components of a, 499
  equal, 499
  inner (scalar) product of, 499
  linearly dependent, 501
  linearly independent, 501
  n-dimensional, 499
  norm (or length) of a, 500
  orthogonal (or perpendicular), 500
  orthogonal (or perpendicular), to a subset of, 500
  product by a scalar, 499
  space(s), 499
    dimension of a, 499
    some theorems on, 501
    spanning (or generating) a, 500
    subspace(s), 500
    orthocomplement of a, 501
  transpose of a, 499
  zero, 499
Venn diagram, 1

W

Wald's lemma for sequential analysis, 383
Weak Law of Large Numbers (WLLNs), 198–200, 210–214
  generalized, 201
    application to Chi-square, Negative Exponential and Poisson distributions of the, 202
Wilcoxon-Mann-Whitney, one-sample test, 498
  two-sample test, 498